C0-FXN-488

CHEMICAL ZOOLOGY

Volume III

Contributors to This Volume

ELIZABETH J. ARTHUR
ALAN F. BIRD
JEAN BIRD
M. B. CHITWOOD
J. DEL CASTILLO
PHILIP DOEZEMA
U. H. M. FAGERLUND
RAGNAR FÄNGE
DONALD FAIRBAIRN
SHIRLEY E. FREEMAN
W. P. FREEMAN
JOHN CARRUTHERS FERGUSON
T. W. GOODWIN
TRYGGVE GUSTAFSON
W. L. NICHOLAS
JOSÉ OLIVER-GONZÁLEZ
IVAN PRATT
W. P. ROGERS
MORTON ROTHSTEIN
RICHARD C. SANBORN
HOWARD J. SAZ
MALCOLM H. SMITH
R. I. SOMMERVILLE
GEORGES UBAGHS
H. R. WALLACE

CHEMICAL ZOOLOGY

Edited by MARCEL FLORKIN
LABORATOIRES DE BIOCHIMIE
INSTITUT LÉON FREDERICQ
LIÈGE, BELGIUM

and

BRADLEY T. SCHEER
DEPARTMENT OF BIOLOGY
UNIVERSITY OF OREGON
EUGENE, OREGON

Volume III

ECHINODERMATA, NEMATODA, AND ACANTHOCEPHALA

ACADEMIC PRESS New York and London 1969

102316
SHEPHERD COLLEGE LIBRARY

COPYRIGHT © 1969, BY ACADEMIC PRESS, INC.
ALL RIGHTS RESERVED.
NO PART OF THIS BOOK MAY BE REPRODUCED IN ANY FORM,
BY PHOTOSTAT, MICROFILM, OR ANY OTHER MEANS, WITHOUT
WRITTEN PERMISSION FROM THE PUBLISHERS.

ACADEMIC PRESS, INC.
111 Fifth Avenue, New York, New York 10003

United Kingdom Edition published by
ACADEMIC PRESS, INC. (LONDON) LTD.
Berkeley Square House, London W.1

LIBRARY OF CONGRESS CATALOG CARD NUMBER: 67-23158

PRINTED IN THE UNITED STATES OF AMERICA

List of Contributors

Numbers in parentheses indicate the pages on which the authors' contributions begin.

ELIZABETH J. ARTHUR* (429), Department of Biological Sciences, Purdue University, Lafayette, Indiana

ALAN F. BIRD (253, 561), C.S.I.R.O., Horticultural Research Station, Glen Osmond, South Australia

JEAN BIRD (253), Mitcham, South Australia

M. B. CHITWOOD** (223), Beltsville Parasitological Laboratory, Animal Disease and Parasite Research Division, Agriculture Research Service, Beltsville, Maryland

J. DEL CASTILLO (521), Department of Pharmacology, School of Medicine, San Juan, Puerto Rico

PHILIP DOEZEMA† (101), Hopkins Marine Station of Stanford University, Pacific Grove, California

U. H. FAGERLUND (123), Fisheries Research Board of Canada, Vancouver Laboratory, Vancouver, British Columbia, Canada

RAGNAR FÄNGE (207, 593), Department of Zoophysiology, University of Gothenburg, Gothenburg, Sweden

DONALD FAIRBAIRN (361), Department of Zoology, University of Massachusetts, Amherst, Massachusetts

SHIRLEY E. FREEMAN (47), Defense Standards Laboratories, Maribyrnong, Victoria, Australia

* Present address: Department of Biology, University of Pittsburgh, Pittsburgh, Pennsylvania.

** Present address: Pindell School Road, Fulton, Maryland.

† Present address: Department of Zoology, University of Washington, Seattle, Washington.

W. P. FREEMAN (47), Department of Experimental Neurology, University of Melbourne, Melbourne, Australia

JOHN CARRUTHERS FERGUSON (71), Florida Presbyterian College, St. Petersburg, Florida

T. W. GOODWIN (135), Department of Biochemistry, University of Liverpool, Liverpool, England

TRYGGVE GUSTAFSON (149), Department of Developmental Physiology, The Wenner-Gren Institute for Experimental Biology, University of Stockholm, Stockholm, Sweden

W. L. NICHOLAS* (289), Department of Zoology, University of Illinois, Urbana, Illinois

JOSÉ OLIVER-GONZÁLEZ (555), Department of Medical Zoology, School of Medicine, San Juan, Puerto Rico

IVAN PRATT (245), Oregon State University, Corvallis, Oregon

W. P. ROGERS (379, 465), Waite Agricultural Research Institute, University of Adelaide, Glen Osmond, South Australia

MORTON ROTHSTEIN (289), Biology Department, State University of New York at Buffalo, Buffalo, New York

RICHARD C. SANBORN (429), Department of Biological Sciences, Purdue University, Purdue Indianapolis Regional Campus, Indianapolis, Indiana

HOWARD J. SAZ (329), Department of Pathobiology, The Johns Hopkins University School of Hygiene and Public Health, Baltimore, Maryland

MALCOLM H. SMITH (501), Department of Biochemistry, University of Sheffield, Sheffield, England

R. I. SOMMERVILLE (465), Department of Zoology, University of Adelaide, Adelaide, South Australia

GEORGES UBAGHS (3), Department of Paleontology, University of Liége, Liége, Belgium

H. R. WALLACE (561), C.S.I.R.O., Division of Horticultural Research, Glen Osmond, South Australia

* Present address: Department of Zoology, The Australian National University, Canberra, A.C.T. Australia.

QP
514
F5.28
v.3

Preface

Zoology is currently undergoing a period of transition in which chemical knowledge is progressively integrated with the more classic knowledge of morphology and systematics. Biochemical studies of species, as well as of higher taxa, open new disciplines to the zoologist and offer new viewpoints in considering problems of structure, function, development, evolution, and ecology. The biochemist has considerable opportunities for broadening his sphere of investigation because of the enormous selection of animal species available for study from which a great variety of compounds can be obtained and reactions observed. There are abundant prospects for fruitful collaboration between the biochemist and zoologist in studies in which the characteristics of the animal and the biochemical constituents and processes interact in significant ways.

Very often the initial obstacle in undertaking investigations in new fields is the complexity and scattered character of the literature. This treatise is aimed primarily at making it possible for zoologists and chemists, who have a limited knowledge of the literature in fields other than their own, to gain a valid impression of the present state of knowledge in chemistry and zoology and an introduction to the existing literature. Thus, we have invited research workers who have contributed significantly to problems involving combined chemical and zoological approaches to summarize the knowledge in their specific disciplines of interest and competence. The authors have been encouraged to be critical and synthetic and to include mention of gaps in knowledge as well as the established information.

The treatise is arranged by phyla, an arrangement which seemed most suitable for presenting chemical information of zoological significance and for bringing to the attention of chemists those aspects of biochemical diversity of greatest potential interest. Each section, dealing with a major phylum, is introduced by a discussion of the biology and systematics of the group. This is followed by chapters dealing with various aspects of the biochemistry of the group. In general, the authors of individual chapters have been given full freedom, within the limitations of space, to develop their assigned topic. We thought that in this way the reader would have the advantage of the author's personal experience in and attitude toward his field, and that this would more than compensate for any unevenness in coverage that might result.

We are grateful to Professor K. M. Wilbur for his help in the early planning of this treatise, to the authors for their cooperation and patience, and to the staff of Academic Press for their careful work.

November, 1968

Marcel Florkin
Liège, Belgium

Bradley T. Scheer
Eugene, Oregon

Contents

Chapter 7. **Fertilization and Development**

TRYGGVE GUSTAFSON

Chapter 8. **Pharmacology of Echinoderms**

RAGNAR FÄNGE

Section II: NEMATODA AND ACANTHOCEPHALA

Chapter 1. **The Systematics and Biology of Some Parasitic Nematodes**

M. B. CHITWOOD

Chapter 2. The Biology of the Acanthocephala

Ivan Pratt

Chapter 3. Skeletal Structures and Integument of Acanthocephala and Nematoda

Alan F. Bird and Jean Bird

Chapter 4. Culture Methods and Nutrition of Nematodes and Acanthocephala

Morton Rothstein and W. L. Nicholas

Chapter 5. Carbohydrate and Energy Metabolism of Nematodes and Acanthocephala

Howard J. Saz

Chapter 6. **Lipid Components and Metabolism of Acanthocephala and Nematoda**

DONALD FAIRBAIRN

Chapter 7. **Nitrogenous Components and Their Metabolism: Acanthocephala and Nematoda**

W. P. ROGERS

Chapter 8. **Osmotic and Ionic Regulation in Nematodes**

ELIZABETH J. ARTHUR AND RICHARD C. SANBORN

Chapter 9. **Chemical Aspects of Growth and Development**

W. P. ROGERS AND R. I. SOMMERVILLE

Contents of Other Volumes

Volume I: PROTOZOA

Volume II

Section I: PORIFERA

Section II: COELENTERATA, CTENOPHORA

Section III: PLATYHELMINTHES, MESOZOA

Volume IV: ANNELIDA, ECHIURA, SIPUNCULA

CHEMICAL ZOOLOGY

Volume III

SECTION I

ECHINODERMATA

CHAPTER 1

General Characteristics of the Echinoderms

Georges Ubaghs[1]

I. Introduction

The echinoderms make up one of the principal branches of the animal kingdom and one of the most distinctive. They are represented in present-day nature by five classes; the Crinoidea (sea lilies, feather stars), the Holothuroidea (sea cucumbers), the Echinoidea (sea urchins, sand dollars, etc.), the Asteroidea (sea stars or starfish) and the Ophiuroidea (brittle stars); these last two classes are sometimes considered subclasses, in which case they are grouped into another class, the Stelleroidea. In past times, many other classes existed, which all disappeared before the end of the Paleozoic era.

The echinoderms are sharply distinguished from other groups by their organization. Nonetheless, they are not entirely isolated; their first larval stages show such a striking resemblance with the *Tornaria* larva of the Enteropneusta that these two groups are generally considered to be of common ancestry. However, the metamorphoses at the end larval stages of echinoderms and the distinctiveness of their final organization

[1] Translated by Mark Tepfer, University of Oregon, Eugene, Oregon.

prove that they have evolved greatly from any common ancestry with another branch. The existence, since the lowest Cambrian, of several distinct classes of well-developed echinoderms proves the extreme age of their origins.

Echinoderms are still varied and numerous today (6200 species were listed in 1964). They play an important role in marine ecology, they are found at all latitudes, on all types of sea-bottom, and at all depths, from the littoral zone to the deeps. Except for some rare pelagic species, all belong to the benthos, either motile or sessile. They consume great quantities of detritus and plankton, some of them being active predators, notably capable of controlling the production of certain mollusks. The larva as well as the adults make up a considerable food supply for other animals; some species are edible. The study of their distribution has proved very useful to zoogeographers, and they are among the most characteristic animals of various animal communities.

The eggs of echinoderms have played an important role in the history of cytology and biology in general. They have been useful in the study of problems in fertilization, germinal localization, polyspermy, parthenogenesis, etc., as well as endless research in experimental embryology. Finally, let us remember that the starfish, brittle stars, and holothurians are excellent material for experiments in regeneration and the physiology of the nervous system.

Excellent general outlines have been published by Cuénot (1948), Hyman (1955), A. M. Clark (1962), Nichols (1962), and Kaestner (1963). In this introductory chapter we will limit ourselves to the most important aspects of the morphology, embryology, phylogeny, and systematics of these organisms.

II. Distinguishing Characteristics

The echinoderms are deuterostomous marine invertebrates, in which the coelom is typically enterocoelous and whose bilaterally symmetrical swimming larva are ultimately transformed through their metamorphoses into organisms with five rays. The bilateral symmetry of the larva is always disrupted by an asymmetry that is shown throughout the ontogenesis of the larva by the atrophy of certain structures on the right side of the body and by the compensatory development of organs on the left side. Three characteristics are particularly typical of this phylum: (1) the middle coelom, called the hydrocoel, develops into a water-vascular system, that is to say into a system of sacs and canals that produce a series of hollow tentacles on the body surface; (2) the mesenchyme secretes an internal skeleton (endoskeleton), usually highly developed, composed of distinct plates or ossicles and made up of a network of calcareous crystals and an organic matrix; (3) the organization is more or less completely influenced by radial symmetry, usually pentamerous, which is most apparent in the skeleton, the nervous system, the circulatory system, and the system of canals and sinuses that develop from the middle and posterior coeloms.

Being organisms with three cell layers (triploblastic) and being provided with a vast general cavity, the echinoderms are clearly above the coelenterates on the zoological scale. The first naturalists, misled by the radial symmetry of both phyla, grouped them in one phylum, the Radiata or Zoophyta. Nonetheless, in many ways the echinoderms show a low level of organization. They have neither head nor brain. Their nervous system is integrated enough to allow rather complex movements, but it is rather primitive and their sensory organs are poorly developed. Even though their lacunar or hemal system, in particular the axial organ, seems locally capable of pulsating or contracting, there is no true heart and the vessels that constitute this system are simple cavities, without true walls. Only rarely does respiration take place in specialized organs; it is usually accomplished by simple exchanges across thin walls of various systems of organs. The echinoderms have no excretory organs; ameboid cells excrete the insoluble wastes, while soluble wastes pass directly through the walls into the surrounding medium. Finally, reproductive products are emitted directly into the water.

The body of the echinoderms, except that of some fossil classes, is typically divided in sectors by five zones or rays, called *ambulacra,* which radiate around the mouth and which are distinguished by the series of hollow tentacles, the *podia* (sing. *podion*) or "tube feet" that are aligned along either side of the axes. These characteristic tentacles are part of the water-vascular system which, we have already said, develops from the middle coelom or hydrocoel of the larva. During metamorphosis, the hydrocoel, after having encircled the esophagus, sends out five tubular evaginations which develop tentacles as they grow. These evaginations are the first buds of the rays. Parallel to the rays, and possibly under their inductive influence, the skeleton, the nervous system, the lacunar (circulatory) system and the entire sinus of coelomic origin organize in the same manner. This is in fact one of the most remarkable of the distinctive traits of the echinoderms: the unity of plan exhibited by organs of such diverse origins and functions. According to the arrangement of these evaginations and the formations that accompany them, two fundamental modes of growth can be distinguished, and to a certain extent these growth patterns condition the general form of the body. These patterns are: the radial pattern, which is characteris-

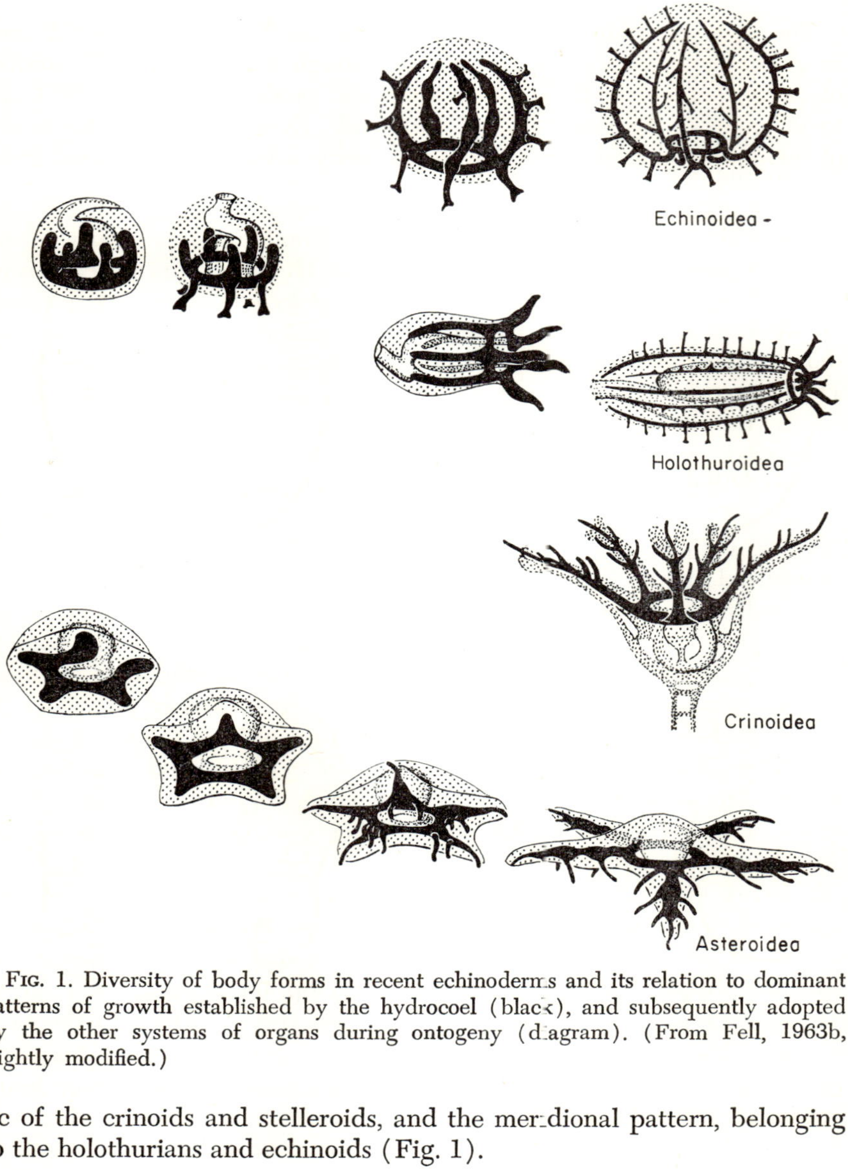

FIG. 1. Diversity of body forms in recent echinoderms and its relation to dominant patterns of growth established by the hydrocoel (black), and subsequently adopted by the other systems of organs during ontogeny (diagram). (From Fell, 1963b, slightly modified.)

tic of the crinoids and stelleroids, and the meridional pattern, belonging to the holothurians and echinoids (Fig. 1).

III. Ontogenesis

To better understand the organization of the echinoderms and their place in the animal kingdom, it is useful to consider the principal stages of their ontogenesis.

The eggs of echinoderms are generally small (0.75 mm. to 0.2 mm. in diameter) and poor in metaplasm, though a relatively high number of species (especially the incubative species) produce larger ones (0.3 mm. to 0.5 mm. in diameter), filled with yolk. Even before the beginning of their development they can show polarity, asymmetry, germinal local-

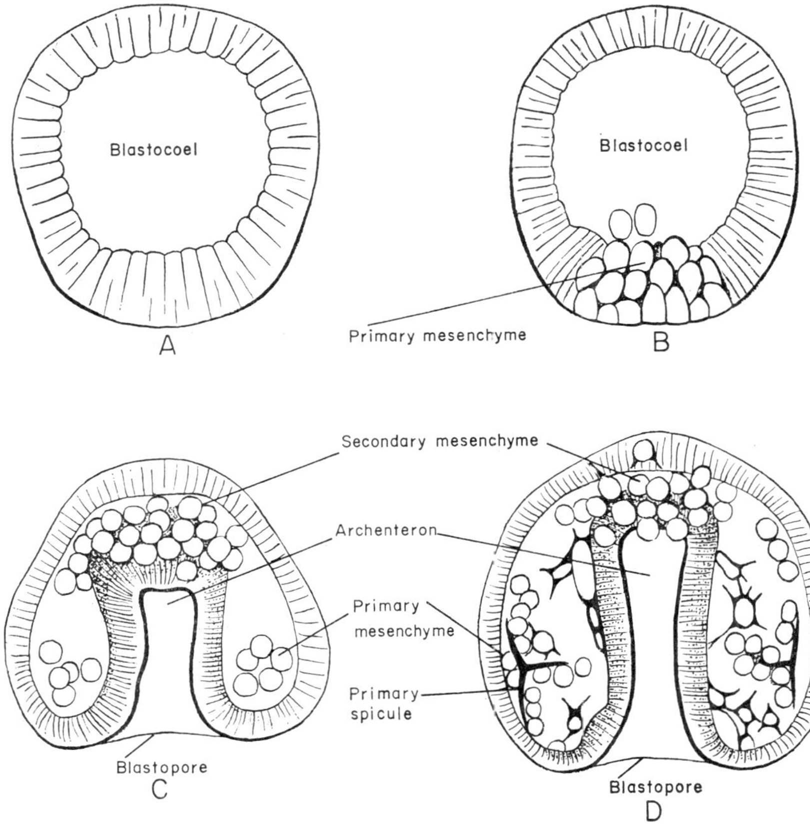

FIG. 2. Optical sections through the coeloblastula (A) and three stages (B–D) in the formation of the gastrula and of the mesenchyme of *Paracentrotus lividus* (echinoid). (After Boveri, 1901.)

izations, and gradients of morphogenetic potentiality that decrease from the poles.

In principal, the cleavage is complete, of a radial nature, and finishes with the formation of a coeloblastula covered with long flagella (Fig. 2A). Cells can begin to emigrate from the vegetal pole of the blastocoel to begin the formation of a true mesenchyme (Fig. 2B). The coeloblastula rapidly transforms itself by invagination into a gastrula whose tubular archenteron opens to the exterior by the blastopore (Fig. 2C–D). The cells that will form the mesoderm detach themselves from the apical region of the archenteron. Generally, at this stage or at the blastula stage, the embryo abandons its ovular coverings and becomes a larva.

The eggs that are small and poor in metaplasm develop into swimming larvae, with a mouth and more or less developed projections (larval arms); these larvae nourish themselves on plankton (planktotrophic larvae). The larva is transformed into the final organism through metamorphoses; therefore the development is indirect. On the contrary, the eggs that are rich in yolk produce modified larvae, less active or even completely lacking a free phase, whose development is more or less condensed and who are nourished only from their own reserves (lecithotrophic larvae); the existence of a species has been reported whose larval stage has been completely suppressed and whose development is therefore direct. Since generally the lecithotrophic larvae have vestiges of the organization of planktotrophic larvae, direct development is generally considered a secondary modification of indirect development.

Even though incontestable convergence phenomena have been reported between the larvae of different classes, and certain classes have larvae of several types, in general each class has its particular type of larva, designated by its own name: the barrel-shaped larva or vitellaria of the crinoids, auricularia of the holothurians, bipinnaria of the asteroids, ophiopluteus of the ophiuroids and echinopluteus of the echinoids. Certain of these larvae have skeletons which are reabsorbed later.

All of these larvae, except the vitellaria, show a group of common characteristics that define what is called the dipleurula larva of the echinoderms (Fig. 3). They are bilaterally symmetrical and show on one of their sides a saddle-shaped depression which is bordered by a more or less contoured ciliate ectoderm band, extending to the ends of the arms. The digestive tract, which comes directly from the archenteron, and whose origin is therefore endodermal, opens on the ventral side with a newly formed mouth, and an anus which is the old blastopore or which opens where the blastopore used to be. The digestive tract is arch-shaped and divides into an esophagus, a gastric expansion and an intestine. The mouth is at the bottom of the saddle-shaped depression mentioned above and therefore is in a ventral position, but the anus, also ventral, stays outside this depression. The part of the body of the larva which is anterior to the mouth, that is to say the preoral lobe, encloses the nervous elements and sometimes a tuft of sensory hairs. On either side of the digestive tract are concentrated the coelomic buds, of which the most anterior on the left side opens on the exterior by a microscopic dorsal pore, the hydropore. There is no nephridial organ.

The characteristics of the dipleurula larva are very important phylogenetically. The bilateral symmetry indicates first of all that the echinoderms are fundamentally bilateralia. Second, the transformation of the blastopore into an anus proves that they are deuterostomous. Finally, the startling resemblance of this larva with the *Tornaria* larva of the Enteropneusta can only be explained by a common ancestry for the

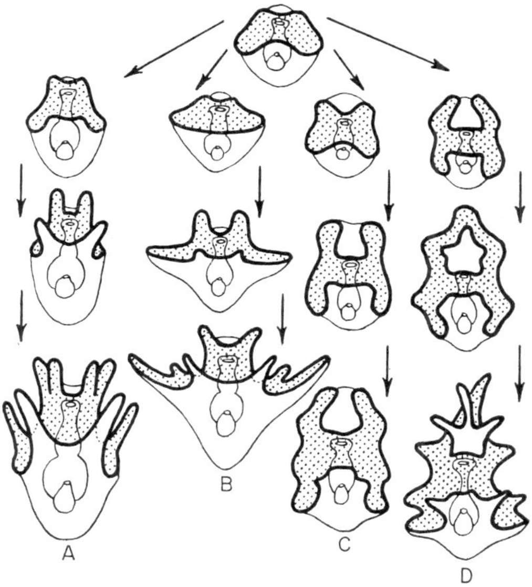

FIG.. 3. Development of echinoderm larvae from Dipleurula (diagram). (A) *Echinopluteus* (echinoid); (B) *Ophiopluteus* (ophiuroid); (C) *Auricularia* (holothuroid); (D) *Bipinnaria* (asteroid). (From Dawydoff, 1948, p. 299, after Müller.)

echinoderms and the latter.[2] This resemblance includes the general appearance, the structural organization, the form of the digestive tract, the lack of excretory organs, and, as we shall see, the formation and development of the coelom.

The coelom is generally derived from the top of the archenteron, from which it is separated in the form of one or sometimes two closed sacs (Fig. 4). When, as is ordinarily the case, only one sac forms, it divides into two vesicles symmetrically placed on either side of the archenteron. Then the two vesicles lengthen and divide into three suc-

[2] Nonetheless see Fell (1948), who contests the phylogenetic significance of the resemblances between the Tornaria and Dipleurula larvae.

cessive coelomic compartments: an anterior compartment (axocoel), a middle compartment (hydrocoel), and a posterior compartment (somatocoel). The enterocoelous origin[3] of the coelom and its division into three compartments confirm the systematic position of the echino-

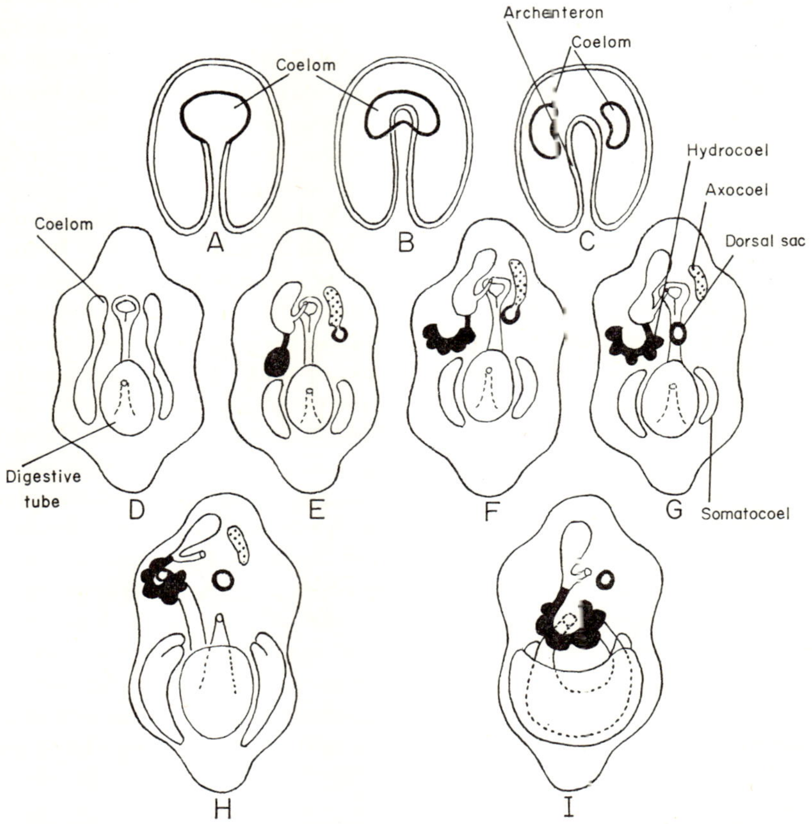

FIG. 4. Formation and first evolution of coeloms during gastrulation (A,B,C), larval stage (D–G) and metamorphoses (H,I) of echinoderms (diagram). Black: left hydrocoel (water-vascular system); stippled: right axocoel. (From Dawydoff, 1948, p. 287.)

derms in the deuterostomes; like the hemichordates and the pogonophores, they are organisms with three pairs of coelomic pouches.

[3] The fact that in certain echinoderms with direct development the coelom is formed by schizocoely in no way decreases the phylogenetic importance of the usual way in which this group of cavities develops.

From this point the originality of the organization of the echinoderms begins to appear. As a matter of fact, the division of the coelomic compartments is rarely accomplished in the same way on the left as on the right. Normally the division of the right antimere is incomplete: the right hydrocoel does not become individualized, or is only represented by a transitory bud, and the right axocoel, lacking in the holothurians and crinoids, is always less developed than the left. Moreover, the left axocoel and the left hydrocoel stay in direct communication or, as in the crinoids, are connected later by a canal. Also, the left axocoel (but not the right) opens to the exterior by a pore, placed on the left of the dorsal face. So, in the arrangement and development of the larval coeloms we discover one of the most basic characteristics of the echinoderms: their asymmetry, which manifests itself from the origin of the coelomic arrangement or is sometimes even apparent in the egg. Asymmetry influences all subsequent ontogenetic processes and is therefore of primordial importance. Nonetheless, this is only a secondary characteristic, since there are larvae with reversed symmetry or which show equally perfect divisions on the right and left sides, which proves that the morphogenetic potentialities are fundamentally bilateral.

Whether its larva becomes fixed (crinoids, certain asteroids) or stays free (other echinoderms), the echinoderm undergoes metamorphoses through which it becomes more and more asymmetrical and finally acquires an entirely new radial symmetry.[4] So one observes: (1) a very marked tendency, from the beginning of these transformations, to concentrate the formation of the imago on the left side of the larva, which radically modifies the plan of bilateral symmetry; (2) the reduction or disappearance of the anterior coelomic sacs of the right antimere; (3) the preponderance of the growth of the larva on the left side as compared to the right; (4) the crescent and finally the ring shape taken by the left hydrocoelom and the budding from this ring of five primary tentacles which are, as we have already seen, the buds of the five rays of the adult organism.

These metamorphoses are unique to the echinoderms. They show that the echinoderms have followed an evolutionary path uniquely their own and explain the importance of the gap that separates the echinoderms from other groups, between which there are no intermediaries.

Having retraced the principal stages of development, let us now pass on to an examination of the principal characteristics of the final organism.

[4] Remember that in species that develop directly the embryo changes directly into an adult and that lecithotrophic larvae generally go through abridged metamorphoses. But in any case the final result is the same.

IV. Symmetry

We have just seen that the fundamental symmetry of the echinoderms is bilateral, but that it is altered by a very precocious asymmetry on which is grafted the radial symmetry of the adult. The appearance of asymmetry certainly represents a primary phenomenon of the ontogenesis of the echinoderms. Asymmetry is apparent during the entire embryonic course, and the metamorphoses stabilize it. No doubt it bears a great phylogenetic significance, since even though fossil echinoderms

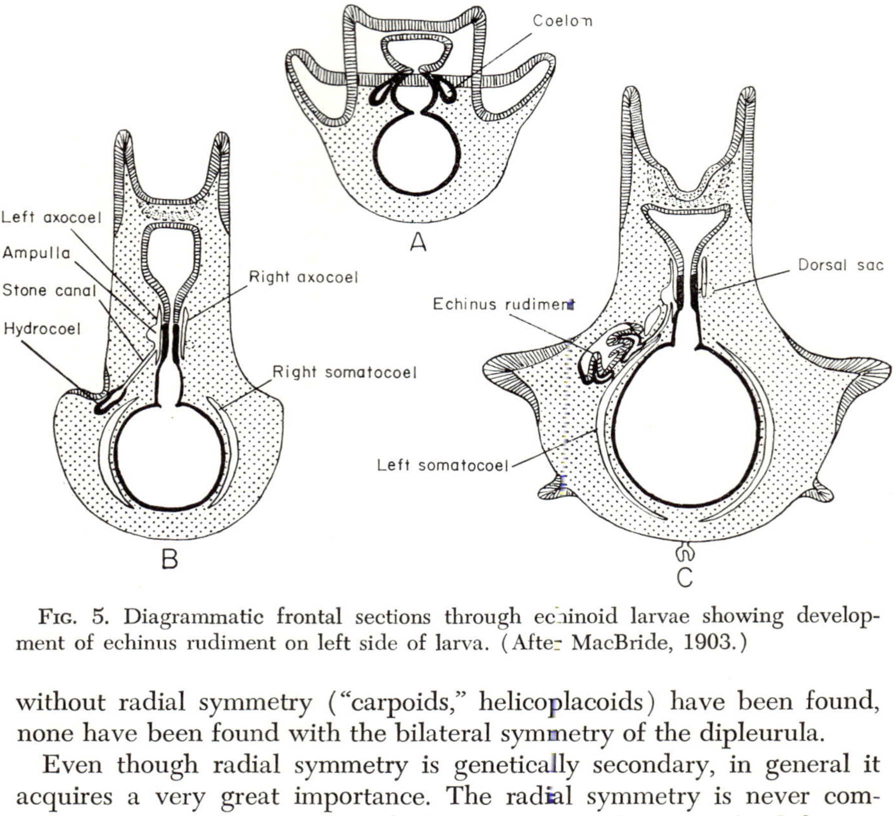

FIG. 5. Diagrammatic frontal sections through echinoid larvae showing development of echinus rudiment on left side of larva. (After MacBride, 1903.)

without radial symmetry ("carpoids," helicoplacoids) have been found, none have been found with the bilateral symmetry of the dipleurula.

Even though radial symmetry is genetically secondary, in general it acquires a very great importance. The radial symmetry is never complete; some structures, such as the hydropore, which permit the definition of certain systems of bilateral symmetry escape it. But these arrangements have no connection with the original scheme of bilaterality. In fact, the principal axis of the imago and the longitudinal axis of the larva never coincide; they can even be nearly perpendicular (asteroids, echinoids) (Fig. 5). In certain forms, the spatangoid sea urchins for example, the radial symmetry is in turn dominated by bilateral symmetry; but this is an adaptive phenomenon, a secondary acquisition, without deep morphological significance.

V. Skeleton

By its histological makeup and its crystallographic properties, the skeleton of the echinoderms makes up one of the most important diagnostic traits of the branch. It is always internal and of mesodermal origin. It is therefore an endoskeleton. In some cases, for example the

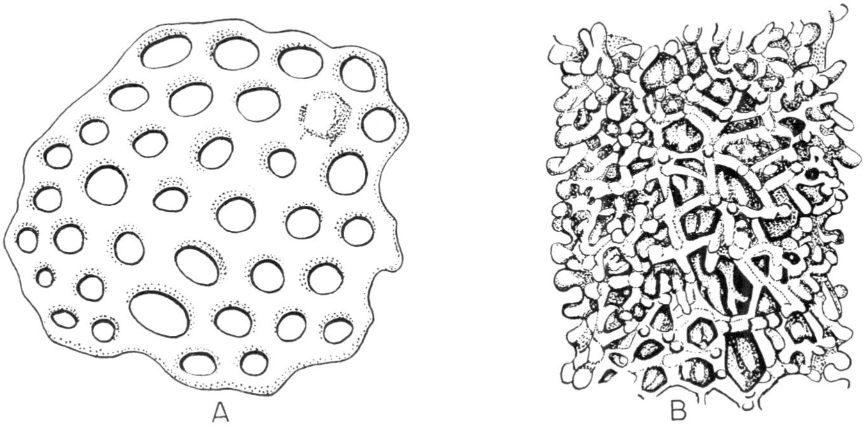

FIG. 6. Microstructure of echinoderm endoskeleton. A. Two-dimensional framework, holothurian ossicle (*Echinocucumis hispida* (Barrett)) ×100. (After Ostergren, 1938.) B. Three-dimensional framework, part of radial piece of the calcareous ring of the holothurian *Caudina chilensis* (J. Müller) ×400. (After Hozawa, 1928.)

spines of sea urchins, it is only covered by a thin epidermal layer, which can even disappear locally by abrasion.

The skeleton is made up of many distinct elements: plates, various ossicles, spicules, spines, microscopic pincers (pedicellaria). These elements may be isolated in the dermis or on the contrary may be contiguous and form varied structures, supple or rigid, extremely rich in functional specialization.

The skeleton of the echinoderms (except in rare cases) is always made up of a calcareous network, whose pores can be more than 50% of the total volume. In the living organism these pores are filled with connective tissue. In its most simple form (larval skeletons, many spicules of holothurians), this network is two-dimensional (Fig. 6A), but in most units it is three-dimensional (Fig. 6B). Generally, all skeletal parts start with a calcareous granule which changes rapidly into a three-cleft spicule, covered by a thin protoplasmic mantle. The spicule grows, bifurcates at its extremities, and its ramifications fuse at their points of contact. In this manner is constructed the reticulated structure in whose interstices are lodged the calcigenous cells which are joined into a syncytium. In the initial stages of calcification the crystallographic properties control the form of the spicules, but in later stages the morphology of skeletal units is determined by the configuration of the organic matter that surrounds the calcite.

The mineral part of the skeleton is composed of a magnesium calcite (71–95% calcium carbonate and from 3–15% magnesium carbonate), with very small quantities of calcium sulfate, calcium phosphate, aluminum salts, iron salts and various trace elements. The magnesium content is controlled by at least three factors: the temperature of the seawater, the position in the skeleton, and the inherited characteristics of the individual (Raup, 1966).

The examination under polarized light of a thin section taken from any skeletal unit of an echinoderm reveals that it always behaves (with a very few exceptions) like a single crystal. However, X-rays seem to show that in reality skeletal parts are an aggregate of elementary crystals whose optical axes are almost perfectly aligned (Garrido and Blanco, 1947; Nissen, 1963).

In relationship to morphology, crystal units show a certain orientation. For example, in the corona of the echinoids, the optical axes of the plates are generally either perpendicular or nearly tangent to the surface of the plates. However, homologous parts sometimes show different orientations depending on the species under consideration, a diversity from which Raup (1962a,b) has drawn useful taxonomic and phylogenetic criteria.

VI. Nervous System

The rather primitive nervous system of the echinoderms consists of three localized networks at different levels of the body and all three are affected by radial symmetry. Two of these three networks—the ectoneural and the hyponeural—are found under the oral face and their distribution is more or less like that of the middle coelom, that is to say the water-vascular system. The third or entoneural network occupies an aboral position and follows—except in the crinoids—the sinuses of the posterior coelom.

The ectoneural network, essentially sensory and in close relation with the subepithelial plexus that covers nearly the whole body, is made up of a ring or pentagon around the esophagus and five radial nerves that run the length of the ambulacra. In the crinoids and asteroids, the radial nerves are situated immediately under the epidermis that carpets the bottom of the ambulacral grooves. In the other classes, which have no ambulacral grooves, the nerves are found under the epineural canals that probably represent invaginated grooves (Fig. 7). The ectoneural network, in all extant echinoderms except the crinoids, is the most developed and the most important from the functional point of view.

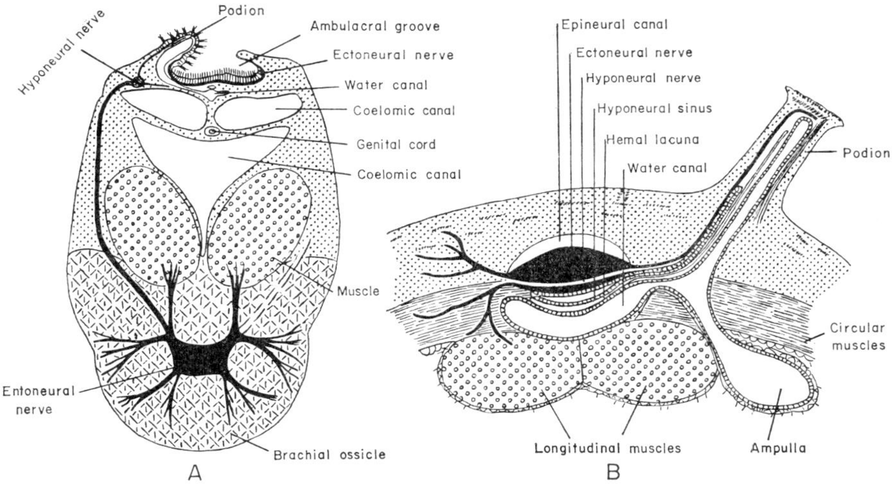

FIG. 7. Diagrammatic sections through ambulacra of a crinoid (A) and a holothuroid (B).

The hyponeural network is found in the adoral wall of the sinuses coming from the posterior coelom and like the latter is made up of a periesophageal ring and radial nerves. It is generally separated from the ectoneural system only by a thin connective layer. It innervates the muscles of the body wall and, in the crinoids, various internal structures. It is lacking in the sea urchins, except in those that have a chewing apparatus whose numerous muscles it controls.

The entoneural sensory-motor network seems to exist in all echinoderms except the holothurians and asteroids. It is found in the walls of the posterior coelom. In the sea urchins and ophiuroids, it is in the aboral sinus ring and passes from there to the gonads. It makes up the principal network of the crinoids, which helps to distinguish the crinoids from the other echinoderms. In the crinoids it forms a sort of ganglionic cup, the chambered organ, enclosing a coelomic cavity derived from the posterior coelom and divided into five chambers by connective tissue membranes; this mass, which coordinates swimming and walking movements, presents the characteristics and exercises the functions of a nervous center. From its sheath powerful nerves go to the arms and stem (or the fixative organs if the stem is lacking in the adult stage).

The sensory organs are poorly developed. Other than a general sensitivity to touch, light, odors, and water quality, the echinoderms may possess tactile organs, light receptors, photogenic organs, and organs of balance.

VII. Digestive Tube

The digestive tube is essentially derived from the digestive tube of the larva. Straight and affected by radial symmetry in the stelleroids, in the other echinoderms it consists of a more or less elongated tube, twisted into a spiral in the crinoids, the holothurians and young echinoids; this character is partially masked in the adult echinoids, because generally their tube goes through a considerable elongation. Its principal divisions—esophagus, stomach, intestine and rectum—are usually only distinguished by slight histological differentiation. Various appendices are attached, of which the most remarkable are the respiratory trees of the holothurians—ramified tubes, ending in thin-walled vesicles, and opening on the terminal cloaca of the intestine. They fill with water pumped by the anus and contract alternatively several times a minute.

The mouth generally occupies the center of the oral face; toward it converge the ambulacra. The anus—which is lacking in the ophiuroids and certain asteroids—is in more varied positions: placed on the adoral face of the crinoids, it is ordinarily lateral (but never apical) in the fossil forms which lived attached to the sea floor. It is aboral in the holothurians, the regular echinoids, and most of the asteroids; it migrates toward one of the interrays and can even become adoral in the irregular sea urchins.

VIII. Coelom

In all the echinoderms the characters of the coelom are essentially alike. Apart from the general cavity, the coelom contains a whole group of cavities and sinuses carpeted with an endothelium of mesodermal origins. The coelom, which is typically formed—as we have already seen—by enterocoely, divides during its development into three pairs of compartments (axocoel, hydrocoel, somatocoel), each element of which has its own destiny.

The left axocoel is not affected by radial symmetry. Remember that in the beginning it is connected to the left hydrocoel (except in the crinoids, where the communication is established later) by a canal, which will become the stone canal of the adult. On the other hand, when hardly formed, it puts forth a tubular diverticulum, the hydroporic canal, which opens on the exterior through the hydropore. Later the stone canal and the hydroporic canal are placed end to end, so that by this positioning, only a very small axocoelous bulb remains, situated on the path of the definitive stone canal (Fig. 5B).

But that is not the end of the role of the left axocoel. It still furnishes an important cavity, the axial sinus, which extends more or less parallel to the axis of the body in the interray that contains the hydropore. This sinus doesn't exist in the echinoids and is said to disappear in the holothurians during ontogenesis. This sinus encloses the axial organ (which we shall describe later) and, in the stelleroids, the already mentioned stone canal with which it is connected at its aboral extremity. In the asteroids, at its adoral extremity, it opens on the internal part of the circumoral ringlike sinus which is just above the periesophageal nervous ring and which is divided into an inner ring (arising, as we have just seen, from the axial sinus) and an external ring (hyponeural ring, arising from the left somatocoel as described later) by an oblique partition (Fig. 8).

Of the right axocoel, when it exists (the crinoids and holothurians have none), only a small cavity remains, the dorsal sac or madreporic vesicle, which in the asteroids and echinoids caps the aboral extremity of the axial organ. Thought to be closed, it communicates nonetheless with the general cavity by a cleft, at least in the sea urchins (Boolootian and Campbell, 1964). In the ophiuroids, the right axocoel forms the right portion of the axial sinus, divided in two by the axial organ itself, and in the echinoids (having no axial sinus) it seems to constitute, apart from the dorsal sac that we have just seen, the lumen of the axial organ itself. The close physiological and morphological relationship between the dorsal sac and the axial organ—central part of the circulatory system—make it generally considered homologous to the heart vesicle of the Enteropneusta and Pterobranchia.

While the right hydrocoel, if it appears, stays rudimentary and is quickly obliterated, the left hydrocoel plays a considerable role in ontogenesis since it furnishes the whole of the water-vascular system except the hydropore and the hydroporic canal (Fig. 8). The central part of this system is the oral water ring, which is connected with the left axocoelous bulb by a canal, the stone canal. The latter, except in the crinoids and most holothurians, opens to the exterior through a ciliated funnel (hydropore) or through multiple pores in a special skeletal plate,

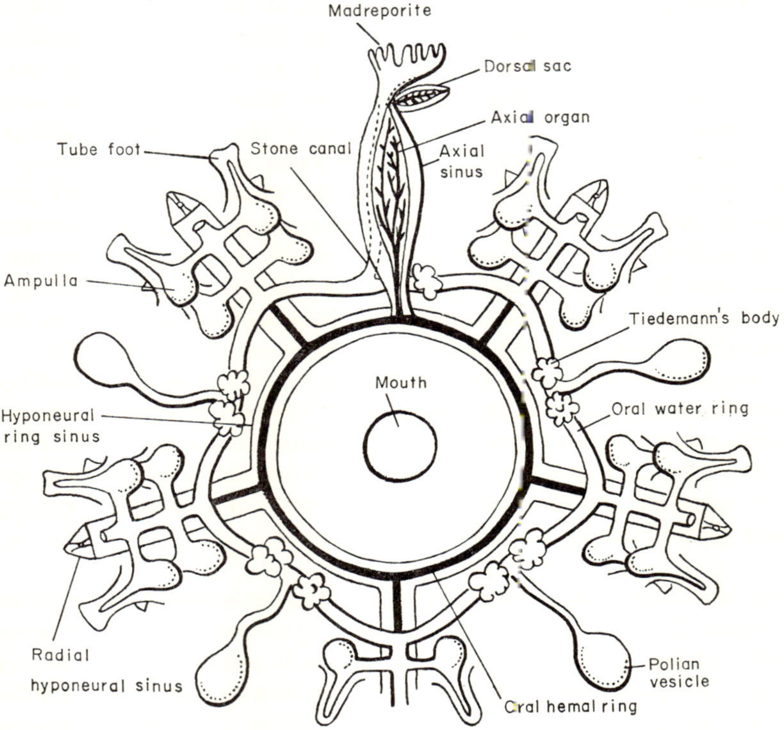

FIG. 8. Organization of water-vascular, hemal and hyponeural sinus systems in asteroid (diagram). (From Ubaghs, 1968a.)

the madreporite, on whose internal face is generally found a collecting sac, attached to the dorsal sac.

From the oral ring a long radial water canal extends into each ray; the canal is terminated by a caecum or small tentacle. Along its entire length, this canal puts forth short branches on both sides that end in hollow evaginations of the body wall, the podia or tube feet, which altogether make up the ambulacrum. These tube feet, in spite of a histological structure that is fundamentally the same in the whole group, show a remarkable polymorphism related to the diversity of their functions. Other than helping respiration, they are one of the exit routes of waste-carrying phagocytes. They can also capture food, provide locomotion, dig holes, receive various sensations, etc. When tube feet are used to move the animal, as is the case in numerous asteroids, echinoids, and holothurians, they end in suction cups which allow the animal to adhere to the substratum or the objects on which it is moving. In these three classes, they are each provided with an internal bulb (ampulla) which by its contractions controls the pressure of the coelomic fluid in the corresponding foot. When there is no bulb, as in the crinoids and ophiuroids, other mechanisms (division of the radial canal by muscular fibers or sphincters, differentiation of the muscles of the podia into two antagonistic systems) fill the same necessary function. The podia are therefore hydraulic organs and the whole water-vascular system seems to have no role other than to provide for their functioning.

The oral ring can put forth directly special podia that surround the mouth and have a nourishing or sensory function; examples are the various types of oral tentacles of the holothurians, the oral or labial podia of the crinoids and the buccal podia of the ophiuroids.

In the interrays the oral ring also carries (except in the crinoids and echinoids) one or several large pediculate, contractile vesicles, the polian vesicles, which seem to serve as reservoirs for the internal fluid. Finally, the oral ring also receives the tubules from the Tiedemann's bodies of the asteroids and the cavities of the spongy organs of the echinoids, organs whose function is not known for certain.

The two somatocoels, on either side of the archenteron in the embryo, are displaced during development in such a manner that the left becomes parallel to the hydrocoelic ring to which it adheres, while the right occupies an aboral position; a principal mesentery, oriented perpendicularly to the axis of the imago, separates them (Fig. 4I). Both curve in as they grow and their horns tend to come together until they form, on the medial-dorsal line, in the interray that contains the hydropore, two superimposed additional mesenteries perpendicular to the principal mesentery. The two additional mesenteries play a much more important role in ontogenesis than the principal mesentery: they become the seat of the axial complex, which is made up of the axial organ, the dorsal sac, the madreporite, the stone canal, and the genital rachis.

The two somatocoels produce the general cavity of the adult, a vast free space, except in the crinoids where it is invaded by connective membranes and fibers. This cavity penetrates deeply into the rays of the asteroids, slightly in the arms of the ophiuroids, and by four canals in the arms of the crinoids where it also occupies the axial conduit of the stem. Certain portions can be more or less isolated by membranes, forming distinct cavities, such as the peripharyngial cavity of the sea urchins which covers the jaw apparatus or the perianal cavity of the holothurians.

In addition, the larval somatocoel gives birth to a whole group of sinuses that form a distinct system, variously called hyponeural, perihemal, pseudohemal or perilacunar. We have already seen that in the imago the somatocoel is attached to the hydrocoelic ring. The wall of the somatocoel buds five diverticula, in the interray, which pass on the dorsal face of the ring, and each of them divides into two branches which meet the branches coming from the neighboring diverticula and

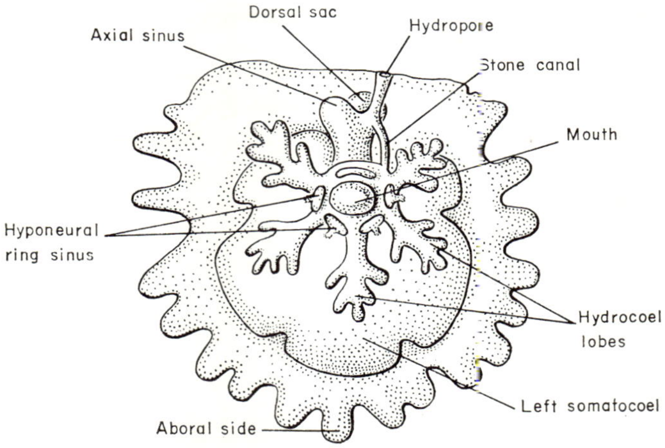

FIG. 9. *Astropecten aurantiacus,* immediately after metamorphosis, oral view, showing formation of the hyponeural ring sinus as pouches from the somatocoel. Diameter, about 0.5 mm. (After Horstadius, 1939.)

fuse with them to form a periesophageal ring (Fig. 9). (Remember that in the asteroids this ring constitutes the external part of the circumoral ring.) From this ring extend five radial sinuses whose small lateral branches are borne toward the podia and in the asteroids, toward marginal sinuses that follow the edges of the ambulacral sinuses (Fig. 10). In the adult this system is lodged between the ectoneural system and the ambulacral system. Its sinuses accompany or surround the circulatory system, so that they are often considered as a second circulatory system. In reality these systems have very different origins, since the one is coelomic and the other blastocoelic.

In the echinoids and stelleroids the left somatocoel also produces, this time from the aboral side, the genital coelom, which consists of a perianal ring and five interradial pouches containing the gonads (Fig.

10); these are connected with a peduncle to the genital rachis which is enclosed in the perianal ring. The rachis is transitory in the echinoids, but permanent in the stelleroids. Note also that the perianal ring, which loses all relation with the left somatocoel, comes secondarily into communication with the axial sinus. In the crinoids the definitive genital bud is closely connected with the left horn of the right somatocoel (and not the left as in the other echinoderms); the genital rachis which grows from there toward the bottom is enclosed in a coelomic tube

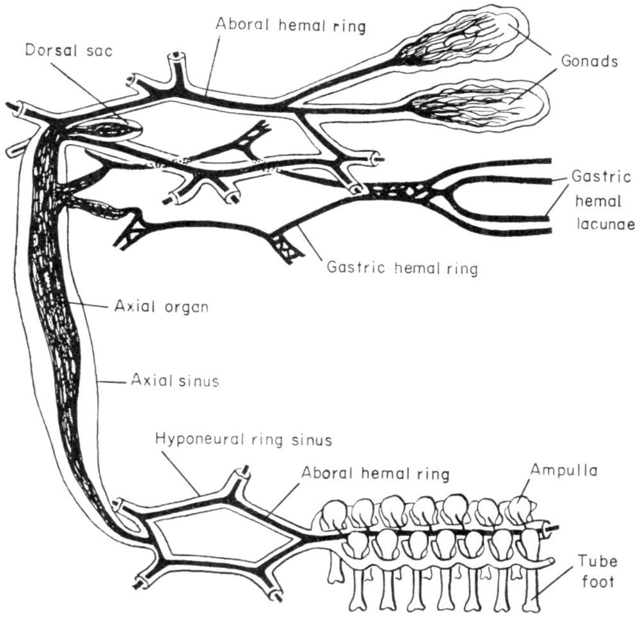

FIG. 10. Scheme of axial organ and of hemal and hyponeural systems in asteroid. (From Ubaghs, 1968a, modified from Cuénot.)

of the same origin. As for the holothurians, they have no aboral genital coelom. Their single gonad is born in the principal mesentery on the left side in the neighborhood of the hydroporic canal.

IX. Lacunar Hemal System

This system, whose cavities represent blastocoelic spaces, is composed of cavities and sinuses hollowed in the mesenchyme. These channels have no true walls and no endothelial sheathing. Closely associated with the other cavity systems, often even surrounded by coelomic sinuses, these channels are divided into absorbent lacunae spread in a network over a great part of the digestive tract, an adoral network and an aboral network. The adoral network follows all the canals of the ambulacral system and is found between the latter and the hyponeural sinus network, with which it is closely associated (Fig. 10). The hemal system accompanies the genital rachis to the interior of the genital coelom and penetrates into the gonads. The two networks are connected to each other and to the absorbent lacunae of the digestive tract by lacunae and numerous capillaries, which penetrate the axial organ.

X. Axial Organ

The axial organ (also called axial gland, brown gland, ovoid organ, dorsal organ, heart, etc.) exists in all the echinoderms except the holo-

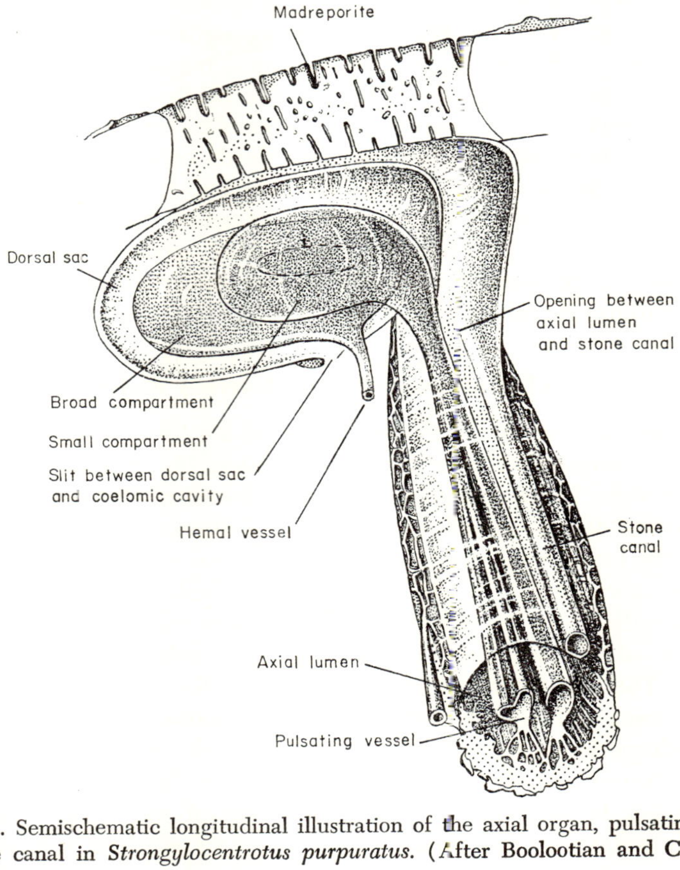

FIG. 11. Semischematic longitudinal illustration of the axial organ, pulsating vessel and stone canal in *Strongylocentrotus purpuratus*. (After Boolootian and Campbell, 1964.)

thurians, who are thought not to have any or to have only a rudimentary one. It is an elongated body, soft, generally brown or reddish, covered by the peritoneum and made of lacunar tissue. It is derived, not from the true walls of the neighboring coeloms, but from a concentration of connective formations in a longitudinal fold of the wall of the axial sinus (left axocoel). In the asteroids and echinoids its aboral portion (reduced to a finger-shaped process placed directly under the madreporite) is enclosed in a separate coelomic cavity, the dorsal sac, derived from the right axocoel. In the ophiuroids the right half is also enveloped by the right axocoel.

The function of the axial organ has been the root of many controversies. Recently, Boolootian and Campbell (1964) have shown that the axial organ of at least certain echinoids is provided with an internal pulsing vessel and that the aboral portion is composed of two contractile chambers: one, opaque, is in relationship with the pulsing vessel; the other, larger and thinner, is placed around the internal surface of the dorsal sac that envelops it (Fig. 11). The rhythmic contractions of the pulsing vessel and the successive contractions of the two contractile chambers could circulate the coelomic fluids coming from the general cavity through the hemal system. Nonetheless, it would seem doubtful that it acts as a true circulatory system (Burton, 1964). The contractile and pulsatile portions observed could have another role: to mix the fluids and to promote the mixing of the materials of diverse origins that are found there.

XI. Coelomocytes and Internal Fluids

The internal cavities of the echinoderms are filled with a fluid having an ionic composition near that of seawater, but rich in free cells or coelomocytes. These coelomocytes can be found in all tissues, all organs, all cavities of the hemal system and all the coelomic compartments except the hyponeural sinuses. Their abundance, their diversity—a recent revision recognizes 18 different types (Endean, 1966)—and their ubiquity allow one to suppose that they play an important role in the economy of the organism, even though this role is most often poorly known. The phagocytes ingest foreign particles, destroy dead or damaged tissue, and transport them to the exterior by escaping through the madreporite or through the wall of the podia, the branchial papillae, the respiratory papules, the digestive tract, etc. On the other hand, they don't seem to intervene in the elimination of nitrogenous substances (mainly ammonia and urea) which, because of their solubility, pass directly to the exterior via the respiratory surface. Other coelomocytes seem to play a role in the fabrication of pigments, in the synthesis of collagen, in the transport of oxygen, the formation of clots, and the secretion of certain deposits of calcium carbonate. Finally, it is possible that certain coelomocytes help in digestion, the transportation and conservation of nutritive material, even though in a general manner this role seems rather to be vested in the coelomic fluid itself.

XII. Reproductive System

The first bud of the gonads appears under the aboral face of the young stelleroid or echinoid, in the secondary mesentery, under the form of genital cells in close relation with the peritoneum of the adjacent left somatocoel. This first bud is transformed into a strand or genital rachis, in which the genital cells are surrounded by connective tissue. The left somatocoel produces a vesicle which caps the genital rachis, separates itself from the parent coelom and becomes the genital coelom. The rachis bifurcates at its aboral end in two branches that diverge and rejoin to form a perianal genital ring contained in the coelomic ring which accompanies it in its development. This double ring receives hemal cavities from the axial organ and buds five diverticula in the interrays into which emigrate the gonocytes; they are the future adult gonads. The genital coelom remains, but the genital ring disappears in the echinoids while it remains in the stelleroids after the formation of the gonads.

In the holothurians the single gonad originates in the dorsal mesentery, formed by the horns of the left somatocoel, in the neighborhood of the stone canal. It keeps this position in the adult. In the crinoids, a primary gonad appears in the larva in the principal mesentery, in the same interray that contains the anus and hydropore. This primary gonad disappears, to be replaced by a secondary bud situated in the aboral vertical (secondary) mesentery against the axial organ. This bud builds a genital rachis, destined to disappear in the calyx but from which the gonocytes emigrate later to the arms and appendices, the pinnules, the final location of the gonads.

XIII. Position in the Zoological System

Since the Lower Cambrian, the echinoderms have been represented by several classes (Fig. 12). Some, like the edrioasteroids, seem to have all the essential characteristics of the group, including pentaradial symmetry. Others, such as the Helicoplacoidea or the "Carpoidea," whose oldest representatives are from the Middle Cambrian, though they do

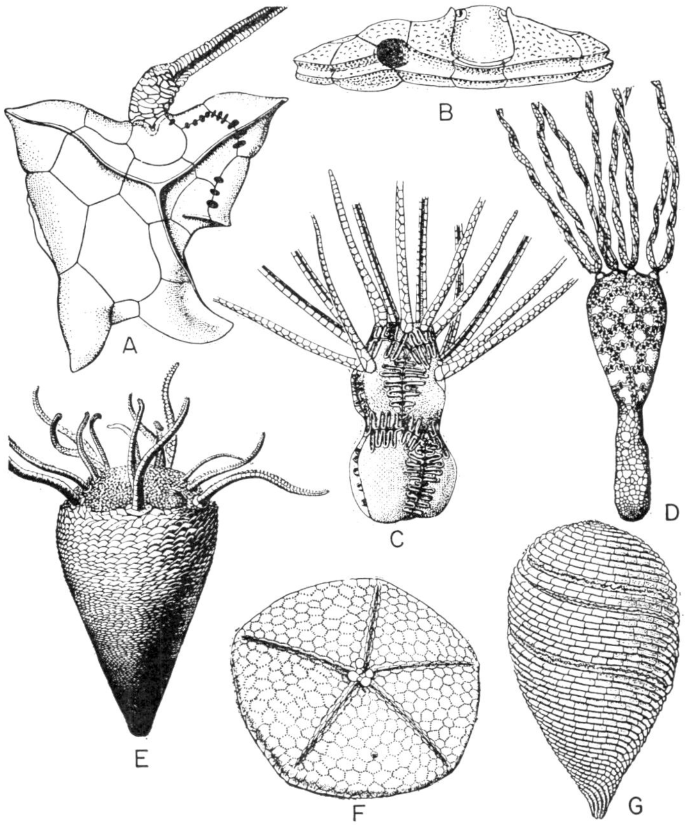

FIG. 12. Some Lower and Middle Cambrian echinoderms. A. *Ceratocystis perneri* Jaekel ("carpoid" Stylophora, M. Camb.), ×1.5. (From Ubaghs, 1967.) B. *Trochocystites bohemicus* Barrande ("carpoid" Homostelea, M. Camb.), ×1.5. (From Ubaghs, 1968d.) C. *Lichenoides priscus* Barrande (Eocrinoidea, M. Camb.), ×2. (From Ubaghs, 1968b.) D. *Gogia spiralis* Robison (Eocrinoidea, M. Camb.), ×2. (From Robison, 1965.) E. *Lepidocystis wanneri* Foerste (Lepidocystoidea, L. Camb.), ×1.1. (From Durham, 1968.) F. *Stromatocystites pentagularis* Pompeckj (Edrioasteroidea, M. Camb.), ×2. (From Durham, 1967, after Pompeckj.) G. *Helicoplacus curtisi* Durham & Caster (Helicoplacoidea, M. Camb.), ×2. (From Durham and Caster, 1963.)

not show any trace of radial symmetry and represent, perhaps, delayed evidence of a preradiate radiation, are nonetheless, by their asymmetry, the nature of their skeleton, and the presence of an ambulacrum, authentic echinoderms. So paleontology does not directly show the origins of the group, whose ancestral forms must have lived in such a distant past that it seems illusory to hope to find the trace. On the other hand, embryology and comparative anatomy demonstrate that the echinoderms are not as isolated in the animal kingdom as it might seem at first glance.

So, the essentials for the solution of the problem of their affinities are the following points: (1) retention of the blastopore as the anus or the piercing of the latter at the location of the blastopore; (2) enterocoelous origin of the coelom; (3) division of the coelom into three successive pairs of pouches; (4) presence of hollow tentacles in each of which a diverticulum of the middle coelom penetrates; (5) existence of a dorsal pore that puts the anterior coelom in communication with the exterior; (6) production by the anterior coelom of a dorsal vesicle which enters into a close relation with the hemal system; (7) finally, the extraordinary similarity between the *Dipleurula* and *Tornaria* larvae. These characteristics prove that the echinoderms are deuterostomes and that they are most closely related to the Hemichorda (Enteropneusta and Pterobranchia).[5] The resemblances between the ontogeneses of these two groups are so numerous and so profound that they cannot be fortuitous or the results of simple convergences. The resemblances suggest the existence of real ties of a common ancestry, while the precise nature of these ties is beyond our knowledge. Some have thought to make a taxonomic expression of these relationships by uniting the echinoderms and the Hemichorda in a single unit, the Ambulacraria (Metschnikoff) or the Coelomopora (Grobben).

Following Bateson, the origin of the chordates has often been sought in relation to the echinoderms. This hypothesis, admitted by many zoologists, but rejected by others, only makes sense if it is taken in its general significance. It would not justify the direct comparisons sometimes attempted between certain fossil echinoderms like the carpoids on the one hand, and *Amphioxus* or the cyclostomes on the other, since these comparisons would be based on simple analogies of form or structure.

XIV. Relationships among the Classes of Echinoderms

Paleontology is not very helpful in the search for the origins of the phylum and it does no better with the problem of the relationships among the classes. In fact, since their appearance in the Lower Paleozoic, the classes show their peculiar architectures. By consequence, their origins are obscure and it is evident that the great diversification of the echinoderms was already accomplished before the dawn of the Cambrian age.

[5] This relationship, contested by Fell (1948), is admitted by most zoologists.

However, in one case, that of the asteroids and ophiuroids, paleontology shows that these two classes, today quite separate, converge in the direction of a common ancestor, probably represented by the Somasteroidea—a group first found in fossil state and then living (Fell, 1962). It also weakens the hypothesis of a common ancestry between the ophiuroids and the echinoids, which is based on the convergence of certain biochemical and embryological characteristics. But in fact, it does not furnish any decisive argument on the origin of the Somasteroidea. The latter, by their ciliary-mucous feeding method and the similarity offered by their arms with the arms of the crinoids, suggested to Fell (1963a) that they could have branched from the crinoids. This hypothesis, built on structural and functional analogies, hardly seem convincing. It raises problems of embryological order, anatomy and biochemistry (Bolker, 1967), and is contrary to the available paleontological evidence. The old opinion that the asteroids came from the edrioasteroids because the ambulacral grooves of these two classes show certain similarities is no longer acceptable. The ancestors of the stelleroids, like those of the other classes, are still to be discovered.

It is generally agreed that during their phylogenesis, the echinoderms passed through a sessile phase and that their attachment was accomplished by the ancestral form of the preoral lobe. We have looked in this sessile nature for the explanation of their essential characteristics, that is to say, the preponderant influence of the left side during ontogenesis and the radial structure that affected the organization so deeply. If such a phase truly existed, it must necessarily have been long before the Cambrian, since free forms can be found from the base of the system; some completely lack radial symmetry (Helicoplacoidea), and some have radial symmetry (Edrioasteroidea). Therefore, the existence of this phase cannot be proved by fossil evidence, and if one supposes, as has been often done, that all echinoderms are descended from sessile fossil groups such as the cystoids or the eocrinoids, this amounts to formulating hypotheses without any paleontological or anatomical basis.

The real situation seems both more simple and more complex. Before the Upper Ordovician all the classes of echinoderms, recent or fossil, had already made their appearance. They are certainly much older, for as soon as they appear on the stratigraphic scale, they all behave like autonomic units, distinct, specialized, and often already diversified. The representation of the phylogenesis of the echinoderms can only take the appearance of a series of parallel lines, whose original connections remain conjectural (Fig. 13).

102316

SHEPHERD COLLEGE LIBRARY

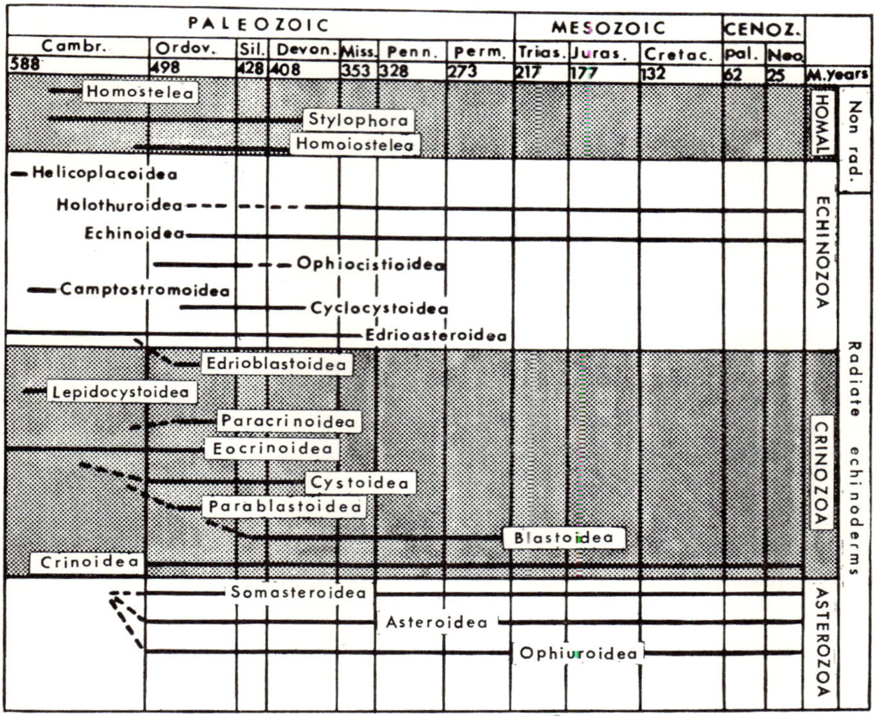

FIG. 13. Phylogeny and classification of Echinodermata.

XV. Classification

The attempt has often been made to group the classes of echinoderms in more comprehensive taxonomic units. Various essays have been made in that direction. The most widespread consists of dividing the phylum into two subphyla, the *Pelmatozoa*, or echinoderms attached during their entire existence or most of it, and the *Eleutherozoa*, or freely motile echinoderms. In the present day, the former are represented by the crinoids, the second by the holothurians, the echinoids, the asteroids, and the ophiuroids. It is certainly true that the crinoids differ more from the other echinoderms than the latter differ among themselves. By many characteristics—the orientation of their oral face toward the surface of the sea, the aboral position of their brachial skeleton, the peculiar development of their entoneural nervous system, their sinusary coelomic system—they reveal that they belong to an entirely distinct evolutionary line. But nothing proves that the other fossil classes (Blastoidea, Cystoidea, Edrioasteroidea, Eocrinoidea, Paracrinoidea, etc.) whose members were sessile, as are modern crinoids, belong to the same phyletic group as the crinoids, nor that all free forms are necessarily related. It has become more and more obvious that such a grouping into free echinoderms and attached echinoderms expresses less the real genetic affinities than structural and functional analogies caused by the adoption of similar ways of life.

Recently Fell (1965) proposed the recognition of four subphyla based on the types of symmetry and the preferential orientations of the growth axes. They are: (1) the Homalozoa, whose body is fundamentally asymmetrical and who shows no trace of radial symmetry; (2) the Crinozoa, with typical radial symmetry, a generally globoid body, with growth axes partly meridional and partly divergent as rays, and with nutritive appendices diverging upward and to the exterior; (3) the Asterozoa, with very pronounced radial symmetry, divergent growth axes forming projecting rays or arms and a star-shaped body; (4) the Echinozoa, with typically radial symmetry, a predominately meridional growth pattern and a globoid, cylindrical, or discoidal body, lacking arms and nutritive appendices.

This division of the classes of echinoderms into four large units probably expresses the real affinities of the classes better than the division into Pelmatozoa and Eleutherozoa. Nonetheless this is only a hypothetical division, since we don't know the origin of any of the classes very precisely. It is, moreover, a construction based on predetermined concepts rather than a comparison hierarchized progressively from the organisms themselves. Nonetheless, this attempt at grouping the classes is the most complete and the most comprehensive that has been presented to date. It is for this reason that we adopt it here.

[The classes are those recognized by the *Treatise on Invertebrate Paleontology* (Moore, 1968). Those marked by an asterisk are extinct.]

Subphylum Homalozoa
- * Class Homostelea (Middle Cambrian)
- * Class Stylophora (Middle Cambrian—Middle Devonian)
- * Class Homoiostelea (Upper Cambrian—Lower Devonian)

Subphylum Crinozoa
- * Class Eocrinoidea (Lower Cambrian—Middle Ordovician)
- * Class Lepidocystoidea (Lower Cambrian)
- * Class Paracrinoidea (Middle Ordovician)
- * Class Cystoidea (Lower Ordovician—Upper Devonian)
- * Class Edrioblastoidea (Middle Ordovician)
- * Class Blastoidea (Silurian—Permian)

* Class Parablastoidea (Ordovician)
Class Crinoidea (Lower Ordovician—Recent)
Subphylum Asterozoa
Class Stelleroidea
Subclass Somasteroidea (Lower Ordovician—Recent)
Subclass Asteroidea (Lower Ordovician—Recent)
Subclass Ophiuroidea (Lower Ordovician—Recent)
Subphylum Echinozoa
* Class Helicoplacoidea (Lower Cambrian)
* Class Camptostromoidea (Lower Cambrian)
* Class Edrioasteroidea (Lower Cambrian—Lower Carboniferous)
* Class Cyclocystoidea (Middle Ordovician—Middle Devonian)
* Class Ophiocistioidea (Lower Ordovician—Upper Silurian, ? Middle Devonian)
Class Echinoidea (Middle Ordovician—Recent)
Class Holothuroidea (Middle Ordovician—Recent)

XVI. Systematics

Phylum Echinodermata
Subphylum I. Homalozoa

The Homalozoa (Fig. 14) are extinct echinoderms, having depressed and asymmetrical bodies, without radial symmetry. This last characteristic opposes them to almost all the other echinoderms. Their relations to the other classes are unknown. They do not seem to have had any descendants. Perhaps they belong to a very ancient offshoot of the group composed of several branches that were able to separate themselves from the central trunk before the phylum acquired radial symmetry, that is to say before the beginning of the Cambrian era. Three groups can be distinguished, previously grouped into one class, the Carpoidea or Heterostelea, but now it would seem preferable to make a separate class of each of them, because of the profound differences between their respective organizations and also because of our total ignorance of their reciprocal phylogenetic relationships. These three classes are: the Homostelea (Fig. 14A), the Stylophora (Fig. 14B), and the Homoiostelea (Fig. 14C).

Subphylum II. Crinozoa

The Crinozoa (Figs. 15, 16) are typically radial echinoderms whose viscera are enclosed in a theca made of plates. They are attached, with rare exceptions, either directly to the substratum or by means of a stalk. The mouth occupies the pole opposite to the point of fixation or insertion

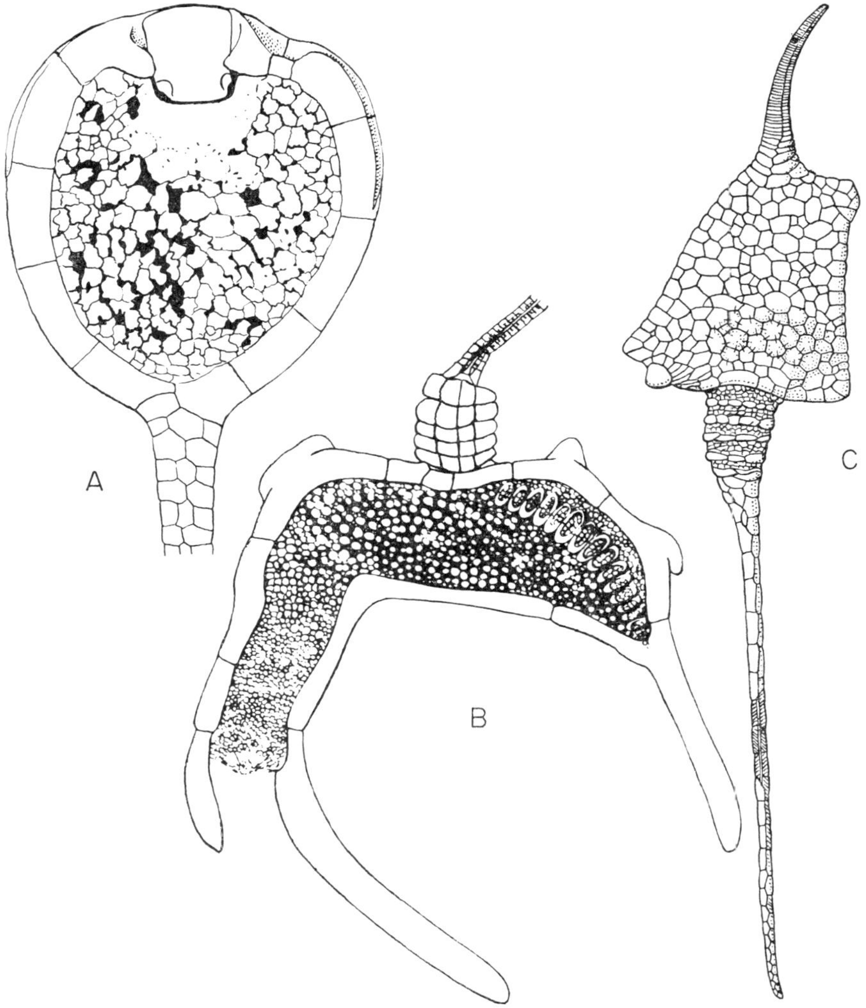

FIG. 14. Representatives of Homalozoa. A. *Trochocystites bohemicus* Barrande (Homostelea, M. Camb.), upper side, ×2.25. (From Ubaghs, 1968d.) B. *Cothurnocystis elizae* Bather (Stylophora, Up. Ordovician), upper side, ×2. (From Ubaghs, 1968c.) C. *Dendrocystites sedgwicki* Barrande (Homoiostelea, M. Ordovician), ×0.67. (From Bather, 1913.)

of the stalk; the anus occupies an adoral or lateral position, but is never aboral. They have arms which enclose extensions of the body cavity, or articulated appendices outside the theca, the brachioles. The skeleton of the arms prolongs the one that forms the aboral part of the theca. The skeleton of the brachioles is confined to these organs and has little

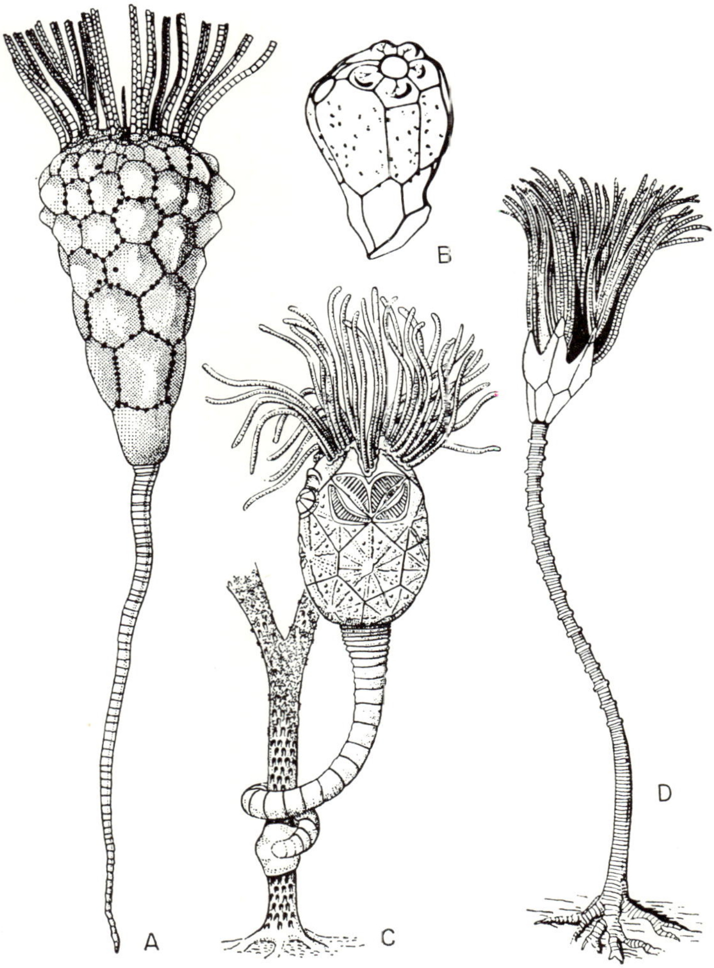

FIG. 15. Representatives of fossil Crinozoa. A. *Rhopalocystis destombesi* Ubaghs (Eocrinoidea, L. Ordovician), ×3. (From Ubaghs, 1963). B. *Eucystis angelini* Regnéll (Cystoidea Diploporita, U. Ordovician or L. Silurian), about ×2. (From Regnéll, 1945). C. *Lepadocystis moorei* Meek (Cystoidea Rhombifera, U. Ordovician), about ×2. (From Kesling and Mintz, 1961). D. *Orophocrinus fusiformis* Wachsmuth & Springer (Blastoidea, L. Carboniferous), about ×1. (From Cuénot, 1948, after Bather.)

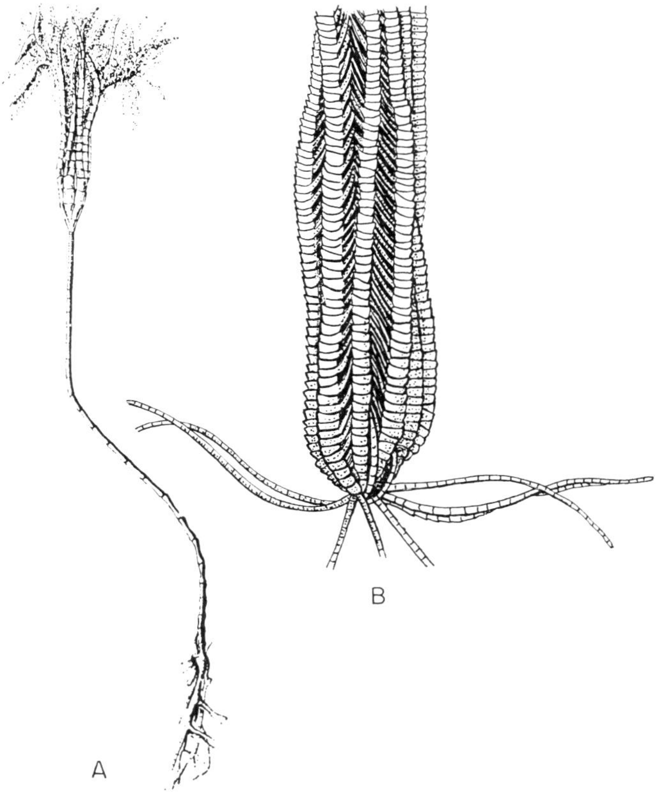

FIG. 16. Representatives of Crinoidea. A. A stalked species, *Ilycrinus carpenteri* Dan. and Kor., about ×1. (After Danielssen, 1892.) B. An unstalked species, *Homalometra denticulata* (P. H. Carpenter), ×2. (After A. H. Clark, 1918.)

connection with the skeleton of the theca. Both arms and brachioles gather food that is suspended in the water and send it along ciliated grooves to the mouth.

Fossil classes

Up to eight classes of Crinozoa (see above: classification) have been defined, all of which belong to Paleozoic formations, except for the Crinoidea. The extinct classes, which it is not useful to describe in detail here, were characterized by their thecae, which are closed except for a few orifices, and by the presence of brachioles (and not true arms) (Fig. 15). The brachioles are arranged along or at the end of the food grooves, which may be long or very short, and which, depending on the form under consideration, run to the surface of the theca or are inserted in the wall of the theca, or even sometimes become internal and are transformed into tunnels. In most of these classes, of which many have only one genus or very few genera, pores, slits, or folds in or between the plates of the theca seem to have provided for respiratory exchanges between the coelomic fluid and the exterior. Radial symmetry is evident sometimes in almost the entire skeleton, sometimes only in the food grooves; radial symmetry is completely lacking in some specialized forms that have lost several rays. The number of their rays, depending on the form under consideration, varies from 2 to 5, which suggests that the pentamerous organization had not yet been definitely established. Figure 15 shows a typical example of each of the principal classes: the Eocrinoidea (Fig. 15A), the Cystoidea (Fig. 15B, C), and the Blastoidea (Fig. 15D).

Class Crinoidea

The crinoids, living and extinct, are characterized by true arms, a theca that is clearly divided into an aboral capsule situated under the roots of the arms and an adoral arch that spreads between the arms, and by a fundamental pentamerous symmetry (Fig. 16). In principle there is a stalk, which may be reduced to a short cylindrical base, as in the recent *Holopus,* or even completely rejected at an early stage of development, as in the comatulids (feather stars). The crinoids live on plankton and debris suspended in the water, which are captured by the tentacles (podia) in the food grooves which are spread to the ends of the arms and their small lateral branches, the pinnules. The radial ambulacral canal runs in an open groove on the adoral side of the arms and pinnules (Fig. 7A). The arms are sometimes simple, but generally they are divided into two or more branches, which can cause an enormous extension of the "fishing area" of the animal. The genital glands, in all recent crinoids, are found in the arms or more generally in certain of their pinnules.

During the Paleozoic era the crinoids filled the seas with thousands of species. At the end of that era, three of the recognized subclasses disappeared, but the fourth, the Articulata, developed considerably, starting in the Triassic and becoming particularly important in the Jurassic. At the present time, the crinoids are represented by some 75 stalked species and about 540 stalkless ones. They live at all depths, from the littoral zone to the greatest ocean depths.

Subphylum III. Asterozoa

The Asterozoa (Figs. 17–19), like the Crinozoa and the Echinozoa, belong to the group of radial echinoderms. They are free (even though their larvae are sometimes temporarily fixed) and generally have star-shaped bodies, usually with five arms or rays. The arms appear after metamorphosis as expansions of the body that grow in a horizontal plane and enclose elongations of the organs and cavities of the body. In all of them (except most of the Ophiuroids) the digestive tract, affected by radial symmetry, has ceca that penetrate deeply into the arms (Fig. 18B). Neither the Crinozoa nor the Echinozoa have this peculiarity. The mouth opens on the lower side (turned toward the substratum), the anus (when it exists) opens on the opposite side. The axial skeleton of the arms is formed of two series of ossicles (ambulacral plates) which act as a roof or sheathing for the radial ambulacral canal.

FIG. 17. Representative of a fossil somasteroid: *Villebrunaster thorali* Spencer (L. Ordovician), adoral face, ×2. (After Spencer, 1951.)

The proximal ossicles of the axial skeleton are changed into buccal parts or a jaw apparatus.

As a group, the Asterozoa are both diversified and compact. The sharply defined limits that separate the modern ophiuroids and asteroids become blurred when the Paleozoic forms are brought into consideration. Thus many authors, and especially paleontologists, only recognize one class, the Stelleroidea, which they divide into three subclasses: the Somasteroidea, the Asteroidea, and the Ophiuroidea.

Class Stelleroidea

Subclass 1. Somasteroidea

The Somasteroidea (Fig. 17) are apparently the most fundamental and archaic type of Asterozoa. Known in the fossil state in Paleozoic formations, they seem to be represented at the present day by the genus

Platasterias (Fell, 1962). In certain traits, especially their feeding habits, they are similar to the crinoids. Each of their ambulacral plates carries on the outside a transverse series of elongated ossicles, arranged end to end in such a way as to form rods between which run channels

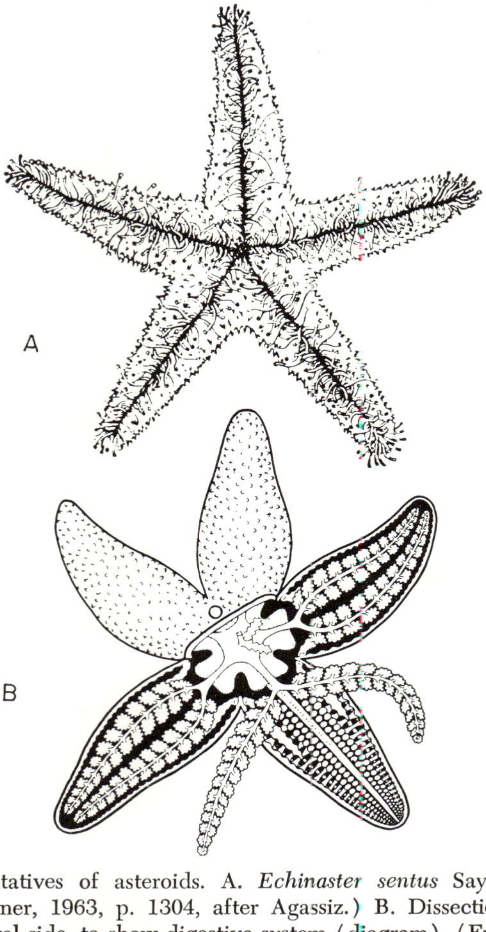

FIG. 18. Representatives of asteroids. A. *Echinaster sentus* Say (adoral face), ×0.64. (From Kaestner, 1963, p. 1304, after Agassiz.) B. Dissection of a starfish, viewed from the aboral side, to show digestive system (diagram). (From Buchsbaum, 1948.)

that are open on the adoral face, and are protected by covering platelets or movable spines. These channels lead to a radial groove, which is generally quite shallow, but can in some cases deepen temporarily when the ambulacral ossicles, which are by nature only slightly raised, become erect. Such a structure, by its pinnate nature, resembles to a certain degree a crinoid arm with pinnules. But the resemblance is superficial, since the transverse channels are between and not on the rods, while the ambulacral structures remain confined to the axial part of the arms

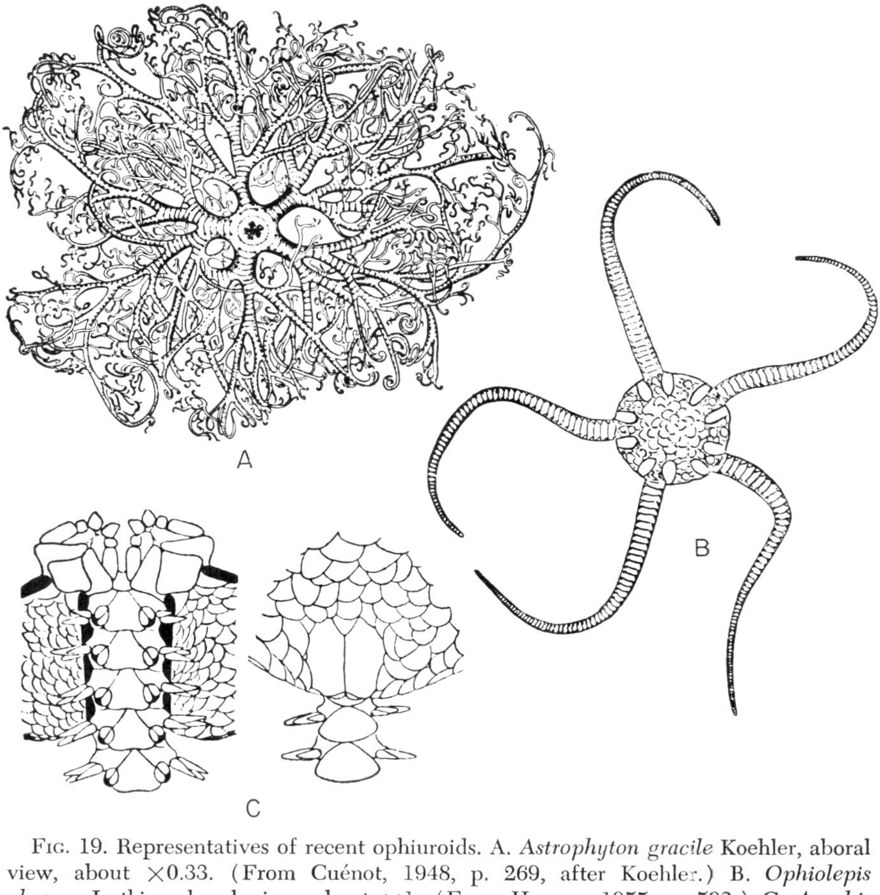

FIG. 19. Representatives of recent ophiuroids. A. *Astrophyton gracile* Koehler, aboral view, about ×0.33. (From Cuénot, 1948, p. 269, after Koehler.) B. *Ophiolepis elegans* Lutkin, aboral view, about ×1. (From Hyman, 1955, p. 592.) C. *Amphipholis* squamata (D. Chiaje), part of adoral side and of aboral side, ×12. (From Mortensen, 1927.)

without sending any elongation toward the transverse channels. The latter, unlike the pinnules of the crinoids, do not have the structure of "miniature arms." The ambulacral feet are in a double row the length of the axis of each arm; they are attached to small cups on the ambulacral plates; they do not have terminal suckers, but in some species have internal bulbs. *Platasterias* has no anus.

The transition from the somasteroids to the other subclasses of the Stelleroidea is by way of the primitive Asteroidea and Ophiuroidea, living and fossil. Their arms have maintained in a more or less altered state the ancestral pinnate arrangement. Though, in these forms, the longitudinal growth axes predominate, unequivocal traces of a transverse system of gradients still exist. Thus, the Asteroidea and Ophiuroidea can be shown as two diverging branches coming from a common somasteroid stock.

Subclass 2. Asteroidea

The Asteroidea (Fig. 18) exhibit the following distinguishing characteristics: relatively wide arms, of variable length (almost none in some species, considerable in others), arms with a large internal cavity, generally little distinguished from the central disc that connects them; distinct ambulacral plates that are arranged to form a deep open groove in which are placed the radial ambulacral canal and the structures that accompany it; locomotory ambulacra with internal ampullae and usually terminal suckers; external gills (papulae) on the aboral face; spines and granules generally distributed in great number on all the body surface; an anus is generally present.

Around 1200 species of asteroids are found in all regions of the oceans. Certain species principally eat particles and use ciliary mucous mechanisms to bring them to the mouth. Many others are predators and vigorously attack mollusks and other organisms, including other species of starfish.

Subclass Ophiuroidea

The ophiuroids (Fig. 19) are easily recognized from their typically cylindrical, flexible arms, clearly separate from the central disc, which, thanks to their powerful muscles, are used for locomotion. The small sensorial ambulacral feet are not used in moving the organism. They have neither suckers nor internal ampullae. In all modern forms the ambulacral plates are fused by pairs to form brachial ossicles, called vertebrae, which are jointed, and the ambulacral grooves are transformed into internal canals; in most paleozoic forms, the ambulacral plates are distinct and the ambulacral grooves are not closed by specialized plates as is the case with almost all modern ophiuroids. Respiration takes place in branchial pouches in the disc, near the base of the arms. There is no anus.

The importance of the ophiuroids in the biology of the sea is far from negligible. The number of species (more than 1900) is considerable. Their distribution is almost universal and the very high number of individuals makes them a dominant and characteristic element of many marine communities.

Subphylum IV. Echinozoa

The subphylum Echinozoa is the least clearly defined. It groups classes that are very different by appearance, organization, and way of life, and whose phyletic relationships are far from being elucidated. Fundamentally, they are forms that are globoid or cylindrical which never have arms. The growth gradients, at least in the most representative types, are meridional; in any case they are never divergent as in the Asterozoa.

Fossil classes

There are five fossil classes, all of which are limited to Paleozoic formations. They contain a small number of genera and species, except the Edrioasteroidea which acquired a certain importance. We will only consider the principal classes here.

The Helicoplacoidea, from the lowest Cambrian of the western United States, are among the strangest of the echinoderms (Fig. 12G). Their spindle-shaped body is enveloped in a flexible and apparently "expansible" test, whose plates are arranged in spiral. The mouth opens at one extremity of the body and the anus (unobserved) probably at the opposite end. Only a single ambulacrum exists, sometimes with a short branch; it is inserted in the body wall. There is no symmetry properly speaking, and since there is only one ambulacrum it is hardly conceivable that the organization is derived from a radial structure, transformed by torsion. The presence of a single ambulacrum makes it equally probable that one of the original hydrocoels is reduced or suppressed, which suggests a pronounced asymmetry of organization. To sum up, in spite of their unusual appearance, the Helicoplacoidea indicate a level of evolution comparable to that of the Homalozoa.

If the Helicoplacoidea were not attached, the Edrioasteroidea (Lower Cambrian—Lower Carboniferous) were mostly attached and, like the crinoids, held their adoral face uppermost (Fig. 12F). But they had neither stem nor arms nor any exothecal appendices of any sort. Their ambulacra, five in number, were inserted in the wall of the capsule or theca which enveloped the visceral mass; they radiated from around the mouth and stayed essentially restricted to the adoral face. The anus opened between two ambulacra on the adoral face. A third pore, the hydropore, was generally present between the mouth and the anus.

The Ophiocistioidea (Lower Ordovician—Upper Silurian, ?Middle Devonian) were free echinoderms that rested on their adoral face. In the center of the adoral face opened the mouth which was surrounded by a flexible peristome, similar to that of the sea urchins, and armed with five jaws. The rest of this face was divided into five very wide ambulacra, separated by narrow interambulacra. The ambulacra did not spread to the aboral face. The most remarkable characteristic of this group is undoubtedly the presence of giant ambulacral feet, which had a complete scaly skeletal covering. These ambulacral feet grew from the junctions between the plates of the ambulacral area; they must have been attached to a radial ambulacral canal situated under the plates that formed the ambulacral zones of the test and which ended, at the edge of the test, in a single tentacle of the same size as the adjacent ambulacral feet. The anus was situated on the aboral face, either in a lateral or marginal position. The Ophiocistioidea, of which only a few forms are known, show curious resemblances with the early growth stages of a very young Echinoidea.

Class Echinoidea

The Echinoidea are free echinoderms whose visceral mass is entirely enclosed in a test made up of columns of plates arranged along the meridians (Fig. 20). Apart from a double circle of apical plates, columns of ambulacral plates and columns of interambulacral plates can be distinguished. The ambulacral feet are provided with internal ampullae and ordinarily a terminal sucker. There is no ambulacral groove (except in some forms of the Ordovician) and the radial ambulacral canal, originally enclosed between the ambulacral plates, is situated under the ambulacral plates. The spines, always present and extremely diversified, are mobile, moved by muscles and attached to protuberances of the test. They are essentially for locomotion and defense. A very complicated and characteristic chewing apparatus exists in most echinoids, except for some very specialized forms. The mouth is always situated on the face turned toward the substratum; it is ordinarily central. The anus, primitively apical, has migrated toward the side, and even to the adoral face of the so-called irregular sea urchins. Finally, the echinoids are unique, with the asteroids, in having pedicellaria.

The echinoids have been known since the Ordovician. About 800 species are counted in the modern seas, spread from the intertidal zone to depths of over 7000 meters, and from the polar regions to the equator. The number of known fossil species is over 7500.

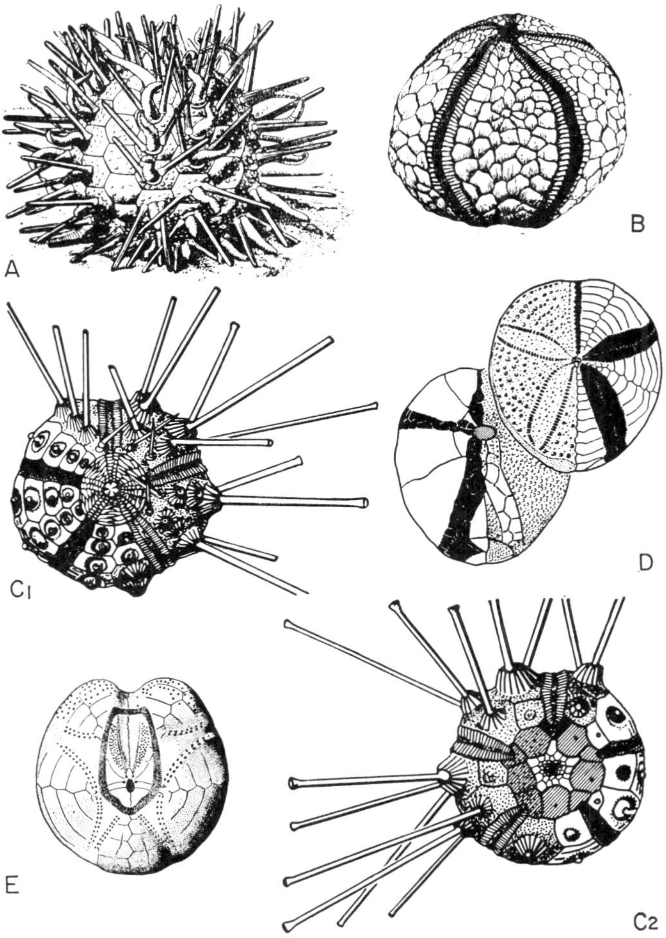

FIG. 20. Representatives of fossil and Recent echinoids. A. *Bothriocidaris eichwaldi* Mannil, Ordovician, ×2. (After Durham, 1966.) B. *Aulechinus grayae* Bather & Spencer, Ordovician, ×1. (After MacBride and Spencer, 1938.) C. *Stereocidaris tubifera* Mortensen, Recent, C_1 adoral face, C_2 aboral face, ×0.5. (After Fischer, 1952.) D. *Eupatagus mooreanus* Pilsbry, Eocene, ×0.5. (After Fischer, 1952, p. 700.) E. *Echinocardium cordatum* (Pennant), Recent, ×0.5. (After Cuénot, 1948.)

Class Holothuroidea

The holothurians are free echinoderms, with fundamentally cylindrical bodies, elongated along the oral-aboral axis (Fig. 21). The body wall, usually thick and leathery, has powerful longitudinal and circular muscles. The skeleton, which is sometimes nonexistent, usually consists of

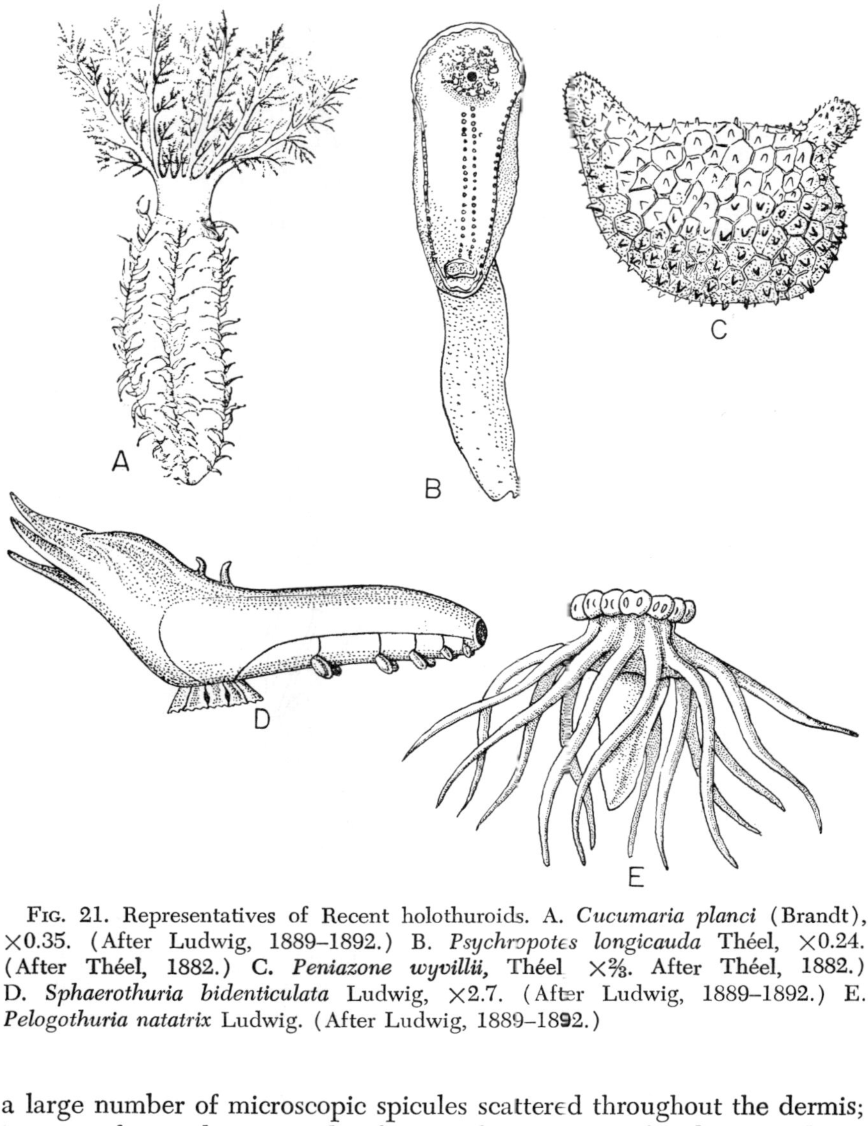

FIG. 21. Representatives of Recent holothuroids. A. *Cucumaria planci* (Brandt), ×0.35. (After Ludwig, 1889–1892.) B. *Psychropotes longicauda* Théel, ×0.24. (After Théel, 1882.) C. *Peniazone wyvillii,* Théel ×⅔. After Théel, 1882.) D. *Sphaerothuria bidenticulata* Ludwig, ×2.7. (After Ludwig, 1889–1892.) E. *Pelogothuria natatrix* Ludwig. (After Ludwig, 1889–1892.)

a large number of microscopic spicules scattered throughout the dermis; in some forms these spicules fuse to form a test of imbricate plates, but these never have a regular arrangement as in the other classes. The mouth opens at or near one end of the body, the anus at the opposite end. A circle of tentacles, which are only modified podia, surround the mouth to capture food. The ambulacral grooves are totally lacking; they are replaced by internal epineural canals. The podia, variously arranged along the ambulacra, sometimes spread to the interambulacral zones; they are absent in some forms. They are usually provided with internal ampullae and terminal suckers. The madreporite is generally internal. They have a single gonad, placed in the same interray as the madreporite. Their symmetry, essentially pentamerous, is altered by a secondary bilateral symmetry. The animal, which moves with the mouth forward and the oral-aboral axis parallel to the bottom, usually shows a differentiated "ventral" face and "dorsal" face. The first has three ambulacra (trivium) and the second has only two (bivium). The podia of the trivium function like those of the starfish and sea urchins, but those of the bivium are often atrophied or modified. A peculiar respiratory organ exists in three of the five orders of Holothuroidea; it consists of a pair of highly ramified hollow tubes that come from the cloaca. These tubes, called respiratory trees, are alternately filled and emptied of seawater pumped by the anus; they are true water lungs.

The Holothuroidea have often been connected with the Echinoidea because of the fact that they have no arms, that their radial ambulacral canals are in a meridian position, and that these canals are situated under the skeleton of the body wall. But their very simple larvae, their single gonad and, in the most primitive forms, the position of the madreporite between the mouth and genital pore suggest a phylogenetic position far from the sea urchins. Their origin remains completely mysterious.

The Holothuroidea, generally benthonic, crawl or rest on the bottom or on algae, or squeeze themselves into the mud or in cracks in rocks. Most species are sedentary or slightly active, but some are able to swim for short periods; some very rare forms have become pelagic (Fig. 21E). Particularly abundant in shallow tropical waters, they can be found at all depths, from the oceanic deeps where they make up a very high percentage of the biomass. The number of modern species is more than 700.

REFERENCES

Bather, F. A. (1913). *Trans. Roy. Soc. Edinburgh* **49**, Part 2, 359–529.

Bather, F. A., Goodrich, E. S., and Gregory, J. W. (1900). *In* "A Treatise on Zoology" (E. R. Lankester, ed.), Vol. 3, pp. 1–344. Adam & Charles Black, London.

Bolker, H. L. (1967). *Nature* **213**, 904–905.

Boolootian, R. A., and Campbell, J. L. (1964). *Science* **145**, 173–175.

Boveri, T. (1901). *Zool. Jahrb. Abt. Anat. Ontog. Tiere.* **14**, 630–653.

Burton, M. P. M. (1964). *Nature* **204**, 1218.

Buchsbaum, R. (1948). "Animals without Backbones," p. 303. Univ. of Chicago Press, Chicago, Illinois.

Clark, A. H. (1918). *Siboga Expeditie* **42B**, 1–300.
Clark, A. H. (1915–1950). "A Monograph of the Existing Crinoids," Vol. I, Part 1 (1915); Part 2 (1921); Part 3 (1931); Part 4a (1951); Part 4b (1947); Part 4c (1950).
Clark, A. H. (1900). *U.S. Natl. Museum, Bull.* **82.**
Clark, A. M. (1962). "Starfishes and their Relations," pp. 1–119. British Museum (Natural History), London.
Cuénot, L. (1948). *In* "Traité de Zoologie" (P.-P. Grassé, ed.), Vol. 11, pp. 1–275. Masson, Paris.
Danielssen, D. C. (1892). Crinoidea. *Norske Nordhavs Expedition, 1876–1878* **21,** pp. 1–28.
Dawydoff, C. (1948). *In* "Traité de Zoologie" (P.-P. Grassé, ed.), Vol. 11, pp. 277–363. Masson, Paris.
Durham, J. W. (1966). *Biol. Rev.* **41,** 368–391.
Durham, J. W. (1967). *Yale Sci. Mag.* **39,** No. 2, 24–28.
Durham, J. W. (1968). *In* "Treatise on Invertebrate Paleontology" (R. C. Moore, ed.), Part S, pp. 631–634. Geol. Soc. Am. and Univ. of Kansas.
Durham, J. W., and Caster, K. E. (1963). *Science* **140,** 820–822.
Endean, R. (1966). *In* "Physiology of Echinodermata" (R. A. Boolootian, ed.), pp. 301–328. Wiley (Interscience), New York.
Fell, H. B. (1948). *Biol. Rev.* **23,** 81–107.
Fell, H. B. (1962). *Science* **136,** 633–636.
Fell, H. B. (1963a). *Phil. Trans. Roy. Soc. London* **B246,** 381–485.
Fell, H. B. (1963b). *Smithsonian Inst., Ann. Rept.,* pp. 457–490.
Fell, H. B. (1965). *Breviora* **219,** 1–17.
Fischer, A. G. (1952). *In* "Invertebrate Fossils" (R. C. Moore, C. G. Lalicker, and A. G. Fischer, eds.), p. 678. McGraw-Hill, New York.
Garrido, J., and Blanco, J. (1947). *Compt. Rend.* **224,** 485.
Hörstadius, S. (1939). *Publ. Staz. Zool. Napoli* **17,** 212–312.
Hozawa, S. (1928). *Sci. Rept. Tohoku Impe. Univ., Fourth Ser.* **3,** 361–378.
Hyman, L. H. (1955). "The Invertebrates," Vol. IV, pp. 1–763. McGraw-Hill, New York.
Kaestner, A. (1963). "Lehrbuch der speziellen Zoologie," Part 1, No. 5, pp. 981–1423. Fischer, Jena.
Kesling, R. V., and Mintz, L. W. (1961). *Contrib. Museum Paleontol. Univ. Mich.* **17,** 123–148.
Ludwig, H. (1889–1892). Die Seewalzen. *Bronn's Klassen* **2,** Sect. 3, 1–460.
MacBride, E. W. (1903). *Phil. Trans. Roy. Soc. London* **B195,** 285–327.
MacBride, E. W., and Spencer, W. K. (1938). *Phil. Trans. Roy. Soc. London* **B229,** No. 558, 91–136.
Moore, R. C., ed. (1968). "Treatise on Invertebrate Paleontology," Part S, pp. i–xxix, 1–650, Univ. of Kansas Press, Lawrence, Kansas.
Mortensen, T. (1921). "Studies of the Development and Larval Forms of Echinoderms," pp. 1–261. G. E. C. Gad, Copenhagen.
Mortensen, T. (1927). "Handbook of the Echinoderms of the British Isles," pp. 1–471. Oxford Univ. Press, London and New York.
Mortensen, T. (1928–1951). "Monograph of the Echinoidea," 15 vols. C. A. Reitzel, Copenhagen.
Nichols, D. (1962). "Echinoderms," pp. 1–200. Hutchinson, London.
Nissen, H. (1963). *Neues Jahrb. Geol. Paleontol. Abhandl.* **117,** 230–234.

Phillip, G. M. (1965). *Nature* **208**, 766–768.
Raup, D. M. (1962a). *J. Paleontol.* **36**, 793–810.
Raup, D. M. (1962b). *Systematic Zoology* **11**, 97–108.
Raup, D. M. (1966). *In* "Physiology of the Echinodermata" (R. A. Boolootian, ed.), pp. 379–395. Wiley (Interscience), New York.
Regnéll, G. (1945). *Medd. Lunds Geol.-Mineral. Inst.* **108**, 1–255.
Robison, R. A. (1965). *J. Paleontol.* **39**, 355–364.
Spencer, W. K. (1951). *Phil. Trans., Roy. Soc. London* **B235**, No. 623, 87–129.
Théel, H. (1882). "Report on the Holothuroidea," Part I. Report on the Scientific Results of the Voyage of H.M.S. "Challenger," Zoology, Vol. 4, Part 13. London.
Ubaghs, G. (1963). *Notes Service Geol. Maroc* **23**, No. 172, 25–44.
Ubaghs, G. (1967). *Univ. Kansas Paleontol. Contrib.*, Paper 22, pp. 1–16.
Ubaghs, G. (1968a). *In* "Treatise on Invertebrate Paleontology" (R. C. Moore, ed.), Part S, pp. 3–60, Geol. Soc. Am. and Univ. of Kansas.
Ubaghs, G. (1968b). *In* "Treatise on Invertebrate Paleontology" (R. C. Moore, ed.), Part S, pp. 455–495, Geol. Soc. Am. and Univ. of Kansas.
Ubaghs, G. (1968c). *In* "Treatise on Invertebrate Paleontology" (R. C. Moore, ed.), Part S, pp. 495–565, Geol. Soc. Am. and Univ. of Kansas.
Ubaghs, G. (1968d). *In* "Treatise on Invertebrate Paleontology" (R. C. Moore, ed.), Part S, pp. 565–581, Geol. Soc. Am. and Univ. of Kansas.

CHAPTER 2

Ionic Patterns

Shirley E. Freeman[1] *and W. P. Freeman*

I. Introduction

Muscles of animals of the phylum Echinodermata share the ability to accumulate K^+ and partially exclude Na^+ with most animal and plant tissues. Steinbach (1962) has commented on the widespread nature of this ability, which may almost be considered a criterion of life. It is of interest phylogenetically to study ionic patterns at different evolutionary levels; in contrast to vertebrate muscle, holothuroidean muscle appears to be the only invertebrate muscle which has been subjected to a systematic investigation.

This work was commenced by Steinbach (1937, 1940) and was confirmed and extended by the present authors and co-workers (W. P. Freeman and Simon, 1964; Simon *et al.*, 1964a,b). Steinbach (1937, 1940) used the longitudinal muscle bands of *Thyone briareus*. We have studied the histology and ionic patterns of the closely related holothuroidean *Stichopus mollis;* a limited investigation of *Paracaudina australis* was also carried out.

[1] Formerly Shirley E. Simon. Present Address: Defence Standards Laboratories, Maribyrnong, Victoria, Australia.

II. Histology

The early work of Steinbach (1937) on the distribution of K^+ and Cl^- in *Thyone* muscle rested on Hall's (1927) histological interpretation of the structure of the retractor muscles of *Cucumaria miniata.* Hall suggested that these muscles were composed of giant cells consisting of a number of radially grouped fibrils. Olson (1938) concluded that Hall had been wrong in her suggestion, and showed that the "fibrils" were themselves the muscle cells, which were regularly arranged in a matrix of connective tissue. Steinbach (1940) reinterpreted his data in line with Olson's finding, with a considerable alteration in emphasis. The histology of *Stichopus mollis* (W. P. Freeman and Simon, 1964) and *Paracaudina australis* appears similar to that of *Thyone* as demonstrated by Olson (1938).

A. Structure

The arrangement of the longitudinal muscle bands in *Stichopus* and *Paracaudina* may be seen in Figs. 1 and 2. There are five muscle bands

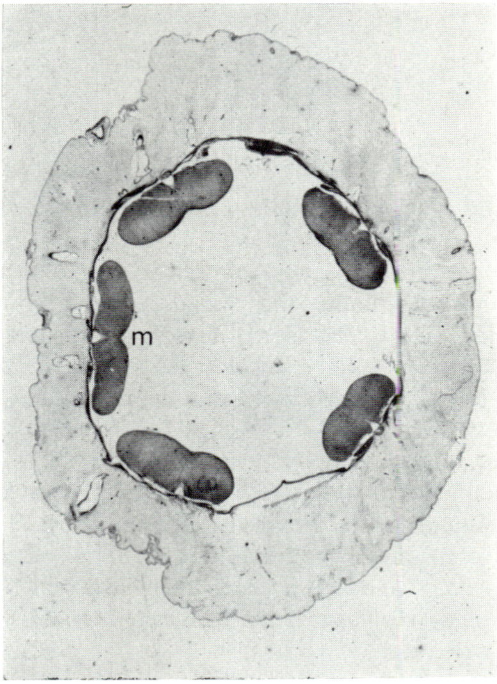

FIG. 1. Transverse section through a holothuroidean, *Stichopus mollis,* showing the five longitudinal muscle bands (m) attached to the body wall. The viscera have been removed. ×4.

in *Stichopus,* which run along the body wall from the calcareous ring to the anus; in *Paracaudina* each muscle is longitudinally divided into two, so that there are ten muscle bands. In both species the muscles are covered by a coelomic lining epithelium; beneath this there is a thin but dense layer of connective tissue, which thins abruptly to form a diffuse matrix between the muscle cells.

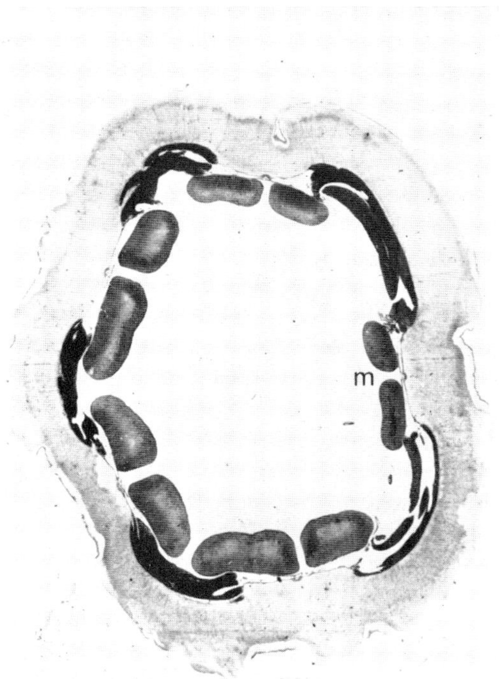

FIG. 2. Transverse section through a holothuroidean, *Paracaudina australis,* showing the five pairs of longitudinal muscles (m). ×3.

The muscle cells are arranged in groups of from two to ten cells, which appear as rosettes in transverse section (see Figs. 3 and 5). The muscle cells appear to be devoid of striations. They are mononucleate; the nucleus is placed external to the contractile elements of the cell (Figs. 4, 7, and 8). The innervation of individual muscle cells is shown in Fig. 5; the nerve may be seen passing across the section before branching and sending fine branches to the muscle cells.

The nerves appear similar in size to those demonstrated in *Astropecten irregularis* (Pennant) by Smith (1950), and may be compared with electron micrographs of the innervation of the lantern retractor muscle of *Echinus esculentus* (Cobb and Laverack, 1966).

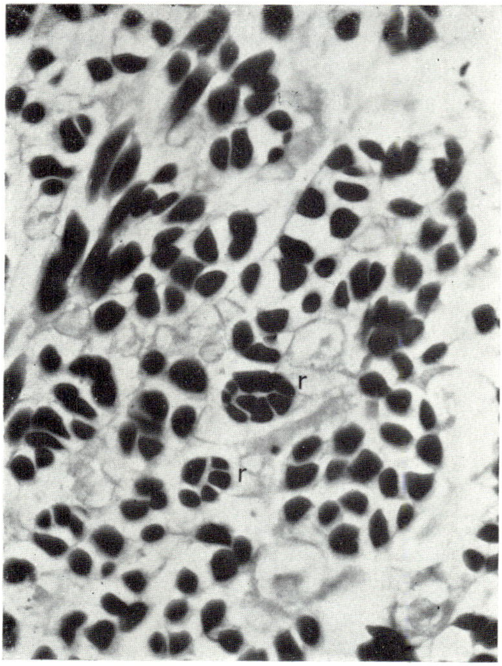

FIG. 3. Transverse section through the muscle showing the individual muscle fibers arranged in rosettes. *Paracaudina australis.* ×280.

B. DIMENSIONS OF MUSCLE CELLS

The length of the muscle cells was determined in isolated single fibers, which were prepared by a maceration procedure (W. P. Freeman and Simon, 1964). The fibers were measured under low power, using a calibrated eyepiece grid. Only obviously intact, tapered fibers were measured. The mean length was 352 ± 3 μ (S.E. of 125 observations); the range in fiber length was from 240 to 555 μ. The distribution of fiber length was not significantly different from a normal distribution. Fiber diameter was determined using transverse sections which were photographed under high magnification with a calibrated eyepiece grid. The precautions taken to avoid compression of the section and ensure random sampling are outlined in W. P. Freeman and Simon (1964). Table I shows the results of these measurements.

These measurements were also carried out on sections of *Paracaudina.* We noted, however, that fiber diameter varied from one part of the muscle to another. Photographs which were taken of areas adjacent to the body wall showed smaller fibers than more central areas, while

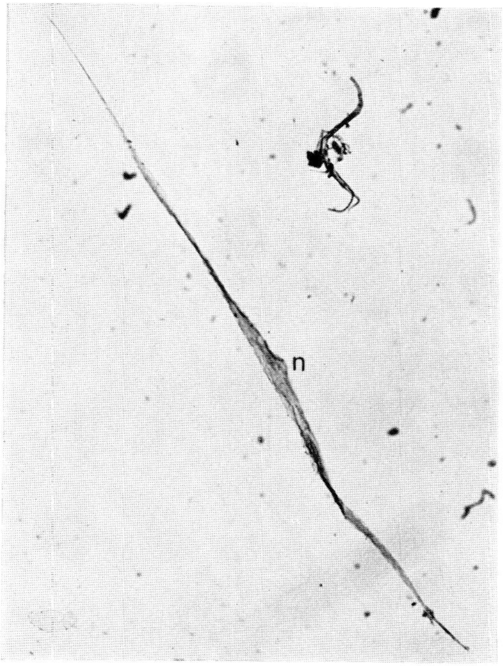

FIG. 4. Isolated single muscle cell 480 μ in length, prepared by teasing, showing external nucleus (n). *Stichopus mollis*. ×220.

areas close to the lumenal border showed a scattered distribution of very large cells. Measurements taken in central areas gave a mean fiber diameter of 4.4 μ, which is close to the value found for *Stichopus*. However, a figure as low as 2.2 μ was found for the mean diameter adjacent

TABLE I

THE DIAMETER OF HOLOTHUROIDEAN MUSCLE FIBERS[a]

Fixative	Fiber diameter (μ)[b]			
	Animal 1		Animal 2	
Brazil	3.4 ± 0.1(26)	3.3 ± 0.1(28)	3.8 ± 0.2(15)	3.4 ± 0.1(22)
Marine bouin	3.1 ± 0.1(24)	2.9 ± 0.1(14)	4.6 ± 0.4(11)	4.9 ± 0.2(10)
Absolute alcohol	3.7 ± 0.2(25)	3.4 ± 0.2(23)	3.6 ± 0.1(24)	4.2 ± 0.2(10)

[a] From table 1, W. P. Freeman and Simon (1964, p. 27).

[b] Figures given are average values ±S.E. The figures in parentheses are the number of observations.

to the body wall, and as high as 9.1 μ close to the lumenal border. The extreme range in fiber diameters was from 1.7 μ to 22.0 μ.

It may be seen in *Stichopus mollis* that the fiber length exceeds the diameter approximately 100-fold. The muscles are both smaller in diameter and shorter than those of the lantern retractor muscle of *Echinus esculentus,* as described by Cobb and Laverack (1966). The fibers of the latter muscles are similar in size to the largest fibers we measured

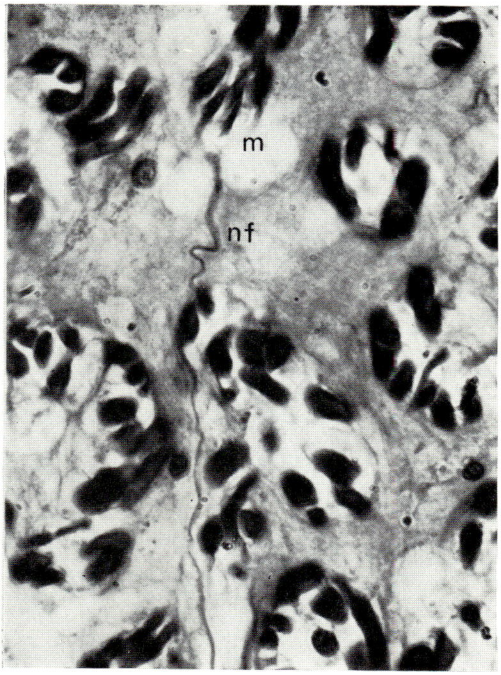

FIG. 5. Transverse section through the muscle showing a nerve fiber (nf) dividing and sending branches to the muscle cells (m). *Stichopus mollis.* ×725.

in *Paracaudina. Echinus* muscle also differs in the placement of the nucleus. We noted that the nucleus in *Stichopus* was placed symmetrically with regard to the long axis of the muscle cell, whereas Cobb and Laverack noted a more proximal placement with regard to the lantern. Figures 4, 7, and 8 illustrate the placement of the nucleus in *Stichopus.*

III. Dimensions of the Extracellular Space

Since any attempt to determine the distribution of ions across the cell membrane must take into account the relative volumes of the intra-

and extracellular water, we have approached this problem in several ways.

First, we have measured the volume of the extracellular space, using planimeter measurements of photomicrographs of transverse sections. Second, the space available to ^{131}I-labeled albumin was determined (Simon *et al.*, 1964a). Third, use was made of a recently developed technique (S. E. Freeman *et al.*, 1966) whereby the distribution of lissamine-rhodamine-200 labeled human serum albumin may be studied under the fluorescence microscope.

The measurement of the cell area in photographs of high power views of transverse sections was found to present some difficulties. Apart from the inherent error in the planimetry of small areas, the number of cells per unit area was found to vary widely from one part of a section to another. Areas of low cell density gave values for the extracellular space as high as 77%; densely packed areas gave values several orders of magnitude lower.

The use of ^{131}I-labeled albumin as an index of the extracellular water has proved a reliable and reproducible method in a number of muscles (S. E. Freeman *et al.*, 1966). The albumin space in holothuroidean muscle showed some scatter, and was in general smaller than was indicated by planimetric methods. Average values of 30% were found in one group of muscles (Simon *et al.*, 1964a), although values as high as 50% were occasionally obtained. The difference between the two methods was previously ascribed to the fact that the muscles were routinely blotted before counting the radioactivity. This procedure was shown to remove part of the extracellular fluid, because of the spongelike texture of the muscle. Recently, however, in studying frozen sections of *Stichopus mollis*, it has been noted that the muscle cells are rather more densely packed than was found in paraffin sections, regardless of the method of fixation.

While the use of ^{131}I-labeled albumin was a convenient index of the size of the extracellular compartment, it was not possible to consider it precise, since certain areas of the extracellular space might have been inaccessible to such a large molecule. High concentrations of mucopolysaccharide in the extracellular space could cause a partial exclusion of albumin (Ogston and Phelps, 1960), leading to an underestimation of the size of this compartment. However, minimal quantities only of such material could be demonstrated with alcian blue (W. P. Freeman and Simon, 1964).

There was, however, a possibility that the connective tissue holding groups of muscle cells in rosettes might have been sufficiently dense to exclude albumin. To check this point the distribution of lissamine-rhodamine labeled albumin was studied in frozen sections. This technique (S. E. Freeman *et al.*, 1966) permits of the precise localization of fluorescence in the tissue, without translocation during processing. A difficulty was encountered in transferring the technique from amphibian to marine tissue, because of the high salt concentration of seawater. As was described previously (S. E. Freeman *et al.*, 1966), the

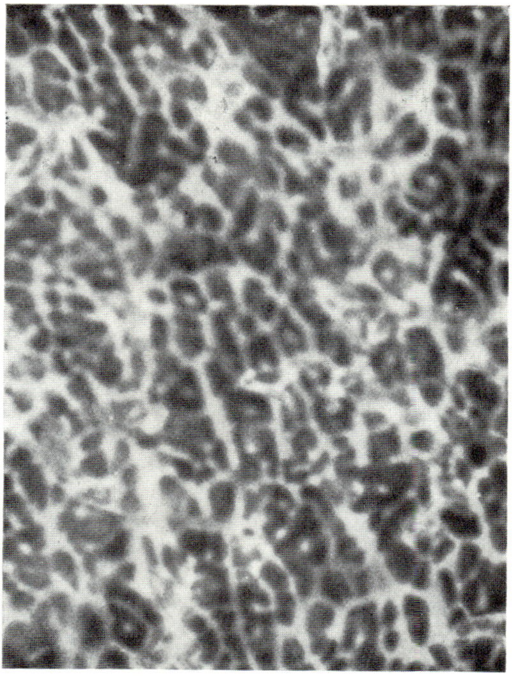

FIG. 6. Transverse section of fluorochrome-albumin treated muscle showing the brightly fluorescent connective tissue and the dark, nonfluorescent muscle cells. *Stichopus mollis.* ×136.

human albumin was conjugated with the fluorescent dye in Na_2CO_3 solution. This was then neutralized with HCl, and other salts added to make an artificial seawater (Dakin, 1960). It was found that the albumin slowly salted out of this solution. Consequently it was not possible to prepare concentrated solutions of fluorescent albumin for equilibration with holothuroidean muscle, and the sections obtained were not as brilliantly fluorescent as were obtained for vertebrate tissue.

Figure 6 shows a transverse section of holothuroidean muscle, which was equilibrated for 18 hours with fluorescent albumin prior to freezing and sectioning. It may be seen that the fluorescent albumin has penetrated all parts of the muscle, with the exception of the muscle cells. These appear as small dark areas in the photomicrograph; they have retained the typical rosette arrangement seen in paraffin sections. It was not possible to detect any difference in the intensity of fluoresence within a rosette compared to the connective tissue matrix. Consequently

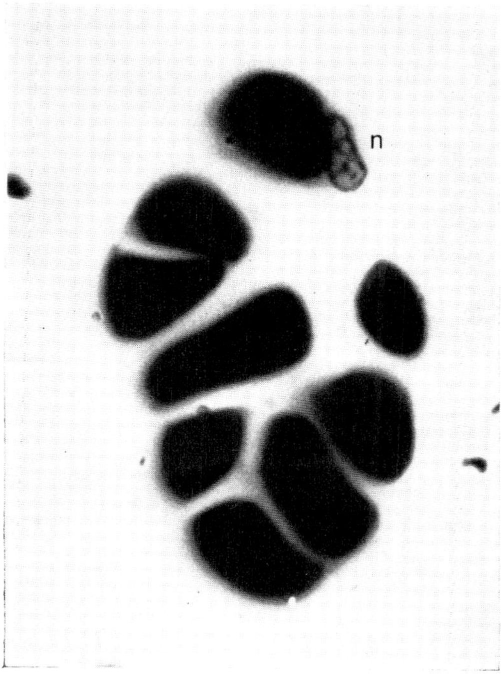

FIG. 7. Transverse section of a rosette of muscle cells showing an external nucleus (n). *Paracaudina australis.* ×2000.

we must assume that albumin can penetrate all the extracellular water of holothuroidean muscle, and offers a precise index of it.

IV. Kinetics of Albumin Diffusion

A study has been carried out of the kinetics of albumin diffusion in holothuroidean muscle (Simon *et al.*, 1964a). After removal from the animal, pieces of longitudinal muscle were equilibrated in filtered seawater containing ^{131}I-labeled albumin. The rate of loss of tracer from the muscle was then determined by serial transfer of it through a series of pots of inactive seawater. The radioactivity was counted, and an integrated curve was plotted of the amount, Q, remaining in the muscle at any time t. These curves were drawn on semilogarithmic coordinates; a typical curve may be seen in Figure 9.

Since it appears from the previous discussion that albumin is contained exclusively in the extracellular compartment, one would expect its rate of loss to approximate the form of the diffusion curve for a cylinder of the same dimensions as the muscle. As may be seen in Fig. 1, the

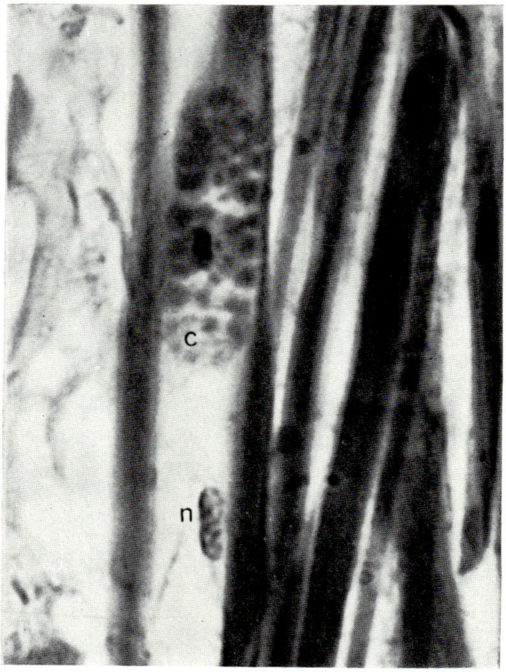

FIG. 8. Longitudinal section through the muscle showing the lateral placement of the nucleus of a muscle cell (n) and a large wandering coelombocyte (c). *Paracaudina australis.* ×450.

muscle is in fact elliptical in section. After removal from the animal it tended to become slightly more cylindrical. Measurements of the two diameters were made, and an average value used in calculating the thoeretical curve for diffusion from a cylinder. The measurements were not capable of great precision, since the muscles contracted on handling, and consequently would have varied in diameter from time to time during the efflux experiments. In order to overcome these practical difficulties all muscles were measured, and a group mean of the major and minor axis diameters determined. The theoretical curve for diffusion from a cylinder was then calculated, using this figure and the diffusion coefficient for human albumin in free solution.

The mathematical treatment of the problem has been set out elsewhere (Simon *et al.*, 1964a); it was derived from equation 5.22 of Crank's *Mathematics of Diffusion* (1956). Figure 10 illustrates the form of the theoretical curve compared to three experimental albumin efflux curves.

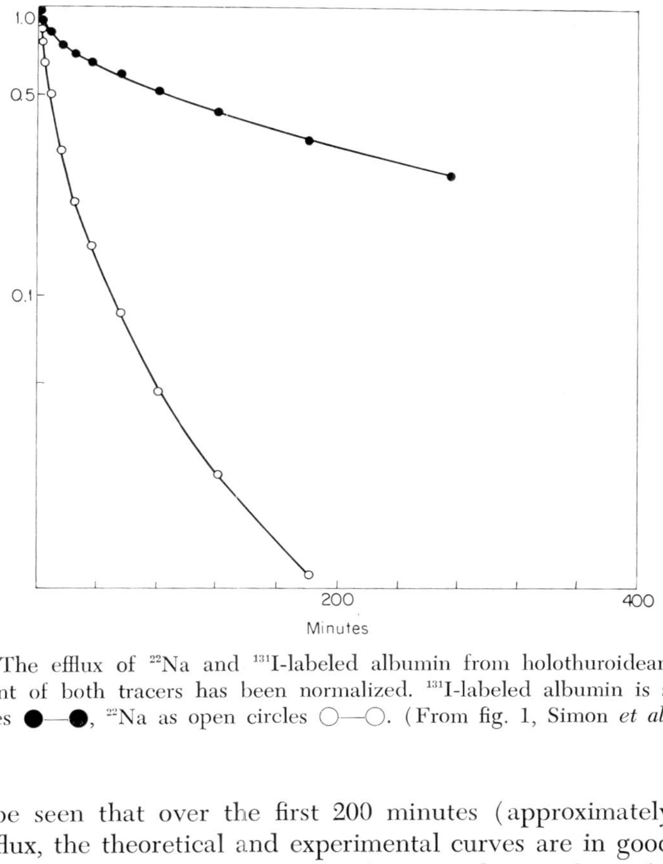

FIG. 9. The efflux of ^{22}Na and ^{131}I-labeled albumin from holothuroidean muscle. The content of both tracers has been normalized. ^{131}I-labeled albumin is shown as solid circles ●—●, ^{22}Na as open circles ○—○. (From fig. 1, Simon *et al.*, 1964a, p. 79.)

It may be seen that over the first 200 minutes (approximately 70%) of the efflux, the theoretical and experimental curves are in good agreement. At longer times the experimental curves deviate from theory, as they never straighten. The last 30% of the albumin leaves the muscle at a slower rate than predicted from the cylindrical model. It is not possible to ascribe the deviation of the experimental curves to a uniform increase in path length or decrease in diffusion coefficient, since the theoretical curve straightens after the loss of 50% of material.

It is noteworthy that the albumin efflux from other types of muscle show similar deviations from theory, when compared with the appropriate mathematical model. In the sartorius muscle of a toad (S. E. Freeman *et al.*, 1966) the experimental curve could not be approximated by either the theoretical curve for a plane sheet or a cylinder, but showed a late-time deviation rather similar to that found with holothuroidean muscle.

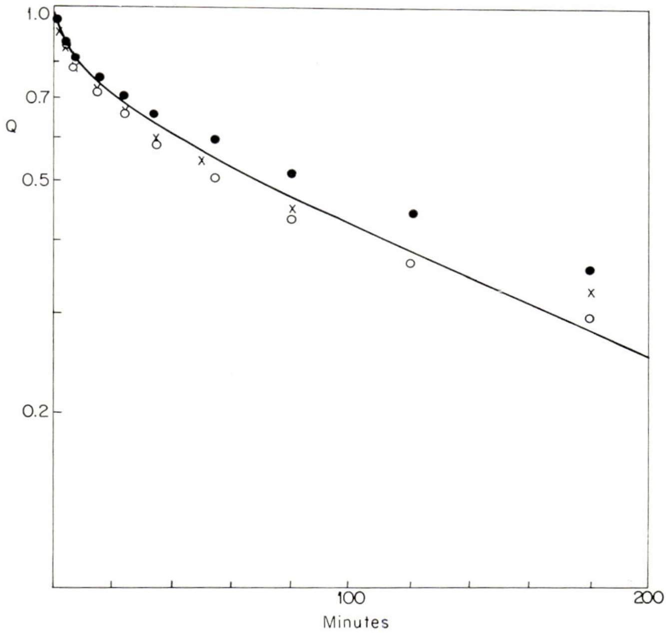

FIG. 10. A comparison of three experimental ^{131}I-labeled albumin efflux curves with the theoretical curve for diffusion from a cylinder of radius 0.2 cm. D = 5.8 $\times 10^{-7}$ cm.^{2} sec.$^{-1}$. The theoretical curve is shown as a solid line. (From fig. 2, Simon *et al.*, 1964a, p. 80.)

A study was made of the magnitude of the factors which could account for this deviation in toad muscle, and it seems reasonable to consider albumin efflux in holothuroidean muscle in the light of this study. It was concluded that the extracellular space of the toad sartorius is not homogeneous, but contains regions where albumin is sequestered so as to have an increased path length for diffusion, or a decreased diffusion coefficient, or both. An estimate was made with an analog computer of both the amount of albumin that was subjected to this sequestration, and the alteration in path length or diffusion coefficient for this amount of albumin which would cause the mathematical model to conform to the theoretical curve. From the computer solution it was found that approximately 10% of the albumin leaves the sartorius muscle at a rate consistent with a 100-fold decrease in diffusion coefficient, or a 10-fold increase in path length.

Taking into account the fact that the computer solution for albumin diffusion in the sartorius was based on a plane sheet model, whereas albumin diffusion in holothuroidean muscle is more properly described by a cylindrical model, it is likely that the sequestration of albumin is similar in both preparations.

The factors which cause the sequestration may, however, differ somewhat, because of the marked histological differences between the two muscles. It was suggested (S. E. Freeman *et al.*, 1966) that albumin diffusion in the sartorius may be impeded by very dense connective tissue, such as occurs in the adventitia of blood vessels and nerve sheaths. Structures such as these are absent in the holothuroidean. Further, the collagen network per se is unlikely to offer resistance to the diffusion of macromolecules (Ogston and Sherman, 1961).

A factor which may, however, impede diffusion in the holothuroidean is the acid mucopolysaccharide which was demonstrated histochemically to occur in association with the reticular fibers. Although histochemical techniques do not afford a quantitative estimate of this substance, the amount present appeared to be very small. Because of this it was thought unlikely that it would either be sufficient in quantity to accumulate cation (Manery, 1961) or exclude albumin (Ogston and Sherman, 1961) to a significant extent. It is, however, possible that very small amounts of mucopolysaccharide distributed through the extracellular water might, because of the extensive branching of the long molecules, afford resistance to albumin diffusion in certain areas.

V. Ionic Content

A. Steady-State Ionic Levels

The ionic content of the muscle was determined by analyzing the muscle strips for Na^+, K^+ and Cl^- immediately after removal from the animal. The *in vivo* content of these ions, and the effect of soaking the muscle strips in seawater are shown in Table II. The muscle strips were soaked in filtered seawater for 5 or 18 hours; as there was no difference between the two groups, the results have been pooled. Changes in Na^+ and Cl^- content on soaking were insignificant; there was a slight loss of K^+. The results shown in Table II have been expressed on a wet weight basis; the water content of the muscle was 84.0 ± 0.2% (S.E. of 10 observations). The intracellular ionic content of the muscle strips was estimated concurrently with the extracellular space. The muscles were immersed in seawater containing ^{131}I-labeled albumin for 5 hours, they were then blotted and weighed and the radioactivity due to ^{131}I. counted. They were then analyzed for total ionic

TABLE II

THE IONIC CONTENT OF HOLOTHUROIDEAN MUSCLE[a]

	Na^+ mM/kg.	K^+ mM/kg.	Cl^- mM/kg
In vivo	321 ± 12(8)	84.9 ± 2.4(8)	293 ± 21(8)
Soaked in seawater	346 ± 8(32)	72.7 ± 2.3(32)	307 ± 12(23)

[a] From table 1, Simon *et al.* (1964a, p. 78).

[b] Figures shown are ±S.E. The number of observations is shown in parentheses.

content. The intracellular ionic levels, as calculated from these data, were as follows: Na^+ 241 ± 9 mM/kg.; K^+ 100 ± 3 mM/kg.; Cl^- 214 ± 46 mM/kg. Each figure is ±S.E. of 20 observations. Weight changes on soaking were random and small in magnitude.

B. THE EFFECT OF INCREASE IN EXTERNAL K^+ ON THE IONIC CONTENT OF MUSCLE

Experiments were carried out to determine the effect of increases in external K^+ on the ionic content of holothuroidean muscle. Boyle and Conway (1941) were able to demonstrate increases in intracellular K^+ and Cl^- in frog skeletal muscle in response to external increases in a manner consistent with a Donnan equilibrium for these ions across the membrane. Robertson (1957) was unable to duplicate these findings in marine invertebrate muscle. Holothuroidean muscle also yielded results which were inconsistent with Donnan postulates.

The experiments were carried out in two ways; first solid KCl was added to seawater, resulting in a hyperosmotic solution. Second, the Na^+ level of an artificial seawater was reduced as the K^+ level was increased, the Cl^- level remaining constant. ^{131}I labeled albumin was added to all solutions, so that the extracellular space was determined at the same time as the ionic content.

The results of the first series of experiments are shown in Table III. Weight changes on soaking the muscles in the hyperosmotic solutions were very variable. The weight loss could not be related to the KCl level of the seawater, nor would any two pieces of muscle necessarily show a similar weight loss at any given KCl level. Average values for weight losses are shown in Table III; the ionic contents were calculated on the final weight.

There was no obvious explanation for the variability of the weight changes; it was, however, noted that high K^+ solutions caused a permanent contracture of the muscle. Were the contracture associated with a volume change, then variations in the amount of shortening might account for the variation in weight change. This suggestion received some confirmation from the finding that the tension developed by the muscle in response to high K^+ solutions showed a similar variability.

TABLE III

THE EFFECT OF HIGH KCl SEAWATER ON THE IONIC CONTENT OF HOLOTHUROIDEAN MUSCLE[a]

K_o^+ (mM/l)	Cl_o^- (mM/l)	Weight change (%)	K_i^+ (mM/kg)	K_i^+/K_o^+	Cl_i^- (mM/kg)	Cl_o^-/Cl_i^-	Na_i^+ (mM/kg)	Na_o^+/Na_i^+
10	580	−11.1	118 ± 4	11.8	230 ± 19	2.52	239 ± 18	2.32
40	610	−15.9	130 ± 5	3.25	265 ± 17	2.30	235 ± 16	2.36
80	650	−23.0	150 ± 4	1.87	277 ± 10	2.34	278 ± 19	2.00
120	690	−21.2	166 ± 3	1.38	258 ± 12	2.68	242 ± 9	2.30
180	760	−31.0	222 ± 4	1.23	280 ± 9	2.71	261 ± 11	2.13
240	810	−26.6	280 ± 4	1.16	367 ± 39	2.21	291 ± 26	1.91

[a] From table 1, Simon *et al.* (1964b, p. 90).
[b] Figures for ionic content are ±S.E. of six observations.

It may be noted from Table III that increases in external K^+ brought about increases in the intracellular level of this ion in an approximately linear manner. The slope of the curve of K_i^+ against K_o^+ was, however, less than would be expected from the Donnan theory; further, the ratio of K_i^+/K_o^+ was at no time equal to the ratio Cl_o^-/Cl_i^-.

The Na^+ level of the muscles tended to rise slightly in high K^+ solutions; the ratio of Na_o^+/Na_i^+ was very similar to the Cl^- ratio. This finding recalls the results of Simon *et al.* (1957), using the sartorius muscle of the toad.

In the second series of experiments carried out, the Na^+ level of the seawater was lowered as the K^+ level was increased, and consequently both the tonicity and the Cl^- level remained constant. Weight changes were again erratic; however, there was a consistent decrease in volume in high K^+ solutions, despite the fact that the solutions were isoosmotic with seawater. This finding is at variance with results obtained with amphibian muscle, where there is a consistent volume increase in high K^+ — low Na^+ solutions (Simon *et al.*, 1957).

The effect of variation in the external K^+ level on the intracellular ionic content may be seen in Table IV. It was of interest that there was no significant change in intracellular K^+ content at zero external K^+. The only statistically significant increase in K^+ content was found

TABLE IV

THE EFFECT OF VARIATION IN THE K^+ AND NA^+ CONTENT OF SEAWATER ON THE IONIC CONTENT OF HOLOTHUROIDEAN MUSCLE[a]

K_o^+ (m*M*/liter)	Na_o^+ (m*M*/liter)	Weight change (%)	K_i^+ (m*M*/kg.)	K_i^+/K_o^+	Cl_i^- (m*M*/kg.)	Cl_o^-/Cl_i^-	Na_i^+ (m*M*/kg.)	Na_o^+/Na_i^+
0	556	− 5.5	104 ± 7	∞	242 ± 14	2.36	235 ± 11	2.37
10	556	− 6.0	95.5 ± 4	9.55	257 ± 18	2.26	270 ± 7	2.06
40	516	−13.0	98.0 ± 3	2.45	246 ± 9	2.36	233 ± 8	2.22
120	436	−20.0	212 ± 24	1.77	291 ± 30	2.00	188 ± 24	2.32
240	316	−15.5	115 ± 16	0.48	172 ± 24	3.37	160 ± 16	1.97

[a] From table 2, Simon *et al.* (1964b, p. 93).

[b] Figures for ionic content are ±S.E. of six observations.

at 120 m*M* K^+ in the seawater. As in the previous experiment the ratio of K_i^+/K_o^+ was at no time equal to the ratio of Cl_o^-/Cl_i^-. The Na^+ content of the tissue fell as the external Na^+ level was lowered; the ratio of Na_o^+/Na_i^+ remained constant.

VI. Ionic Exchange in Holothuroidean Muscle

The steady-state exchange of Na^+, K^+ and Cl^- was determined in holothuroidean muscle using the radioactive isotopes ^{22}Na, ^{42}K, and ^{36}Cl. The methods used have been outlined previously (Simon *et al.*, 1964a,b); it suffices to comment here that the kinetics of ^{22}Na exchange were studied concurrently with the exchange of ^{131}I-labeled albumin, using a double isotope technique which involved an analysis of the energy spectra of the isotopes.

A. ^{22}NA EXCHANGE

The efflux of ^{22}Na was followed in muscles which had been equilibrated in seawater containing ^{22}Na and ^{131}I-labeled albumin for periods of 4 or 18 hours. Similar data were obtained in both instances.

As was described in Section IV, the amount of radioactivity remaining in the muscle was plotted against time on semilogarithmic coordinates, after normalizing the initial content of the tracer. Figure 9 shows the efflux curves of both substances.

It was found that the curve of ^{22}Na efflux never straightened; it was therefore not possible to divide it into a fast and a slow component, as has been customary with Na^+ efflux curves of other muscle preparations, (Harris and Burn, 1949; Keynes, 1954; Hodgkin and Horowicz, 1959). The time of 90% loss of activity from the muscle was taken as

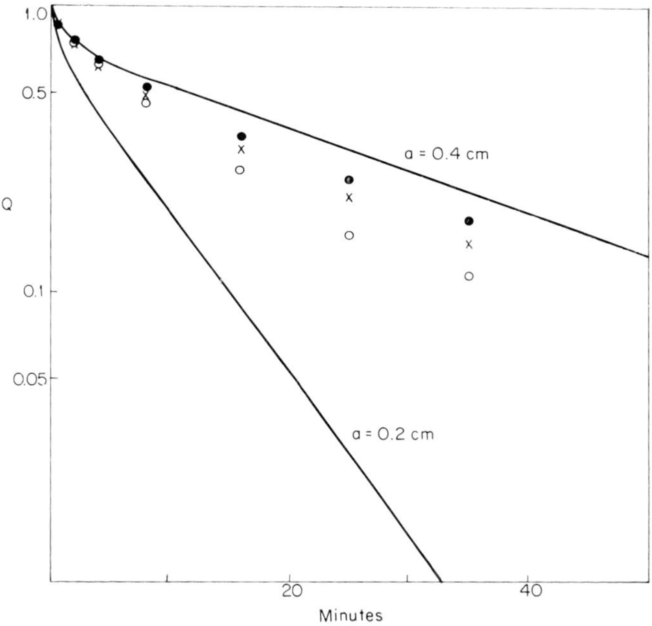

FIG. 11. A comparison of three ^{22}Na efflux curves with the theoretical diffusion curves for cylinders of 0.2 and 0.4 cm. radius (solid lines). $D = 1.48 \times 10^{-5}$ cm.^{2} sec.$^{-1}$. (From fig. 3, Simon *et al*, 1964a, p. 81.)

a measure of the rate of loss of Na^+. The average $T_{9/10}$ of the Na^+ efflux was 37.6 minutes, and the range was from 27 to 50 minutes (11 experiments).

The theoretical curve for diffusion from a cylinder was calculated in the same way as was done for albumin, using the mean axis diameter as the radius, and the diffusion coefficient for Na^+ in a solution of the appropriate salinity.

Figure 11 illustrates the first 50 minutes of three ^{22}Na efflux curves compared with the theoretical diffusion curve calculated for the observed radius (0.2 cm.) and twice the observed radius. The experimental points fall between the two theoretical curves. It is of interest that in no instance did the experimental efflux curve approximate the theoretical curve for the measured radius of the cylinder. This cannot be ascribed to an error in the estimate of the true radius, since the curves shown were obtained from the same muscles as the albumin curves shown in Fig. 10, and are consequently subject to the same errors as these curves.

The Na^+ efflux curves cannot be divided into a fast and a slow component, because they never straighten. They refer therefore to loss of tracer from both intra- and extracellular compartments. As may be seen from Fig. 11, approximately 90% of Na^+ leaves the muscle at a rate which would be consistent with an increase in path length of less than twice the measured radius. On the basis of the albumin space, which was measured concurrently with the Na^+ efflux, approximately 36% of the total Na^+ would be intracellular. It must be assumed therefore that the membrane does not impose a severe restriction on the diffusion of Na^+ in this muscle.

B. ^{36}Cl Exchange

The exchange of ^{36}Cl was followed in a way similar to that used with ^{22}Na. However, since the isotope is a beta-emitter, it was not possible to estimate the extracellular space with ^{131}I-labeled albumin at the same time as the Cl^- efflux was carried out. It was found that the normalized curve of Cl^- efflux was closely similar to the curve of Na^+ efflux. The mean $T_{9/10}$ was 37.0 minutes, and the range was from 30 to 43 minutes (5 experiments).

Once again the experimental curves were compared with the theoretical curve for diffusion from a cylinder of the measured radius and twice the measured radius. As was found with the Na^+ efflux curves, the experimental curves fell between the two theoretical curves, in a manner indistinguishable from that found with Na^+.

Although we were unable to make a concurrent estimate of the extracellular space with the Cl^- efflux, it may be seen from Table II that the total Cl^- space is similar in magnitude to the total Na^+ space. Therefore a similar proportion of the Cl^- efflux must be intracellular, and again it must be assumed that the cell membrane does not impose a severe restriction to the diffusion of Cl^-.

C. ^{42}K Exchange

The entry of ^{42}K into pieces of muscle was followed by immersing the tissue for increasing time intervals in seawater containing ^{42}K, and counting the radioactivity gained by the muscle. It was found that after an initial rapid entry the rate slowed down considerably. Figure 12 shows the semilogarithmic plot of the entry of ^{42}K into muscles from solutions containing 10, 25, and 50 mM/liter K^+. It will be recalled that Australian seawater contains 10 mM/liter K^+, therefore the lowest level may be considered physiological. It will be seen from Fig. 12 that exchange with cell K^+ was incomplete over the time shown in all

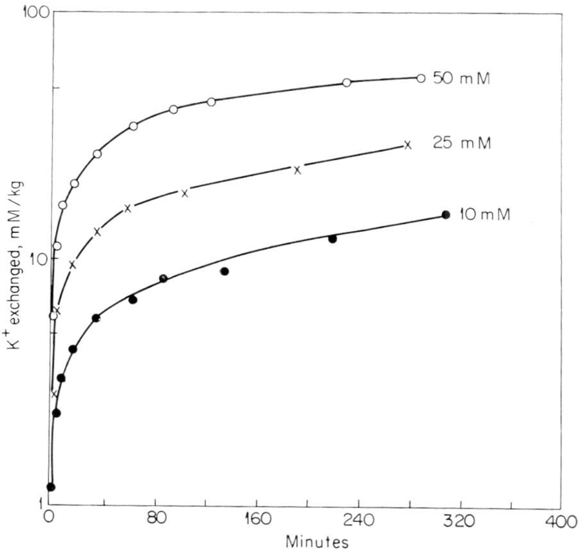

FIG. 12. The rate of entry of ^{42}K into holothuroidean muscle. The influx solution contained 10, 25, or 50 mM/liter KCl, as shown in the figure. The ordinate is mM/kg. K^+ exchanged, plotted on a logarithmic scale. (From fig. 5, Simon *et al.*, 1964b, p. 94.)

instances. In other experiments exchange was followed for longer periods. At 10 mM/liter external K^+ exchange was still incomplete at 13 hours; there was no evidence that exchange had ceased at this time, so it cannot be said that there is a "nonexchanging" fraction of cell K^+.

As has been found with other muscle preparations, the rate of exchange of cell K^+ was markedly hastened by increasing the level of K^+ in the influx solution. At 50 mM/liter external K^+ exchange was complete at 13 hours. At levels of external K^+ that were higher than in seawater there was some net entry of K^+, so that such experiments cannot be considered strictly as exchanges at the steady state.

The rate of loss of ^{42}K was followed by serial transfer of the muscle through pots of inactive seawater, after equilibration periods of either

4 or 13 hours. The equilibration solutions contained either 10 or 50 mM/liter K^+. It is clear from the influx data that the specific activity of the tissue would only be equal to that of the equilibration solution when the muscles were equilibrated for 13 hours in 50 mM/liter K^+. Should cell K^+ be nonhomogeneous, then short periods of equilibration would lead to efflux from unevenly labeled fractions.

Efflux curves have been plotted on semilogarithmic coordinates; in all instances the curve was found to straighten after approximately 40

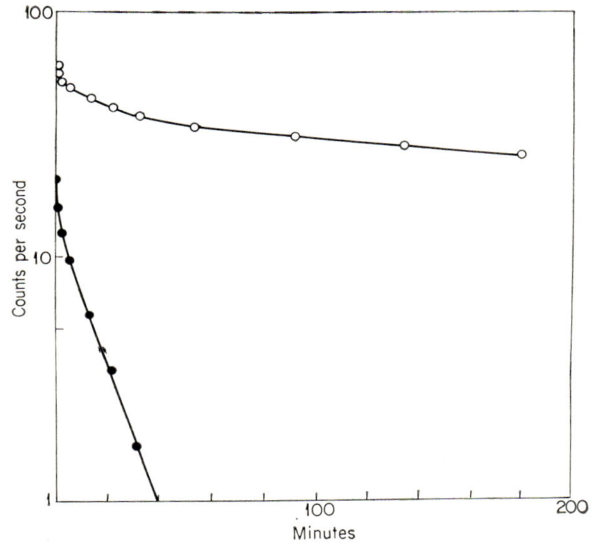

FIG. 13. The rate of loss of ^{42}K from holothuroidean muscle. The open circles ○—○ show the amount of radiactivity remaining in the muscle, plotted on a logarithmic scale. The solid circles ●—● show the fast component of efflux, obtained by subtracting the straight line extrapolate of the slow component from the observed points (see text). (From fig. 7, Simon *et al.*, 1964b, p. 95.)

minutes. Figure 13 illustrates a typical efflux curve; in this experiment both the equilibration and washout solutions contained 10 mM/liter K^+. Consequently, save for the incomplete exchange of tracer with cell K^+, the experimental conditions were close to the physiological situation.

The efflux curves were analyzed according to the procedure of Harris and Burn (1949). The method assumes a model comprising a slow rate-limiting step represented by a single exponential term, and a fast diffusion limited step, represented by an exponential series.

The quantity of K^+ exchanging in the fast component was found to be independent of the time of equilibration, with equilibration solutions of either 10 or 50 mM/liter K^+. In the first instance the fast fraction contained 10.1 ± 0.6 mM/kg. K^+ (S.E. of 20 observations). As the K^+ level of the external medium was 10 mM/liter this means that if the K^+ in this fraction is at the concentration of the external medium, then the total muscle water is available to this fraction. When the muscles were equilibrated in a 50 mM/liter K^+ solution, the fast fraction of the efflux contained 44.3 ± 2.3 mM/kg. K^+ (S.E. of 8 observations).

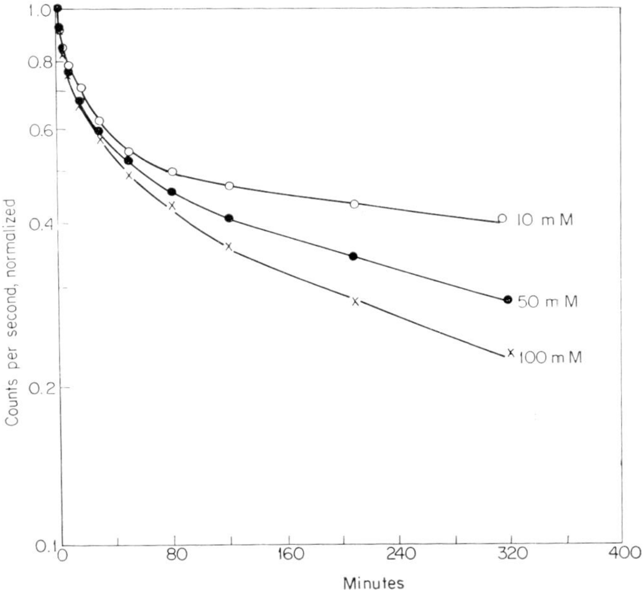

FIG. 14. The rate of loss of ^{42}K from holothuroidean muscle into seawater containing 10, 50, or 100 mM/liter KCl. (From fig. 9, Simon *et al.*, 1964b, p. 97.)

Thus in this instance 88% of the total muscle water is available to the fast fraction.

It would seem from these data that either part of the fast fraction is accumulated in the extracellular space, or part of this fraction must be intracellular.

The subtraction of the extrapolated slow component from the observed points of the efflux curve leaves a function which itself straightens after the loss of from 40–50% of the material in it. The normalized fast component was compared with the theoretical curve for diffusion from a cylinder, in the same way as was done with the Na^+ and Cl^- curves. It was found that the experimental curves fell between the curve calculated for the measured radius and the curve calculated for twice the measured radius, in a way closely similar to that found for Na^+ and Cl^-.

The quantity of material in the slow component of the efflux was dependent both upon the time of equilibration and the K^+ level of the equilibration solution. As would be expected from the influx data, slow component ^{42}K was increased both by increased time of equilibration and by raising the K^+ level of the equilibration solution. However, the rate of loss of slow component ^{42}K was independent of the time of equilibration (and hence the quantity of K^+ exchanged) and of the K^+ level of the equilibration solution. It was, however, sensitive to K^+ level of the washout solution. The half-time of slow component K^+ loss in muscles which were equilibrated in 10 m*M*/liter K^+ solution was 308 ± 38 minutes (S.E. of 11 observations). The size of the standard error indicates the scatter in results, which was as great between pieces of muscle from one animal as it was between animals.

The half time of the slow component was significantly decreased by efflux into solutions with a raised K^+ level, and increased by efflux into K^+-free solutions. This effect is illustrated in Fig. 14.

VII. Discussion

The pattern of ionic distribution in holothuroidean muscle shows certain differences from that which has been established in vertebrate preparations. These differences may result partly from the histological structure of the tissue, and may in part reflect basic differences in discriminatory mechanisms.

The study of the distribution and kinetics of exchange of albumin in the extracellular space indicate that this substance offers a reliable estimate of the extracellular water. Thus, concurrent measurements of ionic content show that Na^+ and Cl^- are partly contained in the intracellular water. The rate of exchange of these ions across the cell membrane appears, however, to be rapid. The efflux curves were found never to straighten, so it was not possible to divide the efflux into fast and slow components, as has been done with vertebrate muscle. The exchange of tracer is slower than would be expected if diffusion alone were involved; one must assume that the ions are impeded in some way when diffusing out of the muscle. The apparent increase in path length or decrease in diffusion coefficient is not uniform. The first 90% of both Cl^- and Na^+ efflux leaves the muscle at a rate which would be consistent with an increase of less than twice the estimated length; the last 10% in both instances leaves more slowly. It is difficult to envisage how, in the light of this evidence, the cell membrane could offer significant resistance to the passage of these ions. It must be realized however, that the surface-to-volume ratio of the muscle cells is very great. Thus, even were the ionic flux across the membrane relatively low, the total rate of exchange could be high, because of the large surface available for diffusion.

The accumulation of K^+ in the cell shows certain quantitative differences from the pattern found in vertebrate skeletal muscle. Alteration of the external K^+ level led to changes in the internal ionic content which were incompatible with the Donnan postulate that $(K_o^+)(Cl_o^-) = (K_i^+)(Cl_i^-)$. On the contrary, it was found that alteration of the external Na^+ or Cl^- level led to proportionate changes in the intracellular content of both ions; the ratio Na_o^+/Na_i^+ remained constant, and approximately equal to the ratio Cl_o^-/Cl_i^-. It is possible to explain the small increases in K_i^+ in response to external increases in K^+ level if one assumes Cl^- is excluded from some part of the cell water. Such an exclusion would restrict K^+ entry, for lack of a companion anion. It is not possible on present evidence to suggest where such a region might be located.

It is not possible to assume that K^+ has difficulty in penetrating any part of the cell water since first, exchange with ^{42}K is complete under suitable experimental conditions, and second, the fast component of K^+ exchange has the dimensions of the total cell water. The exchange of K^+ across the cell membrane proceeds at a rate which is closely similar to that of Na^+ and Cl^-; it would seem that in the resting muscle the membrane is equally permeable to all three ions.

Some information as to the state of the intracellular K^+ can be obtained from the kinetic experiments. The efflux data have been analyzed according to the procedure of Harris and Burn (1949), and the quantity of material in the fast and slow components obtained by a straight line extrapolation to the zero time axis. The correction suggested by Huxley (1960) was not applied, since the ratio of the half time of the slow component to the half time of the exponential portion of the fast component was of the order of 30. Under these circumstances a straight-line extrapolation would lead to a negligible error. However, interpretation of the data in terms of a model in which diffusion is limited by a resistive membrane presents obvious difficulties.

K^+ retention could be due to a specific accumulation, resulting in a decreased diffusion coefficient for this ion. If such an accumulation were due to binding sites distributed homogeneously throughout the cytoplasm then the slow component should be represented by an exponential series, rather than a single exponential term, and should have the same form as the diffusion curve shown in Fig. 10. The experimental curves cannot be fitted by such a curve, so that this simple model is unacceptable. Were one, however, to assume that the binding sites are situated in a nonhomogeneous medium, and that intersite diffusion is fast compared to the exchange process, then the experimental data could be explained.

The apparently high permeability of the cell membrane to the three ions we have studied presents problems for the interpretation of bioelectric phenomena. Intracellular potentials cannot be recorded accurately in this preparation, because of the small size of the cells. It seems likely, however, that the resting potential across the membrane amounts to some 50–60 mV. It is difficult to see how the current during an action potential could be carried by Na^+, because of the high resting permeability to this ion. One must conclude either that the kinetic data are misleading, because of the large surface-to-volume ratio of the cells, or that the inward current during an action potential may be carried by a divalent cation.

REFERENCES

Boyle, P. J., and Conway, E. J. (1941). *J. Physiol. (London)* **100,** 1.

Cobb, J. L. S., and Laverack, M. S. (1966). *Proc. Roy. Soc.* **B164,** 651.

Crank, J. (1956). "The Mathematics of Diffusion." Oxford Univ. Press (Clarendon), London and New York.

Dakin, W. J. (1960). "Australian Sea Shores." Angus & Robertson, Sydney, Australia.

Freeman, S. E., Freeman, W. P., and Dewhurst, D. J. (1966). *Australian J. Exptl. Biol. Med. Sci.* **44,** 539.

Freeman, W. P., and Simon, S. E. (1964). *J. Cellular Comp. Physiol.* **63,** 25.

Hall, A. R. (1927). *Publ. Puget Sound Biol. Sta., Univ. Wash.* **5,** 205.

Harris, E. J., and Burn, G. P. (1949). *Trans. Faraday Soc.* **45,** 508.

Hodgkin, A. L., and Horowicz, P. (1959). *J. Physiol. (London)* **145,** 405.

Huxley, A. F. (1960). Appendix 2, Part 1A, p. 163. *In* "Mineral Metabolism" (C. L. Comar and F. Bronner, eds.), Academic Press, New York.

Keynes, R. D. (1954). *Proc. Roy. Soc.* **B142,** 359.

Manery, J. F. (1961). *In* "Mineral Metabolism" (C. L. Comar and F. Bronner, eds.), Vol. I, Part B, p. 551. Academic Press, New York.

Ogston, A. K., and Phelps, C. F. (1960). *J. Physiol. (London)* **78,** 827.

Ogston, A. K., and Sherman, T. F. (1961). *J. Physiol. (London)* **156,** 67.

Olson, M. (1938). *Biol. Bull.* **74,** 342.

Robertson, J. D. (1957). *In* "Invertebrate Physiology" (B. Scheer, ed.), p. 229. Univ. of Oregon Publ., Eugene, Oregon.

Simon, S. E., Shaw, F. H., Bennett, S., and Muller, M. (1957). *J. Gen. Physiol.* **40,** 753.

Simon, S. E., Muller, J., and Dewhurst, D. J. (1964a). *J. Cellular Comp. Physiol.* **63,** 77.

Simon, S. E., Edwards, S., and Dewhurst, D. J. (1964b). *J. Cellular Comp. Physiol.* **63,** 89.

Smith, J. E. (1950). *Phil. Trans. Roy. Soc. London Ser. B* **234,** 521.

Steinbach, H. B. (1937). *J. Cellular Comp. Physiol.* **9,** 429.

Steinbach, H. B. (1940). *J. Cellular Comp. Physiol.* **15,** 1.

Steinbach, H. B. (1962). *Perspectives Biol. Med.* **5,** 338.

CHAPTER 3

Feeding, Digestion, and Nutrition in Echinodermata

John Carruthers Ferguson

I. Introduction

While the phylum Echinodermata is one of the most distinct and easily defined in the animal kingdom, these animals possess a great variety of feeding and nutritional mechanisms. It is perhaps this property, more than any other, which has been responsible for the success of the group from the Cambrian period to the present time. The nutritional physiology of relatively few of the approximately 4500 living species has been studied, however, and in most cases our understanding even of these is very incomplete. Considering these factors, and in full awareness of the difficulty of making meaningful generalizations about the widely divergent groups, the following discussion will attempt to integrate the relevant information that is available on the nutrition of these forms.

II. General Nutritive Requirements

A. Diversity of Diet

Most echinoderms will utilize almost any source of nutrients compatible with their feeding apparatus and digestive capabilities. This is particularly true for many starfishes. Wells *et al.* (1961), for example, recovered 91 different species of invertebrates representing 5 phyla from the stomachs of 124 specimens of *Astropecten articulatus.* Other workers have observed the various species of *Asterias* and *Pisaster* to feed on numerous kinds of mollusks, crustaceans (especially barnacles), polychaets, ascidians, moribund fish, and other forms (Hunt, 1925; Chidester, 1929; Galtstoff and Loosanoff, 1939; Feder, 1959; L. S. Smith, 1961; Mauzey, 1966).

The diets of sea urchins are likewise often quite varied. Fuji (1962) was able to feed 12 different types of algae to *Strongylocentrotus intermedius.* The purple sea urchin, *Arbacia punctulata,* will eat algae, coral polyps, mussels, sand dollars, dead fish, etc. (Harvey, 1956). Comparable diversity has been reported for *Echinus esculentus* (Stott, 1955) and *Psammechinus miliaris* (Eichelbaum, 1910) and doubtless occurs in numerous other species. The antarctic urchin *Sterechinus neumayeri* subsists mainly on diatoms, but will also eat red algae and seal feces (Pearse and Giese, 1966).

Many brittlestars probably are also able to subsist off of a multiplicity of food sources. While these forms are usually scavengers and ciliary-mucoid feeders, Fontaine (1965) has shown that *Ophiocomina nigra,* at least, uses a whole assortment of both microphagous and macrophagous mechanisms to collect practically any food material that is within its size range.

Cannibalistic behavior is occasionally exhibited by starfish, urchins, and other echinoderms (Coe, 1912; Galtstoff and Loosanoff, 1939; Harvey, 1956; Hancock, 1957). This phenomenon is seen most frequently in crowded aquarium specimens provided with inadequate food, and then generally only moribund individuals suffer. Even when a number of animals are clumped closely together in a community attack against some prey, those immobilized near the bottom are not likely to be harmed provided they are healthy.

Although many echinoderms are able to subsist on a great variety of foods, they may still exhibit marked preferences when provided a selection. This selectivity need not be due entirely to their own behavior, but may in part result from the effectiveness of the defensive reactions of the various prey species (cf. Feder, 1963; Hancock, 1965). Dietary preferences may also change with size, age, season, or physiological state. Small specimens of *Asterias rubens*, for example, exist mainly on barnacles, while larger ones prefer mussels and other shellfish (Hancock, 1955). Likewise, young *Asterias forbesi* prey upon oyster spat while larger individuals select, in order of preference, soft-shell clams, oysters, mussels, and hard shell clams. In general, smaller or damaged bivalves are more readily eaten (Galtstoff and Loosanoff, 1939). Vevers (1949) has noted that there is an optimum relationship between the size of a starfish and its food source. This relationship must be maintained if the starfish population is to remain large and homogeneous.

The preferred diets of sea urchins also change with size and age. Kawamura (1964) and Kawamura and Taki (1965) have demonstrated that the proportion of calcareous algae devoured by *Strongylocentrotus intermedius* is inversely proportional to the size of the urchins. Such changes are also correlated with slight variations in the geographical location of the different size urchin populations.

B. Quantity of Food

The enormous appetites of many echinoderms are legend. Sea urchins in particular may feed almost constantly if conditions are suitable. *Strongylocentrotus intermedius* will ingest up to 6% of its own weight in algae in a day (Fuji, 1962). *Echinus esculentus* can clear the barnacles and algal film off of nearly 6 $cm.^2$ of surface in an hour (Forster, 1959). Other regular urchins also have large digestive capacities (cf. Mortensen, 1938; Moore *et al.*, 1963a; Boolootian and Lasker, 1964; Lewis, 1964; Lawrence *et al.*, 1965; Moore and McPherson, 1965). Heart urchins and related forms ingest great quantities of material, but much of it is inert substrate (Buchanan, 1966). In contrast, sand dollars appear to have relatively sparse appetites (Hyman, 1958; Goodbody, 1960; Culver, 1961).

The greatest food capacities are attributed to the predaceous asteroids. In one instance, two specimens of *Asterias forbesi* were observed to devour 25 year-old oysters in 3½ days (Galtstoff and Loosanoff, 1939). The "smallest" size *Asterias rubens* can eat up to 200 barnacles in 1 week, while larger ones will dispose of two or three adult mussels (Hancock, 1955). Numerous other accounts also attest to the voraciousness of these and other starfish (*viz.* Mead, 1899; Hunt, 1925; Needler, 1941; Ino *et al.*, 1955; Wells *et al.*, 1961; Fenchel, 1965).

The feeding capacities of the remaining echinoderms are more difficult to evaluate and probably vary considerably from species to species.

Buchanan (1964), for example, has determined that the brittlestar, *Amphiura filiformis,* grows rapidly and has a high metabolic rate and food requirement, while *Amphiura chiajei,* on the other hand, is slow growing, long-lived, and has a low metabolic rate. Likewise, Tanaka (1958) points out that some sea cucumbers feed continuously while others exhibit diurnal activity.

The maximum feeding rates that have been observed in various instances for different species do not necessarily reflect their true nutritional requirements over an extended period of time. Small, young specimens are almost always more ravenous than larger, older ones (Galtstoff and Loosanoff, 1939; Kuznetzof, 1946; Fuji, 1962; Moore and McPherson, 1965). As a result of seasonal temperature fluctuations, reproductive cycles, and other factors, many species feed very little if at all for several months of the year. *Asterias forbesi* in Long Island Sound is most active from August to December. Cold weather in the winter initially inhibits its feeding but inhibition continues throughout the spring until after spawning occurs in the middle of the summer (Galtstoff and Loosanoff, 1939). Only 5% of the specimens of *Pisaster ochraceus* are found feeding in January and February as compared to 80% in July and August (Mauzey, 1966). Likewise, Moore (1937) has reported that *Echinus esculentus* only feeds during the summer and autumn, at which time digested pigment is deposited in the test. Less drastic seasonal variations in feeding have been reported for a number of other forms (Hancock, 1957; Tanaka, 1958; Fuji, 1962; Lawrence *et al.*, 1965; McPherson, 1965).

C. Survival in the Absence of Food

It is obvious from the study of seasonal feeding cycles that some echinoderms can endure rather long periods without apparent sources of nutrients. *Asterias forbesi* under normal circumstances assimilates many times as much food as is necessary for keeping it in good condition (Coe, 1912). Mead (1899) has shown that the rate of growth of this species is directly dependent on the amount of food available. Thus, excess food is, to a large extent, transformed into excess growth with little additional benefit to the organism. Vevers (1949) established a number of *Asterias rubens* in a box with practically no food for 17¼ months. Several specimens survived this period although all lost weight. Galtstoff and Loosanoff (1939) starved *A. forbesi* for 2 months without apparent harmful effects. Specimens of *Echinaster echinophorus* have been kept for over 2 months in tanks of artificial seawater changed twice daily. Over this period the animals slowly lost weight and deteriorated (personal observations). Lasker and Boolootian (1960) starved the sea urchin, *Strongylocentrotus purpuratus,* for 18 days and noted that the gonads were partially resorbed during this period of restricted food intake.

III. Mechanisms for Collecting and Ingesting Nutrients

A. Soluble Nutrients

One of the greatest revolutions in the understanding of the nutrition of echinoderms has come from the recent appreciation of the role of soluble nutrients in the economy of these forms. The first sound evidence that echinoderms, and many other invertebrates, could utilize the dissolved organic materials found in seawater was provided by Stephens and Schinske (1961). They demonstrated the uptake over 1-day periods of a high percentage (33–97) of dissolved glycine by *Asterias forbesi, Henricia sanguinolenta, Thyone briareus,* and *Leptosynapta inhaerens.* More recently *Ophiactis simplex, Ophiactis arenosa,* and *Ophionereis annulata* were also added to the list (Stephens and Virkar, 1965, 1966).

Ferguson (1963, 1967a,b, unpublished observation) has confirmed this phenomenon in *Asterias forbesi, Asterias vulgaris, Henricia sanguinolenta, Echinaster echinophorus,* and *Lytechinus variegatus.* Each of these forms has demonstrated the ability to absorb significant quantities of a variety of amino acids and glucose from low concentrations in its surrounding medium. Moreover, most of these nutrients are assimilated directly into the epidermal cells, and do not seem to move on into the internal tissues. It appears that there is a significant degree of dependency of these organisms on exogenous dissolved nutrients for the maintenance of their superficial tissues, while ingested food materials serve primarily for the support of their internal structures.

The pedicellaria found on many species of echinoderms probably function in conjunction with this assimilation mechanism. Small organisms, trapped in these structures, either autolyze or are digested by epidermal glandular secretions, and serve to enrich the fluid medium in the immediate proximity of the epidermis. The presence of different types of glands on the pedicellaria, spines, tube feet, and other areas has been noted and described by many workers (Hamann, 1884, 1885, 1887, 1889; Fujiwara, 1935; J. E. Smith, 1937; Tortonese, 1947; Pérès, 1950; Nichols, 1959a, 1961; Buchanan, 1962, 1963; Mendes *et al.,* 1963; Fontaine, 1963, 1964, 1965; Chaet, 1965; Chaet and Philpott, 1960, 1961, 1964). While the secretions of these glands doubtless serve a variety of functions, some of them have been reported as definitely toxic and possibly proteolytic.

Migratory amebocytes may also be important in epidermal digestion. According to Pequignat (Pequignat, 1966a,b; Pequignat and Tiffon, 1967) these, together with epidermal secretions, can "predigest" a variety of materials on the surface of the animals. The digestive products are then absorbed directly through the epidermis or carried into the gut. Pequignat has observed such digestion in a number of forms, including *Echinocardium cardium, Psammechinus miliaris, Asterias rubens,* and *Ophiothrix fragilis. P. miliaris* makes an especially interesting case. In its normal feeding position a "digestive pouch" is formed between its oral surface and the substrate. Large quantities of "spherulocytes" move out through the branchial tufts into this chamber and digest the organic matter on the substrate, permitting the partially solubilized products to be sucked up into the gut. The lantern teeth aid in the reduction of hard material. More fascinating still is the conjecture that the shells and other debris which this species (as several other urchins) collects on itself serve a nutritional function. The organic matter which naturally accumulates on these objects (algae, bacteria, *etc.*) is digested by the spherulocytes and absorbed by the epidermis.

B. Particulate Materials

The principal source of food for most echinoderms is most probably particulate matter of various kinds, including suspended plankton, small bottom organisms, detritus, and plant and animal debris. Numerous specific adaptations have been developed by the different species for collecting these materials.

1. Suspension Feeding

The crinoids, more so than the other groups, have developed great dependence of this source of nutrients. Unfortunately, knowledge of the physiology of these organisms is largely limited to the genus, *Antedon,* which is probably not very representative of the class as a whole. In these animals feeding takes place with the arms held up in a bowllike position, and food particles are entrapped in mucus "shot out" by papillae on the tube feet of the pinnules and the arms. The tube feet "flick" the food, with an additional "shot" of mucus, into ciliated grooves which carry it to the mouth (Nichols, 1960).

Several species of holothuroids are also suspension feeders. *Cucumaria curata* and *Psolus chitonoides,* as examples, live on rocks and other exposed areas and entrap floating organisms, particularly diatoms and small pelagic crustaceans (MacGinitie and MacGinitie, 1949; Brumbaugh, 1965). These food organisms are captured by retractile adhesive papillae located at the tip of each tentacle branch, and are passed to the mouth by the inward bending of the tentacles (Brumbaugh, 1965).

Some ophiuroids feed in a somewhat similar manner. *Amphiura filiformis* holds up its arms into the current and entraps suspended matter and phytoplankton (Buchanan, 1964). *Ophiocomina nigra* secretes a net of mucus threads between its arm spines. The material that becomes entangled in this meshwork is formed into boluses and is transported to the mouth by the tube feet (Vevers, 1956; Fontaine, 1965). Basket-stars spread out their numerous branching arms like a gigantic fan to seize planktonic particles and pass them to the mouth (Cuénot, 1948; Schroeder, 1965).

Reports of true suspension feeding among the regular echinoids are rare, although scavenging activity of pedicellaria has often been seen. Considering the diversity of this group, and the fact that a number of species have been shown to produce integumentary secretions (Hamann, 1887; Nichols, 1961), it is probable that feeding comparable to the suspension feeding of ophiuroids has been overlooked. As will be discussed later, there is evidence that some sand dollars subsist on nannoplankton.

In asteroids suspension feeding is widespread, although not always discernible from other forms of ciliary-mucoid feeding. Ciliary surface currents have been demonstrated in a number of species, and among other functions probably have a nutritional basis (Gemmil, 1915; Gislén, 1924; Budington, 1942). The greatest dependence on suspension feeding occurs in such forms as *Henricia, Porania, Linkia,* and *Echinaster,* which have well-developed Tiedemann pouches—very efficient flagellary pumps lying at the bases of the digestive glands (Anderson, 1960, 1962). The effectiveness of these may be seen in Gemmil's demonstration (1915) that *Porania* can maintain itself exclusively on particulate matter. More recently, Rasmussen (1965) has reported high efficiencies of phytoplankton utilization by *Henricia. Henricia* and *Echinaster* are often found in association with sponges (Vasserot, 1961) and appear to be benefitting from their currents in a kind of "energy commensalism" (Rasmussen, 1965). On a sea table specimens of *Echinaster* always congregate in areas of rapidly flowing water, probably demonstrating a behavior pattern related to the same purpose (Ferguson, unpublished observation).

2. *Particle-Deposit Feeding*

Particulate matter naturally collects on the sea floor, and this detritus makes an ideal habitat for numerous smaller organisms. Compositely, this material serves as a rich source of nutriment to many echinoderms.

The sea cucumber, *Cucumaria pseudopopulifera,* grows in a distinctive U-shape that enables it to lie in the bottom and extend its tentacles over the sea floor to entrap material in mucus strings, which are fed into its mouth (MacGinitie and MacGinitie, 1949). *Thyone* assumes a somewhat similar position and mops its tentacles over the bottom. Particles are grasped in the finer tentacular branches and trapped in mucus. Each tentacle, in turn, bends toward the mouth and scrapes off the adhering particles. The stomach contents are found to consist mainly of protozoans, small worms, crustaceans and other animals, together with diatoms, unicellular and filamentous algae, and detritus (Coe, 1912).

Brittlestars are especially well adapted to particle-deposit feeding, as has been demonstrated for several species, including *Amphiura chiajei* (Buchanan, 1964), *Ophiocomina nigra* (Vevers, 1956; Fontaine, 1964), *Ophiothrix fragilis* (Vevers, 1956), *Ophiothrix spiculata,* and *Ophiura leutkeni* (Austin, 1966). *Amphiodia barbarae,* a long-armed form that normally dwells several inches deep in mud, extends its arms to the surface where they sweep up detritus. The food is transferred to the mouth via ciliary-mucus tracts (MacGinitie and MacGinitie, 1949).

Many regular urchins are also efficient scavengers. Specimens of *Lytechinus variegatus* kept on a sea table scoop up loose material like vacuum cleaners and redeposit much of it as compact fecal pellets. They will also use their lanterns to rasp off brown algae growing on the exposed surfaces. Mortensen (1938) has noted that a number of deep-sea urchins survive mainly on plant debris, often of terrestial origin, which has sunk to the bottom.

A few irregular urchins also make use of the detritus on the sea floor, in addition to relying heavily on in-bottom feeding. *Echinocardium cordatum,* for example, is normally found lying several inches deep in sand, but it can extend its "funnel-building" tube feet to the surface to collect food materials (Buchanan, 1966). The digestive tracts of *Echinocardium flavescens, Spatangus purpureus, Brissopsis lyrifera,* and *Schizaster fragilis* are found to contain many materials from the sea floor including diatoms, foraminiferans, worms, coelenterates, mollusks, and echinoderms (Eichelbaum, 1910).

As previously indicated, ciliary-mucoid feeding is widespread among asteroids and is used to collect both suspended and sedentary food particles. The cushion star, *Patiria miniata,* is particularly well adapted for the utilization of the latter class of nutrients. It has an enormous stomach which is extruded over the bottom. Almost anything it covers is rapidly digested by enzymes supplied from the digestive glands via gutters in the stomach wall. Digested products are swept up into the digestive glands by flagella on the ridges between the gutters (MacGinitie and MacGinitie, 1949; Anderson, 1959; Araki, 1965). *Asterina gibbosa* and possibly other stars feed in a similar manner. A more direct approach is used by some species of *Astropecten* which may completely engorge themselves with detritus (Ferguson, unpublished observation).

3. *In-bottom Feeding*

A number of echinoderms collect their nutriment from particulate matter found within the sea bottom. Many holothuroids are especially well suited for this type of existence. *Caudina* is noted for its ability to burrow through mud and sand while swallowing large quantities of the substrate (MacGinitie and MacGinitie, 1949). *Stichopus japonicus* also ingests a great deal of bottom material, but apparently is capable of some selection. The nitrogenous content of its initial food was measured as 0.11% while the adjacent sea bottom contained only 0.03% nitrogen (Tanaka, 1958). A similar degree of selectivity has been seen in *Leptosynapta inhaerens,* whose stomach contents include (in addition to much sand) detritus, diatoms, ostracods, small mollusks, and crustacean appendages (Sanders *et al.,* 1962).

The irregular echinoids are another group that is particularly well adapted to in-bottom feeding. Heart urchins, such as *Echinocardium* and *Moira,* live buried in sandy bottoms and utilize a remarkedly diversified array of tube feet to maintain their burrows, collect food, and aid in respiration (Nichols, 1959a,b,c). Food particles trapped in mucus on "feeding" tube feet are scraped off in the mouth by means of delicate spines called the oral rakers. Food particles are also collected in mucus tracts in the ambulacral furrows and funneled toward the mouth by means of a "mucus rope," which is picked up by "oral" tube feet. Large quantities of substrate are passed through the gut. The principal food materials appear to be foraminiferans, diatoms, and other small organisms (Chesher, 1963; Buchanan, 1966). *Meoma ventricosa,* a large tropical spatanganoid, engorges itself with vast amounts of coralline sand and subsists on the foraminifera and other organic matter it contains (Hyman, 1955; Ferguson, unpublished observation).

The feeding of sand dollars is less well understood. The MacGinities (1949) reported that cilia on the aboral spine plates of *Dendraster excentricus* cause food particles to become entrapped in strings of mucus behind the spines. The mucus strings are carried to the posterior edge from which they move in special converging tracts to the mouth. Hyman (1958) failed to observe any feeding behavior of *Mellita quinqueisperforata* but did note that the gut contents consisted solely of nannoplankton-like organisms. Culver (1961) also failed in direct attempts to observe feeding in this species, but was able to demonstrate spasmodic uptake into the gut of large numbers of ^{32}P-labeled *Chlamydomonas.* From these observations it would seem that this species must depend at least in part, on suspension feeding. True in-bottom feeding has been described, however, for *Mellita sexiesperforata*. Goodbody (1960) states that this form collects fine food particles from the sand by the sorting action of spines on the aboral surface. These are carried to the oral region by means of ciliary-mucoid currents. The podia play an accessory role in collecting food.

C. Food Masses

Echinoderms, of course, are not limited to a diet of particulate food. A number of predaceous and herbivorous species survive almost exclusively on larger food masses which must be masticated and ingested, swallowed whole, or digested externally.

1. Internal Digestion

Many ophiuroids, including *Ophiocomina nigra, Ophiocomina fragilis,* and *Ophioderma panamensis,* can seize carrion or living organisms with their arms and pass the food mass to the mouth. Some mastication takes place by means of the teeth and digestion is completed within the stomach (MacGinitie and MacGinitie, 1949; Vevers, 1956; Fontaine, 1965).

In regular echinoids the food is ground up externally by the alveolar teeth of the complex Aristotle's lantern, and only smaller particles are admitted to the digestive tract. The chief food of many urchins, including *Strongylocentrotus intermedius, Diadema antillarum, Tripneustes esculentus, Lytechinus variegatus,* and *Paracentrotus lividus,* is usually algae of various sorts (Kitching and Ebling, 1961; Fuji, 1962; Lewis, 1964; Moore and McPherson, 1965). Several other species are primarily carnivorous. *Psammechinus miliaris* will attack many live animals, and on occasion even penetrate and devour oysters and other bivalves (Hancock, 1957). The intestinal contents of *Echinus esculentus* include fragments of shells, barnacles, and serpulid worms—all well worked over by the teeth (Chadwick, 1900). Particles of mussels and gastropods are seen in the fecal pellets of *Stomopneustes variolaris* (Giese *et al.*, 1964).

Some starfish, notably the various species of *Astropecten* and *Luidia,* frequently ingest their prey whole (MacGinitie and MacGinitie, 1949; Kisch, 1957; Wells *et al.*, 1961; Macne and Kalk, 1962). Swallowing is probably accomplished by partial eversion and retraction of the cardiac stomach. Mollusks that can seal themselves tight in their shells can survive for many days in the stomach cavity; those with less tight shells or a lower tolerence to anoxia succumb more rapidly (Christensen,

1962). Multirayed starfishes, such as *Pycnopodia,* also ingest large objects, especially sea urchins, through their dilated oral openings.

2. External Digestion

Because of the economic implications, considerable attention has been attracted to the notorious ability of certain asteroids to seize and open bivalves, and to digest them in their own shells by means of a bulbous inverted cardiac stomach. Among the forms known to feed in this manner are *Asterias, Pisaster, Marthasterias, Evasterias,* and *Coscinasterias.* Considerable controversy has arisen over how these animals are able to open the bivalves. Christensen (1957) has reviewed the history of these arguments. The most current view, based on the experiments of Feder (1955), Burnett (1955, 1960), Lavoie (1956), and Christensen (1957) is that the starfish can exert enough force with its tube feet to overcome the resistance of its prey, but in actuality seldom needs to, as its very thin cardiac stomach can penetrate through minor openings between the supposedly tightly closed valves. Once a small portion of the stomach has entered the mantle cavity, enzymes are released whose toxic effects overcome any further resistance on the part of the mollusk.

The stomach is everted by opening the mouth, and relaxing muscles in the stomach wall and retractor harness. The muscles of the body wall then contract and by hydrostatic pressure force the stomach out of the mouth as fluid-filled vesicles. Retraction occurs mainly through the actions of the stomach's own muscles, with the retractor harness possibly giving a final tug (Cuénot, 1887; Anderson, 1954).

Enzymes are produced by the digestive glands and passed to the exterior via grooves in the stomach wall. Digested products are taken up into the body by flagellary currents maintained over the remaining surface of the stomach (Anderson, 1953, 1954, 1960).

IV. Factors Affecting Food Selection

A. Chemosensitivity to Food

In working with *Asterias forbesi,* Galtstoff and Loosanoff (1939), and Loosanoff and Shipley (1947), were unable to demonstrate a significant chemosensitivity of this species to food, and concluded that starfish usually locate their food by coming in contact with it while crawling over the bottom. Nelson (1946), however, points out that *A. forbesi* will appear to travel several miles to seek out new beds of seed oysters. Carthy (1958) states, apparently on the basis of work by Henschel, that carbohydrates fail to initiate a feeding reaction in *Asterias rubens,* but that peptides and amino acids will. Likewise, Araki (1965) notes that proteins and protein derivatives are strongly stimulating to *Patiria miniata,* while protein hydrolyzates, amino acids, and carbohydrates have considerably less effect. These latter reports verify observations by the MacGinities (1949) and Fenchel (1965) of positive reactions to food offered starfish and brittlestars in aquaria. Milligan (1916) also observed a positive response with the urchin, *Psammechinus miliaris,* which will move directly toward food placed up to 19 inches away. It is most likely, then, that these organisms can sense the presence of food (particularly injured animals releasing body juices) but their reactions to such stimuli vary under different conditions and are subject to other uncertainties.

B. Biological Associations

While most echinoderms feed independently of the influence of other animals, cases have been documented in which the selection of food is, at least in part, tied in with another organism which does not serve as an important source of nutrition. An excellent example of such behavior has been presented by Hancock (1957), who demonstrated that the urchin, *Psammechinus miliaris,* will attack adult oysters only when they contain *Cliona* or *Polydora.*

Reference has already been made to the even more intriguing case of the feeding relationship of *Henricia* (and doubtless *Echinaster* and other forms) and sponges. Apparently, these starfish only slightly injure the sponges on which they are usually found. They benefit from the association by extracting minute food organisms from the currents produced by their hosts (Rasmussen, 1965).

There are, of course, many other symbiotic relationships between echinoderms and various organisms, but a discussion of these is beyond the scope of this treatise.

C. Environmental Effects on Feeding

1. Temperature

Environmental temperature is a most important factor in controlling the geographical range and seasonal cycles of many species. Temperature may also have a direct effect on the feeding behavior and rate of growth, although relatively few observations have been made of this property. Galtstoff and Loosanoff (1939) found that at low winter temperatures of 2°–3°C. *Asterias forbesi* remained inactive and would not feed. Likewise, considerable resistance to feeding may be encountered in this species at temperatures much above 20°C (personal observations). Fuji (1962) noted that the intake of algae by *Strongylocentrotus intermedius* was greatest at a temperatue of 10°C. Moore *et al.* (1963a), and Moore and McPherson (1965) have observed that the maximum rate of feeding of *Tripneustes esculentus* and *Lytechinus variegatus* occurs at the mean-ambient temperature of the locality and shifts with the seasons. However, maximum growth in these species is greater at lower temperatures (Moore *et al.*, 1963a,b).

2. *Salinity*

Even though there are many observations of the restrictive effect of salinity on the distribution of echinoderms (*viz.* G. F. M. Smith, 1940; Ursin, 1960), there have been practically no attempts to measure the influence of this factor on their nutrition. In one study, however, Loosanoff (1945) observed that *Asterias forbesi* will feed normally even at its lowest limit of salinity tolerance (18 parts per thousand).

3. *Substrate*

Like many other marine organisms, most echinoderms are adapted to a limited range of substrates and are normally found only in areas in which these occur. Feeding in some species, such as the burrowing heart urchins or ciliary-mucoid feeding brittlestars, is impossible if the animal is removed from its normal habitat. In other cases, the condition of the substrate would seem to have little obvious influence on the feeding mechanism. But even in these forms, the substrate may exert an effect on the feeding behavior. In at least one instance, a starfish (*Asterias rubens*) has been conditioned with food to respond to different surfaces (Sokolov, 1961). It is, therefore, possible that in nature specimens "learn" to associate various types of bottoms with specific sources of food. It is common practice among some workers to place sand, gravel, or rocks in aquaria containing starfish as "they seem to do better" under such conditions. Objective studies are needed to evaluate the true significance of this factor.

4. *Current*

The nutrition of suspension-particle feeding echinoderms is strongly dependent on natural currents, since these forms often lack means of maintaining sufficient circulation of the seawater themselves. Many other echinoderms are also commonly found in regions of flowing water, particularly in the littoral zone. While the factors that attract the many species to these areas are probably quite complex, a positive response to current has been observed in several cases and doubtless is often of prime importance. Allusion has already been made to this property in *Echinaster*. In another study, Moore and McPherson (1965) placed *Lytechinus variegatus* in troughs of running seawater and observed that increasing the flow from 2.7 to 5 liters per minute more than tripled their feeding rate. These investigators were further able to show that this effect was due solely to the current and not nutrients dissolved in the water.

5. Light

Light is another factor that can strongly influence the behavior of echinoderms. Some species are secretive while others demonstrate a distinctly positive phototaxis. It is somewhat difficult to evaluate these reactions solely in terms of nutrition, since light can also influence the behavior of many of the food organisms as well as possible predators. Tanaka (1958) has noted that sea cucumbers vary considerably in their responses to illumination, with some feeding only during the daytime and others feeding day and night. Likewise, the starfishes *Astropecten polyacanthus* and *Luidia sarsi* exhibit diurnal rhythms in feeding activity, and these appear to be controlled by light (Mori and Matutani, 1952; Fenchel, 1965). Other species, including *Asterias forbesi,* can be found feeding at any time. The urchin, *Diadema antillarum,* feeds mainly in the afternoon and early evening (Lewis, 1964). *Lytechinus variegatus* and *Arbacia punctulata* are more variable, exhibiting selective responses to different kinds of light (Sharp and Gray, 1962). Most of these responses seem to have little relationship to nutrition. *L. variegatus,* like a number of other urchins, heaps shells on itself during daylight, and this has been interpreted by these last workers as a negative phototaxis. As has been pointed out, the heaping behavior may be more significant as a nutrient collecting mechanism.

V. Digestive Mechanisms

A. Functional Types of Digestive Tracts

While epidermal assimilation of certain types of nutrient materials in echinoderms is now well established, the bulk of the food collected by these organisms is reduced and assimilated by the specialized organs of their digestive systems.

Two basic types of digestive systems are found in the phylum—a tubular tract, characteristic of crinoids, echinoids, and holothuroids, and a saclike gut, occurring in ophiuroids and some asteroids. In functional terms, the tubular tract appears to represent primarily an adaptation for continuous feeding rather than, as in vertebrates, a mechanism for the systematic degradation of the food materials. It has a particular advantage in its efficiency for compacting feces, a process of less importance to the more active forms which possess the saclike gut.

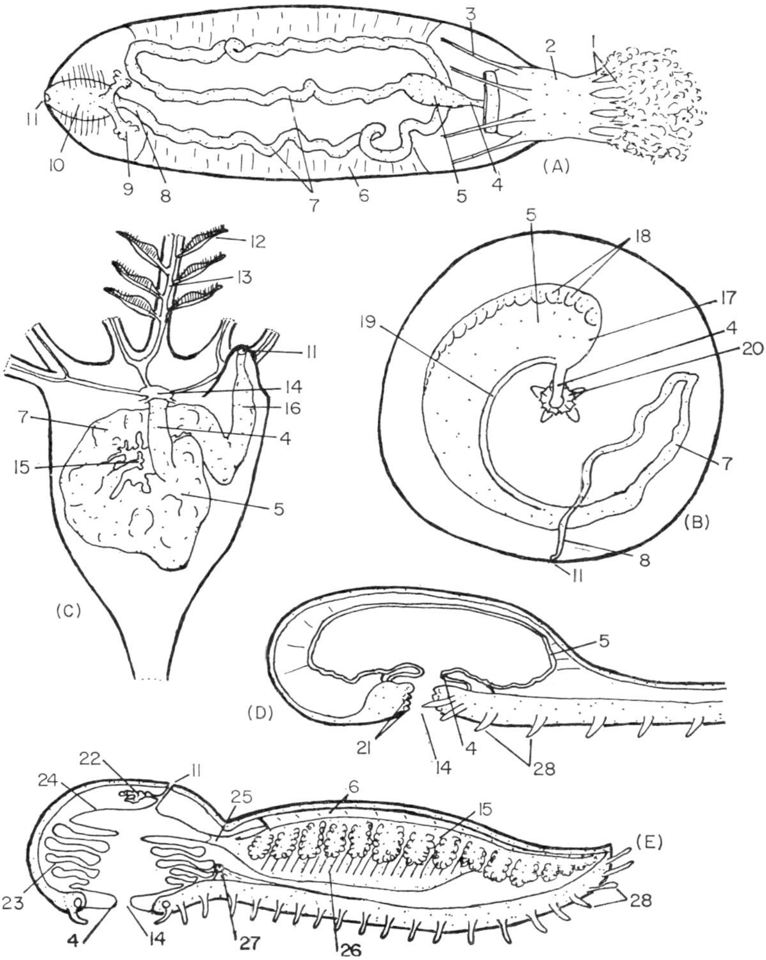

FIG. 1. Diagrammatic representations of the major features of the digestive systems of different types of echinoderms. A., sea cucumber; B., sand dollar; C., sea lily; D., brittlestar; E., starfish: 1, tentacles; 2, introvert; 3, retractor muscle; 4, esophagus; 5, stomach; 6, mesentery; 7, intestine; 8, rectum; 9, respiratory tree; 10, cloaca; 11, anus; 12, pinnule; 13, food groove; 14, mouth; 15, digestive diverticulum; 16, anal sac; 17, cecum; 18, festoons; 20, lantern; 21, teeth; 22, rectal cecum; 23, cardiac stomach; 24, pyloric stomach; 25, pyloric duct; 26, Tiedemann's pouch; 27, retractor harness; 28, tube feet.

Figure 1 illustrates diagrammatically some of the major variations in the structure of echinoderm digestive tracts. It is impossible for any such single illustration to fairly portray the great diversity of structure that exists within many of the individual classes. Neither do the published sources create a true appreciation of the diversities of microscopic and gross structure that occur. Detailed study has shown repeatedly that even slight variations in the morphology of different species may signal marked differences in function. In other respects, a number of the accounts in the literature are quite excellent. Some of the most important descriptive sources include; Tiedemann (1816), Hamann (1884, 1885, 1887, 1889), Cuénot (1887, 1891, 1948), Ludwig (1899), Schneider (1902), Chadwick (1900, 1907, 1923), Tennent and Keiller (1911), Hayashi (1935), Hyman (1955), Anderson (1953, 1954, 1959, 1960), Stott (1955, 1957), and Fuji (1961). In addition, there are numerous more limited reports, many of which are listed in the bibliography of this chapter.

B. Regional Specializations of Digestive Tracts

1. Teeth

Functional teeth are found in only two classes—Ophiuroidea and Echinoidea. In the former, they consist of vertical rows of preoral spinelike projections mounted in five pairs of "half-jaws" operated by a series of muscles. They generally appear to be adapted for selecting and grinding up small food particles.

The teeth of echinoids are incorporated into a most complex masticatory apparatus, the Aristotle's lantern. In regular urchins, this consists of 40 calcareous plates and 6 sets of muscles. Unlike the teeth of ophiuroids, this structure is adapted for rasping up bits of food and, in some cases, grinding burrows into rocks and other substrates. They have even been reported to have damaged steel pilings (Irwin, 1953). The tooth structure is similar to that of vertebrates although chemically distinct (MacGregor *et al.*, 1956). As each tooth wears away it is continually replaced by new growth on the other end (Holland, 1965). Cobb and Laverack (1966a,b,c) recently have restudied the structure and function of the lantern of *Echinus esculentus,* and described the maintenance of its complex rhythmic activities. In irregular urchins the lantern is reduced and usually does not function as a chewing organ.

2. Esophagus

The esophagus is the only portion of the digestive tract that is clearly present in all echinoderms. In asteroids and ophiuroids it may occur as little more than a short connecting duct between the oral opening and the stomach. In starfish, at least, it is rich in specialized cells with elongate nuclei, and in mucous cells (Anderson, 1954). In the other classes the esophagus is a very conspicuous organ and is often modified into a muscular pharynx. Generally it has important functions in collecting strings of food particles and propelling them into the stomach or intestine. In regular urchins the esophagus contains many unicellular glands rich in acid mucopolysaccharides as well as an abundance of mucous cells (Stott, 1955; Fuji, 1961; Holland and Nimitz, 1964). It can serve as a site for both digestion and absorption (Farmanfarmaian and Phillips, 1962). Irregular urchins lack typical gland cells but possess, instead, distinctive glandular crypts (Koehler, 1883; Hyman, 1955).

3. *Stomachic Enlargements*

In most forms the esophagus sooner or later enlarges into a region generally referred to as the stomach or stomach-intestine. In ophiuroids the stomach represents the bulk of the digestive system. It has the form of a festooned chamber and serves for the digestion, assimilation, and probably storage of the ingested food products. In spite of its apparent multitude of functions, Hamann (1889) reports little cellular specialization of its lining. There appear to be no modern studies of its structure or biochemical nature.

The stomach of asteroids is a very remarkable organ which, as has already been indicated, takes on a variety of functions, including the collection of food. It is divided into a major chamber, the cardiac stomach, and a smaller region, the pyloric stomach. The latter part functions primarily as a distributor to the digestive glands. In most asteroids the cardiac stomach is considerably larger than the space it occupies and is, therefore, normally seen formed into many folds. When everted, it can expand into often voluminous vesicles. It is well endowed with muscles and also firmly attached to various types of retractor harnesses (Anderson, 1954, 1959, 1960). The lining is rich in mucous and other types of secretory cells, but in most cases does not produce significant quantities of enzymes (Sawano, 1936; Anderson, 1954, 1959; Araki, 1965).

In holothuroids and crinoids the stomach occurs as a variously differentiated region of the first of usually three loops of the gut. In *Thyone* it is a well-demarcated structure and is clearly divided into an anterior cardiac and posterior pyloric portion. In many other sea cucumbers the differentiation of this region of the gut is less conspicuous and the term "stomach" is reserved for only the more posterior portion, the initial area being called the "crop" (Stott, 1957). The entire gut in these forms is encompassed in an elaborately developed hemal network. A battery of enzymes has been reported from the stomach of *Holothuria* (Oomen, 1926) which apparently come from fine granules in the epithelial lining (Stott, 1957), or from the numerous amebocytes which migrate between the stomach and the hemal plexus (Oomen, 1926). Absorption of dyes and other types of materials by the migrating amebocytes has also been reported by both these investigators.

The stomach of urchins is not clearly set off from the intestine. It sometimes begins as a rather prominent cecum, and may be slightly festooned along its outer edges. Secretory granules, apparently enzymes, are abundant in the lining cells. The lumen often contains many kinds of amebocytes, which appear to aid in digestion and possibly transport (Stott, 1955; Fuji, 1961). Lipid and glycogen stores have been noted, especially in the more posterior portions (Lasker and Giese, 1954; Fuji, 1961).

4. *Siphon*

In most echinoids the inner margin of the stomach is formed into a separate muscular tube, the siphon. Unlike much of the rest of the gut, it apparently is unciliated and never contains food materials (Stott, 1955). This structure probably functions as a peristaltic pump, which helps draw food particles down the esophagus and expel feces from the hindgut without disturbing the digestive actions taking place in the stomach. The internal passage of water is doubtless also an aid to respiration.

5. *Digestive Diverticula*

As previously mentioned, many echinoids possess a cecum located at the juncture of the esophagus and the stomach. This normally has few functions of its own distinct from those of the remainder of the stomach. Crinoids may also have varying numbers of cecal outpocketings of unknown significance.

It is only in the asteroids that digestive diverticula ("digestive glands," "pyloric ceca," "hepatic ceca") form truly significant portions of the digestive systems, and are, in fact, generally the largest organs of the body. Two of these structures occur in each ray suspended by paired mesenteries. They develop as outpocketings of the pyloric stomach and, in spite of their massive appearance, they retain the simple structure of the rest of the digestive system—pseudostratified epithelia underlain by thin layers of connective tissue, muscle, nerve, and peritoneum. They remain connected to the pyloric stomach by a simple, tubular duct, as in *Asterias,* or by more complex Tiedemann's pouches (*Henricia, Echinaster, Astropecten, etc.*). The Tiedemann's pouches generally are formed by the adhesion of lateral parts of the epithelia of the expanded duct to create parallel ciliated tubes which by their divergence from the main duct produce a powerful suction (Anderson, 1960, 1961). Zymogen and mucous cells are located in many regions of the ceca (Anderson, 1953), and these organs have been shown to be the principal source of the digestive enzymes (Sawano, 1936; Anderson, 1959). They also function in the absorption and storage of food materials (Irving, 1926; Farmanfarmaian *et al.*, 1958; Anderson, 1953; Ferguson, 1964a).

Most starfish also possess a set of small diverticula from the intestine, the so-called "rectal ceca." Their structure is considerably different from that of the pyloric ceca, and their true function is not clear. Some food absorption has been reported in those of *Patiria miniata* by Araki (1965).

6. *Intestine*

An intestine is not found in ophiuroids and in several families of asteroids; in other asteroids it is very reduced. The intestine of the remaining classes, however, is a large and conspicucus structure making up most of the remaining portions of the three major loops of the gut. The physiological significance of the loops, and the possible implication of countercurrent distribution, has never been evaluated beyond the obvious accommodation of increased length such an arrangement provides. The cells of the intestines of echinoids and holothuroids are usually of a more uniform nature than those of the preceding gut regions, and they generally lack significant secretory activity (Stott, 1955, 1957; Fuji, 1961; Lewis, 1964). It is probable that any digestion is restricted to the anterior portions except that which might occur through the actions of the numerous microorganisms which are usually found in the intestinal contents (Lasker and Giese, 1954).

Little absorption appears to take place in the intestine, particularly in the more posterior regions. Oomen (1926) and Schreiber (1929, 1930, 1932 believed this was due to the impermeability of the gut wall in this region. They and many others (Kawamoto, 1927; Stott, 1955, 1957; Fuji, 1961; Boolootian and Lasker, 1964) report that digestive products can, however, be removed by amebocytes migrating through the gut wall into the coelomic cavity or hemal plexus. In any case, most of the more easily assimilated nutrients are absorbed in the stomach, leaving little to be taken up in the intestine except possibly the more inert particles to which the amebocytes might be more likely attracted. Thus, the primary function of the intestine appears to be conduction and consolidation of feces.

7. *Terminal Specializations*

In those echinoderms which possess a significant intestine, there is usually also some kind of distinctive specialization of the terminal end of the digestive tract. A rectum is distinguishable in many species and doubtless functions to expel the feces. In holothuroids the rectum is followed by a muscular cloaca which gives off the respiratory trees and maintains their ventilation. Crinoids generally possess an "anal sac" which probably also has a respiratory function (Yakovlev, 1944). None of these structures has been shown to be otherwise significant in nutrition.

C. Movement of Nutrient Materials

The ingestion of food materials in many species of echinoderms takes place by means of ciliary-mucoid currents (or "flagellary," as some authors prefer). This action may be aided by movements of the peristome, introvert, or lantern. In some individuals structural modification such as the Tiedemann's pouches exist to enhance the ciliary action. The digestive tracts are often lined with cilia, which can be important in moving and mixing the gut contents (cf. Gemmil, 1915; Irving, 1924; Budington, 1942). In forms with tubular tracts, the muscle layers of the gut appear to have the primary responsibility for moving the food materials. Strong peristaltic motions have been repeatedly observed in the intestines of both sea urchins and sea cucumbers (Olmsted, 1917; Schreiber, 1930; Stott, 1955; D'Agostino and Farmanfarmaian, 1960; Fuji, 1961; Prosser *et al.*, 1965). By regulation of the hydrostatic pressure of the perivisceral coelom, the body wall may also be used to control the movement of different regions of the gut in certain cases (Gemmil, 1915; Anderson, 1954).

D. Enzyme Systems

As one might expect, echinoderms ordinarily are capable of producing a variety of extracellular enzymes sufficient to reduce the bulk of their food into an easily assimilated form. These enzymes are almost always produced by secretory cells interspersed in the same general regions in which absorption takes place, and there is little specialization of different regions of the gut for the digestion of the various classes of nutrients. The enzymes are all effective at neutral or slightly acid pH values and major deviations from this range probably do not normally occur.

These facts have been established through the efforts of numerous investigators including: Chapeaux (1893), Stone (1897), Scott (1901), Roaf (1908), Crozier (1918), Van der Heyde (1922), Irving (1926),

Oomen (1926), Weese (1926), Sawano (1936), Lasker and Giese (1954), Stott (1955), Eppley and Lasker (1959), Lasker and Boolootian (1960), Fuji (1961), Farmanfarmaian and Phillips (1962), Lewis (1964), and Araki (1965). Unfortunately, many of these workers fail to agree on specific points, and the methods used by a number of the earlier ones must be viewed with suspicion. As an example of the complications that are faced in such studies, Lasker and Giese (1954) showed that the pH of the freshly opened gut of *Strongylocentrotus purpuratus* is 7.2 to 7.3, but that it will fall rapidly to lower levels in a brief time. These same investigators failed to observe an agar-digesting enzyme that was present, probably because of heat denaturization (Farmanfarmaian and Phillips, 1962).

There is some evidence for a slight lowering of the pH in the initial regions of the gut of urchins. Stott (1955) recorded average values for *Echinus* gut as: mouth, 5.6; esophagus, 6.0; stomach, 6.3; junction, 6.7; and intestine, 6.8. While these values might be too low for the reasons indicated by Lasker and Geise, they do suggest a trend. Acid secretions (basophilic acid-mucopolysaccharides) have been reported in the esophagus by Stott and in the same region of *Strongylocentrotus* by Fuji (1961) and Holland and Nimitz (1964). These acids may be particularly significant in dissolving carbonates, especially those of calcareous algae. They may be quite important for reef-eroding species such as *Diadema antillarum* (Lewis, 1964).

There is also a slight fall in the pH of the gut of some sea cucumbers. Tanaka (1958) notes that during the 12 hours in which digestion is active in *Stichopus japonicus*, the pH drops from 6.1 to 5.6. Similarly, Crozier (1918) reported a pH of 4.8 to 5.5 for feeding animals as compared with 5.2 to 6.5 for fasted specimens. It is probable that these lower pH values are also due to acid mucopolysaccharides, as these substances have been reported present in the esophageal cells of *Cucumaria curata* (Brumbaugh, 1965).

Changes in acidity appear to be less significant in starfish. Irving (1926) records the pH of the fluid in the vicinity of the stomach of *Patiria miniata* at 7.3 to 7.5, while the contents of the pyloric ceca give a reaction of 6.6 to 7.0.

Only a little is known about the biochemical characteristics of the various enzymes produced by these forms. The proteases are quite powerful in all specimens that have been examined. Even di- and tripeptides are rapidly hydrolyzed (Sawano, 1936). Amylases (β-glucosidases) also have been widely found and can reduce starch to individual reducing sugars (Lasker and Giese, 1954; Tanaka, 1958; Araki, 1965). Galactosidases have been reported for *Strongylocentrotus* by Lasker and Giese

(1954), and alginases by Eppley and Lasker (1959), Lasker and Boolootian (1960), and Farmanfarmaian and Phillips (1962). Oshima (1931) was the first to record alginase activity in echinoderms using *Sphaerechinus pulcherinus;* he could not observe it in starfish.

The presence of true lipases in echinoderms is open to doubt. Several investigators have reported them in various groups, including Asteroidea (Stone, 1897), Echinodea (Lewis, 1964), and Holothuroidea (Clerc, 1904; Oomen, 1926). Many others failed to detect such ability. As suggested by Chapeaux (1893), fats might be digested intracellularly by amebocytes migrating in and out of the digestive cavity. There can be little doubt that this class of compounds is utilized by these organisms, although much of their own lipid is synthesized *de novo* (Allen and Giese, 1966).

E. Role of Microorganisms

Large numbers of bacteria, protozoa, and other microorganisms dwell in the intestinal regions of echinoderms. Lynch (1929, 1930) for example, has described several species of ciliates usually found in *Strongylocentrotus.* These ciliates appear to feed primarily on bacteria, which in turn obtain their nutrition from ingested algal material. The bacteria may be so numerous in the guts of sea urchins that they form mucilaginous envelopes around the algal pellets (Lasker and Giese, 1954; Farmanfarmaian and Phillips, 1962). There is some doubt, however, as to the significance of these microorganisms to the nutrition of the organisms. Certainly they do not appear to be essential, as their numbers may be drastically reduced without ill effects in a rapidly feeding urchin or in an urchin fed algae which have been kept in strong solutions of antibiotics (Farmanfarmaian and Phillips, 1962). In addition, they are found most abundantly in the intestine, where absorption, except possibly by amebocytes, appears to be limited. Nevertheless, there is still the very high probability that they are a factor in the ability of sea urchins to increase their digestive efficiencies when food is sparse. They might also contribute essential micronutrients in some cases.

F. Amebocytic Digestion

The possible contribution of certain types of amebocytes to digestion has already been mentioned. Large numbers of amebocytes containing spherule inclusions were first noted in the lumen of sea cucumbers by Frenzel (1892). In the following year, Chapeaux (1893) described digestion of fat droplets by the amebocytes of *Asterias glacialis.* Enriques (1902), Oomen (1926), and Schreiber (1930, 1932) all reported that amebocytes contributed enzymes to the stomach of holothuroids, the source of the enzymes being parts of the hemal system. After observations on *Echinus* and *Holothuria,* Stott (1955, 1957) challenged the view of these workers, but maintained the opinion that amebocytes do participate in intracellular digestion. Fuji (1962) supported Stott with his own observations on *Strongylocentrotus.* More recently, Pequignat (1966a) has reported the release of enzymes by migrating spherule-coelomocytes externally on several different echinoderms in a process he refers to as "skin digestion."

There can be no question that large numbers of coelomocytes invade the digestive organs of various holothuroids, echinoids, and possibly other echinoderms. Many of these are phagocytic. They pick up food particles, particularly the more inert ones, and either carry them off or break them down. As the mortality of these cells appears to be high, it is reasonable to assume that they can alter, through their autolysis, the digestive climate of the gut. More evidence, however, is needed in order to define their exact role in the physiology of these forms.

VI. Fate of the Digested Products

A. Nutrient Absorption

While the principal locations in which nutrients are absorbed are known, and have been indicated in the previous sections, the mechanisms by which these materials are taken up are, to a large extent, still unresolved. At least three processes might be significant: (1) simple diffusion of soluble digestive products into or through the gut epithelia, (2) active absorption and transport of these products by the gut cells, and (3) removal of digestive products by migrating amebocytes.

1. *Simple Diffusion*

The gut contents, particularly of carnivorous species, often contain high concentrations of soluble nutrients. The coelomic fluid surrounding the gut, on the other hand, invariably is very deficient in free organic substances (Lang and MacLeod, 1920; Myers, 1920; Delaunay, 1931; Hilts and Giese, 1949; Giordano *et al.*, 1950; Lasker and Giese, 1954; Bennett and Giese, 1955; Tanaka, 1958; Greenfield *et al.*, 1958; Giese *et al.*, 1959; Vasu and Giese, 1966; Ferguson, 1962a, 1964a; Lange, 1964). There may exist, then, a steep concentration gradient between the two compartments, and it seems only reasonable that some nutrients should enter the body fluids by diffusion along this gradient. A fairly rapid movement of ^{14}C-labeled nutrients into the coelom of *Strongylocentrotus purpuratus* after feeding has been described by Farmanfarmaian and Phillips (1962) and Boolootian and Lasker (1964), suggesting such a process. An active mechanism, however, would still have to exist in order to remove the nutrients from the coelomic fluids. Vasu and Giese (1966) observed a slight rise in the coelomic fluid nitrogen after feeding in *Pisaster ochraceous*, although Ferguson (1964a) could not detect such a change in *Asterias forbesi*. In both forms, the nitrogen content of the fluid was very variable.

In some forms, such as *Stichopus japonicus*, the gut contents are very low in organic materials (Tanaka, 1958). These animals are still able to absorb the nutrients with great efficiency even though the concentration gradient in these cases would be slight. Thus, while simple diffusion might possibly account for some absorption of the digestive products, it cannot by itself be responsible for all of it.

2. *Active Absorption*

Direct evidence for the active absorption of food products by the guts of echinoderms has never been obtained. In one study D'Agostino and Farmanfarmaian (1960) did demonstrate that *in vitro* the movement of glucose from the luminal to the coelomic side of *Leptosynapta inhaerens* gut could be inhibited by $MgCl_2$. It is uncertain, however, how this effect relates to the transport mechanism, as Lawrence *et al.* (1967) could observe little transport in *Stichopus paravimonsis*. Studies by Ferguson (1962b, 1964b, 1966a,b, 1967a,b, 1968) indicate that it is a common property of most starfish tissues to be able to take up dissolved nutrients from very dilute incubative solutions. The uptake mechanism exhibits competitive inhibition within various classes of substances, but is usually insensitive to the traditional metabolic inhibitors (which also fail to reduce the respiratory rate of isolated organs). It is this absorption by the cells of the body that appears to maintain the low concentrations of organic substances in the body fluids. Coupled with an apparent slow "leaking" of nutrients from the cells and an active circulation of the body fluids (cf. Irving, 1924; Budington, 1942), it is probably also responsible for much of the transport and exchange of metabolites between the various tissues of the body.

3. *Amebocytic Transport*

The phagocytic property of amebocytes has led a great many workers to believe that some types of these cells have a significant role in collecting food material from the gut and transporting it through the body. This view was probably first suggested by Cuénot (1891), although later discarded by him. It was further advanced by Chapeaux (1893), Enriques (1902), Oomen (1926), Kindred (1924, 1926), Schreiber (1930, 1932), Liebman (1950), Stott (1955, 1957), and Boolootian and

Lasker 1964). All of these investigators were primarily interested in holothuroids or echinoids which generally have a greater diversity of amebocyte types than do the other classes, and more complex hemal systems to serve as pathways for their distribution.

Whether in fact these amebocytes are significant agents for the absorption and distribution of nutrients cannot be considered resolved. There is little doubt that they can absorb dissolved substances just as other cells do, and can also carry phagocytized particles throughout the body. These functions, however, need not represent nutrient translocation. Hilgard *et al.* (1967) related the specific uptake of foreign proteins by *Strongylocentrotus purpuratus* amebocytes to an immunological response. In any case, the actions of amebocytes appear to be of secondary importance to the more direct distribution of nutrients by the circulation of the coelomic fluids.

B. Storage

The nutrients absorbed into the body need not be utilized immediately, but may be deposited in storage depots located in various regions. These storage depots appear to be most pronounced in forms which exhibit markedly discontinuous feeding or reproduction, and are less conspicuous in other species. As the digestive tract and the gonads are usually the most massive organs of the body, it is in them that the major food reserves are normally found.

In starfish, lipid stores, together with some glycogen and mucoprotein, are located in regions of the pyloric ceca (Anderson, 1953; Greenfield *et al.*, 1958; Ferguson, 1964a; Pearse, 1965) or in the Tiedemann's pouches associated with them (Anderson, 1963). These reserves may be seen to disappear after 6 to 8 weeks of starvation (Anderson, 1953). In the annual periodicity of some species, the gonads regularly appear to grow at the expense of food reserves in the pyloric ceca (Greenfield *et al.*, 1958; Farmanfarmaian and Phillips, 1962; Mauzey, 1966). During prolonged periods of starvation growth ceases and all the organs may atrophy to sustain the animal (Mead, 1899; Galtstoff and Loosanoff, 1939; Greenfield *et al.*, 1958). Holland *et al.* (1967) noted also that the coelomic nitrogen concentration of *Strongylocentrotus purpuratus* decreased during starvation.

The site of storage in sea urchins has been subject to some controversy. While the primary reserves of *Strongylocentrotus* are most likely located in the esophagus and stomach (Fuji, 1961; Farmanfarmaian and Phillips, 1962; Lawrence *et al.*, 1965), the gonads may reach considerable size and thus represent a major reservoir of material that can be utilized in time of starvation (Hilts and Giese, 1949; Giese *et al.*, 1959). Moore

(1937) has shown that the gonads of *Echinus esculentus* maintain the animal during the early winter. In the tropical urchin, *Stomopneustes variolaris,* the gonads appear to be the major store although both gut and body wall possess significant reserves in some individuals (Giese *et al.,* 1964). In all cases, lipid is the primary storage material with carbohydrate being of secondary importance.

Very little is known about nutrient storage in the other classes. As they generally exhibit more continuous feeding schedules, they perhaps are less dependent on specialized depots and rely more on a general build-up of reserves throughout the body. It is likely that some seasonal accumulation would occur in the gonads of annually reproducing species.

REFERENCES

Allen, W. V., and Giese, A. C. (1966). *Comp. Biochem. Physiol.* **17,** 23.
Anderson, J. M. (1953). *Biol. Bull.* **105,** 47.
Anderson, J. M. (1954). *Biol. Bull.* **107,** 157.
Anderson, J. M. (1959). *Biol. Bull.* **117,** 185.
Anderson, J. M. (1960). *Biol. Bull.* **119,** 371.
Anderson, J. M. (1961). *Am. Zoologist* **1,** 338.
Anderson, J. M. (1962). *Am. Zoologist* **2,** 387.
Anderson, J. M. (1963). *Proc. 16th Intern. Congr. Zool., Washington, D.C.* **3,** 124.
Araki, G. S. (1965). *Dissertation Abstr.* **23,** 4306.
Austin, W. C. (1966). Doctoral Dissertation, Stanford University.
Bennett, J., and Giese, A. C. (1955). *Biol. Bull.* **109,** 226.
Boolootian, R. A., and Lasker, R. (1964). *Comp. Biochem. Physiol.* **11,** 273.
Brumbaugh, J. H. (1965). *Dissertation Abstr.* **26,** 552.
Buchanan, J. B. (1962). *Proc. Zool. Soc. London* **138,** 645.
Buchanan, J. B. (1963). *Proc. Zool. Soc. London* **141,** 251.
Buchanan, J. B. (1964). *J. Marine Biol. Assoc. U.K.* **44,** 565.
Buchanan, J. B. (1966). *J. Marine Biol. Assoc. U.K.* **46,** 97.
Budington, R. A. (1942). *Biol. Bull.* **83,** 438.
Burnett, A. L. (1955). *Biol. Bull.* **109,** 355.
Burnett, A. L. (1960). *Ecology* **41,** 583.
Carthy, J. D. (1958). "An Introduction to the Behavior of Invertebrates." Allen & Unwin, London.
Chadwick, H. C. (1900). *Liverpool Marine Biol. Comm.* **3,** 1.
Chadwick, H. C. (1907). *Liverpool Marine Biol. Comm.* **15,** 1.
Chadwick, H. C. (1923). *Liverpool Marine Biol. Comm.* **25,** 1.
Chaet, A. B. (1965). *Ann. N.Y. Acad. Sci.* **118,** 921.
Chaet, A. B., and Philpott, D. E. (1960). *Biol. Bull.* **119,** 308.
Chaet, A. B., and Philpott, D. E. (1961). *Biol. Bull.* **121,** 373.
Chaet, A. B., and Philpott, D. E. (1964). *J. Ultrastruct. Res.* **11,** 345.
Chapeaux, M. (1893). *Bull. Acad. Roy. Med. Belg.* [3] **26,** 227.
Chesher, R. H. (1963). *Bull. Marine Sci. Gulf Caribbean* **13,** 549.
Chidester, F. E. (1929). *Science* **70,** 1818.
Christensen, A. M. (1957). *Limnol. Oceanog.* **2,** 180.
Christensen, A. M. (1962). *Proc. Constal. Shallow Water Res. Conf., 1961,* Tallahassee, p. 69.

Clerc, A. (1904). *Compt. Rend. Soc. Biol.* **56**, 748.
Cobb, J. L. S., and Laverack, M. S. (1966a). *Proc. Roy. Soc.* **B164**, 624.
Cobb, J. L. S., and Laverack, M. S. (1966b). *Proc. Roy. Soc.* **B164**, 641.
Cobb, J. L. S., and Laverack, M. S. (1966c). *Proc. Roy. Soc.* **B164**, 651.
Coe, W. R. (1912). *Conn., State Geol. Nat. Hist. Surv. Bull.* **19**, 1.
Crozier, W. J. (1918). *J. Exptl. Zool.* **26**, 379.
Cuénot, L. (1887). *Arch. Zool. Exptl. Gen.* **5**, No. 2, 1.
Cuénot, L. (1891). *Arch. Biol.* **11**, 313.
Cuénot, L. (1948). *In* "Traité de Zoologie" (P.-P. Grassé, ed.), Vol. II, pp. 1–363. Masson, Paris.
Culver, S. (1961). Master's Thesis, Duke University.
D'Agostino, A. S., and Farmanfarmaian, A. (1960). *Biol. Bull.* **119**, 301.
Delaunay, H. (1931). *Biol. Rev.* **6**, 265.
Eichelbaum, E. (1910). *Wiss. Meeresuntersuch.* **11**, 187.
Enriques, P. (1902). *Arch. Zool. Ital.* **1**, 1.
Eppley, R. W., and Lasker, R. (1959). *Science* **129**, 214.
Farmanfarmaian, A., and Phillips, J. H. (1962). *Biol. Bull.* **123**, 105.
Farmanfarmaian, A., Giese, A. C., Boolootian, R. A., and Bennett, J. (1958). *J. Exptl. Zool.* **138**, 355.
Feder, H. M. (1955). *Ecology* **36**, 765.
Feder, H. M. (1959). *Ecology* **40**, 271.
Feder, H. M. (1963). *Ecology* **44**, 505.
Fenchel, T. (1965). *Ophelia* **2**, 223.
Ferguson, J. C. (1962a). *Am. Zoologist* **2**, 522.
Ferguson, J. C. (1962b). *Biol. Bull.* **123**, 482.
Ferguson, J. C. (1963). *Am. Zoologist* **3**, 524.
Ferguson, J. C. (1964a). *Biol. Bull.* **126**, 33.
Ferguson, J. C. (1964b). *Biol. Bull.* **126**, 391.
Ferguson, J. C. (1966a). *Comp. Biochem. Physiol.* **19**, 259.
Ferguson, J. C. (1966b). *Am. Zoologist* **6**, 529.
Ferguson, J. C. (1967a). *Biol. Bull.* **132**, 161.
Ferguson, J. C. (1967b). *Biol. Bull.* **133**, 317.
Ferguson, J. C. (1968). *Comp. Biochem. Physiol.* **24**, 921.
Fontaine, A. R. (1963). *Proc. 16th. Intern. Congr. Zool., Washington, D.C.* **1**, 87.
Fontaine, A. R. (1964). *J. Marine Biol. Assoc. U.K.* **44**, 145.
Fontaine, A. R. (1965). *J. Marine Biol. Assoc. U.K.* **45**, 373.
Forster, G. R. (1959). *J. Marine Biol. Assoc. U.K.* **88**, 361.
Frenzel, J. (1892). *Arch. Anat. Physiol., Physiol. Abt.* p. 81.
Fuji, A. (1961). *Bull. Fac. Fisheries, Hokkaido Univ.* **11**, 195.
Fuji, A. (1962). *Japan. J. Ecol.* **12**, 181.
Fujiwara, T. (1935). *Annotationes Zool. Japon.* **15**, 62.
Galtsoff, P. S., and Loosanoff, V. L. (1939). *U.S. Bur. Fisheries, Bull.* **49**, 75.
Gemmil, S. F. (1915). *Proc. Zool. Soc. London* **1**, 1.
Giese, A. C., Greenfield, L., Huang, H., Farmanfarmaian, A., Boolootian, R., and Lasker, R. (1959). *Biol. Bull.* **116**, 49.
Giese, A. C., Krishnaswamy, S., Vasu, B. S., and Lawrence, J. (1964). *Comp. Biochem. Physiol.* **13**, 367.
Giordano, M. F., Harper, H. A., and Filice, F. P. (1950). *Wasmann J. Biol.* **8**, 1929.
Gislén, T. (1924). *Zool. Bidr. Uppsala* **9**, 262.
Goodbody, I. (1960). *Biol. Bull.* **119**, 80.

Greenfield, L., Giese, A. C., Farmanfarmaian, A., and Boolootian, R. A. (1958). *J. Exptl. Zool.* **139**, 507.
Hamann, O. (1884). "Beiträge zur Histologie der Echinodermen," Vol. 1. Fischer, Jena.
Hamann, O. (1885). "Beiträge zur Histologie der Echinodermen," Vol. 2. Fischer, Jena.
Hamann, O. (1887). "Beiträge zur Histologie der Echinodermen," Vol. 3. Fischer, Jena.
Hamann, O. (1889). "Beiträge zur Histologie der Echinodermen," Vol. 4. Fischer, Jena.
Hancock, D. A. (1955). *J. Marine Biol. Assoc. U.K.* **34**, 313.
Hancock, D. A. (1957). *Proc. Zool. Soc. London* **129**, 255.
Hancock, D. A. (1965). *Ophelia* **2**, 253.
Harvey, E. B. (1956). "The American Arbacia and Other Sea Urchins." Princeton Univ. Press, Princeton, New Jersey.
Hayashi, R., (1935). *J. Fac. Sci., Hokkaido Univ., Ser. VI* **4**, 1.
Hilgard, H. R., Hinds, W. E., and Phillips, J. H. (1967). *Comp. Biochem. Physiol.* **23**, 815.
Hilts, S. V., and Giese, A. C. (1949). *Anat. Record* **105**, 620.
Holland, N. D. (1965). *J. Exptl. Zool.* **158**, 275.
Holland, N. D., and Nimitz A. (1964). *Biol. Bull.* **127**, 280.
Holland, L. Z., Giese, A. C., and Phillips, J. H. (1967). *Comp. Biochem. Physiol.* **21**, 361.
Hunt, O. D. (1925). *J. Marine Biol. Assoc. U.K.* **13**, 560.
Hyman, L. H. (1955). "The Invertebrates," Vol. 4. McGraw-Hill, New York.
Hyman, L. H. (1958). *Biol. Bull.* **114**, 54.
Ino, T., Sagara, J., Jamado, S., and Tamakawa, M. (1955). *Bull. Japan. Soc. Sci. Fisheries* **21**, 32.
Irving, L. (1924). *J. Exptl. Zool.* **41**, 115.
Irving, L. (1926). *J. Gen. Physiol.* **10**, 343.
Irwin, M. (1953). *Science* **118**, 307.
Kawamoto, N. (1927). *Sci. Rept. Tohoku Univ., Fourth Ser.* **2**, 239.
Kawamura, K. (1964). *Sci. Rept. Hokkaido Fisheries Expt. Sta.* **2**, 39.
Kawamura, K., and Taki, J. (1965). *Sci. Rept. Hokkaido Fisheries Expt. Sta.* **4**, 22.
Kindred, J. E. (1924). *Biol. Bull.* **46**, 228.
Kindred, J. E. (1926). *Biol. Bull.* **50**, 147.
Kisch, B. S. (1957). *Bull. Centre Etude Rech. Sci. (Biarritz)* **2**, 9.
Kitching, J. A., and Ebling, F. J. (1961). *J. Animal Ecol.* **30**, 373.
Koehler, R. (1883). *Ann. Musee. Hist. Nat. Marseilles, Zool.* **1**, Mem. 3.
Kuznetzof, V. V. (1946). *Bull. Acad. Sci. USSR, Ser. Biol.* **4**, 431.
Lang, R. S., and MacLeod, J. J. R. (1920). *Quart. J. Exptl. Physiol.* **12**, 331.
Lange, R. (1964). *Comp. Biochem. Physiol.* **13**, 205.
Lasker, R., and Boolootian, R. A. (1960). *Nature* **188**, 1130.
Lasker, R., and Giese, A. C. (1954). *Biol. Bull.* **106**, 328.
Lavoie, M. E. (1956). *Dissertation Abstr.* **16**, 1534.
Lawrence, J. M., Lawrence, A. L., and Holland, N. D. (1965). *Nature* **205**, 1238.
Lawrence, D. C., Lawrence, A. L., Greer, M. L., and Mailman, D. (1967). *Comp. Biochem. Physiol.* **20**, 619.
Lewis, J. B. (1964). *Cana. J. Zool.* **42**, 549.
Liebman, E. (1950). *Biol. Bull.* **98**, 46.
Loosanoff, V. L. (1945). *Trans. Conn. Acad. Arts Sci.* **36**, 813.
Loosanoff, V. L., and Shipley, D. D. (1947). *Anat. Record* **99**, 87.
Ludwig, H. (1899). *Bronn's Klassen* **2**, Sect. 3.
Lynch, J. E. (1929). *Univ. Calif. (Berkeley) Publ. Zool.* **33**, 27.

Lynch, J. E. (1930). *Univ. Calif. (Berkeley) Publ. Zoo.* **33**, 307.
MacGinitie, G. E., and MacGinitie, N. (1949). "Natural History of Marine Animals." McGraw-Hill, New York.
MacGregor, A. B., MacInnes, D. G., and Marsland, E. A. (1956). *Proc. Zool. Soc. London* **127**, 573.
Macne, W., and Kalk, M. (1962). *J. Animal Ecol.* **31**, 93.
McPherson, B. F. (1965). *Bull. Marine Sci.* **15**, 228.
Mauzey, K. P. (1966). *Biol. Bull.* **131**, 125.
Mead, A. D. (1899). *Bull. U.S. Fish. Comm.* **19**, 203.
Mendes, E. G., Abbud, L., and Umiji, S. (1963). *Science* **139**, 408.
Milligan, N. H. (1916). *Zoologist* **20**, 81.
Moore, H. B. (1937). *J. Marine Biol. Assoc. U.K.* **21**, 711.
Moore, H. B., and McPherson, B. F. (1965). *Bull. Marine Sci.* **15**, 855.
Moore, H. B., Jutare, T., Bauer, J. C., and Jones, J. A. (1963a). *Bull. Marine Sci. Gulf Caribbean* **13**, 23.
Moore, H. B., Jutare, T., Jones, J. A., McPherson, B. F., and Roper, F. E. (1963b). *Bull. Marine Sci. Gulf Caribbean* **13**, 267.
Mori, S., and Matutani, K. (1952). *Publ. Seto Marine Biol. Lab.* **2**, 213.
Mortensen, T. (1938). *Annotationes Zool. Japon.* **17**, 225.
Myers, R. G. (1920). *J. Biol. Chem.* **41**, 119.
Needler, A. W. H. (1941). *Bull. Fisheries Res. Board Can.* **60**, 1.
Nelson, J. R. (1946). *Bull. Bingham Oceanog. Collection* **9**, 31.
Nichols, D. (1959a). *Quart. J. Microscop. Sci.* **100**, 73.
Nichols, D. (1959b). *Phil. Trans. Roy. Soc. London* **B242**, 347.
Nichols, D. (1959c). *Systematics Assoc. Publ.* **3**, 61.
Nichols, D. (1960). *Quart. J. Microscop. Sci.* **101**, 105.
Nichols, D. (1961). *Quart. J. Microscop. Sci.* **102**, 157.
Olmsted, J. (1917). *J. Exptl. Zool.* **24**, 333.
Oomen, H. (1926). *Pubbl. Staz. Zool. Napoli* **7**, 215.
Oshima, K. (1931). *J. Agr. Chem. Soc. Japan* **7**, 332.
Pearse, J. S. (1965). *Biol. Antarctic Seas II, Antarctic Res. Ser.* **5**, 39.
Pearse, J. S., and Giese, A. C. (1966). *Biol. Bull.* **130**, 387.
Pequignat, E. (1966a). *Nature* **210**, 397.
Pequignat, E. (1966b). *Bull. Soc. Linneenne Normandie* [10] **7**, 222.
Pequignat, E., and Tiffon, Y. (1967). *Compt. Rend.* **264**, 3014.
Pérès, J. (1950). *Arch. Zool. Exptl. Gen.* **86**, 118.
Prosser, C. L., Nystrom, R. A., and Nagae, N. (1965). *Comp. Biochem. Physiol.* **14**, 53.
Rasmussen, B. N. (1965). *Medd. Danmarks Fiskeri-og Hauundersogelser* **4**, 157.
Roaf, H. E. (1908). *Biochem. J.* **3**, 462.
Sanders, H. L., Goudsmit, E. M., Mills, E. L., and Hampson, G. E. (1962). *Limnol. Oceanog.* **7**, 63.
Sawano, E. (1936). *Sci. Rept. Tokyo Bunrika Daigaku, Sect. B* p. 179.
Schneider, K. C. (1902). "Lehrbuch der Vergleichenden Histologie der Tiere." Fischer, Jena.
Schreiber, B. (1929). *Boll. Soc. Ital. Biol. Sper.* **4**, 1187.
Schreiber, B. (1930). *Publ. Staz. Zool. Napoli* **10**, 235.
Schreiber, B. (1932). *Arch. Zool. Ital.* **16**, 865.
Schroeder, R. E. (1965). "Something Rich and Strange." Harper, New York.
Scott, F. J. (1901). *Contrib. Can. Biol. Fisheries* **1**, 49.

Sharp, D. T., and Gray, I. E. (1962). *Ecology* **43**, 309.
Smith, G. F. M. (1940). *J. Fisheries Res. Board Can.* **5**, 84.
Smith, J. E. (1937). *J. Marine Biol. Assoc. U.K.* **22**, 345.
Smith, L. S. (1961). *Behaviour* **18**, 148.
Sokolov, V. A. (1961). *Tr. Murmanskogo Morsk. Biol. Inst., Akad. Nauk SSSR* **3**, 49.
Stephens, G. C., and Schinske, R. A. (1961). *Limnol. Oceanog.* **6**, 175.
Stephens, G. C., and Virkar, R. A. (1965). *Am. Zoologist* **5**, 661.
Stephens, G. C., and Virkar, R. A. (1966). *Biol. Bull.* **131**, 172.
Stone, E. (1897). *Am. Naturalist* **31**, 1036.
Stott, F. C. (1955). *Proc. Zool. Soc. London* **125**, 63.
Stott, F. C. (1957). *Proc. Zool. Soc. London* **129**, 129.
Tanaka, Y. (1958). *Bull. Fac. Fisheries, Hokkaido Univ.* **9**, 14.
Tennent, D. H., and Keiller, V. H. (1911). *Papers Tortugas Lab. Carnegie Inst. Wash.* **3**, 111.
Tiedemann, F. (1816). "Anatomie der Röhren-Holothurie des pomeranzfarbingen Seesterns und Stein-seeigels." Thomann, Landshut.
Tortonese, E. (1947). *Monitore Zool. Ital.* **56**, 73.
Ursin, E. (1960). *Medd. Danmarks Fiskeri-og Hauundersogelser* **2**, 204.
Van der Heyde, H. C. (1922). Academical Dissertation, Univ. of Amsterdam.
Vasserot, J. (1961). *Bull. Soc. Zool. France* **86**, 796.
Vasu, B. S. and Giese, A. C. (1966). *Comp. Biochem. Physiol.* **19**, 351.
Vevers, H. G. (1949). *J. Marine Biol. Assoc. U.K.* **28**, 165.
Vevers, H. G. (1956). *Proc. Zool. Soc. London* **126**, 484.
Weese, A. D. (1926). *Publ. Puget Sound Biol. Sta., Univ. Wash.* **5**, 165.
Wells, H. W., Wells, M. J., and Gray, I. E. (1961). *Biol. Bull.* **120**, 265.
Yakovlev, N. N. (1944). *Compt. Rend. Acad. Sci. USSR* **44**, 796.

CHAPTER 4

Carbohydrates and Carbohydrate Metabolism of Echinoderms

Philip Doezema[1]

I. Introduction

To contribute information to comparative biochemistry is one of the primary purposes for this author's study of carbohydrate metabolism in echinoderms. Comparative biochemistry is a point of view rather than a separate field of study; a point of view which attempts to relate the results of biochemical research with a given organism to its unique position in both an ecological and an evolutionary sense, and, in turn, to relate these findings to the basic unity of living organisms.

All of the important early work on the biochemistry and physiology of the echinoderms is reviewed by Hyman (1955). A large body of biochemical information about echinoderms has been accumulated in the course of investigations on the development of sea urchin eggs. Much of the significance of this work is related to its developmental context, but a detailed discussion of biochemical embryology is beyond

[1] The work done in this laboratory was supported by Grants 1-F1-GM-23-159-03 and 3-T01-GM-647-05 from the U.S. Public Health Service.

the scope of this chapter. Metabolism of eggs and embryos has been reviewed elsewhere (Rothschild, 1956; Monroy, 1965) and also by Gustafson in this volume.

This chapter is intended to be a discussion of carbohydrate metabolism from the point of view of adult echinoderms. References to the biochemistry of development will be made only for comparative purposes and where information from adult animals is incomplete. The content reflects the author's interest in glycogen metabolism of sea urchins and the fragmentary state of knowledge in the biochemistry of echinoderms.

II. Mucopolysaccharides

A. Definitions and General Aspects

A mucopolysaccharide is often defined as a hexosamine-containing polysaccharide, although there is disagreement on this point (Briacombe and Webber, 1964). The mucopolysaccharides may be conveniently divided into two groups, the neutral and acid mucopolysaccharides; the latter contain carboxyl or sulfate groups which give an acid reaction while the former do not. In many instances the mucus and connective tissue substances which contain mucopolysaccharides have been studied only by histochemical techniques so that the above definitions are difficult to apply with certainty. The periodic acid-Schiff (PAS) procedure is widely used for the histochemical detection of neutral mucopolysaccharides. Glycogen is also stained by the PAS procedure but can be specifically removed by amylase. The acidic mucopolysaccharides, which as a rule do not react to the PAS procedure, do stain with toluidine blue, azure A and alcian blue. The procedures and criteria for histochemical detection of mucopolysaccharide substances are discussed more completely by Pearse (1961). The histochemistry of polysaccharides of invertebrates is the subject of a review by Gabe (1962) which includes a section on echinoderms.

The functions of mucopolysaccharide substances are highly diverse and in many instances completely unknown. The epithelial mucous secretions of the appendages are often adhesive and serve to trap food particles or assist in locomotion by providing adhesion between the walking surface and the substrate. Epithelial secretions of the alimentary tract probably lubricate the fragile tissues of the gut against abrasion by food particles. Mucopolysaccharide substances in connective tissue serve as binding cements between the connective tissue fibers. Examples of substances that have these functions will be given in the following paragraphs.

B. Mucopolysaccharide Substances in the Five Echinoderm Classes

Mucous gland cells were detected histochemically in the skin of the crinoid, *Antedon bifida* (Dimelow, 1958, cited by Giese, 1966b). *Antedon mediterranea* has mucous cells in the epidermis which are especially abundant near the food grooves. The esophagus contains clearly defined mucous cells, whereas in the remainder of the gut mucous substances are present but discrete mucous gland cells are not found. The chambers of the chambered organ contain free spherules of a material which contains much sulfated acid mucopolysaccharide (Holland, 1966).

Many ophiuroids are well supplied with mucous gland cells of several types located on the body surface, in the arm plates and spines, and on the podia (Buchanan, 1963; Austin, 1966). The secretions of these gland cells consist of several histochemically distinguishable mucopolysaccharide substances, those of the spines being important in trapping food while the glands on the podia secrete material for capturing and transferring food and also for locomotion (Austin, 1966; Fontaine, 1964). Large quantities of a mucous secretion, which is probably largely the product of a single type of integumentary gland, can be obtained from *Ophiocomina nigra* by mechanical irritation. After partial purification the mucous contained protein, amino sugar, and sulfate, and also stained metachromatically with toluidine blue (Fontaine, 1964).

The epidermis of the tentacles in the holothurian *Cucumaria curata* has numerous gland cells as does the epithelium of the anterior part of the gut, the peribuccal lip, the pharynx, and the stomach. The papillae on the tips of the tentacles secrete a special adhesive mucous which may have a role in capturing food (Brumbaugh, 1964). The staining reactions of these cells suggest neutral-, acid-, and sulfated acid mucopolysaccharides. An acid mucopolysaccharide material has been demonstrated histochemically in the longitudinal muscle of the holothurian *Stichopus mollis* (Freeman and Simon, 1964). Mucoprotein substances have been obtained by water extraction from the tissues of two holothurians, *Stichopus japonicus* and *Cucumaria japonica*. The carbohydrate portion of the material from *C. japonica* was further purified and separated into two fractions, one of which was a polyfucose sulfate; the other contained galactosamine, glucuronic acid, and sulfate in the ratio of 1:1:1 (Motohiro, 1960a,b,c,). Fibers of the cuverian tubules of *Holothuria leucospilota* consist of a collagenlike material that is associated with a mucopolysaccharide (Endean, 1957).

The anterior portions of the digestive tract of echinoids have numerous unicellular gland cells (Fuji, 1961), (Holland and Nimitz, 1964). The secretions of the mucous glands in *Strongylocentrotus purpuratus* are primarily acidic. Some of the cells incorporate sulfate-^{35}S and the label moves toward the lumen of the gut after 18 to 24 hours (Holland and Nimitz, 1964).

Partial chemical characterization of the jelly coat substance from eggs of several species of sea urchins has shown that it is an acidic protein-carbohydrate complex containing sulfate. Fucose is the primary carbohydrate constituent of the jelly coat substances of *Arbacia lixula, Pseudocentrotus depressus, Echinocardium cordatum, Strongylocentrotus droebachiensis,* and *Hemicentrotus pulcherrimus* (Monroy *et al.*, 1954; Minganti and Vasseur 1959; Vasseur, 1952; Vasseur and Immers, 1949; Nakano and Ohashi, 1954). L-Galactose is the only sugar present in the jelly coat of *Echinus esculentus* eggs (Vasseur, 1950), fructose is the sole carbohydrate constituent of the jelly from *Echinarachnius parma* (Bishop, 1951), and galactose is present in the jelly substance from the eggs of *Strongylocentrotus purpuratus* (Tyler, 1949).

An acid polysaccharide which is different from the jelly substance is released from the cortical granules of the sea urchin upon fertilization. This substance, which contains protein, aldose, hexosamine, and sulfate is thought to account for the reduction in pH of the medium by sea urchin eggs at fertilization. The polysaccharide from *Pseudocentrotus depressus* contains fucose, glucose, and hexosamine, while that from *Anthocidaris crassispina* contains only fucose, galactose, and hexosamine; the ratio of aldose to hexosamine to sulfate is 3:1:3 in both species (Ishihara, 1964). Cortical granule material isolated from unfertilized eggs of *Arbacia* contains protein, galactose, fucose, mannose, glucosamine, and galactosamine (Faust *et al.*, 1959). Immers (1961) studied the incorporation of sulfate-^{35}S into sea urchin ovaries autoradiographically by injecting labeled sulfate into the body cavity of several species of sea urchins. Sulfate was initially incorporated uniformly throughout the cytoplasm of the oocytes and then gradually moved outward to the cortex.

The jelly coat of starfish eggs is very similar to that of sea urchins. The carbohydrate portion of the jelly coat of *Asterias amurensis* eggs contains fucose, galactose, and glucosamine (Muramatsu, 1965). In a series of papers on the digestive systems of sea stars Anderson (1953, 1954, 1959, 1960) describes the location and staining properties of mucous gland cells from several species. The tube feet of starfish secrete adhesive mucus which is probably associated with locomotion (Defretin, 1952). In an ultrastructural study, Chaet and Philpott (1964) found small ellipsoidal packets in the terminal disc of the tube feet of *Asterias forbesi* which are filled with a fibrous material that stains with PAS and alcian blue. The asteroid *Pteraster tesselatus* when disturbed secretes large amounts of a mucous which is probably irritating to other animals (Rodenhouse and Guberlet, 1946, cited by Ward, 1965).

The sialic or neuraminic acids, condensation products of pyruvic acid and substituted mannosamine, are components of many mucoproteins and mucopolysaccharides that do not contain uronic acid (Gottschalk, 1960). As a rule the sialic acid residue occupies a terminal position in some oligo- or polysaccharide prosthetic group attached to protein or lipid (Gottschalk, 1960).

Sialic acid was demonstrated in the viscera of three asteroids, two holothurians, and one crinoid; from the viscera, eggs, and sperm of three echinoids; and from two whole ophiuroids (Warren, 1963). In each instance it proved to be an *N*-glycolylneuraminic acid; however, there is little information about the substance to which the sialic acid is bound. About 25% of the sialic acid in *Arbacia* eggs is soluble in lipid solvents, suggesting that at least this much is bound to glycolipid (Warren and Hathaway, 1960). About 55% of the sialic acid of *Paracentrotus lividus* eggs is found in the 2000 g particulate fraction but very little was found in the jelly layer (Perlman *et al.*, 1959). Extracts of ovary and viscera of *Arbacia punctulata* were shown to catalyze the following reaction (Warren, 1963):

$$\begin{array}{l} N\text{-acetyl-D-mannosamine-6-phosphate} + \text{phosphoenolpyruvate} + H_2O \\ \quad \rightarrow N\text{-acetyl-neuraminic acid-9-phosphate} + \text{phosphate} \end{array}$$

This reaction had previously been shown to occur in mammalian tissue (Roseman *et al.*, 1961; Warren and Felsenfeld, 1961).

III. Steroid Glycoside Toxins from Echinoderms

The steroid glycosides or saponins consist of a steroid structure (aglycone) to which is attached a chain of several monosaccharides. Holothurin, a toxin from sea cucumbers, was recognized as a saponin by Nigrelli *et al.* (1955) who showed that holothurin from *Actinopyga agassizi* could be hydrolyzed and fractionated into a mixture of several steroid aglycones and a group of sugars. Subsequent communications showed that there were four steroid aglycones, each of which had a side chain of quinovose, 3-*O*-methyl glucose, glucose, and xylose in that order (Chanley *et al.*, 1959, 1960). The xylose is attached to the steroid portion of the molecule, presumably by the hydroxyl group at C_3. Toxic saponins are found in many other holothurians (Yamanouchi, 1955; Nigrelli and Jakowska, 1960) and in some starfish (Hashimoto and Yasumoto, 1960; Rio *et al.*, 1963). Although saponins are well-known plant products, these are the first reports of such substances from animals (Nigrelli *et al.*, 1955).

IV. Structure and Distribution of Glycogen

A. Definition and General Aspects

Glycogen is a branched polysaccharide containing glucose units only and having 1–4 and 1–6 α-glucosidic linkages in a ratio that varies between 12 and 18 to 1. Glycogen is the reserve polysaccharide in nearly all animal cells (Benazzi-Lentati, 1941; Gabe, 1962) and in some plants and bacteria. In mammals, at least, glycogen has a central role in carbohydrate metabolism as the primary short-term energy reserve. The presence of glycogen in echinoderms is well documented (Moore *et al.*, 1913; Stott, 1931; Giese *et al.*, 1958; Greenfield *et al.*, 1958; Giese, 1966a,b). Glycogen from *Arbacia* eggs has been characterized as to optical rotation, hydrolysis to reducing sugar, and its staining reaction with iodine (Hutchens *et al.*, 1941). However, the nature of the material isolated from echinoderms by the standard techniques for glycogen has never been fully investigated so that many authors have hesitated to use the term glycogen without reservation (Giese, 1966a; Mohri, 1957; Spikes, 1949; Rothschild and Cleland, 1952).

B. Structure of Glycogen from *Strongylocentrotus purpuratus*

From the following it will be evident that glycogen from the purple sea urchin fits the above definition. Glycogen was obtained from whole gonads of *S. purpuratus* by extraction with cold 10% trichloracetic acid and precipitated with ethanol (Stetten *et al.*, 1956). After being twice reprecipitated from water the material contained less than 0.05% nitrogen. This material was converted quantitatively to glucose after hydrolysis in 1 *N* HCl for 3 hours as determined by Nelson's reducing sugar method (Ashwell, 1957), glucose oxidase (Worthington Glucostat reagent), and paper chromatography (Becker and Shefner, 1964; Jermyn and Isherwood, 1949). The presence of 1–4 and 1–6 bonds was confirmed by partial acid hydrolysis (Wolfrom *et al.*, 1951) which yielded only two disaccharides. They were identified by paper chromatography as maltose (4-α-glucosyl-glucose) and isomaltose (6-α-glucosyl-glucose). The results of a chain length analysis and several other standard procedures for characterization of glycogen are given in Table I together with representative data from other sources for comparison. It can be seen from the table that *S. purpuratus* glycogen is very similar to glycogens that have been isolated from a wide variety of sources. Recent work has

TABLE I

PROPERTIES OF GLYCOGENS[a]

Source	Optical rotation	Average chain length	β-Amylase limit (%)	Exterior chain length	Interior chain length	Iodine complex (mμ)	Reference[b]
S. purpuratus	190	13	47	9–10	2–3	405–410	4
Rabbit liver	188–200	11–18	31–60	6–11	3–5	450–490	1,2
Mytilus	192–196	9–13	40–46	6–9	2–4	420–435	1,2
Ascaris	196	12	48–50	7–9	2–4	420–440	2
Tetrahymena	195	13	44	(9)	(3)		1
Arthrobacter	183	8–9	54	6	2	420–440	3

[a] Average chain length of *Strongylocentrotus purpuratus* glycogen was determined by oxidation with sodium metal periodate (Hassid and Abraham, 1957). The exterior and interior chain lengths were calculated from the following relationships (Ghosh and Preiss, 1965). ECL = (number of glucose units removed by β-amylase) + 2.5; average chain length = $ECL + ICL + 1$.

[b] References: (1) Manners (1957); (2) Kjolberg *et al.* (1963); (3) Ghosh and Preiss (1965); (4) Doezema (1967).

shown that native glycogen exists in particulate form and can be isolated quantitatively by water extraction and high speed centrifugation (Barber *et al.*, 1965; Mordoh *et al.*, 1965; Bueding and Orrell, 1964). A large fraction of the glycogen from sea urchin tissue by the first steps of the procedure of Bueding and Orrell was sedimentable at 20,000 g in 1 hour, which shows its particulate nature (Doezema, 1967).

C. Glycogen Content of Echinoderm Body Components

The results of an extensive survey of the glycogen content of the body components of echinoderms are included in a paper by Giese (1966b) from which the data in Table II are taken. From this it may be said that the highest values for glycogen content are to be found in the sea urchins and sand dollars and, of the organs, the digestive tract and the gonads contain the most glycogen. Histochemical tests indicate that the glycogen of the digestive tract and diverticula is found in the tall columnar epithelium, which lines the alimentary tract and accounts for most of its thickness (Anderson, 1953; Fuji, 1961; Austin, 1966). Fuji (1961) states that glycogen is never found in the rectum of *Strongylocentrotus intermedius*. This is certainly not true of either *S. purpuratus* or *S. franciscanus*, as will be evident from Fig. 2 and the following paragraph. The lack of glycogen staining in the rectum of *S. intermedius* may be an artifact of histological procedure.

The digestive tract of *S. purpuratus* is typical of regular echinoids in that it consists of pharynx, esophagus, and an intestine which makes two circuits of the coelomic cavity, one in a clockwise direction as viewed from the oral side, and then doubles back on itself to form the second circuit which terminates in the anus (Fig. 1). The two circuits are functionally different with respect to retention of food, enzymic activity, incorporation of algal food (Farmanfarmaian and Phillips,

TABLE II

GLYCOGEN CONTENT OF ECHINODERM BODY COMPONENTS[a]

Body Component	Glycogen % dry wt.	Body Component	Glycogen % dry wt.
Pisaster ochraceus (asteroidea)		*Dendraster excentricus* (echinoidea)	
Testis	0.20–0.43	Testis-gut	2.4
Ovary	0.25–0.45	Ovary-gut	7.9
Stomach	1.50	Body wall	0.69
Ceca	0.60–2.1		
Body wall	0.11	*Parastichopus californicus* (holothuroidea)	
		Gonad	0.36
Patiria miniata (asteroidea)		Gut	0.51
Testis	2.24	Resp. tree	0.71
Ovary	0.57	Long. muscle	1.71
Body wall	0.15	Body wall	1.13
Stomach	1.11	Arms	0.16
Ceca	1.06		
		Ophioderma panamensis (ophiuroidea)	
Strongylocentrotus purpuratus (echinoidea)		Gonad	1.02
Testis	2.02–7.30	Gut	1.17
Ovary	3.90–5.50	Body wall	0.13
Gut	3.10		
Body wall	0.09	*Florometra perplexa* (crinoidea)	
Lantern (whole)	0.22	Arms and pinnae	0.035
Lantern muscle	1.6	Viscera	0.009
		Calyx	0.031
		Cirri	0.009

[a] From Giese (1966b).

1962), and histological structure (Fuji, 1961). Glycogen is not uniformly distributed along the intestine, instead most of the glycogen is found in the hindgut. Figure 2 shows representative results from a survey of glycogen distribution along the gut of *S. purpuratus*. Glycogen content of the foregut seldom exceeds 5% (dry weight), but large deposits of glycogen, when present, are always found in the hind gut (Doezema, 1967). Preliminary results indicate that a similar pattern of glycogen storage is found in the intestine of *S. franciscanus* (Doezema, 1967).

D. Gonads, Glycogen, and Reproductive Cycles

The ovaries and testes of many echinoderms go through an annual growth cycle which terminates in a more or less synchronized spawning into the sea where fertilization takes place. During the growth cycle a large amount of nutrient material, especially lipid, is accumulated by the gonad (Giese, 1959, 1966a). The roles of lipid and glycogen in the growth of gonads and gametes have been studied by examining lipid and glycogen levels in the gonads of echinoderms in relation to the yearly cycle. Sudden increases in gonad weight and glycogen content indicate periods of intense metabolic activity in the addition of storage

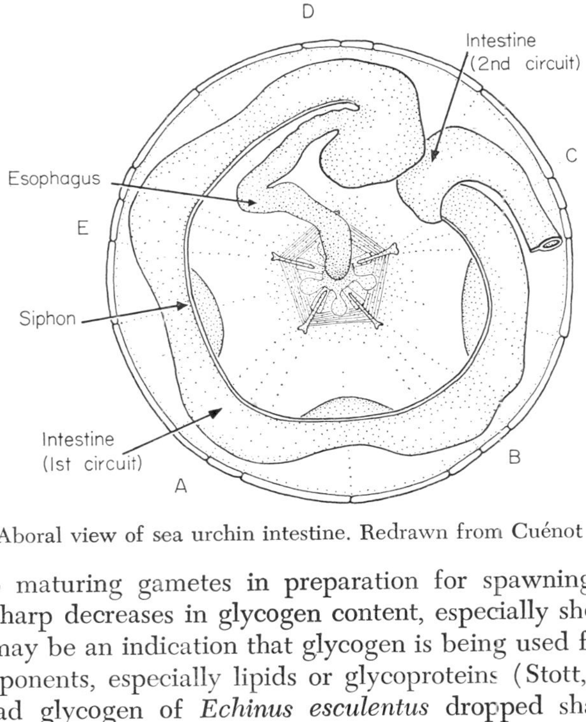

Fig. 1. Aboral view of sea urchin intestine. Redrawn from Cuénot (1948).

products to maturing gametes in preparation for spawning (Giese *et al.* 1958). Sharp decreases in glycogen content, especially shortly before spawning, may be an indication that glycogen is being used for synthesis of egg components, especially lipids or glycoproteins (Stott, 1931).

The gonad glycogen of *Echinus esculentus* dropped sharply about the time the gametes were maturing and began to rise during and after spawning (Stott, 1931). In a more complete study in which data were taken for a complete year, the glycogen content of *S. purpuratus* gonads dropped during November and December prior to spawning in January (Giese *et al.*, 1958). Although transformation of nutrient material in connection with gamete maturation was suggested, there were large fluctuations at other times of the year. Correlation of gonad glycogen

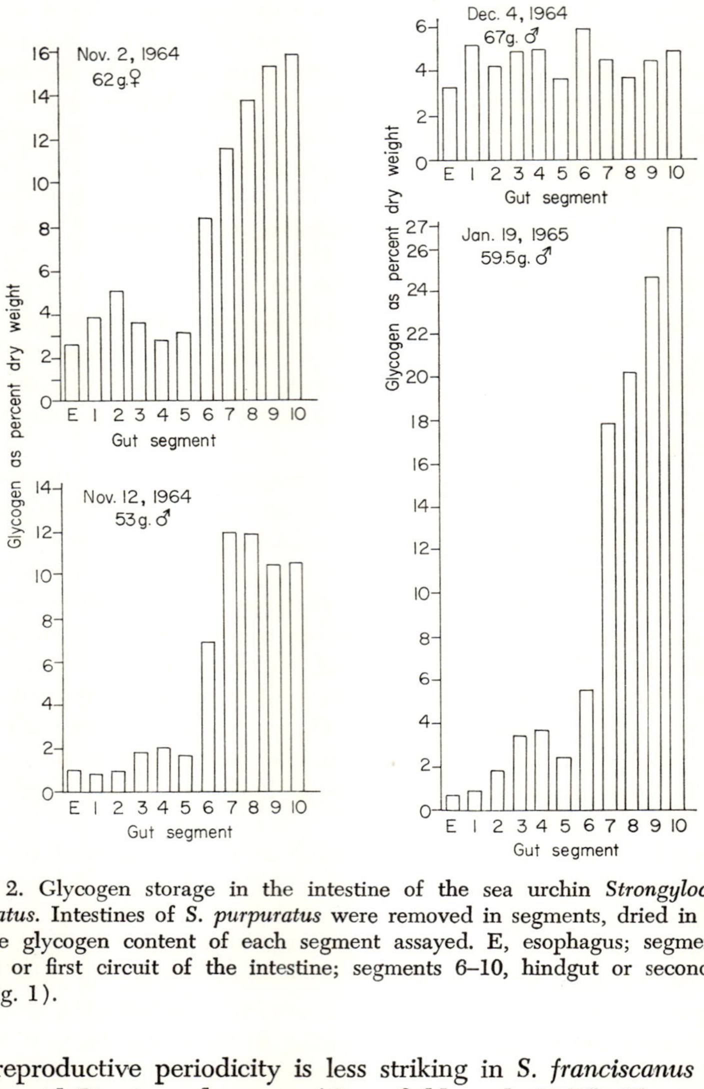

FIG. 2. Glycogen storage in the intestine of the sea urchin *Strongylocentrotus purpuratus*. Intestines of *S. purpuratus* were removed in segments, dried in acetone, and the glycogen content of each segment assayed. E, esophagus; segments 1–5, foregut or first circuit of the intestine; segments 6–10, hindgut or second circuit (see Fig. 1).

with reproductive periodicity is less striking in *S. franciscanus* and in the asteroid *Pisaster ochraceous* (Greenfield *et al.*, 1958). Glycogen content in *Pisaster* gonads is about 1% at most (Greenfield *et al.*, 1958) whereas in the sea urchins, *E. esculentus*, *S. purpuratus*, and *S. franciscanus*, the gonad glycogen may amount to more than 10% of dry weight (Moore *et al.*, 1913; Stott, 1931; Giese *et al.*, 1958; Giese, 1966b). Much of the glycogen of the gonads is found in the gametes themselves. Örstrom and Lindberg (1940) report glycogen content of *P. lividus* eggs to be about 3.6% dry weight. *E. esculentus* eggs have about 1.5% dry weight of glycogen and *Arbacia* has about 3% glycogen in the eggs (Cleland and Rothschild, 1952a; Hutchens *et al.,* 1941). Glycogen is also present in sea urchin sperm (Spikes, 1949; Mohri, 1957). From data given by Rothschild and Cleland (1952), a value of 1.1% dry weight of glycogen can be calculated for sperm of *E. esculentus.*

V. Physiology and Metabolism of Glycogen

A. Glycogen as Nutritional Reserve

The data that are presently available indicate that glycogen stored in the gut of echinoderms is available as nutritional reserve while that of the gonads is not. The rate of utilization of glycogen during starvation seems to be low, but that may be a property of heterothermic animals in general.

The data for gonads are relatively scarce and must be treated with caution because of the changes which are taking place with respect to the reproductive cycle. Twenty days of starvation did not affect the glycogen content of *E. esculentus* gonads (Moore *et al.,* 1913). Glycogen was 3.5% of dry weight in gonads of *S. franciscanus* that had been starved for 60 days and 3.8% in animals that had starved 90 days, showing no decrease (each figure is the mean of 10 animals), although animals that had been fed *Macrocystis* deposited glycogen in the gonads during this time (A. L. Lawrence, 1966). This is in agreement with recent results which show that lipid stores in the gonad decline only slightly during 60 days of starvation whereas gut lipid is readily used (J. M. Lawrence *et al.,* 1966).

Figure 3 shows that marked depletion of the glycogen stores in the gut of *S. purpuratus* is noticeable in a matter of weeks. This is taken as evidence that the glycogen of the intestine is a storage product that is readily available for use during starvation (Doezema, 1967). Anderson (1953) states that during prolonged starvation glycogen in storage cells of the pyloric ceca of *Asterias forbesi* is used first and that the disappearance of lipid and a glycoprotein which is stable to diastase follows.

Utilization of glycogen in the echinoderms seems almost insignificant when compared to that of a mammal. A rat which has about 14% glycogen (dry weight) in the liver will lose over 90% of this glycogen reserve in 24 hours (Stimpson, 1965). The difference in rate of glycogen utilization between the rat and the sea urchin may be accounted for by the fact that oxygen consumption in the rat is on the order of 100 times greater than it is in the sea urchin (Altman and Dittmer, 1964; Giese *et al.,* 1966). In some heterothermic vertebrates glycogen disappearance is as slow as or slower than in the echinoderms. In livers of young goldfish, which contain about as much glycogen as rat livers, only 50% of the glycogen was used during 8 days of starvation and older goldfish lost no liver glycogen after prolonged fasting (Stimpson, 1965). Livers of starved and estivating African lungfish lost only half of the stored glycogen during a period of 6 months (Janssens, 1964). Oxygen con-

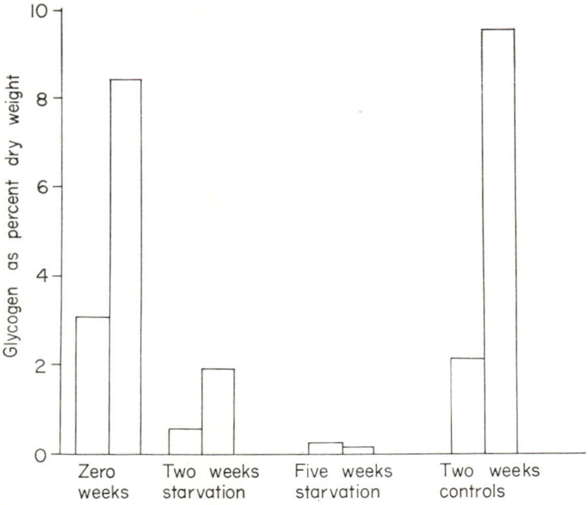

FIG. 3. Effect of starvation on gut glycogen of *S. purpuratus*. Glycogen analyses were performed separately on fore- and hind-gut. The left bar of each pair represents foregut, the other hind-gut. There were six animals in the time zero and 5-week samples and four animals in each of the 2-week samples.

sumption in goldfish and lungfish is only slightly greater than for sea urchins (Altman and Dittmer, 1964).

B. SYNTHESIS OF GLYCOGEN IN THE INTESTINE OF *S. purpuratus*

Since it has been shown that glycogen storage is located primarily in the second circuit of the intestine (Fig. 2), it would seem probable that this could be accounted for in one of two ways: either glycogen is broken down more rapidly in the first circuit of the intestine, or it is synthesized more rapidly in the second circuit. The starvation experiment eliminates the first hypothesis. Figure 3 shows that for at least the first 2 weeks of starvation the rate of glycogen loss from the two circuits was about equal.

Experiments were then undertaken to test the second hypothesis that localized synthesis accounts for the distribution of glycogen in the gut.

Concurrent measurements were made of glycogen content and glycogen synthesis from radioactive glucose in relation to position along the gut using the system of ten divisions that was used in the glycogen distribution study. Figure 4 shows the results of an experiment in which the total amount of glycogen and the amount of radioactivity in the glycogen is plotted for each of ten segments plus the esophagus. The experiment was performed *in vitro* so that a point on the synthesis curve of Figure

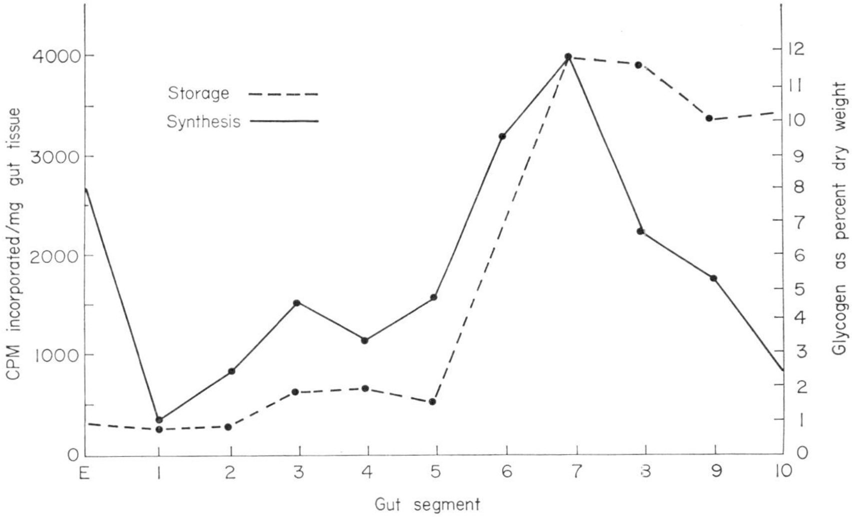

FIG. 4. Glycogen storage and synthesis in the intestine of the sea urchin. Previously washed segments were placed separately in 1 ml. of artificial seawater containing 0.5 μ curies/ml. of radioactive glucose at about 0.1 *M* concentration. The segments were incubated at 15° for 6 hours and then dropped into acetone, after which the glycogen was purified and sampled for glycogen assays and radioactive counting. Most of the radioactivity in the solution had been removed by 1 hour.

4 represents only the ability of that segment to absorb glucose from a standard solution and incorporate it into glycogen, because all effects of rates of passage of materials along the alimentary tract are eliminated. The results show that the rate of glycogen synthesis is much greater in the second circuit of the intestine. The sharp increase of glycogen synthesis correlates well with the increase of glycogen content at segment 6, the one immediately posterior to the sharp bend in the intestine (Fig. 1).

The coelomic cavities of a number of animals were injected with radioactive glucose. The pattern of incorporation of radioactivity into glycogen in these experiments failed to show any prominent peaks; the incorporation was nearly uniform along the entire intestine.

The results of these two experiments may have some significance in the digestive physiology of the sea urchin. In the *in vitro* experiment both lumen and coelomic surfaces of the intestinal segments were exposed to uniform concentrations of radioactive sugar solution. Where the radioactive glucose was injected only the coelomic surface was exposed. Since incorporation was uniform when only the coelomic surface was exposed, the peak of incorporation noted in the first experiment must be due to enhanced uptake by a small region of the lumen surface or digestive epithelium. This region of enhanced uptake is just posterior to the junction between the first and second circuits of the intestine where glycogen content begins to increase and it is also just here that the siphon enters the gut.

The siphon is a small tube which runs along the inner border of the intestine from the junction of the esophagus to a point just before the beginning of the second circuit, by-passing the first circuit (Fig. 1). A current of water passes through the siphon in an oral to anal direction. If a dye solution is placed into the water surrounding an urchin, color appears in the esophagus, the siphon, and the second circuit of the intestine while the first circuit remains free of color (Cuénot, 1948). One function proposed for the siphon is to divert water which has been swallowed with food to the posterior circuit of the intestine with the result that the digestive enzymes in the anterior portions of the gut are not diluted (Cuénot, 1948). It is well established that sea urchins have the ability to digest a wide variety of polysaccharides (Hyman, 1955). Enzymes which release reducing sugar from agar and fragments of the alga *Iridaea* are far more abundant in the esophagus and the most anterior parts of the intestine (Farmanfarmaian and Phillips, 1962). It is likely that since digestive activity is very high in the anterior digestive tract, fluid which enters the siphon would be quite rich in dissolved sugars and oligosaccharides. If this is so, then the area of high incorporation of sugar demonstrated in Fig. 4 would represent a modification for the uptake of these dissolved substances which are deposited there by the siphon. This hypothesis is further supported by the observation that the hind-gut, which begins just posteriorly to the siphon, is the primary location of glycogen storage.

Farmanfarmaian and Phillips (1962) and Boolootian and Lasker (1964) studied uptake of nutrients by feeding algae which had been labeled by exposure to $NaH^{14}CO_3$ to *S. purpuratus* and measured the distribution of total radioactivity during digestion and absorption. At one day after feeding most of the radioactivity in the intestine had been incorporated into the first circuit. Incorporation of algal radioactivity into other intestinal tissue components probably obscured the carbohydrate labeling pattern since glycogen constitutes only about 5% of the dry weight. Lipid and protein constitute a much larger proportion of the intestinal tissue.

C. Pathways of Carbohydrate Metabolism

Glycogen phosphorylase activity has been demonstrated in extracts of gut and jaw muscles of *Strongylocentrotus purpuratus* (Doezema,

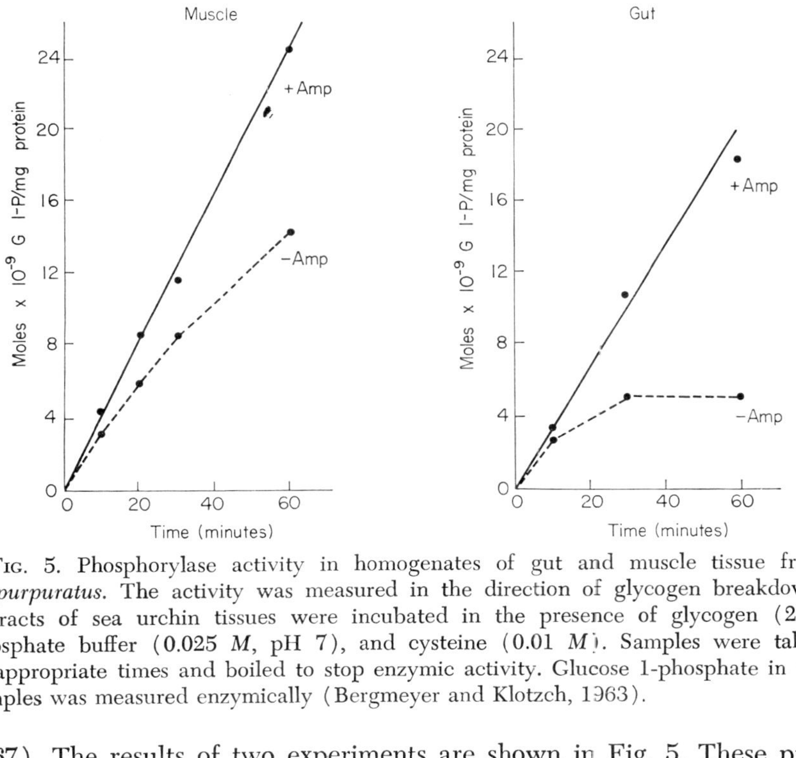

FIG. 5. Phosphorylase activity in homogenates of gut and muscle tissue from *S. purpuratus*. The activity was measured in the direction of glycogen breakdown. Extracts of sea urchin tissues were incubated in the presence of glycogen (2%), phosphate buffer (0.025 *M*, pH 7), and cysteine (0.01 *M*). Samples were taken at appropriate times and boiled to stop enzymic activity. Glucose 1-phosphate in the samples was measured enzymically (Bergmeyer and Klotzch, 1963).

1967). The results of two experiments are shown in Fig. 5. These preliminary results show that sea urchin phosphorylase may be similar to mammalian muscle phosphorylase. The enzyme from rabbit muscle exists in two interconvertible forms, phosphorylase a and phosphorylase b. The transformations b → a and a → b are catalyzed by a specific kinase and phosphatase respectively (Krebs *et al.*, 1964):

$$2 \text{ phosphorylase b} + 4 \text{ ATP} \rightarrow \text{phosphorylase a} + 4 \text{ ADP}$$
$$\text{phosphorylase a} + 4 \text{ H}_2\text{O} \rightarrow 2 \text{ phosphorylase b} + 4 \text{ P}_i$$

Phosphorylase b is inactive without AMP. AMP activates the b form, enhances the activity of the a form, and inhibits the activity of the phosphatase. Phosphorylase from S. *purpuratus* intestine and muscle behaves in a manner consistent with this scheme. In the presence of AMP the activity is stable and maximal, but in the absence of AMP activity is initially somewhat lower and falls off immediately. Phosphorylase is also present in S. *purpuratus* eggs (Epel, 1966), but the low level of hexose monophosphates in unfertilized sea urchin eggs suggests that it may be inactive or in some way blocked before fertilization (Aketa, 1964). Bergami, Mansour, and Scarano (Mansour, 1966) demonstrated phosphorylase activity in eggs and muscle of *Sphaerechinus granularis.* There was only slight activation by AMP in the presence of fluoride and no difference in activity between homogenates of fertilized and unfertilized eggs.

Hexokinase has been studied in eggs and embryos of *Arbacia* (Krahl *et al.,* 1954) and in the gut and jaw muscles of S. *purpuratus* (Doezema, 1967). Aketa (1957) reported that reducing sugar was absent in eggs of several sea urchins that he examined. Glucose and other free reducing sugars, the substrates of hexokinase, are present in the body fluid of S. *purpuratus* and S. *franciscanus* at levels of 0.1–0.2 μmoles/ml. (Bennett and Giese, 1955; Lasker and Giese, 1954; J. M. Lawrence *et al.,* 1966). Ferguson (1964a,b) showed that free sugars of the body fluid are available to the tissues by diffusion and this would also be true of eggs prior to spawning. Phosphoglucomutase activity is present in extracts of jaw muscles and gut of S. *purpuratus* (Table III) and is slightly more active than either phosphorylase or hexokinase (Doezema, 1967).

The glycolytic pathway in echinoderms has been studied primarily in sea urchin eggs. There is evidence for most of the enzymes of the glycolytic pathway in eggs (Cleland and Rothschild, 1952; Ycas, 1954; Ishihara, 1957) and several of the corresponding glycolytic esters have been detected (Aketa *et al.,* 1964). Lactic acid accumulates under anaerobic conditions in unfertilized eggs (Cleland and Rothschild, 1952; Aketa, 1957) and shortly after fertilization, eggs of several urchins show a transient appearance of lactic acid (Aketa, 1963, 1964).

Good evidence exists for most of the enzymes of the hexose monophosphate pathway in eggs (Krahl *et al.,* 1955; Bäckstrom, 1959, 1963; Isono and Ishida, 1962). Experiments with glucose-^{14}C labeled in specific carbon atoms are consistent with the idea that both the glycolytic-tricarboxylic acid cycle and the hexose monophosphate pathways of glucose oxidation are operating concurrently in sea urchin eggs and embryos. Position 1 of the glucose molecule contributes to respiratory CO_2 to a much greater extent than does either the 2 or the 6 position (Krahl, 1956; Bäckstrom *et al.*, 1960).

The hexose monophosphate pathway may also be quite important in tissues of adult sea urchins. Glucose 6-phosphate dehydrogenase and 6-phosphogluconate dehydrogenase were high relative to hexokinase and phosphorylase in extracts of gut and muscle of *S. purpuratus* (Table III) (Doezema, 1967). It must be remembered, however, that estimates

TABLE III

ACTIVITIES OF ENZYMES FROM SEA URCHIN TISSUES

Enzyme	Source	Activity μM/mg. protein/minute	Number of observations	Ref.[a]
Glucose 6-phosphate dehydrogenase	Gut (*S. purp.*)	0.013 –0.023	4	1
	Muscle (*S. purp.*)	0.002 –0.004	2	1
	Eggs	0.043 –0.075		3
6-Phosphogluconate dehydrogenase	Gut (*S. purp.*)	0.009	1	1
	Muscle (*S. purp.*)	0.001 –0.002	2	1
	Eggs (*Arbacia*)	0.021 –0.028		3
Phosphoglucomutase	Gut (*S. purp.*)	0.004 –0.008	4	1
	Muscle (*S. purp.*)	0.003 –0.006	2	1
Hexokinase	Gut (*S. purp.*)	0.0006–0.0012	3	1
	Muscle (*S. purp.*)	0.001	2	1
	Eggs (*Arbacia*)	0.006 –0.017		2
Phosphorylase	Gut (*S. purp.*)	0.0003	1	1
	Muscle (*S. purp.*)	0.0004	1	1

[a] References: (1) Doezema (1967); (2) Krahl *et al.* (1954); (3) Krahl *et al.* (1955).

of G 6-P dehydrogenase activities in crude extracts tend to be slightly high because of contaminating 6-phosphogluconate dehydrogenase activity (Glock and McLean, 1953).

Many sea urchins are herbivorous, feeding mainly on algae (Boolootian and Lasker, 1964; Lasker and Giese, 1954; Farmanfarmaian and Phillips, 1962). *S. purpuratus* and *S. franciscanus* show a distinct preference for the brown alga, *Macrocystis,* and also consume large amounts of *Laminaria* and *Gigartina,* a brown and a red respectively (Leighton, 1966). *S. purpuratus* is able to digest algal fragments (Farmanfarmaian and Phillips, 1962; Boolootian and Lasker, 1964) and many algal polysaccharides (Epley and Lasker, 1959; Lasker and Giese, 1954; Huang and Giese, 1958; Murakami, 1956, 1960). The brown algae contain large amounts of laminarin (β,1-3 glucan), fucoidin (polyfucose sulfate), alginic acid (polymannuronic acid), and free mannitol. Red algae contain the starchlike glucose polymer, floridean starch, and sulfated galactose polymers such as agar and carrageenin (Peat and Turvey, 1965). After feeding on the red alga, *Iridaea,* free galactose appears in the perivisceral fluid of *S. purpuratus.* Much of this galactose probably originates as galactosyl glycerol which is abundant in the tissues of *Iridaea* (Farmanfarmaian and Phillips, 1962). Free mannitol appears in the perivisceral fluid of *S. purpuratus* after a meal of *Macrocystis,* a brown alga (Boolootian and Lasker, 1964).

In order to extend the understanding of how the sea urchin metabolizes food material, a study was made of the ability of gut segments *in vitro* to use some of the sugars and related substances of the algae as precursors of glycogen. Radioactive glucose, galactose, mannose, mannitol, and fucose were tested. Radioactivity was not incorporated into glycogen from fucose, but all the other sugars were good glycogen precursors. After hydrolysis of the radioactive glycogen, paper chromatograms showed only one radioactive peak over the glucose spot (Doezema, 1967).

The experiments with radioactive sugars imply certain pathways of metabolism. In vertebrates glucose is incorporated into glycogen by the following sequence (Leloir *et al.,* 1959):

$$\text{G 1-P} + \text{UTP} \rightleftharpoons \text{UDPG} + \text{PP}$$
$$\text{UDPG} + (\text{glucose})_n \rightarrow (\text{glucose})_{n+1} + \text{UDP}$$

The other sugars must undergo molecular rearrangement to form glucose and then be incorporated into glycogen. The following sequence found in bacteria, yeast, and vertebrates is responsible for the epimerization of galactose (Kalckar, 1965):

$$\text{Gal 1-P} + \text{UDPG} \rightleftharpoons \text{UDPGal} + \text{G 1-P}$$
$$\text{UDPGal} \rightleftharpoons \text{UDPG}$$
$$\text{UDPG} + \text{PP} \rightleftharpoons \text{G 1-P} + \text{UTP}$$

Mannose and mannitol enter the glucose pool by way of fructose 6-P. Mannose is phosphorylated at the 6 position and transformed to fructose 6-P by phosphomannose isomerase which has been studied in yeast and rabbit muscle (Slein, 1950; Noltman and Bruns, 1958). One possible pathway for mannitol is oxidation at the 2 position to form fructose by an NAD-linked polyol dehydrogenase. Such an enzyme has been studied in *Acetobacter;* however, the polyol dehydrogenase of the rat and mouse has specificity such that sorbitol is oxidized to fructose but mannitol is not a suitable substrate (McCorkindale and Edson, 1954). Alternatively, mannitol could be phosphorylated by mannitol kinase and oxidized by mannitol 1-P dehydrogenase, which is the pathway in *E. coli* (Klungsøyr, 1966). L-Fucose appears not to enter the glucose metabolism of the sea urchin. The same is true for rats and mice. Fucose does not increase the glycogen content of fasted mice (Shull and Miller, 1960), and radioactive fucose is not incorporated into glycogen in rats but does appear in glycoprotein (Coffey *et al.*, 1964).

VI. Conclusions

An overall view of the mucopolysaccharide substances from echinoderms suggests similarities to substance found in the vertebrates. The mucoid substances isolated from the Japanese holothurians is identical in its constituents to chondroitin sulfate. The constituents of the jelly coat substances from sea urchin eggs, which probably determine the specificity of sperm to egg interaction are quite reminiscent of the mammalian blood group substances which are certainly involved in biological specificity.

Insofar as it has been studied, the comparative biochemistry of carbohydrate metabolism in echinoderms has suggested more unity with the rest of the animal kingdom than diversity. One possible exception is the use of mannitol by a sea urchin which seems to be rather unique in the animal kingdom; however, this ability is almost certainly related to the fact that mannitol is very abundant in the natural food of this urchin. One might expect to find other fundamental differences in the regulation of carbohydrate metabolism. Further investigation of phosphorylase and UDPG-glycogen glucosyl transferase, both of which exist in multiple forms in the vertebrates, would be important for comparative purposes. Even more interesting is the prospect of hormonal regulation in the echinoderms in reference to the recent paper by Wilson and Falkmer (1965) reporting the presence of insulin in the pyloric ceca of the starfish *Pisaster ochraceus*. Experiments testing the effects of this material on the incorporation of glucose into glycogen or alternatively, its effects on the reducing sugars in the perivisceral fluid, would go far in assessing the possible physiological significance of starfish insulin.

REFERENCES

Aketa, K. (1957). *Embryologia Nagoya* **3**, 367–278.

Aketa, K. (1963). *Exptl. Cell Res.* **30**, 93–97.

Aketa, K. (1964). *Exptl. Cell Res.* **34**, 192–194.

Aketa, K., Bianchetti, R., Marré, E., and Monroy, A. (1964). *Biochim. Biophys. Acta* **86**, 211–215.

Altman, P. I., and Dittmer, D. S. (1964). "Biology Data Book." Federation of American Societies for Experimental Biology.

Anderson, J. M. (1953). *Biol. Bull.* **105**, 47–61.

Anderson, J. M. (1954). *Biol. Bull.* **107**, 157–173.

Anderson, J. M. (1959). *Biol. Bull.* **117**, 185–201.

Anderson, J. M. (1960). *Biol. Bull.* **119**, 371–398.
Ashwell, G. (1957). *In* "Methods in Enzymology" (S. P. Colowick and N. O. Kaplan, eds.), Vol. 3, pp. 73–105. Academic Press, New York.
Austin, W. C. (1966). Doctoral Dissertation, Stanford University.
Bäckstrom, S. (1959). *Exptl. Cell Res.* **18**, 347–356.
Bäckstrom, S. (1963). *Exptl. Cell Res.* **32**, 566–569.
Bäckstrom, S., Hultin, K., and Hultin, T. (1960). *Exptl. Cell Res.* **19**, 634–636.
Barber, A. A., Harris, W. W., and Padilla, G. M. (1965). *J. Cell Biol.* **27**, 281–292.
Becker, M. J., and Shefner, A. M. (1964). *Nature* **202**, 803–804.
Benazzi-Lentati, G. (1941). *Arch. Zool. Ital.* **29**, Suppl.; *Attualita Zool.* **5**, 35–69.
Bennett, J., and Giese, A. C. (1955). *Biol. Bull.* **109**, 226–247.
Bergmeyer, H. U., and Klotzch, H. (1963). *In* "Methods of Enzymatic Analysis" (H. U. Bergmeyer, ed.), pp. 131–133. Academic Press, New York.
Bishop, D. W. (1951). *Biol. Bull.* **101**, 215–216.
Boolootian, R. A., and Lasker, R. (1964). *Comp. Biochem. Physiol.* **11**, 273–289.
Briacombe, J. S., and Webber, J. M. (1964). "Mucopolysaccharides." Elsevier, Amsterdam.
Brumbaugh, J. H. (1964). Doctoral Dissertation, Stanford University.
Buchanan, J. B. (1963). *Proc. Zool. Soc. London* **141**, 251–259.
Bueding, E., and Orrell, S. (1964). *J. Biol. Chem.* **239**, 4018–4020.
Chaet, A. B., and Philpott, D. E. (1964). *J. Ultrastruct. Res.* **11**, 354–362.
Chanley, J. D., Ledeen, R., Wax, J., Nigrelli, R. F., and Sobotka, H. (1959). *J. Am. Chem. Soc.* **81**, 5180–5183.
Chanley, J. D., Perlstein, J., Nigrelli, R. F., and Sobotka, H. (1960). *Ann. N.Y. Acad. Sci.* **90**, 902–905.
Cleland, K. W., and Lord Rothschild. (1952). *J. Exptl. Biol.* **29**, 285–294.
Coffey, J. W., Miller, N., and Sellinger, O. Z. (1964). *J. Biol. Chem.* **239**, 4011–4017.
Cuénot, L. (1948). *In* "Traité de Zoologie" (P. P. Grassé, ed.), Vol. 11, pp. 3–363. Masson, Paris.
Defretin, R. (1952). *Compt. Rend.* **234**, 1806–1808.
Dimelow, E. J. (1958). Ph.D. Dissertation, University of Reading.
Doezema, P. (1967). Ph.D. Thesis, Stanford University.
Endean, R. (1957). *Quart. J. Microscop. Sci.* **98**, 455–478.
Epel, D. E. (1966). Personal communication.
Epley, R. W., and Lasker, R. (1959). *Science* **129**, 214–215.
Farmanfarmaian, A., and Phillips, J. H. (1962). *Biol. Bull.* **123**, 105–120.
Faust, R. G., Jones, R. F., and Parpert, A. (1959). *Biol. Bull.* **117**, 394.
Ferguson, J. C. (1964a). *Biol. Bull.* **126**, 33–53.
Ferguson, J. C. (1964b). *Biol. Bull.* **126**, 391–406.
Fontaine, A. R. (1964). *J. Marine Biol. Assoc. U.K.* **44**, 145–162.
Freeman, W. P., and Simon, S. E. (1964). *J. Cellular Comp. Physiol.* **63**, 25–38.
Fuji, A. (1961). *Bull. Fac. Fisheries, Hokkaido Univ.* **11**, 195–202.
Gabe, M. (1962). *In* "Handbuch der Histochemie" (W. Graumann and K. Neumann, eds.), Vol. 2, Part 1, pp. 95–356. Fisher, Jena.
Ghosh, H. P., and Preiss, J. (1965). *Biochim. Biophys. Acta* **104**, 274–277.
Giese, A. C. (1959). *Ann. Rev. Physiol.* **21**, 547–576.
Giese, A. C. (1966a). *Physiol. Rev.* **46**, 244–298.
Giese, A. C. (1966b). *In* "Physiology of the Echinodermata" (R. A. Boolootian, ed.), pp. 757–796. Wiley (Interscience), New York.

Giese, A. C., Greenfield, L., Huang, H., Farmanfarmaian, A., Boolootian, R., and Lasker, R. (1958). *Biol. Bull.* **116**, 49–58.
Giese, A. C., Farmanfarmaian, A., Hilden, S., and Doezema, P. (1966). *Biol. Bull.* **130**, 192–201.
Glock, G. E., and McLean, P. (1953). *Biochem. J.* **55**, 400–408.
Gottschalk, A. (1960). "The Chemistry and Biology of Sialic Acids and Related Substances. Cambridge Univ. Press, London and New York.
Greenfield, L., Giese, A. C., Farmanfarmaian, A., and Boolootian, R. A. (1958). *J. Exptl. Zool.* **139**, 507–524.
Hashimoto, Y., and Yasumoto, T. (1960). *Bull. Japan. Soc. Fisheries* **26**, 1132–1138.
Hassid, W. Z., and Abraham, S. (1957). *In* "Methods in Enzymology" (S. P. Colowick and N. O. Kaplan, eds.), Vol. 3, pp. 34–50. Academic Press, New York.
Holland, N. D. (1966). Personal communication.
Holland, N. D., and Nimitz, S. A. (1964). *Biol. Bull.* **127**, 280–293.
Huang, H, and Giese, A. C. (1958). *Science* **127**, 475.
Hutchens, J. O., Keltch, A. K., Krahl, M. E., and Clowes, G H. A. (1941). *J. Gen. Physiol.* **25**, 717–731.
Hyman, L. H. (1955). "The Invertebrates," Vol. IV. McGraw-Hill, New York.
Immers, J. (1961). *Exptl. Cell Res.* **24**, 356–378.
Ishihara, K. (1957). *J. Fac. Sci., Univ. Tokyo, Sect. IV* **8**, 71–93.
Ishihara, K. (1964). *Exptl. Cell Res.* **36**, 354–367.
Isono, N., and Ishida, J. (1962). *J. Fac. Sci., Univ. Tokyo, Sect. IV* **9**, 357–368.
Janssens, P. A. (1964). *Comp. Biochem. Physiol.* **11**, 105–117.
Jermyn, M. A., and Isherwood, F. A. (1949). *Biochem. J.* **44**, 402–407.
Kalckar, H. M. (1965). *Science* **150**, 305–313.
Kjölberg, O., Manners, D. J., and Wright, A. (1963). *Comp. Biochem. Physiol.* **8**, 353–365.
Klungsøyr, L. (1966). *Biochim. Biophys. Acta* **122**, 361–364.
Krahl, M. E. (1956). *Biochim. Biophys. Acta* **20**, 27–32.
Krahl, M. E., Keltch, A. K., Walters, C. P., and Clowes, G. H. A. (1954). *J. Gen. Physiol.* **38**, 31–39.
Krahl, M. E., Keltch, A. K., Walters, C. P., and Clowes, G. H. A. (1955). *J. Gen. Physiol.* **38**, 431–439.
Krebs, E. G., Gonzales, C., Posner, J. B., Love, D. S., Bratvold, G. E., and Fischer, E. H. (1964). *Ciba Found. Symp., Control Glycogen Metab.* pp. 200–210.
Lasker, R., and Giese, A. C. (1954). *Biol. Bull.* **106**, 328–340.
Lawrence, A. L. (1966). Unpublished data.
Lawrence, J. M., Lawrence, A. L., and Giese, A. C. (1966). *Physiol. Zool.* **39**, 281–290.
Leighton, D. L. (1966). *Pacific Sci.* **20**, 104–113.
Leloir, L. F., Olavarria, J. M., Goldemberg, S. H., and Carminatti, H. (1959). *Arch. Biochem. Biophys.* **81**, 508–520.
McCorkindale, J., and Edson, N. L. (1954). *Biochem. J.* **57**, 518–523.
Manners, D. J. (1957). *Advan. Carbohydrate Chem.* **12**, 261–298.
Mansour, T. E. (1966). Personal communication.
Minganti, A., and Vasseur, E. (1959). *Acta Embryolo. Morphol. Exptl.* **2**, 195–203.
Mohri, H. (1957). *J. Fac. Sci., Univ. Tokyo, Sect. IV* **8**, 51–64.
Monroy, A. (1965). "Chemistry and Physiology of Fertilization." Holt, New York.
Monroy, A., Tosi, L., Giardina, G., and Maggio, R. (1954). *Biol. Bull.* **106**, 169–177.

Moore, B., Whiteley, E., and Adams, A. (1913). *Biochem. J.* **7,** 127–141.
Mordoh, J., Leloir, L. F., and Krisman, C. R. (1965). *Proc. Natl. Acad. Sci. U.S.* **53,** 86–91.
Motohiro, T. (1960a). *Bull. Japan. Soc. Sci. Fisheries* **26,** 1171–1174.
Motohiro, T. (1960b). *Bull. Japan. Soc. Sci. Fisheries* **26,** 1175–1178.
Motohiro, T. (1960c). *Bull. Japan. Soc. Sci. Fisheries* **26,** 1179–1182.
Murakami, S. (1956). *Sci. Rept. Saitama Univ.* **B2,** 191–204.
Murakami, S. (1960). *Sci. Rept. Saitama Univ.* **B3,** 261–266.
Muramatsu, T. (1965). *J. Biochem.* **57,** 223–225.
Nakano, E., and Ohashi, S. (1954). *Embryologia (Nagoya)* **2,** 81–86.
Nigrelli, R. F., and Jakowska, S. (1960). *Ann. N.Y. Acad. Sci.* **90,** 884–892.
Nigrelli, R. F., Chanley, J. D., Kohn, S. K., and Sobotka, H. (1955). *Zoologica* **40,** 47–48.
Noltman, E., and Bruns, F. H. (1958). *Biochem. Z.* **330,** 514–520.
Örstrom, A., and Lindberg, O. (1940). *Enzymologia* **8,** 367–384.
Pearse, A. G. E. (1961). "Histochemistry." Churchill, London.
Peat, S., and Turvey, J. R. (1965). *Chem. Org. Naturstoffe* **23,** 1–45.
Perlman, P., Bostrom, H., and Vestermark, A. (1959). *Exptl. Cell. Res.* **17,** 439–446.
Rio, G. J., Ruggeri, G. D., Martin, S. J., Stempien, F., Jr., and Nigrelli, R. F. (1963). *Am. Zoologist* **3,** 554–555.
Rodenhouse, I., and Guberlet, J. (1946). *Univ. Wash. Publ. Biol.* **12,** 23–47.
Roseman, S., Jourdian, G. W., Watson, D., and Rood, R. (1961). *Proc. Natl. Acad. Sci. U.S.* **47,** 958–961.
Rothschild, Lord. (1956). "Fertilization." Wiley, New York.
Rothschild, Lord, and Cleland, K. W. (1952). *J. Exptl. Biol.* **29,** 66–71.
Shull, K. H., and Miller, O. N. (1960). *J. Biol. Chem.* **235,** 551–553.
Slein, M. W. (1950). *J. Biol. Chem.* **186,** 753–761.
Spikes, J. D. (1949). *Am. Naturalist* **383,** 285–298.
Stetten, M. R., Katzen, H. M., and Stetten, D., Jr. (1956). *J. Biol. Chem.* **222,** 587–599.
Stimpson, J. H. (1965). *Comp. Biochem. Physiol.* **15,** 187–197.
Stott, F. C. (1931). *J. Exptl. Biol.* **8,** 133–150.
Tyler, A. (1949). *Am. Naturalist* **83,** 195–219.
Vasseur, E. (1950). *Acta Chem. Scand.* **4,** 1144–1145.
Vasseur, E. (1952). *Acta Chem. Scand.* **6,** 376.
Vasseur, E., and Immers, J. (1949). *Arkiv Kemi* **1,** 39.
Ward, J. A. (1965). *J. Exptl. Zool.* **158,** 365–371.
Warren, L. (1963). *Comp. Biochem. Physiol.* **10,** 153–171.
Warren, L., and Felsenfeld, H. (1961). *Biochem. Biophys. Res. Commun.* **5,** 185–190.
Warren, L., and Hathaway, R. (1960). *Biol. Bull.* **119,** 354.
Wilson, S., and Falkmer, S. (1965). *Can. J. Biochem.* **43,** 1615–1624.
Wolfrom, M. L., Lassettre, E. N., and O'Neill, A. N. (1951). *J. Am. Chem. Soc.* **73,** 595–599.
Yamanouchi, T. (1955). *Publ. Seto Marine Biol. Lab.* **4,** 183–203.
Yčas, M. (1954). *J. Exptl. Biol.* **31,** 208–217.

CHAPTER 5

Lipid Metabolism

U. H. M. Fagerlund

I. Introduction

A considerable store of knowledge pertaining to lipids in echinoderms is accumulating. A large part of this has been achieved through the work of Bergmann (1962) and Giese (1966a,b), and members of their groups. Investigations have centered around the quantitative determination of lipid components. Much of the earlier work is of limited value since contents vary a great deal with gonad development and nutritional state of the animal, and information about these and other factors has not always been supplied. Little is known about the interrelationships between lipids in this phylum and whether the same metabolic pathways exist as in mammalian tissue. Since the death of Dr. Werner Bergmann in 1959, rather few reports have appeared on marine sterols, particularly on those of echinoderms. Much of the earlier work on attempts to identify individual sterol components will have to be revised, since the difficulty in resolving mixtures from marine invertebrates with conventional techniques has resulted in unreliable data. In this field, as well as in that of fatty acids of invertebrates, application of newer analytical tools, such as gas chromatography and mass spectrometry will bring new and interesting results.

II. Lipids

A. Distribution and Storage

Two comprehensive reviews of data on lipid contents of echinoderms have appeared, both with the focus on sterol composition. Toyama

(1958) listed invertebrates investigated by Japanese workers, which included representatives of all five classes of echinoderms now living. The content of total lipids, glycerides, unsaponifiables and sterols was given and also the iodine number of the glycerides. Bergmann (1962) reviewed most of the literature on the lipid content of marine invertebrates, including some unpublished data. Data on total fat, unsaponifiables, and sterols were furnished. Rather few investigations have been made of the distribution of lipids in various components of the body and the changes taking place in these during starvation, development, and sexual cycle. Giese (1966a,b) has recently reviewed these problems. The author discussed exhaustively work carried out with echinoderms by his group, some not previously published, and by others. Therefore, only a few of the most pertinent studies will be dealt with here.

The gross chemical composition of the flesh of the sea cucumber *Stichopus japonicus* was determined by Tanikawa *et al.* (1955) during a 12-month period. Peaks in the fat content were recorded during February, June, and November. Greenfield *et al.* (1958) gave a detailed account of the monthly biochemical changes, including those of lipids, in gonads and hepatic ceca of two starfish and a sea urchin. There was a substantial drop in lipid content from the gravid to the spent ovary. Chemical analyses of the Antarctic sea urchin *Sterechinus neumayeri* were carried out by Pearse and Giese (1966) over an 11-month period, but no seasonal changes in organic constituents of the gonad were found. Giese (1961, 1966b) reported the nonprotein N, protein, glycogen, and lipid content in various components and organs of some representatives of the five classes of echinoderms. The observations indicated that the amount of lipid is closely related to the nutritional state of the animal. An analysis of the lipids in various components of a sea urchin *Somopneustes variolaris* was undertaken by Giese *et al.* (1964). Phospholipids were predominant in the gut while the major portion of gonad lipids was contained in the fraction consisting of fatty acids, triglycerides, and sterols.

The lipid content of the body fluid was determined in a few echinoderms by Lawrence (1965) and an attempt was made to establish whether the supply of lipid is maintained in the fluid during starvation. Two starfish and two chitons retained some lipid in the body fluid after prolonged starvation. Some further aspects of starvation may be mentioned. The gonad of *Echinus esculentus* declined in size during the winter months when food was scarce, but not when food was abundant (Moore, 1935). In *Allocentrotus fragilis* (deep-sea urchin), lipid was lost from the gut and body wall during starvation and after a month the gonad had decreased in size (Giese, 1961). Lawrence *et al.* (1966)

observed that neutral lipids decreased much more than polar ones in *Strongylocentrotus purpuratus* during starvation.

B. GLYCERYL ETHERS AND PHOSPHOLIPIDS

Alkyl ethers of glycerol are widespread among marine invertebrates and elasmobranch fishes. The most common are batyl alcohol (α-octadecyl-glyceryl ether) and selachyl alcohol (α-oleyl-glyceryl ether). Batyl alcohol was first isolated from invertebrates by Dorée (1909) during studies of the unsaponifiable fraction of starfish lipids. Because of its characteristics, which in many ways resemble those of sterols, batyl alcohol was found in fractions containing these compounds. It was not until 1943 that Matsumoto *et al.* and Bergmann and Stansbury (1943) proved independently that the material isolated from starfish was identical with batyl alcohol.

The α-glyceryl ether content in the unsaponifiable fractions of lipids of several marine animals, including four echinoderms, *Asterias forbesi, Arbacia punctulata, Echinarachnius* sp. and *Thyone biareus* was determined by Karnovsky and Brumm (1955). They found the starfish diverticulum to be a particularly rich source of α-glyceryl ethers. In young animals of 2-inch length no glyceryl ethers were detected when the whole animal was extracted, but in older animals the ethers accumulated with age in the diverticulum. Karnovsky *et al.* (1955) correlated histochemical studies with chemical determination of phosphorus, α-glyceryl ethers, and aldehydogenic glycerides in the diverticulum of the starfish *Asterias forbesi.* From the phosphorus values the phosphatide content in the fat of diverticulum of two animals was calculated to be 40.8 and 37.6%. Lipid aldehydogenic compounds were detected not only in the phosphatide fraction (plasmalogens) but also in the acetone-soluble neutral fat. The aldehydogenic compounds were believed to be glycol acetals.

The nature of the glycerol-aldehyde linkage in plasmalogens was established by Rapport *et al.* (1957) working with bovine muscle to be that of an α,β-unsaturated glyceryl ether rather than an acetal as had been believed earlier. Eichberg *et al.* (1961) confirmed the presence of neutral plasmalogens in the diverticulum of *Asterias forbesi.* They proved that the plasmalogen of the starfish also has a double bond in the α,β-position of the fatty aldehyde moiety and that the aldehyde is attached in the α-position of glycerol. On the basis of evidence obtained by chromatography of lipids it was suggested that three categories of glycerol-containing lipids were present, namely: (1) conventional neutral triglycerides and glycerophosphatides; (2) glyceryl ethers saturated in the α,β-position to the oxygen bridge (e.g., batyl alcohol) existing as fatty acid diesters, and the corresponding phosphatides; and (3) fatty acid diesters of glyceryl-alkenyl ether (neutral plasmalogens), and the well-known phosphatide plasmalogens. The authors also suggested that lipids of the batyl alcohol type in starfish might be derived metabolically from the plasmalogens.

Prompted by a report by Bergmann and Landowne (1958) that the sea anemone *Anthopleura elegantissima* contained plasmalogens of the acetal type, Rapport and Alonzo (1960) investigated 11 species of invertebrates, including three echinoderms. They found the plasmalogen to be mostly of the α,β-unsaturated type. The quantity of phosphatidal choline (plasmalogen of phosphatidyl choline; naming suggested by Rapport, M. M.) in all 11 animals was much less than the quantity of phosphatidal ethanolamine in analogy with mammalian tissues, but contrary to Bergmann and Landowne's observation of *A. elegantissima*, in which only choline plasmalogen was present. The marine tissues studied were all shown to be as rich a source of plasmalogens as mammalian brain.

A paper chromatographic analysis of lipids of 46 species of invertebrates, including an ophiuroid and an echinoid was undertaken by Hack *et al.* (1962). A basic phosphatide pattern was established with phosphatidal ethanolamine (plasmalogen) as a major component in agreement with Rapport and Alonzo (1960). Lecithin (phosphatidyl choline) was present in all animals, but the corresponding choline plasmalogen was only a minor constituent. Also generally present in small amounts were a monophosphoinositide and a polyglyceryl phosphatide (cardiolipin). The two echinoderms contained two inositide components, one resembling diphosphoinositide. The high sphingomyelin content in *Anthopleura* reported by Bergmann and Landowne (1958) was not matched in these animals.

Gilbertson and Karnovsky (1963) presented results to complement previously reported investigations by the same group (Eichberg *et al.*, 1961) on the lipids of starfish diverticulum. Phosphatides were removed from total lipids by flocculating with acetone to obtain neutral lipids. The presence in the neutral lipids of glyceryl alkenyl ethers in the form of diesters, as previously indicated, was established. Of total lipid 72.5% was neutral, 16.1% of this fraction was glyceryl ethers and 31.9 mole% of these were of the α,β-unsaturated type. It was also established that glyceryl alkyl ethers actually exist in nature as diesters.

Todd and Rizzi (1964) separated the lipids of whole starfish *Asterias forbesi* by acetone extraction and then obtained 66% acetone-soluble (neutral) material and 34% phospholipids. Each fraction was then saponified. The unsaponifiable portion of the neutral fraction contained 36–42%

glyceryl ethers, the phospholipid unsaponifiable 27%. Using kerosene-impregnated paper the authors were able to separate naturally occurring glyceryl ethers from other compounds and from each other, except selachyl alcohol and chimyl alcohol when present, the mixture of which first had to be hydrogenated before separation was achieved. No selachyl alcohol was found however. The glyceryl ethers obtained from the unsaponifiable portion of phospholipid and neutral fractions consisted of 35% and 55% respectively of batyl alcohol, the rest was chimyl alcohol. No unsaturated glyceryl ethers were found.

The phospholipids of sea urchin spermatozoa were investigated by Mohri (1961). The major components of spermatozoa from *Anthocidaris crassispina* were phosphatidyl ethanolamine and ethanolamine plasmalogen, which together made up 50–80% of the total phospholipids. Two minor inositol-containing compounds were found. Phosphatidyl choline constituted 10–30% of the total. Spermatozoa from *Hemicentrotus pulcherrimus* also contained phosphatidyl ethanolamine and ethanolamine plasmalogen as major components together with a minor inositol-containing phospholipid and phosphatidyl choline as 40% of the total phospholipids. A small amount of sphingomyelin was also found.

The lipid content of the developing embryo of the sea urchin *Paracentrotus lividus* was also studied by Mohri (1964b). The content of total lipids did not change significantly up to the mesenchyme blastula stage, but began to decrease with the onset of gastrulation. Phospholipids showed the same changes, while total nitrogen continued to decrease slightly until the gastrula stage and then began to increase. The predominant components of lipids were triglycerides. Isono (1965) used thin layer and paper chromatography to separate the phospholipids of eggs of the sea urchins *Hemicentrotus pulcherrimus* and *Anthocidaris crassispina*. Phosphatidyl choline and phosphatidyl ethanolamine were the largest components. Phosphatidyl serine, lysophosphatidyl choline and sphingomyelin were also present, but no inositol-containing phospholipids were found. No marked qualitative changes in phospholipids were observed during early development. The hydrolyzates of phospholipid fractions contained some nitrogenous compounds other than choline, ethanolamine, and serine.

C. Fatty Acids of Glycerides

Fats of marine invertebrates, as those of fishes, contain highly complex mixtures of saturated and unsaturated fatty acids bound to glycerol. Toyama (1958) listed the iodine values of glyceride fatty acids of a large number of invertebrates. The values in 14 echinoderms ranged from 84 (*Coscinasterias acutespina*) to 202 (*Clupeaster japonicus*). The unsaturated fatty acid composition was generally similar to that of fish oils. Variations were observed not only from species to species, but also from animal to animal within the same species, depending on location and season.

With the aid of gas-liquid chromatography, a thorough investigation of the composition of fatty acids of a starfish *Pisaster ochraceus* was made by Rodegker and Nevenzel (1964) and the composition was compared to that of two mollusks, a barnacle and a mussel, upon which starfish are known to feed. Considerable differences were found between the fatty acid composition of starfish and that of mussels and barnacles, which feed on detritus and plankton. The authors reported the triglyceride, phospholipid, and total lipid content of the starfish. Fatty acid composition was then determined in these fractions. Finally, fatty acids were analyzed in the lipids from three components of the animal, namely hepatic cecum, male gonad, and arm. The major constituent of *Pisaster* was 7-eicosaenoic acid (20:1), which comprised 18–29% of lipids. Quite marked differences in fatty acid composition were noted between the triglyceride and phospholipid fractions of hepatic cecum. A similar finding in mussels was made by Lovern (1938). On the other hand, differences were also considerable when the fatty acid content was compared in the phospholipid fractions from cecum, gonad, and skeletal musculature of the arm. The results contribute to the existing evidence which indicates that ingested lipids are broken down in invertebrates before being absorbed (deposited).

An investigation has been made of the fatty acid composition of marine animals taken at various depths up to 4400 meters (Lewis, 1966). The content of medium chain saturated and long chain polyunsaturated acids decreased with increasing depth while that of oleic acid increased. The authors postulated that the reason was the paucity of plant matter as food. The benthic, mud-feeding holothurian *Scotoplanes theeli* had a fatty acid composition, which was markedly different from that of a pelagic species. This was attributed to the bacterial composition of the diet, which indicated a subsidiary food chain on the ocean floor.

D. Lipogenesis

Evidence for biogenesis of lipids of some echinoderms has been secured with the help of radioactive tracer compounds. Karnovsky and Brumm (1955) exposed pieces of starfish (*Asterias forbesi*) diverticulum to acetate-^{14}C and α-glycerol-^{14}C. The carbon of both acetate and glycerol was incorporated into the fatty acids of neutral fat and phosphatide, although specific activity was much higher with acetate as substrate. The fatty acids of phosphatides were about twice as active as those of the neutral fat fraction. Activity was significantly incorporated into the alkyl moiety of glycerol ethers from both labeled substrates. Labeled glycerol was incorporated as such into glyceryl ethers, but acetate was not incorporated into glycerol.

Mohri (1964b) followed the incorporation of ^{14}C-labeled acetate and glycerol in the embryo of the sea urchin *Paracentrotus lividus* during different stages of development. The results showed that lipid metabolism is intensified during embryonic development, particularly in the later phase. In unfertilized eggs the incorporation of labeled acetate was only slight but increased with progressive development. Most of the label was incorporated into fatty acids. The specific activity of the phosphatidyl ethanolamine fraction was much higher than that of the phosphatidyl choline fraction. As was found in starfish by Karnovsky and Brumm (1955), incorporation of glycerol-^{14}C was similar to that of acetate-^{14}C but occurred to a much lesser extent. The incorporation of glycerol into total lipids in contrast to that of acetate decreased somewhat after fertilization and reached a maximum at the pluteus stage. The highest specific activity was found in the phosphatidyl ethanolamine fraction, as was also the case when acetate was the substrate.

Incorporation of ^{14}C-labeled acetate and glucose into various compartments of the starfish *Pisaster ochraceus* was measured by Allen (1965), and Allen and Giese (1966). Predictably, hepatic cecal tissue incorporated acetate-1-^{14}C into lipids more rapidly than uniformly labeled glucose-^{14}C. Incorporation of acetate was decreased markedly with length of time after the last feeding. Addition of glucose to the incubation medium stimulated lipogenesis in cecal tissue from starved animals. Tissue from starved and fed animals oxidized acetate-1-^{14}C at the same rate, as indicated by radioactivity of respired CO_2. The rate of lipid synthesis in the hepatic cecum was highest in the fall, when fat deposition was highest. In the gonads the highest rate of synthesis occurred during the early phase of gonad maturation. Polar lipids predominated in testis, body wall, cardiac, and pyloric stomach. The high concentration of neutral lipids (70% of total lipids) in the hepatic cecum indicated the storage function of this organ. When tissue samples were incubated with acetate-1-^{14}C and the lipids fractionated into polar and neutral material, it was found that the rate of incorporation was the same into both fractions in ovarian and hepatic cecal tissue incubations, but about 3–5 times greater in neutral fractions than in polar ones in body wall and testis incubations. The specific activity of total lipids of body fluids was equal to that of the neutral fraction of hepatic cecal lipids, which was taken as an indication that lipid is transported by the body fluids.

Ellingsboe and Karnovsky (1965) exposed hepatic cecal tissue from starfish briefly to acetate-^{14}C and then incubated the tissue with unlabeled acetate for various intervals up to 6 hours. Lipids were separated on silica gel columns into phosphatides and neutral lipids. In contrast to the findings of Allen and Geise (1966), who reported equal incorporation into polar and neutral fractions, the specific activity of fatty acids of phosphatides was about 4 times that of fatty acids of neutral lipids. The values for neutral and phosphatide alkyl ethers increased with time. In the neutral fraction the specific activity of alkenyl ether glycerides rose to 40 times the values for alkyl ether glycerides, which points to alkenyl ethers as precursors of alkyl ethers in this system.

E. Phosphatide Utilization by Sea Urchin Spermatozoa

Lardy and Phillips (1941) observed that bull spermatozoa in sugar-free Ringer solution were able to move only if oxygen was present and then a decrease in phospholipid content took place. Rothschild and Cleland (1952) found that 5.5% of the dry weight of spermatozoa of the sea urchin *Echinus esculentus* consisted of phospholipids, while seminal plasma contained only a very small amount. They demonstrated that the ether-soluble phospholipids disappeared during aerobic incubation of the spermatozoa and that the O_2 uptake observed coincided fairly well with the theoretical oxygen requirement for oxidation of the disappeared phospholipids. It was concluded that oxidative breakdown of phospholipids, located in the middle piece of the spermatozoon, is the principal source of energy required for movement. Their results were later confirmed by Mohri (1957a) who studied several species of Japanese sea urchins. The same author (1957b) also demonstrated the presence of a fatty acid oxidizing system and later (1962) found a stimulatory effect of certain tricarboxylic acid cycle intermediates on oxygen uptake, which suggested that this cycle is operative. Maggio and Monroy (1955) had found that, when spermatozoa of the sea urchin *Arbacia lixula* were incubated with lipovitelline (the lipoprotein of hen's egg yolk) ether-extractable phospholipids were initially liberated and later de-esterified to ether-insoluble phosphatides (glyceryl-phosphorylcholine). Mohri (1959) found that phosphatidyl choline and phosphatidyl ethanolamine are hydrolyzed by homogenates of *Anthocidaris crassispina* with liberation of two molecules of fatty acid from one molecule of phospholipid ester. Phospholipase C was not present in spermatozoa of sea urchin.

An alternative source of energy, involving carbohydrate metabolism through a hexose monophosphate shunt was indicated in spermatozoa of *Arbacia lixula* and *Paracentrotus lividus* (Ghiretti and D'Amelio, 1956; Scarano and Rocca, 1959). However, the amount of carbohydrate initially present was found by Rothschild and Cleland (1952) and Mohri (1957a) to be insufficient to account for the intense respiration of the spermatozoa after dilution with seawater. Carbohydrate, therefore, could not be the only substrate metabolized. In order to prove that endogenous phospholipids are being utilized even by the sea urchins, in which the operation of the hexosemonophosphate shunt had been suggested, Mohri (1964a) investigated these animals together with the sea urchin *Sphaerechinus granularis*. Even in these animals the amount of endogenous phospholipids utilized accounted for most of the oxygen uptake measured. Recently, Afzelius and Mohri (1966) studied the changes taking place during respiration in the mitochondria of aged spermatozoa of the sea urchin *Brissopsis lyrifera*. When potassium hydroxide was present to remove respiratory carbon dioxide, a reduction of cristae of mitochondria occurred. This change was accompanied by a decrease of the amount of endogenous phospholipids.

Very little is known about which phosphatides are involved in motility of spermatozoa. Mohri (1961), who fractionated phospholipids from spermatozoa from *Hemicentrotus pulcherrimus* into cephalin and cephalin plasmalogen, inositol phospholipid and lecithin, found after aerobic incubation a considerable reduction in all phospholipid fractions.

F. Sterols

Several general reviews of investigations of sterols in marine invertebrates have appeared, of which the latest ones will be mentioned here. Bergmann (1952) prepared a comprehensive list of naturally occurring sterols and grouped them in relation to optical rotation. Idler (1958) listed newer information on sterols of marine invertebrates. Toyama (1958) reviewed work done by Japanese investigators including 14 species of *Echinodermata*. Kita and Toyama (1960) made a survey of 25 invertebrates in Japanese waters, including 5 species of echinoderms. Ultraviolet and infrared absorption of unfractionated sterols was determined in order to detect the presence of conjugated double bonds at C-5 and C-7 and terminal methylene groups. The sterol mixture of the echinoid *Heliocidaris crassispina* had a UV absorption peak at 255 mμ. A comprehensive review and discussion of available information on sterols of marine invertebrates was prepared by Bergmann (1962). A short summary will be given here of the characteristics of echinoderm sterols and reference will be included to the sparse new information which has appeared since the preparation of Bergmann's review.

The sterols of asteroids are entirely $^{7}\Delta$-unsaturated, which is shown by their low optical rotations. A variety of sterols containing 27, 28, and 29 carbon atoms have been found in this class. It is noteworthy that the sterols of the common starfish *Pisaster ochraceus* have been found to contain 7,24(28)-ergostadien-3β-ol (19% of total sterols) (Fagerlund and Idler, 1959) a sterol which differs from 24-methylene cholesterol, which is very common in mollusks on which starfish feed, in that it contains a double bond at C-7 rather than at C-5.

The ophiuroids contain predominantly $^5\Delta$-sterols in complex mixtures, reminiscent of plant sterols. C-7 unsaturated sterols appear to be entirely absent.

Our knowledge of crinoid sterols is rather scanty, but it appears that the sterols are $^5\Delta$-unsaturated. Preliminary investigations of the sterols of an unidentified crinoid in our laboratory (Fagerlund and Idler, 1958) revealed the presence of absorption bands in the infrared spectrum at 890 and 975 cm^{-1} characteristic of a terminal methylene group and a C-22(23) double bond respectively. Gas chromatographic analysis (Fagerlund, 1962) of the crude sterol mixture showed the presence of a major peak with the mobility of a monounsaturated $^5\Delta$-sterol and in addition four minor peaks, of which one had the mobility of 24-methylene cholesterol. The gas chromatograms resembled strikingly those of bivalves, although the peak corresponding to a monounsaturated $^5\Delta$-sterol was stronger in the crinoid.

The holothurians are outstanding because their content of sterols is appreciably lower than that of other echinoderms. Consequently, a large amount of starting material is needed in order to obtain a sufficient amount of sterol for investigation. This is probably one of the reasons why very little is known about the sterols of this class. The properties of those so far investigated have been characteristic of $^7\Delta$-sterols.

The sterols of echinoids are C-5 unsaturated and cholesterol is often a major component. Bergmann and Domsky (1960) isolated by chromatographic methods from the slate pencil sea urchin (systematic name not given) an unidentified sterol believed to be C-7 unsaturated. Two echinoids have been investigated in our laboratory (Fagerlund and Idler, 1958). The *p*-phenylazobenzoyl esters of the sterols of the sand dollar, *Echinarachnius parma,* when chromatographed on a SiO_2-Celite column, separated into two components. The minor one, comprising 11% of total sterols, contained 22.5% of a $\Delta^{5,7}$-sterol. The IR spectrum gave evidence for the presence of a terminal methylene group. The major component had a melting point characteristic of cholesterol. The sterol mixture of the other echinoid investigated by us, namely the sea urchin *Strongylocentrotus franciscanus,* separated chromatographically into three components. A minor one (10% of total sterols) showed evidence for a terminal methylene group and a 22(23)-double bond in the IR spectrum. The major component which made up 83.1% of total sterols had a melting point characteristic of cholesterol. Gas chromatographic analysis (Fagerlund, 1962) of the sterol mixtures of the two echnoids yielded one major peak with the mobility of cholesterol and in addition three minor components, of which the largest had the mobility of 24-methylene cholesterol. As with the crinoid sterols investigated in this laboratory, the mobilities of all components coincided with those of corresponding peaks in the gas chromatograms of mollusk sterols, the only difference being the relative magnitude of component peaks.

Very little is known about the pathways of sterol biosynthesis and metabolism in echinoderms. Fagerlund and Idler (1960) have shown that a starfish *Pisaster ochraceus* efficiently transforms the C-5 double bond of ingested cholesterol to the C-7 position.

On the basis of the observed efficiency of this conversion and because the sterols of starfish are mostly C-7 unsaturated, while the major portion of the sterols of ingested mollusks are C-5 unsaturated, it may be assumed that the starfish converts not only 24-methylene cholesterol but the bulk of exogenous $^{5}\Delta$-sterols to the corresponding $^{7}\Delta$-sterols. Two starfish, *Pisaster ochraceus* (Allen, 1965) and *Paracentrotus lividus* (Salaque *et al.*, 1966), have been found to incorporate labeled acetate into unsaponifiable matter, but the sterol fraction was not labeled (Salaque *et al.*, 1966) after several recrystallizations when the animals were extracted 16 hours after the labeled acetate was injected.

Available evidence indicates that qualitative similarities exist between the sterol makeup of many invertebrate phyla although quantitative differences often obscure these facts. Whether the similarities are caused by the presence of similar enzyme systems for sterol transformation or by similarity in the pattern of absorption and deposition of exogenous sterols remains to be ascertained. This knowledge will be required in order to properly evaluate the theory advanced by Bregmann (1962) that sterol composition is a reflection of the position of an animal on the evolutionary scale.

REFERENCES

Afzelius, B. A., and Mohri, H. (1966). *Exptl. Cell Res.* **42**, 10.

Allen, W. V. (1965). *Dissertation Abstr.* **26**, 454.

Allen, W. V., and Giese, A. C. (1966). *Comp. Biochem. Physiol.* **17**, 23.

Bergmann, W. (1952). *In* "Progress in the Chemistry of Fats and Other Lipids" (R. T. Holman, W. O. Lundberg, and T. Malkin, eds.), Vol. 1, pp. 18–69. Pergamon Press, Oxford.

Bergmann, W. (1962). *Comp. Biochem.* **3**, 103–162.

Bergmann, W., and Domsky, I. I. (1960). *Ann. N.Y. Acad. Sci.* **90**, Art. 3, 906.

Bergmann, W., and Landowne, R. A. (1958). *J. Org. Chem.* **23**, 1241.

Bergmann, W., and Stansbury, H. A., Jr. (1943). *J. Org. Chem.* **8**, 283.

Dorée, C. (1909). *Biochem. J.* **4**, 72.

Eichberg, J., Gilbertson, J. R., and Karnovsky, M. L. (1961). *J. Biol. Chem.* **236**, PC15.

Ellingsboe, J., and Karnovsky, M. L. (1965). *Federation Proc.* **24**, 477.

Fagerlund, U. H. M. (1962). Unpublished data.
Fagerlund, U. H. M., and Idler, D. R. (1959). *J. Am. Chem. Soc.* **81**, 401.
Fagerlund, U. H. M., and Idler, D. R. (1960). *Can. J. Biochem. Physiol.* **38**, 997.
Fagerlund, U. H. M., and Idler, D. R. (1958). Unpublished data.
Ghiretti, F., and D'Amelio, V. (1956). *Exptl. Cell Res.* **10**, 734.
Giese, A. C. (1961). *Biol. Bull.* **121**, 141.
Giese, A. C. (1966a). *Physiol. Rev.* **46**, 244.
Giese, A. C. (1966b). *In* "Physiology of the Echinodermata" (R. A. Boolootian, ed.), pp. 757–796. Wiley (Interscience), New York.
Giese, A. C., Krishnaswamy, S., Vasu, B. S., and Lawrence, J. (1964). *Comp. Biochem. Physiol.* **13**, 367.
Gilbertson, J. R., and Karnovsky, M. L. (1963). *J. Biol. Chem.* **238**, 893.
Greenfield, L., Giese, A. C. Farmanfarmaian, A., and Boolootian, R. A. (1958). *J. Exptl. Zool.* **139**, 507.
Hack, M. H., Gussin, A. E., and Lowe, M. E. (1962). *Comp. Biochem. Physiol.* **5**, 217.
Idler, D. R. (1958). Quoted by H. L. A. Tarr. *Ann. Rev. Biochem.* **27**, 225.
Isono, Y. (1965). *Sci. Papers Coll. Gen. Educ., Univ. Tokyo* **15**, 87.
Karnovsky, M. L., and Brumm, A. F. (1955). *J. Biol. Chem.* **216**, 689.
Karnovsky, M. L., Jeffrey, S. S., Thompson, M. S., and Deane, H. W. (1955). *J. Biophys. Biochem. Cytol.* **1**, 173.
Kita, M., and Toyama, Y. (1960). *Nippon Kagaku Zasshi* **81**, 485.
Lardy, H. A., and Phillips, P. H. (1941). *Am. J. Physiol.* **133**, 602.
Lawrence, J. M. (1965). Ph.D. Dissertation, Stanford University.
Lawrence, J. M., Lawrence, A. L., and Giese, A. C. (1966). *Physiol. Zool.* **39**, 281.
Lewis, R. W. (1966). *Dissertation Abstr.* **26**, 5012.
Lovern, J. A. (1938). *Biochem. J.* **32**, 1214.
Maggio, R., and Monroy, A. (1955). *Exptl. Cell Res.* **8**, 240.
Matsumoto, T., Yajima, M., and Toyama, Y. (1943). *J. Chem. Soc. Japan, Pure Chem. Sect.* **63**, 1203.
Mohri, H. (1957a). *J. Fac. Sci., Univ. Tokyo, Sect. IV* **8**, 51.
Mohri, H. (1957b). *Annotationes Zool. Japon.* **30**, 181.
Mohri, H. (1959). *Sci. Papers Coll. Gen. Educ., Univ. Tokyo* **9**, 269.
Mohri, H. (1961). *Sci. Papers Coll. Gen. Educ., Univ. Tokyo* **11**, 109.
Mohri, H. (1962). *Embroyologia (Nagoya)* **7**, 252.
Mohri, H. (1964a). *Pubbl. Staz. Zool. Napoli* **34**, 53.
Mohri, H. (1964b). *Biol. Bull.* **126**, 440.
Moore, H. B. (1935). *J. Marine Biol. Assoc. U.K.* **20**, 109.
Pearse, J. S., and Giese, A. C. (1966). *Biol. Bull.* **130**, 387.
Rapport, M. M., and Alonzo, N. F. (1960). *J. Biol. Chem.* **235**, 1953.
Rapport, M. M., Lerner, B., Alonzo, N., and Franzl, R. E. (1957). *J. Biol. Chem.* **225**, 859.
Rodegker, W., and Nevenzel, J. C. (1964). *Comp. Biochem. Physiol.* **11**, 53.
Rothschild, Lord, and Cleland, K. W. (1952). *J. Exptl. Biol.* **29**, 66.
Salaque, A., Barbier, M., and Lederer, E. (1966). *Comp. Biochem. Physiol.* **19**, 45.
Scarano, E., and Rocca, E. (1959). *Pubbl. Staz. Zool. Napoli* **31**, 209.
Tanikawa, E., Akiba, M., and Yoshitani, S. (1955). *Bull. Fac. Fisheries, Hokkaido Univ.* **5**, 341.
Todd, D., and Rizzi, G. P. (1964). *Proc. Soc. Exptl. Biol. Med.* **115**, 218.
Toyama, Y. (1958). *Fette, Seifen, Anstrichmittel* **60**, 909.

CHAPTER 6

Pigments in Echinodermata

T. W. Goodwin

I. Crinoidea

A. Anthraquinones

The Australian crinoids *Comatula pectinata* (L) and *C. cratera* (A. H. Clark) contain the 6-methyl and 6-8-dimethylethers of rhodocomatulin (I) (Sutherland and Wells, 1959, 1967), 8-butyryl,4,5,7-tetrahy-

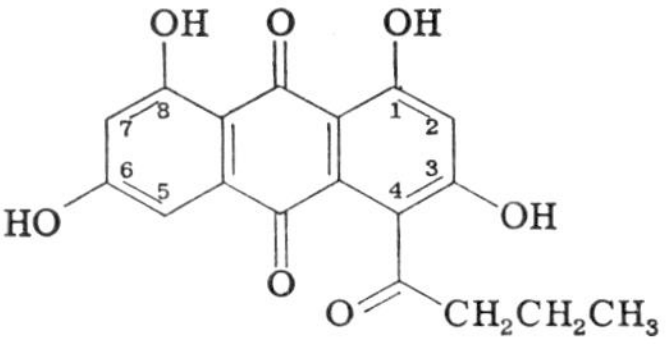

(I)

droxy-2-methoxyanthraquinone is also present (Powells *et al.*, 1967). Hydroxyanthraquinones present in *Ptilometra australis* are 1,6,8-trihydroxy-3(1-hydroxypropyl)-anthraquinone, 1,6,8-trihydroxy-3-(2-hydroxypropyl) anthraquinone and 1,6,8-trihydroxy-3-propylanthraquinone-2-carboxylic acid (Powells and Sutherland, 1967). Only the acidic derivative is present in *Tropiometra afra.*

A red quinoid pigment from the feather star *Antedon bifida* (Dimelow, 1958) has not been characterized fully although its low pK_a value, 3.8, suggests a naphthoquinone rather than an anthraquinone (Thomson, 1962).

B. Carotenoids

Neither Lederer (1935a) nor Karrer and Solmssen (1935) could find carotenoids in the sea lily *Antedon rosacea* (*bifida*), but Lönnberg (1931) had earlier reported the presence of unidentified carotenoids in *A. petasus,* and Dimelow (1958) made similar observations on *Antedon bifida.*

II. Echinoidea

A. Naphthoquinones

The red pigment in the perivisceral fluid of several sea urchins was first called echinochrome by MacMunn in 1885. Echinochrome (II) was

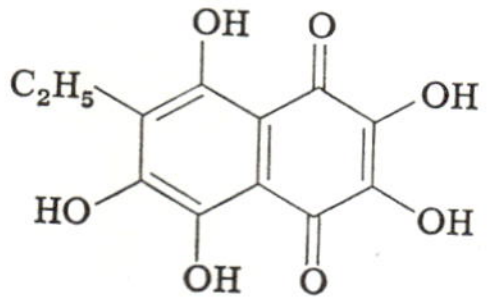

(II)

first crystallized by McLendon (1912) and fully characterized by Kuhn and Wallenfels (1943); it was synthesized by Gauhe and Wallenfels (1943) and later by Anderson *et al.* (1965) by a different route. It is now known as echinochrome A following the nomenclature proposals of Goodwin *et al.* (1951). It was also suggested that the pigments of the calcareous parts of echinoids (the test and the spines) should be called spinochromes while those from fleshy parts should be called echinochromes (Goodwin and Srisukh, 1950). Later investigations have shown that the first proposal is valid, but that echinochromes are often found in the calcareous parts in small amounts. It was also suggested that the pigments should be named alphabetically in chronological order of isolation; this proposal has not always been followed very seriously.

The various fully characterized spinochromes which have been isolated are summarized with their structures and synonyms in Table I. The distribution of spinochromes and echinochromes is given in Table II. Apart from the species listed in Table II, the sand dollar, *Dendraster excentricus,* contains small amounts of echinochrome-like pigments (Fox and Scheer, 1941) while *D. laevis* contains none (Goodwin and Fox,

TABLE I

NATURALLY OCCURRING SPINOCHROMES IN ECHINOIDEA

Pigment	Structure	Synonyms	References
Spinochrome A	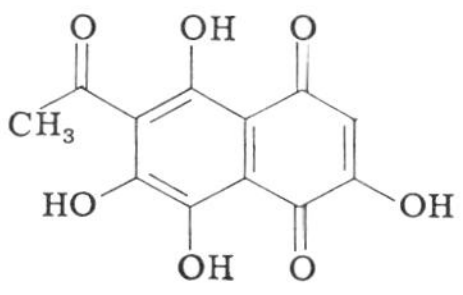	Spinochromes M, Mg, Aka_2, P	Glaser and Lederer (1938); Goodwin and Srisukh (1950); Kuroda and Ohshima (1940); Kuroda and Okajima (1951, 1953); Chang *et al.* (1964, a, b); Minchilli and Musajo (1940); Lederer (1952)
Spinochrome B	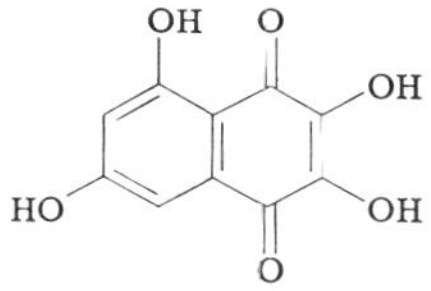	Spinochromes N, P_1, B, M_2	Goodwin *et al.* (1951); Minchilli and Musajo (1942); Kuroda and Okajima (1953, 1960); Kuroda and Ohshima (1940); Glaser and Lederer (1938); Goodwin and Srisukh (1950); Gough and Sutherland (1964)
Spinochrome C	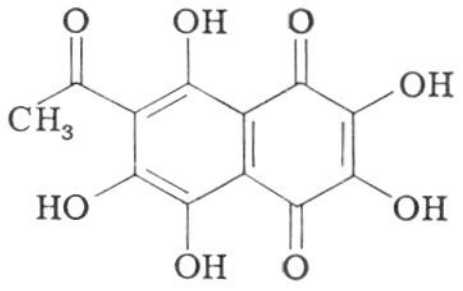	Spinone A, Spinochromes F, F_1, M_3	Glaser and Lederer (1938); Lederer (1952); Kuhn and Wallenfels (1941); Kuroda and Okajima (1951, 1960, 1964)
Spinochrome D	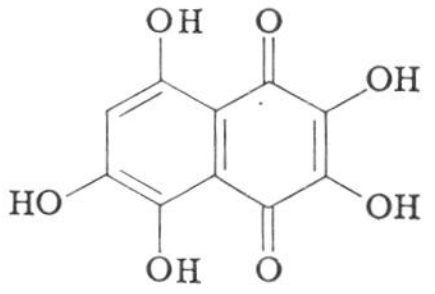	Spinochromes Aka, Aka_1	Kuroda and Okajima (1951, 1953); Anderson *et al.* (1965)
Spinochrome E	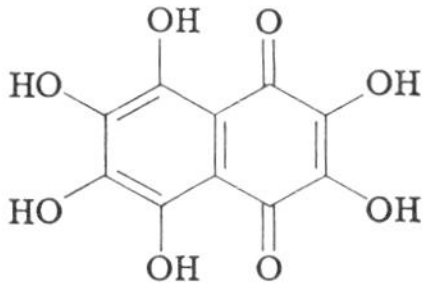		Yoshida (1959); Smith and Thomson (1961)
Spinochrome G	?	? Spinochrome B	Glaser and Lederer (1938)

TABLE II

DISTRIBUTION OF NAPHTHOQUINONES IN ECHINOIDEA

	Echino-chrome	Spinochrome						References
	A	A	B	C	D	E	G	
Anthocidaris crassisipina		+	+	+				Kuroda and Oshima (1940); Kuroda and Koyasu (1944); Anderson and Thomson[a]
Arbacia incisa	+			+				Anderson and Thomson[a]
Arbacia lixula	+			+				Lederer (1952); Kuhn and Wallenfels (1941)
Arabacia pustulosa	+			+				Lederer (1952)
Centrostephanus nitidus	+							Anderson and Thomson[a]
Colobocentratus atratus		+						Chang *et al.* (1964a,b)
Dendraster excentricus	+				+	+		Goodwin and Fox (1955); Anderson and Thomson[a]
Diadema antillarum	+					+		Millott (1957); Anderson and Thomson[a]
Diadema paucispinum		+						Chang *et al.* (1964a,b)
Diadema setosum	+							MacMunn (1885); Nishibori (1959)
Echinometra oblonga		+						Chang *et al.* (1964a,b)
Echinothrix calamaris		+						Chang *et al.* (1964a,b)
Echinothrix diadema		+						Chang *et al.* (1964a,b)
Echinus acutus	+	+	+	+		+		Anderson and Thomson[a]
Echinus esculentus	+	+	+			+		Lederer (1940); Goodwin and Srisukh (1950); Anderson and Thomson[a]
Eucidaris thouarsii	+	+	+	+		+		Anderson and Thomson[a]
Hemicentrotus pulcherrimus		+	+	+		+		Kuroda and Iwakura (1942); Yamasu and Kawaguti (1955); Kuroda and Okajima (1962); Anderson and Thomson[a]
Heterocentrotus mammilatus				+				Kuroda and Ohshima (1940); Kuroda and Okajima (1960); Anderson and Thomson[a]
Paracentrotus depressus		+			+			Kuroda and Ohshima (1940) Kuroda and Okajima (1960); Anderson and Thomson[a]
Paracentrotus lividus (N. Atlantic)		+	+					Glaser and Lederer (1939); Goodwin and Srisukh (1950)
Paracentrotus lividus (Mediterranean)	+	+	+	+		+	+	Lederer (1952); Minchilli and Musajo (1940, 1942)
Psammechinus miliaris		+	+	+		+		Yoshida (1959); Smith and Thomson (1961); Anderson and Thomson[a]

TABLE II (*Continued*)

	Echinochrome	Spinochrome						References
	A	A	B	C	D	E	G	
Scaphechinus mirabilis	+	+						Nishibori (1957); Yamasu and Kawaguti (1955)
Sphaerechinus granularis	+	+		+				Anderson and Thomson[a]
Sterechinus neumayari		+						Anderson and Thomson[a]
Stomopneustes variolaris	+							Nishibori (1961)
Strongylocentrotus franciscanus	+	+	+	+		+		Anderson and Thomson[a]
Strongylocentrotus purpuratus	+							Tyler (1939); Thomson and Anderson[a]
Tripneustes gratilla		+						Chang *et al.* (1964a,b)

[a] Unpublished work quoted by Fox and Hopkins (1966).

1955). Traces are also found in the very pale-colored urchin *Lytechinus pictus* (Fox and Scheer, 1941). Two pigments, echinochrome B and C, have been obtained from *Arbacia lividus,* they are very similar to echinochrome A and have not yet been fully characterized (Kuhn and Wallenfels, 1941).

The characteristic colors of the spines and tests of sea urchins are, except in the case of black specimens which contain melanin (see p. 141), due to the calcium salts of spinochromes. It is a difference in

TABLE III

DISTRIBUTION OF SPINOCHROMES A AND B IN SPINES OF COLOR VARIANTS OF *Paracentrotus Lividus* (NORTH ATLANTIC)

Color variant	Amount (mg./gm.)(fresh wt)	
	Spinochrome A	Spinochrome B
Violet	0.52	0.39
Olive green	0.17	1.07

amounts of the component pigments which produces color variants in the same species. For example, violet and olive-green specimens of *Paracentrotus lividus* contain varying amounts of spinochromes A and B (Goodwin and Srisukh, 1950) [Table (III)] together with small amounts of echinochrome A (Lederer, 1952); the brown specimens of the same species collected in the Mediterranean contain a more complex mixture of echinochrome A and spinochromes A, B, C, E and G (Lederer, 1952).

Echinochrome A is widely distributed within the animals proper but tends to accumulate in the eggs and ovaries and be absent from the testes (Glaser and Lederer, 1939). It is present in the perivisceral fluid attached to protein (Kuhn and Wallenfels, 1943) and some is present in the reduced form. This led to the early suggestion by MacMunn that it was a respiratory pigment, but this could not be confirmed (Cannon, 1927). Claims that echinochrome function in reproduction has also been doubted (Cornman, 1941), but they may well play an important part in photoreception in these animals (Millott and Yoshida, 1957).

B. Carotenoids

Echinenone was first isolated from the gonads of *Paracentrotus* (*Strongylocentrotus*) *lividus* (Lederer, 1935b) and it is present in the gonads of *Echinus esculentus* and in the gonads and perivisceral fluid of *Echinocardium cordatum* (Goodwin and Srisukh, 1951). Its structure, 4-oxo-β-carotene (III), was elucidated by Goodwin and Taha (1950); it is iden-

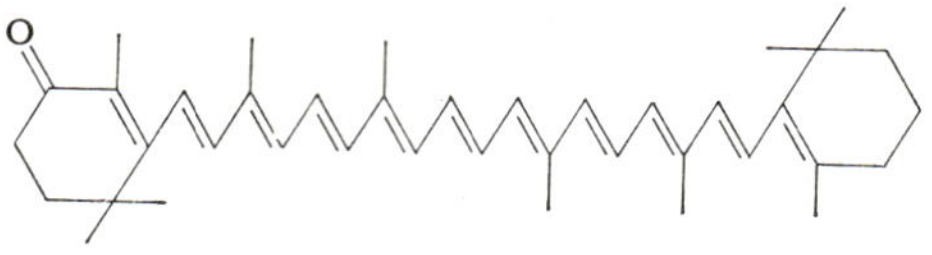

(III)

tical with myoxanthin from the alga *Oscillatoria rubescens* and aphanin from the alga *Aphanizomenon flosaquae* (Hertzberg and Jensen, 1966). It is now known to be widely distributed in nature. It has been reported in the following other echinoderms: *Lytechinus pictus, Dendraster excentricus, Strongylocentricus purpuratus,* and *S. franciscanus* (Fox and Scheer, 1941), and in *Sphaerechinus granularis* (de Nicola and Goodwin, 1954a). β-Carotene is also invariably present along with echinenone, and together they constitute the bulk of the carotenoids present in sea urchin male and female gonads. A number of more strongly adsorbed xanthophylls are also present, but with the exception of pentaxanthin, they have not been completely identified. Lederer crystallized pentaxanthin ($C_{40}H_{56 \text{ or } 58}O_5$, m.p. 209°–210°) from *P. lividus* in 1935 but its exact structure remains undetermined. Fox and Scheer (1941) give exhaustive data on all the specimens they examined. The eggs of *P. lividus* contain two unique carotenoids which have not been fully identified and which have been named paracentrotin A and B. They represent some 40–50% of the total pigments in the eggs and about 10% in the ovaries. They are characterized by unexpected absorption spectral changes on saponification. The change [e.g., in paracentrotin from 488 mμ. to 452 mμ.

(main max.)] is not reversible. There is no paracentrotin of either type in the testes (de Nicola and Goodwin, 1954b). The presence of astaxanthin (see Section IV, A) in *Peronella japonica* (Kawaguti and Yamasu, 1957) represents the first report of this pigment in the Echinoidea.

C. Melanins

Millott and Jacobson (1953) showed that the large epidermal melanophores of *Diadema antillarum* contain granules which, from chemical and histochemical tests, probably consist of a melanin. If the coelomic fluid is exposed to air the amebocytes are thrown down as aggregates which rapidly blacken to form what appears to be melanin.

D. Heme Pigments

A heme pigment is reported in the eggs of *Arbacia punctata* (Ball and Meyerhof, 1940).

III. Holothuroidea

A. Quinones

Only one species has been examined so far: Namakochrome (IV) from *Polycheira rufescens* is a monomethyl ether of spinochrome E

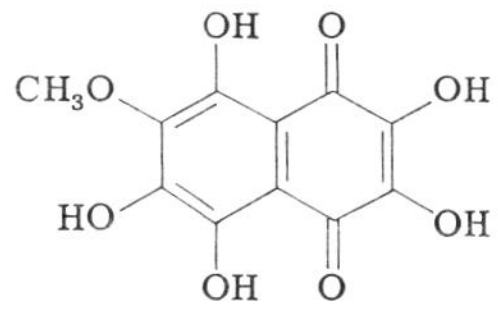

(IV)

(Table I) (i) (Mukai, 1958, 1960). This is the only known report of a naphthoquinone pigment outside the class Echinoidea.

B. Carotenoids

"Lipochromes" were observed in holothurians as early as the end of the nineteenth century (see Fox, 1953). Lönnberg and Hellström (1931) found carotenoids in ovaries of *Mesothuria intestinalis* and *Cucumaria lactea*, and in the tissues of *Psolus phantalus*, *C. elongata*, *Phyllophorus pellucidus*, and *Thyone fusus*. They found none in the red integument of *Stichopus tremulus* but Fox and Scheer (1941) found small amounts of carotenes (including probably echinenone) and xanthophylls in the whole bodies of *S. parvimensis*. Manunta (1943) reported carotenoids in the lungs, mesentery, intestine, and gonads of *Holothuria forskali*, *H. tubulosa*, and *H. polii*. The highest concentrations were found in the gonads of the first two species and in the intestine of the last. Fox and Hopkins (1966) have found carotenes and xanthophylls, including astaxanthin, in *C. lubrica*. Carotenoproteins are said to occur in the gonads of *Holothuria* spp. and *Cucumaria* spp. (Toumanoff, 1927).

C. Melanins

The black pigment in the large epidermal melanophores of *Thyone briareus* is melanin (Millott, 1950) as is the black pigment in the body wall of *Holothuria forskali* (Millott, 1953). The amebocytes of the latter contain a phenolic system similar to that in the sea urchin *Diadema artillarum* (p. 141). The pigment shed from the body wall of *H. grisea* seems to be melanin (Millott, 1954).

D. Porphyrins

Some holothurians, e.g., *Caudina*, *Cucumaria miniata*, and *Thyonella* spp. contain porphyrins as part of their respiratory hemoglobins (Prosser and Brown, 1961; Manwell and Baker, 1963).

IV. Asteroidea

A. Carotenoids

Early work on the lipochromes of asteroids has been summarized by Goodwin (1952) and Fox (1953). Astaxanthin (V) or its oxidative

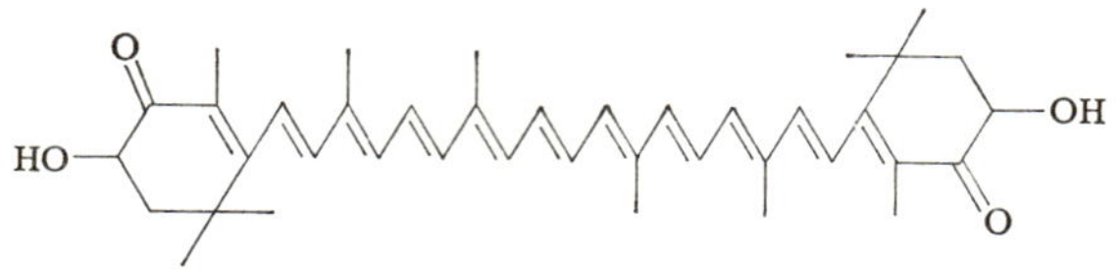

(V)

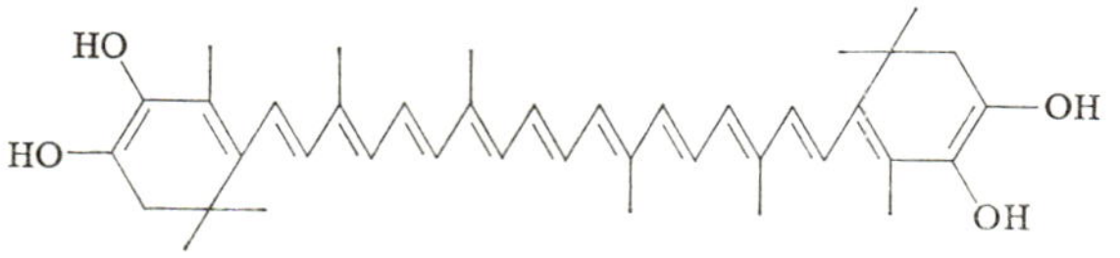

(VI)

artifact astacin (VI) has been obtained from almost all specimens examined during the past 20 to 30 years, viz., *Ophidiaster ophidianus* (Karrer and Benz, 1934); *Asterias rubens* (von Euler and Hellström,

1934; Vevers, 1952); *Crossaster (Solaster) papposus, S. endica, Porania pulvillus* (von Euler and Hellström, 1934); *Echinaster sepositus* (Karrer and Solmssen, 1935; de Nicola and Goodwin, 1954a); *Astropecten californicus, Patiria miniata, Pisaster ochraceous, P. giganteus* (Fox and Scheer, 1941); *Asterina panceri* (de Nicola, 1956); *Marthasterias glacialis* (Vevers and Millott, 1957; de Nicola, 1959a) and *Astopecten auranticus* (de Nicola, 1958). Astaxanthin is, however, absent from a number of color variants of *Henricia leviuscula* (Fox and Hopkins, 1966). β-Carotene is always present, but the relative amount of carotenes to xanthophylls is very low (Fox and Scheer, 1941). The more recent investigations have revealed the presence of a number of pigments with properties which suggest that they are biosynthetic intermediates between β-carotene and astaxanthin. The structure of none of these has yet been unequivocally determined but two from *O. ophidianus, Asterina panceri, Astropecten aurantiacus,* and *M. glacialis* are reported to be 3-hydroxy-4′-oxo-β-carotene (asteroidenone) and 3,3′-hydroxy-4′-oxo-β-carotene (hydroxyasteroidenone) (de Nicola, 1959b). Unresolved mixtures of carotenoids have been reported in *Aphelasterias japonica, Asterias amurensis, Asterina pectinifera, Dipsacaster pretiosa,* and *Solaster borealis* (Nishibori, 1954).

Astaxanthin occurs in many asteroids as chromoproteins—for example in *Porania pulvillus* as a violet blue-red pigment and as a deep blue pigment in *Asterias rubens* (von Euler and Hellström, 1934). A full list is given by Cheeseman *et al.* (1967).

B. Porphyrins

MacMunn (1886) reported the presence of porphyrins in the skin of *Uraster rubens* and other brown asteroids. Much more recently Kennedy and Vevers (1953a,b, 1954) found chlorocruoroporphyrin (VII)

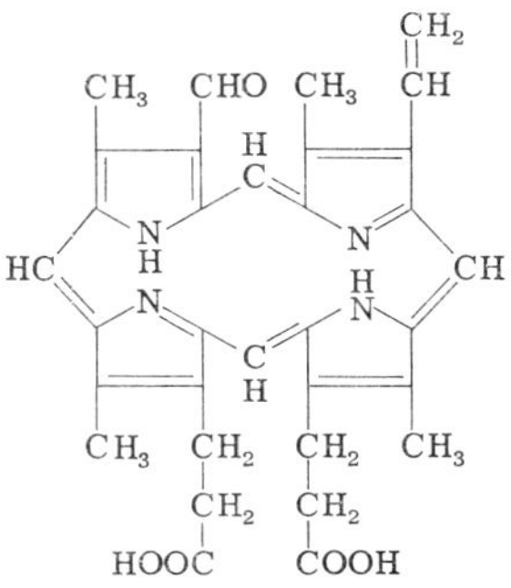

(VII)

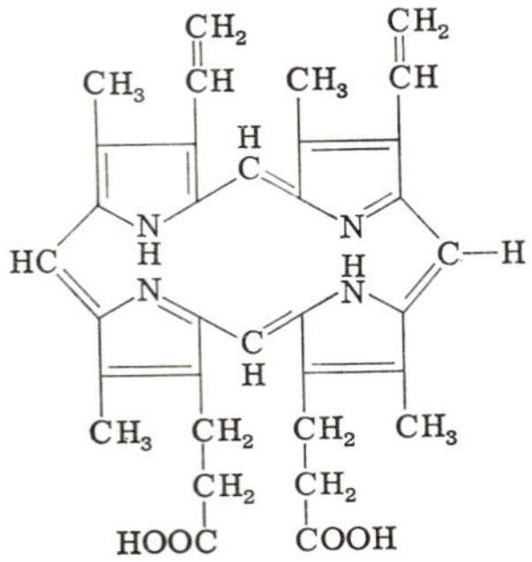

(VIII)

and protoporphyrin (VIII) in *Astropecten irregularis* and *Luidia ciliaris* and protoporphyrin from *Asterias rubens*.

V. Ophiuroidea

A. Carotenoids

Carotenoids have been detected in a number of ophiuroids by Lönnberg (1935) but the only detailed survey has been that of Fox and Scheer (1941). They examined *Ophiopteris papillosa, O. spiculata,* and *Ophiothrix rudis* and found xanthophylls, both free and esterified, present and carotenes absent. Pectenoxanthin, first obtained from *Pecten maximus* by Lederer (1934) was detected, as was a pigment similar to taraxanthin or dinoxanthin; other unstable unidentified pigments were also present. Pectenoxanthin has now been shown to be identical with the acetylenic carotenoid alloxanthin (IX) (Mallams *et al.*, 1967; Campbell *et al.*, 1967).

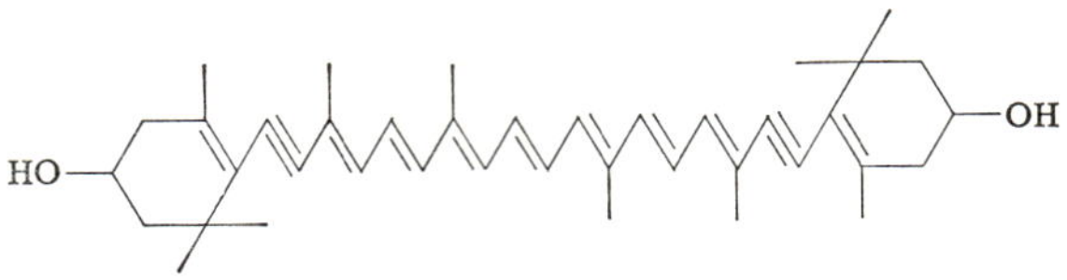

(IX)

Similar observations were made on *Ophiocomina nigra* (Fontaine, 1962). There is a very old report that *Ophioderma longicauda* contains a brown carotenoprotein (de Mérejkowsky, 1883).

B. Melanins

Melanin has been reported in the skin of some ophiuroids (Cloudsley-Thompson, 1946; Fontaine, 1962).

C. Porphyrins

No porphyrins are in the integument of *Ophiocomina nigra* (Kennedy and Vevers, 1954) but breakdown products of chlorophylls are found in the intestinal tract of various Ophiuroidea (Fox and Scheer, 1941; Fontaine, 1962).

VI. Comment

The time is ripe for a large-scale study of the carotenoids of the Echinodermata. The new knowledge of the chemistry of the carotenoids combined with the newer methods of separation and analysis (e.g., thin-layer chromatography and mass spectrometry) should give much clearer information on the biosynthetic sequence in the Asteroidea and on the nature of the complex xanthophylls encountered in other classes.

ACKNOWLEDGMENT

The author thanks Professor R. H. Thomson for valuable information concerning the present status of the naphthoquinone pigments in the Echinoidea.

REFERENCES

Anderson, H., Smith, J., and Thomson, R. H. (1965). *J. Chem. Soc.* p. 2141.
Ball, E. G., and Meyerhof, B. (1940). *J. Biol. Chem.* **134**, 483.
Campbell, S. A., Mallams, A. K., Waight, E. S., Weedon, B. C. L., Barbier, M., Lederer, E., and Salaque, A. (1967). *Chem. Commun.* p. 941.
Cannon, R. K. (1927). *Biochem. J.* **21**, 184.
Chang, C. W., Moore, R. E., and Scheuer, P. J. (1964a). *J. Am. Chem. Soc.* **86**, 2959.
Chang, C. W., Moore, R. E., and Scheuer, P. J. (1964b). *Tetrahedron Letters* p. 3557.
Cheeseman, D. F., Lee, W. L., and Zagalsky, P. F. (1967). *Biol. Rev.* **42**, 131.
Cloudsley-Thompson, J. L. (1946). *Nature* **157**, 342.
Cornman, I. (1941). *Biol. Bull.* **80**, 202.
de Mérejkowsky, C. (1883). *Bull. Soc. Zool.* **8**, 81.
de Nicola, M. G. (1956). *Exptl. Cell Res.* **10**, 441.
de Nicola, M. G. (1958). *Ric. Sci.* Suppl. **28**, 2076.
de Nicola, M. G. (1959a). *Ric. Sci.* Suppl. **29**, 78.
de Nicola, M. G. (1959b). *Boll. Sed. Accad. Gioenia Sci. Nat. Catania* [4] **4**, 201.
de Nicola, M. G., and Goodwin, T. W. (1954a). *Pubbl. Staz. Zool. Napoli* **25**, 145.
de Nicola, M. G., and Goodwin, T. W. (1954b). *Exptl. Cell Res.* **7**, 23.
Dimelow, E. J. (1958). *Nature* **182**, 812.
Fontaine, A. R. (1962). *J. Marine Biol. Assoc. U.K.* **42**, 1, 9, and 33.
Fox, D. L. (1953). "Animal Biochromes and Structural Colours." Cambridge Univ. Press, London and New York.
Fox, D. L., and Hopkins, T. S. (1966). *In* "Comparative Biochemistry of Pigments" Chapter 12.
Fox, D. L., and Scheer, B. T. (1941). *Biol. Bull.* **80**, 441.

Gauhe, A., Wallenfels, K. (1943). *Chem. Ber.* **76**, 325.
Glaser, R., and Lederer, E. (1938). *Compt. Rend.* **207**, 254.
Glaser, R., and Lederer, E. (1939). *Compt. Rend.* **208**, 1939.
Goodwin, T. W. (1952). "Comparative Biochemistry of the Carotenoids." Chapman & Hall, London.
Goodwin, T. W., and Fox, D. L. (1955). *Experientia* **11**, 270.
Goodwin, T. W., and Srisukh, S. (1950). *Biochem. J.* **47**, 69.
Goodwin, T. W., and Srisukh, S. (1951). *Nature* **167**, 358.
Goodwin, T. W., and Taha, M. M. (1950). *Biochem. J.* **47**, 244.
Goodwin, T. W., Lederer, E., and Musajo, L. (1951). *Experientia* **7**, 375.
Gough, J., and Sutherland, M. D. (1964). *Tetrahedron Letters* p. 269.
Hertzberg, S., and Jensen, S. L. (1966). *Phytochemistry* **5**, 554 and 565.
Karrer, P., and Benz, F. (1934). *Helv. Chim. Acta* **17**, 412.
Karrer, P., and Solmssen, U. (1935). *Helv. Chim. Acta* **18**, 915.
Kawaguti, S., and Yamasu, T. (1957). *Biol. J. Okayama Univ.* **1**, 249.
Kennedy, G. Y., and Vevers, H. G. (1953a). *Nature* **141**, 81.
Kennedy, G. Y., and Vevers, H. G. (1953b). *J. Marine Biol. Assoc. U.K.* **32**, 235.
Kennedy, G. Y., and Vevers, H. G. (1954). *J. Marine Biol. Assoc. U.K.* **33**, 663.
Kuhn, R., and Wallenfels, K. (1941). *Chem. Ber.* **74**, 1594.
Kuhn, R., and Wallenfels, K. (1943). *Chem. Ber.* **76**, 323.
Kuroda, C., and Iwakura, H. (1942). *Proc. Imp. Acad. (Tokyo)* **18**, 74.
Kuroda, C., and Koyasu, K. (1944). *Proc. Imp. Acad. (Tokyo)* **20**, 23.
Kuroda, C., and Ohshima, H. (1940). *Proc. Imp. Acad. (Tokyo)* **16**, 214.
Kuroda, C., and Okajima, M. (1951). *Proc. Japan Acad.* **27**, 343.
Kuroda, C., and Okajima, M. (1953). *Proc. Japan Acad.* **29**, 27.
Kuroda, C., and Okajima, M. (1960). *Proc. Japan Acad.* **36**, 424.
Kuroda, C., and Okajima, M. (1962). *Proc. Japan Acad.* **38**, 353.
Kuroda, C., and Okajima, M. (1964). *Proc. Japan Acad.* **40**, 836.
Lederer, E. (1934). *Compt. Rend. Soc. Biol.* **116**, 150; **117**, 411.
Lederer, E. (1935a). "Les carotenoides des animaux." Hermann, Paris.
Lederer, E. (1935b). *Compt. Rend.* **201**, 300.
Lederer, E. (1940). *Biol. Rev. Cambridge Phil. Soc.* **15**, 273.
Lederer, E. (1952). *Biochim. Biophys. Acta* **9**, 92.
Lönnberg, E. (1931). *Arkiv Zool.* **22A**, No. 14.
Lönnberg, E. (1935). *Arkiv Zool.* **28B**, No. 8.
Lönnberg, E., and Hellström, H. (1931). *Arkiv Zool.* **23A**, 1.
MacMunn, C. A. (1885). *Quart. J. Microscop. Sci.* **25**, 469.
MacMunn, C. A. (1886). *J. Physiol. (London)* **7**, 240.
McLendon, J. F. (1912). *J. Biol. Chem.* **11**, 435.
Mallams, A. K., Waight, E. S., Weedon, B. C. L., Chapman, D. J., Haxo, F. T., Goodwin, T. W., and Thomas, D. M. (1967). *Chem. Commun.* p. 301.
Manunta, C. (1943). *Rend. Ist. Lombardo Sci. Lettere* **B76**, 19.
Manwell, C., and Baker, C. M. A. (1963). *Comp. Biochem. Physiol.* **10**, 39.
Millott, N. (1950). *Biol. Bull.* **99**, 343.
Millott, N. (1953). *J. Marine Biol. Assoc. U.K.* **31**, 529.
Millott, N. (1954). *Bull. Marine Gulf Caribbean* **3**, 305.
Millott, N. (1957). *Proc. Zool. Soc. London* **129**, 263.
Millott, N., and Jacobson, F. W. (1953). *Proc. Roy. Soc.* **B141**, 231.
Millott, N., and Yoshida, M. (1957). *J. Exptl. Biol.* **34**, 394.
Minchilli, M., and Musajo, L. (1940). *Gazz. Chim. Ital.* **70**, 287.

Minchilli, M., and Musajo, L. (1942). *Boll. Sci. Fac. Chim. Ind. Univ. Bologna* **3**, 113.

Mukai, T. (1958). *Mem. Fac. Sci., Kyushu Univ.* **3**, 29.

Mukai, T. (1960). *Bull. Chem. Soc. Japan* **23**, 453 and 1234.

Nishibori, K. (1954). *Bull. Japan. Soc. Sci. Fisheries* **20**, 36.

Nishibori, K. (1957). *Bull. Japan Soc. Sci. Fisheries* **22**, 708.

Nishibori, K. (1959). *Nature* **184**, 1234.

Nishibori, K. (1961). *Nature* **192**, 1293.

Powells, V. H., and Sutherland, M. D. (1967). *Australian J. Chem.* **20**, 535.

Powells, V. H., Sutherland, M. D., and Wells, J. W. (1967). *Australian J. Chem.* **20**, 535.

Prosser, C. L., and Brown, F. A., Jr. (1961). "Comparative Animal Physiology," p. 201. Saunders, Philadelphia, Pennsylvania.

Smith, J., and Thomson, R. H. (1961). *J. Chem. Soc.* p. 1008.

Sutherland, M. D., and Wells, J. W. (1959). *Chem. Ind. (London)* p. 291.

Sutherland, M. D., and Wells, J. W. (1967). *Australian J. Chem.* **20**, 515.

Thomson, R. H. (1962). *Comp. Biochem.* **3**, Part A, 631.

Toumanoff, C. (1927). *Compt. Rend. Soc. Biol.* **96**, 1393.

Tyler, A. (1939). *Proc. Natl. Acad. Sci. U.S.* **25**, 523.

Vevers, H. G. (1952). *J. Marine Biol. Assoc. U.K.* **30**, 569.

Vevers, H. G., and Millott, N. (1957). *Proc. Zool. Soc. London* **129**, 75.

von Euler, H., and Hellström, H. (1934). *Z. Physiol. Chem.* **223**, 89.

Yamasu, T., and Kawaguti, S. (1955). *Biol. J. Okayama Univ.* **2**, 62.

Yoshida, M. (1959). *J. Marine Biol. Assoc. U.K.* **38**, 455.

CHAPTER 7

Fertilization and Development

Tryggve Gustafson

I. Introduction

Gametes and embryos of echinoderms are classical material in developmental biology, cytology, and to some extent in biochemistry. Many of the basic problems of fertilization and cell division have been elucidated by studies of echinoderm eggs. The theory of the early developmental interaction between cells and of embryonic determination is also to a great extent based upon experiments with echinoderm eggs. In fact, the double gradient interactions are not restricted to this group. The possibility of affecting the developmental trend of echinoderm eggs by means of various chemical agents has made them a useful object for studies of the biochemical background of differentiation. Since large amounts of more or less synchronously developing sea urchin embryos may be obtained by artificial fertilization, and they are easy to keep in rather dense cultures, the sea urchin material has been particularly useful for such investigations. This material has therefore also been used by biochemists with more general interest in the mechanisms of protein synthesis and its regulation. Unfortunately, the echinoderm egg is not an ideal material for studies on chromosomes. Important investigations of sea urchin hybrids and of the regional and temporal changes in the nuclear stainability, however, have already shed some light upon the role of the genes and the regulation of genetic activities during sea urchin development. Finally, since it is possible to define some of the cellular activities underlying morphogenesis, it seems possible to trace the causal chain from the genes, or at least from the biochemical activities they control, up to morphogenesis. The possibilities of keeping sea urchins in aquaria and making them shed by means of various stimuli, and even of sending living sea urchins over wide distances with little harm to the animals, has opened the possibility of using sea urchins in every biological laboratory. To some extent, other echinoderms can be handled in the same way; this is important since premature generalizations from sea urchin material should be avoided. This is evident already from comparisons between the sea urchin egg and that of *Asterias*.

As this review covers a very extensive field and the space is very restricted, the author has to confine it to a few main problems without any penetrating analysis.

II. Formation of the Egg

The ovaries of the echinoderms have a wide lumen lined with a rather flat epithelium. According to one view, the oogonia develop from scattered cells within this epithelium. Observations on a sea urchin, on the other hand, indicate that the germinal epithelium covers the outside of the ovary, and that the oogonia pass through its wall into the lumen, where they undergo some divisions and develop into ovocytes, which after an initial growth align along the wall (Tennent and Ito, 1941). In any case, the contact of the growing ovocyte with the wall gradually narrows into a stalk and is finally lost, after which the ovocytes come to lie freely within the lumen.

In various echinoderms, some cells join the growing ovocytes and form a kind of follicle around it, although this is more or less rudimentary and disappears sooner or later. The role of the follicle cells in the nutrition of the ovocytes appears to be unknown (cf. however, Fig. 15 in Runnström, 1966). In the sea urchin, ameboid cells appear to play a role in this connection. Such cells phagocytize unspawned ovocytes and join up into a syncytium that is rich in inclusions. Pseudopods from the syncytium later on join the growing ovocytes. Since the amount of inclusions decreases as the ovocytes grow, the material piled up is probably transferred to the ovocytes (Lindahl, 1932, 1941). In *Antedon,* abortive oogonia play the role of nurse cells (L. A. Harvey, 1931). For further details about echinoderm oogenesis see Raven (1961).

Detailed references to the abundant literature on the chemical composition of the echinoderm egg can be found in Harvey's bibliography (E. B. Harvey, 1956), in Raven's monograph on oogenesis (Raven, 1961), in Monroy's book on the chemistry and physiology of fertilization (Monroy, 1965a) and in Brachet (1960). The author will therefore restrict himself to a few general remarks.

The amount of material piled up during oogenesis of the echinoderms need not be excessive, as the egg generally develops into a free-swimming larva that soon begins to feed actively, in the sea urchins after about 2 or 3 days. Where development is more or less direct—examples of this can be found in all classes of echinoderms—the egg is heavily loaded with yolk, and may be quite large. Exceptions from these rules, however, occur. For further information about this see Dawydoff (1948) and Hyman (1955).

Part of the material piled up may be considered as fuels and raw materials in anabolic reactions during early development. One of the fuels is a glycogenlike carbohydrate, but lipids, amino acids, and proteins also play a role (Hayes, 1938; Lindberg, 1943, 1945; Lindberg and Ernster, 1948; Lindahl and Öhman, 1938; Öhman, 1940, 1944; Hutchens *et al.,* 1942; E. B. Harvey, 1956; Tuft, 1953; Monroy and Maggio, 1964, Monroy, 1965a; Isono, 1963; see Gustafson, 1965). An important source of raw material for protein synthesis is no doubt the yolk granules (Kavanau, 1953, 1954; Perlmann and Kaltenbach, 1957; cf. Monroy and Maggio, 1964). Pure yolk has not been isolated, however, and information about its chemical composition largely derives from histochemical work (cf. Monné and Slautterback, 1950). Another raw material is the free amino acids which are abundant—particularly glycine which also plays an osmotic role (Örström, 1941; Kavanau, 1953, 1954).

The rapid increase of DNA during cleavage stage creates demands for easily available building blocks for DNA within the cytoplasm, as in the amphibians, a demand that is evidently covered to a certain extent (Hoff-Jörgensen, 1954; Elson *et al.,* 1954; Agrell and Persson, 1956; Whiteley and Baltzer, 1958; Gray, 1965; Lövtrup, 1966; Kavanau, 1958; see also Monroy and Maggio, 1964, and Brachet, 1960). As the genetic material during early cleavage is most involved in its own reproduction, and only exists in a condensed state at fertilization and during mitosis, there is restricted time for the formation of messenger RNA during this period. Hence one can expect that the cells contain a certain amount of performed messenger RNA protected in some way from degradation. There is, in fact, evidence that messenger RNA together with ribosomes are present within the unfertilized egg, but in an inactive form (cf. Gross *et al.,* 1965; Nemer, 1965; Section III,N). The role of the cytoplasmic DNA as informational molecules should also be considered (cf. Brachet, 1960; Monroy, 1965a,b).

A great number of enzymes required during early development are available already in the unfertilized egg, many of them in an inactive state, or at least structurally separated from their substrates (see Runnström, 1949, 1966; Monroy, 1965a,b; Section III,L,M). The egg also contains macromolecules required for the first mitotic division, that is often completed within an hour after fertilization and is foreshadowed within a few minutes after the entrance of the sperm. The occurrence of such material has also been demonstrated, although an essential component has to be supplied by the sperm (Section III,O).

Apart from the material necessary for early development, the egg is supplied with compounds and structures that play a special role for the interaction between the sperm and the egg, for the protection of the egg against the entrance of supernumerary spermatozoa, or as a protection against chemical and mechanical influences. The jelly coat and the vitelline membrane surrounding the egg and the cortical granules are examples of this (Section III,B and G). Apart from these structures, the echinoderm egg is equipped with organelles of the type common to eggs and unspecialized cells in general. The ultrastructural features of these have been described by Afzelius and others (see Raven, 1961; Monroy, 1965a; Runnström, 1966). An important tool in studies of the cellular inclusions is the centrifuge microscope of Harvey, by means of which most of the cytoplasmic inclusions can be stratified within the cell according to their relative specific weight, a stratification that may finally lead to a subdivision of the egg into a number of fragments, enriched in one or another of the different structural components (see E. B. Harvey, 1956).

The ovocyte nucleus or germinal vesicle contains one or several nucleoli. For detailed references about its growth, structure, chemical composition, see, e.g., Raven (1961) and Brachet (1960). The nucleoplasm in *Asterias* and sea urchins exhibits a strong reaction of SH groups as in many other animals. The nucleic acid and protein synthesis of the ovocyte nucleus of *Asterias* has been studied radioautographically by Ficq (e.g., 1955, 1961) and direct biochemical studies have also been performed on nucleoli from the same material by Vincent, Balthus, and collaborators (see, e.g., Vincent, 1965). The ovocyte eventually undergoes nuclear maturation, when some of the nuclear constituents are mixed with the cytoplasm.

In the later phase of ovocyte development, various signs of metabolic arrest can be noticed. In *Asterias* there is a gradual decrease in the nucleic acid and protein synthesis, both in the cytoplasm and the nucleus (Ficq, 1955, 1961). The rate of respiration, on the other hand, appears to be unaltered (Horowitz, 1965; cf. however Borei, 1948; Borei and Lybing, 1949). In the sea urchin the metabolic arrest does not appear until nuclear maturation, and is reflected by the intensity of protein synthesis (Monroy and Maggio, 1964), as well as changes in oxygen consumption (Lindahl and Holter, 1941; cf. Maggio and Monroy, 1959). Another feature of this period is that the cortical granules become incorporated in the cortex of the egg, a process that has been described and discussed in detail by Runnström (1966).

An additional feature of the late development of the female gamete is that it becomes capable of a normal reaction with a sperm. To some extent this process must be causally interlinked with the metabolic and ultrastructural changes just mentioned, and the whole set of phenomena is often denoted cytoplasmic maturation. The nature of the maturation process has recently been discussed by Runnström (1966). Although both the *Asterias* and sea urchin eggs are metabolically arrested at the time of shedding and are capable of reacting with the sperm, there is a marked difference between them with respect to the stage of nuclear maturation at the time of sperm entrance.

In forms such as *Asterias,* there is an arrest in nuclear maturation, which may be a reflection of metabolic arrest of the cell (Monroy, 1965a). In this case fertilization precedes nuclear maturation and evokes it. It is interesting to note that the cytoplasm in this case is not immediately in a state that allows the migration of the male nucleus toward the egg pronucleus and the swelling of the male nucleus. One may therefore conclude that some factor essential for the final steps in cytoplasmic maturation is provided by the germinal vesicle. This conclusion is particularly well founded by the experiments of Delage (1901) in his work on fertilization of nucleated and anucleate cell fragments.

In forms such as the sea urchins, on the other hand, nuclear maturation precedes the entrance of the sperm, and the cytoplasm could therefore be expected to be well prepared to allow the sperm nucleus to migrate and swell. It is important to note, however, that even in this case the egg may remain in some kind of underripeness for some time, i.e., in a state where it is refractory to spermatozoa, or at least responds in a more or less defective way (cf. Runnström *et al.*, 1959; Runnström, 1966). A complete maturation can, however, be induced by various agents such as amino acids and periodate. To some extent such agents bring about their effects by swelling and degrading the jelly coat, which may be too tough to allow the penetration of spermatozoa, or by elevation of a smooth fertilization membrane all over the egg surface. According to various observations, however, direct effects on the cell proper by such agents are also involved (cf. Runnström *et al.*, 1959; Runnström, 1966).

III. Fertilization

A. General Considerations

The process of fertilization not only means a combination of genetic information of two haploid gamete nuclei, it also involves a triggering of a series of arrested activities in the female gamete, leading to the onset of mitotic activity, including DNA synthesis, and also an activation of protein synthesis, which sooner or later becomes directed by the genetic information combined in the zygote. Fertilization involves a series of complex structural and biochemical reaction chains, where earlier links are prerequisite for later ones. The reaction chains are also interconnected and to a great extent mutually dependent. Still some of the individual events or sets of events may be experimentally dissociated or interrupted, leading to more or less partial activation. The activating effect of the sperm can be mimicked by means of various chemical agents so that parthenogenetic activation is elicited. Apart from showing that the egg itself is capable of carrying out some initial reactions of development, such experiments make it easier to study the physiology of fertilization, i.e., to define the individual reactions and their causal relations.

The echinoderm egg, particularly the sea urchin egg, is a quite favorable material for such studies, and many of the major discoveries and basic concepts in this field are based upon investigations of this material. The restricted space does not allow a discussion of the early fundamental contributions of Hertwig, Flemming, Boveri and others. One may, however, mention the fertilization theory of F. R. Lillie (1913, 1914, 1919), the concept that fertilization involves a local cytolysis of the egg (Loeb, 1913), the observation on the activation on the changes in respiratory activity of the egg at fertilization (Warburg, 1910), and the concept that fertilization involves a removal of a structural barrier or an inhibitor and hence allows the bringing together of some enzymes with the corresponding substrates (Runnström, 1933, 1949). The emphasis on the analogy between fertilization and nerve and muscle stimulation by R. S. Lillie (1916) is a further example of the generalizations based upon observations on sea urchin eggs. The list on classical observations and concepts can be considerably extended. Also, in recent years the sea urchin egg has played a basic role in studies on the molecular basis of initiation of various synthetic processes, and thus in attempts to apply the concepts of microbial and molecular biology to higher organisms (Section III,N).

Several important events and structures of importance in fertilization were discovered and studied in other materials, which sometimes are more suitable for a particular purpose. Therefore, a discussion of the physiology of fertilization should not be confined to the echinoderm egg. This should also be avoided as a protection from making premature generalizations. However, the purpose of this review and the restricted space available does not allow extended discussion. The historical aspects of the development of current concepts must also to a great extent be omitted. This often means that credit will not be given to the classical contributors to the field, whereas a series of investigators extending their work will be quoted. For more detailed information on all the aspects mentioned, the reader should consult the numerous reviews on fertilization, e.g., J. C. Dan (1956), F. R. Lillie (1919), Loeb (1913), Metz (1957), Monroy (1957, 1965a,b), Rothschild (1956), Runnström (1949), 1956b, 1966), Runnström *et al.* (1959), Tyler (1941, 1948, 1960, 1963), and Perlmann (1959).

B. The Jelly Coat

As most animal eggs, the echinoderm egg is surrounded by a coat, in this case not a tough membrane but a gelatinous layer, the jelly coat. The chemistry of this coat has been studied by Tyler, Vasseur, and others (for references, see Monroy, 1965a; Runnström *et al.*, 1959; Vasseur, 1954). The coat is made up of a rather homogeneous substance consisting of a mucopolysaccharide with highly asymmetrical molecules, with the molecular weight in *Arbacia* of about 30,000. The polysaccharide component, about 80% of the jelly dry weight, is strongly acid, heparinelike, compound with one sulfate molecule per monosaccharide residue, but devoid of amino groups. The monosaccharide residues show a great variation between different species, and may comprise mannose, galactose, glucose, and fucose, the last one occurring rather regularly (see Chapter 4, Vol. III). The additional 20% of the dry weight consists of amino acids which presumably form peptides alternating with the monosaccharide sequences. The variation in the carbohydrate composition is undoubtedly a main reason for the species specificity of some reactions between the sperm and the egg, although the arrangement of the nitrogenous components probably increases the variability. The jelly coat can be rather compact or quite swollen, particularly when the egg has reached complete cytoplasmic maturity. The swollen state, which can also be induced by various amino acids, proteins, and alkali, certainly favors the penetration of the sperm. The jelly may finally also go into solution to a varying extent, and forms an essential part of so-called egg water. This dissolution can be provoked by acid treatment, which makes it possible to investigate some aspects of fertilization.

C. The Fertilizin Hypothesis

On its way to the egg surface, the sperm generally penetrates the jelly coat, although there is a micropylar channel. The jelly coat has therefore played an important role in the development of various concepts about fertilization. According to the fertilizin hypothesis of F. R. Lillie (1913, 1914, 1919), the jelly contains a factor secreted by the egg up to the time of fertilization and denoted as fertilizin. The fertilizin molecule was assumed to contain spermophile as well as ovophile groups, i.e., groups complementary to such in the sperm surface, or potentially complementary to such in the egg surface. The initial step in fertilization was assumed to involve a specific reaction between the egg spermophile groups and the sperm. A structural chain of the ovophile groups would then become able to react with the egg, which thereby would receive its primary incitement to activation. The main further assumptions in this elaborate hypothesis are outlined in Fig. 2-2 in Monroy (1965a).

The fertilizin theory has been considerably modified by Tyler and collaborators (cf. Tyler, 1948, 1956, 1960; Monroy, 1965a), who consider fertilizin identical with the jelly itself. When the sperm enters the jelly, a number of the specific groups of the sperm surface, located at its head and estimated to occur at a number of 10^5 but still covering only 10% of the surface, react with the jelly and become blocked, Since the molecules of the jelly have more than one spermophile group, a jelly solution may lead to an agglutination of spermatozoa. This process is not considered to play an essential positive role under physiological conditions, but has attracted much attention in studies on the specificity of gamete interaction.

Agglutination is reversible, evidently due to a splitting of the molecules, but the surface of the liberated sperm is still blocked by the attached residue, and a similar blocking may be induced by nonagglutinating split products or otherwise altered jelly. The blocking is considered to protect the egg against the successful attachment of too many competent spermatozoa and hence against polyspermic fertilization. Furthermore, it has been hypothesized that the sperm surface is somehow modified by the jelly so as to increase the efficiency of a sperm that still has been unblocked groups. The jelly may, for instance, facilitate the peeling off of the sperm cell membrane during sperm penetration (Monroy, 1965a), and no doubt facilitates the release of the acrosomal reaction (Section III,F). In any case, it is claimed that jelly-free eggs are less easily fertilized than normal eggs. This conclusion, already criticized by Loeb, was later contraindicated by Hagström (B. E. Hagström and Hagström, 1954a) on the basis of his method of studying the rate of fertilization, which is increased rather than decreased if the jelly is removed in a gentle way (see also Runnström *et al.*, 1959). In Hagström's view the function of the jelly would be confined to the protection of the egg against polyspermy, an effect that could be ascribed to a mechanical filter function of the jelly, as well as to its ability to release a premature acrosomal reaction (Section III,F). Furthermore, Hagström considers the possibility that the jelly may be involved in the prevention of cross fertilization, since foreign spermatozoa may have a reduced ability to dissolve it, a conclusion in line with results of Isaka *et al.* (1966).

In spite of the results just mentioned, credit must be given to the fertilizin theory in that it has forwarded the idea that complementary groups in the gametes are essential for the reactions between the sperm and the egg. However, identical groups in the surfaces of the two gametes must also be considered in this connection (Runnström *et al.*, 1959). Immunological work by Tyler, Metz, and Perlmann and collaborators has made it possible to locate and in certain respects define some of the groups that may be involved (for references, see Monroy, 1965a; Perlmann, 1959; Runnström *et al.*, 1959; Runnström, 1966). The sperm may thus be rendered incapable of fertilization by means of antibodies to groups at its surface, apparently to a certain extent species-specific ones. Four different antigens have also been located in the egg surface layers; one of them was considered to be a sperm receptor. Antibodies binding to another group appeared to bring about a certain activation. Further work with ferritin-labeled antibodies has been carried out to map the distribution of surface antigens (Baxandall *et al.,* 1964a,b; Baxandall, 1966).

D. Other Factors in the Interaction Between the Gametes

A number of other results indicate various interactions between the gametes before they have met (for references, see Monroy, 1965a; Runnström *et al.,* 1959). It is at present impossible to try to discuss them even superficially. However, another activity attributed to the fertilizin is, according to Lillie, that of stimulation of sperm motility, an effect that is at least evident in work with aged spermatozoa of low motility. The jelly solution also prolongs the motility of spermatozoa in suspension. The increased motility can be correlated with an increased O_2 consumption, and controversial results on this point seem to be due to the use of spermatozoa in different physiological states. The mechanism of these activations has been suggested to involve the binding of certain toxic ions by the jelly. The activating principle may not be identical with the agglutinating one, as fractionating experiments have shown that the jelly consists of more than one fraction, of which only the main one is agglutinating. One has therefore to consider that some of the factors in the jelly are not carbohydrates, and that they may just be contaminations, i.e., compounds derived from the egg proper. The principle that activates respiration may also have this origin.

An additional property sometimes ascribed to the components outside the egg proper in chemotaxis, a field critically reviewed by Rothschild (1956). There is evidently a pitfall in many of the investigations in this field, at least with regard to the echinoderms, since the jelly traps spermatozoa, giving the impression that they have been attracted. It is therefore not tempting to discuss the elaborate hypotheses in this field, particularly since many attempts to reproduce the experiments underlying them have been negative.

E. Penetration of the Jelly by the Sperm

The penetration of the coat around the egg by the sperm is often made possible, or at least facilitated, by lytic principles given off by the sperm or present in the seminal fluid, particularly in cases where the coat is tough and a functional micropylar channel is lacking. The occurrence of such principles in echinoderms has, however, been a matter of some dispute (see Monroy, 1965a; Runnström *et al.,* 1959).

The presence of a jelly depolymerase in the sea urchin sperm was suggested by Monroy and Ruffo (1947; cf. Vasseur, 1951), a view supported by Messina (1954). The lytic principle may be located at the sperm surface (Brookbank, 1958) and has been considered to be a sulfatase (Hathway and Metz, 1961). That a specific lytic interaction of some kind is involved has also been suggested by Hagström (see Runnström *et al.*, 1959), who considered the possibility that the rate of cross fertilization may be reduced partly due to the decreased ability of foreign sperms to dissolve the jelly. The occurrence of a lytic principle with a certain species specificity has recently been demonstrated in sea urchin spermatozoa by Isaka *et al.* (1966). One may at present only speculate about the exact location of the lytic principle suggested, e.g., about its release from the acrosomal region.

According to some other results, the lytic activity of a sperm suspension may simply be due to the CO_2 produced by the spermatozoa, by their mechanical effect or—as concluded by Monroy and Tosi (1952) and Monroy *et al.* (1954; cf. Vasseur, 1951)—due to an adsorption of the jelly substance to the surface of the sperm. It is as yet unknown to what extent such effects play a role in the penetration of the jelly substance *in situ*. One should finally consider the suggestion of J. C. Dan (1960) that the penetration of the jelly by the sperm is due to the force of the ejecting filament (see the following).

F. Sperm Attachment and Penetration into the Egg

In various groups of animals, particularly in those with a gelatinous coat, the acrosome region of the sperm undergoes a drastic change, the acrosome reaction, which often involves the ejection of a filament, the acrosome filament. The first observation of this filament was made on the starfish by Fol (1878–1879). It must be noted, however, that this author, like R. Chambers (1930), considered the filament to be formed not by the sperm but by the female gamete. That the filament is actually a derivative of the sperm was shown by Popa (1927) in the sea urchin *Arbacia*. The general importance of the acrosome reaction was then gradually uncovered, initially due to studies on starfish and sea urchin material by Dan and collaborators. For references to papers on this reaction see papers and reviews by J. C. Dan (1956), Colwin and Colwin (1962, 1964), and Monroy (1965a). In recent years the acrosomal reaction has been studied by means of the electron microscope, a work initiated by Dan in the sea urchin and extended in a beautiful series of investigations by Colwin and Colwin. Most of the details of the reaction have been elucidated on nonechinoderm species such as the annelid *Hydroides*, the enteropneust *Saccoglossus*, both studied by the Colwins, and the crab *Callinectes* (Brown, 1966). The reaction in these forms is very complicated. There are indications, however, that some of the findings are also applicable to some extent to the echinoderms.

The acrosome reaction in the sea urchin can be triggered by a great number of agents, among which the jelly is one of the most active.

It was in fact egg-water, i.e., seawater containing some dissolved jelly, that in Popa's studies brought about the ejection of the acrosome filament. A premature release of the filament may contribute to the low fertilizing capacity of spermatozoa that once have been agglutinated by a jelly solution. Under natural conditions, when the concentration of jelly in the seawater around the egg is negligible, the acrosomal reaction may be triggered rather early during the passage of the sperm through the jelly coat (J. C. Dan, 1956). According to Afzelius and Murray (1957), the reaction is elicited when the sperm comes in contact with the egg surface. In any case, a too-early release of the reaction would reduce the chances for a contact between the filament and the egg surface. This, together with the mechanical filter action of the jelly may serve as a first device of protection of the egg against polyspermic fertilization (see Runnström *et al.*, 1959).

According to the interpretation of Popa, the acrosomal reaction involves the release of some substance of importance for the activation of the egg. The source of such a substance may be the acrosomal granule, the disappearance of which is associated with the appearance of the filament (see J. C. Dan, 1956). It has sometimes been suggested that such an agent already contributes to a lysis of the jelly, but the exact location of the jelly lysin is so far unknown. Furthermore, the secretion of the acrosomal granule may not occur until the sperm has attached to the egg (cf. Afzelius and Murray, 1957).

The attachment of the sea urchin sperm to the egg plasma membrane is initially weak, and allows the sperm to rotate around its axis, but the attachment is gradually consolidated (see Runnström *et al.*, 1959). The mechanism of bringing the sperm close to the egg, when the acrosome filament is ejected early as in *Asterias*, and its engulfment have been a matter of divergent opinions, and even recent data are to some extent controversial (see Monroy, 1965a). The idea that in *Asterias* the filament actually pulls the head of the sperm toward the egg surface was suggested by Fol and favored by Chambers. The so-called cone of attachment, first described by Fol and appearing at the site of the attachment of the filament, has also been considered to contract and carry the sperm inward. (We must again emphasize that according to the original interpretation the filament was formed by the egg.) That the filament somehow plays a role in this connection cannot be doubted, but the fact that it can be seen fully extended within the cytoplasm is not in line with the idea that it fulfills its function by contracting. Likewise, the idea about the role of the cone of attachment appears to have lost its general validity, since in various animals, including some sea urchins, the formation of the cone is delayed until after the engulfment of the sperm.

In cases where the cone appears early, such as in *Asterias* and the holothurians *Thyone* and *Holothuria,* it has been observed that cone formation involves the extension of the egg plasma membrane that creeps up and around the acrosomal filament. In other cases the filament must also find its way into the egg somehow. Some details of the early sperm penetration in the sea urchin have been described by Afzelius and Murray (1957). When the filament has been burrowed in the egg, the plasma membranes of the egg and the sperm head proper fuse, after which the conjoined membranes break down (see Monroy, 1965a). A funnel is thereby formed, through which the naked sperm nucleus together with a centriole and the mitochondria enter the egg plasm, where also remnants of the filament have been observed in some cases. The fate of the sperm plasma membrane after the sperm has been emptied is variously described. It may remain outside the egg, or may be incorporated as a patch within the egg plasma membrane.

The detachment of the sperm plasma membrane is no doubt essential for a successful fertilization. This idea is supported by the fact that spermatozoa entering through a damaged cortex (Runnström and Kriszat, 1952b), or by injection with a micropipet (Hiramoto, 1962a), neither activate the egg, nor undergo a swelling and form no aster.

As the sperm plasma membrane is peeled off, its role in fertilization is apparently restricted to the establishment of a fusion of the gametes. The peeling off has been visualized as facilitated by the attachment of some jelly substance to the sperm head (cf. Monroy, 1965a). The establishment of the contact between the plasma membranes of the two gametes is likely to depend on some specificity of the membranes at the molecular level. This concept has been somewhat extended by Tyler (see Tyler, 1960; cf. Perlmann, 1959) in the sense that the reaction between complementary groups of the two membranes would trigger the release of some processes of basic importance for the activation of the egg. Quite naturally, the discussion of such hypotheses has been based upon the use of immunological tools, i.e., antibodies specific to groups in the two gametes (see Section II,C). Although this type of work may be considered to represent one of the most promising approaches in the physiology of activation, much further work is required before a more detailed and coherent picture of the reactions concerned can be presented.

G. The Cortical Reaction

The reaction with the sperm not only means the introduction of genetical material into the female gamete, but also the activation of a great number of reactions within the cytoplasm. As mentioned earlier, this is a main result of the studies of parthogenetic activation. The cytoplasmic changes start off with changes in the superficial layers of the egg, the cortical reaction that spreads as a wave from the point of attachment of the sperm. This reaction has been subjected to a great number of investigations, particularly in the sea urchin (for references and details, see Monroy, 1965a, and the extensive review of Runnström, 1966).

According to polarization microscopical investigations, the cortex of the unfertilized sea urchin egg is made up of a series of alternating lipid and protein layers, or of several layers of looped protein molecules with the axis of the micelles parallel to the egg surface and covered by an outer lipid layer. Fertilization is rapidly followed by a shift in the dark-field appearance of the cortex from yellow-red to silver-white and a change in its birefringence. These changes have been considered to reflect an alteration of the dispersion of its lipids (Runnström, 1928b) or disarrangements of the texture of its lipid and protein layers (see Monroy, 1965a; Runnström, 1966).

The molecular rearrangements of the cortex are paralleled by other alterations, partly revealed by the light and electron microscopes. At the surface of the unfertilized egg a so-called vitelline membrane covering the egg plasma membrane may be revealed. Another feature of the superficial layer is the occurrence of cortical granules, generally described as lying below the plasma membrane. The ultrastructure of these granules has been studied by several authors, among which Afzelius (1956) gave the most comprehensive data covering a number of different species; see Runnström (1966) who had also studied their genesis. The granules (Fig. 1), about 0.8μ in diameter, are bounded by a membrane that encloses lamellae and extralamellar bodies of quite characteristics and species-specific morphology and distribution and contains sulfated polysaccharides.

At fertilization an acid is liberated from the egg (Section III,L). This acid is thought to break hydrogen bonds between the vitelline membrane and the plasma membrane (Isaka and Aikawa, 1963a,b). The detached vitelline membrane immediately elevates and expands. The elevation may be assisted by the osmotic activity of sulfated polysaccharides released from the egg surface (cf. Monné and Hårde, 1951; Immers, 1961). The expansion probably involves an uncoiling of protein molecules. As both the acid and the polysaccharides may be derived from the cortical granules, the whole process may reflect an opening of the cortical granules. However, the detachment and beginning elevation somewhat precedes the opening of the granules (see Runnström, 1966). In any case, the process described is an initial step in the formation of the fertilization membrane.

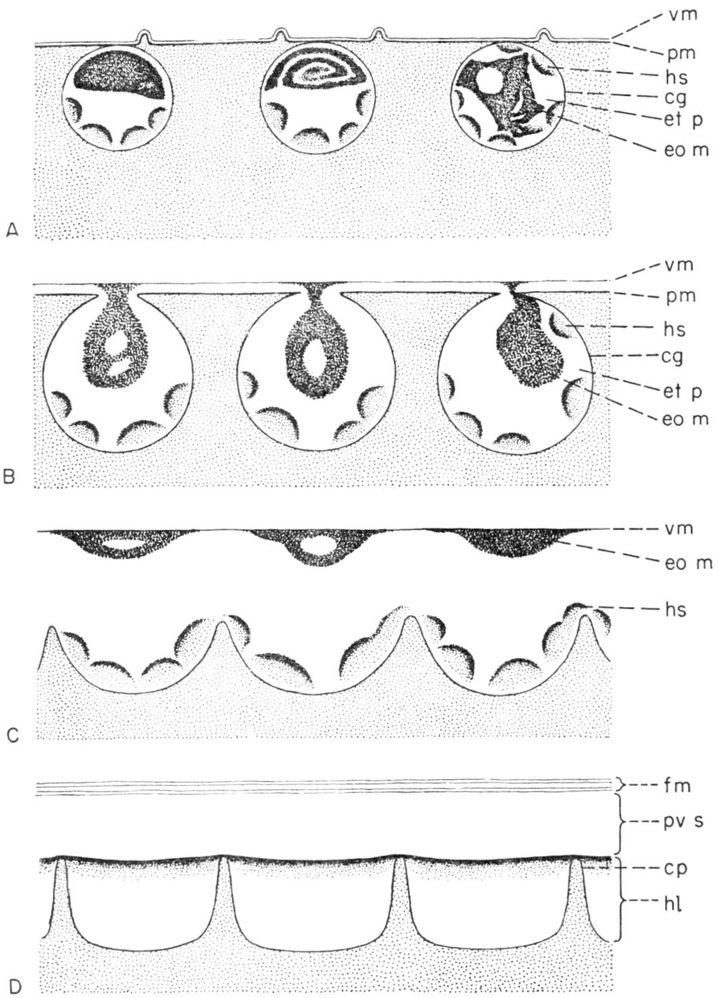

FIG. 1. Diagram of the formation of the fertilization membrane in the sea urchin. A. Surface of the unfertilized egg with vitelline membrane (vm) and cortical granules (cg) with two main types of inclusions. B. Explosion of the cortical granules and elevation of the vitelline membrane. C. Spreading of the material of the cortical granules, the lamellar inclusions (db) on the vitelline membrane, the other inclusions (hs) on the egg surface. D. Fertilization membrane (fm) formed, separated from the surface of the hyaline membrane (hl) by the perivitelline space (pv.s.). (From Endo, 1961a; cf. Monroy, 1965a.)

The cortical granules have long been known to be expelled during this process, a process that has been studied by means of the electron microscope (Afzelius, 1956; Endo, 1961a,b; Runnström, 1966; Wolpert and Mercer, 1961). According to most of the investigators, each granule undergoes an opening at the sites where its limiting membrane is in contact with the egg plasma membrane (Fig. 1). This would mean that the limiting membranes become incorporated into the plasma membrane, which from this time will therefore consist of a mosaic of membrane elements of quite different origin (Monroy, 1965a). The lamellar material is extruded and joins the elevating vitelline membrane as an expanded sheet and attaches by numerous threadlike processes. This process is apparently accompanied by a molecular transformation of the material concerned, as can be concluded from observations on cases where the incorporation is incomplete and a transformation can be observed in the space below the vitelline membrane (see Runnström, 1966). The changes of the granules and the elevation of the fertilization membrane is supposed to be an energy-requiring process as inferred by the effects of uncouplers of phosphorylation (R. Okazaki, 1956; see also Runnström, 1966).

According to the characteristics of birefringence, the fertilization membrane so formed consists of elongated elements parallel to the surface. These elements are eventually linked to each other, partly as a consequence of the oxidation of SH groups (see Runnström, 1966), and is no longer easily removed when the eggs are passed through a narrow-mouthed pipet. After treatment with thioglycollate membrane, however, loses its toughness and becomes elastic and sensitive to proteolytic enzymes.

The extralamellar bodies detach from the wall of the opened cortical granules and form a sheet on top of the villi of the egg surface (K. Dan and Ono, 1954). This is the boundary of the hyaline layer, which during early cleavage and later development forms a layer around the blastomere cluster and the embryo. The villi are gradually retracted and the opened urn-shaped granules open still more and probably deliver more of its contents.

In the starfish the events are rather similar to those described above (see Monroy, 1965a).

H. Changes in Permeability

As could be expected, the structural alterations of the egg surface at fertilization are accompanied by changes in its permeability to various substances (see Monroy, 1965a,b; Runnström, 1966). A number of investigations on sea urchin material thus reveal a transient increase in water permeability (Ishikawa, 1954) and marked changes in the permeability for phosphate (Lindberg, 1950), amino acids (Hultin and others; see Monroy, 1965a,b), etc. The increased traffic through the membrane partly reflects alterations in the ionic distribution within the egg due to changes in ion-binding sites or in compartmentalization, as shown by the increased exchangeability of potassium ions [cf. the classical concept of Heilbrunn (1937) and Mazia (1937), about a liberation of calcium ions within the cortex]. The marked transient release of potassium may therefore have a dual background (see Monroy, 1965a,b). The increased phosphate permeability (Lindberg, 1950), however, probably reflects an enhancement of the efficiency of an ATP-synthesizing system of the cortex, a system that apparently is not preformed and ready to act within the cortical granules, but rather formed *de novo* during an energy-requiring process (Litchfield and Whiteley, 1959; Whiteley and Chambers, 1960). An interesting idea is that the release of potassium ions and perhaps some of the other changes, reflects a similarity between fertilization and nerve stimulation (R. S. Lillie, 1916). The validity of this concept is supported by some observations of the membrane potential of some eggs (see Monroy, 1965a,b). A transient decrease in this potential has been revealed in the sea urchin as well as in *Asterias,* in the latter case clearly correlated with the elevation of the fertilization membrane. Changes in membrane resistance and capacitance of echinoderm eggs have also been recorded, which is in agreement with the changes in water permeability and surface area (see Monroy, 1965a).

I. The Polyspermy-Preventing Mechanisms

An additional expression of the surface alterations at fertilization is the appearance of a barrier protecting the egg against the reaction with additional spermatozoa, i.e., against polyspermic fertilization. We may recall that already the jelly coat seems to be involved in this protection (Section III,E). It is also quite evident that the hardened fertilization membrane must act as an efficient barrier. The question is, however, how polyspermy is efficiently prevented before the membrane has formed, a question that is particularly burning with regard to eggs from which the jelly coat has been removed.

The first detailed concept concerning the ability of the egg surface to prevent polyspermy was formulated by F. R. Lillie as an integral part of his fertilizin theory. According to this, a block against the entrance of additional sperm spreads as a wave from the point of attachment of the first successful sperm (cf. Just, 1919) in the same way as the ultrastructural changes described by later authors. The idea about a spreading block has been generally accepted, even if the underlying mechanism is still debated. The close relationship between the wave of polyspermy prevention and the structural ones is suggested by the fact that they may all be incomplete when fertilization is defective (see below). That the retraction at fertilization of microvilli of the egg surface is involved may be considered in the case of *Asterias*, but this mechanism cannot be of general significance (see Monroy, 1965a).

According to the concept of Rothschild and Swann, there is not one, but two spreading blocks (see Monroy, 1965a,b). An initial wave, completed within a matter of seconds and coinciding with the ultrastructural changes, is connected with a partial block against polyspermy. A slower wave within a minute, coinciding with the breakdown of the cortical granules and associated phenomena, renders the egg completely protected against further spermatozoa. A number of experiments have led to the conclusion that the slow wave of polyspermy prevention actually depends on the breakdown of the cortical granules. The problem is then to define the nature of this dependency.

As the breakdown of the cortical granules is involved in the formation of the fertilization membrane, one could assume that the block to polyspermy merely reflects the formation of a tough membrane barrier. To some extent this assumption is correct, since a mechnical removal of the fertilization membrane in certain sea urchin species makes them accessible to refertilization. Furthermore, the block does not appear if the formation of the membrane is prevented in various ways (see below). But in many species the egg surface itself is capable of stopping the entrance of further spermatozoa (see Monroy, 1965a) and hence must have a barrier of its own. In such cases it has been suggested by Hagström and co-workers that another derivative of the cortical granules, namely the hyaline membrane, serves as the barrier (see Runnström *et al.*, 1959). An argument for this is that refertilization may occur if eggs devoid of the fertilization membrane are treated with Ca-Mg-free seawater, in which the hyaline membrane is known to be dissolved (cf. Ishida and Nakano, 1947, 1950; Sugiyama, 1951; B. E. Hagström and Hagström, 1954b). Furthermore, a number of agents known to inhibit the breakdown of the cortical granules and membrane elevation, and hence also the formation of the hyaline membrane, also inhibit the development of the barrier (cf. R. Okazaki, 1956; Whitely and Chambers, 1960; and the group of Runnström, particularly Hagström; see Runnström *et al.*, 1959; Monroy, 1965a). It must be emphasized that, as far as is known, neither of these agents is effective if the eggs are exposed to them after the hyaline membrane is formed. Possibly the proteolytic enzyme inhibitors just mentioned block the transient proteolytic activity that appears at fertilization, and is supposed to be involved in the cortical reaction (see Runnström *et al.*, 1959; Section III,L). That ovocytes cannot resist polyspermy is in line with all this, since their cortical granules have as yet not accumulated in the cell cortex.

The occurrence of a rapid wave of changes partially protecting the egg against polyspermy is not in line with the results of Ginsburg (1961), and B. Hagström and Allen (1956) are also doubtful about its existence. However, fertilization may involve a series of waves along the egg surface (cf. Runnström, 1966).

As in so many other experiments with complex biological systems, it may be difficult to state that an agent applied interferes primarily with only one structure or function. One may, for instance, suppose that some of the treatments that interfere with the breakdown of the cortical granules, or with the hyaline membrane after it is formed, may also interfere at some other level at the egg surface. The deficiency of the ovocyte in function may also be due to some structural deficiency of the cell surface, other than the incomplete incorporation of the cortical granules (cf. Monroy, 1965a). Furthermore, a correlation between two phenomena does not automatically mean that one is the cause of the other, for instance, the breakdown of the cortical granules for the polyspermy block. Both changes may quite as well depend on the same initial change. However, the role of the breakdown of the cortical granules for the development of the block is attractive, and appears to be the only constructive hypothesis that is both concrete and adequately supported. For further information on the problem concerned, the readers are referred to the original papers referred to in the reviews of Runnström *et al.* (1959) and Monroy (1965a).

J. The Protection Against Cross-Fertilization

The problem of how the rate of cross-fertilization is reduced is a rather controversial one. The possible role of the jelly coat (Section III,E) and the vitelline membrane has been discussed. Removal of this membrane by means of trypsin certainly increases the rate of cross-fertilization (Hultin, 1948a,b), but the evident conclusion from this has been disputed (B. E. Hagström, 1959; cf. however, Runnström, 1966). The view that foreign sperm do not induce a cortical reaction has also been abandoned. In fact, ovocytes which do not respond with a normal cortical reaction still show some species specificity in their reaction to spermatozoa. Certain observations that the swelling of the sperm nucleus in the cytoplasm may be defective suggest that the protective mechanism to some extent is associated with the cytoplasm (cf. B. E. Hagström, 1959; B. E. Hagström and Lönning, 1961). One may argue that this is due to a defective peeling off of the sperm plasma membrane at the egg surface, and hence a defective reaction between the plasma membranes of the two gametes. It seems likely, however, that the protection against cross-fertilization in the sea urchin occurs at different levels, as in many other animals (see Runnström, 1966).

K. The Cortical Reaction and the General Activation of the Egg

The reactions of the egg cortex are accompanied by alterations of the egg as a whole, as shown by the reactions of the nuclei and the onset of cell division (Section III,O), as well as by dramatic changes of the metabolic activities of the cell (Section III,L–N). The casual relation between the processes at the different levels has been considered by various workers. According to some experiments, an inhibition of the reaction of the cortical granules is followed by a defective nuclear and cytoplasmic activation (Allen and Hagström, 1955), whereas in other experiments this correlation is not found (Motomura, 1954; Osano, 1960). One may argue that in the first case the inhibition is brought about by warm seawater that could have a direct effect on the cortex as well as on the interior of the egg, whereas the second case involves a brief treatment with chemical agents that only affect the superficial layers. This would mean that activation involves at least two independent processes. On the other hand, one may argue that the reactions of the cortical granules may just be an indicator of the cortical change at the molecular level, and hence that it is not identical with the self-propagating wave in the cortex. This conclusion is held by Sugiyama (1956) on the basis of a series of experiments which show that some agents locally applied to the egg surface cause local reaction of the cortical granules, whereas other agents probably elicit a self-propagating wave all over the surface, not indicated by a reaction of the cortical granules. This would mean that a chemical disturbance of the indicator reaction in the experiments of Motomura and Osani may permit some subtle cortical events of importance for the deeper layer, whereas the heat treatment would suppress these events and therefore activation as a whole. In conclusion: "It may be surmized that the interaction with the spermatozoa originates a reaction (or rather a series of reactions) in the surface layers of the egg which is responsible both for the breakdown of the cortical granules *and* the activation of the nuclear-cytoplasmic system" (Monroy, 1965a).

L. General Basis of Metabolic Changes at Fertilization

The cytoplasm of the unfertilized egg is under some kind of metabolic inhibition (Section II). Theoretically, this may be brought about in many different ways, by a structural separation of enzymes, by the accumulation of metabolic products that bring about allosteric inhibition of enzyme activities, by the presence of inhibitors with some other mode of action, by the lack of cofactors or substrates, etc. A classical hypothesis of Runnström (1933) is that fertilization involves the removal of a structural barrier between cytochrome *c* and cytochrome oxidase. In a later paper he suggested that the effective attachment of the spermatozoon brings about the release of a 'kinase" from an inhibitor. The "kinase" would then bring some other key enzymes into active state (Runnström, 1949, 1966). Various ways of testing this hypothesis have been tried with the assumption that the inhibitor might be a mucopolysaccharide (Runnström and Hagström, 1955; Runnström and Kriszat, 1950). One must, however, consider that there may be several inhibitors (Runnström, 1966). Even the change from underripeness to full cytoplasmic maturity seems to involve the elimination of certain inhibitions (Runnström and Kriszat, 1950). An interesting discussion of the mechanism of activation which departs from Loeb's hypothesis about surface cytolysis as a basis of activation and the observation that lysophosphatides bring about an activation (Öhman, 1944), has been presented by Monroy (1956, 1965a), who suggests that egg lipases are activated by sperm lipases.

A most dramatic change immediately upon fertilization of the sea urchin egg is the transient production of an acid that displaces CO_2 from seawater (Runnström, 1933; and others; see Runnström, 1966). That this acid has a metabolic origin appears likely, and the phenomenon may therefore serve as a model for other changes at fertilization. However, neither lactic acid nor Krebs cycle intermediates seem to be involved (Runnström, 1933; Ycas, 1954), although it must be admitted that lactic acid shows a transient accumulation in certain sea urchins investigated (Aketa, 1957, 1964). Furthermore, hydrogen ions may be liberated as a consequence of the activation of a nicotinamide-adenine dinucleotidase (Fujii and Ohnishi, 1963). On the other hand, Runnström and Immers (1956) regarded acid formation as the consequence of a release and splitting of sulfated mucopolysaccharides. Mehl and Swann (1961) considered that part of the CO_2 displacement at least is due to the appearance of very strong acidic groups within the cytoplasm, concomitant with the exposure of basic groups. In fact Aketa also (1963) favors the view that the acid concerned is very strong. Since there is a simultaneous release of sulfate groups, he considers that the sulfuric acid is formed in connection with the splitting of sulfated polysaccharides of the exploding cortical granules. A similar idea is suggested by Bäckström (1956), who considers that polysaccharide components split off from the cytoplasmic histone-ribonucleic acid complex are partly extruded.

Fertilization also involves a sudden increase in the activity of various enzymes such as aldolase, glucose 6-phosphate dehydrogenase, 6-phosphogluconate dehydrogenase, ribonuclease, cytochrome oxidase, two different ATPases and, hypothetically, also a nicatinamide-adenine dinucleotidase, a kinase responsible for NAD-NADP interconversion, and phosphorylase (see Monroy, 1965a,b; Runnström, 1966). A most interesting phenomenon is the transient, sequential activation of some proteolytic enzymes (Lundblad, 1954; see also Runnström, 1966). Some of these increases may be due to the release of the enzymes from a bound state, e.g., aldolase from the egg surface and ribonuclease from certain granules. Such a release may be triggered by the release of potassium ions or calcium from a bound state or from certain compartments (see Monroy, 1965a,b). The liberation—or possibly an active inflow—of calcium ions would furthermore activate one of the proteolytic enzymes and an ATPase. One should also consider that the activation of proteolytic enzymes can start a chain reaction, which may include the release of a great number of components and remove inhibitors (cf. Runnström, 1966). The increase in the cytochrome oxidase, on the other hand, is apparently due to the removal of some inhibitor of low molecular weight, the activity of which depends on the availability of SH groups (cf. Monroy, 1965a,b).

Changes of this kind may account for the hydrolytic processes and the metabolic revolution that is a characteristic feature of fertilization. It is of course difficult to present a coherent picture of this. The following is only a rather brief illustration of some main metabolic changes and of some attempts to account for them. For further information see, e.g., Runnström *et al.* (1959), Runnström (1949, 1956b, 1966), and Monroy (1965a,b).

M. Changes in Respiration

The classical observation of Warburg (1910) that fertilization of the sea urchin egg involves a drastic increase in respiration has been confirmed (Yasumasu and Nakano, 1963; Ohnishi and Sugiyama, 1963; cf. however, Borei, 1948). This rise coincides with the formation of the fertilization membrane and the fertilization acid. Such a rise is by no means typical for fertilization in general. It rather seems that fertilization involves a restoration of the respiration to the rate of normally growing cells (cf. Whitaker, 1933; Monroy, 1965a). A transient increase, however, occurs in *Asterias* (Horowitz, 1965). The first respiratory rise in the sea urchin egg is to a certain extent reversed so that a transient peak in oxygen consumption occurs (Laser and Rothschild, 1939; Ohnishi and Sugiyama, 1963), but the respiratory rate remains at a higher level than in the unfertilized egg and shows a gradual increase.

The concept of Runnström (1933) and the removal of the cytochrome oxidase inhibitor (see Monroy, 1965a) are important in this connection. It is furthermore clear that during the subsequent rise there is an increasing saturation of the cytochrome oxidase (see Runnström, 1949, 1956b; Monroy, 1965a,b). The problem, however, requires further consideration.

According to one concept, an essential step in the release of respiration is the removal of a block that prevents the utilization of a glycogenlike polysaccharide piled up in the egg (cf. Öström and Lindberg, 1940; Lindberg, 1943, 1945; Monroy and Vittorelli, 1960). The removal of this block may involve an activation of a phosphorylase (Aketa *et al.*, 1964), which would lead to an increase in the concentration of hexose phosphates and other intermediates in carbohydrate metabolism so that their concentration, low before fertilization, no longer limits respiration (cf. Monroy and Vitorelli, 1960; Aketa *et al.*, 1964). That the released carbohydrate metabolism mainly follows the shunt pathway (for references, see Monroy, 1965a) may not only be due to the increased activity of the shunt enzymes mentioned earlier, but also to an activation of a kinase involved in the interconversion of NAD to NADP (cf. Epel, 1964a,b; Krane and Crane, 1960).

Another concept is that one of the limiting factors of respiration in the unfertilized egg is the shortage of phosphate acceptors (Immers and Runnström, 1960). Actually, the concentration of ADP is here very low, whereas that of ATP is high (Rossi, see Monroy, 1965a), a situation that is markedly changed at fertilization. Part of the ATP-ADP interconversion at this event may be connected with the NAD-NADP interconversion, that was found to occur a few seconds before the respiratory rise (Epel, 1964b). It should be noted, however, that according to Epel and Iverson (1965), the concentrations of ATP, ADP, and AMP do not change significantly at the time of maximal respiration. This may indicate that the availability of small molecules as substrates may be the limiting factor, as suggested earlier and substantiated by Epel, who emphasizes that the limiting substrates may be different in different species.

Another factor that may contribute to the respiratory rise is that the association of mitochondria with each other and with lipid granules is reversed at fertilization (Monné and Hårde, 1951; Pasteels *et al.*

1959). Observations on the capacity of the mitochondria to undergo swelling and on their ultrastructural features also indicate that they undergo a change (cf. Monroy, 1965a).

The change in the pyridine nucleotides at fertilization involve an increased level of NADPH as well as an increased NADPH/NADP ratio (Aketa *et al.,* 1964; Epel, 1965). The causal relation between these changes and the activation of various synthetic processes is suggestive (cf. Epel, 1965), as NADPH is the reductant in many biosynthetic processes, e.g., in the formation of deoxyribosides and lipids, as well as in protein disulfide reduction. The changes concerned may also play a role, if only indirect, in the activation of protein synthesis (Epel, 1965).

N. Changes in Protein and Nucleic Acid Metabolism

Protein synthesis in the unfertilized egg is strongly reduced. One of the most prominent changes at fertilization is the marked increase in the rate of protein synthesis that was first demonstrated by Hultin (1950c). In the sea urchin, this occurs almost at once. In *Asterias* it does not begin until after the formation of the second polar body (Monroy and Tolis, 1964). The mechanism of the increase in the sea urchin has been studied by several investigators, apart from Hultin, also by the groups of Monroy, Tyler, by Nemer, Wilt, Gross and Cousineau, etc., and their collaborators. Detailed references about this can be found in the reviews of Grant (1965) and Monroy (1965a,b).

The increase in protein synthesis is revealed by the increased incorporation of labeled amino acids. Theoretically this increase may be due to a number of factors. An increase in the amino acid penetration undoubtedly occurs, but is not the single cause of the increased incorporation, as an increase also may be seen when the pool has been labeled before fertilization. Furthermore, the differences in incorporation can also be demonstrated in experiments with homogenates of unfertilized and fertilized eggs. The considerable size of the pool, the lively amino acid interconversions within the egg, as well as the *de novo* synthesis of amino acids from glucose appear to rule out the possibility that changes in the availability of amino acids are responsible for the increase in protein synthesis, although, it may be noted, there is an increase in the pool at fertilization (see Runnström, 1966). Most of this indicates that the actual machinery for protein synthesis is somehow blocked before fertilization. The release of this block is evidently not due to an increased activation of amino acids but rather to a change at the ribosomal level. Experiments on the effects of a synthetic messenger added to ribosomes from the unfertilized egg suggested that the deficiency is due to a lack of messenger. On the other hand, ribosomes from this source were incapable of reacting with messenger from the fertilized egg or from liver, which sheds some doubt upon the interpretation that the messenger is failing in the unfertilized egg. A number of experiments have actually shown that messenger is already present before fertilization. Actinomycin-C, which stops messenger production, does not interfere with the early protein synthesis, and artificially activated anucleate egg halves have an active protein synthesis in spite of the lack of nuclear information. Finally, RNA from the unfertilized egg is able to activate ribosomes from fertilized eggs as well as from liver. The deficiency of the unfertilized egg must therefore be one involving the state of the messenger in the intact egg or the state of the ribosomes. The deficiency concerned has been further elucidated by means of ultracentrifuge experiments, which have shown that most of the ribosomes of the unfertilized eggs occur as monosomes, and that fertilization involves a marked increase in the proportion of polysomes, a change that continues as development proceeds. Pulse labeling experiments with amino acids indicate that the incorporation is highest in the area of the polysomes.

An interesting observation, which may be related to the ribosomal changes, is that cytoplasmic histones disappear in connection with fertilization (see Bäckström, 1966). On the basis of the observations of his own and other workers, Bäckström suggests that these proteins to a great extent are bound to the ribosomal RNA and somehow interfere with the ribosomal activity. Also the messenger RNA may be kept more or less unavailable and protected from degradation due to a binding to cytoplasmic histones until fertilization (cf. Vincent, 1965). The disappearance of the histones appears to be related to a removal of polysaccharides that mask the stainability of the histones in the unfertilized egg (cf. Section III,L). The inactivating role of the histones is in line with the observation that trypsin treatment, which is known to attack histones, brings about a release of the amino acid incorporation of ribosomes from unfertilized eggs (Maggio *et al.*, 1965). Under natural conditions, such a removal could be due to the transient proteolytic activity at fertilization detected by Lundblad (see Lundblad, 1954; Runnström, 1966).

There is also some RNA synthesis soon after fertilization of the sea urchin (see Monroy, 1965a). However, the reservoir of ribosomes of the unfertilized egg largely serves the need of the embryo during early development (Nemer, 1965). This would mean that an important fraction of the RNA formed is messenger, which agrees with various observations (Nemer, 1963, 1965; Wilt, 1963; Gross *et al.*, 1965). However, the immediate postfertilization increase in protein synthesis cannot be due to this (see Gross *et al.*, 1965; Nemer, 1965). For further information about RNA synthesis in echinoderm development, see Grant (1965).

Fertilization also involves an onset of DNA synthesis, which appears to start even before the pronuclei have fused (Simmel and Karnofsky, 1961). The change in the hydrogen carrier system mentioned above may be involved in this activation. The idea has also been advanced that the DNA synthesis is released as a result of enzymic changes that lead to changes in the deoxyribonucleotide pool (for references, see Monroy, 1965a,b). Such a concept is well in line with the conclusion that certain enzymes involved in the reduction of ribonucleotides in other systems undergo an allosteric change in activity under the influence of deoxyribonucleotides (e.g., see Larsson and Reichard, 1966). Other mechanisms suggested by Mazia and collaborators involve an activation of enzymes required for the utilization of the deoxyribonucleotides in DNA synthesis (for references, see Grant, 1965). The role of the cytoplasmic DNA and RNA as a raw material in nuclear DNA synthesis should also be considered in this connection (Sections II and IV,E,H).

For further information about masked messenger RNA and cytoplasmic DNA see Tyler (1967).

O. Physicochemical Changes of the Egg Proteins and Cleavage of the Egg

Fertilization of the sea urchin egg involves a series of prominent changes in its physical properties (see Runnström *et al.*, 1959; Runnström, 1966). To a great extent these must be due to the alteration of certain proteins. A classical observation in this connection is the decrease in solubility in 1 *M* KCl of certain proteins upon fertilization (Mirsky, 1936). Some of the changes appear to be reminiscent of alterations that proteins undergo upon denaturation (see Monroy, 1965a). Changes in the proportion of SH and SS groups of a protein fraction have also been observed by Sakai and Dan (see Monroy, 1965a; Runnström, 1966). Runnström has discussed some of these changes, particularly on the basis of the results of Sakai (1960a,b, 1962a,b), and put them in relation to the activation of proteolytic enzymes (Lundblad, 1954) and the release of calcium (Heilbrunn, 1937; Mazia, 1937) or its uptake at fertilization (Runnström, 1966). The dramatic effects of calcium ions on the state of the egg proteins have been studied by the addition of calcium to homogenates (Hultin, 1949, 1950a,b; Gross and Pearl, 1950; Gross, 1954, 1956). The changes appear to involve the system isolated by Sakai that can undergo contraction and relaxation and therefore may account for the rounding up of the egg at fertilization, and presumably also the phase separation within the egg. Changes of this kind are considered to play an important role in mitosis.

According to an old concept, first formulated by Boveri and Wilson, but recently supported by a beautiful series of experiments by Hiramoto (1962a,b), the sperm supplies the egg with a centrosome as a substitute for the egg centrosome that is lost during maturation. The reason why the egg, although fully capable of forming asters with centrioles, still needs a supply by the sperm is provoking. It seems as if the cytoplasm undergoes some change at maturation essential for the maintenance and formation of the structures concerned.

After its entrance into the egg the sperm nucleus begins to migrate toward the egg nucleus, which itself moves to the center of the egg. In *Asterias*, however, the movements first occur after the second polar body has formed. The mechanism for these movements appears to be less obvious than first considered by E. L. Chambers (1939). Another feature of the pronuclei is their swelling. Although coinciding in time, migration and swelling appear to be rather independent. The factor responsible for swelling of the egg nucleus thus appears to be a diffusible agent that can pass through nonactivated cytoplasm, whereas nuclear migration requires the presence of an activated cytoplasm (Allen, 1954; Allen and Hagsträm, 1955). For further references and discussion about these problems see Monroy (1965a).

The sea urchin egg has played an important role in the analysis of the problem of cell division. Due to the restricted space available, reference should only be given to some reviews of the literature covering these problems. The formation of the mitotic spindle has been reviewed by Mazia (e.g., Mazia, 1961), who was able to isolate spindles from dividing sea urchin eggs and hence made them available for biochemical investigations. The problem of plasma cleavage has been reviewed by Wolpert (e.g., Wolpert, 1960), who favors the so-called astral relaxation theory, which incorporates features common to other theories, especially Marsland's most recent version of the cortical gel contraction theory (1956). One of the important postulates of this theory is that cleavage is initiated by relaxation of the membrane in the polar regions, which expand in area and allow the furrow region to constrict and divide the egg in two cells. Another postulate is that this relaxation of the membrane is due to its closeness to an aster. The localization of a relaxing system during cleavage has recently been elucidated by means of an immunological method (Kinoshita and Yazaki, 1968).

IV. Development

A. Polarity of the Egg

The point of attachment of the growing ovocyte is at the vegetal pole. At the animal pole, where the polar bodies later on appear, a cone of cytoplasm pierces the jelly coat and produces the so-called micropyle (see Raven, 1961). The mature egg of *Echinocardium* keeps an animal protrusion (Gustafson, 1952). In *Paracentrotus* another indicator of polarity may be seen. During maturation or shortly afterward, the cortical pigment often concentrates in a subequatorial band (Boveri, 1901a). The pigment band is more or less paralleled by a ring that in the dark-field stands out, owing to its yellow color (Runnström, 1928a). Observations on the development of giant polar bodies indicate that the animal-vegetal polarity is established already at the time of maturation (Gustafson, 1946). Some echinoderm eggs also show signs of a dorsoventral (oral-aboral) polarity before fertilization. Eggs of *Echinocardium* treated with various SH compounds or some other agents develop two concavities of different size, the larger of which presumably corresponds to the ventral side (Gustafson, 1952). One evidence for this is that surface-active agents, which make the concavities disappear or prevent their formation, interfere with the dorsoventral development of the larva (Gustafson and Sävhagen, 1950). Centrifugation experiments (see below) also favor the concept that dorsoventrality is predetermined, although not fixed, before fertilization. The physiological polarity is finally reflected by the formation of the fertilization membrane, which often starts at the site of the larger concavity (Runnström, 1962).

The origin of the polarity is an unsolved problem. One may hypothesize that it reflects the polarity of the cells in the germ epithelium, or that it is induced by the supply of material to the ovocyte when it is attached to the wall, etc. The problem is of course particularly difficult with regard to the dorsoventrality of the egg. The nature of the polar differences is another obscure problem (see Gustafson, 1965; Runnström, 1966). Experiments on the effect of removal of part of the inner cytoplasm of unfertilized sea urchin eggs support the view that the animal-vegetal polarity is somehow related to the cortex (Hörstadius *et al.*, 1950). This conclusion was already drawn by experiments by Runnström, Lindahl, and Pease (see Gustafson, 1965) with various sea urchins, where the inclusions of the egg were stratified by centrifugation. The same conclusion is also applicable to the dorsoventral polarity, which is not completely changed by the stratification of the cytoplasmic components. The occurrence of the yellow ring mentioned above, as well as the effects of surface-active agents, suggest that the polarity is somehow related to the architecture of the lipid-protein layers of the cortex.

B. General Principles of Determination

On the basis of vital staining, the main regions of the sea urchin larva have been projected on the egg (see Hörstadius, 1939). By interpolation, a more detailed map may be obtained (Figs. 2, 3). The sea urchin larva as well as those of other echinoderms is highly regulative, however. The polarity only determines the sequential order of the organ rudiments along the animal-vegetal axis. Their relative size, on the other

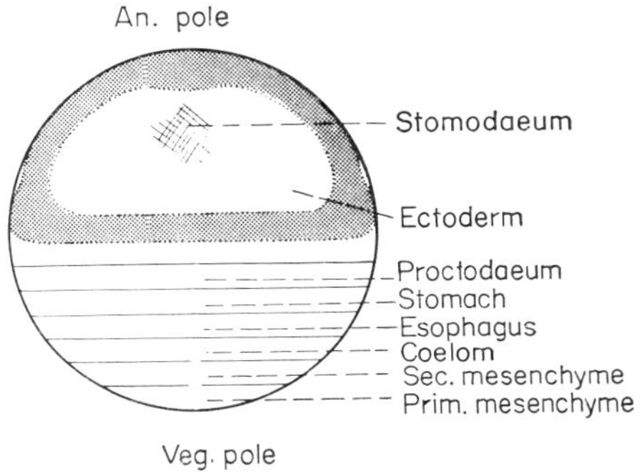

FIG. 2. A map of the presumptive regions of the egg. All the zones indicated have not yet been mapped by vital staining technique, and the diagram should be considered as an extrapolation from available results. The relative proportions of zones 1–6 are overscaled for convenience. The stippled zones correspond to thickened regions of the ectoderm but must be considered as entirely schematic.

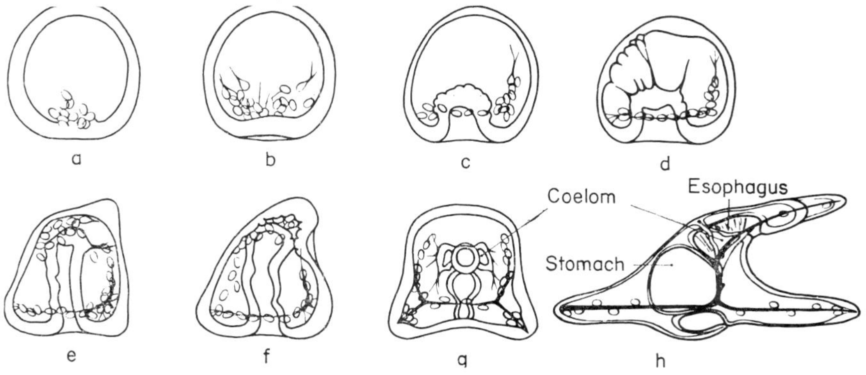

FIG. 3. Diagram of the development of the sea urchin larva (*Psammechinus miliaris*) from the mesenchyme blastula stage (a) to an early pluteus stage (h); (c–f and h) larvae in profile (g) larva seen from the ventral side.

hand, depends on the competition between an animal and a vegetal principle, the so-called morphogenetic gradients, which can be symbolized as in Fig. 4 (see Hörstadius, 1939). As the principles are antagonistic, the animal one may suppress the vegetal (animalization) or vice versa (vegetalization). The interactions are evidently mediated by certain substances. In case of extreme animalization, the whole larva develops into an ectodermal vesicle of a structure similar to the most animal region of a normal larva. An extremely vegetalized larva, on the other hand, may be practically devoid of an ectoderm. All transitions between these two types may be obtained (Fig. 4). The outcome of the competition between the two gradients can be changed by the addition or the removal of animal or vegetal material, or by treatment of the material with certain chemical agents (Section IV,E). It is important to note that development does not depend on the total amount of material present. A normal larva may develop from egg fragments of suitable

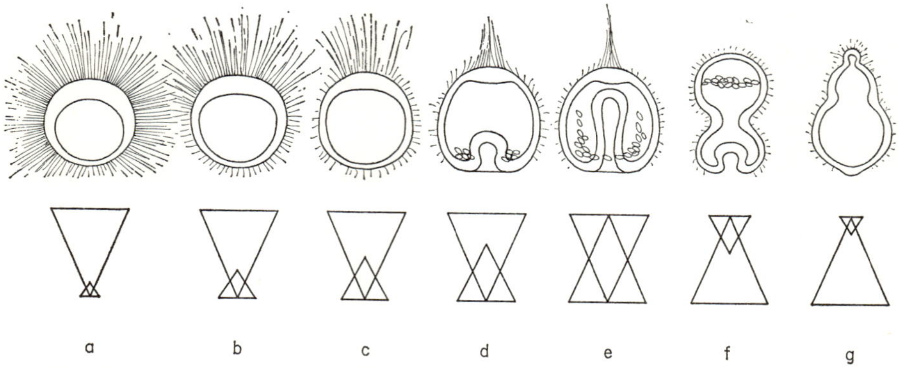

FIG. 4. Diagram of sea urchin larvae of about the same age but varying in the animal and the vegetal tendencies. (e) Normal gastrula, (a–d) larvae animalized to a varying extent, (f–g) larvae vegetalized to a different extent; in (g) the ectoderm is reduced to a small knob. The hypothetical double gradient system corresponding to each larva is indicated.

animal and vegetal strength combined during early cleavage stages. Furthermore, animal and vegetal halves may be normalized, provided they are submitted to proper chemical treatment (see Hörstadius, 1939; von Ubisch, 1929). For further details of such experiments see the reviews of Hörstadius (1939) and Lindahl (1942). The results have beautifully supported the double gradient concept of Runnström (1928a,b, and c). A theoretical treatment of such a concept has been presented by Apter (1966).

The dorsoventral polarity has sometimes been considered to represent a separate gradient (Czihak, 1961). However, it may merely reflect a different steepness of the animal and vegetal gradients along the ventral and dorsal side. One may in fact suggest that the ventral side is characterized by rather strong animal properties, an idea which is in line with morphogenetic considerations (see Gustafson and Wolpert, 1963, 1968). For further discussion of this problem see Gustafson (1965).

This description of the dynamics of determination is only a first approximation. There are also other interactions between the different levels within the eggs. These interactions may be described as feedback phenomena. An example of this is that the development of the animal plate inhibits the formation of a similar structure close to it (Lindahl, 1936). Another example is that the development of primary mesenchyme cells at the vegetal pole leads to an inhibition of the formation of primary mesenchyme in the adjacent zone (Boveri, 1901b). Upon removal of part of the animal or vegetal material, the structures concerned are formed by the adjacent material, which is no longer inhibited by the more polar region. Other examples of a similar kind have been discussed by Gustafson (1965).

C. Chemical Mediators of Interaction

Various attempts have been made to extract the animal and vegetal control substances. Recent experiments by Hörstadius *et al.* (1967) have shown that a potent animalizing agent can be isolated from unfertilized eggs. The agent is a rather small molecule, presumably a small peptide. Encouraging results have also been reported by the same authors concerning the isolation of a substance with vegetalizing properties. Although these substances are present even in the unfertilized egg, one may suppose that more of them are produced during development (G. Czihak, 1966; Section IV,E).

The question may be asked to what extent the mediators of the feedback interactions are identical with the animal and vegetal substances, which are involved in the competition between the two gradients (Gustafson, 1965). That both types of interactions may be mediated by the same type of agents may be in line with the concept of a gradual transition from the animal to the vegetal pole in the morphogenetic properties of the cells (Section IV,L). Such a transition may reduce the need of a series of feedback mediators, each characteristic of a particular zone. The observations of Czihak should also be considered in this connection (Section IV,E).

D. Progress of Determination

Experiments on the effect of brief lithium treatment indicate that the determination of the relative size of the ectoderm starts at the 16-cell stage and is completed within about 3 hours (Lindahl, 1942; Bäckström and Gustafson, 1953).

Another way to shed light upon the progress of determination is to treat isolated animal halves with lithium or to combine them with micromeres at different stages of development (see Hörstadius, 1939). Such experiments have shown that the animal material retains its morphogenetic plasticity for a rather long period of time, and that the determination proceeds from the vegetal toward the animal pole.

After the period of determination, the vegetal agent is still produced, although it is not capable of effecting the determination of the larva. It must be remembered, however, that the vegetalizing effect of lithium may be considerably enhanced if a treatment started early is prolonged after this period (Hörstadius, 1936a). Lithium may increase the production of the vegetal agent (Hörstadius, 1936a,b) and hence its radius of action, and also directly retard some processes in the animal region so that a larger section of it becomes affected by the vegetal control substances (see Gustafson, 1965).

E. Chemical Interference with Development

Among the chemical agents that suppress the animal development in the sea urchin, i.e., bring about vegetalization, the lithium ion is the best known (Herbst, 1892; Hörstadius, 1939; Lindahl, 1936, 1942; Gustafson, 1950, 1954, 1965; Bäckström and Gustafson, 1953; see Ranzi, 1962; Runnström, 1928c). Other vegetalizing agents are *o*-, *m*-, and *p*-tyrosine (Fudge-Mastrangelo, 1966), chloramphenicol (Hörstadius, 1963; Lallier, 1961), phenazone (Lallier, 1959) and often 2,4-dinitrophenol (Hörstadius, 1953). A well-known animalizing agent is thiocyanate if applied for some hours before fertilization (Lindahl, 1936, 1942; Runnström and Thörnblom, 1963). Other animalizing agents are *o*-iodosobenzoic acid (Runnström and Kriszat, 1952a; Bäckström, 1953), 2-thio-5-methylcytosine (Gustafson and Hörstadius, 1956), thiomalic acid (Lallier, 1952), oxidized lipoic acid (Wolfson and Fry, 1965; Runnström, 1956a), zinc ions (Lallier, 1955), sulfate-free seawater (see Lindahl, 1936, 1942; Runnström *et al.*, 1964) glucose, and other sugars (Hörstadius, 1959), trypsin (Runnström and Immers, 1966) etc. Various natural amino acids, including glutamine, reduced glutathione, structural analogs of amino acids, purines and pyrimidines, ascorbic acid and glucoascorbic acid have also a certain animalizing or vegetalizing effect (for references, see Gustafson, 1965; Lallier, 1964). The effect of one agent may be counteracted and even overcompensated by a subsequent treatment with another agent (see, e.g., Lindahl, 1936, 1942; Gustafson and Hörstadius, 1956). Synergic effects of various agents and lithium have also been described (e.g., Gustafson and Hörstadius, 1955, 1957; see also Gustafson, 1965). For further examples see Lallier (1964).

Theoretically, the agents that bring about animalization and vegetalization could bring about their effects in a variety of ways, for instance, by interfering with the structural and biochemical polarity of the egg before fertilization or during cleavage, by interfering with the production of animal and vegetal substances that mediate the interactions between the different zones, by affecting the metabolic processes influenced by these substances, by affecting the transcription of the genetic message, or the utilization of the messenger produced during protein synthesis. It must be emphasized, however, that most of the agents concerned exert their specific morphogenetic effects only if the treatment occurs during early cleavage stages, when the fate of the development of the zones of the egg is determined, at least to a certain extent (Section IV,D).

One exception to this rule is thiocyanate, the animalizing effect of which is mainly obtained if the eggs are treated before fertilization (see Lindahl, 1942). Glucose and other sugars are also exceptional, since they affect development if they are applied after hatching (Hörstadius, 1959). In this case one may suggest that the agent does not interfere with the process of primary determination, but rather with the actual process of morphogenesis, more precisely in the establishment of increased contact between the cells in the animal region. Actinomycin-C and mitomycin-C also affect development of certain regions if applied at an advanced stage (Markman, 1963). These effects, which are confined to the differentiation of restricted regions of the mesendoderm and therefore cannot be described as vegetalization or animalization, show that the transcription of the genetic information occurs at different times in the different regions.

A concept favored by the author is that the period of early cleavage and blastula formation is a period when the region governed by the animal processes undergoes a preparation for a sudden major burst in protein synthesis that starts at the onset of gastrulation (see Gustafson, 1950, 1954, 1965). According to this concept, one could say that development involves a process of amplification in the sense that the intensity of the preparatory processes determines the intensity during the following anabolic burst. The micromeres have been assumed to inhibit the preparatory processes in their vicinity, an action that is enhanced by lithium (cf. Gustafson and Hasselberg, 1951; Gustafson and Lenicque, 1952; Lenicque *et al.*, 1953). An animal-vegetal gradient is therefore established, both in the preparatory processes and in the following burst. The inhibitory action of the micromere descendants is gradually overcome, and anabolic activities can therefore also start in the vegetal region. To some extent this interpretation of the role of the micromeres has been adopted by Runnström (1957, 1961), according to the terminology of whom the micromeres "moderate" the anabolic processes in the rest of the germ, a moderation that is expressed by an inhibition of anabolic processes. According to this interpretation, the size of the ectoderm would correspond to an area where the preparatory processes have been intense at any early stage. If the vegetal material is removed or the egg is treated with potent animalizing agents, it may extend over the whole larva.

It is difficult to make a coherent picture of the preparatory processes and to localize the target of the normal morphogenetic substances and the chemical agents that interfere with the double gradient system. When considering this problem one should realize, however, that the preparatory processes presumably involve some protein synthesis, as labeling experiments show an increase in amino acid incorporation even during blastula formation, and since certain enzymes such as glucose 6-phosphate dehydrogenase show a rise during this period (Section IV,F).

Some of the chemical agents that favor the vegetal development are analogs of amino acids and building blocks of nucleic acids (Gustafson and Hörstadius, 1955), but a number of natural amino acids also have a certain effect (Gustafson and Hörstadius, 1957). It must be admitted that the effects are often rather weak. In combination with weak lithium solutions, that in themselves are quite ineffective, vegetalization is clearly potentiated, however. This is for instance the case with D-leucine. As a general tentative conclusion of such experiments, it was suggested that any kind of treatment which may disturb the anabolic functions of the egg, e.g., by creating an inbalance in the pool of normal building blocks etc., weakens the animal development.

A most interesting contribution in line with this is that vegetalization may be produced by treatment with weak concentrations of D- and L-*p*-tyrosine, *o*-tyrosine, and *m*-tyrosine during cleavage (Fudge-Mastrangelo, 1966). The tyrosines are synergistic with lithium and 2,4-dinitrophenol but, unlike these agents, the tyrosines do not affect respiration. On the other hand, they have a marked depressive effect on the incorporation of C-labeled amino acids. Since their own incorporation is negligible and their effect can be completely annulled by a full complement of amino acids, Fudge-Mastrangelo suggested that the tyrosines limit the endogenous supply of amino acids available for synthesis of proteins, associated with the elaboration of animal structures, by inhibiting the release of storage proteins.

According to the results of Ranzi and collaborators (see Ranzi, 1962), a common denominator of the action of vegetalizing and animalizing substances is that the former tend to induce a fibrous transformation of proteins, whereas the latter brings about a depolymerization. An effect on the state of protein molecules has also been advocated by Ling (1962). With the work of Fudge-Mastrangelo in mind, one may suggest that the coarsening effect of lithium may retard the utilization of the yolk, whereas the animalizing agents have the opposite effect.

The mode of action of the agents concerned may be different, however, and more complicated. Since some animalizing agents tend to oxidize SH groups, one could speculate that the S-S compounds formed may serve as hydrogen acceptors that favor the shunt activity (cf. Gustafson, 1965). It has also been suggested that the effect of lithium on the phosphatides of cellular membranes is a key to its morphogenetic effects (cf. Raven, 1958, 1961).

When considering the dynamics of determination Agrell (1953, 1956) emphasized the occurrence of a vegetal-animal gradient in mitotic rate during early development. Other investigators (B. E. Hagström, 1963; B. E. Hagström and Lönning, 1963, 1964) claim that the rate of cleavage is higher in the animal than in the vegetal region, and that all animalizing agents studied by them bring about an acceleration of the rate of division, whereas vegetalizing agents bring about a decrease. Although conflicting, such observations underline the importance of the length of the interphase periods, and the time at which cleavage has proceeded far enough to allow a slowing down of the rate of cleavage. Broadly speaking, one may suggest that the production of messenger RNA from interphase nuclei could be favored in the animal region during early development, a concept that apparently fits with the idea proposed by Agrell.

That the nuclear activities play a role at a rather early stage in animal development is shown by the observations that actinomycin-C and mitomycin-C treatment at an advanced blastula stage tends to weaken the animal trend of development of animal halves, as well as the development of the ectodermal part of isolated vegetal halves (Markman and Runnström, 1963). Treatment with actinomycin-C during restricted periods of more advanced stages, on the other hand, interferes with the development of various parts of the mesendoderm, different regions depending on the time of treatment (Markman, 1963). On the whole, it would seem that messenger RNA is formed at different times in different regions, and that it starts in the animal one. The dark-field appearance and stainability of the nuclei in the animal and vegetal region is in agreement with this (Section IV,J). The authors therefore consider that the normal control substances interfere at the genetic level (cf. also Czihak, 1966). In excellent agreement with the general picture is also the fact that chloramphenicol, which is known to interfere with the utilization of the messenger, brings about vegetalization. One may in this connection consider that lithium is known to interfere with the state of nucleic acids in solution (for references, see Gustafson, 1950). The problem seems to require some further considerations, however.

The morphogenetic action of actinomycin on whole larvae has been studied by Gross and Cousineau (1963). According to their results, the development of the ectoderm is less affected than that of the mesendoderm. One possible reason is that the ectoderm is a less sensitive indicator of certain morphogenetic disturbances than the mesendoderm, which is characterized by dramatic morphological changes at gastrulation. One may, however, suggest that early development, including the development of the ectoderm, is more dependent on messenger preformed within the egg than is the development of the mesendoderm. In this connection one should consider the fact that after hybridization the disturbances are mainly restricted to the vegetal region (Chen *et al.*, 1960), and that the paternal antigens do not appear until gastrulation (Harding *et al.*, 1954). One should finally consider that, in spite of all, the actinomycin may also affect the vegetal cells during early cleavage stages; see the following.

The micromeres begin to form RNA as soon as they appear, possibly due to an early retardation in the rate of division of these tiny cells and a corresponding prolongation of their interphase periods (see Agrell, 1958; Czihak, 1965). In any case, the production of RNA occurs in the nuclei and is inhibited by actinomycin, as shown by Czihak (1966) who considers that it plays a role both as an inducer of endodermal development and in the feedback interactions within the vegetal material.

That 8-azaguanin incorporation into the RNA of the micromeres deletes their endoderm-inducing capability, but leaves their own differentiation into primary mesenchyme unaffected (Czihak, 1966), indicates that the messenger production related to their own future development occurs later. On the whole, the observations and interpretations neither contradict the idea that the animal trend of development favors an early onset of protein synthesis (either at the expense of preformed messenger RNA or of messenger formed *de novo* within the animal region) nor that the vegetal material counteracts this (e.g., as a consequence of a reprogramming of the ribosomes or by a hybridization with the DNA). The specificity of the endodermal development (Section K), which can be considered as a new trend in the blastula, can be due to the type of messenger produced in the vegetal region. This would fit with the original ideas of Runnström about the double gradient system. One should consider, however, that the interactions between the animal and vegetal region are mutual, and that an early development of the animal region inhibits the animal trend of development in adjacent regions (cf. Lindahl, 1936), perhaps in the way suggested by Weiss (1947) and Rose (1957; cf. Gustafson and Lenicque, 1952).

F. Overall Changes in Respiratory Activity and Protein Synthesis

The first detailed investigation of the overall respiratory changes of the developing sea urchin egg was performed by Lindahl (1939a, 1942), according to whom it may be described by the curve in Fig. 5. A similar

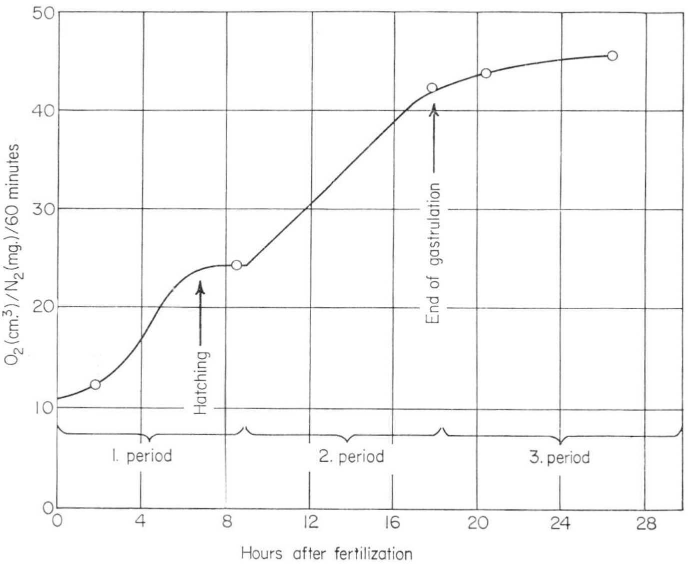

FIG. 5. Changes in the rate of respiration during larval development of the sea urchin *Paracentrotus lividus*. (From Lindahl, 1939a.)

curve can evidently be obtained with *Asterias* (Borei and Lybing, 1949), and is by no means unique for echinoderms. Studies by means of the Cartesian diver have shown that the early rise of the curve has a microstructure with marked fluctuations that can be correlated with the mitotic cycles, which in early stages are synchronous to some extent (Holter and Zeuthen, 1957).

The respiration of the newly fertilized egg has the respiratory quotient 0.7. If this respiration makes up a constant fraction even during cleavage and the early respiratory rise reflects a new, increasing fraction, the respiratory quotient of this fraction would be 1.0, indicating that it corresponds to a combustion of carbohydrates (Lindahl and Öhman, 1938).

For various reasons (cf. Hörstadius and Gustafson, 1947) one was led to think that the carbohydrate combustion concerned essentially proceeds along the hexose monophosphate shunt pathway. This suggestion is well in line with various metabolic studies (Lindberg and Ernster, 1948; Keltch *et al.*, 1956; Krahl, 1956; Ghiretti and D'Amelio, 1956; Isono, 1962, 1963; etc.). The changes in activity of glucose 6-phosphate dehydrogenase and 6-phosphogluconate dehydrogenase (Fig. 6) (Bäckström, 1959a, 1963a) and the fate of variously labeled glucoses (Bäckström *et al.*, 1960) also fits with this, and also data of Isono (see Gustafson, 1965) on the oxidation of carbohydrates and lipids. The strong fixation of CO_2 during cleavage stages (Hultin and Wessel, 1952),

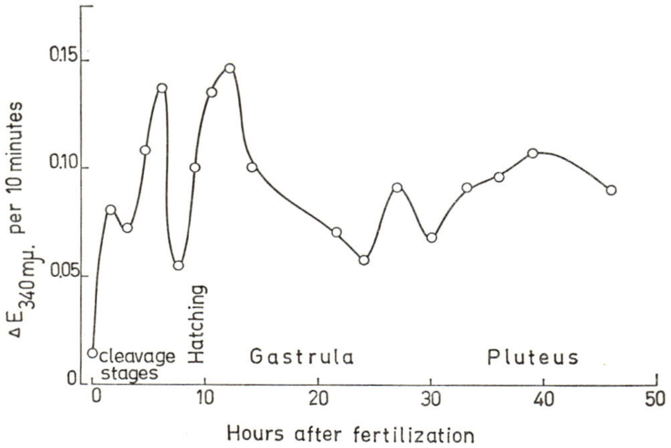

FIG. 6. Changes in the activity of 6-phosphogluconate dehydrogenase during larval development of the sea urchin *Paracentrotus lividus*. (From Bäckström, 1963a.)

which indicates a lively formation of reduced nicotinamide-adenine dinucleotide (NADPH), has also been used as an argument.

However, the interruption of the early respiratory rise is correlated with a decrease in the shunt enzymes, and the subsequent steady rise in respiration occurs when the enzymes concerned show a new peak in activity and a series of small fluctuations (Fig. 6). In order to account for the respiratory changes after the first rise, one therefore has to consider other factors that may determine the respiratory activity.

Some light may be shed on the problem by studies on the mitochondrial population. This remains rather constant during cleavage. Nevertheless, a certain rise in mitochondrial stainability has been described during this period (Gustafson and Lenicque, 1955). This indicates that the mitochondria undergo a certain change even during this early period, a conclusion supported by their incorporation of labeled amino acids (Guidice and Monroy, 1958; Nakano *et al.,* 1958; Nakano and Monroy, 1958). The most marked changes in the mitochondrial population, however, start at the onset of gastrulation. The evidence for this is multifold. There is thus a strong rise in the activity of succinodehydrogenase and malic dehydrogenase at this time (Gustafson and Hasselberg, 1951). Studies on the effect of CO in the respiratory activity are furthermore in line with a rise in cytochrome oxidase (Lindahl, 1939b), although direct determinations of this activity indicate that the activity in homogenates is influenced by various factors, for instance, by inhibitors (cf. Deutsch and Gustafson, 1952; Maggio and Monroy, 1959; Maggio *et al.,* 1960). That mitochondria show a strong increase in activity at the onset of gastrulation is also supported by the work of Hultin (1953a,b) and with Isono's studies on the oxidation of lipids; see also Black (1964). Mitochondrial counts on vital stained material (Gustafson and Lenicque, 1952), as well as counts on homogenates (Shaver, 1956), give direct evidence for the concept that the number of mitochondria increases at the time concerned. Furthermore, electron microscopical investigations (Berg *et al.,* 1962; Berg and Long, 1964) indicate that the mitochondria also change their appearance in the early gastrula, a conclusion also drawn by Shaver. One may therefore conclude that the second rise in respiration to a great extent reflects an increase in mitochondrial activities, overlapped by some fluctuations in the hexose monophosphate shunt.

Another approach to the problem of respiratory control has been to study the effects of 2,4-dinitrophenol on the respiration of whole larvae and animal or vegetal halves (Immers and Runnström, 1960; De Vincentiis *et al.,* 1966; De Vincentiis and Runnström, 1968). This agent brings about a marked increase of the respiratory activity, and changes the general shape of the respiratory curve so that a distinct minimum can be seen at the time of the plateau in untreated larvae. As dinitrophenol brings about a release of the respiratory control to the same extent as the addition of the phosphate acceptor ADP (adenosine diphosphate), the conclusion was drawn that the respiratory intensity is controlled by the rate of consumption of ATP (adenosine triphosphate), which to a great extent reflects the rate of protein synthesis. As pointed out by the authors, this is in good agreement with the studies of Kavanau (1953, 1954) on the changes in the free amino acids and proteins in the embryo. The occurrence of two main waves in protein synthesis during development is also in agreement with studies on the incorporation of labeled amino acids in developing sea urchin eggs (cf. Monroy, 1963; Monroy and Maggio, 1964). The second rise is furthermore in line with the appearance of new antigens at the onset of gastrulation (Perlmann and Gustafson, 1948; Perlmann, 1953; and others; see Monroy and Maggio, 1964). That respiration is to a great extent controlled by the ATP consumption, however, by no means contradicts the suggestion that the formation of shunt and mitochondrial enzymes is a basic requirement for the respiratory changes. As pointed out by the authors, the uncoupled phosphorylation in fact reveals the "total amount" of respiratory enzymes.

G. Gradients in Respiratory Activity

As the extent of the first respiratory rise varies with the size of the developing ectoderm (Lindahl, 1939a, 1942), one might expect that a considerable fraction of this rise is located in the animal half of the larva and plays a role in its development into an ectoderm. This could mean that the first peak in shunt enzyme activity would be concentrated in the animal region. One may accordingly speculate that the second peak corresponds to the vegetal region.

Information about the distribution of the dehydrogenase activities can be obtained from studies on the reduction of various stains in vitally or supravitally stained larvae performed by various authors (Child, 1936, 1941a; Hörstadius, 1952, 1955; Lallier, 1958; Bäckström, 1959b), Fig. 7. These observations are in line with the general picture in the preceding paragraph. There is thus an animal-vegetal gradient in reducing capability in early cleavage stages, whereas an opposite gradient appears in the late blastula. Furthermore, the extension of the first (animal-vegetal) reduction gradient appears to vary in proportion to the size of the future ectoderm, as shown by studies on normal, animalized, and vegetalized whole larvae, and animal halves with a varying number of implanted micromeres. The second (vegetal-animal) reduction gradient shows a similar relation to the extent of the future mesendoderm in such material (Hörstadius, 1952, 1955).

The reduction gradients may to a great extent reflect the distribution of the shunt activity, as the rate of terazolium autoreduction in early development closely resembles the changes in the activity of the shunt enzymes (Bäckström, 1963b). This does not necessarily mean that the rate of reduction exactly reflects local differences in the amount of the shunt enzymes. In fact, the early rise in glucose 6-phosphate dehydrogenase activity is not markedly affected by animalization and vegetalization (Bäckström, 1959a). To overcome this difficulty, Bäckström drew attention to the fact that the shunt activity reflects the rate of reoxidation of reduced nicotinamide-adenine dinucleotide phosphate (NADPH). One important pathway for this reoxidation involves the respiratory chain. This pathway is slow, but can be accelerated if lactate is supplied, since lactic dehydrogenase mediates the transfer of hydrogen from the reduced NADPH to NAD, which serves as a mediator to the respiratory chain. The observation that lactate, pyruvate, and propanediol phosphate (which can be considered as a potential lactate) have an animalizing

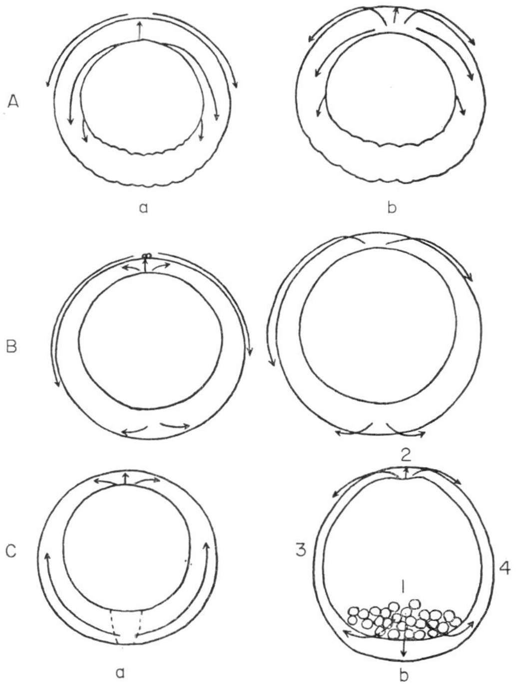

FIG. 7. Reduction gradients of various sea urchin species in different stages of development. Intensity of reduction falls off in the direction of the arrows. A. Early blastula (a, *Strongylocentrotus;* b, *Dendraster*). B. Late blastula (*Patiria*); reduction more rapid on one side, presumably the ventral side. C. Late blastula (a) and mesenchyme blastula (b); both of *Strongylocentrotus.* (From Child, 1941a.)

tendency (Hörstadius and Gustafson, 1947) fits with the concept that a gradient of animal-vegetal shunt activity plays a role in the determination of the size of the ectoderm and suggests that the reoxidation of NADPH is a rate-limiting factor. Experiments of Bäckström (1959b) in which he studied the tetrazolium reduction gradients in larvae kept in buffer solution to which he added NADP, NAD, and substrates for NADP- and NAD-dependent dehydrogenases, are in line with this. The experiments showed that the rate-limiting factor is not the total amount of NADP present, but that the reduction can be mediated by NAD. The rate-limiting factor could therefore be a dinucleotide transhydrogenase or a diaphorase, but also other steps in the respiratory chain. Observations on the early regional variations of the Nadi reaction might be consistent with this idea (Czihak, 1963; cf. Child, 1941b, 1944). It is, however, possible that some of the animalizing agents favor the shunt by increasing the role of another pathway for the hydrogen (Gustafson, 1965).

The second (vegetal-animal) reduction gradient might also to some extent reflect a local increase in the shunt activity, this time in the mesendodermal region. In fact it coincides with the appearance of the second peak in shunt activity. The reduction becomes affected by the addition of NAD from the time of gastrulation (Bäckström, 1959b). Furthermore, the primary mesenchyme is characterized by a strong NAD diaphorase activity (Czihak, 1962b).

The reducing activity of the animal region appears to be proportional to the subsequent rise in mitochondrial enzymes (cf. Gustafson and Hasselberg, 1951). This suggests that the shunt plays a role in the later anabolic activities in the larvae. If so, the first reduction gradient could correspond to a subsequent animal-vegetal gradient in anabolic processes in the blastula and early gastrula, whereas the second reduction gradient could correspond to anabolic processes that start in the vegetal region of the late blastula and mesendoderm. (For further discussion about the reduction in gradients see Section IV,G.)

H. Overall Nucleic Acid Synthesis

As the hexose monophosphate shunt plays a role in the formation of nucleotides, one may hypothesize that it prepares the cells for nucleic acid synthesis. The importance of ribose phosphates in the purine metabolism in early sea urchin development was in fact suggested by Hultin (1953d). It may of course play a dominating role in DNA synthesis, which is one of the most conspicuous processes during cleavage (Brachet, 1937, 1960) that far exceeds the RNA synthesis, and which starts even before the pronuclei have fused (Simmel and Karnofsky, 1961). The literature on DNA synthesis in sea urchin development has been extensively reviewed, e.g., by Monroy and Maggio (1964) and Grant (1965). As the egg contains some DNA precursors (see Section II; Monroy and Maggio, 1964; Tyler, 1967), one could suggest that the shunt is not a limiting factor for DNA synthesis in early cleavage stages, and this is in agreement with studies of Karnofsky and Basch (1960), Nemer (1962), and others, even if the precursors may not meet the requirements for more than a few cleavages.

Studies on the changes in the RNA content during early development have given somewhat conflicting results (Elson *et al.*, 1954; Bäckström, 1959c; Agrell and Persson, 1956; see also Monroy and Maggio, 1964). This may indicate that new RNA mostly forms from cytoplasmic RNA, the decrease of which tends to mask the synthesis. This is in line with the studies by Markman (1957) on the cytoplasmic pyronine stainability and the observations of Tocco *et al.* (see Monroy and Maggio, 1964) that the increase in nuclear RNA corresponds to a decrease in cytoplasmic RNA. There is, however, evidence for a *de novo* synthesis of ribonucleotides and RNA from studies on incorporation of labeled precursors such as ammonium chloride, glycine, acetate, formate, carbonate, adenine, and phosphate (for references, see Maggio and Monroy, 1964; Gustafson, 1965). It is particularly interesting that the incorporation of carbonate into guanine and adenine of RNA was marked during the first respiratory rise (Hultin, 1953c,d). As the formation of ribosomal RNA does not occur until later (Nemer, 1965), one may speculate that the RNA to some extent is messenger RNA. This is in agreement with the observations of Ficq, Aiello, and Scarano and of Tocco *et al.* (both mentioned by Monroy and Maggio, 1964). For further information on the messenger RNA problem in relation to sea urchin development see Tyler (1967).

I. Gradients in RNA Synthesis and Nuclear Activity

The RNA synthesis during early development has been studied by means of labeled adenine. Such experiments show an increase in nucleic acid synthesis during two phases (Markman, 1960). Experiments with ribonuclease-treated material have made it possible, however, to obtain a general picture of the synthesis of RNA, and when the isotope labeling and ribonuclease digestion are carried out in combination with radioautography, information may also be obtained about the regional intensities of the RNA synthesis (Markman, 1961a). On the whole there is a good correlation between the reduction gradients and the incorporation of adenine into RNA, although an RNA synthesis in the micromere descendents in the most vegetal region starts before the vegetal-animal reduction gradient becomes distinct (Agrell, 1958; Czihak, 1965). Studies on the incorporation into animal and vegetal halves by means of radioautography, as well as Geiger counter readings of the total adenine incorporation, are in agreement with the general view about a close relation between the reduction gradients and RNA synthesis (cf. Markman, 1961b).

Experiments by means of the Hale ferric acetate reagent in combination with ribonuclease indicate that the nuclei in the animal region of the mesenchyme blastula contain stainable RNA, in contrast to those of the vegetal region. In animalized larvae the region with this stainability is larger than in normal larvae in proportion to the increased size of the ectoderm (Immers, 1956a,b; Markman, 1957; Immers *et al.*, 1967).

Observations on the dark-field appearance of the nuclei of the early sea urchin larva fit with the concept that the nuclei of the animal region are among the first to take part in a transcription process (Runnström, 1961; Markman and Runnström, 1963; Immers *et al.*, 1967). The Hale reagent can also stain DNA and may thus give important information about the state of the genetic material itself. The DNA in the early blastula is poorly stained. In the mesenchyme blastula and gastrula the stainability appears in the nuclei of the animal region, whereas those in the vegetal region are still poorly stained. The low stainability indicates that most of the DNA is in a condensed and repressed state, e.g., due to the presence of histones. A stainability can be induced in all cells by treatment of the material before staining with an alkaline phosphate buffer. It is interesting that the longer development proceeds, the less alkaline need the buffer be in order to induce stainability. This suggests that the genetic material undergoes a gradual change until it becomes capable of serving as a template in a DNA-dependent RNA synthesis.

As a general, tentative conclusion the shunt gradient corresponds to a gradient in nuclear nucleic acid synthesis, which may include messenger RNA as well as ribosomal RNA. The probem remains of how the nuclei become switched on and become involved in this synthesis.

J. Gradients in Protein Synthesis

The gradients in RNA synthesis are apparently paralleled by gradients in protein synthesis, as shown by radioautographic studies together with Geiger counter measurements of the incorporation of labeled leucine into animal and vegetal halves isolated at the 16-cell stage (Markman, 1961b). This conclusion appears to be contradicted by the results of Berg (1965). In a recent paper by Markman (1967), the original results have been nicely confirmed, however, and the divergencies in the results between different authors have been explained in an apparently satisfactory way. Furthermore, the cytoplasmic basic proteins, which may be considered to suppress the ribosomal activity, decrease in concentration in the late blastula and gastrula, and this decrease starts in the ectoderm (Bäckström, 1965).

The gradients in leucine incorporation and the disappearance of cytoplasmic basic proteins can be expected to be paralleled by gradients in the intensity of enzyme synthesis. The marked rise in the activity of mitochondrial enzymes at the onset of gastrulation (Section F) was in fact suggested to consist of two main fractions, first an increase in the animal region, then one in the vegetal (Gustafson and Hasselberg, 1951). Observations on vital stained material by Gustafson and Lenicque (1952) are in line with this. The first peak in the number of stainable mitochondria appeared in the animal region, a second peak or rather a series of peaks appeared in the vegetal material. The extent of the region in which the stainable mitochondria first appeared was proportional to the size of the ectoderm, as shown by studies of animalized and vegetalized whole larva. In addition, implantation of micromeres in animal halves isolate at the 16-cell stage brings about an inhibition of mitochondrial development in the vegetal region of the animal half (Lenicque *et al.*, 1953). These results have not been repeated by Shaver

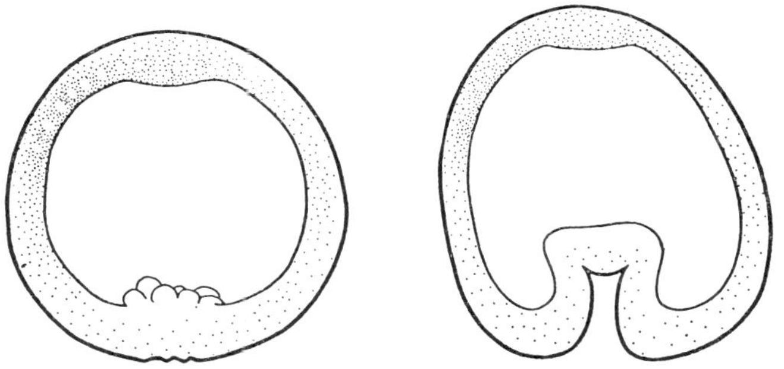

FIG. 8. Distribution of Janus green stainable mitochondria in early development of *Psammechinus miliaris*. 1. Early mesenchyme blastula. 2. Early gastrula seen in profile. (From Czihak, 1962b.)

(1956), presumably since the stain Nile blue sulfate is difficultly soluble in seawater of high salinity, or since the lipid droplets in the sea urchin studied might have interfered with the mitochondrial counts. Cytological studies with the light microscope of fixed material (Mori *et al.*, 1958) and electron microscopical investigations by Berg *et al.* (1962) and Berg and Long (1964) have also questioned the validity of the result. The mitochondrial gradient within the ectoderm has, however, been confirmed by studies by Czihak (1962b) with Janus green vital staining (Fig. 8). From this study it is also clear that the mitochondrial population has its maximal density at the oral region of the ectoderm.

Apart from what has been said above, an explanation for the conflicting observations is that the stainability reveals an activity of functional mitochondria. The stainability is apparently due to an active transport mechanism that concentrates the stain into the mitochondria, where it presumably precipitates as a calcium salt. In support of this, Strecker and Gustafson (in preparation) have shown that various inhibitors of respiration and oxidative phosphorylation such as amytal, 2,4-dinitrophenol, and pentachlorophenol, inhibit the uptake of the stain. It is therefore quite likely that the mitochondrial gradients reflect gradients in particularly active mitochondria, and that the nonstainable or weakly stainable mitochondrial precursors of Gustafson and Lenicque correspond to mitochondria that are still rather incomplete with respect to their capability for oxidative phosphorylation (for further information see Gustafson, 1965).

One would anticipate that animal-vegetal mitochondrial gradient should reveal itself by differences in the respiratory activity of animal and vegetal halves, even in early stages when the mitochondrial population appears to undergo some changes. Early observations, however, failed to show any difference in oxygen consumption between the two halves (Holter and Lindahl, 1940). Later studies of Immers and Runnström (1960) De Vincentiis *et al.* (1966), and De Vincentiis and Runnström (1967) on the effect of 2,4-dinitrophenol explained this lack of correspondence in a very interesting way (cf. Section IV,F), which fits nicely with the concept that there is both a gradient in the mitochondria with respect to their number or content of respiratory enzymes and in their ability to carry out oxidative phosphorylation. For further discussion about problems as well as about the relation between the animal-vegetal and the dorso-ventral polarity of the larva see Gustafson (1965).

K. Biochemical Specificity of the Animal and Vegetal Development

According to the results presented in the previous sections, the animal-vegetal or rather ventroanimal-vegetal gradient appears to favor an early onset of protein synthesis. The opposite gradient appears to be characterized by essentially the same sequence of metabolic events as the animal, although their start is delayed (see further Section IV,E). It is evident, however, that the animal and vegetal region also show qualitative biochemical differences. In the vegetal region some cells lay down a skeleton, an activity that must require particular enzymic activities, presumably also the formation of some kind of collagen-like material (see below). Other cells produce a red pigment (see Monroy and Maggio, 1964). The vegetal region has also a stronger alkaline phosphatase activity than the ectoderm (Evola-Maltese, 1957; Hsiao and Fujii, 1963), a difference that appears at the time of gastrulation when the total activity increases (Mazia *et al.*, 1948; Gustafson and Hasselberg, 1950;

Flickinger, 1957). Other biochemical characteristics have been described by Czihak (1962a) and Bäckström (1957).

An important line in the discussion of the metabolic differences between the animal and vegetal region departs from the observation of Herbst (1904) on the effects of the lack of sulfate on sea urchin development. Such a lack brings about a peculiar type of animalization characterized by radial symmetry and a decreased capability to serve as a substrate for the primary mesenchyme cell pseudopods. There is also a depression of protein synthesis. The primary mesenchyme cells also show a reduced capability of migration, and may even be destroyed. The whole vegetal region may in fact show signs of damage. To some extent the animalization may be a consequence of this damage. A peculiar result is, however, that isolated animal halves show an increased extent of animalization, whereas the vegetal halves become still more vegetalized (see Runnström *et al.*, 1964). It thus appears that the sulfate plays an important role for the animal as well as for the vegetal halves, but that to some extent the function of sulfate is different in the two halves.

The first analysis of the role of sulfate was undertaken by Lindahl (see Lindahl, 1942), who suggested that the vegetal region in the mesenchyme blastula is a site of strong degradation of proteins. That a strong degradation of yolk occurs in the primary mesenchyme is quite evident, since it soon becomes devoid of yolk granules. Lindahl suggested that the sulfate is needed for the detoxification of aromatic waste products formed in the vegetal region as a result of a combustion of aromatic amino acids, and that the morphological effects of the lack of sulfate can be ascribed to a vegetal intoxication. On the basis of this, Ten Cate (1952) suggested that the oxidation of aromatic amino acids is connected with the formation of a collagenlike protein in the mesenchyme cells, which might serve as a matrix for the skeleton. This suggestion has been discussed by Gustafson (1965). Among the observations that favor this idea is that ascorbic acid, which elsewhere is involved in collagen formation, shows a strong increase in the advanced gastrula, and that an insoluble ascorbic acid complex is localized particularly in the vegetal area (Bäckström, 1956, 1957). Furthermore, ascorbic acid enhances the vegetalizing action of lithium ions, whereas glucoascorbic acid enhances the animalization of animal halves. The pigment production in some of the secondary mesenchyme cells may also be related to the metabolism of aromatic amino acids. The abundant formation of unmasked mucopolysaccharides in the vegetal region may also be looked upon as an aspect of formation of connective tissue substances in this region (cf. Runnström *et al.*, 1964).

Later work by Runnström and collaborators has focused the attention on the role of the sulfate in the formation of acid polysaccharides, which to a great extent are sulfated (see Runnström *et al.*, 1964; Hörstadius *et al.*, 1966; Inners and Runnström, 1965). The formation and distribution of such polysaccharides has been studied by Immers by means of the ferric-acetate reagent of Hale. Radioautographic studies by Immers showed that such polysacharides are formed to about the same extent in the animal as in the vegetal region. It is clear, however, that they are linked to a certain extent to proteins which mask their stainability by the Hale reagent, and that this masking varies between the different regions. Some of the morphogenetic effects of sulfate deficiency may be related to the role of the polysaccharides at the cell surfaces. The effect is supposed to vary, depending on the cells in which the polysaccharides occur, and the extent to which they are associated with proteins. In the animal region, particularly in the acron, the polysaccharides are assumed to increase the adhesion between the cells and hence their joining up into an animal plate. Furthermore, the polysaccharides in certain zones of the ectoderm play a role in the attachment of the primary mesenchyme. In the vegetal region, where the cell adhesion is weaker, the polysaccharides still play an important role for the maintenance of the integrity of the structures. Polysaccharides covering the primary mesenchyme cells thus appear to play a role for their motility, including their formation of pseudopods. The same applies to the pseudopod-forming cells at the tip of the archenteron during gastrulation, as well as when they are released as secondary mesenchyme cells. This picture is well in line with the analysis of the cellular basis of morphogenesis carried out by means of time-lapse cinematography by Gustafson and Wolpert (reviews, 1963, 1968) and the observations of Motomura (1960) and K. Okazaki *et al.* (1962). The formation of the sulfated mucopolysaccharides apparently depends upon the formation of messenger RNA, e.g., it is involved in the formation of appropriate enzymes in sulfate metabolism and the protein moieties of the complexes. Treatment with actinomycin-C at appropriate time hence mimics of effects of sulfate deprivation with respect to the behavior of the primary mesenchyme cells.

Runnström and collaborators also emphasize that the sulfate and the sulfated polysaccharides are involved in the control of the nuclear activities and nucleic acid and protein anabolism, and therefore play an even more fundamental role in the cellular activities. On the basis of a variety of evidences, it has been suggested that sulfated polysaccharides serve as inhibitors for nucleases and proteases. One may also speculate about their role in the removal of histones from the genetic material, which would mean that they act as derepressors. As the amino acid incorporation in animal halves is higher than in the vegetal region, even after sulfate deprivation, it was suggested that the sulfated polysaccharides are more important for the anabolic reactions in the vegetal than in the animal region.

Another approach to the problem of regional biochemical differentiation is the one presented by Willmer in his inspiring book *Cytology and Evolution* (Willmer, 1960), in which he points out that cells may be grouped in a few basic cell types and discusses some of their metabolic characteristics. The classification of Willmer is also applicable to the sea urchin embryo, in which the cells in the most animal region can be classified as epitheliocytes with strong adhesive properties, the most vegetal cells as amebocytes or amebo-mechanocytes and the cells in the zones between (see Fig. 2) can be considered as transitional forms (see Gustafson and Wolpert, 1963, 1968). On the basis of various data from the literature Lövtrup (1966) has made an attempt to apply the general ideas of Willmer to the biochemical development of the sea urchin larva and classed the enzymes as amebocytic, epitheliocytic, and mechanocytic.

L. Biochemical Activities and Morphogenesis

In order to shed some light upon the way in which the metabolic activities of the cells are translated into morphogenetical activities, detailed time-lapse cinematographic studies of the behavior of the individual cells has been carried out (reviewed in Gustafson, 1961, 1964; Gustafson and Wolpert, 1963, 1968). From such studies it seems that one can arrange the morphogenetic behavior of the cells along a spectrum with a high adhesion between the cells at one end and a very low adhesion at the other, and with all transitions in between. The animal plate of the gastrula is thus characterized by a high cellular adhesion. A high adhesion also characterizes the ciliated band, which extends from the animal plate and surrounds the ventral side (cf. Fig. 2), and which in strongly animalized animal larvae forms a plate continuous with the animal plate. The adhesion of the rest of the ectoderm is somewhat lower than in the regions just mentioned. At the vegetal pole of the late blastula the adhesion is very low, as indicated by the entrance of the primary mesenchyme cells into the blastocoel. A rather low adhesion is also characteristic of the early endomesoderm, although it varies from zone to zone. The decrease in adhesion can be correlated with a pulsatory activity of the cells and ability to form pseudopods. These two latter activities seem to be closely related, since cells with a strong pulsatory activity later on have a lively pseudopodal activity.

The ability to form pseudopods, however, seems to decrease more rapidly in the direction of the animal pole than the pulsatory activity.

The time course of the changes in morphogenetic activity can be correlated with the time pattern of the main biochemical activities. The increase in adhesion starts in the ectodermal region, particularly in the regions of the ciliated band and the animal plate, as can be concluded from the gradual appearance of a dorsoventral shape of the larva. A new phase in the morphogenetic activities characterized by a decrease in adhesion, pulsations, and pseudopod formation starts at the vegetal pole and spreads upward from zone to zone (cf. Fig. 2). These activities play an important role in the release of the primary mesenchyme, in the process of invagination, and in coelom formation. In further development, however, the adhesion apparently increases within the entomesoderm.

One could summarize these observations by saying that the animal development is characterized by the enhancement of the epitheliocytic properties by the cells. In contrast, the cells in the vegetal regions acquire amebocytic and mechanocytic properties to a varying extent, and form free mesenchyme cells, the "mesothel" of the coelom, and the more epithelial structure of the endoderm. This pattern seems to be a rather general feature of developing embryos, as emphasized by Willmer (1960).

It has been considered that the changes in cellular behavior to a great extent reflect changes in the cell surfaces (Gustafson and Wolpert, 1963, 1968). If one tries to discuss these changes in terms of their relation to the metabolic gradients, one has to consider that the processes in the animal and vegetal region are similar to some extent, although their time pattern is different, and that qualitative differences are superimposed upon the quantitative ones. Our present knowledge, it must be admitted, is very fragmentary. However, it has been suggested that an increased adhesion between the cells is related to the intensity of their oxidative phosphorylation. This fits with the observation that uncoupling agents tend to decrease the cellular contacts. The decrease in the cellular contacts in the vegetal region at the onset of gastrulation may be due to a slow development of the energy-supplying mechanism, and to the onset of anabolic processes that consume ATP required for the maintenance of cell contacts. One may also consider the role of the acid polysaccharides of the cell surfaces (Section IV,K). The increased cell contacts may thus become stabilized by the appearance of certain polysaccharide-protein complexes, whereas the decrease in adhesion between the cells may be facilitated by their excretion of certain types of polysaccharides that tend to separate them (see Gustafson, 1965).

The relation between low adhesion and the pulsatory and pseudopodal activity may also reflect changes in the cell surface (cf. Immers, 1961). The ability of the pseudopodal tips to attach at the contact points between the ectoderm cells (Gustafson, 1963) could indicate that the high adhesiveness at these points can be utilized by other cells. It is possible that the ability of the mesenchyme to attach is due to the presence of mucopolysaccharides in certain zones of the ectoderm (see Motomura, 1960; K. Okazaki *et al.*, 1962; and the work of Runnström and collaborators, Section K).

The further development of the larva results in a shape of increasing complexity. It seems, however, that the morphogenetic activities of the cells that bring about these changes are of the same type as during gastrulation (cf. Gustafson and Wolpert, 1963, 1968).

REFERENCES

Afzelius, B. A. (1956). *Exptl. Cell Res.* **10**, 257.
Afzelius, B. A., and Murray, A. (1957). *Exptl. Cell Res.* **12**, 325.
Agrell, I. (1953). *Arkiv Zool.* [2] **6**, 13.
Agrell, I. (1956). *In* "Bertil Hanström Zoological Papers in Honor of His Sixty-Fifth Birthday" (K. G. Wingstrand, ed.), p. 27. Zooolgical Institute, University of Lund, Lund, Sweden.
Agrell, I. (1958). *Arkiv Zool.* [2] **11**, 435.
Agrell, I., and Persson, H. (1956). *Nature* **178**, 1398.
Aketa, K. (1957). *Embryologia (Nagoya)* **3**, 267.
Aketa, K. (1963). *Exptl. Cell Res.* **30**, 93.
Aketa, K. (1964). *Exptl. Cell Res.* **34**, 192.
Aketa, K., Bianchetti, R., Marré, E., and Monroy, A. (1964). *Biochim. Biophys. Acta* **86**, 211.
Allen, R. D. (1954). *Exptl. Cell Res.* **6**, 403.
Allen, R. D., and Hagström, B. (1955). *Exptl. Cell Res.* **9**, 157.
Apter, M. J. (1966). "Cybernetics and Development." Pergamon Press, Oxford.
Bäckström, S. (1953). *Arkiv Zool.* [2] **4**, 485.
Bäckström, S. (1956). *Exptl. Cell Res.* **11**, 322.
Bäckström, S. (1957). *Exptl. Cell Res.* **13**, 333.
Bäckström, S. (1959a). *Exptl. Cell Res.* **18**, 347.
Bäskström, S. (1959b). *Exptl. Cell Res.* **18**, 357.
Bäckström, S. (1959c). *Arkiv Zool.* [2] **12**, 339.
Bäckström, S. (1963a). *Exptl. Cell Res.* **32**, 566.
Bäckström, S. (1963b). *Acta Embryol. Morphol. Exptl.* **6**, 235.
Bäckström, S. (1965). *Acta Embryol. Morphol. Exptl.* **8**, 20.
Bäckström, S. (1966). *Acta Embryol. Morphol. Exptl.* **6**, 235.
Bäckström, S., and Gustafson, T. (1953). *Arkiv Zool.* [2] **4**, 485.
Bäckström, S., Hultin, K., and Hultin, T. (1960). *Exptl. Cell Res.* **19**, 634.
Baxandall, J. (1966). *J. Ultrastruct. Res.* **16**, 158.
Baxandall, J., Perlmann, P., and Afzelius, B. A. (1964a). *J. Cell Biol.* **23**, 609.
Baxandall, J., Perlmann, P., and Afzelius, B. A. (1964b). *J. Cell Biol.* **23**, 629.
Berg, W. E. (1965). *Exptl. Cell Res.* **40**, 469.

Berg, W. E., and Long, N. D. (1964). *Exptl. Cell Res.* **33**, 422.
Berg, W. E., Taylor, D. A., and Humphreys, W. J. (1962). *Develop. Biol.* **4**, 165.
Black, R. E. (1964). *Exptl. Cell Res.* **33**, 361.
Borei, H. (1948). *Biol. Bull.* **95**, 124.
Borei, H., and Lybing, S. (1949). *Nature* **163**, 451.
Boveri, T. (1901a). *Zool. Jahrb., Abt. Anat. Ontog. Tiere* **14**, 630.
Boveri, T. (1901b). *Verhandl. Physik.-Med. Ges. Wurzburg* [N.S.] **34**, 145.
Brachet, J. (1937). *Arch. Biol.* (*Liege*) **48**, 529.
Brachet, J. (1960). "The Biochemistry of Development." Pergamon Press, Oxford.
Brookbank, J. W. (1958). *Biol. Bull.* **115**, 74.
Brown, G. G. (1966). *J. Ultrastruct. Res.* **14**, 425.
Chambers, E. L. (1939). *J. Exptl. Biol.* **16**, 409.
Chambers, R. (1930). *Biol. Bull.* **58**, 344.
Chen, P. S., Baltzer, F., and Zeller, C. (1960). *In* "Symposium on Germ Cells and Development" (S. Ranzi, ed.), p. 506. Baselli, Rome.
Child, C. M. (1936). *Arch. Entwicklungsmech. Organ.* **135**, 426.
Child, C. M. (1941a). "Patterns and Problems of Development." Univ. of Chicago Press, Chicago, Illinois.
Child, C. M. (1941b). *Proc. Natl. Acad. Sci. U.S.* **27**, 523.
Child, C. M. (1944). *Physiol. Zool.* **17**, 129.
Colwin, A. L., and Colwin, L. H. (1964). *In* "Cellular Membranes in Development" (M. Locke, ed.), p. 233. Academic Press, New York.
Czihak, G. (1961). *Arch. Entwicklungsmech. Organ.* **153**, 353.
Czihak, G. (1962a). *Fortschr. Zool.* **14**, 238.
Czihak, G. (1962b). *Arch. Entwichlungsmech. Organ.* **154**, 29.
Czihak, G. (1963). *Arch. Entwicklungsmech. Organ.* **154**, 272.
Czihak, G. (1965). *Arch. Entwicklungsmech. Organ.* **156**, 504.
Czihak, G. (1966). *Arch. Entwicklungsmech. Organ.* **157**, 199.
Dan, J. C. (1956). *Intern. Rev. Cytol.* **5**, 365.
Dan, J. C. (1960). *Exptl. Cell Res.* **19**, 13.
Dan, K., and Ono, T. (1954). *Embryologia* (*Nagoya*) **2**, 87.
Dawydoff, C. (1948). *In* "Traité de Zoologie" (P.-P. Grassé, ed.), Vol. II, p. 277. Masson, Paris.
Delage, Y. (1901). *Arch. Zool. Exptl. Gen.* [3] **9**, 285.
Deutsch, H. F., and Gustafson, T. (1952). *Arkiv Kemi* **4**, 221.
De Vincentiis, M., and Runnström, J. (1967). *Exptl. Cell Res.* **45**, 681.
De Vincentiis, M., Hörstadius, S., and Runnström, J. (1966). *Exptl. Cell Res.* **41**, 535.
Elson, D., Gustafson, T., and Chargaff, E. (1954). *J. Biol. Chem.* **209**, 285.
Endo, Y. (1961a). *Exptl. Cell Res.* **25**, 383.
Endo, Y. (1961b). *Exptl. Cell Res.* **25**, 518.
Epel, D. (1964a). *Biochem. Biophys. Res. Commun.* **17**, 62.
Epel, D. (1964b). *Biochem. Biophys. Res. Commun.* **17**, 69.
Epel, D., and Iverson, R. M. (1965). *In* "Control of energy metabolism" (B. Chance, R. W. Eastbrook, and Y. R. Williams, eds.) p. 267. Academic Press, New York.
Evola-Maltese, C. (1957). *Acta Embryol. Morphol. Exptl.* **1**, 98.
Ficq, A. (1955). *Arch. Biol.* **66**, 509.
Ficq, A. (1961). *Exptl. Cell Res.* **23**, 427.
Flickinger, R. A. (1957). *Biol. Bull.* **112**, 21.
Fol, H. (1878–1879). *Mem. Soc. Phys. Geneve* **26**, 89.
Fujii, T., and Ohnishi, T. (1963). *J. Fac. Sci., Univ. Tokyo, Sect. IV* **9**, 333.

Ghiretti, F., and D'Amelio, V. (1956). *Exptl. Cell Res.* **10**, 734.
Ginsburg, A. S. (1961). *Dokl. Akad. Nauk SSSR* **152**, 501.
Grant, P. (1965). *In* "The Biochemistry of Animal Development" (R. Weber, ed.), Vol. 1, p. 483. Academic Press, New York.
Gross, P. R. (1954). *Biol. Bull.* **107**, 364.
Gross, P. R. (1956). *J. Cellular Comp. Physiol.* **47**, 429.
Gross, P. R., and Cousineau, G. H. (1963). *Exptl. Cell Res.* **33**, 368.
Gross, P. R., and Pearl, W. (1950). *J. Cellular Comp. Physiol.* **52**, 147.
Gross, P. R., Malkin, L. I., and Mayer, W. A. (1965). *In* "Molecular and Cellular Aspects of Development" (E. H. Bell, ed.), p. 316. Harper, New York.
Guidance, G., and Monroy, A. (1958). *Acta Embryol. Morphol. Exptl.* **2**, 58.
Gustafson, T. (1946). *Arkiv Zool.* **38A**, No. 4.
Gustafson, T. (1950). *Rev. Suisse Zool.* **57**, Suppl. 1, 77.
Gustafson, T. (1952). *Arkiv Zool.* [2] **3**, 273.
Gustafson, T. (1954). *Intern. Rev. Cytol.* **3**, 277.
Gustafson, T. (1961). *In* "Biological Structure and Function" (T. W. Goodwin and O. Lindberg, eds.), Vol. 2, Part 3, p. 497. Academic Press, New York.
Gustafson, T. (1963). *Exptl. Cell Res.* **32**, 570.
Gustafson, T. (1964). *In* "Primitive Motile Systems in Cell Biology" (R. D. Allen and N. Kamiya, eds.), p. 333. Academic Press, New York.
Gustafson, T. (1965). *In* "The Biochemistry of Animal Development" (R. Weber, ed.), Vol. 1, Part 2, p. 140. Academic Press, New York.
Gustafson, T., and Hasselberg, I. (1950). *Exptl. Cell Res.* **1**, 371.
Gustafson, T., and Hasselberg, I. (1951). *Exptl. Cell Res.* **2**, 642.
Gustafson, T., and Hörstadius, S. (1955). *Exptl. Cell Res.* Suppl. 3, 170.
Gustafson, T., and Hörstadius, S. (1956). *Zool. Anz.* **156**, 102.
Gustafson, T., and Hörstadius, S. (1957). *Pubbl. Staz. Zool. Napoli* **29**, 407.
Gustafson, T., and Lenicque, P. (1952). *Exptl. Cell Res.* **3**, 251.
Gustafson, T., and Lenicque, P. (1955). *Exptl. Cell Res.* **8**, 114.
Gustafson, T., and Sävhagen, R. (1950). *Arkiv Zool.* **42A**, No. 10.
Gustafson, T., and Wolpert, L. (1963). *Intern. Rev. Cytol.* **15**, 139.
Gustafson, T., and Wolpert, L. (1967). *Biol. Rev.* **42**, 447.
Hagström, B., and Allen, R. D. (1956). *Exptl. Cell Res.* **10**, 14.
Hagström, B. E. (1959). *Arkiv Zool.* [2] **12**, 127.
Hagström, B. E. (1963). *Biol. Bull.* **124**, 55.
Hagström, B. E., and Hagström, B. (1954a). *Exptl. Cell Res.* **6**, 479.
Hagström, B. E., and Hagström, B. (1954b). *Exptl. Cell Res.* **6**, 491.
Hagström, B. E., and Lönning, S. (1961). *Sarsia* **4**, 5.
Hagström, B. E., and Lönning, S. (1963). *Arkiv Zool.* [2] **15**, 377.
Hagström, B. E., and Lönning, S. (1964). *Sarsia* **15**, 17.
Harding, C. V., Harding, D., and Perlmann, P. (1954). *Exptl. Cell Res.* **6**, 202.
Harvey, E. B. (1956). "The American Arbacia and other Sea Urchins." Princeton Univ. Press, Princeton, New Jersey.
Harvey, L. A. (1931). *Proc. Roy. Soc.* **B107**, 417.
Hathaway, R. R., and Metz, C. B. (1961). *Biol. Bull.* **120**, 360.
Hayes, F. R. (1938). *Biol. Bull.* **74**, 267.
Heilbrunn, L. V. (1937). "An Outline of General Physiology." Saunders, Philadelphia, Pennsylvania.
Herbst, C. (1892). *Z. Wiss. Zool.* **55**, 446.
Herbst, C. (1904). *Arch. Entwicklungsmech. Organ.* **17**, 306.

Hiramoto, Y. (1962a). *Exptl. Cell Res.* **27**, 416.
Hiramoto, Y. (1962b). *Exptl. Cell Res.* **28**, 323.
Hoff-Jörgensen, E. (1954). *Proc. Symp. Colston Res. Soc.* **7**, 79.
Holter, H., and Lindahl, P. E. (1940). *Compt. Rend. Trav. Lab. Carlsberg, Ser. Chim.* **23**, 249.
Holter, H., and Zeuthen, E. (1957). *Pubbl. Staz. Zool. Napoli* **29**, 285.
Horowitz, B. A. (1965). *Exptl. Cell Res.* **38**, 620.
Hörstadius, S. (1936a). *Arch. Entwicklungsmech. Organ.* **135**, 1.
Hörstadius, S. (1936b). *Arch. Entwicklungsmech. Organ.* **135**, 40.
Hörstadius, S. (1939). *Biol. Rev. Cambridge Phil. Soc.* **14**, 132.
Hörstadius, S. (1952). *J. Exptl. Zool.* **120**, 421.
Hörstadius, S. (1953). *J. Embryol. Exptl. Morphol.* **1**, 327.
Hörstadius, S. (1955). *J. Exptl. Zool.* **129**, 249.
Hörstadius, S. (1959). *J. Exptl. Zool.* **142**, 141.
Hörstadius, S. (1963). *Develop. Biol.* **7**, 144.
Hörstadius, S., and Gustafson, T. (1947). *Zool. Bidr. Uppsala* **25**, 271.
Hörstadius, S., Lorch, I. J., and Danielli, J. F. (1950). *Exptl. Cell Res.* **1**, 188.
Hörstadius, S., Immers, J., and Runnström, J. (1966). *Exptl. Cell Res.* **43**, 444.
Hörstadius, S., Josefsson, L., and Runnström, J. (1967). *Develop. Biol.* **16**, 189.
Hsiao, S. C., and Fujii, W. K. (1963). *Exptl. Cell Res.* **32**, 217.
Hultin, T. (1948a). *Arkiv Zool.* **40A**, 12.
Hultin, T. (1948b). *Arkiv. Zool.* **40A**, 20.
Hultin, T. (1949). *Arkiv Kemi, Mineral. Geol.* **26A**, No. 27.
Hultin, T. (1950a). *Exptl. Cell Res.* **1**, 159.
Hultin, T. (1950b). *Exptl. Cell Res.* **1**, 277.
Hultin, T. (1950c). *Exptl. Cell Res.* **1**, 599.
Hultin, T. (1953a). *Arkiv Kemi* **5**, 543.
Hultin, T. (1953b). *Arkiv Kemi* **5**, 559.
Hultin, T. (1953c). *Arkiv Kemi* **6**, 195.
Hultin, T. (1953d). *Arch. Neerla. Zool.* **1**, Suppl., 76.
Hultin, T. (1953e). Doctoral Dissertation, Emil Kihlströms Tryckeri AB., Stockholm.
Hultin, T., and Wessel, G. (1952). *Exptl. Cell Res.* **3**, 613.
Hutchens, J. O., Keltch, A. K., Krahl, M. E., and Clowes, G. H. A. (1942). *J. Gen. Physiol.* **25**, 717.
Hyman, L. H. (1955). "The Invertebrates," Vol. 4. McGraw-Hill, New York.
Immers, J. (1956a). *Arkiv Zool.* [2] **9**, 367.
Immers, J. (1956b). *Exptl. Cell Res.* **10**, 546.
Immers, J. (1961). *Arkiv Zool.* [2] **13**, 299.
Immers, J., and Runnström, J. (1960). *Develop. Biol.* **2**, 90.
Immers, J., and Runnström, J. (1965). *J. Embryol. Exptl. Morphol.* **14**, 289.
Immers, J., Markman, B., and Runnström, J. (1967). *Exptl. Cell Res.* **47**, 425.
Isaka, S., and Aikawa, T. (1963a). *Exptl. Cell Res.* **30**, 139.
Isaka, S., and Aikawa, T. (1963b). *Exptl. Cell Res.* **30**, 150.
Isaka, S., Kanatani, H., and Suzuki, N. (1966). *Exptl. Cell Res.* **44**, 66.
Ishida, J., and Nakano, E. (1947). *Dobutsugaku Zasshi* **57**, 117.
Ishida, J., and Nakano, E. (1950). *Annotationes Zool. Japon.* **23**, 43.
Ishikawa, M. (1954). *Embryologia (Nagoya)* **2**, 57.
Isono, N. (1962). *J. Fac. Sci., Tokyo Univ., Sect. IV* **9**, 357 and 369.
Isono, N. (1963). *Annotationes Zool. Japon.* **36**, 126.
Just, E. E. (1919). *Biol. Bull.* **36**, 1.

Karnofsky, D. A., and Basch, R. S. (1960). *J. Biophys. Biochem. Cytol.* **7**, 61.
Kavanau, J. L. (1953). *J. Exptl. Zool.* **122**, 285.
Kavanau, J. L. (1954). *Exptl. Cell Res.* **7**, 530.
Kavanau, J. L. (1958). "Chemical Basis of Development." Johns Hopkins Press, Baltimore, Maryland.
Keltch, A. K., Krahl, M. E., and Clowes, G. H. A. (1956). *J. Gen. Physiol.* **40**, 27.
Kinoshita, S., and Yazaki, I. (1967). *Exptl. Cell Res.* **47**, 449.
Krahl, M. E. (1956). *Biochim. Biophys. Acta* **20**, 27.
Krane, S. M., and Crane, R. K. (1960). *Biochim. Biophys. Acta* **43**, 369.
Lallier, R. (1952). *Experientia* **8**, 271.
Lallier, R. (1955). *Arch. Biol.* (*Liege*) **66**, 75.
Lallier, R. (1958). *Experientia* **14**, 309.
Lallier, R. (1959). *Compt. Rend.* **248**, 1416.
Lallier, R. (1961). *Compt. Rend.* **253**, 3060.
Lallier, R. (1964). *Advan. Morphogenesis* **3**, 147.
Larsson, A., and Reichard, P. (1966). *J. Biol. Chem.* **241**, 2540.
Lenicque, P., Hörstadius, S., and Gustafson, T. (1953). *Exptl. Cell Res.* **5**, 400.
Lillie, F. R. (1913). *J. Exptl. Zool.* **14**, 515.
Lillie, F. R. (1914). *J. Exptl. Zool.* **16**, 523.
Lillie, F. R. (1919). "Problems of Fertilization." Univ. of Chicago Press, Chicago, Illinois.
Lillie, R. S. (1916). *Am. J. Physiol.* **40**, 249.
Lindahl, P. E. (1932). *Arch. Entwicklungsmech. Organ.* **140**, 168.
Lindahl, P. E. (1936). *Acta Zool.* (*Stockholm*) **17**, 179.
Lindahl, P. E. (1939a). *Z. Vergleich. Physiol.* **27**, 136.
Lindahl, P. E. (1939b). *Z. Vergleich. Physiol.* **27**, 233.
Lindahl, P. E. (1941). *Acta Zool.* (*Stockholm*) **22**, 101.
Lindahl, P. E. (1942). *Quart. Rev. Biol.* **17**, 213.
Lindahl, P. E., and Holter, H. (1941). *Compt. Rend. Trav. Lab. Carlsberg, Ser. Chim.* **24**, 49.
Lindahl, P. E., and Öhman, L. O. (1938). *Biol. Zentr.* **58**, 179.
Lindberg, O. (1943). *Arkiv Kemi, Mineral. Geol.* **16A**, No. 15.
Lindberg, O. (1945). *Arkiv Kemi, Mineral. Geol.* **20B**, No. 1.
Lindberg, O. (1950). *Exptl. Cell Res.* **1**, 105.
Lindberg, O., and Ernster, L. (1948). *Biochim. Biophys. Acta* **2**, 471.
Ling, G. N. (1962). "A Physical Theory of the Living State: The Association-induction Hypothesis." Ginn (Blaisdell), Boston, Massachusetts.
Litchfield, J. B., and Whiteley, A. H. (1959). *Biol. Bull.* **117**, 133.
Loeb, J. (1913). "Artificial Parthenogenesis and Fertilization." Univ. of Chicago Press, Chicago, Illinois.
Lövtrup, S. (1966). *Bull. Suisse Acad. Med. Sci.* **22**, 201.
Lundblad, G. (1954). *Arkiv Kemi* **7**, 169.
Maggio, R., and Monroy, A. (1959). *Nature* **184**, 68.
Maggio, R., Aiello, F., and Monroy, A. (1960). *Nature* **188**, 1195.
Maggio, R., Monroy, A., Rinaldi, A. M., and Vittorelli, M. L. (1965). *Compt. Rend.* **260**, 1293.
Markman, B. (1957). *Exptl. Cell Res.* **12**, 424.
Markman, B. (1960). *Exptl. Cell Res.* **23**, 197.

Markman, B. (1961a). *Exptl. Cell Res.* **23**, 118.
Markman, B. (1961b). *Exptl. Cell Res.* **25**, 224.
Markman, B. (1963). *Arkiv. Zool.* [2] **16**, 207.
Markman, B. (1967). *Exptl. Cell. Res.* **46**, 1.
Markman, B., and Runnström, J. (1963). *Exptl. Cell Res.* **31**, 615.
Marsland, D. A. (1956). *Intern. Rev. Cytol.* **5**, 199.
Fudge-Mastrangelo, M. (1966). *J. Exptl. Zool.* **161**, 109.
Mazia, D. (1937). *J. Cellular Comp. Physiol.* **10**, 291.
Mazia, D. (1961). *In* "The Cell" (J. Brachet and A. E. Mirsky, eds.), Vol. 3, p. 77. Academic Press, New York.
Mazia, D., Blumenthal, G., and Benson, E. (1948). *Biol. Bull.* **95**, 250.
Mehl, J. W., and Swann, M. M. (1961). *Exptl. Cell Res.* **22**, 233.
Messina, L. (1954). *Pubbl. Staz. Zool. Napoli* **25**, 454.
Metz, C. B. (1957). *In* "The Beginnings of Embryonic Development" (A. Tyler, R. C. von Borstel, and C. B. Metz, eds.), p. 23. Am. Assoc. Advance Sci., Washington, D.C.
Mirsky, A. E. (1936). *Science* **84**, 333.
Monné, L., and Hårde, S. (1951). *Arkiv Zool.* [2] **1**, 487.
Monné, L., and Slautterback, D. B. (1950). *Exptl. Cell Res.* **1**, 477.
Monroy, A. (1956). *Exptl. Cell Res.* **10**, 320.
Monroy, A. (1957). *Intern. Rev. Cytol.* **6**, 107.
Monroy, A. (1963). *Ann. Biol. Animale, Biochim., Biophys.* **2**, 289.
Monroy, A. (1965a). "Chemistry and Physiology of Fertilization." Holt, New York.
Monroy, A. (1965b). *In* "The Biochemistry of Animal Development" (R. Weber, ed.), Vol. 1, Part 1, p. 73. Academic Press, New York.
Monroy, A., and Maggio, R. (1964). *Advan. Morphogenesis* **3**, 95.
Monroy, A., and Ruffo, A. (1947). *Nature* **159**, 603.
Monroy, A., and Tolis, H. (1964). *Biol. Bull.* **126**, 456.
Monroy, A., and Tosi, L. (1952). *Experientia* **8**, 393.
Monroy, A., and Vittorelli, M. L. (1960). *Experientia* **16**, 56.
Monroy, A., Tosi, L., Giardina, G., and Maggio, R. (1954). *Biol. Bull.* **106**, 169.
Mori, S., Takashima, Y., Yano, K., Kojo, T., Hashimoto, K., and Fujitani N. (1958). *Bull. Exptl. Biol. (Japan)* **8**, 65.
Motomura, I. (1954). *Sci. Rept. Tohoku Univ., Fourth Ser.* **20**, 213.
Motomura, I. (1960). *Bull. Marine Biol. Sta. Asamushi, Tohoku Univ.* **10**, 165.
Nakano, E., and Monroy, A. (1958). *Exptl. Cell Res.* **14**, 236.
Nakano, E., Guidice, G., and Monroy, A. (1958). *Experientia* **14**, 11.
Nemer, M. (1962). *J. Biol. Chem.* **237**, 143.
Nemer, M. (1963). *Proc. Natl. Acad. Sci. U.S.* **50**, 230.
Nemer, M. (1965). *In* "Molecular and Cellular Aspects of Development" (E. Bell, ed.), p. 323. Harper, New York.
Öhman, L. O. (1940). *Arkiv Zool.* **32A**, No. 15.
Öhman, L. O. (1944). *Arkiv Zool.* **36A**, No. 7.
Ohnishi, T., and Sugiyama, M. (1963). *Embryologia (Nagoya)* **8**, 79.
Okazaki, K., Fukushi, T., and Dan, K. (1962). *Acta Embryol. Morphol. Exptl.* **5**, 17.
Okazaki, R. (1956). *Exptl. Cell Res.* **10**, 476.
Örström, Å. (1941). *Arkiv Kemi, Mineral. Geol.* **15A**, No. 1.
Örström, Å., and Lindberg, O. (1940). *Enzymologia* **8**, 367.
Osano, K. (1960). *Sci. Rept. Tohoku Univ., Fourth Ser.* **26**, 77.

Pasteels, J. J., Castiaux, P., and Vandermeerssche, G. (1959). *Arch. Biol. (Liege)* **69**, 627.

Perlmann, P. (1953). *Exptl. Cell Res.* **5**, 394.

Perlmann, P. (1959). *Experientia* **15**, 41.

Perlmann, P., and Gustafson, T. (1948). *Experientia* **4**, 481.

Perlmann, P., and Kaltenbach, J. C. (1957). *Exptl. Cell Res.* **12**, 185.

Popa, G. T. (1927). *Biol. Bull.* **52**, 238.

Ranzi, S. (1962). *Advan. Morphogenesis* **2**, 211.

Raven, C. P. (1958). "Morphogenesis, The Analysis of Molluscan Development." Pergamon Press, Oxford.

Raven, C. P. (1961). "Oogenesis." Macmillan, New York.

Rose, S. M. (1957). *Biol. Rev. Cambridge Phil. Soc.* **32**, 351.

Rothschild, Lord. (1956). "Fertilization." Methuen, London.

Runnström, J. (1928a). *Arch. Entwicklungsmech. Organ.* **113**, 556.

Runnström, J. (1928b). *Protoplasma* **10**, 106.

Runnström, J. (1928c). *Acta Zool. (Stockholm)* **9**, 365.

Runnström, J. (1933). *Biochem. Z.* **258**, 257.

Runnström, J. (1949). *Advan. Enzymol.* **9**, 241.

Runnström, J. (1956a). *Exptl. Cell Res.* **11**, 660.

Runnström, J. (1956b). *Pubbl. Staz. Zool. Napoli* **28**, 315.

Runnström, J. (1957). *Arkiv Zool.* [2] **10**, 523.

Runnström, J. (1961). *Pathol. Biol., Semaine Hop.* **9**, 781.

Runnström, J. (1962). *Exptl. Cell Res.* **27**, 485.

Runnström, J. (1966). *Advan. Morphogenesis* **5**, 222.

Runnström, J., and Hagström, B. E. (1955). *Exptl. Cell Res.* **8**, 1.

Runnström, J., and Immers, J. (1956). *Exptl. Cell Res.* **10**, 354.

Runnström, J., and Immers, J. (1966). *Arch. Biol. (Liege)* **77**, 365.

Runnström, J., and Kriszat, G. (1950). *Exptl. Cell Res.* **1**, 355.

Runnström, J., and Kriszat, G. (1952a). *Exptl. Cell Res.* **3**, 419.

Runnström, J., and Kriszat, G. (1952b). *Exptl. Cell Res.* **3**, 497.

Runnström, J., and Thörnblom, D. (1938). *Acta Biol. Latvica* **8**, 97.

Runnström, J., Hagström, B. E., and Perlmann, P. (1959). *In* "The Cell" (J. Brachet and A. E. Mirsky, eds.), Vol. 1, p. 327. Academic Press, New York.

Runnström, J., Hörstadius, S., Immers, J., and Fudge-Mastrangelo, M. (1964). *Rev. Suisse Zool.* **71**, 24.

Sakai, H. (1960a). *J. Biophys. Biochem. Cytol.* **8**, 603.

Sakai, H. (1960b). *J. Biophys. Biochem. Cytol.* **8**, 609.

Sakai, H. (1962a). *J. Gen. Physiol.* **45, 411.**

Sakai, H. (1962b). *J. Gen. Physiol.* **45**, 427.

Shaver, J. R. (1956). *Exptl. Cell Res.* **11**, 549.

Simmel, E. A., and Karnofsky, D. A. (1961). *J. Biophys. Biochem. Cytol.* **10**, 59.

Strecker, D. D., and Gustafson, T. (in preparation).

Sugiyama, M. (1951). *Biol. Bull.* **101**, 335.

Sugiyama, M. (1956). *Exptl. Cell Res.* **10**, 364.

Ten Cate, G. (1952). *In* "Proceedings of the Symposium on the Biochemical and Structural Basis of Morphogenesis," p. 108. Brill, Leiden.

Tennent, D. H., and Ito, T. (1941). *J. Morphol.* **69**, 347.

Tuft, P. (1953). *Arch. Neerl. Zool. Suppl.* **10**, 59.

Tyler, A. (1941). *Biol. Rev.* **16**, 291.

Tyler, A. (1948). *Physiol. Rev.* **28**, 180.
Tyler, A. (1956). *Exptl. Cell Res.* **7**, 183.
Tyler, A. (1960). *In* "Symposium on Germ Cells and Development" (S. Ranzi, ed.), p. 155. Baselli, Rome.
Tyler, A. (1963). *Am. Zoologist* **3**, 109.
Tyler, A. (1967). *Developmental Biol. Suppl 1*, 170.
Vasseur, E. (1951). *Exptl. Cell Res.* **2**, 144.
Vasseur, E. (1954). "The Chemistry and Physiology of the Jelly Coat of the Sea Urchin Egg." Kilhströmstryck AB., Stockholm.
Vincent, W. S. (1965). *Proc. 11th Intern. Congr. Genet., The Hague, 1963*. Pergamon Press, Oxford.
von Ubisch, L. (1929). *Arch. Entwicklungsmech. Organ* **117**, 80.
Warburg, O. (1910). *Z. Physiol. Chem.* **66**, 305.
Weiss, P. (1947). *Yale J. Biol. Med.* **19**, 235.
Whitaker, D. M. (1933). *J. Gen. Physiol.* **16**, 497.
Whiteley, A. H., and Baltzer, F. (1958). *Pubbl. Staz. Zool. Napoli* **30**, 402.
Whiteley, A. H., and Chambers, E. L. (1960). *In* "Symposium on Germ Cells and Development" (S. Ranzi, ed.), p. 387. Baselli, Rome.
Willmer, E. N. (1960). "Cytology and Evolution." Academic Press, New York.
Wilt, F. H. (1963). *Biochem. Biophys. Res. Commun.* **11**, 447.
Wolfson, N., and Fry, D. S. (1965). *Exptl. Cell Res.* **28**, 66.
Wolpert, L. (1960). *Intern. Rev. Cytol.* **10**, 164.
Wolpert, L., and Mercer, E. H. (1961). *Exptl. Cell Res.* **22**, 45.
Yasumasu, I., and Nakano, E. (1963). *Biol. Bull.* **125**, 182.
Ycas, M. (1954). *J. Exptl. Biol.* **31**, 208.

CHAPTER 8

Pharmacology of Echinoderms

Ragnar Fänge

I. Introduction

Echinoderms are believed to be related to chordates but they are more primitive and lack a fully developed central nervous system, kidney, and real blood circulatory system. This might be one of the reasons why adult echinoderms rarely are used in physiological research. Eggs and embryos, on the other hand, are much appreciated research objects, and an enormous amount of work has involved testing how various substances act on sea urchin eggs before and after fertilization.

This survey brings together information on the effect of drugs on echinoderms and the occurrence of biologically active substances within these animals.

Aspects of echinoderm physiology are treated by Hyman (1955), Hoyle (1957), Prosser and Brown (1961), Smith (1965), and Boolootian (1966). Pharmacology has been surveyed by Alender and Russell (in Boolootian, 1966).

II. Effects of Different Substances on Various Systems

A. General Properties of Muscles

Holothurian muscle fibers are mononucleate, measuring 3–5 μ in diameter and about 350 μ in length (Gay and Simon, 1964). In other echinoderms the structure of the fibers probably is much the same. Although they are nonstriated, the echinoderm muscles in many respects resemble vertebrate skeletal muscles (Riesser, 1933). Ribbonlike nerve endings serve for motor innervation (Hoyle, 1957; Smith, 1965; Pople and Ewer, 1954).

The uptake of oxygen by holothurian muscle is low. In *Stichopus* it is stimulated by glucose and potassium ions (Gay and Simon, 1964). In *Parastichopus* *p*-phenylenediamine (0.01 *M*) or ascorbate (0.02 *M*) increased the rate of oxygen consumption from 10 to 30 μg. to 150–250 μg. per gram wet weight per hour. Succinate stimulated respiration but glucose and cytochrome *c* had no such effect. Mitochondria, cytochrome *c*, and cytochrome oxidase could not be demonstrated. Cell fractionation in 0.25 *M* sucrose showed uptake of oxygen to be limited to the supernatant after centrifugation. Flavin content was low, but a chemically unidentified fluorescent substance was discovered (Mattisson, 1962). Probably the energy metabolism is dependent on glycolysis, as glycogen and glycolytic enzymes are present (Boyland, 1928) Gay and Simon (1964), however, assume metabolism to be based on fat. Ionic composition was investigated by Simon *et al.* (1964a,b; and this volume) in muscles of *Stichopus* (see the Chapter by Freeman and Freeman in this volume).

B. Body Wall and Masticatory Muscles

1. *Holothurian Body Wall Muscles*

In holothurians five muscle bands run along the whole length of the animals. Such bands prepared from *Holothuria nigra* were used by Hill (1926) in studies on viscous elastic properties of muscle. They are very powerful but slow. The speed of contraction is less than 1/100 of that of a frog's striated muscle. They are tetanized by 1–2 stimuli per second. Certain observations, which are difficult on the skeletal muscles of the frog, are possible with the slow holothurian body wall muscles.

a. Acetylcholine and Related Compounds. The sensitivity to acetylcholine is high. Uneserinized preparations from *Stichopus regalis* are contracted by 10^{-8}–10^{-7} gm./ml. (Bacq, 1939). Similar preparations from *Holothuria nigra* respond to acetylcholine in 5×10^{-9} gm./ml. (Welsh, 1954). Ambache and Sawaya (1953) and Sawaya and Ancona Lopez (1959) observed muscles from *Holothuria grisea* to respond to acetylcholine in 10^{-14}–10^{-8} gm./ml. Florey (1956) reports that muscles from *Actinopyga* are contracted by acetylcholine in concentrations down to 10^{-11} gm./ml.

In typical experiments Sawaya and Ancona Lopez (1959) placed 2–3 cm. long pieces of holothurian body wall muscle in a 10-ml. bath of seawater at room temperature. After repeated testing and change of saline, the preparations lengthened more and more. The authors also reported that during the dissection of the living animals the muscle bands shortened from an initial length of about 40 cm. to a length of 2 cm.

Due to their sensitivity to acetylcholine, holothurian muscles can be used for determination of acetylcholine in extracts in the same way as the frog's rectus abdominis and the dorsal muscle of the leech (*Hirudo medicinalis*). The holothurian muscles are insensitive to histamine (Bacq, 1939; Sawaya and Ancona Lopez, 1959), and epinephrine and tyramine (Bacq, 1939). Unfortunately these muscles are sensitive to touch and contract on handling (Simon *et al.*, 1964a).

b. Eserine. This drug increases the sensitivity to acetylcholine about 100 times (Welsh, 1954; *Holothuria nigra*). However, at the same time sensitivity to histamine, epinephrine, tyramine, and adenosine is produced (Bacq, 1935). The eserinized muscle seems to be very sensitive to any change in the composition of the medium. Even the response to electrical stimulation is increased (Bacq, 1939). In contrast to the uneserinized muscle, which relaxes immediately after each assay, the eserine-treated muscle remains contracted for long periods. Because the sensitivity of the holothurian body wall muscle is adequate without eserine, the use of eserine in order to increase the sensitivity to acetylcholine is unnecessary.

c. Further Contracting Substances. Increase of tonus is observed after caffein and strophanthin. Veratrine increases the mechanical response to electrical stimulation and causes double peak contractions (Riesser, 1933). Iriye and Dille (1940) reported that epinephrine (10^{-5}), nicotine (5×10^{-7}), veratrine (10^{-3}), and $BaCl_2$ (2.5×10^{-3}) increase tonus.

d. Cholinolytic Substances. Atropine counteracts acetylcholine and eserine (Iriye and Dille, 1940; Riesser, 1933). A blocking effect by D-tubocurarine has also been observed (Ambache and Sawaya, 1953; compare also Magazanik, 1965).

2. *Holothurian Pharyngeal Muscles*

These muscles insert upon calcareous structures which possibly resemble the lantern ossicles of sea urchins. The contraction by acetylcholine is blocked by curare, but the latter does not affect contraction produced by epinephrine or propagation of impulses in the muscle (Du Buy, 1936a,b). Prosser (1954) showed that action potentials were conducted short distances by nerve action at a rate of 17 cm./sec., but there was no propagation from one end of the muscle to the other. Pople and

Ewer (1954) stimulated a preparation from the sea cucumber *Cucumaria sykion* indirectly via the radial nerve. They observed two kinds of contractions after an unusually long refractory period of about 100 msec.

Magazanik (1965), examining the sensitivity to cholinomimetic and cholinolytic drugs, plotted dose-response curves for the pharyngeal protractor muscle of the sea cucumber, *Cucumaria frondosa.* Nicotinelike and muscarinelike agents caused contraction. Sensitivity to decamethonium and succinylcholine was low. The muscle was more blocked by the mononitrogen compound "mesphenal" than by D-tubocurarine.

3. *Echinoid Lantern Retractor Muscle*

The lantern of Aristotle of regular echinoids is a complex of muscles and calcareous structures used to chew up the food. The lantern retractor muscle of the sea urchins *Echinus esculentus* (Bacq, 1937) and *Parechinus angulosus* (Boltt and Ewer, 1963) responds to acetylcholine and eserine as do holothurian muscles. Epinephrine (10^{-5} M) is inhibitory whereas eserine, nicotine, and curare produce rhythmic activity (Boltt and Ewer, 1963).

C. Coelomic Systems

The physiology of the fluid-filled systems derived from the coelom is poorly known. Probably they function in transport of nutrients, oxygen, and metabolic products, and the water-vascular or ambulacral system is used in locomotion.

1. *Pulsatory Structures*

Coelomic, hemal, and water vascular systems of echinoids communicate within the axial gland. This organ has been ascribed lymph glandlike, endocrine, and heartlike functions. It contains a contractile structure assumed to pump fluid from the coelom into the hemal system. This and other pulsatory vessels of echinoderms contract only a few times per minute and maintain a slow circulation at low pressure within primitive hemal systems (Hyman, 1955). A rhythmically contracting vessel of the holothurian *Stichopus* was studied by Wyman and Lutz (1930) and Prosser and Judson (1952). Epinephrine activated but atropine slowed the rhythmicity. Acetylcholine, tetramine, nicotine, and eserine inhibited. The spontaneous activity of the *Stichopus* vessel probably is myogenic. A propagated action potential appears as a wave of negativity (Prosser and Judson, 1952).

2. *The Ambulacral System*

The tube feet are hydraulically moved by the contractile ampullae. The latter are innervated from the hyponeural tissue. The neuromuscular junctions, as in *Amphioxus,* contain no nerve endings but consist of thin muscular processes approaching the nerve elements (Flood, 1966). Extension of the tube feet is brought about by contraction of the ampullae and simultaneous relaxation of the longitudinal muscles of the feet. Longitudinal tube feet muscles from *Echinus esculentus* are relaxed by nitrite (Fänge, 1948).

3. *Permeability Problems*

The fact that the tube feet of starfishes (*Asterias*) have longitudinal but not circular muscles was utilized by Binyon (1964) in studying the passage of water. Ligated tube feet keep a constant diameter but change length if their fluid volume changes. Shortening indicates that water disappears by ultrafiltration through the walls. Binyon suggests that disappearing water normally is compensated by influx of water due to osmotic hypertonicity upheld by a "simple potassium pump." The ambulacral fluid differs from seawater by a higher concentration of potassium. As a consequence of a continuous loss of water across the tube feet, the ambulacral system may perform excretory functions eliminating soluble waste products with the water. The organic anions fluorescein and chlorphenol red, when introduced into the coelomic cavity of echinoderms, are concentrated by the ambulacral system (Fänge and Krog, 1961).

Henri (1906) and Henri and Lalou (1904) investigated the permeability of the respiratory trees of the holothurian, *Stichopus,* to dyes and other solutes. In spite of conditions favorable for ultrafiltration, no passage of water was observed and the structure proved impermeable to chloride, sugar, sulfate, and urea. By observations of the reoxidation of injected methylene blue, Bertolini (1933) showed diffusion of oxygen from the respiratory trees into the coelomic fluid.

D. VARIOUS ORGANS

1. *Digestive Tract*

The intestine of echinoderms has longitudinal and circular muscles and is capable of peristalsis.

Acetylcholine (10^{-6} gm./ml.) contracts the intestine of holothurians (Beauvallet, 1938) and sea urchins (Prosser *et al.,* 1965). The effect is enhanced by eserine. Acetylcholine likewise activates muscles of starfish digestive glands (Ferguson, 1966). Florey and McLennan (1959) found cholinergic transmission in the esophagus of the sea urchin, *Strongylocentrotus droebachiensis,* to be inhibited by γ-aminobutyric acid.

Contrary to the holothurian pharyngeal retractor muscles, which propagate impulses by nerves relatively rapidly, the intestinal muscles of the sea urchin, *Arbacia,* and the holothurian, *Thyone,* conduct action potentials at the slow rate of 4–6 cm./sec. Conduction is not blocked by D-tubocurarine, procaine, tricaine, or tetracaine. It seems to take place from muscle fiber to muscle fiber (Prosser *et al.,* 1965).

2. Cloaca and Respiratory Trees

The respiratory trees and the cloaca of holothurians make slow rhythmic contractions. The rhythmicity of the cloacal muscles has interested several researchers (see Hyman, 1955, for references). Eserine, nicotine, and histamine increase tonus in the cloaca of *Cucumaria* whereas curare, strychnine, epinephrine, and atropine inhibit (Wyman and Lutz, 1930).

3. Gonadal Walls

Sexual products are ejected by the contraction of muscles forming a sheet around the gonads. In echinoids artificial spawning can be induced by injection into the body cavity of acetylcholine or potassium chloride (Iwata and Fukase, 1964). In starfishes the same effect is produced by extracts of the radial nerve (Unger, 1962; Kanatani, 1964; Chaet, 1966). The extracts also affect maturing of the gonads. The active substance may be a peptide.

III. Pharmacological Effects on Larvae

In embryos of the sea urchin, *Paracentrotus lividus,* a marked increase of cholinesterase activity appears at a stage when the larvae begin to swim by coordinated ciliary activity (Augustinsson and Gustafson, 1949). The enzyme can be demonstrated already at 7 hours after fertilization and the activity reaches a high level at 50 hours. The findings indicate the presence of cholinergic mechanisms in locomotory and coordinatory mehanisms of the larvae. Chudakova and Buznikov (1965) found unfertilized eggs and embryos of the sea urchin, *Strongylocentrotus droebachiensis,* to contain acetylcholine, epinephrine, norepinephrine, and 5-hydroxytryptamine. During cleavage the acetylcholine content changes rhythmically. Cleavage was blocked by nicotine or thalidomide possibly by affecting endogenous cholinergic systems. Buznikov *et al.* (1964) present the hypothesis that 5-hydroxytryptamine acts as a hormonal controller of embryogenesis in Protostomia and Deuterostomia.

Effects of drugs on the digestive tract of sea urchin larvae have recently been studied by Gustafson (1967). In order to allow microscopic observations, he arrests the otherwise rapidly swimming larvae by calcium carbonate crystals precipitated in the meshes of a piece of plankton net. Two kinds of responses are observed: rapid esophageal and slow intestinal. The responses are followed by time-lapse microphotography and by direct microscopy in combination with digital counting. The larval gut is rhythmically active in a characteristic pumping movement. Probably the pumping is necessary for the normal development of the intestinal apparatus. It may also assist in respiration by moving the body fluids, and it may keep up fluid pressure in the primary body cavity by ultrafiltration across the intestinal wall. The pumping activity is very sensitive to external influences. Carbon dioxide stimulates and oxygen depresses the activity. Gustafson has observed marked effects on the gut contractions not only from cholinomimetic drugs but also from tryptamine, 5-hydroxytryptamine, histamine, catecholamines, and reserpine. The results indicate that the sea urchin's larval intestine somewhat unspecifically reacts to many different substances. Some substances, however, are distinctly much more active than others.

IV. Pharmacological Effects on Whole Animals

Greenwood (1890), in a comparative study, showed that the action of nicotine depends on the degree of development of the nervous system. *Asterias* and *Antedon* are paralyzed by 0.05% concentration. In ophiurids

FIG. 1. Schematic drawing showing responses of starfishes to nicotine and strychnine as observed by Moore (1920). (According to own experience on *Asterias rubens* concentrations of 10^{-4} gm./ml. of these substances are efficient).

and crinoids an excitatory phase preceded the paralysis. Thus in echinoderms as in vertebrates, nicotine acts by first stimulating and then depressing synapses. Moore (1920) observed that nicotine causes ventral flexure and strychnine dorsal flexure of the arms of the starfish, *Asterias* (Fig. 1). Strychnine also increases the sensitivity to touch in holothurians (Henri, 1903a). Florey (1951), in a comparative study, concluded that the responses to strychnine, picrotoxin, and pentylenetetrazol vary in parallel to the systematic position of the animals. Deuterostomes react by convulsions while protostomes are more or less insensitive. Strychnine and similar drugs are thought to increase neuronal excitability by selectively blocking inhibition. Eggart and Umrath (1956) assumed that the strychnine effect is due to inhibition of an enzyme that inactivates sensory transmitters. This hypothesis was criticized by Florey (1962).

A tranquilizing effect was observed on *Holothuria tubulosa* after treatment with chlorpromazine and perphenazine (Katona and Wollemann, 1964).

V. Occurrence of Biologically Active Substances

A. Transmitters and Hormones, etc.

Small amounts of acetylcholine or a similar substance were demonstrated in perfusates of electrically stimulated holothurian muscle (Bacq, 1939), and in various echinoderm organs (Unger, 1962; Mendes *et al.*, 1963). Several investigators have observed cholinesterase activity in echinoderms (Bacq, 1937; Augustinsson and Gustafson, 1949; Iwata and Fukase, 1964), and the muscles of echinoderms respond to small doses of acetylcholine. These observations all point toward the function of acetylcholine as a motor transmitter substance in echinoderms.

Echinoderms, in addition to acetylcholine, contain several other physiologically active substances. Östlund (1954) showed the presence of "catechol-4," a compound possibly related to catecholamines. Traces of 5-hydroxytryptamine have been identified (Welsh and Moorhead, 1960; Buznikov *et al.*, 1964). Rather large amounts of histamine occur in the intestine of the sea urchin *Echinaster*, and in *Holothuria* sp. However, the mode of distribution of histamine does not indicate that it has any specific function in echinoderms (Mettrich and Telford, 1963). Various physiologically active substances occur, in extracts of radial nerve of starfish (Unger, 1962; Chaet, 1966). Recently Cottrell (1967) detected dopamine and norepinephrine in the nervous system of *Asterias rubens* and *Echinus esculentus.*

Cells resembling mast cells are present in the holothurian *Parastichopus* (Rollefsen, 1965), but it is not known whether these cells store amines. An enormous increase of the number of coelomic leukocytes is produced in holothurians by injection of histamine (Cornil *et al.*, 1935).

Factors acting on sperms and eggs of echinoderms have been described and some of these have been given the role of hormone acting at the fertilization. However, classical fertilizin theory has been criticized (Hagström, 1956). Estrogen and estradiol, active on mammals, are found in small quantities in starfishes, and insulin has been detected (Barrington, 1964; Boticelli *et al.*, 1961; Wilson and Falkmer, 1965). Any biological functions of these vertebrate hormones in echinoderms have not yet been found.

B. Toxic Factors

1. *Sea Urchins*

The pedicellaria are three-jawed movable pincers by which sea urchins free themselves from settling larvae. Von Uexkull (1899) characterized these structures as "Giftzangen" and reported that a frog's heart is stopped by a single pedicellarian from *Sphaerechinus*. In this sea urchin the pedicellariae measure 10 mm. or more in length, but in other forms they are much smaller.

Henri and Kayalof (1906) injected macerated pedicellariae from *Sphaerechinus* into various animals. Forty pedicellariae killed a rabbit within 2–3 minutes. Crabs, fishes, and lizards were also killed, but echinoderms and frogs were insensitive. Pérès (1949), however, found that frogs are paralyzed by large doses. Toxic effects on man and mice were observed by Fujiwara (1935), if pedicellariae from *Toxopneustes* were placed on the skin.

The pedicellarian toxins, like certain other animal venoms, probably cause hemolysis by catalyzing the formation of lysolecithins. Thus Lévy (1925) found that extracts from the pedicellariae of *Paracentrotus lividus, Psammechinus miliaris,* and *Echinus esculentus* dissolve erythrocytes only in the presence of lecithin. Extracts from pedicellariae of *Tripneustes gratilla* are hemolytic and depress mammalian blood pressure by a direct effect on the heart. The active substance appears to be nondialyzable, resistant to boiling and low pH (C. Alender, cited by Russel, 1965). The toxins are usually thought to be produced by glands of a special type of pedicellariae called "glandular" or "gemmiform." But active extracts have been made also from other types of pedicellariae and from soft tissues of sea urchins (Pérès, 1949). An acetylcholine-like factor occurs in the pedicellariae of *Lytechinus variegatus* (Mendes *et al.,* 1963). Certain species of sea urchins are reported to be dangerous to man (Hyman, 1955; Russel, 1965). In some cases harmful effects are due to the long and sharp spines (Rocha and Fraga, 1962).

2. *Starfishes*

There are several reports of toxicity of starfishes. Extracts of *Asterias* or *Marthasterias* cause almost irreversible contracture of isolated muscle preparations (Fänge, 1963; Feder and Arvidsson 1967). It is possible but not proved that the toxins also produce escape responses in certain animals. For example, the bivalve *Pecten* starts frenetic swimming movements when touched by starfishes. Analogous responses are known from gastropods and coelenterates. Probably many different compounds released from starfishes may be active in evoking escape responses. Feder and co-workers (Feder and Lasker, 1964; Feder and Arvidsson, 1967), by use of Sephadex columns and other methods, partly purified from starfishes (*Pisaster, Marthasterias*) factors acting on mollusks. The substances adsorb to proteins and are thermostable. Their chemical nature was not clarified. Rio *et al.* (1965) and Ruggieri (1965) found extracts from starfishes to immobilize sperms and cytolyze or animalize eggs of sea urchins. The toxins produce a marked eosinophilia in fishes and are hemolytic. Fish heart is arrested in systole. The toxins appear to be saponins chemically related to holothurin.

3. *Sea Cucumbers*

Holothurin is a heat-stable, surface-active steroid saponin contained in the Cuvierian organs and other tissues of the Bahamian sea cucumber *Actinopyga agassizi* (Alender and Russell, in Boolootian, 1966). The same or similar substances are found in many other species of holothurians (Arvy, 1954). The molecular weight of holothurin is about 1150. It is highly toxic to many organisms and has some antitumorous effects (Dennis *et al.*, 1963; Ruggieri and Nigrelli, 1960; Nigrelli and Jakowska, 1960). It has strong hemolytic properties. The properties of holothurin are far better known than those of the toxins found in starfishes and sea urchins. Probably toxins of saponin type are widely distributed in echinoderms. The biological functions of the toxins is not clear, but they may give a passive protection against enemies.

4. *Brittle Stars*

The ophiuroid *Ophiocoma nigra* secretes a mucopolysaccharide with a very low pH. The acid mucus probably is highly offensive to fishes and other marine animals. Like heparin, it has some anticoagulatory properties on vertebrate blood (Fontaine, 1964).

C. Miscellaneous

Ackerman and co-workers identified a large number of low molecular nitrogenous compounds from echinoderms: eledonin, betaininogen, glycine betaine, asterubin, taurine (Ackermann, 1962; Ackermann *et al.*, 1964). Asterubin, a sulfur-containing guanidine derivative, in a dose of 20 mg./kg. increases blood pressure in dogs.

Kutscher and Ackermann many years ago found that vertebrates are rich in creatine, whereas invertebrates contain arginine (Ackermann,

1962). This observation has stimulated discussion on phylogenesis of vertebrates, as later authors discovered that creatine phosphate and arginine phosphate both occur in echinoderms. It is now known that creatine phosphate has a relatively wide distribution in invertebrates. However, as it occurs also in animals lacking the enzyme for its formation, it may be exogenous in some invertebrates (Stephens *et al.*, 1965). All echinoderms have arginine phosphate while echinoids, holothuroids, and ophiuroids also possess creatine phosphate (van Thoai and Roche, 1960).

Homarine, a quaternary ammonium base, occurs in starfishes, sea urchins, sea cucumbers, and in many other marine invertebrates. As it is absent in freshwater animals (Gasteiger *et al.*, 1960), it may, like taurin and certain amino acids, participate in intracellular osmoregulation (Florkin, 1962; Jeuniaux *et al.*, 1962a,b; Lange, 1964). Hultin *et al.* (1954) found homarine, *N*-methylpicolinic acid, in sea urchin embryos.

VI. Summary

Echinoderm muscles probably have a cholinergic innervation. Both nicotinelike and muscarinelike agents produce contraction. The muscles usually have low sensitivity to cathecholamines, 5-hydroxytryptamine, and histamine.

Free-swimming larvae react to many substances, but the specificity of pharmacological receptors is low.

Although echinoderms lack a distinct central nervous system, their responses to strychnine, nicotine, and tranquilizers indicate a high degree of complication of the nervous organization.

Acetylcholine and physiologically active peptides occur in extracts of nervous tissue.

Echinoderms produce toxins probably of saponin nature.

The coelomic systems and the respiratory trees of holothurians are of interest in studies of permeability and transport through biological membranes.

ACKNOWLEDGMENTS

I thank Dr. T. Gustafson, Stockholm, for personal information on unpublished results. This work was supported by the Swedish National Research Council.

REFERENCES

Ackermann, D. (1962). *Ber. Physik. Med. Ges. Wurzburg* **70**, 1.
Ackermann, D. *et al.* (1964). *Z. Physiol. Chem.* **336**, 1.
Ambache, N., and Sawaya, P. (1953). *Physiol. Comp.* **3**, 53.
Arvy, L. (1954). *Compt. Rend.* **239**, 1432.
Augustinsson, K.-B., and Gustafson, T. (1949). *J. Cellular Comp. Physiol.* **34**, 311.
Bacq, Z. M. (1935). *Arch. Intern. Physiol.* **42**, 24.

Bacq, Z. M. (1937). *Arch. Intern. Physiol.* **44,** 174.

Bacq, Z. M. (1939). *Arch. Intern. Physiol.* **49,** 25.

Barrington, E. J. W. (1964). "Hormones and Evolution," p. 58. English Univ. Press, London.

Beauvallet, M. (1938). *Compt. Rend.* **127,** 213.

Bertolini, F. (1933). *Pubbl. Staz. Zool. Napoli* **13,** 1.

Binyon, J. (1964). *J. Marine Biol. Assoc. U.K.* **44,** 577.

Boltt, R. E., and Ewer, D. W. (1963). *J. Exptl. Biol.* **40,** 727.

Boolootian, R. A., ed. (1966). "The Physiology of the Echinodermata." Wiley (Interscience), New York.

Boticelli, C. P. *et al.* (1961). *Proc. Soc. Exptl. Biol. Med.* **106,** 887.

Boyland, E. (1928). *Biochem. J.* **22,** 362.

Buznikov, G. A. *et al.* (1964). *J. Embryol. Exptl. Morphol.* **12,** 563.

Chaet, A. B. (1966). *Biol. Bull.* **130,** 43.

Chudakova, I. V., and Buznikov, G. A. (1965). *Zh. Evolyutsionnoi Biokhim. i Fiziol.* **1,** 557.

Cornil, L. *et al.* (1935). *Compt. Rend. Soc. Biol.* **120,** 162.

Cottrell, G. A. (1967). *Brit. J. Pharmacol.* **29,** 63.

Dennis, T. C. *et al.* (1963). *Toxicol. Appl. Pharmacol.* **5,** 1.

Du Buy, H. G. (1936a). *Biol. Bull.* **71,** 408.

Du Buy, H. G. (1936b). *Am. J. Physiol.* **116,** 22.

Egghart, E., and Umrath, K. (1956). *Z. Vergleich. Physiol.* **39,** 133.

Fänge, R. (1963). *Sarsia* **10,** 19.

Fänge, R. (1918). Unpublished observation.

Fänge, R., and Krog, J. (1961). Unpublished observation.

Feder, H. M., and Arvidsson, J. (1967). *Arkiv. Zool.* **19,** 369–379.

Feder, H. M., and Lasker, R. (1964). *Life Sci.* **3,** 1047.

Ferguson, J. C. (1966). *Comp. Biochem. Physiol.* **19,** 259.

Flood, P. R. (1966). *J. Comp. Neurol.* **126,** 181.

Florey, E. (1951). *Z. Vergleich. Physiol.* **33,** 237.

Florey, E. (1956). *Can. J. Biochem. Physiol.* **34,** 669.

Florey, E. (1962). *In* "Neurochemistry" (Elliot, Page, and Quastel, eds.), p. 673. Thomas, Springfield, Illinois.

Florey, E., and McLennan, H. (1959). *J. Physiol.* (*London*) **145,** 66.

Florkin, M. (1962). *Bull. Acad. Belg. Classe Sci.* **48,** 687.

Fontaine, A. R. (1964). *J. Marine Biol. Assoc. U.K.* **44,** 145.

Fujiwara, T. (1935). *Annotationes Zool. Japon.* **15,** 62.

Gasteiger, E. L. *et al.* (1960). *Ann. N.Y. Acad. Sci.* **90,** 622.

Gay, W. S., and Simon, S. E. (1964). *Comp. Biochem. Physiol.* **11,** 183.

Greenwood, M. (1890). *J. Physiol.* (*London*) **2,** 573.

Gustafson, T. (1967). Personal communication.

Hagström, B. (1956). Doctoral Thesis, Stockholm.

Henri, V. (1903a). *Compt. Rend.* **137,** 669.

Henri, V. (1903b). *Compt. Rend. Soc. Biol.* **55,** 1314.

Henri, V., and Kayalof, M. (1906). *Compt. Rend. Soc. Biol.* **60,** 884.

Henri, V., and Lalou, S. (1904). *J. Physiol. Paris* **6,** 9.

Hill, A. V. (1926). *Proc. Roy. Soc.* **B100,** 108.

Hoyle, G. (1957). "Comparative Physiology of the Nervous Control of Muscular Contraction." Cambridge Univ. Press, London and New York.

Hultin, T. *et al.* (1954). *Arkiv Kemi* **6,** 477.

Hyman, L. H. (1955). "The Invertebrates," Vol. 4, McGraw-Hill, New York.
Iriye, T. T., and Dille, J. M. (1940). *Pharm. Arch.* **10,** 93.
Iwata, K. S., and Fukase, H. (1964). *Biol. J. Okayama Univ.* **10,** 51.
Jeuniaux, C., Bricteux-Gregoire, S., and Florkin, M. (1962a). *Cahiers Biol. Marine* **3,** 107.
Jeuniaux, C., Bricteux-Gregoire, S., and Florkin, M. (1962b). *Arch. Intern. Physiol. Biochim.* **70,** 155.
Kanatani, H. (1964). *Science* **156,** 1177.
Katona, F., and Wollemann, M. (1964). *In* "Comparative Neurochemistry" (D. Richter, ed.), p. 445. Pergamon Press, Oxford.
Lange, R. (1964). *Comp. Biochem. Physiol.* **13,** 205.
Lévy, R. (1925). *Compt. Rend.* **181,** 690.
Magazanik, L. G. (1965). *Zh. Evolyutsionnoi Biokhim. i Fiziol.* **1,** 220; references in *Comp. Biochem. Physiol.* **17** (1966).
Mattisson, A. G. (1962). Doctoral Thesis, Lund, Sweden.
Mendes, E. H. *et al.* (1963). *Science* **139,** 408.
Mettrick, D. F., and Telford, J. M. (1963). *J. Pharm. Pharmaco.* **15,** 694.
Moore, A. R. (1920). *J. Gen. Physiol.* **2,** 201.
Nigrelli, R. F., and Jakowska, S. (1960). *Ann. N.Y. Acad. Sci.* **90,** 884.
Östlund, E. (1954). *Acta Physiol. Scand.* **31,** Suppl. 112, 1.
Pérès, J. M. (1949). *Arch. Zool. Exptl. Gen.* **86,** 118.
Pople, W., and Ewer, D. W. (1954). *J. Exptl. Biol.* **31,** 114.
Prosser, C. L. (1954). *J. Cellular Comp. Physiol.* **44,** 247.
Prosser, C. L., and Brown, F. A., Jr. (1961). "Comparative Animal Physiology." Saunders, Philadelphia, Pennsylvania.
Prosser, C. L., and Judson, C. L. (1952). *Biol. Bull.* **102,** 249.
Prosser, C. L. *et al.* (1965). *Comp. Biochem. Physiol.* **14,** 53.
Riesser, O. (1933). *Arch. Exptl. Pathol. Pharmakol.* **172,** 194.
Rio, G. J. *et al.* (1965). *Toxicon* **3,** 147.
Rocha, G., and Fraga, S. (1962). *Arch. Dermatol.* **85,** 406.
Rollefsen, I. (1965). *Arbok Univ. Bergen, Mat.-Nat. Ser.* **8,** 1.
Ruggieri, G. D. (1965). *Toxicon* **3,** 157.
Ruggieri, G. D., and Nigrelli, R. F. (1960). *Zoologica* **45,** 1.
Russel, F. E. (1965). *Advan. Marine Biol.* **3,** 255.
Sawaya, P., and Ancona Lopez, A. A. (1959). *Univ. Sao Paulo, Fac. Filosof., Cienc. Letras, Bol.* **232;** *Zoologica* **22,** 75.
Simon, S. E. *et al.* (1964a). *J. Cellular Comp. Physiol.* **63,** 77.
Simon, S. E. *et al.* (1964b). *J. Cellular Comp. Physiol.* **63,** 89.
Smith, J. E. (1965). *In* "Structure and Function of the Nervous System of Invertebrates" (T. H. Bullock, and G. A. Horridge, eds.), Vol. 2, p. 1519. Freeman, San Francisco, California.
Stephens, G. C. *et al.* (1965). *Biol. Bull.* **129,** 573.
Unger, H. (1962). *Zool. Jahrb., Physiol. Abt.* **69,** 481.
van Thoai, N., and Roche, J. (1960). *Ann. N.Y. Acad. Sci.* **90,** 923.
von Uexkull, J. (1899). *Z. Biol.* **19,** 334.
Welsh, J. H. (1954). *Nature* **173,** 955.
Welsh, J. H., and Moorhead, M. (1960). *J. Neurochem.* **6,** 146.
Wilson, S., and Falminer, S. (1965). *Can. J. Biochem.* **43,** 1615.
Wyman, L., and Lutz, B. R. (1930). *J. Exptl. Zool.* **57,** 441.

SECTION II

NEMATODA
ACANTHOCEPHALA

CHAPTER 1

The Systematics and Biology of Some Parasitic Nematodes

M. B. Chitwood

The problems of parasitism are far more complex than is generally realized or than can be covered here. This chapter is designed to supply a usable background of the relationships and biology of some parasitic nematodes with emphasis on parasites of laboratory animals. It is not to be considered a definitive treatise, rather it is a very selective coverage designed to correlate relationships based on (1) general morphology (2) histology, (3) life cycles, and (4) site of the adult in the definitive host. Species names are used but rarely. The basic classification on which this skeleton revision is constructed is that of Chitwood (1937). Authors of generic names are given only in the tables but complete citations are given for species names the first time they are used.

All known groups of animals are subject to parasitism and relationships of hosts may often be demonstrated by relationships of their parasites. This fact should be considered in selecting hosts and parasites for *in vivo* studies. Some nematodes may be host-specific and others have little host specificity and some genera may have species in a broad range of hosts. For this reason, classifications based on host relationships are purely artificial. Most adult nematodes tend to be extremely organotropic in the normal host.

Strongylida and Ascarida, parasites of the lumen of the gastrointestinal tract, comprise most of the species used in research for several reasons. These nematodes account for a large share of economic losses in livestock; many species develop readily in laboratory animals; their direct life cycles create few problems; and they are readily obtained. Many nematodes are parasitic in tissue or in the circulatory system, and these parasites may complicate and confuse *in vivo* experimental results. Some confusing data on immune reactions may be due to the presence of unsuspected parasites. The fact that many gastrointestinal parasitic nematodes have a tissue phase, even in the normal host, may also obscure results. A single specimen of rhesus monkey may harbor six or more genera of nematodes in the gastrointestinal tract, another in the lungs, one or more filarial parsites in the tissues, one or more species of cestodes, two or three of trematodes, and some acanthocephalans in the intestine, another species of trematode in the bile duct, still another in the liver, and a few pentastomes in the mesenteries.

Primates normally have many parasites transmissable to man, and some are quite closely related to swine parasites. New World monkeys have some parasites that are most closely related to those of rodents, and murine parasites, *Hymenolepis nana* and *Mastophorus muris* (Gmelin, 1790) Seurat, 1915, can be transmitted to monkeys.

Most of the normal parasites of rats and mice (murine rodents), gerbils and *Rhombomys* (microtine rodents), *Sigmodon* and *Cricetus* (hamsters) (cricetine rodents) are related, although each type of rodent appears to have at least one parasite peculiar to the host. Guinea pigs (*Cavia*) and *Chinchilla,* hystricomorph rodents, demonstrate the effects of geographic separation in having different parasites. Parasites of rabbits and hares are closely related to those of ruminants. The list of parasites reported from rats grows with the rat's invasion of each new habitat, illustrating their susceptibility to many parasites of other rodents.

Classification

The phylum Nematoda (Table I) is divided into two classes, the Secernentea (= Phasmidia) and the Adenophorea (= Aphasmidia). Each class contains both monogenetic and digenetic groups. The first group contains many free-living soil and plant parasitic, and most animal parasitic nematodes, and is characterized by (1) the presence of phasmids, posterior lateral sensory organs, manifested by a simple pore connected to a small tube, (2) excretory duct with cuticular terminal lining; (3) lateral canals usually present; (4) cephalic sensory organs papilloid (rarely setose); (5) amphids, porelike, usually small; (6) somatic pa-

TABLE I
NEMATODA

Class	Order	Suborder	Superfamily
Secernentea (Phasmidia)	Rhabditida	Rhabditina	Rhabdiasoidea Rhabditoidea
	Strongylida	Strongylina	Syngamoidea Ancylostomatoidea Strongyloidea
		Trichostrongylina	Trichostrongyloidea Heligmosomatoidea
		Metastrongylina	Metastrongyloidea Protostrongyloidea Pseudlioidea
	Ascarida	Ascaridina	Ascaridoidea Heterocheiloidea
		Heterakina	Heterakoidea Subuluroidea Aspidoderoidea Cosmocercoidea
		Oxyurina	Oxyuroidea Syphacioidea
	Spirurida	Spirurina	Spiruroidea Thelazioidea Gnathostomatoidea Acuarioidea Physalopteroidea
		Camallanina	Camallanoidea Dracunculoidea
		Filariina	Filarioidea Onchocercoidea
Adenophorea (Aphasmidia)	Trichinellida[a]	Trichinellina	Trichinelloidea
	Dioctophymatida	Dioctophymatina	Dioctophymatoidea Eustrongyloidea

[a] Trichinellidae Ward, 1907, appears to be the oldest valid supergeneric name for this group.

pillae usually absent; (7) males with paired genital papillae; (8) nuclei of hypodermis usually confined to chordal area. Life cycles may be direct or indirect.

Two orders of the Adenophorea are parasitic in a wide range of hosts. These are characterized by (1) absence of phasmids, (2) terminal excretory duct rarely cuticularly lined, (3) males usually without paired caudal papillae, (4) nuclei of hypodermis not confined to chordal areas. Much expansion of Chitwood's 1937 classification has taken place, some necessary and some not. Too many divisions tend to obscure relationships and new taxa based only on host differences are ridiculous. The families discussed herein are mostly restricted to parasites of laboratory animals and their relatives.

The taxonomic progression within the Secernentea is based on the genus *Rhabditis,* which is considered closest to the hypothetical ancestral forms. Rhabditids are usually free-living but some forms are semiparasitic and some completely parasitic with larval forms similar to those of the free-living generation *Rhabdias* and *Strongyloides,* and the first-stage larva of the Strongylina and Trichostrongylina. The intestinal nematodes, *Strongyloides,* may be found in all types of vertebrates except fish. *Rhabdias* occurs only in the lungs of reptiles and amphibians.

I. Rhabditida

Rhabdias and *Strongyloides* (Rhabditida, Rhabdiasoidea) have simple meromyarian-platymyarian musculature (few, flat cells in each sector), require no intermediate host, and each has a free-living heterosexual generation which produces a parasitic generation for which no male is known (exception, *Parastrongyloides*). More than one free-living generation may occur in some forms and parasitic generation of *S. stercoralis* (Bavay, 1876) is known to reproduce *in situ.* Skin penetration by larvae is the usual mode of infection, as it is for many members of the Strongyloidea.

II. Strongylida

The Strongylida contains three suborders, Strongylina, Trichostrongylina, and Metastrongylina; members of the first two have direct life cycles; those of the latter require an intermediate host.

Morphology in the Strongylida is extremely diverse. However, three characteristics are peculiar to the order; namely, (1) the intestine with few multinucleate cells; (2) the strongyloid bursa with paired genital papillae, the free-living larval stages usually retaining the sheaths of the first and/or second molt.

A. Strongylina

The Strongylina (Table II), hookworms, strongyles, cyathastoms, oesophagostoms, syngamids, etc., have highly specialized mouth parts

TABLE II
STRONGYLINA

Superfamily	Family	Genus	Host
Ancylostomatoidea	Ancylostomatidae	*Ancylostoma* (Dubini, 1843), Creplin, 1845	Man, primates, carnivores, mustelids
		Cyclodontostomum Adams, 1933	Rodents
		Galonchus Railliet, 1918	Tigers, leopards
		Gaigeria Railliet and Henry, 1910	Sheep and goats
	Uncinariidae	*Uncinaria* Frolich, 1789	Carnivores
		Arthrostoma Cameron, 1927	Felids
		Monodontus Molin, 1861	Rodents, deer, *Tapirus*
		Necator Stiles, 1903	Man, primates, mustelids, rodents
Syngamoidea	Syngamidae	*Syngamus* Siebold, 1836	Fowl
		Cyathostoma Blanchard, 1849	Fowl
		Mammomonogamus Ryjikov, 1948	Mammals
		Stephanurus Diesing, 1839	Swine
	Deletrocephalidae	*Theileriana* Monnig, 1924	*Procavia*
Strongyloidea	Strongylidae	*Strongylus* Mueller, 1780	Equines
		Ransomus Hall, 1916	Rodents
		Cyathostomum Molin, 1861	Equines
		Globocephalus Molin, 1861	Primates and sulids
	Oesophagostomidae	*Oesophagostomum* Molin, 1861	Man, primates, pigs, cattle, horses, etc.
		Ternidens Railliet and Henry, 1909	Asiatic primates

and the blood feeders usually show some modification of the anterior intestinal cells, with dark pigmentation throughout the gut. The musculature is usually simple meromyarian (containing a few flat or slightly upturned cells). Most of these parasites inhabit the digestive tract as adults and have a direct life cycle. However, *Syngamus* and *Cyathostoma* are tracheal parasites and usually require an intermediate host. The Ancylostomatidae (hookworms) may enter the host by skin penetration, by direct ingestion of the infective larvae, by intrauterine infection, and via the colostrum.

Strongylina parasites of laboratory animals are mostly restricted to a few genera and species of the Ancylostomatoidea (corona radiata absent, oral opening guarded by teeth or cutting plates), to the Strongyloidea (corona radiata present), and Syngamoidea (corona radiata absent, stoma hexagonal, without lips). In the wrong host, some hookworms burrow under the skin and cause "creeping eruption." Three basically different types of developmental cycles occur in the Strongylidae, namely: (1) burrowing of the infective larvae into the intestinal mucosa of the host for a period of development before returning to the intestinal lumen for the final molt; (2) penetration of the intestinal wall and entry into the circulatory system for a developmental period before returning to the intestinal lumen for the final molt; and (3) penetration of the intestinal wall long enough to molt and re-entry into the intestinal lumen.

Ancylostomatoidea consists of morphologically different groups, remarkable differences and similarities occurring in each group. The hookworms (*Ancylostoma, Uncinaria, Necator, Mondontus, etc.*), with their heavily sclerotized mouth parts, parasitize the upper digestive tract of primates, carnivores, mustelids, rodents, ruminants, reptiles, and amphibians.

The Strongyloidea (*Strongylus, Ternidens, Oesophagostomum,* and the *Cyathostomum* group (inhabit the entire intestine in mammals and birds, while *Stephanurus* of swine is adapted to the kidney area.

The Strongylina and Trichostrongylina produce rapidly developing thin-shelled ova that may reach a larval stage during passage from the host body. The larvae usually hatch and continue development in a free, moist environment. Two molts occur, one or both sheaths being retained to form a protective covering until entering a suitable host. Some of these free-living resistant larvae remain infective for years.

Old World primates of the Cercophithecoidea and Hominoidea tend to be parasitized by a wide range of strongyloids, while New World primates have more trichostrongyloid and metastrongyloid parasites. To my knowledge, all records of *Oesophagostomum* infections in New World primates have come from zoo animals; from dealers where conditions are uncertain at best; or from mixed colonies. In Old World primates species of *Oesophagostomum, Ternidens,* and *Globocephalus* are primary parasites but infections of *Ancylostoma duodenale* (Dubini, 1843), Creplin, 1845, and *Necator* spp. may be accidental.

Carnivores also are usually parasitized by strongyloids and metastrongyloids, rarely by trichostrongyloids (in felids and mustelids). On the other hand, rodents are host, with few exceptions (*Ransomus* and *Monodontus*), to trichostrongyloid and metastrongyloid parasites.

B. TRICHOSTRONGYLINA

The Trichostrongylina (Table III) is a large suborder which mostly consists of small, slender bursate nematodes, with platymyarian-meromyarian musculature and usually lacks specialized mouth parts. Trichostrongylin nematodes parasitize all groups of vertebrates, except fish, but the ruminants harbor the great majority of species and most of these are not discussed here. Their life cycles are direct, the first three stages are saprozoic and the third stage enter the host as ensheathed larvae. All members of the suborder fall into one of two superfamilies, the Trichostrongyloidea with amphidelphic females with vulva usually in posterior third of body, and the Heligmosomatoidea with monodelphic females with the vulva near the anus.

The Dictyocaulidae, in lungs of ruminants, is placed with the trichostrongyloids for the following reasons: (1) the life cycle is direct, the larvae develop within the vitelline membrane to an infective stage within the shed cuticles of one or both molts; (2) the musculature is meromyarian-platymyarian; (3) uteri are opposed and the vulva is not near the anus. *Dictyocaulus* is undoubtedly a transitional genus adapting to tissue parasitism, as is *Stephanurus*, which has polymyarian-meromyarian muscles.

Knowledge of tissue-invading nematodes is not all of recent origin; but except for the filarioids, little has been known of their significance until recently. The finding of the adult rat lungworm (*Angiostrongylus*) in the brain of man in Hawaii, and the "meningeal worm" of deer (*Odocoileostrongylus*) throughout the eastern United States, and *Dictyocaulus* in the inguinal canal of deer has raised the question of what is the "normal" site of these parasites. The occurrence of ascarid larvae in the brain of the woodchuck leads to further speculation regarding the wandering of larval nematodes.

C. METASTRONGYLINA

The Metastrongylina (Table IV) is one of the most interesting groups and probably the least understood. All of the metastrongylins require an intermediate host and have meromyarian-coelomyarian musculature; stoma reduced or absent, esophagus without bulb even in first-stage larvae, and bursa well developed, reduced, or absent, with rays somewhat fused. The female is didelphic, usually prodelphic with the vulva

TABLE III
TRICHOSTRONGYLINA

Superfamily	Family	Genus	Host
Trichostron-gyloidea	Dictyocaulidae	*Dictyocaulus*, Railliet and Henry, 1907	Ruminants, equines
		Ollulanus Leuckart, 1865	Cats, fox, swine, rodents
	Trichostron-gylidae	*Trichostrongylus* Looss, 1905	Ruminants, equines, primates, rabbits
		Libyostrongylus Lane, 1923	*Cricetomys* (hamsters)
		Paralibyostrongylus, Ortlepp, 1939	Rodents and primates
		Cooperia Ransom, 1907	Ruminants
		Obeliscoides Graybill, 1924	Rabbits, hares, and hamsters
		Molineus Cameron, 1923	Primates, raccoons, felids, mustelids
		Hyostrongylus Hall, 1921	Swine and *Cavia* (guinea pigs)
		Pithecostrongylus Lubimov, 1930	*Simia satyrus*
		Ornithostrongylus Travassos, 1921	Fowl
		Citellinema Hall, 1916	Sciurids
		Citellinoides Dikmans, 1939	Rodents (*Zapus*)
		Torrestrongylus Vigueras, 1935	Rodents
		Amidostomum Railliet and Henry, 1909	Fowl
		Epomidiostomum Skrjabin, 1915	Fowl
		Pseudamidostomum Boulenger, 1926	Fowl
		Graphidium Railliet and Henry, 1909	Rabbits and hares
		Graphidiodes Cameron, 1923	Hares, spider monkeys, and rodents
		Nochtia Travassos and Vogelsang, 1929	Primates
		Haemonchus Cobb, 1898	Ruminants
		Nematodirus Ransom, 1907	Ruminants and rabbits
		Mecistocirrus Railliet and Henry, 1912	Ruminants
Heligmoso-matoidea	Heligmoso-matidae	*Heligmosomum* Railliet and Henry, 1909	Rodents
		Heligmonina Baylis, 1928	*Cricetomys*
		Nematospiroides Baylis, 1926	Rodents
		Heligmonoides Baylis, 1928	Rodents
		Viannella Travassos, 1920	*Cavia*, lagomorphs, and rodents
		Brevistriata Travassos, 1937	Sciurids
		Heligmodendrium Travassos, 1937	Rodents
		Trichobaylisia Travassos, 1937	*Cricetomys*
		Longistriata Schulze, 1926	Shrews, moles, rabbits, hares, rats, *Cavia* and *Cricetomys*
		Mirandaia Travassos, 1937	Muridae
		Nippostrongylus Lane, 1923	Muridae

behind midbody, usually preanal (exception, Crenosomatinae). The typical strongylidan intestine, with few multinucleate cells, identifies this group with the Strongylida. Members of this group are primarily lungworms, but some invade the central nervous system, muscles, pancreas, and move freely in the lymphatics.

TABLE IV
METASTRONGYLINA

Superfamily	Family	Genus	Host
Metastrongyloidea	Metastrongylidae	*Metastrongylus* Molin, 1861	Swine
		Neometastrongylus Kreis, 1944	Capra
Protostrongyloidea	Protostrongylidae	*Protostrongylus* Kamensky, 1905	Ruminants and rabbits
		Odocoileostrongylus Shul'ts, 1951	Deer and moose
	Angiostrongylidae	*Bronchostrongylus* Cameron, 1931	Felids
		Angiostrongylus Kamensky, 1905	Rodents and man
		Angiocaulus Schulz, 1951	Mustelids
		Gurltia Wolffhugel, 1934	Felids
		Troglostrongylus Vevers, 1922	Felids
		Crenosoma Molin, 1861	Carnivores and insectivores
		Hepatojarakus Yeh, 1955	Rats (inferior vena cava)
Pseudalioidea	Filaroididae	*Filaroides* von Beneden, 1858	Dog, primates, and mustelids
		Metathelazia Skinker, 1931	Felids, primates
		Parafilaroides Dougherty, 1946	Pinnipeds

Basically, there are three distinct types of metastrongylin life cycles in land mammals. The Metastrongyloidea, usually swine parasites, produce eggs which pass out of the host's body before hatching, and develop in the intermediate host, usually earthworms.

The Protostrongyloidea produce unsegmented ova in the host tissue where development to first-stage larvae takes place. The eggs hatch and larvae leave the host, usually via the digestive tract, and seek a mollusk for an intermediate host. This superfamily includes parasites of rabbits, cervids, cats, rats, and several other groups.

The Pseudalioidea produce fully developed first-stage larvae which may be enclosed in a shell (*Metathelazia* Skinker, 1931) or in a thin membrane (*Filaroides* Beneden, 1858). Most descriptions in this group are inadequate, having been made on very poor, often fragmentary specimens, and the few life cycles which have been studied implicate a mollusk intermediate host. Marine mammals have their metastrongylin parasites, the Pseudaliidae.

The first description of a metastrongylin living as an adult in tissues other than the lungs was of *Elaphostrongylus cervi* Cameron, 1931, in the muscles of European deer. The genus *Elaphostrongylus* is not known to occur naturally in North America, although the next description of a similar genus, *Parelaphostrongylus odocoilei* (Hobmaier and Hobmaier, 1934) Boev and Shul'ts, 1950, was of specimens collected from muscles of deer in the western United States. Fenstermacher, in 1934, found metastrongylin nematodes in the eye of moose in Minnesota, but the worms were not identified for many years. Whitlock (1952) described "*Neurofilaria cornellensis*" in the brain of sheep in New York, and the following year Berg, Baudet, and Verwey found "*Protostrongyloides,*" a synonym of *Elaphostrongylus,* in the cranial cavity of a deer in Sweden. In 1955 DeGiusti reported the meningeal worm of deer in Michigan. This nematode, *Odocoileostrongylus tenuis* (Dougherty, 1945) Shul'ts 1951 (= *Pneumostrongylus tenuis* Dougherty, 1945; *Neurofilaria cornellensis* Whitlock, 1952) has been reported from Cervidae throughout the eastern half of North America.

Mackerras (1954) described the life cycle of *Angiostrongylus cantonensis* in the rat where she found migration to the brain, but it was not until Rosen (1962) reported *A. cantonensis* from the brain of man that metastrongylins attained medical significance.

Natural infections of metastrongylin lungworms have not been reported for *Gerbillus, Cavia,* and *Rhombomys.*

III. Ascarida

A. Ascaridina

The Ascarida may be divided into three suborders, listed from least to most complex, the Oxyurina, Heterakina, and Ascaridina. Ascaridina (Table V) is made up of two superfamilies (Ascaridoidea and Heterocheiloidea (= Anisakoidea), both containing parasites of mammals as well as of cold-blooded vertebrates. The latter also contains parasites of birds.

The Ascaridoidea contains the largest of intestinal parasites and is characterized as follows: mouth usually with three lips, interlabia absent, esophagus club-shaped without valvular apparatus, ventriculus (except *Dujardinascaris, Toxocara,* and *Neoascaris*) or intestinal cecum; spicules equal or subequal and many papillae present on male tail, and body musculature polymyarian-coelomyarian. Parasites of mammals, reptiles, and amphibians. Only the mammalian forms are covered here.

TABLE V
ASCARIDINA

Superfamily	Family	Genus	Host
Ascaridoidea	Ascaridae	*Ascaris* Linnaeus, 1758	*Homo*, swine
		Parascaris Yorke and Maplestone, 1926	Equines
		Toxascaris, Leiper, 1907	Felids, canids
		Lagocheilascaris Leiper, 1907	*Felis* (and man)
	Toxocaridae	*Toxocara* Stiles, 1905	Canids, felids
		Neoascaris Travassos, 1927	Cattle
Hetero-cheiloidea	Hetero-cheilidae	*Heterocheilus*, Diesing, 1839	Sirenia
		Anisakis Dujarbin, 1845	Marine mammals
		Porrocaecum Railliet and Henry, 1912	Adults in fowl
		Terranova Leiper and Atkinson, 1914	Marine mammals
		Contracaecum Railliet and Henry, 1912	Marine mammals and fowl

Members of the Heterocheiloidea probably always require at least one intermediate host and many use several transport hosts. Some of these parasites may cause great economic loss in fish and some may be dangerous to man if infected fishes are consumed raw. In Japan, *Anisakis* has been shown to cause tumors in the stomach of man. Deaths of persons who consumed fresh raw marinated herring have been reported in Holland as due to *Porrocaecum,* parasitic in birds. The deaths may have been caused by *Terranova,* a closely related genus parasitic in marine mammals. The Heterocheiloidea are characterized as having unusual modifications of the esophagus, such as an esophageal ventriculus, sometimes with an anteriorly directed cecum (*Porrocaecum* and *Terranova*) and/or ventricular appendage (*Contracaecum* and *Raphidascaris*); interlabia are usually present.

B. Oxyurina

The Oxyurina (Table VI) have simple, direct life cycles, the infective eggs being ingested by the host. This group has platymyarian-meromyarian musculature. They are commonly called pinworms and are

TABLE VI
Oxyurina

Superfamily	Family	Genus	Host
Oxyuroidea	Oxyuridae	*Oxyuris* Rudolphi, 1803	Equines
		Enterobius Leach, 1853	Man—Old World primates
		Trypanoxyuris Vevers, 1923	New World primates
		Lobatorobius Skrjabin and Schik., 1951	*Ateles* and *Lagothrix*—monkeys
		Oxyuronema Kreis, 1932	Howler monkey, *Allouata*
		Passalurus Dujardin, 1845	Rabbits and hares
		Probstmayria Ransom, 1907	Equines, large primates
	Heteroxynematidae	*Heteroxynema* Hall, 1916	Sciurids, rabbits, etc.
		Dentostomella Schulz and Krepkagorskaya, 1932	*Gerbillus* and *Rhombomys*
		Dermatoxys Schneider, 1866	Rabbits and hares
	Aspiculuridae	*Labiostomum* Aktar, 1941	Rodents
		Aspiculurus Schulz, 1924	Rodents
		Eugenuris Schulz, 1948	Rodents
Syphacioidea	Syphaciidae	*Syphacia* Seurat, 1916	Rodents *Mus*, *Gerbillus*, *Apodemus*, squirrels
		Citellina Prendel, 1928	Sciurids and marmots
		Wellcomia Sambon, 1907	Porcupine
		Helminthoxys Freitas, Lent, and Almeida, 1937	Sciurids

somewhat restricted in host groups. No oxyuroid has been reported as a normal parasite of dogs, cats, or guinea pigs, although they occur in many laboratory animals, as well as in a wide variety of other hosts. The genera *Enterobius* and *Probstmayria* occur in higher primates. In New World primates, *Trypanoxyuris*, very closely related to *Entero-*

bius, *Buckleyenterobius*, with a buccal tooth, *Oxyuronema* (closely related to *Oxyuris*), and *Lobatrobius* are commonly found.

Oxyuris equi (Schrank, 1788) infects horses and other equids, while species of *Skrjabinema* infect Cervidae. *Probstmayria*, also parasitic in horses, is unique among mammalian parasites in reproducing continuously in the same host. Transmission between hosts is probably accomplished by food or water contamination.

Rats, mice, and hamsters harbor *Syphacia* and *Aspiculuris;* gerbils, *Syphacia;* and rabbits *Dermatoxys* and *Passalurus*.

C. Heterakina

Heterakina (Table VII) also have a direct life cycle with species occurring in all vertebrate orders and are intermediate between the

TABLE VII
Heterakina

Superfamily	Family	Genus	Host
Heterakoidea	Heterakidae	*Heterakis* Dujardin, 1845	Fowl
		Ganguleterakis Lane, 1914	Mammals (mostly rodents)
		Paraspidodera Travassos, 1914	Guinea pigs
	Aspidoderidae	*Aspidodera* Railliet and Henry, 1912	Marsupials and edentates
		Pseudaspidodera Baylis and Daubney, 1922	Fowl
		Sexansodera Skrjabin and Schikhobalova, 1947	*Dasypus*, *Tolypeutes*, *Tatus*
	Ascarididae	*Ascaridia* Dujardin, 1845	Fowl
Subuluroidea	Parasubuluridae	*Parasubulura* Van den Berghe and Vuylsteke, 1938	Insectivora
	Subuluridae	*Subulura* Molin, 1860	Fowl and primates
		Oxynema Linstow, 1899	Canids, *Petrodromus*, *Vulpes*, and *Xerus*
		Aulonocephalus Chandler, 1935	Fowl and *Suricatta*
	Cruziidae	*Cruzia* Travassos, 1917	Turtles and marsupials
		Pseudocruzia Wolfgang, 1953	Swine

oxyurins and ascarins. The excretory system is closer to that of the oxyurins, the esophagus is similar to those of the atractids of the Oxyuroidea and *Ascaris,* and the musculature is polymyarian-low coelomyarian. The males have a preanal suckerlike disc or modification thereof, usually two equal or subequal spicules, and caudal alae. The females are amphidelphic and approximately the same size as the males. The mouth capsule and a valvular apparatus in the esophageal bulb may be present or absent.

Gallinaceous birds are parasitized by species of *Heterakis, Subulura,* and *Ascaridia,* while water birds are mostly parasitized by heterocheilids. Hawks, eagles, and falcons are parasitized by *Porrocaecum* (heterocheilid), larvae of which may be found in muscles of shrews. Species of *Porrocaecum* and *Subulura* occur in owls. Of the passeriform birds, only robins, magies, pipits, and crows normally harbor *Porrocaecum.*

Species of *Aspidodera* may be found in marsupials and endetates while the closely related *Pseudaspidodera* and *Sexansodera* are parasitic in birds. Members of the Subuluroidea parasitize almost all vertebrates except marine forms.

IV. Spirurida

The order Spirurida contains three suborders, the Spirurina, Camallanina, and Filariina, all of which require an intermediate host in which development to the infective stage takes place.

A. Spirurina

The Spirurina (Table VIII) usually use insects, rarely copepods, for intermediate hosts. These parasites usually inhabit the upper digestive tract. *Spirocerca* normally in the esophagus may lodge in the aorta or stomach. *Gongylonema,* which lives embedded in the mucous membranes of the mouth or esophagus, like most spirurins, produces thick-shelled embryonated eggs which pass out in the feces. These eggs must be ingested by a suitable intermediate host in which growth and development to the third stage and encystment occurs. *Protospirura,* a common rat parasite, may be transmitted by cockroaches to mice and monkeys. New World monkeys harbor natural infections of species of *Rictularia, Physaloptera, Gongylonema,* and *Spirura;* while old World monkeys commonly harbor *Abbreviata, Physaloptera, Streptopharagus,* and *Rictularia.* Cats and dogs harbor *Physaloptera,* and *Spirocerca* is normally parasitic in dogs; *Rictularia* and *Gnathostoma* are found in cats. Few spirurins have been reported as normal parasites of hamsters (*Protospirura*), rabbits, and guinea pigs, but *Gongylonema, Rictularia,* and *Physaloptera* have been reported from gerbils and *Protospirura, Streptopharagus,* and *Physocephalus* from hamsters. *Trichospirura,* a tiny fragile spiruroid, has recently been found in the pancreatic ducts of marmosets.

Thelazia and *Oxyspirura* are parasitic in the eyes. *Thelazia,* parasitic in dogs, deer, goats, and man, is transmitted by flies, as are the species of *Habronema* and *Draschia,* but *Oxyspirura* is transmitted by cockroaches and parasitizes the eyes of fowl. *Geopetitia,* a tissue parasite of birds, is morphologically similar to *Crassicauda,* kidney parasites of whales.

The *Acuariidae,* the *Tetramerinae* (Spiruridae), *Oxyspirura* (Thelaziidae), and *Gongylonema* (Gongylonematidae) contain most of the spirurin parasites of birds. These are parasitic in the digestive tract, some living in the lining of the esophagus, the proventriculus, and some under the lining of the gizzard.

Fishes have their particular group of spiruroids as do marine mammals. Probably the most interesting of the latter is *Crassicauda* in the kidneys of cetaceans. These nematodes may attain a length of 50 cm. or more at least half of which is a fine capillary part extending into the collecting tubules.

The Physalopteridae probably have the widest host range—reptiles, mammals, and birds. *Gongylonema* is not very host-specific, but in the adult stage is confined entirely to mammals and fowl.

B. Camallanina

Of the Dracunculidae (Dracunculoidea) (Table I), only one genus, *Dracunculus,* parasitizes mammals. These are large nematodes with cuticularized thickenings on the head; they exhibit extreme sexual dimorphism and require a copepod intermediate host. The small male has equal spicules and the tail of both sexes is sharply curved. The viviparous female produces an enormous quantity of embryonic larvae, characterized by a very long, pointed tail and large phasmids. When these larvae are ingested by a copepod, the larvae reach the body cavity in a few hours, where, in the course of 5–12 days, two molts take place, after which they become infective to the definitive host. Sexually mature females have been found deep in the tissues of experimentally infected dogs 67 days after infection. Fertilization may take place in the abdominal (or pleural) cavities. After about a year, the female migrates to a subcutaneous site where a blister is formed. Each time the blister comes in contact with water, quantities of larvae are forcefully ejected. *Dracunculus insignis* (Leidy, 1858), Chandler, 1942, occurs in carnivores, raccoons, and mustelids from Texas to New England.

TABLE VIII
SPIRURINA

Superfamily	Family	Genus	Host
Spiruroidea	Spiruridae	*Spirura* Blanchard, 1849	Primates, rodents, insectivores, carnivores
		Chitwoodspirura Chabaud and Rousselot, 1956	*Gorilla*
		Mastophorus Diesing, 1853	Rodents, primates, carnivores
		Protospirura Seurat, 1914	Rodents, primates
		Spiruracerca Erickson, 1938	*Zapus*
	Tetrameridae	*Tetrameres* Creplin, 1846	Fowl
		Microtetrameres Travassos, 1915	Fowl
	Habronematidae	*Habronema* Diesing, 1861	Equines, fowl
		Odontospirura Wehr, 1933	*Rhea americana*
		Parabronema Baylis, 1921	Swine, primates, elephants, *Rhinoceros*
		Draschia Chitwood and Wehr, 1934	Equines
Thelazioidea	Thelaziidae	*Thelazia* Bosc, 1819	Mammals, fowl
		Oxyspirura Drasche, 1897	Fowl
	Spirocercidae	*Spirocerca* Railliet and Henry, 1911	Dogs, horses, goats, man
		Cylicospirura Vevers, 1922	Mammals
		Cyathospirura Baylis, 1934	Mammals
	Ascaropsidae	*Ascarops* Beneden, 1873	Swine, rodents
		Physocephalus Diesing, 1861	Swine, other mammals
		Streptopharagus Blanc, 1912	Primates, gerbils, murines
		Simondsia Cobbold, 1864	Swine, *Bradypus*, *Tapirus*
		Leiuris, Leuckart, 1850	*Bradypus*
		Texicospirura Chitwood and Cordero 1966	Peccary

	Rictulariidae	*Rictularia* Froelich, 1802	Cats, bats, primates, fox, dogs, rodents
		Rictulariodes Hall, 1916	*Lemmus*
		Echinonema Linstow, 1898	*Perameles, Dasyurus*
	Gongylonematidae	*Gongylonema* Molin, 1857	Man, primates, swine, rodents, horses, cows, cervids
Gnathostomatoidea	Gnathostomatidae	*Gnathostoma* Owen, 1836	Mammals
		Tanqua Blanchard, 1904	Reptiles
		Echinocephalus Molin, 1858	Fishes
Acuarioidea	Acuariidae	Numerous genera	Fowl
Physalopteroidea	Physalopteridae	*Physaloptera* Rudolphi, 1819	Reptiles, amphibians, mammals
		Abbreviata Travassos, 1915	Man, primates
		Skrjabinoptera Schul'ts, 1927	Larvae encysted in rodents; adults in reptiles and amphibians
Crassicaudoidea	Crassicaudidae	*Geopetitia* Chabaud, 1951	Fowl
		Crassicauda Leiper and Atkinson, 1914	Marine mammals

C. Filariina

Knowledge of relationships of the filariae (Table IX) and their hosts is gradually emerging, several complete life cycles having been worked out in the past decade. *Brugia, Litomosoides, Dirofilaria,* and *Filaria* have been elucidated.

Many hosts harbor two or more species of filariae; this is particularly

TABLE IX
Filariina

Superfamily	Family	Genus	Host
Filarioidea	Filariidae	*Filaria* Mueller, 1787	Mustelids
		Parafilaria Yorke and Maplestone, 1926	Cattle
Onchocercoidea	Dirofilariidae	*Dirofilaria* Railliet and Henry, 1911	Carnivores, man
		Micipsella Seurat, 1921	Rabbits
		Dirofilariaeformia Lubimov, 1935	Rodents
		Bostrichodera Sandground, 1938	*Choelopus didactylus*
		Tawila Khalil, 1932	Primates
		Cystofilaria Skrjabin and Shikhobalova, 1944	Dogs
		Loa Stiles, 1905	Hominoids, primates
	Onchocercidae	*Onchocerca* Creplin, 1846	Hominoids, ruminants
		Wuchereria Silva Araujo, 1877	Man
		Brugia Buckley, 1958	Primates, dogs, cats
		Eleophora Railliet and Henry, 1912	Horses, sheep, deer, ox, etc.
		Litomosa Yorke and Maplestone, 1926	Bats
		Litomosoides Chandler, 1931	Bats, rodents
		Paralitomosa Nagaty, 1935	Primates
		Edesonfilaria Yeh, 1960	Primates
		Macacanema Schad and Anderson, 1963	Primates
	Dipetalonematidae	*Dipetalonema* Diesing, 1861	Primates, dogs
		Tetrapetalonema Faust, 1935	Primates
		Acanthocheilonema Cobbold, 1870	Dogs
		Migonella Lent and Freitas, 1946	Bats
		Ackertia Vaz, 1934	*Cavia*
		Skrjabinaria Lubimov, 1927	Seals
		Skrjabinofilaria Travassos, 1925	Marsupials
	Setariidae	*Setaria* Viborg, 1795	Equines, ruminants
	Stephanofilariidae	*Stephanofilaria* Ihle and Ihle-Landenburg, 1933	Ruminants

true of birds. Although at least four genera of filariae are known to occur in dogs, only rarely are more than two present in a single host. Separation of *Dirofilaria* from the dipetalonematids is simple; however, ultimately, this latter group must be broken down into several genera. The genera *Tetrapetalonema* and *Acanthocheilonema* consist of small, fragile forms, while *Dipetalonema gracile* represents the large, sturdy dipetalonematids, the most common filaria of New World primates.

Filaria martis, type of the family Filariidae, produces thick-shelled eggs containing an embryo, as do most bird filariae. The esophagus of *Filaria martis* is closer in fine structure to that of the spiruroids.

Members of the Onchocercoidea produce microfilaria. A tentative classification of filariae parasitic in mammals is presented herein and is based on acceptance of the family Dipetalonematidae, as well as that of the Onchocercidae. Failure to accept the separate and equal status of these as families has resulted in chaotic classifications.

Two genera presently included in the Onchocercoidea, but without subfamily designation, are *Edesonfilaria* and *Macacanema.* These nematodes differ from other filarioids in that the esophagus bears some apparent resemblance to that of *Dracunculus.*

V. Adenophorea

The mammalian parasitic Adenophorea is composed of only two orders. But one of these, the Trichinellida (Table X) constitutes a very large segment of parasitic nematodes with a wide assortment of host animals as well as locations within the host. These nematodes usually have direct life cycles, but differ radically in structure and development from all other known nematodes.

The Trichinellida and Dioctophymatida differ from the Secernentea in that the nuclei of the hypodermis are not confined to four chords, but there may be eight or more small chords. Setose papillae may be present on the tegument. Only the Dioctophymatida have esophagi similar to those of the Secernentea. In the trichinelloids, the esophageal glands are reduplicate, forming a stichosome along the esophageal tube. The system which in the Secernentea is termed the excretory system is entirely lacking in the Trichinellida. There is, however, a system peculiar to this group, which may act as excretory system, the bacillary band, which has highly specialized cells and contains tiny pores to the external surface. In *Trichuris* there is a single bacillary band terminating at or near the base of the glandular esophagus. In all the other genera there are two bands which may run the length of the body (*Anatrichosoma*).

TABLE X
ADENOPHOREA–TRICHINELLOIDEA

Superfamily	Family	Genus	Host
Trichinelloidea	Trichinellidae	*Trichinella* Railliet, 1895	Mammals
	Trichuridae	*Trichuris* Roederer, 1761	Large intestine, colon, cecum, rectum of man, carnivores, ruminants
		Trichuroides Ricci, 1949	Bats
	Capillariidae	*Capillaria* Zeder, 1800 (Capillaria)	Esophagus, large intestine, small intestine, urinary bladder in vertebrates
		Capillaria (*Hepaticola*) Hall, 1916	Liver of mammals
		Capillaria (*Eucoleus*) Duj., 1845	Lungs of mammals
		Capillaria (*Thominx*) Duj., 1845	Spicule sheath, spinous
	Trichosomoididae	*Trichosomoides* Railliet, 1895	Urinary bladder of rats (neotenic male in uterus of female)
	Anatrichosomatidae	*Anatrichosoma* Smith and Chitwood, 1954	Subcutaneous lesions, nasal mucosa, rhesus monkeys, man, (males not neotenic)

A single gonad in both sexes and vulva in the female at or near the base of the esophagus is characteristic of the Trichinellida. Biopercu-late eggs containing unsegmented ova are the rule, but *Trichosomoides* and *Anatrichosoma* characteristically produce embryonated eggs and some species of *Capillaria* (usually bat parasites) occasionally produce embryonated eggs.

The Capillariidae parasitize a wide range of hosts as well as the great-est variety of locations in the host. Normally, their life cycle is direct, but transport hosts are not unusual and in some instances may be obligatory.

Numerous attempts have been made to split *Capillaria* into separate genera, viz., *Capillaria* in the digestive tract, *Hepaticola* in the liver, and *Eucoleus* in the lungs. Adequate male characters are present, but no female characters have been correlated with them except for the characteristic eggs of *Hepaticola* and *Eucoleus*.

Trichosomoides crassicauda, Bellingham, 1840, parasitic in the urinary bladder of rats, has a very small neotenic male which spends most of its life in the uterus of the female. The female has a rather flaccid body with the cuticle lacking rigidity.

Anatrichosoma cutaneum (Swift, Boots, and Miller, 1922) and *A. cynamolgi* Smith and Chitwood, 1954 have been found in subcutaneous lesions and in the nasal mucosa of rhesus monkeys. These differ from *Trichosomoides* in that the male is as long as the female and does not live inside her uterus. Both genera produce large, embryonated eggs.

Cystoopsis, a subcutaneous parasite of *Acipenser* (fish), has two rows of stichocytes instead of one and the female has a long thin body and a globular posterior body. The vulva is in the esophageal region. The small male is cylindrical as in other genera.

Trichinella spiralis, though primarily a parasite of swine, man, and rats, has been found in numerous carnivores, mice, rodents, and mon-goose. Experimentally it has been established in monkeys, goats, sheep, guinea pigs, pigeons, and probably others.

VI. Dioctophymatida

This order is composed of only two families, the Dioctophymatidae (Table XI), parasites in the kidneys, rarely the body cavity of mammals; and the Eustrongylidae, parasites of birds. *Dioctophyma renale*, the giant kidney worm of carnivores and mustelids, and *Sobolophyma*, parasitic in shrews, are the sole representatives of the Dioctophymatidae. The life cycle of *D. renale* requires an oligochaete as an intermediate host.

TABLE XI
ADENOPHOREA–DIOCTOPHYMATOIDEA

Superfamily	Family	Genus	Host
Dioctophymatoidea	Dioctophymatidae	*Dioctophyma* Collet-Meygret, 1802	Kidney of dogs and other carnivores (life history Karmanova 1959)
		Sobolophyma Petrow, 1930	Shrews
Eustrongyloidea	Eustrongylididae	*Eustrongylides* Jagerskiold, 1909	Fowl
		Hystrichis Dujardin, 1845	Fowl

Eustrongylides require a fish intermediate host, while *Hystrichis,* also parasitic in birds, may use some marine invertebrates as well as fish. All of these genera may use transport hosts.

REFERENCES

Cameron, T. W. M. (1931). *J. Helminthol.* **9,** 153–156.

Chitwood, B. G. (1937). *In* "Rabot Gel'mint, 30-Year Jubileum K. I. Skrjabin," pp. 69–80. Moscou (Russian transl.).

Chitwood, B. G. (1961). *In* "Principles of Nematology" (G. Thorne, ed.), p. 91. McGraw-Hill, New York.

Chitwood, B. G., and Chitwood, M. B. (1950). "An Introduction to Nematology" (Revised). Monumental Printing Co., Baltimore, Maryland.

DeGiusti, Dominic L. (1955). *J. Parasitol.* **41,** 42.

Dougherty, E. C. (1945). *Parasitology* **36,** 199–208.

Fenstermacher, R. (1934). *Bull. Minn. Agr. Exptl. Sta.* No. 308.

Hobmaier, A., and Hobmaier, M. (1934). *Proc. Soc. Exptl. Biol. Med.* **31,** 509–514.

MacKerras, M. J., and Sandars, D. F. (1954). *Nature* **173,** 956–957.

Mozgovoi, A. A. (1953). *Osnovy Nematodol.* **2,** 616 (Russian text).

Simpson, G. G. (1945). *Bull. Am. Museum Nat. Hist.* **85,** 350 pp.

Skrjabin, K. I., Shikhobalova, N. P., and Sobolev, A. A. (1949). "Spiruraty I Filariaty," Vol. I. P. 519. Moskva and Leningrad (Russian text).

Skrjabin, K. I., Shikhobalova, N. P., and Orlov, I. B. (1957). *Osnovy Nematodol.* **6,** 587 (Russian text).

Weinstein, P. P., Rosen, L., Lagueur, G. L., and Sawyer, T. K. (1962). *J. Parasitol.* **48,** 51–52.

Whitlock, J. H. (1952). *Cornell Vet.* **42,** 125–132.

Yorke, W., and Maplestone, P. A. (1926). "The Nematode Parasites of Vertebrates." London.

CHAPTER 2

The Biology of the Acanthocephala

Ivan Pratt

Redi in 1664 and van Leeuwenhoek in 1665 first wrote of the Acanthocephala; the first names used were Acanthocephalus in 1771 by Koelruether for a fish parasite and Echinorhynchus by F. Mueller in 1776. R. Leuckart in 1862 described the first life cycle, that of *Pomphorhynchus proteus* in freshwater fish. The anatomical and histological studies of Kaiser (1893) and Hamann (1891) brought out the unique characteristics of the hypodermis and the nature of the cells therein and elsewhere in the body. The chief students of the group in this century have been Luhe (1911), Meyer (1931, 1932–1933), and especially Van Cleave, who published many papers on these worms between 1913–1953. Golvan (1958–1962) has published a series of papers on the genera, hosts, geographical distribution, and other topics. An abridged paper in English (Golvan, 1964) including illustrations of the genera, and a key was assembled by Pratt and Merritt and published by Oregon State University Press.

Van Cleave (1941) published a paper outlining the reasons why this group of animals was a separate phylum and since that time it has been accepted as one. He stated that this group had very little in common with the nematodes and listed the following characteristics relating acanthocephalans to cestodes: (1) similarity of external tissues, (2) flattened shape in life, (3) similar excretory systems, (4) similar early embryology, (5) somewhat similar life cycles, and other less evident similarities. However, no one has attempted to include the Acanthocephala into the class Cestoda of Platyhelminthes. The remoteness of the similarities and the unique lacunar sacs of the Acanthocephala make it seem more reasonable to set up a separate phylum, the closest affinities of which are with the cestodes.

Golvan (1958) decided that the Acanthocephala were most closely related to the Priapuloidea and therefore might be included in the phylum or super-phylum Aschelminthes, an idea proposed by Meyer

(1928, 1932–1933). It may be that new evidence gathered with new techniques yet undeveloped or unused on this group will solve the affinities of this group of parasitic worms.

Van Cleave (1948) proposed that the Acanthocephala be divided into two classes, with two orders each with the characteristics given in Table I.

TABLE I
CLASSES OF ACANTHOCEPHALA ACCORDING TO VAN CLEAVE

Class Metacanthocephala Van Cleave, 1948	With 6 or 8 cement gland cells.
Order Palaeacanthocephala Meyer, 1931	Definitive hosts aquatic, few terrestrial; fish, reptiles, birds, mammals.
Order Archiacanthocephala Meyer, 1931	Definitive hosts are terrestrial birds and mammals.
Class Eoacanthocephala Van Cleave, 1936	With 1 syncytial cement gland. Definitive hosts are cold-blooded vertebrates.
Order Neoacanthocephala Van Cleave, 1936	Body wall without spines. Definitive hosts usually fish, rarely amphibians and reptiles.
Order Gyracanthocephala Van Cleave, 1936[a]	Body wall bearing spines. Definitive hosts usually fish, rarely amphibians and reptiles.

[a] Hyman (1951) thought that the single difference of body spines in this order was insufficient difference on which to base an entire order. Golvan (1959) stated that the order probably should be preserved because presence of spines was a primitive character and hence probably a fundamental difference between the Gyracanthocephala and the unarmed and hence more evolved Neoacanthocephala.

Yamaguti (1963) stated that, "The Acanthocephala are gonochoristic Nemathelminthes lacking a digestive tract in all stages of development; the assimilation of nutritive substances takes place thru the body wall." This is an unsupported statement which this helminthologist cannot accept at this time despite the eminence of Professor Yamaguti. Petroschenko (1956, 1958) devised a new classification system parallel to, but not identical with those previously discussed. It may be that some of the new rearrangements he proposed will prove to be correct and be accepted when other students have critically reviewed them.

The widely distributed Eoacanthocephalan, *Neoechinorhynchus rutili* (Mueller, 1780) found in the adult stage in the small intestine of salmonid fish will be used to illustrate the anatomy of an acanthocephalan as well as a life history.

The body is divided into two parts, the forebody (presoma) with proboscis and neck, and the main body. The proboscis is an eversible structure with rows of radially or spirally arranged hooks on its surface. Its size varies in the different genera, and its growth is completed during the larval stages. The neck is of various relative sizes, but usually can enclose the withdrawn proboscis. At the posterior rim of the neck there are two internal sacs, the lemnisci, one on each side of the neck. The cavities of the lemnisci are continuous with the lacunar system of the proboscis, and connect directly with the large ring canal of the neck.

The main body is a short to long sac, that of the female coming to a simple blunt end with a terminal genital pore. The posterior end of the male is a flexible and variably shaped bursa directed posteriad and containing a penis. Males are usually somewhat smaller overall than the females. The trunk or main body is usually distinctly flattened dorsoventrally in life, but is usually round in cross section because the body becomes distended prior to fixation.

There is a cuticle on the surface which covers an unique hypodermis made up of three layers. The outer layer consists of short fibrils lying at right angles to the cuticle. The middle layer has fibers parallel to the surface and arranged in longitudinal, oblique, and radial patterns. The thicker inner layer has radial fibrils, the lacunar system and the characteristic "giant nuclei." The lacunar canals (Fig. 3) which do not have definite walls, branch and anastomose repeatedly. The trunk lacunae are not connected to those in the neck and proboscis. The nuclei go thru a development in each individual from a simple circular or oval shape to a large, ameboid shape. In some species each nucleus fragments into numerous smaller pieces, in others they become dendritic and threadlike. The total number of nuclei in the entire organism is constant in a species. It seems likely that interesting information might be gathered with electron microscope and histochemical studies of the hypodermis and cuticle of the acanthocephalans.

The presoma has the same sort of body wall as the trunk, but thinner. There are no longitudinal main lacunar canals. The inner radial layer of the hypodermis at two points on the posterior rim is prolonged into two sacs, the lemnisci. The lemnisci have a limited number of giant nuclei. It has been suggested that these two sacs are reservoirs for the presomal fluid when this structure is invaginated.

Retractor muscles of the proboscis extend from the summit of the proboscis to the base of the proboscis receptacle. Another set of muscles encircles all the retractor muscles and the lemnisci; they originate near the posterior boundary of the neck and insert on the trunk and are the retractors of the neck.

The nuclei of the proboscis arrive at the adult number very early in the embryological development (Fig. 4) and the proboscis reaches almost adult proportions and is functional early in the invertebrate host. This has an obvious adaptive advantage to an animal which must attach immediately to an intestinal wall after ingestion by the vertebrate host.

The nervous system consists of a ganglion of 80 cells, more or less, embedded in the ventral wall of the proboscis sheath with fibers running to various structures. In the male there is a pair of genital ganglia with fibers to the genital structures. Sense organs are reduced to three in the proboscis, one terminal and a pair of lateral ones. Around the rim of the male bursa there are seven or eight spherical sense organs; all the sense organs may be tactile.

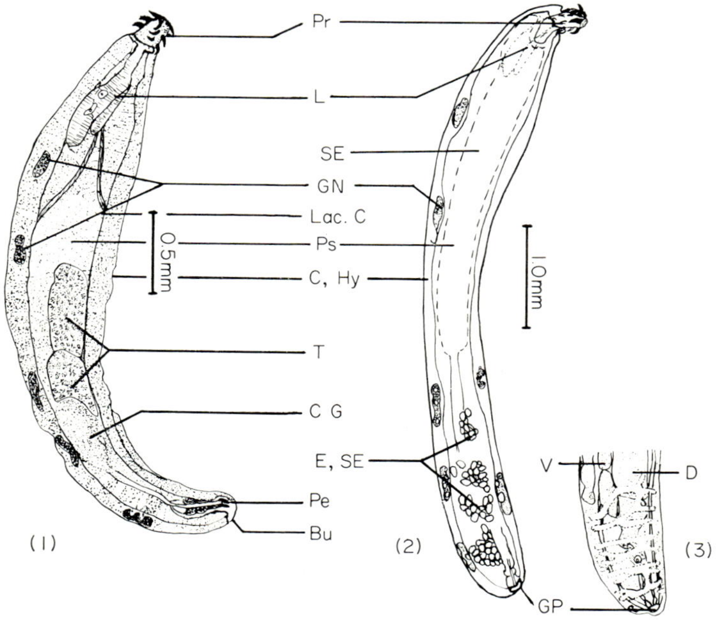

FIG. 1. Male *Neoechinorhynchus rutili* redrawn from Van Cleave, 1950.

FIG. 2. Female *N. rutili* redrawn from Van Cleave, 1950.

FIG. 3. Posterior end of young female *N. rutili* from Van Cleave, 1950. Key to abbreviations on Figs. 1, 2, and 3. Bu, bursa copulatrix; C G, cement gland; C and Hy, cuticle and hypodermis; D, dorsal ligament sac; E and SE, eggs and shelled embryos; GN, giant nuclei of hypodermis; GP, genital pore; L, lemnisci; Lac C, lacunal canals; Pe, penis; Pr, proboscis; Ps, pseudocoel; SE, shelled embryos; T, testes; V, ventral ligament sac.

In the order Archiacanthocephala there are protonephridia with up to 700 flame bulbs. The canals open into the reproductive tubes and thence to the exterior. The members of the other three orders do not have excretory systems.

There is no digestive system in any acanthocephalan nor any systems for respiration or circulation. The body cavity is not lined with a peritoneal epithelium, hence is termed a pseudocoel.

The Acanthocephala are all dioecious. The female has an ovary that was contained within dorsal and ventral ligament sacs extending from the posterior end of the proboscis receptacle or to the adjacent body wall to the posterior part of the body cavity during development. However, the sacs disintegrate before maturity and release the ovary into the pseudocoel where it fragments into separate developing eggs and balls of eggs. The mechanism for discharge of fully developed shelled embryos from the pseudocoel consists of a wide-open uterine bell into which immature eggs and shelled embryos alike may pass. Immature eggs pass back to the pseudocoel, but the shelled embryos are captured by the end of the uterine tube which leads to the muscular uterus. The latter forces the shelled embryos along to the exterior.

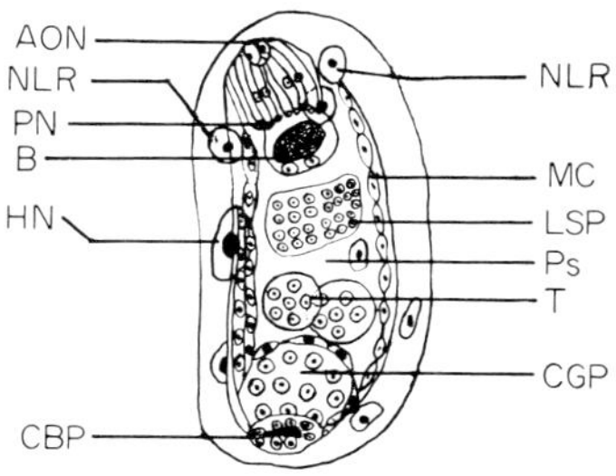

FIG. 4. Acanthella of *N. rutili* 38 to 45 days after entering ostracod. Size 0.305 mm. long by 0.146 mm. wide. (After Merritt and Pratt, 1964.) AON, apical organ nuclei; B, brain; CBP, copulatory bursa primordium; CGP, cement gland primordium; HN, hypodermis nuclei; LSP, ligament sac primordium; MC, muscle cells; NLR, nuclei (3) of lemniscal ring; PN, proboscis nuclei; Ps, pseudocoel; T, testes.

The male reproductive system is anchored to the ligament sac which encloses the two testes anteriorly and attaches to the genital sheath in the area of the cement gland or glands. The two testes are drained by sperm ducts which join to form a seminal vesicle. In the Archiacanthocephala the excretory ducts join at this point and the tube leading through the penis is called an urogenital canal. In the other orders the tube leading from the seminal vesicle through the penis is the sperm canal. In the class Metacanthocephala there are either six or eight cement gland cells, in class Eoacanthocephala there is a syncytial cement gland, sometimes with a cement reservoir. In both classes ducts from these glands join the posterior region of the seminal vesicle, and their product passes along the duct leading to the posterior. The secretion of these glands attaches the male bursa to the posterior end of the female and ensures transfer of sperm to the interior of the female. At the end of copulation the female pore is sealed off to ensure the presence of the sperm inside the female. However, it also acts like a "chastity belt," one that antedates the Crusades by eons of time. The Saefftigen's pouch is an hydrostatic sac to force the bursa and penis out of the posterior end of the body during copulation. The bursa is an eversible, bell-shaped structure which fits into the posterior end of the body except during copulation. It is a part of the body wall. The proximal part of the bursa is a strongly muscular cap, presumably used to grasp the posterior end of the female.

Identification of acanthocephalans is not especially difficult, at least to genera. The proboscis, which is the most necessary structure in preliminary recognitions, is a long or short cylinder with recurved hooks arranged into various patterns that are constant to a degree for orders, families, and genera. In the Archiacanthocephala the proboscis hooks are usually arranged in concentric circles; in the other orders the hooks are in alternating radial rows starting at the anterior tip of the proboscis. This arrangement is termed quincunxial (Fig. 5). A few species have

FIG. 5. Anterior view of proboscis of *N. rutili* showing the quincunxial arrangement of the proboscideal spines. (Redrawn from Golvan, 1959.)

an irregular arrangement of the hooks on the proboscis, some have one or two rows of larger hooks on the ventral surface and others have the dorsal row or rows markedly smaller. The larger hooks have extensions embedded in the proboscis, the smaller ones are sometimes called armature spines and attach to the surface of the proboscis.

The identification of an acanthocephalan to species requires an adult male specimen usually stained and mounted on a slide. One must be able to count and determine the pattern of hooks on the proboscis, the number of cement glands, the number and form of the nuclei in the hypodermis, the main anterior-posterior lacunar canals on the main body and various measurements and size ratios of parts of the body. There are more than 78 genera, and host specificity is not marked at the species level, but genera are fairly closely limited to particular groups of hosts and sometimes to particular continents.

Life histories of acanthocephalans are of two types. The adult or definitive stage of all these worms is found in the small intestine of a vertebrate. In light infections the worms are usually attached near the front end of the small intestine, in heavier infections they may be found also further along the length of the small intestine. Members of the order Archiacanthocephala parasitize terrestrial animals in both stages of the life cycle. The best-known forms are *Moniliformis moniliformis* (Bremser, 1811) and *M. dubius* (Meyer, 1932) parasitizing rats and cockroaches and several other insects, and *Macracanthorhynchus hirudinaceous* (Pallas, 1781) in swine, dogs, and other mammals and scarabaeid beetles, especially the May beetle in the eastern and midwestern United States of America. In both genera 7 or 8 weeks are required to develop infective larvae in the body cavity of the insects and a similar length of time for the intestinal, adult form to mature and lay eggs.

The members of the other orders all conform to an aquatic type life cycle with the larval stages developing in a crustacean. Isopods (*Asellus* sp.), amphipods (*Gammarus* sp.), and ostracods (*Cypria* sp.) are genera of crustaceans known to be intermediate hosts for acanthocephalans.

Neoechinorhynchus rutili is a cosmopolitan species found in cold-water fishes, including trout. The shelled embryos pass out into the intestinal contents of the host fish and thence out with the feces into the water, where they must be ingested by the ostracod, *Cypria turneri* in Oregon waters. Here the acanthor hatches and bores through the digestive tract into the body cavity. The acanthor (shelled embryo before escaping from the shell) of *Macracanthorhynchus hirudinaceous* as drawn by Kates (1943) and redrawn by Van Cleave (1947) possessed anterior larval spines for penetration through the gut wall of the beetles as well as small body spines. However, neither Ward (1940) nor Merritt and Pratt (1964) found spines on the acanthor of *Neoechinorhynchus rutili* nor on *N. cylindratus* (Ward, 1940).

Within 6 to 12 days after penetration of the acanthor of *N. rutili* into the body cavity of *Cypria turneri* it had become transformed into an immobile acanthella, which developed 48 to 57 days before it was infective to the fish host. During this period the reproductive organs and proboscis with spines develop into nearly adult form (Merritt and Pratt, 1964). If an acanthella reaches the definitive host before the time it becomes infective, or if an unsuitable host ingests the acanthella, it penetrates the gut wall and continues to develop on the outside of the gut wall or elsewhere into a stage called a juvenile. The juveniles remain infective to the usual definitive host for some time, but the author has failed to obtain infections in a suitable definitive host with juveniles that appeared to have been waiting a long time in that stage as judged by degree of encapsulation by the host where they were found. The second animal invaded is called the paratenic host or in French *hôte d'attente,* a more self-explanatory name. It would appear that 7 or 8 weeks must pass in the definitive hosts of aquatic acanthocephalans as in the terrestrial ones before eggs are found in the feces of the host. This fact was established by Ward (1940) for *N. cylindratus* and by Reish (1950) for *Parafilicollis* (= *Polymorphus*) *kenti* with the assistance of the author during Reish's vacation.

The acanthocephalans, like the cestodes, absorb all nutrients through the cuticle and hypodermis. The electron microscopy study of this area along with histochemistry and biochemistry of the nutrition of the worm should produce interesting information. The various degrees of specificity of the worm to its hosts offers possibilities of study of the nature of this phenomenon. The mechanism of formation of the giant nuclei and the function intrigue the author. The fertilization, subsequent development, and formation of the embryo shell floating in fluid in the pseudocoel has not been described. These are some of the research problems of general interest arising from the study and use of acanthocephalans by biological investigators.

REFERENCES

Golvan, Y. J. (1958). *Ann. Parasitol.* **33**, 358–602.
Golvan, Y. J. (1959). *Ann. Parasitol.* **34**, 7–52.
Golvan, Y. J. (1960). *Ann. Parasitol.* **35**, 138–165, 350–386, 573–593, and 713–723.
Golvan, Y. J. (1961). *Ann. Parasitol.* **36**, 612–647 and 717–736.
Golvan, Y. J. (1962). *Ann. Parasitol.* **37**, 1–72.
Golvan, Y. J. (1964). "An Illustrated Key to the Genera of Acanthocephala," Oregon State University Lithoprint. OSU Bookstores, Inc., Corvallis, Oregon.
Hamannn, O. (1891). *Zt. Naturwissenschaften* **25**, 113–231.
Hyman, L. (1951). "The Invertebrata," Vol. 3. McGraw-Hill, New York.
Kaiser, J. E. (1893). *Bibliotheca Zool. II,* **7.**
Kates, K. C. (1943). *Am. J. Vet. Res.* **4**, 173–181.
Luhe, M. (1911). In "Die Susswasser Fauna Deutschlands" (A. Brauer, ed.), Vol. 16. (Reprinted by J. Cramer, Weinheim, 1961.)
Merritt, S. V., and Pratt, I. (1964). *J. Parasitol.* **50**, 394–400.
Meyer, A. (1928). *Zool. Jahrb. Abt. Anat. Ontog. Tiere* **50**, 117–218.
Meyer, A. (1932–1933). *In* "Klassen und Ordnungen des Tierreichs" (Bronn, ed.), 4 vols., 2 parts. 4 Bd. 2 Abt., 2 Buch. Leipzig.
Petroschenko, V. I. (1956, 1958). "Acanthocephalans of Domestic and Wild Animals," Vols. 1 and 2 (in Russian).
Reish, D. J. (1950). *J. Parasitol.* **36**, 496.
Van Cleave, H. J. (1941). *Am. Naturalist* **75**, 31–47.
Van Cleave, H. J. (1947). *J. Parasitol.* **33**, 118–125.
Van Cleave, H. J. (1948). *J. Parasitol.* **34**, 1–20.
Ward, H. L. (1940). *Trans. Am. Microscop. Soc.* **59**, 327–347.
Yamaguti, S. (1963). "Systema Helminthum," Vol. 5. Wiley (Interscience), New York.

CHAPTER 3

Skeletal Structures and Integument of Acanthocephala and Nematoda

Alan F. Bird and Jean Bird

I. Introduction

In this chapter the integument or body wall of two quite unrelated invertebrate metazoans is studied from the points of view of structure, chemical composition, and function.

It will become clear that the reason the Acanthocephala and Nematoda are considered together here is not because of many basic similarities that they show but rather because of superficial resemblance which happens to be greater between these two groups than between the Acanthocephala and other groups. Because it contains far fewer species and is a much less adaptable group than the Nematoda, its impact on man is correspondingly much less and so research on this group has been limited.

We consider that the body wall includes the cuticle and its appendages, the tissue from which the cuticle is formed, and the muscles which cause movement of this exoskeleton. In both Acanthocephala and Nematoda the cuticle is formed from an underlying layer which is usually syncytial. However, in structure and function the integuments of these two groups are completely different. The Acanthocephala lacking a gut absorb their food through the body wall while the cuticle of the Nematoda acts as an effective barrier to changes in the chemical environment. Far from being inert, as the term exoskeleton may imply, the integuments of both of these groups of animals are metabolically active structures containing various enzymes.

In the following pages we attempt to relate the physiological functions of these body walls with their structure and composition.

II. Acanthocephala

A. Skeletal Structures

1. *Hooks*

a. Structure and Composition. The proboscis in the Acanthocephala is the main organ of attachment of the parasite to the host. It burrows into the longitudinal muscle layers of the gut wall of the host and a fibrous nodule is developed round it (Pflugfelder, 1956; Nicholas and Hynes, 1958). The proboscis, together with its associated structures, is called the presoma (Van Cleave and Bullock, 1950).

The proboscis of acanthocephalans is armed with recurved hooks which vary in size, number, and arrangement in different species but are similar in shape. The hooks in *Polymorphus* consist of a core of cytoplasm surrounded by hard, nonliving material. They arise from the basement membrane of the proboscis wall and are covered with cuticle (Crompton and Lee, 1965).

Crompton (1963) has shown that protein is present in the hooks and there is some phosphatase activity. However, no carbohydrate or lipoprotein has been demonstrated.

b. Function. As the proboscis is the main organ of attachment of the parasite to the host, the function of the hooks is, presumably, one of adhesion.

2. *Spines*

a. Structure and Composition. Cuticular spines of varying size, appearance, distribution, and extent occur on the trunk of many acanthocephalans. In contrast to the proboscis hooks, in *Polymorphus* the spines are formed from the general body wall and are covered with striped layer and cuticle. Toward the apices of the spines, the striped layer becomes thinner and has fewer canals. The spines are supported by fibers of the felt layer (Crompton and Lee, 1965). Histochemical tests indicate the presence of protein, as in the proboscis hooks and no carbohydrate or lipoprotein has been detected (Crompton, 1963).

b. Function. The trunk spines of Acanthocephala function as secondary organs of attachment and help to keep the parasite in close contact with the intestinal mucosa of the host. Van Cleave (1952) suggested that this intimate contact might be important in the nutrition of the parasite, although Edmonds' experiments (1965) suggest that nutrients do not come from the mucosa itself. Crompton (1963) found that the striped layer of *Polymorphus minutus* was rich in alkaline phosphatase in the region of the trunk spines. This evidence for enzyme activity supports Van Cleave's speculation on the significance which the spines might have in the nutrition of this worm as they must also serve to increase the absorptive area of the parasite.

B. Integument

1. *Cuticle, Hypodermis, and Muscle*

a. Structure and Composition. Recent observations on the ultrastructure of the body wall of *Moniliformis dubius* (Nicholas and Mercer, 1965) and on the ultrastructure (Crompton and Lee, 1965) and histochemistry (Crompton, 1963) of the body wall of *Polymorphus minutus* have shown that the body wall or integument of the Acanthocephala is a unique structure completely different from the integuments of other metazoan parasites, but with close similarities within the group itself.

The Acanthocephala have no alimentary tract and because nutrients are probably absorbed through the body wall, a knowledge of its structure is of particular interest.

The body wall differs slightly in structure in different regions (Crompton and Lee, 1965). The anterior region or presoma contains more spines, less canals, and smaller muscles. It is harder to section and has not been examined in the same detail that the posterior region or metasoma has. The following description applies to the body wall of the main body of the Acanthocephala.

Measurements on fresh frozen sections of adult female *Moniliformis dubius* cut through the middle region and examined under phase contrast (Fig. 1A) show that the body wall, including the muscles, is approximately 300 μ thick. It is covered by a thin, irregular deposition of mucopolysaccharide (Crompton, 1963) which is probably secreted by the worm. The external layer or cuticle consists of an outer plasma membane covering a homogeneous layer (Figs. 1E and 2) which is made up of lipoprotein stabilized by disulfide linkages (Crompton, 1963).

The body wall is pierced by pores (Figs. 1E and 2) which lead to canals in the layer beneath, giving it a striped appearance, and for this reason it is called the striped layer. In *Moniliformis dubius* it is 3.5 μ thick and is clearly visible in sections cut with a cryostat and viewed under phase contrast (Figs. 1B, C and 2). The canals, which are lined with the plasma membrane of the cuticle (Figs. 1E and 2)

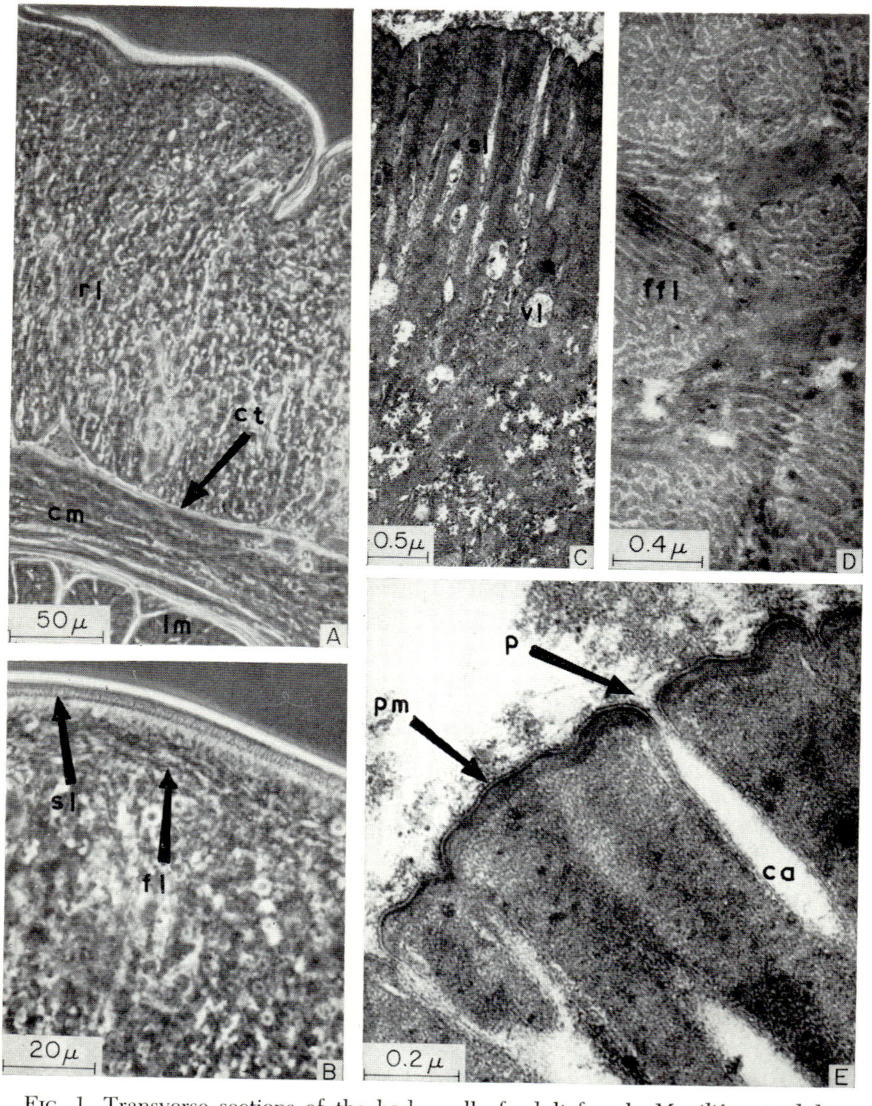

FIG. 1. Transverse sections of the body wall of adult female *Moniliformis dubius*. A. Frozen section examined under phase contrast ×265, showing circular muscle (cm), connective tissue (ct), longitudinal muscle (lm) and the radial layer (rl). B. Frozen section examined under phase contrast ×636, showing the felt layer (fl) and the striped layer (sl). C. Electron micrograph ×15,900, showing the striped layer (sl) and the vesicular layer (vl). D. Electron micrograph ×31,800, showing fibers of the felt layer (ffl). E. Electron micrograph ×63,600, showing a canal (ca), a pore (p), and the plasma membrane (pm).

may open into the vesicles of the vesicular layer beneath (Figs. 1C and 2). This layer merges with the felt layer which consists of masses of disoriented fibers interspersed with lacunae (Figs. 1C, D and 2).

The felt layer in turn merges with the radial layer (Fig. 2) where there are less fibers but larger and more numerous lacunae. This layer contains more mitochondria than the other two and it also contains

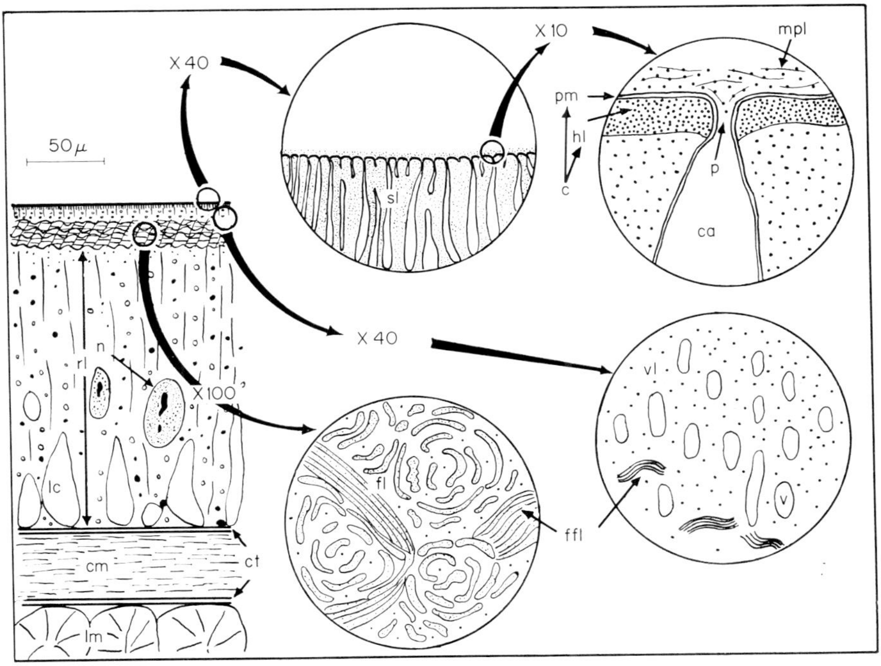

FIG. 2. Diagram illustrating a transverse section of the body wall of *Moniliformis dubius* viewed both under the light and electron microscopes, showing cuticle (c), canal (ca), circular muscle (cm), connective tissue (ct), fibers of felt layer (ffl), felt layer (fl), homogeneous layer (hl), longitudinal muscle (lm), mucopolysaccharide layer (mpl), nucleus (n), pore (p), plasma membrane (pm), radial layer (rl), striped layer (sl), vesicle (v), and vesicular layer (vl).

the nuclei of the body wall (Crompton and Lee, 1965). Histochemical tests have revealed the presence of lipid and polysaccharide in all three of these layers.

A number of enzymes have been detected in the layers of the body wall internal to the cuticle in different species of Acanthocephala. These enzymes, absent in some of the species studied, are alkaline phosphatase, lipase, nonspecific esterase, and leucine aminopeptidase. Their localization in the body wall and its relationship to function will be considered later in this chapter.

The radial layer is separated from a layer of connective tissue, visible under low power phase contrast (Figs. 1A and 2), by a convoluted plasma membrane and a broad basement membrane which is visible under the electron microscope (Nicholas and Mercer, 1965; Crompton and Lee, 1965). No cell walls have been seen in the body wall, which is therefore considered to be a syncytium.

Associated with the body wall and internal to it is a layer of circular muscle (Figs. 1A and 2) separated from a layer of longitudinal muscle by connective tissue.

b. Function. A number of enzymes have been detected by histochemical methods in the body wall of Acanthocephala. Alkaline phosphatase (Bullock, 1958; Crompton, 1963) and leucine aminopeptidase (Crompton, 1963) activity were detected in the striped layer and may be associated with absorption and secretion in this region. Leucine aminopeptidase has also been detected in the radial layer (Crompton, 1963) and nonspecific esterase throughout the body wall beneath the cuticle.

Crompton (1963) suggests that all layers beneath the cuticle are of metabolic importance, the striped layer being associated with absorption and the radial layer, which contains many mitochondria (Crompton and Lee, 1965), being involved in the metabolism of absorbed compounds.

Crompton and Lee (1965) suggest that lipid is absorbed through the canals of the neck, passed through the lemnisci into the body cavity, then into the connective tissue of the metasoma where it enters the radial layer through the convoluted plasma membrane and is utilized.

However, to this date only small molecules have been shown to pass through the integument.

Graff (1964) has demonstrated the absorption of ^{14}C labeled glucose through the body wall of *Moniliformis dubius* and Rothman and Fisher (1964) have shown both in this species and in *Macracanthorhynchus hirudinaceus* that a number of ^{14}C-labeled amino acids are also absorbed through the integument *in vitro*.

Edmonds (1965) using ^{14}C-labeled leucine and ^{32}P-labeled sodium phosphate has shown that *in vivo* these substances pass into *Moniliformis dubius* and are obtained by the worm from the host's gut contents rather than its gut mucosa.

Bryant and Nicholas (1965) have examined the route of glucose oxidation in the body wall of *Moniliformis dubius* and found it similar to that described for other helminths involving the formation of pyruvate by the glycolytic pathway and the carboxylation of pyruvate by malic enzyme to give malate, which was converted to either fumarate and succinate or to oxaloacetate and succinate.

There is some speculation on the means by which food is taken through the body wall of Acanthocephala. It has been tentatively suggested by Nicholas and Mercer (1965) and supported by Edmonds and Dixon (1966) that pinocytosis on the surface of the body wall plays a part in the assimilation of nutrients. However, these observations are probably influenced by the interpretation of electron micrographs which do not show the continuity of the plasma membrane from the surface through the pores and into the canals, which can be demonstrated in transverse section (Figs. 1E and 2). Hence material in the canals of the striped layer has not necessarily been assimilated and may in fact still be outside the body wall.

The hypothesis put forward by Crompton and Lee (1965) that nutrients pass through pores and canals into the worm receives support from observations on the ultrastructure of a number of species of Acanthocephala. It is clear that this invagination of the plasma membrane leads to an increase in the absorptive area of the body wall in much the same way as the absorptive area of the host's intestine is increased by infoldings.

III. Nematoda

A. Skeletal Structures

1. Mouth Parts

a. Teeth. The nematode mouth is terminal and is bordered by six lips which may be fused or otherwise modified. They are lacking altogether in some species. The mouth leads into the buccal cavity (or stoma) which shows great variation in shape, size, and degree of differentiation. In many species of nematodes teeth are developed. Nematode teeth fall into two categories, namely, odontia and onchia. Cobb (1919) defined odontia as teeth arising from the labial region while onchia arise more posteriorly and are more closely associated with the esophagus.

The development of teeth is closely associated with the feeding habits of nematodes. Thus, the zooparasitic nematodes which feed on the intestinal contents of the host and the free-living ones which feed on bacteria do not, in the main have well-developed teeth although they may have small or sometimes large odontia. On the other hand, the zooparasites which feed on the intestinal or respiratory tract mucosa of the host, the predatory free-living forms, and those which feed on plant material usually have large teeth or cutting plates or both. *Ancylostoma*, for example, has two ventro-lateral cutting plates at the entrance to the buccal cavity and in addition to these there may be sharp lancets or teeth. The cutting plates are odontial in origin while the lancets are onchial.

Inglis (1964a) in his detailed study of the head of the marine Enoplida found that in some species, three massive mandibles are formed in addition to the three large onchia which are present. The mandibles are formed from a thickening of the cuticle lining the buccal cavity wall. He considers that the mandible-onchium system must function as a unit although its components are derived from different sources. He also considers that the mandible complex has a skeletal function as well as acting as a gripping organ or jaws.

Other types of toothlike structures which may occur in nematodes are denticles and leaf crowns (corona radiata). Both these types of structure arise in the labial or buccal regions and may, therefore, be considered to be types of odontia. Little is known of the structural and chemical composition of nematode teeth. As they are all cuticular in origin, it seems likely that their composition resembles that of the cuticle of the body wall.

b. Stylets. In all the phytoparasitic nematodes, many of the free-living dorylaims and several zooparasitic species, a special type of feeding mechanism, namely, a stylet or spear, is developed. In the zooparasites which have stylets, the stylet may be present only in the infective stage, but in some species it persists in the adult. Occasionally the stylet is absent in degenerate male phytoparasites but is always present in the larval forms. The stylets in the phytoparasites and the dorylaims are different in origin. In the Tylenchida the stylet is hollow, tripartite, cuticular, and usually has basal knobs or swellings for the attachment of muscles. Allen (1960) considers it to have developed, at least in part, from a collapse and fusion of the walls of the buccal cavity (stoma) and hence it is referred to as a stomatostyl. In the Dorylaimida the spear is different in that its anterior part is developed from a special cell in the anterior part of the esophagus and migrates into place during and after each molt. It is therefore more properly considered to be an onchiostyl since it originates from the same region as do onchia. However, the term odontostyl has been used by several authors, notably Coomans and de Coninck (1963). In dorylaim larvae, the spear tip of the next molt can often be seen developing in the esophagus behind the functional spear. In *Nygolaimus* a subventral tooth is developed in the same way as the spear of the Dorylaimida and is considered to be homologous with it. It is not hollow but it serves the same purpose as the spear.

The tylenchid stomatostyl and the dorylaim onchiostyl are both used in feeding. The stylet is used to puncture the cell wall and the cell contents are withdrawn, usually through the hollow stylet. In the case of *Nygolaimus*, the prey is pierced by the tooth, which is then withdrawn and the contents are sucked into the esophagus.

The endophytoparasitic nematodes use the spear for penetration of the host as well as for feeding once inside the host. Those of the zooparasites which also possess spears also use them for penetration and/or laceration of the host tissue.

2. *Spicules*

Most male nematodes possess copulatory spicules which are lodged in and secreted by spicular pouches. The pouches are formed by cloacal invaginations which have the same histological structure as the cloacal wall. There are usually two spicular pouches and they unite prior to their entrance into the cloaca. The spicules consist of hard scelerotized cuticle with a cytoplasmic core. They appear to be well supplied with nerves, and have been shown to contain esterase (Lee, 1962a; Bird, 1966). Typically there are two spicules although there is only one in some species and they vary greatly in shape and size.

The spicules are often accompanied by an accessory structure, the gubernaculum, which is formed from the wall of the spicular pouch. The gubernaculum is essentially a plate in the groove of which the spicules move (Chitwood and Chitwood, 1950), but in some species the median part projects into the spicular pouch separating the spicules.

In the Strongyloidea another accessory structure, the telamon, is sometimes developed in the ventral and lateral walls of the cloaca. It is scelerotized and immovable. The most important difference between the gubernaculum and the telamon is that the gubernaculum is formed from the spicular pouch while the telamon is formed from the cloacal lining. Both apparently serve to direct the spicules when they are protruded, to prevent their piercing the cloacal wall.

The spicules are thought primarily to spread open the vulva and vagina during copulation and apparently do not transmit the sperm. The order Trichuroidea is unique among the nematodes in that it has a long cloaca (often lined with spines) which is eversible in males and acts as a cirrus. In the Trichuridae (but not in other groups) the cirrus is accompanied by a single copulatory spine.

The strongyloid nematodes are different from all the others in that the males have a conspicuous expanded bursa (reduced in some species) supported by muscular rays, thus differing from the dictophymoid bursa which has muscles but no rays and the bursa of rhabditoids and other forms which consists of cuticular alae. The strongyloid bursa typically has two lateral lobes (each supported by six muscle rays) and a small median lobe with one ray. The rays may branch and divide and the pattern so formed is important taxonomically.

3. *Other Cuticular Structures*

a. Outgrowths. Alae are cuticular projections or extensions either extending throughout the length of the body or confined to the anterior or posterior ends. There are three different types of alae. Longitudinal alae are sublateral solid ridges on top of the lateral cords. They are usually four in number and their shape appears to vary from species to species (Inatomi *et al.*, 1963). Cervical alae are modified lateral alae in the anterior region of some zooparasites. They are composed of the cortical and matrix layers of the cuticle and usually have a hard skeletal support. Caudal alae occur in the posterior region of some males. They apparently serve as clasping organs during copulation. There may be sensory papillae on the caudal alae.

Papillae which are considered to have a sensory function (possibly tactile) occur on the lips of nematodes. Typically, there are three papillae or setae on each of the medial lips and two on each of the lateral lips. The papillae or setae are arranged in two circles of six and an outer circle of four but the basic pattern may be modified by fusion or reduction. They are innervated by six cephalic papillary nerves. In addition to the cephalic papillae, the anterior region often bears a pair of cervical papillae in the vicinity of the nerve ring. They are called deirids and are innervated by lateral nerves.

The genital papillae which occur in males are rounded cuticular elevations which are sometimes on stalks. They may be surrounded by punctations forming rosettes and are sometimes supported by cuticular plates called plectanes. In several families of zooparasites a single large sucker, the preanal sucker, occurs. This may be circular or oval and is supplied with muscle fibers.

Goodey (1951) described a bright, lenslike structure in the vicinity of the excretory pore which he called the hemizonid. The hemizonid, which forms a semicircle on the ventral side of the nematode body is formed by an alteration in the layering of the body wall and lies between the hypodermis and the cuticle. Cephalids, which are similar to hemizonids, occur in the cephalic region of the nematode and extend in a complete circle round the wall (Hirschmann, 1956). In both hemizonids and cephalids part of the middle cuticular layer and the innermost oblique fiber layers are absent (Hirschmann, 1960).

Goodey (1959) observed nerve fibers passing from the circumesophageal nerve ring to the hemizonid as it encircles the ventral half of the body between the cuticle and hypodermis. Caveness (1961) has described a small "mimic" hemizonid on the ventral side of the body, generally at the level of the basal bulb. He has called it the hemizonion and suspects that it, like the hemizonid, functions as a nerve commissure.

Various other cuticular structures, e.g., spines, warts, punctations occur in some nematodes. Punctations, which appear as minute dots on the cuticle, are considered by Inglis (1964a) to be the openings of punctation canals or their derivatives viewed end on. Sometimes the cuticle is swollen in a blisterlike manner and this is termed inflation. It is caused by the separation of the cortical layers from the rest and is largely due to liquefaction of the matrix layer (Hirschmann, 1960).

b. Invaginations. A pair of lateral sensory organs called amphids occurs in the cephalic region of nematodes. Their external openings are cuticular and vary in shape in different nematodes. The external openings connect with amphidial ducts which widen to form amphidial pouches in which are located a group of narrow sensory neurons. The presence of esterase has been reported in the amphidial pouches of several phytoparasitic species (Rohde, 1960; Lee, 1964; Bird, 1966) and it is possible that it may have an active role in the nematodes' sensory systems. It has been suggested that the amphids are chemoreceptors but, because of their size, it is difficult to obtain experimental evidence for this.

Many nematodes possess a pair of cuticular pouches (papillae or platelike scutella in some species) near the posterior end. These are called phasmids and appear to be sensory in nature. Some phasmids have a sensilla which is innervated by lateral caudal nerves and may also have a gland (Hirschmann, 1960).

The buccal cavity, esophagus, rectum, cloaca, and vagina are also lined with cuticle.

B. Integument

1. Cuticle

a. Structure. The nematode cuticle shows great structural diversity in different families and there are often differences, not only between the sexes within a single species, but in thickness and sometimes structure in different regions of the same individual.

The structure of the nematode cuticle has recently been reviewed (Inglis, 1964c; Lee, 1966b). Inglis suggests that the principal layers be named cortical, matrix, and basal, and Lee accepts this terminology. These layers are analogous with the terms cortical, homogeneous, and fiber layers (Bird and Rogers, 1965), although Inglis points out that spiral or oblique fibers do not occur in the cuticle of all nematodes and cites the Chromadorida as an example. All Inglis's observations are based on studies made under the light microscope, but support for his findings concerning the absence of spiral and oblique fibers in the Chromadorida has been supplied by B. D. Watson (1965a) who examined the body wall of *Euchromadora vulgaris* under the electron microscope and was not able to demonstrate a system of diagonal fibers. However, she noted that the basal layer contained collagen and this has been associated with the fiber and other layers of *Ascaris* (Fauré-Fremiet and Garrault, 1944; Chitwood and Chitwood, 1950; Dawson, 1960). Its structure appears hard to resolve and further studies are required before the absence of nondiagonal fibers is established. However, for the sake of uniformity in nomenclature it may be said that the nematode cuticle always consists of at least two major layers—the cortical layer and the basal layer. These layers are nearly always separated by a third layer, the homogeneous or matrix layer, which is absent from the cuticles of adult females of the family Heteroderidae (Wieser, 1953; Ferris and Siegel, 1957; Bird and Rogers, 1965) and the cuticle of the mature larva of *Trichinella spiralis* (Beckett and Boothroyd, 1961).

Research on the structure of nematode cuticles, particularly at the electron microscope level, progresses rapidly and many different species are now being examined. This research serves to highlight the diversity of structure of the nematode cuticle and the extent to which it may digress from the basic pattern. For instance, the cuticle of adult *Nippostrongylus brasiliensis* (Lee, 1965; Jamuar, 1966) has a fluid-filled layer which contains hemoglobin and esterase and is traversed by struts which are supported by fibrils of collagen. Again the cuticles of *Trichuris muris* and *T. vulpis* (Sheffield, 1963) and *Capillaria hepatica* (Wright, 1963) differ from other nematodes in that they contain specialized regions known as bacillary bands which involve the underlying hypodermis. However, the structure of the cuticles of many nematodes parasitic in animals, such as oxyurids, strongylids, and ascarids (Bird, 1958a) as well as that of some nematodes parasitic in plants, such as *Xiphinema index* (Roggen *et al.*, 1967) shows great similarity. All these forms have cuticles with outer and inner cortical layers, a homogeneous layer, and fiber layers. Because this type of cuticle is usually associated with large nematodes which are often readily available for teaching purposes involving both chemical analysis and structural differentiation using the light microscope, we propose to consider the cuticle of *Ascaris lumbricoides* in some detail and compare the cuticles of various other forms with it.

Research on the structure of ascarid cuticles has been going on for more than 120 years. The history of investigation in this field has been summarized (Bird and Deutsch, 1957). Perhaps the most significant single contribution in the early days came from van Bömmel (1895) whose drawings of transverse and longitudinal sections (together with surface preparations) of *Ascaris* cuticle have not been greatly improved by subsequent light microscope studies and whose nomenclature for *Ascaris* cuticle is still used with only minor modification.

The use of the electron microscope has in the past decade led to considerable advances in our knowledge of the structure of cuticle, particularly in small forms of about half a millimeter in length.

The outermost layer of the nematode cuticle consists of a thin osmiophilic layer first detected by Bird and Deutsch (1957) in *Ascaris lumbricoides;* they stated that it was less than 100 mμ. thick. This observation has subsequently been verified in this species by B. D. Watson (1965b) and in *Parascaris equorum* by Hinz (1963) who reported that it was about 45 mμ. thick in this species. It has been suggested by Lee (1965) that this layer may in fact be a triple-layered membrane similar to that first reported by Beckett and Boothroyd (1961) in *Trichinella spiralis* and subsequently by Lee (1965, 1966a) and Jamuar (1966) in the larva and adult of *Nippostrongylus brasiliensis,* by Anya 1966a) in *Aspicularis tetraptera,* by Thust (1966) in the larvae of *Ascaris lumbricoides* and by Roggen *et al.* (1967) in *Xiphinema index.*

It is also visible on the surface of the cuticles of a number of other nematodes in published papers in which its presence has not been reported; for instance, it appears to be present on the surface of the cuticle of adult *Meloidogyne javanica* (Fig. 3H). This three-layered membrane which covers the nematode and lies on the surface of the external cortical layer may be common to all members of the Nematoda and in this respect they may resemble the Acanthocephala (Figs. 1E and 2).

There is considerable variation in the structure of the cortical layers. In some forms such as *Nippostrongylus brasiliensis* (Lee, 1965; Jamuar, 1966) it is only a single homogeneous layer. However, in the majority of forms that have been studied it consists of two layers, external and internal cortical layers. In *Aspicularis tetraptera* (Anya, 1966a) the outer cortex is amorphous and not very electron dense, but in forms such as *Ascaris lumbricoides, Parascaris equorum, Euchromadora vulgaris,* and *Meloidogyne javanica* it is extremely electron dense (Bird, 1957; Bird and Rogers, 1965; Hinz, 1963; B. D. Watson, 1965a,b). In *Meloidogyne* it is present (Figs. 3G,H,I,J,K) in both larval and adult stages as a dense osmiophilic layer.

In *Ascaris lumbricoides* (B. D. Watson, 1965b) the external cortical layer appears amorphous while the internal cortical layer is made up

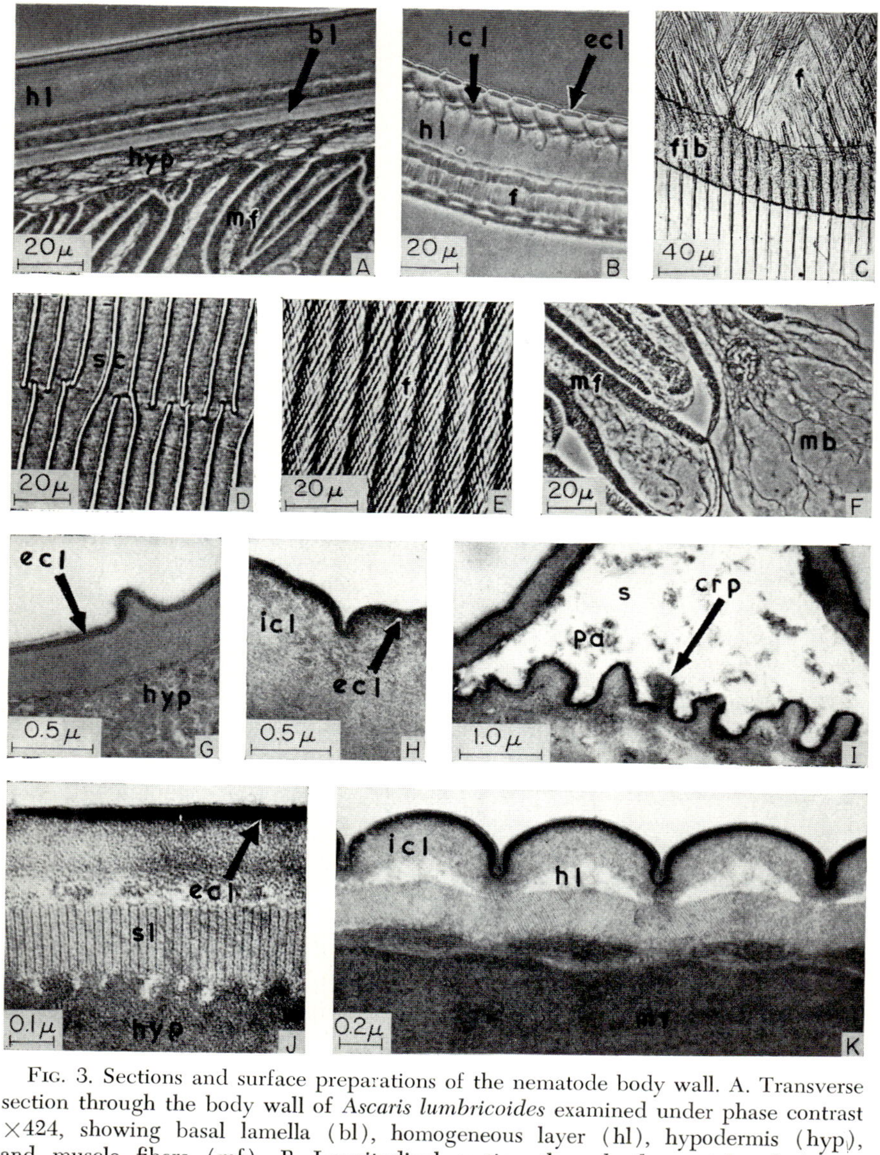

FIG. 3. Sections and surface preparations of the nematode body wall. A. Transverse section through the body wall of *Ascaris lumbricoides* examined under phase contrast ×424, showing basal lamella (bl), homogeneous layer (hl), hypodermis (hyp), and muscle fibers (mf). B. Longitudinal section through the cuticle of *Ascaris lumbricoides* examined under phase contrast ×371, showing external cortical layer (ecl), fiber layers (f), homogeneous layer (hl), and internal cortical layer (icl). C. Surface view of *Ascaris lumbricoides* cuticle after some of the layers had been stripped showing fiber layers (f) and fibrillar layer (fib) ×371. D. Surface view of *Ascaris lumbricoides* cuticle ×371, showing the striations on the surface of the cuticle (sc). E. Surface view of fiber layers (f) in the cuticle of *Ascaris lumbricoides*

of a loose network of fibers 75–100 mμ. in diameter. The transverse grooves on the surface of the cuticle extend inward through the external cortical layer and divide it into a series of rings.

The external cortical layer is often demonstrated as a definite layer by histological methods using the light microscope, and its distinctive composition is discussed later in this chapter.

Under the transverse grooves in ascarids are lines of structures which have been variously referred to as "circular lamellae" by van Bömmel (1895), "*Saftbahnen*" by Toldt (1899), "fibers" by *Goldschmidt* (1904), "strands of condensed material" by Chitwood and Chitwood (1950), "pore canals" by Bird (1958a), Bogoiavlenskii (1958), and B. D. Watson (1965b), "sectioning artifacts" by Hinz (1963) and "a reflection of the plasticity of the matrix layer" by Inglis (1964c).

Whether or not these are pore canals has not yet been clearly resolved. They may function both as skeletal supports and channels through which material may travel to the surface layers of the cuticle. Structures with similar functions but which are morphologically different may occur throughout the Nematoda.

Experimental evidence is needed before the function of these structures can be established. In particular they should be studied during growth of the cuticle. B. D. Watson (1965b) has done this with *Ascaris lumbricoides*. This nematode is only about 1–2 cm. in length after its final molt and over a period of several months it grows up to 30 cm. Watson measured changes in thickness of the whole cuticle as the worm grew and found that thickness of the cuticle is directly proportional

---

×477. F. Transverse section through somatic muscle of *Ascaris lumbricoides* examined under phase contrast ×318, showing belly of muscle (mb) and muscle fibers (mf). G. Longitudinal section through the body wall of a larva of *Meloidogyne javanica* viewed under the electron microscope ×21,200, showing the external cortical layer (ecl) and hypodermis (hyp). H. Longitudinal section through the cuticle of an adult female *Meloidogyne javanica* viewed under the electron microscope ×21,200, showing the external cortical layer (ecl) and internal cortical layer (icl). I. Transverse section through the cuticle of a larva of *Meloidogyne javanica* in the process of molting viewed under the electron microscope ×10,600, showing areas of resorption in the cortex (crp), particles (pa), and the space between the cuticles (s). J. Transverse section through the body wall of an infective larva of *Meloidogyne javanica* viewed under the electron microscope ×63,600, showing the external cortical layer (ecl), the hypodermis (hyp), and the striped layer (sl). K. Longitudinal section through the body wall of an infective larva of *Meloidogyne javanica* viewed under the electron microscope ×29,150, showing the homogeneous layer (hl), the internal cortical layer (icl), and muscle fibers (mf). Figures 3B, C, D and E after Bird and Deutsch (1957) *Parasitology;* reproduced by kind permission of Cambridge University Press. Figures 3G, H and I after Bird and Rogers (1965) *Nematologica;* reproduced by kind permission of E. J. Brill, Leiden.

to the length of the worm. She also measured increases in width of individual layers during this period and found that the homogeneous layer increases in thickness more rapidly than the fiber layers and that the fiber layers grow more rapidly than the cortex. Watson found that in young forms the pore canals extend into the fiber layers. When viewed in tangential sections under the electron microscope after staining with uranyl acetate, these canals have dark walls enclosing a light lumen. These observations appear to be the only ones that have been made on cuticular growth in ascarids. Both molting and growth have been studied in *Meloidogyne javanica* (Bird and Rogers, 1965) and this will be discussed later.

Electron microscope studies of the homogeneous layer in ascarids have led to conflicting reports. B. D. Watson (1965b) states that it is amorphous, an observation that is in keeping with the report of Bird and Deutsch (1957) that is does not show birefringence when viewed under the polarizing microscope. Bird and Deutsch (1957) reported radial striations in the homogeneous layer, a finding which was not in keeping with the absence of any marked birefringence in this area and which, in view of Watson's recent investigation and the considerable advances in embedding techniques which have taken place over the last decade, is perhaps open to criticism. This layer in *Parascaris equorum* was viewed under the electron microscope by Hinz (1963) who reported the presence of fine striations running parallel to the surface. Taking into account phase contrast and polarizing microscope studies on fresh frozen material (Bird and Deutsch, 1957) and recent electron microscope studies (B. D. Watson, 1965b), this layer has perhaps been best described by Lee (1966b) as "rather plastic and homogeneous."

The boundary layer which lies between the homogeneous layer and the fiber layers is about 1 μ thick in *Ascaris lumbricoides* (B. D. Watson, 1965b). It is visible in sections of the cuticle examined under phase contrast, exhibits birefringence, and is visible in stained sections (Bird and Deutsch, 1957). Both the homogeneous layer and boundary layer can be seen in Figs. 3A and B.

The fiber layers of *Ascaris* cuticle are readily demonstrated under the light microscope (Figs. 4, 3A, B, C and E). Their presence has been repeatedly verified by a large number of workers in many different species. Under the electron microscope it can be shown that these layers are made up of irregularly shaped fibers which consist of bundles of tightly packed fibrils each less than 10 mμ. in diameter (B. D. Watson, 1965b). These fibrils do not show the banding visible in vertebrate collagen fibrils under the electron microscope but X-ray diffraction studies

(Fauré-Fremiet and Garrault, 1944; Picken *et al.*, 1947) and various chemical tests (Chitwood, 1936; Bird, 1957; Dawson, 1960) indicate its close affinities.

Each fiber layer consists of parallel strands which run in spirals at about 75° to the longitudinal axis of the worm. The middle of the three

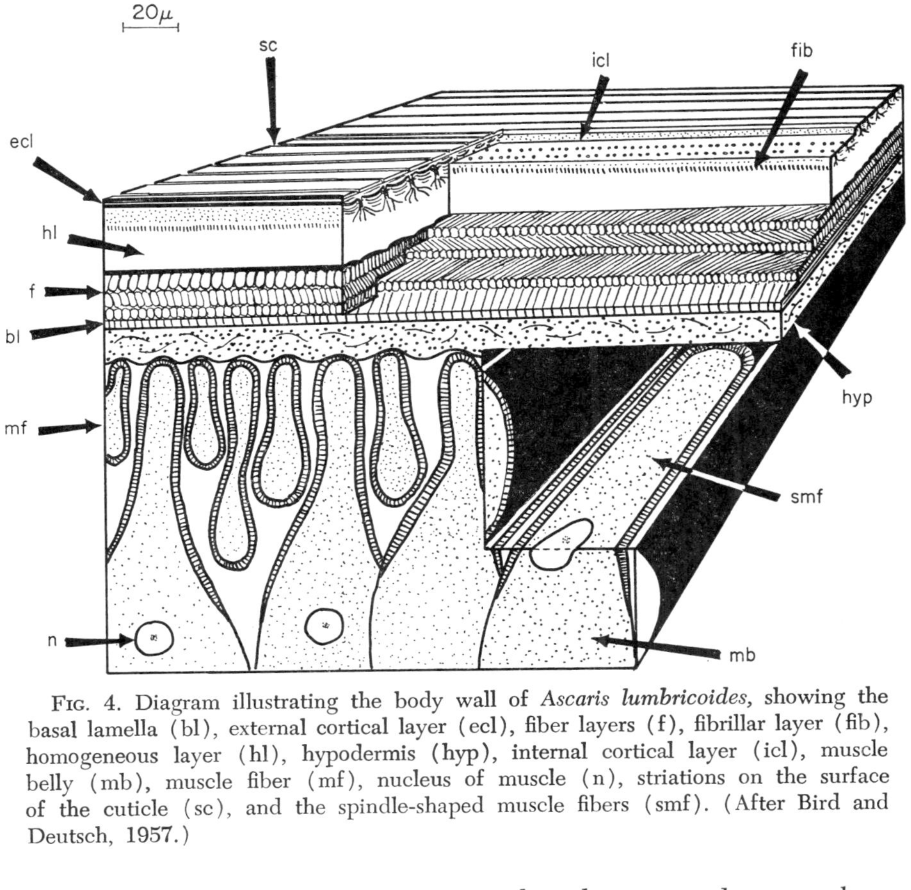

FIG. 4. Diagram illustrating the body wall of *Ascaris lumbricoides*, showing the basal lamella (bl), external cortical layer (ecl), fiber layers (f), fibrillar layer (fib), homogeneous layer (hl), hypodermis (hyp), internal cortical layer (icl), muscle belly (mb), muscle fiber (mf), nucleus of muscle (n), striations on the surface of the cuticle (sc), and the spindle-shaped muscle fibers (smf). (After Bird and Deutsch, 1957.)

layers runs in the opposite direction to the other two and crosses them at an angle of 135° to form a mobile lattice which provides for an anisometric expansion and contraction under the action of the longitudinal muscles (Harris and Crofton, 1957).

A layer known as the basal lamella separates the fiber layers from the hypodermis (Figs. 4, 3A and B). It consists of fine fibrils similar to those of the fiber layers (Hinz, 1963; B. D. Watson, 1965b) and it merges with the hypodermis beneath. A similar state of affairs appears to exist in *Meloidogyne javanica* where the fiber layers appear to merge with the hypodermis (Bird and Rogers, 1965). However, Watson (1965b) reports the presence of a plasma membrane in *Ascaris lumbricoides* which separates the basal lamella from the hypodermis. Similarly in *Xiphinema index* (Roggen *et al.*, 1967) and the larva of *Nippostrongylus brasiliensis* (Lee, 1966a) the cuticle is separated from the hypodermis by a plasma membrane. However, similar structures in *Parascaris equorum* are thought by Hinz (1963) to be membranes of the hypodermal endoplasmic reticulum.

Comparative studies of the nematode cuticle have been recently done by Inglis (1964b,c). All the material used was examined under the light microscope so that all observations and conclusions were limited to this level of resolution. Nevertheless many different species of nematodes have been examined in these studies and a number of interesting points have been raised.

Inglis (1964c) interprets all modifications of the nematode cuticle as being based on and built around a system of punctation canals which evolved "as a method of allowing the cuticle to grow by the incorporation of additional materials from processes of the hypodermis." In his studies on various species of ascarids Inglis (1964b) regards the punctation canals as inwardly running strands from the fibrillar layer which meet with outwardly running strands from the boundary layer. He considers that the functional demands of forms such as *Ascaris, Enoplus, Mermis,* and *Dorylaimus* are met by the development of spiral layers of fibers which are absent in the Chromadorida. Thus this basketwork of fibers does not occur in all nematodes and generalization on its role in nematode locomotion must be treated with caution.

It is interesting to note that the larval cuticles of many different parasitic species show a basically similar pattern (Inatomi *et al.*, 1963; Eckert and Schwarz, 1965; Lee, 1966a; Jamuar, 1966; Bird, unpublished observations). This consists broadly of cortical, homogeneous, striped, and basal layers (Figs. 3J, K and 5).

The striped or striated layer is common to larvae of *Meloidogyne javanica* and *Tylenchulus semipenetrans* (Bird, unpublished), *Haemonchus contortus, Trichostrongylus colubriformis, Cooperia punctata, Bunostomum trigonocephalum, Strongyloides ransomi* (Eckert and Schwarz, 1965), *Necator americanus, Ancylostoma duodenale, Ancylostoma caninum, Trichostrongylus orientalis, Strongyloides stercoralis, Strongyloides fulleborni* and *Strongyloides ratti* (Inatomi *et al.*, 1963), and *Nippostrongylus brasiliensis* (Inatomi *et al.*, 1963; Lee, 1965; Jamuar, 1966). In *N. brasiliensis* Jamuar (1966) reports that this layer, when viewed under the electron miscroscope in sagittal section, consists of two sets of parallel dense lines running at right angles. In both longitudinal and transverse sections these are the vertical stripes and from a sagittal section Jamuar (1966) has calculated that their anterior-posterior periodicity is 18 mμ., while the lateral periodicity is 27 mμ. The distance between the stripes probably varies according to methods of preparation. For instance, Lee (1966a) reports that it is 15 mμ. when viewed in longitudinal section in the same species. It varies a little from species to species as Inatomi and his co-workers have shown that

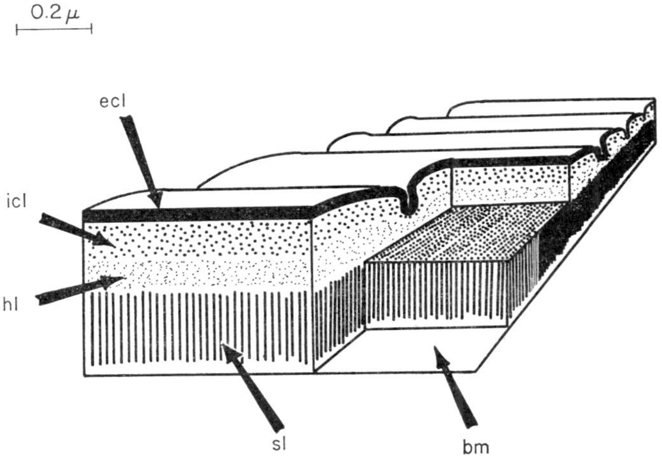

FIG. 5. Diagram illustrating the cuticle of a typical second-stage infective larva, showing the basement membrane (bm), external cortical layer (ecl), homogeneous layer (hl), internal cortical layer (icl), and the striped layer (sl). (After Inatomi *et. al.,* 1963.)

in the eight species that they studied it varied from 9 mμ. to 15 mμ. In transverse sections of larvae of *Meloidogyne javanica* it is about 18 mμ. (Fig. 3J).

The function of this striped layer or striated layer as Lee (1966a) calls it, has been discussed by Inatomi *et al.* (1963) who consider that it may be concerned with molting. On the other hand, Lee (1966a) considers that its regular spacings suggest that it is composed of a protein with very close linkages between the molecules, resulting in a resistant layer which would protect the larva from fluctuations in its environment.

In view of the fact that reports of these structures so far are associated with nematodes which are subjected to fluctuating environments, it would appear that Lee's speculations are the more probable.

b. Composition. Since the early days of investigation into the composition of nematode cuticle, it has been established that its solids consist predominantly of protein and, unlike arthropod cuticle, it does not contain chitin (Lassaigne, 1843; Sukataschoff, 1899; Flury, 1912; Magath, 1919; Mueller, 1928). In later years traces of lipid and carbohydrate have also been detected in small quantities in the cuticles of various species of nematodes (Chitwood, 1936, 1938; Trim, 1949; Bird, 1956, 1957, 1958a; Bird and Rogers, 1956; Simmonds, 1958; Monné, 1959; M. R. Watson and Silvester, 1959; Mackintosh, 1960; Lee, 1962b; Beames, 1964; Anya, 1964, 1966a).

The cuticle of *Ascaris lumbricoides* contains about 75% water (Fairbairn, 1957). It is not stiff and unyielding and, in fact, Harris and Crofton (1957) report changes in length of 10–15% for pressure differences within the normal physiological range. Quantitative estimation of the amino acid content of *Ascaris lumbricoides* cuticles using the methods of Moore and Stein (1951) have been done (Bird, 1957; M. R. Watson and Silvester, 1959). The results of these separate analyses are shown in Table I. From this table it can be seen that, with the exception of valine, there is close agreement in the quantities of amino acids present. Hydroxyproline, which was not verified on the column (Bird, 1957), was nevertheless identified by means of paper chromatography using specific reagents. On the basis of these experiments it was concluded that the whole cuticles contain proteins which are related to the vertebrate collagens. For instance, they contain hydroxyprolyl residues, have a high content of glycyl residues, and have an infrared absorption band at 3330 $cm.^{-1}$ However, in many respects their amino acid composition is at variance with typical collagen proteins. For instance, hydroxylysyl residues have not been detected.

The individual layers of a nematode cuticle were first studied by Chitwood (1936) using enzymes to separate and digest the layers of *Ascaris lumbricoides* cuticle. Since this earlier work the layers in a number of nematode cuticles have been examined using X-ray diffraction, electron microscopy, chromatography, electrophoresis, and histochemistry.

A thin osmiophilic superficial layer has been recorded from a number of nematodes and can only be seen under the electron microscope. In *Ascaris lumbricoides* it is less than a tenth of a micron thick and may occur commonly throughout the Nematoda as a three-layered membrane, as suggested by Lee (1965). Chemical evidence for the lipid content of this outermost layer has been obtained (Chitwood, 1938; Trim, 1949; Bird, 1957).

The composition of the external cortical layer has received considerable attention over the past decade. Because of its sulfur content it has been considered to be a keratin (Flury, 1912; Chitwood, 1936). Further support for this view has come from Savel (1955) who resolved the amino acids obtained by acid hydrolysis of the pepsin-resistant part of *Ascaris lumbricoides* cuticle and compared the histidine–lysine–arginine ratio that he obtained with that obtained for keratins. Further

TABLE I
AMINO ACID COMPOSITION OF *Ascaris lumbricoides* CUTICLE GIVEN AS GRAMS OF AMINO ACID NITROGEN[a]

Amino acid	After Bird (1957)	After M. R. Watson and Silvester (1959)
Alanine	5.6	6.0
Amide N	—	3.39
Ammonia	4.3	—
Arginine	11.2	12.5
Aspartic acid	5.2	5.4
Glutamic acid	5.9	6.0
Glycine	18.1	20.4
Histidine	1.94	2.09
Hydroxylysine	—	0.0
Hydroxyproline	—	1.58
Isoleucine	1.44	1.25
Leucine	1.94	1.55
Lysine	7.7	6.2
Methionine	1.00	0.77
Phenylalanine	0.88	0.67
Proline	22.1	22.7
Serine	1.78	1.82
Threonine	1.44	1.47
Tyrosine	0.50	0.24
Valine	4.25	1.29
Unknown	1.51	—
Total	96.8	95.3
Method of hydrolysis	80% formic acid –11.7 *N* HCl (1:1)	6 *N* HCl at 110° for 24 hours
Method of analysis	Stein and Moore (1950)	Moore and Stein (1951)

[a]Values per 100 gm of total nitrogen.

evidence for keratin has been provided by the histochemical detection of disulfide and sulfydryl groups in this layer (Carbonell and Apitz, 1961; Anya, 1966a). The fact that hydroxyproline has not been detected in this layer indicates that it does not resemble a typical vertebrate collagen (Savel, 1955; Bird, 1957).

However, it has been shown that this layer gives a collagen-type X-ray diffraction pattern (Fauré-Fremiet and Garrault, 1944), is soluble in hot dilute alkali (Fairbairn, 1957) and is not completely dissolved by 0.5% thioglycollate (Bird, 1957, 1958b; Anya, 1966a). There is considerable variation in histidine–lysine–arginine ratios between different sources of keratin and even the figures obtained for similar keratins vary with different investigators (Crewther *et al.*, 1965) so that this ratio is not of much significance in the determination of different types of protein.

Because of the small size of the majority of nematodes and the thinness of the external cortical layer, histochemical methods, which are relatively nonspecific, have been used and, as has been mentioned above, both disulfide and sulfydryl groups have been demonstrated in this layer in *Ascaris lumbricoides* by Carbonell and Apitz (1961) and Anya (1966a) and also in the same region in *Aspiculuris tetraptera* and *Syphacia obvelata* by Anya (1966a). This corroborates the earlier reports of sulfur-containing protein and the later demonstration of sulfur-containing amino acids in hydrolyzates of this layer (Bird, 1957), although these have also been detected in hydrolyzates of the fiber layers (Bird, 1957). However, both Carbonell and Apitz (1961) and Anya (1966a) report the presence of sulfydryl groups in this and other layers, so that these groups are not necessarily related to the resistance of the external cortical layer to various solvents. Nevertheless, disulfide groups appear to be peculiar to the external cortical layer and are probably associated with its resistant properties.

In addition, there is evidence for quinone tanning in this layer (Brown, 1950; Bird, 1957, 1958b; Monné, 1960) where both catechol oxidase and polyphenols have been detected by histochemical methods. However, Anya (1966a) has not been able to repeat these experiments in *Ascaris lumbricoides, Syphacia obvelata,* and *Aspiculuris tetraptera,* nor Lee (1965) in *Nippostrongylus brasiliensis.* Nevertheless this enzyme, together with its substrate, has been detected in the adult female cuticle of both *Heterodera rostochiensis* (Ellenby, 1946) and *Meloidogyne javanica* (Bird, 1958b), where its activity can be inhibited by copper poisons such as diethyldithiocarbamate. In addition, Monné (1960) has reported the presence of polyphenols in the cuticles of several species of nematodes. He divides these into free polyphenols which are present in the external cortical layer and bound polyphenols which can be unmasked by treatment with hot methanolic HCl so that their presence can be detected in other layers of the cuticle. Using this technique, Ellenby (1963) has reported the presence of polyphenols in the innermost parts of the cuticle of female *Heterodera rostochiensis.* In addition Monné (1959) reports quinone tanning in the larval cuticle of *Dictyocaulus viviparus* using histochemical techniques. But quinone tanning was not detected in the cuticles of larvae of *Haemonchus contortus,*

Nippostrongylus muris, and *Trichostrongylus* spp. by Bird and Rogers (1956) using various solvents and paper chromatographic methods. However, histochemical methods were not used.

From these observations it appears that quinone tanning occurs in the external cortical layers of various species of nematodes and as more sensitive techniques are developed it may in fact be shown that this is commonly found throughout the Nematoda.

It is interesting at this juncture to consider the observations of Hughes (1959), who studied the thin colorless cuticle of the mite *Acarus siro* and came to the conclusion that this cuticle is tanned through sulfhydryl groups reacting with quinones as the protein produced from this arthropod's epidermis contains sulfhydryl groups, whereas that in insects does not. He suggests that "a basic tanning mechanism reacting with epidermally produced proteins of cuticle might well be expected to produce different types of tanned cuticle, dependent on the type of cuticle with which it reacted." It seems plausible to assume on the evidence available that similar reactions may occur in the external cortical layers of nematode cuticle.

Almost a quarter of the fresh weight of *Ascaris lumbricoides* cuticle is made up of water-soluble proteins (Chitwood, 1936; Bird, 1957) and these appear to come from the fiber layers. They pass into solution at 38° in borate buffer (pH 8.8) within 72 hours (Bird, 1957), and can be shown by precipitation tests, electrophoresis, and ultracentrifugation to consist of at least five proteins of a similar fairly low molecular weight. Two of these are precipitated between 55° and 60° and three are heat stable.

From the reports of various workers on the chemical composition of the innermost layers of the nematode cuticle (Chitwood, 1936; Fauré-Fremiet and Garrault, 1944; Picken *et al.*, 1947; Bird, 1954, 1957; Bird and Rogers, 1956; Simmonds, 1958; M. R. Watson and Silvester, 1959; Dawson, 1960; Anya, 1966a) it is clear that proteins which have the characteristics of collagens exist. That is, as already mentioned, they give a collagen type X-ray diffraction pattern, contain hydroxyproline, have a high content of glycyl residues, and in some cases fibers exist which show collagenlike banding under the electron microscope (Jamuar *et al.*, 1964; Lee, 1965; Jamuar, 1966).

Furthermore, it has been reported that collagenase hydrolyzes both the fourth larval stage cuticular sheath of *Nippostrongylus muris* (Simmonds, 1958) and all layers of the cuticle except the external cortical layer in *Ascaris lumbricoides* (Dawson, 1960).

Generally speaking then, it appears that the proteins which make up the cuticle other than the external cortical layer have many affinities with the collagens.

In the past there has been a tendency to compare and name the proteins obtained from nematode cuticles after those that have been investigated in birds and mammals. It is clear that to attempt to compare the external cortical layer of nematodes with vertebrate keratins because both contain disulfide groups and lack hydroxyproline is open to criticism. A similar line of argument holds for comparison between nematode cuticle and vertebrate collagens although here the evidence for similarity both physical and chemical is stronger. It may be better to regard the proteins which make up the nematode cuticle as being unique to the phylum with characteristics of their own which may become more pronounced as the methods of investigation become more sophisticated.

The hypothesis has been put forward (Bird, 1957) that the cuticle of a nematode such as *Ascaris* is in a state of constant metabolic activity which is increased at times, such as during molting. Recently, some evidence has accumulated which lends support to this hypothesis. In addition to catechol oxidase which has been mentioned above, several other enzymes have been found in nematode cuticles. Esterase (Lee, 1961, 1962a, 1965; Anya, 1966a), adenosine triphosphatase, and acid phosphatase (Anya, 1966b) have been demonstrated in the cuticles of several nematodes. Hemoglobin (Lee, 1965), ribonucleic acid, and ascorbic acid (Anya, 1966b) have also been detected. The presence of ribonucleic acid in the cuticle suggests that protein synthesis might occur in this layer. As has been indicated above, the homogeneous layer of *Ascaris* cuticle consists largely of a collagen-type protein. Ascorbic acid has been shown to be associated with collagen fibrogenesis and it has been suggested (Bartoš, 1965) that it is required for the hydroxylation of proline residues. Since adenine triphosphatase and acid phosphatase participate in the formation of some structural proteins, it seems reasonable to suppose (as Anya, 1966b, suggests) that the cuticle of nematodes might participate, at least to some extent, in the synthesis of some of the proteins of which it is composed.

Many of the chemicals reported to occur in nematode cuticles are components of oxidation-reduction systems, namely, sulfhydryl and disulfide forms of certain compounds, the catechol-*o*-quinone system and the ascorbic acid-dehydroascorbic acid system. All these may function as electron carriers in biological oxidations. Thus it is hardly feasible to regard the cuticles of nematodes as inert exoskeletons; on the contrary, they appear to be associated with considerable biochemical activity.

2. *Hypodermis*

a. Structure. The hypodermis lies between the cuticle and the somatic muscle. A plasma membrane separating the hypodermis from the cuticle has been reported in several species of nematodes although it may not be of universal occurrence. Chitwood and Chitwood (1950) described the hypodermis of the simplest forms of nematodes as being cellular and this has more recently been substantiated by Maggenti (1961) and B. D. Watson (1965a) who found cells in the hypodermis of the free-living nematodes *Plectus* and *Euchromadora vulgaris* respectively. In the more complex forms however, the hypodermis is usually syncytial, particularly in adult worms.

Characteristically the nematode hypodermis is a thin layer which is thickened in the dorsal, ventral, and lateral positions to form four hypodermal cords which project into the pseudocoelomic cavity and divide the somatic muscles into four quadrants. The hypodermis may be covered by a basement membrane (Hinz, 1963; Lee, 1966a). In some nematodes, such as the Mermithidae, four subsidiary cords occur in a submedian position between the four main cords. In some species the hypodermal cords are poorly developed in the anterior region but become more apparent in the region of the esophageal bulb or nerve ring. The lateral cords usually become more developed than the dorsal and ventral ones and commonly run the whole length of the body, frequently showing on the surface of the body as pale, lateral lines. The dorsal and ventral cords tend to disappear posteriorly but this is not necessarily so since, in some species, either the ventral or dorsal cord, or both are more prominent in the tail region.

The hypodermal nuclei are usually confined to the cords but the intercordal hypodermis may be nucleate (Chitwood and Chitwood, 1950; Hinz, 1963). The subsidiary cords are usually anucleate and the dorsal cord often is also. Typically there are three rows of nuclei in the lateral cords and from none to three rows in the median cords (Hyman, 1951) but these numbers are subject to variation between different species and in different regions of the same nematode. The hypodermis, particularly in the region of the lateral cords, contains mitochondria and endoplasmic reticulum. Ribosomes (Bird and Rogers, 1965; B. D. Watson, 1965b) and abundant endoplasmic reticulum (B. D. Watson, 1965b) have been observed in the hypodermis of larvae prior to molting. Fibers passing through the hypodermis and attaching the muscle to the cuticle were reported by Goldschmidt (1909) and subsequently by B. D. Watson (1965b) and Lee (1965). Unicellular glands frequently occur in the lateral cords of some free-living nematodes and in many instances open through sublateral pores in the cuticle. The hypodermal cords contain the longitudinal body nerves and in some nematodes the lateral excretory canals.

In the cephalic region the cuticle lining the stoma is surrounded by a thin layer of hypodermis, and Wright (1965) has reported thin hypodermal connections between this stomatic hypodermis which underlies the cuticle in *Xiphinema index.* The stomatic hypodermis and the somatic hypodermis which has been reported in several genera probably corresponds with the arcade described by Chitwood and Chitwood (1950). Elsea (1951) found spindle-shaped arcade cells in the band of hypodermis surrounding the digestive tract at the stomatal level in female *Meloidogyne hapla.*

In at least two genera of the superfamily Trichuroidea, specialized regions of the cuticle and the underlying hypodermis occur. These are the bacillary bands which are confined to the lateral position in the esophageal region in *Trichuris,* but extend the length of the body and may be dorsal, ventral, or lateral in *Capillaria.* Recently, Wright (1963) and Sheffield (1963) have studied the bacillary bands in *Capillaria* and *Trichuris,* respectively, at the electron microscope level. Wright has described hypodermal gland and hypodermal nongland cells in *Capillaria* while Sheffield has found cells corresponding with the gland cells (which he calls bacillary cells) in *Trichuris.* In both genera, the hypodermal cords in the bacillary band region are a single cell in thickness and are elongate and perpendicular to the cuticle. In *Trichuris* the basal end of the gland cell lies on a thick basement membrane while in *Capillaria* it is ensheathed by nongland cells which lie on a thin basement membrane. In both genera, the gland cell is located under a pore in the cuticle and at its apical end the cell membrane is deeply infolded to form a system of lamellae which increases the surface area of the cell. The nucleus is usually located near the center of the cell and mitochondria, endoplasmic reticulum, and Golgi apparatus are present. Several hypotheses regarding the function of hypodermal gland cells have been presented but to date no particular function can be assigned to them.

b. Composition. Little work appears to have been done on the composition of the hypodermis of nematodes, perhaps because it is difficult to separate from the cuticle and muscles. In addition to glycogen and fat which have been detected in the hypodermal cords of several genera (von Kemnitz, 1912; Fairbairn, 1957; Hinz, 1963; Anya, 1964), Roggen *et al.* (1967) have found protein-containing vacuoles and leucine aminopeptidase in *Xiphinema index. Ascaris lumbricoides* hypodermis also contains leucine aminopeptidase (Lee, 1962b), hemoglobin (Smith and Lee, 1963), and esterase as a "cap" on the fat globules (Lee, 1962a).

c. Function. It is generally accepted that the cuticle is secreted by the hypodermis, and changes associated with cuticle secretion and molting which occur in the hypodermis will be discussed below. A further role for the hypodermis has been suggested by Roggen (1966), namely, that it is involved in osmoregulation in *Xiphinema index* and possibly other species.

3. *Muscle*

a. Structure and Composition. The musculature of the body wall of nematodes is unique in the animal kingdom. Instead of having branches sent to them from the nervous system, the muscle cells themselves send processes or arms (Fig. 6B) to the longitudinal nerves or the nerve ring.

The history of investigation into nematode musculature has been comprehensively reviewed by De Bell (1965) who draws attention to the credit which must go to Schneider (1860, 1863) whose work, done over a hundred years ago, has only recently been established with the aid of the electron microscope and the intracellular microelectrode.

Because of its convenience, Schneider's terminology for muscle is still in use, although no longer valid as a means of taxonomic separation of nematode genera. Somatic muscle is made up of a single layer of more or less spindle-shaped cells running longitudinally beneath the hypodermis. This longitudinal muscle layer is usually divided into a number of rows between the cords. There may not be any rows or only two, in which case the structure is of the holomyarian type. Usually there are two to five rows of muscle cells between the cords and the condition is known as meromyarian (Fig. 6A). If there is a large number of muscle rows in each field, as in *Ascaris* (Figs. 3A, 4 and 6B), it is the polymyarian type. Of course there are transitions between these various types and the number of muscle rows may even vary in different parts of the same worm.

A muscle cell consists essentially of a contractile fibrillar portion and a noncontractile protoplasmic portion which contains the nucleus and reserves of glycogen and fat.

Platymyarian muscle cells are flat, with the fibrillar region limited to a basal zone running parallel with the hypodermis (Fig. 6A). If the protoplasmic zone bulges into the pseudocoel with the fibrillar zone extending up its sides (Figs. 3F and 6B), it is called a coelomyarian muscle cell.

Most small nematodes and larvae have a meromyarian muscle arrangement and cells of the platymyarian type.

In recent years the ultrastructure of this type of musculature has been studied in several nematodes, namely *Nippostrongylus brasiliensis* (Lee, 1966a; Jamuar, 1966), *Euchromadora vulgaris* (B. D. Watson, 1965a), and *Xiphinema index* (Wright, 1965; Roggen *et al.*, 1967). A similar type of musculature is found in *Meloidogyne javanica* and *Tylenchulus*

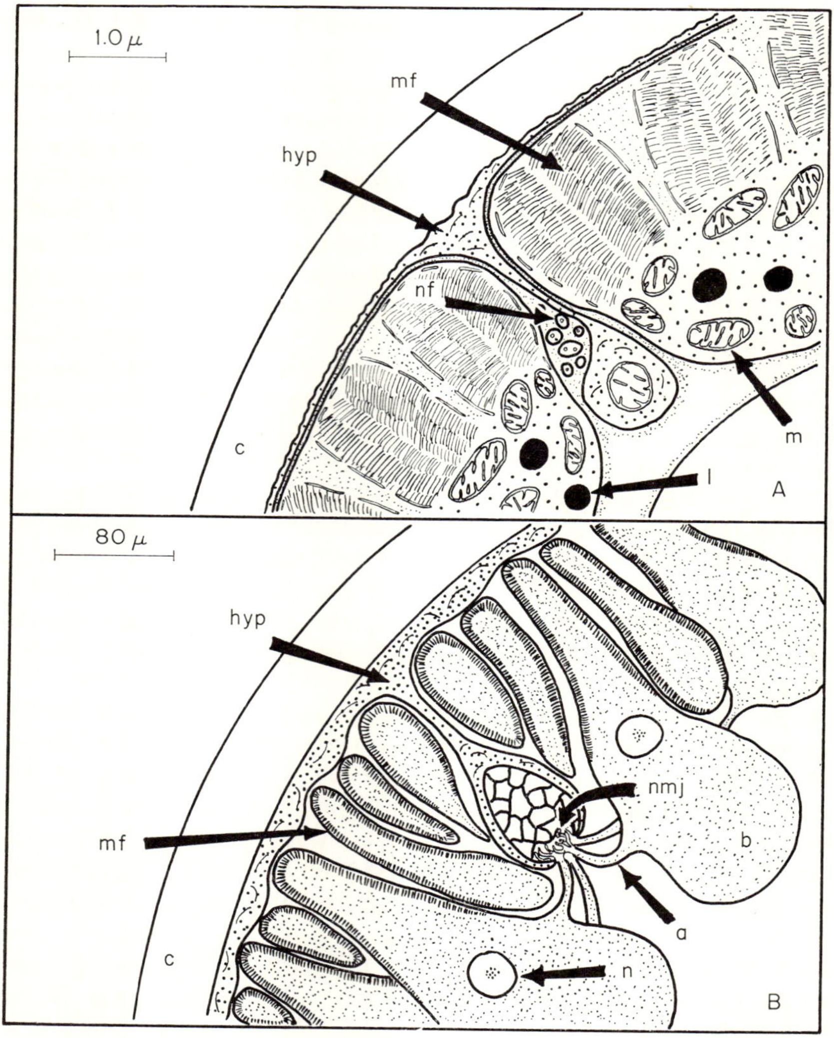

FIG. 6. A. Diagram illustrating platymyarian muscle cells, showing cuticle (c), hypodermis (hyp), lipid body (l), mitochondrion (m), muscle fiber (mf), and nerve fiber (nf). (After Lee, 1966a.) B. Diagram illustrating coelomyarian muscle cells, showing innervation processes or arms (a), belly (b), cuticle (c), hypodermis (hyp), muscle fibers (mf), nucleus of muscle cell (n), and neuromuscular junctions (nmj). (After Rosenbluth, 1965b.)

semipenetrans and the platymyarian muscle cell is probably found in the majority of nematodes (Fig. 6A).

Two types of myofilaments have been reported in the fibrillar portion of the muscle cell of the larva of *Nippostrongylus brasiliensis*. Lee reports that they are 22 mμ. and 6 mμ. in diameter and Jamuar that they are 24 mμ. and 9 mμ., while in *Xiphinema index* Roggen and his coworkers report that they are 26 mμ. and 5–6 mμ. in diameter. In *Euchromadora vulgaris* the myofilaments are 20–25 mμ. and 5–7 mμ. in diameter. In some cases each of the large filaments is surrounded by a number of the smaller filaments. Lee (1966a) reports that in the larvae of *Nippostrongylus brasiliensis* the thicker filaments, which probably contain myosin, are surrounded by about twelve thin filaments, which probably contain actin. In *Xiphinema index* some regions of the fibrillar portion of the muscle cell contain only thick myofilaments, some only thin, and some have both sizes.

The noncontractile portion of the platymyarian muscle cell (Fig. 6A) contains large mitochondria with numerous cristae, ribosomes, endoplasmic reticulum, glycogen granules, and lipid bodies. The whole of the muscle cell is covered by a basement membrane. These muscles do not appear to be either cross or obliquely striated. In *Euchromadora vulgaris* they have a direct connection with the cuticle (B. D. Watson, 1965a).

Large nematodes like *Ascaris* all have a polymyarian muscle arrangement and cells of the coelomyarian type (Figs. 3A, F, 4 and 6B). Wright (1964, 1966) has examined the muscle systems in a number of polymyarian nematodes and the ultrastructure of the somatic muscles of *Ascaris* and *Parascaris* has been examined recently (Hinz, 1963; Reger, 1964; B. D. Watson, 1965b; Rosenbluth, 1965a,b). Rosenbluth's electron micrographs have led to a better understanding of the coelomyarian muscle cell. This cell consists of three components, namely the fiber, which is contractile, the belly, which is essentially a sac of glycogen, and the arm, which extends to a nerve cord and establishes a myoneural junction (Fig. 6B). Rosenbluth considers that the muscle fibers of the fiber component are made up of two types of myofilaments which interdigitate and form H, A, and I bands as has been described in striated muscle (Hanson and Huxley, 1955).

The thick filaments are about 23 mμ. in diameter and are surrounded by 10 to 12 thin filaments which are about 8 mμ. in diameter. Watson considers that they are 30 mμ. and 5–7 mμ. in diameter respectively, while Reger reports that they are 17.5 mμ. and 9 mμ. in diameter. Only one type of myofilament has been reported in *Capillaria hepatica* (Wright, 1964) and this is about 25 mμ. in diameter. Generally speaking, it seems that the two types of myofilaments found in the contractile part of coelomyarian muscle cells are similar to those found in the contractile part of platymyarian muscle cells.

Over the years there has been considerable discussion as to whether the muscles of *Ascaris* should be considered as smooth or striated. Rosenbluth (1965a) considers that this confusion arose because it is neither smooth nor striated but is intermediate, having characteristics of both, although being more closely related to striated. He calls it obliquely striated and with the aid of complex three-dimensional diagrams explains how the pronounced stagger of the myofilaments results in the obliquity of the striations.

One of the differences that arises between the coelomyarian muscle cell of the large ascarids and the platymyarian muscle cell of the small nematodes is that the mitochondria in the former are located in the sarcoplasmic core which lies between the U-shaped contractile fibers, wheras in the platymyarian muscle cell the mitochondria lie in the noncontractile portion. However, the sarcoplasmic core continues into the muscle belly and a few mitochondria are found here together with a nucleus and a large number of glycogen granules which are depleted during starvation. The belly gives rise to one or in some cases two (Wright, 1965) tubular arms or innnervation processes which extend to nerve cords where they break up into several processes which make synaptic contact with nerve fibers (Fig. 6B).

Wright (1965) has recently examined the musculature of eight polymyarian nematodes and a single meromyarian nematode and found that in all polymyarian types there is interconnection between the muscle cells by means of cytoplasmic bridges. These are absent in the meromyarian nematode. Wright suggests that where there are few contractile elements to be coordinated, as in the meromyarian-platymyarian nematodes, coordination of muscle action is accomplished within the nervous system, but in the polymyarian-coelomyarian nematodes the cytoplasmic bridges have a significant role in the coordination of movement and probably operate as pathways by which contraction impulses are spread from cell to cell.

b. Neuromuscular Physiology. Because of its size and availability, practically all the work done on the physiology of nematode muscle has been done with *Ascaris*. Using *in vitro* nerve-muscle preparations Baldwin and Moyle (1949), Norton and De Beer (1957) and Mellanby (1955) found that *Ascaris* resembled vertebrate skeletal muscle in its response to drugs such as acetylcholine which produce contracture and to drugs which block this response. Further evidence to support this theory of cholinergic transmission has been provided by a number of workers (Jarman, 1959; De Bell *et al.*, 1963; del Castillo *et al.*, 1963, 1964a,b; Goodwin and Vaughn Williams, 1963). Using recording microelectrodes they have measured the transmembrane potential from *Ascaris* muscle bellies under normal conditions and under the influence of drugs.

It is thought (De Bell, 1965) that the nervous system in *Ascaris* serves only to modulate, at its site of origin, an activity which is initiated and exhibited continuously by the muscle cell membrane. This is achieved by two different neurohumors which are liberated from the longitudinal nerves. One of them (probably acetylcholine) accelerates membrane activity and the other inhibits it.

4. *Molting*

In this chapter molting will be considered in relation to the cuticle and hypodermis. Only passing reference will be made to the involvement of other structures as these are discussed in detail in Chapter 9, Section 7.

The cuticle is formed from the hypodermis so that the stimuli for molting must trigger off the changes which occur in this structure prior to the onset of molting. Briefly, it has been suggested (Rogers and Sommerville, 1963) that these stimuli are similar to those found in the Insecta and Amphibia, namely that under certain conditions neurosecretory cells are stimulated to produce secretions which activate glands which produce hormones that initiate molting. Some evidence is accumulating to support this hypothesis (Sommerville, 1957; Davey, 1964, 1966; Davey and Sau, 1967; Rogers, 1965; Fisher, 1966; Roggen *et al.*, 1967).

In the Nematoda descriptions of molting in which the structural changes associated with breakdown of the old and formation of the new cuticle are limited to only a few genera, notably *Meloidogyne javanica* (Bird, 1959; Bird and Rogers, 1965), *Phocanema decipiens* (Davey, 1965), and *Ascaris lumbricoides* (Thust, 1966). *M. javanica* is less than a millimeter in length while *P. decipiens* grows to a length of 60 mm. However, the molting stages of *M. javanica* are readily obtainable from the host plant and so this process can be studied under natural conditions whereas *P. decipiens,* which occurs as an adult worm in various species of seals, has to be cultured. The larva, which occurs in the muscles of cod, was grown in a medium containing antibiotics at 35°. In this medium the cuticle of the worm grew from a thickness of about 5 μ to about 30 μ after 3 weeks. This rate of development was thought to be similar to that which occurs *in vivo.* At the onset of molting the nuclei and nucleoli of the hypodermis enlarge, mitochondria multiply, and the cytoplasm becomes vacuolated. Davey (1965) also reports that muscle cells are associated with molting as lipid is secreted from the muscles into the hypodermis during this stage. The cuticle is formed by condensation at the outer edge of the hypodermis. Three layers are formed and the outer cuticle is cast as soon as the second layer of the new cuticle is formed.

It has been reported by Roggen *et al.* (1967) that cyclical fluctuations of the enzyme leucine aminopeptidase occur in the hypodermis of *Xiphinema index* and these are related to molting. The concentration of the enzyme increases during the period between molts and reaches a peak just before molting. During the molt the concentration of the enzyme diminishes and simultaneously the cuticle is loosened from the hypodermis and a new cuticle is formed. Davey and Sau (1967) have reported that molting in *Phocanema* is similarly dependent on this enzyme, which is synthesized by the excretory gland under the control of neurosecretory cells in the dorsal and ventral ganglia.

In *M. javanica* the hypodermis becomes granular and increases in width at the start of molting. When viewed under the electron microscope these granules resemble ribosomes. Histochemical tests at this stage reveal the presence of protein and ribonucleic acid. The innermost layer of the cuticle starts to separate from the hypodermis and an interrupted osmiophilic line, the future external cortical layer of the new cuticle, starts to form on the surface of the hypodermis. Other layers are formed underneath this and the new cuticle becomes convoluted and eventually a stage is reached when both cuticles are of similar thickness with large spaces in between (Fig. 3I). These spaces become filled with particles which appear to be associated with the breakdown and resorption of the innermost layers of the outer cuticle. This is thought to take place through regions in the newly formed cuticle where the external cortical layer has not been formed (Bird and Rogers, 1965). Eventually only the external cortical layer of the outermost cuticle remains and this may often be seen as a loose thin layer around parts of the cuticle of adult female specimens since in this genus they are sedentary parasites and it cannot be rubbed off. Similarly, in *Ascaris lumbricoides* larvae (Thust, 1966; Crandall and Arean, 1967), all that remains of the molted cuticle is the external cortical layer.

In *Meloidogyne,* as in *Ascaris* and *Phoconema,* the cuticle continues to grow and increase in thickness after the final molt. The process described above involving resorption of the innermost layers of the outer cuticle before the cuticle is shed may only be a specialized process common to the family Heteroderidae and related to their sedentary habit.

In the infective larvae of a large number of species parasitic in animals, it is clear that this resorption of the cuticle does not take place. Both Inatomi *et al.* (1963) and Eckert and Schwarz (1965) have measured the thickness of the old and the new cuticle in these species and found that in all cases the old cuticle or sheath is thicker. However, in the sheath nematode *Hemicycliophora,* a nematode in which the shed cuticle of the last molt is similarly retained around the adult as a sheath, sections viewed under the electron microscope (Bird, unpublished) show that this outer cuticle is thinner than the inner cuticle and that there are particles in the space between the cuticles which resemble those that have been described in *M. javanica.* These particles resemble the granules described by Noble-Nesbitt (1963) which appear to contain molting enzymes which digest the soft parts of the old cuticle in the Collembolan *Podura aquatica* and it is tempting to assume that the particles described in *M. javanica* have a similar function.

IV. Conclusion

Clearly, the body walls of members of both the Acanthocephala and Nematoda are complex structures which show considerable morphological diversity. Nevertheless, within each of these groups there are basic structures common to the entire group whose functions in many cases are not entirely known.

In the Acanthocephala, perhaps the most interesting functional problem concerns the mechanisms by means of which food passes through the body wall into the animal. Electron microscope studies support the hypothesis that these nutrients pass through pores on the surface and it is clear that the absorptive surface is considerably increased by this system of canals and vesicles. Histochemical tests have demonstrated considerable metabolic activity in the body wall.

In the Nematoda the ultrastructure and histochemistry of body wall structures such as the external cortical layer and fiber layers of the cuticle, the two types of myofilaments in the musculature and some of the components of the hypodermis are more or less similar throughout the group.

Many parasitic nematodes have become adapted to particular environments and have often developed or lost certain structures in the course of their evolution, which makes them atypical of the group in some respects. These forms are often used for teaching purposes because of their size and availability.

A good example of the basic uniformity of structure which exists at the ultrastructural level in nematodes is found in the cuticle of free-living infective larvae of a number of parasites of both animals and plants.

Perhaps the most interesting functional problems of nematode body walls are those concerned with cuticle formation, growth and permeability, molting, osmoregulation, neurosecretion, and neuromusclar physiology. Studies of problems of this nature will probably further substantiate the hypothesis that the cuticle itself is an active living structure.

REFERENCES

Allen, M. W. (1960). *In* "Nematology" (J. N. Sasser and W. R. Jenkins, eds.), pp. 136–139. Univ. of North Carolina Press, Chapel Hill, North Carolina.
Anya, A. O. (1964). *Parasitology* **54**, 555.
Anya, A. O. (1966a). *Parasitology* **56**, 179.
Anya, A. O. (1966b). *Nature* **209**, 827.
Baldwin, E., and Moyle, V. (1949). *Brit. J. Pharmacol.* **4**, 145.
Bartoš, F. (1965). *Nature* **204**, 1104
Beames, C. G., Jr. (1964). *Exptl. Parasitol.* **15**, 387.
Beckett, E. B., and Boothroyd, B. (1961). *Ann. Trop. Med. Parasitol.* **55**, 116.
Bird, A. F. (1954). *Nature* **174**, 362.
Bird, A. F. (1956). *Exptl. Parasitol.* **5**, 350.
Bird, A. F. (1957). *Exptl. Parasitol.* **6**, 383.
Bird, A. F. (1958a). *Parasitology* **48**, 32.
Bird, A. F. (1958b). *Nematologica* **3**, 205.
Bird, A. F. (1959). *Nematologica* **4**, 31.
Bird, A. F. (1966). *Nematologica* **12**, 359.
Bird, A. F., and Deutsch, K. (1957). *Parasitology* **42**, 319.
Bird, A. F., and Rogers, W. P. (1956). *Exptl. Parasitol.* **5**, 449.
Bird, A. F., and Rogers, G. E. (1965). *Nematologica* **11**, 224.
Bogoiavlenskii, Iu. K. (1958). *Biophysics (USSR) (English Transl.)* **3**, 449.
Brown, C. H. (1950). *Nature* **165**, 275.
Bryant, C., and Nicholas, W. L. (1965). *Comp. Biochem. Physiol.* **15**, 103.
Bullock, W. L. (1958). *Exptl. Parasitol.* **7**, 51.
Carbonell, L. M., and Apitz, C. R. (1961). *Exptl. Parasitol.* **10**, 263.
Caveness, F. E. (1961). *Proc. Helminthol. Soc. Wash., D.C.* **28**, 169.
Chitwood, B. G. (1936). *Proc. Helminthol. Soc. Wash., D.C.* **3**, 39.
Chitwood, B. G. (1938). *Proc. Helminthol. Soc. Wash., D.C.* **5**, 68.
Chitwood, B. G., and Chitwood, M. B. (1950). "An Introduction to Nematology." Monumental Printing Co., Baltimore, Maryland.
Cobb, N. A. (1919). *Contrib. Sci. Nematol.* **8**, 213.
Coomans, A., and de Coninck, L. (1963). *Nematologica* **9**, 85.
Crandall, C. A., and Arean, V. M. (1967). *J. Parasitol.* **53**, 105.
Crewther, W. G., Fraser, R. D. B., Lennox, F. G., and Lindley, H. (1965). *Advan. Protein Chem.* **20**, 191.
Crompton, D. W. T. (1963). *Parasitology* **53**, 663.
Crompton, D. W. T., and Lee, D. L. (1965). *Parasitology* **55**, 357.
Davey, K. G. (1964). *Can. J. Zool.* **42**, 731.
Davey, K. G. (1965). *Can. J. Zool.* **43**, 997.
Davey, K. G. (1966). *Am. Zoologist* **6**, 243.
Davey, K. G., and Sau, P. K. (1967). *Nature* **214**, 737.
Dawson, B. (1960). *Nature* **187**, 779.
De Bell, J. T. (1965). *Quart. Rev. Biol.* **40**, 233.
De Bell, J. T., del Castillo, J., and Sanchez, V. (1963). *J. Cellular Comp. Physiol.* **62**, 159.

del Castillo, J., de Mello, W. C., and Morales, T. A. (1963). *Arch. Intern. Physiol. Biochim.* **71**, 741.
del Castillo, J., de Mello, W. C., and Morales, T. A. (1964a). *Brit. J. Pharmacol.* **22**, 463.
del Castillo, J., de Mello, W. C., and Morales, T. A. (1964b). *Experientia* **20**, 141.
Eckert, J., and Schwarz, R. (1965). *Z. Parasitenk.* **26**, 116.
Edmonds, S. J. (1965). *Parasitology* **55**, 337.
Edmonds, S. J., and Dixon, B. R. (1966). *Nature* **209**, 99.
Ellenby, C. (1946). *Nature* **157**, 302.
Ellenby, C. (1963). *Proc. 16th Intern. Congr. Zool. Washington, D.C., 1963* Vol. 1, 139.
Elsea, J. R. (1951). *Proc. Helminthol. Soc. Wash., D.C.* **18**, 53.
Fairbairn, D. (1957). *Exptl. Parasitol.* **6**, 491.
Fauré-Fremiet, E., and Garrault, H. (1944). *Bull. Biol. France Belg.* **78**, 206.
Ferris, V. R., and Siegel, B. M. (1957). *Nematologica* **2**, 16.
Fisher, J. M. (1966). *Australian J. Biol. Sci.* **19**, 1073.
Flury, F. (1912). *Arch. Exptl. Pathol. Pharmakol.* **67**, 275.
Goldschmidt, R. (1904). *Zool. Anz.* **28**, 259.
Goldschmidt, R. (1909). *Arch. Zellforsch.* **4**, 81.
Goodey, J. B. (1951). *J. Helminthol.* **25**, 33.
Goodey, J. B. (1959). *Nematologica* **4**, 157.
Goodwin, L. G., and Vaughn Williams, E. M. (1963). *J. Physiol. (London)* **168**, 857.
Graff, D. J. (1964). *J. Parasitol.* **50**, 230.
Hanson, J., and Huxley, H. E. (1955). *Symp. Soc. Exptl. Biol.* **9**, 228–264.
Harris, J. E., and Crofton, H. D. (1957). *J. Exptl. Biol.* **34**, 116.
Hinz, E. (1963). *Protoplasma* **56**, 202.
Hirschmann, H. (1956). *Proc. Helminthol. Soc. Wash., D.C.* **23**, 140.
Hirschmann, H. (1960). *In* "Nematology" (J. N. Sasser and W. R. Jenkins, eds.), pp. 130–135. Univ. of North Carolina Press, Chapel Hill, North Carolina.
Hughes, T. E. (1959). *J. Exptl. Biol.* **36**, 363.
Hyman, L. H. (1951). *In* "The Invertebrates" (E. J. Boell, ed.), Vol. III, p. 217. McGraw-Hill, New York.
Inatomi, S., Sakumoto, D., Itano, K., and Tanaka, H. (1963). *Japan. J. Parasitol.* **12**, 16.
Inglis, W. G. (1964a). *Bull. Brit. Museum, Zool.* **11**, 266.
Inglis, W. G. (1964b). *Bull. Soc. Zool. France* **89**, 317.
Inglis, W. G. (1964c). *Proc. Zool. Soc. London* **143**, 465.
Jamuar, M. P. (1966). *J. Parasitol.* **52**, 209.
Jamuar, M. P., Swift, H., and Lewert, R. M. (1964). *J. Parasitol.* **50**, 36.
Jarman, M. (1959). *Nature* **184**, 1244.
Lassaigne, M. (1843). *Compt. Rend.* **16**, 1087.
Lee, D. L. (1961). *Nature* **192**, 283.
Lee, D. L. (1962a). *Parasitology* **52**, 241.
Lee, D. L. (1962b). *Parasitology* **52**, 533.
Lee, D. L. (1964). *Proc. Helminthol. Soc. Wash., D.C.* **31**, 285.
Lee, D. L. (1965). *Parasitology* **55**, 173.
Lee, D. L. (1966a). *Parasitology* **56**, 127.
Lee, D. L. (1966b). *Advan. Parasitol.* **4**, 187.

Mackintosh, G. M. (1960). *Nematologica* **5**, 158.
Magath, T. B. (1919). *Trans. Am. Microscop. Soc.* **38**, 49.
Maggenti, A. R. (1961). *Proc. Helminthol. Soc. Wash., D.C.*, **28**, 118.
Mellanby, H. (1955). *Parasitology* **45**, 287.
Monné, L. (1959). *Arkiv. Zool.* [2] **12**, 343.
Monné, L. (1960). *Arkiv. Zool.* [2] **13**, 287.
Moore, S., and Stein, W. H. (1951). *J. Biol. Chem.* **192**, 663.
Mueller, J. F. (1928). *Z. Zellforsch. Mikroskop. Anat.* **8**, 361.
Nicholas, W. L., and Hynes, H. B. N. (1958). *Ann. Trop. Med. Parasitol.* **52**, 36.
Nicholas, W. L., and Mercer, E. H. (1965). *Quart. J. Microscop. Sci.* **106**, 137.
Noble-Nesbitt, J. (1963). *Quart. J. Microscop. Sci.* **104**, 369.
Norton, S., and De Beer, E. J. (1957). *Am. J. Trop. Med. Hyg.* **6**, 898.
Pflugfelder, O. (1956). *Z. Parasitenk.* **17**, 371.
Picken, L. E. A., Pryor, M. G. M., and Swann, M. M. (1947). *Nature* **159**, 434.
Reger, J. F. (1964). *J. Ultrastruct. Res.* **10**, 48.
Rogers, W. P. (1965). *Comp. Biochem. Physiol.* **14**, 311.
Rogers, W. P., and Sommerville, R. I. (1963). *Advan. Parasitol.* **1**, 109.
Roggen, D. R. (1966). *Nematologica* **12**, 289.
Roggen, D. R., Raski, D. J., and Jones, N. O. (1967). *Nematologica* **13**, 1.
Rohde, R. A. (1960). *Proc. Helminthol. Soc. Wash., D.C.* **27**, 121.
Rosenbluth, J. (1965a). *J. Cell Biol.* **25**, 495.
Rosenbluth, J. (1965b), *J. Cell Biol.* **26**, 579.
Rothman, A. H., and Fisher, F. M. (1964). *J. Parasitol.* **50**, 410.
Savel. J. (1955). *Rev. Pathol. Comparee Hyg. Gen.* **55**, 213.
Schneider, A. (1860). *Arch. Anat. Physiol. Leipzig*, p. 224.
Schneider, A. (1863), *Arch. Anat. Physiol. Leipzig*, p. 1.
Sheffield, H. G. (1963). *J. Parasitol.* **49**, 998.
Simmonds. R. A. (1958). *Exptl. Parasitol.* **7**, 14.
Smith, M. H., and Lee, D. L. (1963). *Proc. Roy. Soc.* **B157**, 234.
Sommerville, R. I. (1957). *Exptl. Parasitol.* **6**, 18.
Stein, W. H., and Moore, S. (1950). *Cold Spring Harbor Symp. Quant. Biol.* **14**, 179.
Sukataschoff, B. (1899). *Z. Wiss. Zool.* **66**, 377.
Thust, R. (1966). *Zool. Anz.* **177**, 411.
Toldt, K. (1899). *Arb. Zool. Inst. Univ. Wien* **11**, 289.
Trim, A. R. (1949). *Parasitology* **39**, 281.
van Bömmel, A. (1895). *Arb. Zool. Inst. Wurzburg* **10**, 191.
Van Cleave, H. J. (1952). *Exptl. Parasitol.* **1**, 305.
Van Cleave, H. J., and Bullock, W. L. (1950). *Trans. Am. Microscop. Soc.* **69**, 288.
von Kemnitz, G. (1912). *Arch. Zellforsch.* **7**, 463.
Watson, B. D. (1965a). *Quart. J. Microscop. Sci.* **106**, 75.
Watson, B. D. (1965b). *Quart. J. Microscop. Sci.* **106**, 83.
Watson, M. R., and Silvester, N. R. (1959). *Biochem. J.* **71**, 579.
Wieser, W. (1953). *Medd. Vaxtskyddanst. Stockholm* **65**, 1.
Wright, K. A. (1963). *J. Morphol.* **112**, 233.
Wright, K. A. (1964). *Can. J. Zool.* **42**, 483.
Wright, K. A. (1965). *Can. J. Zool.* **43**, 689.
Wright, K. A. (1966). *Can. J. Zool.* 44, 329.

CHAPTER 4

Culture Methods and Nutrition of Nematodes and Acanthocephala[1]

Morton Rothstein and W. L. Nicholas

I. Culture Methods and Nutrition of Free-Living Nematodes

A. Introduction

In order to study properly the nutrition and culture of nematodes, it is desirable to establish the organisms in axenic culture. Only in this way can the metabolic abilities of the nematodes be separated from those of coexisting and interacting organisms. One may settle for a monoaxenic culture, but the best way to attain this is to obtain axenic nematodes and then add the second organism or tissue, for example, alfalfa callus tissue for plant parasitic nematodes (Krusberg, 1961). This chapter will devote itself, in the main, to recent work on the culture and nutrition

[1] Part I dealing with free-living nematodes, was prepared by M. Rothstein. Parts II and III, dealing with parasitic nematodes and Acanthocephala were prepared by W. L. Nicholas.

of nematodes, free-living and parasitic, and will refer only in passing to work already thoroughly reviewed (Dougherty *et al.*, 1959; Nicholas, *et al.*, 1959; Dougherty, 1960).

The study of axenic cultivation of invertebrates is at present in a rudimentary state. Compared with cultivation of bacteria or protozoa, the relationship is that of an infant to a well-developed youth. The reasons for this situation are quite obvious though not readily excusable. Historically, the development of bacteriology played a starring role in the development of Western medicine. The rapid evolution of biochemistry, the widespread use of microbiological assay, and more recently, the demand for specialized culture conditions for genetic studies, have advanced techniques for pure culture of bacteria and protozoa as a natural consequence. On the other hand, development of axenic culture methods for invertebrates had no such impetus. Indeed, use of organisms for purposes of taxonomy or experiments for biology students seldom require axenic conditions. A further inhibition of development of axenic culture of invertebrates arises from technical difficulties involved, over and above those found in the culture of microorganisms. Axenization, feeding habits (e.g., requirement for particulate foods), liquid vs. solid support, oxygen tension, light vs. dark, osmotic regulation, special requirements of larval or other stages, mating, and hatching factors are some of the more obvious considerations.

Of the invertebrates (excluding insects), nematodes drew early attention because of the large number of parasitic species and their obvious relation to health and agriculture. It is quite apparent that parasitic species, with their complicated life cycles, demand complex and ever-changing environmental and nutritional conditions and thus are less than ideal organisms with which to begin studies in axenic cultivation. Nonetheless, because of the importance of parasitic nematodes many studies have been undertaken with the object of axenic cultivation of these organisms and considerable progress has been made. However, it is the free-living nematodes with their simpler requirements which have provided successful axenic cultures of invertebrate species, generation after generation.

B. Axenization of Nematodes

Historically, the first axenization of a nematode was carried out by Zimmerman (1921) on the vinegar eel, *Turbatrix aceti.* He washed the nematodes with hydrogen peroxide for 10 minutes, twice a day, rinsing them each time with sterile water. After 10 days of treatment, the organisms were free of bacteria. In 1956, Nicholas performed the whole procedure in a covered Seitz filter using a clarifying grade filter pad, so that successive solutions could readily be drawn off. Mortality was heavy, but large numbers of worms could be used, thus assuring numerous axenized survivors. Nicholas and McEntegart (1957) also devised a method for axenization of nematodes in which gravid females were placed in a wire gauze boat (20 μ pore size) and submerged for 10 minutes in 10 volumes percent H_2O_2. The boat containing the killed females was washed and immersed in a solution of penicillin (1000 μ/ml.) and chloramphenicol (250 μ/ml.) in a sterile centrifuge tube. After 24 hours (*Caenorhabditis briggsae*) or 48 hours (*Rhabditis maupasi* and *Rhabditis axei*), the newly hatched larvae were recovered by centrifugation, the dead females being retained in the gauze boat.

Mechanical disruption of adult worms in the presence of 1.0% merthiolate has also been used to obtain axenic larvae of *Panagrellus redivivus* (Cryan *et al.*, 1963). The larvae were subsequently washed in antibiotic solutions (penicillin and streptomycin).

Simple washing procedures have been used by Krusberg (1961) in preparing six species of plant parasitic nematodes for large-scale growth on plant tissue cultures. Five or six baths of malachite green (20 p.p.m.) and streptomycin (1000 p.p.m.) for a period of 3–4 hours yielded organisms which grew on alfalfa callus in agar for protracted periods, apparently without the presence of bacteria. On the other hand, brief exposure to a sequence of bacteriostatic agents for purposes of surface sterilization apparently left small numbers of viable bacteria on or in the nematodes (Myers and Krusberg, 1965). The difference may lie in the time of exposure to antibiotics.

In general, many small free-living nematodes are bacteria feeders; intact microorganisms cannot get into the gut because of the limited size of the esophagus. Since this is true for many nematodes, bacterial contamination is limited to the surface of the organisms, which explains the relative ease of axenization. For example, Nicholas axenized a species of *Acrobeloides* by simple washing with sterile water and with 0.1% merthiolate (Nicholas, 1962). Of particular help is the resistance of soil-inhabiting nematodes to the antibiotics. For example, *Rhabditis pellio* in a solution of 1,000,000 units of penicillin and 100 mg. of streptomycin/ml., survived for several days (Dougherty and Calhoun, 1949). *C. briggsae* was reported more resistant to penicillin than is the mouse (Gochnauer and McCoy, 1954). *T. aceti* exhibits sensitivity to streptomycin at 1000 units/ml. over a 24-hour period (18% survival vs. 75% for controls). Merthiolate, at a 1:1000 dilution under the same conditions, left no survivors (Ells, 1955). One problem in the axenization of free-living nematodes which may offer difficulty is in the elimination of fungi, molds, and bacteria resistant to antibiotics. For this reason, short exposure to such agents as merthiolate or nystatin may be advisable.

Two procedures have been reported for the more or less mechanized axenization of large numbers of nematodes. Ells (1955) axenized *T. aceti* by washing the organisms on a fritted glass funnel of medium porosity permitting bacteria, but not worms, to pass through. Washing solutions of antibiotics (5000 units of dihydrostreptomycin and 500 units of penicillin per milliliter) were passed through a fine porosity filter into the worms in the medium porosity funnel. After 24 hours, the nematodes were rinsed and washed for 1 hour in 1:1000 merthiolate. After final rinsing, the nematodes were free of bacteria. The solutions used were those selected by Dougherty (1949).

Another large-scale procedure which was effective for a number of nematode species was recently reported by Cryan (1963). Solutions containing antibiotic are dripped slowly onto the concentrated nematodes in a funnel lined with lens paper. The nematodes migrate through the lens paper and drop into a separatory funnel where they collect above the stopcock. Opening of the stopcock permits the worms to drop into a second separatory funnel containing glass beads. The nematodes migrate through the beads, thus leaving behind any debris, and they can be tapped off at the lower stopcock. By this procedure, *Haemonchus contortus, Trichostrongylus colubriformis, Strongyloides papillosus, Rhabditis pellio, Panagrellus redivivus,* and *Caenorhabditis briggsae* have been axenized. Nystatin was added to control molds. The procedure is a simple one and appears to yield excellent results. It might be difficult, however, to use with *T. aceti* since this organism generally swims to the top of solutions and exhibits negative geotropism.

C. Growth and Nutrition

As previously mentioned, the first reported axenic cultivation of a free-living nematode was carried out by Zimmerman (1921). He reported that *Turbatrix aceti,* the vinegar eel, would grow with continued reproduction in a bacteria-free solution of peptone, glucose, lecithin, sodium chloride, and a yeast extract. More recently, Ells (1955) failed to obtain reproduction with Zimmerman's medium, possibly because of the differences in the "lecithin" used. However, Ells (1955) reported that addition to the medium of an ether extract of fresh rat liver yielded unspecified rates of growth and reproduction. These rates were reported to be slower than in medium contaminated by bacteria which apparently serve as a food source for the nematode. In fact, Nicholas (1956), in a careful attempt to duplicate the work, was unable to demonstrate sustained growth of *T. aceti* with either the medium of Zimmerman or of Ells. As Nicholas points out, in neither earlier experiment was serial subculture attempted, so that the true capacity of the earlier media to support growth was never determined. Ability of organisms to grow in subculture is an important criterion of the suitability of any medium. In this laboratory also, *T. aceti* fails to grow in subculture if an ether extract of beef liver is used as a supplement.

The successful culture of metazoa under axenic conditions, generation after generation, has been limited to a number of nematodes, to certain insects, and to a single annelid, *Enchytraeus fragmentosus* (Dougherty and Solberg, 1961). For nematodes (and the annelid, *E. fragmentosus*) tissue extracts (from liver or chick embryo) are necessary additions to even the most complex basal media. By use of a liver protein fraction (LPF), *C. briggsae* has been maintained in continuous serial subculture since 1954; the same liver fraction, or chick embryo extract (CEE) are effective in the axenic cultivation of a number of free-living nematodes. In fact, a method for continuous culture of nematodes on a thin film of medium has recently been reported (Hansen and Cryan, 1966a). The achievement of a completely defined medium which will support serial axenic cultures for indefinite periods has perhaps been most closely approached for *Drosophila* (Sang, 1962; see also a review by Rodriguez, 1966) and for the free-living nematode, *C. briggsae*. In the latter case, purification of the required tissue supplement (Sayre *et al.*, 1961, 1967) led to the isolation of a semipurified protein from mammalian liver which when added to the basal medium, supports the continued reproduction of *C. briggsae*. Thus, although the nature of the action of the "growth factor" and its absolute structure are as yet unknown, the medium is approaching a "defined" status.

Chaudhuri (1964) reports in his thesis, the growth of axenic nematodes (*Diplogasteroides* and *Diplogaster nudicapitatus*) in a medium composed of casein, gelatin, thiamine, and buffer supplemented by rabbit serum. The work includes detailed studies of the effect of various parameters such as pH, temperature, area of the container, etc. on growth and motility of the organisms.

1. Assay Methods

The initial work which led eventually to the successful axenic culture of a variety of nematodes was performed on *C. briggsae*. An effective bioassay was developed by Dougherty *et al.* (1959) to test the ability of medium components to support reproduction in this organism. The method involves the sterile transfer of adult nematodes through a series of dishes (Bureau of Plant Industry watch glasses or BPI dishes) containing potassium phosphate buffer. The washed adult nematodes are left in the final dish, where they lay eggs which subsequently hatch. The newly hatched larvae can then be transferred individually to small test tubes containing the media to be tested. The time required for the freshly hatched larvae to mature and produce new larvae (F_1 time) is considered a measure of the efficacy of the medium. A variation of the procedure is to wash an egg mass (rather than adult worms) through the BPI dishes. Hatching is allowed to take place in the final dish. This method has been used to test nutrients for *C. briggsae,* two strains of *C. elegans* and *Acrobeloides buetschlii* (Nicholas, 1962). The method is eminently suitable for self-fertilizing hermaphrodites such as *C. briggsae*. However, it also has been used to study axenic cultivation of *Panagrellus redivivus* which is bisexual. Ten larvae were used in each test solution to assure the presence of both sexes. Total population rather than regeneration time was used as the criterion for evaluating the culture media, since the requirement for copulation was an uncontrollable factor (Cryan *et al.,* 1963).

A more straightforward procedure is the direct counting of nematodes on a slide. The cultures to be counted are swirled and the nematodes quickly picked up in an appropriately sized pipet. The organisms are transferred to a test tube for dilution, and are once more swirled, picked up, and run out into the slot in a Scott counting slide (0.1 or 0.2 ml.). It is advisable to use a blow-out pipet, and to use the entire contents each time. Since most nematodes sink, they tend to accumulate at the bottom and it is not possible to dispense serial aliquots from a single pipet.

A dissecting microscope (15–30 $\times$ magnification) is excellent for counting. *T. aceti* swims too rapidly for easy counting and should first be chilled in an ice bath. *R. anomala* (= *maupasi*) is often so sluggish as to make it difficult to distinguish dead organisms from live ones. After preincubation of the organisms in 0.01% neutral red (20 minutes), the dead organisms become pink; the live ones remain essentially colorless.

Direct counting of cultures has several advantages over the larval assay method described above. It requires no special skill in recognizing gravid nematodes, or in the semimicro manipulations which have to be carried out under aseptic conditions. A considerable number of experiments can be carried out at one time, whereas the larval procedure would require the tedious isolation of large numbers of washed adults and the individual handling of many larvae. In addition, the initial stimulation of growth rates of *C. briggsae* brought about by addition of protein would not have been observed (see Section D; Rothstein and Cook, 1966) by the larval techniques. Moreover, the procedure is difficult to use for bisexual species. On the other hand, the larval technique gives an absolute answer as to the ability of a medium to permit reproduction and does this without the need for repeated subculture. The method is faster than direct counting and one can use the F_1 time (in hermaphroditic species) as a criterion for evaluating the effectiveness of growth media (Lower *et al.*, 1966).

2. *Tissue Supplements to the Basal Medium*

All efforts to obtain reproduction of *C. briggsae* in defined media failed unless certain tissue extracts were added. The supernatant liquid from a crude aqueous liver extract, centrifuged at 14,000–20,000 *g.* and sterilized by Seitz filtration showed growth-promoting activity, but the fraction precipitated at between 40–60% saturation with ammonium sulfate yielded a very active preparation (LPF-C). The material in this fraction is water soluble, stable (in terms of nematode growth response) at 4° (more than 1 year) and at freezing temperature (more than 3 years). Repeated freezing and thawing of the samples tends to cause precipitation. Autoclaving of the initial liver extract (ALE) resulted in loss of growth-promoting activity, but this result was variable. At the same time, it was found the 10% chick embryo extract (CEE), produced by homogenizing 10-day-old embryos in an equivalent amount of distilled water (w./v.), also resulted in good growth of nematodes, but only if the vitamin concentration of the basal medium (which sufficed when using LPF-C) was raised, or the protein-free supernatant of autoclaved liver was added. Growth-promoting activity also appears to be present in plasma fractions, and is certainly present in bacteria, as *C. briggsae* grows well in monoxenic culture with *E. coli*. Preparations of heat-killed bacteria were inactive, but extracts of broken cells appeared to retain growth-promoting activity. Most of the above work was done by Dougherty and co-workers, and has been thoroughly reviewed by Nicholas *et al.* (1959). Use of LPF-C or CEE also yields an excellent growth response with *T. aceti* (Nicholas, 1956).

From the above results it appears that a heat-sensitive substance, either a protein or a protein-bound material, is a requirement for growth and reproduction of *C. briggsae*. The material, or an adequate substitute, seems to be available from several biological sources—various liver preparations, bacteria, plasma, or chick embryos.

In 1961, Sayre *et al.* reported an effective crude substitute for LPF-C. This was a "heated liver extract" (HLE) which is prepared by heating with stirring, a 50% w./v. liver homogenate (sheep, lamb, horse) to 53° for 6 minutes, removing the precipitate by centrifugation, and sterilizing the supernatant solution by Seitz filtration. The resulting liquid (HLE) when added to the basal medium (EM) supports growth (see below) of *C. briggsae, T. aceti, R. anomala, Acrobeloides buetschlii,* and *P. redivivus.* HLE may be prepared in 250-ml. batches by heating the liver homogenate in 600- or 800-ml. beakers.

D. Basal Medium

Dougherty *et al.* (1959) after a number of trials and substitutions developed several complex basal media (i.e., media excluding the protein supplement) which were revised by Sayre *et al.* (1963) into the following mixture, designated EM 1. This medium has been the basis, with various alterations, for much of the recent work in nematode nutrition, and for attempts to obtain a completely defined medium.

EM certainly appears to have considerable redundancy in its vitamin complement. Nicholas (1962) simplified the medium (see Table I) by removing ten components and replacing calcium pantothenate with D-pantothenic acid. With this basal medium (EM2), growth and populations of *A. buetschlii* were good using 20% CEE as supplement, through six serial subcultures. No direct comparison could be made with EM 1, since in the latter case, 10% instead of 20% CEE was used. It should be noted that *A. buetschlii* grows readily in CEE in water at strengths of 100%, 20%, 10% and it also can be grown in 10% LPF in water and in 30% HLE in water. Obviously, with crude supplements at high level, the function of EM is negligible. It is only when limiting amounts of supplement, free of other nutrients from the liver extract are added, that meaningful manipulations of the basal medium can be made. Thus, changes in EM are reflected in total population number and F_1 time at low levels of HLE. Whether or not these changes are affecting the "balance" of components or supplying a nutritional requirement cannot as yet be determined. Tomlinson and Rothstein (1962) replaced the amino acids in EM with various protein hydrolyzates, and the vitamins by yeast extract. While this is a tendency away from a defined medium, there is a considerable simplification of preparation for the simple purposes of growing axenic nematodes for biochemical study. The amino acids in EM were replaced respectively with 4% w./v. enzyme digest of casein, acid-hydrolyzed casein, enzyme digest of lactalbumin, and soy-peptone. The latter yielded an F_1 time of 4.5 days (HLE supplement, 10% by volume), the enzymic casein and lactalbumin, 7 days, and casein hydrolyzate, 10–12 days. Figure 1 shows the population curves of *C. briggsae* using soy-peptone medium, peptone-yeast extract, and EM respectively.

Obviously, soy peptone is somewhat superior to EM in providing

TABLE I
BASAL MEDIUM FOR AXENIC CULTIVATION OF NEMATODES[a]

Component	mg./liter	Component	mg./liter
L-Arginine	1,300	*N*-Acetylglucosamine	15
L-Histidine	377	DL-Alanine	15
L-Isoleucine	1,148	*p*-Aminobenzoic acid[b]	7.5
L-Leucine	1,918	Biotin[b]	3.75
L-Lysine	1,372	Cyanocobalamine	3.75
L-Methionine	518	Ca folinate	3.75
L-Phenylalanine	1,070	Niacin[b]	7.5
L-Threonine	956	Niacinamide	7.5
L-Tryptophane	245	Pantetheine	3.75
L-Valine	1,360	Ca pantothenate	7.5
L-Alanine	1,860	Pteroylglutamic acid[b]	7.5
L-Aspartic acid	2,160	Pyridoxal·HCl	3.75
L-Cysteine	26	Pyridoxamine·2HCl	3.75
L-Glutamate·Na	733	Pyridoxin·HCl	7.5
Glycine	962	Riboflavin-5-PO_4·Na	7.5
L-Proline	870	Thiamine·Cl·HCl	7.5
L-Serine	1,051	DL-Thioctic acid[b]	3.75
L-Tyrosine	363	NaOH	150
α,ϵ-Diaminopimelic acid	137	KH_2PO_4	1,225.5
L-Glutamine	1,950	KOH	925.5
Adenylic acid	347	Mg·H·citrate·$5H_2O$	915
Cytidylic acid	323	$CaCl_2 \cdot 2H_2O$	220
Guanylic acid	363	K_3·citrate	486
Uridylic acid	324	$Fe(NH_4)_2(SO_4)_2 \cdot 6H_2O$	58.8
Thymine	126	$MnCl_2 \cdot 4H_2O$	22.2
Glucose	1,315	$ZnCl_2$	10.2
Glutathione	408	$CuCl_2 \cdot 2H_2O$	6.5
Choline·Cl	885	Triethanolamine[b]	6.5
i-Inositol·2H2O	648		

pH about 5.6

[a] From Sayre *et al.*, 1963.
[b] The vitamins were first made up as concentrated stock solutions in 5% w./v. triethanolamine.

higher populations of *C. briggsae*. Yeast extract is inferior to the vitamin mix.

1. *Vitamin Requirements in the Basal Medium*

Before nutritional requirements can be completely resolved, the nature of the growth factor must be determined. Work on this problem is under way, but in the meantime, some insight into other requirements of various free-living nematodes has been obtained even in the presence of crude supplements. Nicholas *et al.* (1962) have demonstrated that a 1:50 dilution of the B-vitamin fraction of a medium supplemented with 10% CEE prevented larvae of *C. briggsae* from reproducing. Under various conditions, a requirement could be demonstrated for thiamine, riboflavin, folic acid, panthothenate, niacinamide, and pyridoxine. Biotin was shown to be required by lack of reproduction of *C. briggsae* in the presence of avidin, a biotin inhibitor (Nicholas and Jantunen, 1963).

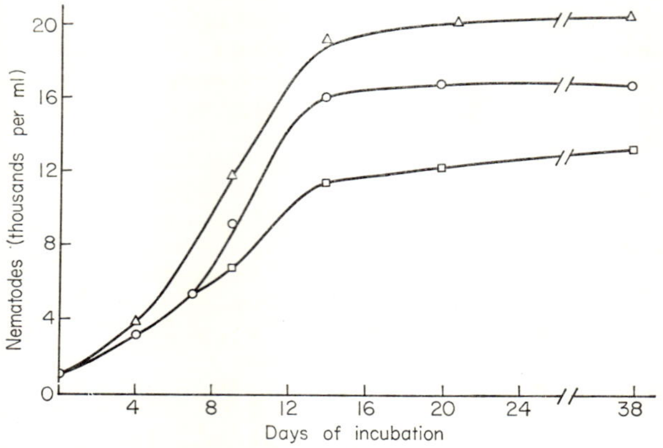

FIG. 1. Growth of *Caenorhabditis briggsae* at 20° in three different media. Δ—Δ Peptone medium. ○—○ Defined medium. □—□ Peptone-yeast extract medium. The worms were incubated in 50-ml. Erlenmeyer flasks containing 5 ml. of medium.

However, it is possible though unlikely that the avidin works *in situ* and is thus limiting use of internally synthesized biotin.

2. *Physical Conditions*

The nematodes under discussion can grow well in liquid media. The depth of the medium may be limiting for some organisms (deep cultures are deleterious to *C. briggsae*) (Tomlinson and Rothstein, 1962), but there seems to be no other limitation on the size or shape of the containers. In the author's laboratory, *T. aceti* can be readily grown in still culture in 2800-ml. Fernbach flasks containing 300-ml. of medium. Erlenmeyer flasks (50 ml.) containing 5 ml. of medium are used routinely in the culture of *C. briggsae, T. aceti,* and *P. redivivus.* A "Morton" closure, or cotton stopper should be used in order to admit a continuous supply of oxygen. Use of shakers may be complicated by precipitation of protein from the tissue supplement. A device for the continuous rather than batch production of axenic nematodes has recently been reported (Hansen and Cryan, 1966a).

E. Comparative Requirements for Culture of *C. briggsae, P. redivivus,* and *T. aceti*

Both *P. redivivus* and *C. briggsae* grow well in serial subculture in peptone medium supplemented with HLE, but *T. aceti* requires the addition of acetic acid (Rothstein and Cook, 1966) (Fig. 2). The effect is not one of pH, as HCl will not replace the acetic acid (Fig. 3). *Turbatrix aceti* will not grow when autoclaved Bacto-liver or autoclaved whole liver homogenate is used as a supplement. The report that *T. aceti* would grow if autoclaved HLE was used (Rothstein and Cook, 1966) could not be substantiated upon repeating the experiments (Rothstein, unpublished results). Of the three free-living nematodes under discussion, only *P. redivivus* (see below) can be grown through repeated subcultures when certain liver preparations are autoclaved before use as supplements to the medium.

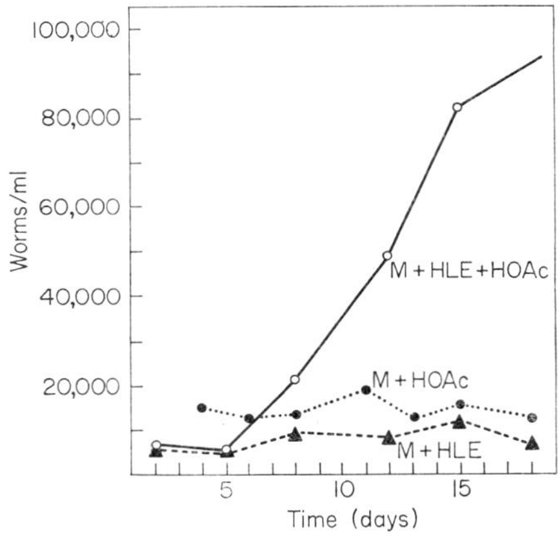

FIG. 2. Growth of *T. aceti* at 30° in soy-peptone basal medium with various additions. M = soy-peptone basal medium.

Addition of autoclaved casein to the supplemented medium has a stimulating effect on growth rate (Fig. 3) Adding acetic acid to medium supplemented with HLE until the pH is 3.4 results in a copious precipitate. The precipitate after washing with acetic acid retained normal growth-supporting ability but the supernatant solution was inactive, as demonstrated by repeated subculture of *T. aceti*.

T. aceti was found to grow in serial subculture over a period of several months at 30°. At 34° after an initial population increase, the organisms did poorly.

P. redivivus was found to yield increasing populations with decreasing pH (peptone medium; HLE) attaining populations of 24,000 worms/ml. at pH 3.4 in 17 days (Table II). This organism, alone of those tested, will grow in serial subculture when using autoclaved Bacto-liver as the supplement. It will not grow on the liver preparation marketed by General Biochemicals. In the presence of autoclaved Bacto-liver, the most favorable pH, in terms of population density, is pH 5.9 (53,000 worms/ml. in 17 days) (Table II) (Rothstein and Cook, 1966).

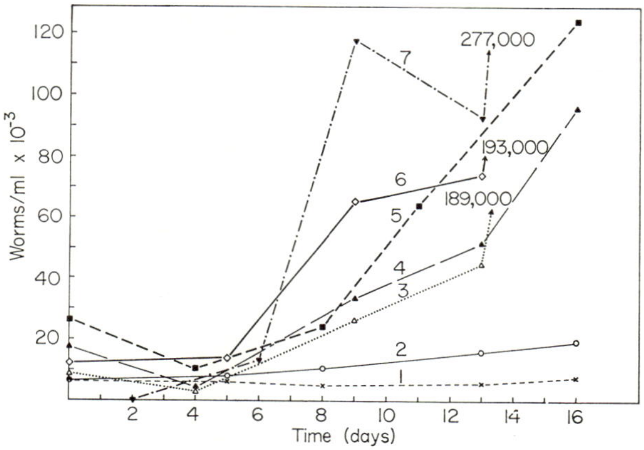

FIG. 3. Growth of *T. aceti* at 30° in soy-peptone basal medium to which the following additions were made: 1. pH adjusted to 3.4 with HCl; 2. Ethanol (2.5% final concentration); 3. Autoclaved casein (100 mg.) + 0.1 ml. HLE; 4. Autoclaved casein (25 mg.) + 0.1 ml. HLE; 5. Autoclaved HLE (0.7 ml.); 6. HLE (0.7 ml.) (standard medium); 7. Autoclaved casein (100 mg.) + 0.2 ml. HLE. Figures represent worm populations at 15 days. The sudden large increase in numbers probably reflects an additional reproductive cycle. The media in 3, 4, 5, 6, and 7 all contained acetic acid.

When using HLE as supplement, increasing amounts of precipitate form with decreasing pH. It is therefore tempting to propose that the growth-promoting material in HLE is more readily available as a precipitate than when in solution. No variable precipitation effect would be seen with autoclaved Bacto-liver, which is already completely precipitated. Thus, the most favorable pH for *P. redivivus* is 5.9 with Bacto-liver, but with HLE, a pH of 3.4 appears best. The latter condition yields extra precipitate which may be assumed to overcome the effects of the very acidic pH. *P. redivivus* is also the only nematode so far tested which grows well in repeated subculture when the medium is supplemented with autoclaved whole liver homogenate. As is the case for *T. aceti,* only the precipitate and not the supernatant solution supports growth in serial subcultures of *P. redivivus* (Rothstein, unpublished).

A further interesting note in the culture of *P. redivivus* is the effect of adding acetic acid to the medium. If acetic acid is added (1%) to peptone—HLE medium (the pH falls to 5.9), a lag of some 12 days occurs before normal growth occurs. Subculture of these organisms into fresh medium containing 1% acetic acid again results in a lag period. However, subculture from the first acetic acid-containing medium into

TABLE II
GROWTH OF *P. redivivus* AT VARIOUS pH VALUES[a]
NEMATODE POPULATION[b]

pH	HLE	Autoclaved Bacto-liver[c]
3.4	24,000 (17)	28,000 (21)
3.8	15,000 (18)	31,800 (17)
4.2	14,250 (18)	32,200 (17)
5.9	7,600 (17)	53,000 (17)
6.8	15,500 (17)[d]	29,500 (22)
8.6	3,450 (17)	8,850 (25)

[a] All experiments were carried through three or more serial subcultures except those run at pH 6.8 and 8.6. Results were essentially unchanged in the subcultures.

[b] Figures in parentheses represent the number of days at which the worm population was counted.

[c] 100 mg. Bacto-liver autoclaved in the medium for 7 minutes (fast exhaust), instead of the usual HLE supplement.

[d] Results from a single flask. The duplicate became contaminated. The apparent stimulation at this pH was noted in a second experiment in which the population was 9800 worms/ml. after 18 days. However, over 30% of the organisms were nonmotile.

normal medium adjusted to pH 5.9 with HCl resulted in immediate growth of the organism.

Temperatures above 25° had a deleterious effect on long-term growth of *P. redivivus* (Rothstein and Cook, 1966). Similar results were noted by Winkler and Pramer (1961) at 28°. Hansen and Cryan (1966b) reported that an elevated temperature (28°) and protein supplement deficiency raised the ratio of males to females in *P. redivivus.*

C. briggsae, like *T. aceti,* but unlike *P. redivivus,* will not tolerate autoclaving of any of the tested liver supplements. Added autoclaved proteins such as casein, soy protein, egg albumin (or heated HLE) stimulate initial population growth though they do not support growth in serial subcultures (Rothstein and Cook, 1966). As with *T. aceti,* protein exerts a sparing effect on HLE. That is, larger populations were obtained with less HLE through the three serial subcultures tested.

F. The Nature of the Growth Factor

It was first noted that the LPF fraction became more active, i.e., more rapid reproduction of *C. briggsae* was attained, if the material was frozen in the presence of a number of compounds such as thymine and urea (Sayre *et al.,* 1963). Treatment of LPF on columns of hydroxylapatite yielded a purified "growth factor" which permitted reproduction of *C. briggsae* (Sayre *et al.,* 1961) and *P. redivivus* (Cryan *et al.,* 1963). This purified material also showed a pattern of freeze activation. Subsequently, the "growth control factor" (GCF, purified LPF) was found to be activated in a number of ways, including freezing in dilute solution, addition of 4% w./v. of the sucrose polymer FICOLL (Pharmacia), or heating at 37° for 24 hours (Hansen *et al.,* 1964; Buecher *et al.,* 1966). Activated growth factor alone does not support reproduction in *C. briggsae.* Basal medium must be added. Of course basal medium without growth factor is also inactive.

In the above study, Hansen *et al.* (1964) showed that transfer from supplemented to basal medium essentially stopped maturation of fourth-stage larvae of *C. briggsae.* Since this is the case, it would be interesting to determine at what stage of nematode development autoclaved protein stimulates reproduction in *C. briggsae* (Rothstein and Cook, 1966). Hansen *et al.* (1964) also compared the size and other characteristics of *C. briggsae* grown in axenic culture (GCF-supplemented medium) and in monoxenic association with *E. coli.* However, differences in size should be interpreted with caution, as the monoxenic cultures were grown on agar slants, and the axenic cultures in liquid medium. Nematodes are known to swell or shrink depending on the tonicity of the medium.

Recently Sayre *et al.* (1967) have studied in great detail the purification of the "growth control factor" for *C. briggsae.* The material precipitated from lamb liver homogenate (40–60% ammonium sulfate saturation) was chromatographed on columns of hydroxylapatite, and the fraction eluted with 0.5 *M* phosphate buffer was used for subsequent studies. This material after rechromatography on hydroxylapatite yielded seven protein peaks, six of which were tested and showed biological activity (i.e., when added to basal medium, the fractions permitted maturation of larvae of *C. briggsae*). All of the peaks, and the original sample, had essentially the same amino acid composition. Sayre *et al.* (1967) suggest that "activation" may really constitute a reaggregation of smaller subunits into a biologically active form of protein. Thus, material stored at 40° (inactive) had a different elution pattern (hydroxylapatite column) from material stored frozen (biologically active).

Since crude HLE does not require activation and purified CGF does, one may speculate further on the nature of the "activation." One possibility is that the protein or protein-bound material must be made available in particulate form. All of the "activators" perhaps "aggregate" the growth factor, to the extent of making it available to the nematodes in particulate form. Thus, HLE which is a crude preparation, ages and always provides a precipitate. There is evidence that *P. redivivus* grows best when HLE is precipitated by lowering the pH of the medium, although in other circumstances, the low pH itself is unfavorable (see Table II). Furthermore, *P. redivivus* will grow in medium supplemented with autoclaved whole liver homogenate. After autoclaving, the washed precipitated protein is effective as a supplement; the supernatant liquid is not. Similarly, when the acetic acid is added to the medium for *T. aceti*, a heavy precipitate forms from the added HLE. This precipitate supports growth of *T. aceti* in serial subcultures; the supernatant solution does not. If these results can be translated to *C. briggsae*, then precipitated material would seem of importance, and the very solubility of "unactivated" GCF may militate against the nematodes receiving an adequate concentration of the growth material. Solubility could also explain the loss of growth activity observed when the purified growth factor is digested with proteolytic enzymes (Sayre *et al.*, 1967). In this respect, Pertel (1964) has reported briefly that a Seitz-filtered water extract of Bacto-liver will permit repeated subculture of *C. briggsae*. The solution is reported to be "clear." However, in our hands a precipitate gradually appears in the extract. Moreover, the yield of organisms grown with this supplement is small (W. L. Nicholas, private communication). It is of interest to note that CEE also precipitates slowly on standing. Removal of the precipitate reduces the potency of the preparation.

Whatever its nature, either as a protein, or a protein-bound material, HLE provides a necessary growth substance for nearly all the free-living nematodes so far tested. Nonetheless, there appear to be other factors which can replace it. For example, CEE rather than LPF appears to be a requirement for *C. elegans* under certain conditions (see Table III).

The growth factor for *C. briggsae* and *T. aceti* is heat labile. Nonetheless, autoclaved whole liver homogenate, but not autoclaved HLE, serves for the growth of *P. redivivus*. One must conclude that the material discarded during the preparation of HLE contains the heat-stable material utilized by *P. redivivus.* Moreover, autoclaved Bacto-liver (Difco) serves as the supplement for the latter organism but the liver product marketed by General Biochemicals does not. Table III summarizes these results.

These results are quite unsatisfying. One must assume that there exist at least two growth factors. One is heat labile and is satisfactory for growth of most free-living nematodes, and the other is heat stable. If it is a question of a single protein of very specific structure, it would be surprising to find that material in such diverse sources as in several varieties of liver, in a large number of mammalian organs (Sayre *et al.*, 1967), in chick embryo, and in bacteria. In addition, the ability of *P. redivivus* to utilize autoclaved liver preparations for growth and reproduction militates against such an interpretation. It appears probable that all of the factors (if they are multiple) are related. A possible explanation might be that a single factor is involved which after autoclaving becomes protein-bound in such a way as to become available or unavailable to the various nematodes, depending on the proteolytic enzymes in each species. However, it should be noted that when autoclaved casein is added to the cultures of *C. briggsae, T. aceti,* or *P. redivivus,* the precipitate is visibly diminished with time. Thus 50 mg. of autoclaved casein in 5 ml. of medium disappears completely within 2 weeks after inoculation with *T. aceti.* There is no doubt that the nematodes can "eat" or at least, dissolve particulate protein in the me-

TABLE III
EFFECTIVENESS OF VARIOUS TISSUE SUPPLEMENTS IN SUPPORTING CONTINUED GROWTH OF NEMATODES

Supplement	Growth[a]	No growth
HLE	Ta, Pr, Cb, Ab, Ra	*C. elegans* (Bergerac strain)[b]
LPF	Ta, Pr, Cb, Ab, Ra	*C. elegans* (Bergerac strain)[b]
Autoclaved HLE		Pr, Cb, Ta
Autoclaved whole liver homogenate	Pr	Ta, Cb
Autoclaved Bacto-liver	Pr	Ta, Cb
CEE	Ta, Rb, Ab, Pr	

[a] Ta = *T. aceti;* Pr = *P. redivivus;* Cb = *C. briggsae;* Ab = *A. buetschlii;* Ra = *Rhabditis anomala.*

[b] Requires CEE. An early type of basal medium was used which supported growth in the Bristol strain of *C. elegans.* The Bergerac strain has not been tested with liver fractions using EM as a basal medium (Hansen *et al.*, 1960). Direct comparisons with the other nematodes should therefore be made with caution.

dium. An alternative idea is the possibility that "A" is the true growth factor (heat labile), but that only *P. redivivus* has the metabolic ability to convert "B," a related but heat-stable material to "A." Acceptance of this alternative is also antithetical to the idea of a specific protein structure. The above comments are speculative, and the possibility remains that the growth factor is indeed involved with a specific protein structure.

A problem which should also be discussed is the possibility of a sterol or fatty acid requirement. The basal "EM-type" medium contains neither, but such a material may well be bound to the protein supplement. Unquestionably, these components are present in the crude HLE and Sayre *et al.*, reported (1967) that small amounts of lipoidal materials could be extracted from purified growth factor preparations. Extraction of these with organic solvents resulted in a loss of activity both of the protein and the extracts. Possibly, protein denaturization caused the loss of activity, but it is also possible that the lipids became unavailable when not attached to the appropriate "carrier."

All insects tested have a sterol requirement, and in our laboratory we have been unable to show the conversion of acetate-2-^{14}C to labeled cholesterol by *T. aceti, C. briggsae,* or *P. redivivus.* Bacteria such as *Escherichia coli,* the normal food of these nematodes, contain no detectable sterols. Thus, if a sterol is present in these nematodes, it must be one other than cholesterol, and therefore, the latter is not a dietary requirement per se. Indeed, we have been unable to replace HLE with casein and cholesterol. The possibility exists, of course, that a cholesterol precursor further along the pathway than acetate can be utilized for sterol synthesis by the nematodes. Linoleic and linolenic acids are "essential" for continued reproduction in certain insects as well as being required nutrients for mammals. Arachidonate has been identified in *P. redivivus* (Sivapalan and Jenkins, 1966) and unless this organism can synthesize polyunsaturated fatty acids, this material must be of dietary origin (the organisms were not raised in a defined medium such as EM). No information is available regarding abilities of nematodes to synthesize fatty acids.

In summary, it may be said that several species of free-living nematodes can be cultured indefinitely under axenic conditions. The ideal of a completely defined medium has been closely approached, though many mysteries remain, centered particularly about the requirement for a complex growth factor or supplement. Nonetheless, the knowledge of culture and nutrition of free-living nematodes has reached a stage more than sufficient to make these organisms of great value for biochemical as well as zoological studies.

Addendum

Two papers of importance to the nutrition of nematodes have recently been reported. It has been shown that *T. aceti* can synthesize polyunsaturated acids *de novo* from acetate-2-^{14}C (Rothstein and Götz, (1968). It has also been shown that *C. briggsae* has an absolute sterol requirement for growth and reproduction (Hieb and Rothstein, 1968). In the latter work, it was shown that *C. briggsae* could be subcultured indefinitely in a medium consisting of *E. coli* cells in buffer, provided that sterols were added. In the light of these findings, it is obvious that the purified growth factor (Sayre *et al.*, 1967) must contain a sterol. Indeed, the factor has been reported to contain cholesterol (Hieb and Sandman, 1967). Since eucaryotic cells lack polyunsaturated fatty acids as well as sterols, it is clear that *C. briggsae,* as well as *T. aceti* (Rothstein and Götz, 1968) can synthesize its requirements for polyunsaturated fatty acids. The discussion in the section above should be considered in the light of these new findings.

II. Culture Methods and Nutrition of Parasitic Nematodes

A. Introduction

Under this heading, attention is directed to experiments in which parasitic nematodes have been successfully grown in culture. Emphasis is placed on axenic cultures, but some monoxenic culture systems will also be considered. This restriction excludes many experiments on their physiology, biochemistry, and chemotherapy, in which the nematode has been maintained briefly *in vitro*. Some of these experiments will, no doubt, be referred to elsewhere in this volume, and many are cited by von Brand (1966) in his textbook on parasite biochemistry.

A comprehensive review of the development of axenic cultures of helminths has recently been published by Silverman (1965a) in *Advances in Parasitology*. However, some important advances have been made in the field since Silverman's review appeared, and to minimize unnecessary duplication, this review will concentrate on the culture of those nematodes with which significant advances continue to be made. No attempt has been made to list all the literature on axenic culture.

Though the development of techniques for the axenic culture of nematodes has proved difficult and often frustrating, efforts to establish satisfactory cultures continue to be made because these offer fresh avenues for the investigation of many aspects of nematode biology. Their value has been discussed by Weinstein (1958, 1966) and some of the problems which have already been investigated with axenic cultures will be touched upon in this article. Though *in vitro* methods, including axenic culture, offer many advantages to the research worker, there are undoubtedly problems in parasitology which must be solved by *in vivo* experiments (Read, 1966).

The life cycles of parasitic nematodes are complex and show great variety. Some knowledge of these life cycles must be presumed in order to keep this article to a reasonable length (see Chapter 1).

B. Axenic Culture

1. *Neoaplectana*

Interest was first drawn to *Neoaplectana glaseri* because of its potential value in controlling the larvae of the Japanese beetle *Popillia japonica.* Glaser (1931) reported its mass culture on meat infusion agar overlaid with live yeast, though cultures died out unless supplemented with extracts of bovine ovarian tissue or dried beetles (Glaser, 1940a).

More significant was the discovery that *N. glaseri* and a related species, *N. chresima* could be cultured axenically through many generations on sterile vertebrate tissues on agar or in saline (Glaser, 1940b; Glaser *et al.,* 1942). Sodium hypochlorite was employed to sterilize the nematodes used to initiate cultures.

In nature, susceptible insects are infected by ingesting ensheathed third-stage larvae which survive for long periods in moist soil. The ingested larvae develop initially in the insect's gut, later rupturing the gut and killing the insect. The dioecius ovoviviparous adults give rise to numerous progeny in the putrifying cadaver. A number of other species of *Neoaplectana* which attack insects are known. Poinar and Thomas (1966) have shown, using axenic insects (*Galleria mellonella*), pure cultures of bacteria, and axenic and monoxenic worms, that the infective larvae of one of these species (referred to as DD-136) transmits a pathogenic bacterium, *Achromobacter nematophilus,* to the susceptible host. In the absence of this or certain other bacteria, the worm kills the insect but does not reproduce. The relationship between *N. glaseri* and bacteria in nature is not known, but it retains its capacity to kill insects after many years in axenic culture (Stoll, 1956; Jackson, 1962a). The axenic culture of *N. glaseri* was greatly refined by Stoll (1953, 1959). It can conveniently be maintained by serial subculture in test tubes on sterile rabbit kidney supported by nutrient agar slopes. The third-stage larvae, required for subcultures, conveniently climb the walls of the tube, leaving other stages and material behind. *N. glaseri* can also be grown in axenic liquid cultures when supplied with an extract of certain vertebrate tissues. Rabbit liver is a good source, and the extract,

RLE, gives the highest yield of worms when prepared from pregnant rabbits (Stoll, 1954, 1959). RLE gives good cultures when mixed with dextrose meat infusion broth, one part to nine of broth.

Jackson (1962a,b) reported the development of the third-stage larva, derived from rabbit kidney cultures, to the adult with limited reproduction in a chemically defined medium, based on a medium developed by Trager for *Leishmania* (a protozoan parasite). Jackson and Siddiqui (1965) and Jackson (1966) found that two other species, *N. dutkii* and *N. carpocapsae,* both insect parasites, could be cultivated on excised sterile mammalian tissue on agar, but that they did not develop as well as *N. glaseri* in either a liquid medium (heat-sterilized meat infusion, peptone broth) or in the defined medium.

2. *Strongylida*

Nematodes of this order include many important parasites of domestic animals and man. The adults of most species occur in the gastrointestinal or respiratory system of vertebrates and pass their eggs or early larval stages with the host's feces. The first and second larval stages usually feed on bacteria in the host feces. Development does not proceed beyond the third, so-called ensheathed larva, which retains the cuticle or sheath of the second larval stage, unless the nematode invades a susceptible host. In some species the larvae can penetrate the unbroken skin of potential hosts and reach the alimentary canal via the circulatory and respiratory systems, but in many the third-stage larva must be ingested. Though it will be convenient to consider the culture of these worms under a single heading, it must be emphasized that the adult worms, even from related species, may inhabit habitats in the host that are physiologically very different, and that the early parasitic stages usually penetrate the mucosal surface of the gut, some making extensive migrations through the host's tissues before reaching their final location.

The free-living larvae of many species can be cultured in fecal material, preferably mixed with some moistened inert carrier such as charcoal or sphagnum moss. Many species can also be cultivated on pure cultures of bacteria, as originally reported by McCoy (1929a,b) and Lapage (1933), but not on heat-killed bacteria. The eggs, which hatch in saline, can be axenized by brief treatment in toxic sterilizing agents. Sodium hypochlorite, in the form of domestic bleaching solutions, is the most convenient. Starting with eggs axenized in this manner, a number of species have been cultured axenically through the free-living phase (see Table IV). All the media used have included complex poorly defined organic materials, and, with one exception (Lawrence, 1948), heat-labile extracts have been required for good yields of third-stage larvae.

Strongyloides, though belonging to a different order, is included in the table.

The axenic culture of *Nippostrongylus brasiliensis* (= *N. muris*) from the egg to the sexually mature adult stage by Weinstein and Jones (1956a,b, 1957a, 1959) was a major advance, even though the worms, which were stunted, failed to copulate. Adult worms were cultured both from the third larval stage (i.e., the infective stage), obtained from

TABLE IV
AXENIC CULTURES OF FREE-LIVING STAGES OF CERTAIN GASTROINTESTINAL PARASITES OF VERTEBRATES

Species	Host	Heat-labile constituent	Authors
Haemonchus contortus	Sheep	Rabbit kidney	Glaser and Stoll (1938a)
Ancylostoma braziliense	Dog	None required	Lawrence (1948)
Ancylostoma caninum	Dog	Chick embryo homogenate or rat liver extract	Weinstein (1953)
Ancylostoma duodenale	Man	Rat liver extract	Weinstein (1953)
Necator americanus	Man	Chick embryo homogenate	Weinstein (1954)
Nippostrongylus braziliensis (= *N. muris*)	Rat	Chick embryo homogenate	Weinstein (1954)
Nippostrongylus braziliensis (= *N. muris*)	Rat	Chick embryo homogenate	Weinstein and Jones (1956a)
Nippostrongylus braziliensis (= *N. muris*)	Rat	Chick embryo homogenate and/or serum	Weinstein and Jones (1957a)
Nippostrongylus braziliensis (= *N. muris*)	Rat	Chick embryo homogenate	Weinstein and Jones (1959)
Nematospiroides dubius	Rat	Chick embryo homogenate	Weinstein and Jones (1957a)
Strongyloides ratti	Rat	Chick embryo homogenate	Weinstein and Jones (1957b)
Strongyloides species	Monkey	Chick embryo homogenate	Weinstein and Jones (1957b)
Cooperia punctata	Bovine	Chick embryo homogenate	Leland (1962)

charcoal/feces cultures and sterilized with sodium hypochlorite, and from the egg (also sterilized with hypochlorite), completing both free-living and parasitic stages in axenic culture. A variety of complex media were used, but all contained an unheated chick embryo homogenate, CEE. The simplest medium for the parasitic stages consisted of CEE, Tyrode solution, and rat or human serum. Free-living stages cultured in CEE media began to develop through the parasitic stages when transferred to a higher temperature, 37.5°, and higher levels of CEE. The problem of obtaining copulation and fertilization of *N. braziliensis* in culture has not yet been solved with worms cultured from the third larval stage. Worms taken directly from the rat's intestine, axenized with antibiotics, and transferred to culture media have produced small numbers of fertile eggs, which hatched (though the larvae survived for only a few days), but copulation was not observed *in vitro* (Weinstein and Jones, 1959). Leland (1963) has also cultured third-stage larvae of *N. braziliensis* to the egg-laying stage.

Jones and Weinstein (1957) reported the axenic cultivation of *Nematospiroides dubius* (from the mouse) by similar techniques to immature adults. Growth from the egg to the third larval stage was completed in a CEE/liver extract medium. When transferred to a CEE/rat serum medium, the larvae developed to the early adult stage.

Sommerville and Weinstein (1964) studied the reproduction of *N. dubius in vivo* and in culture. When immature worms were taken from the mouse intestine, sterilized with antibiotics, and transferred to axenic culture, stunted adult worms developed, which copulated. Males produced few or no spermatazoa and the females only very few eggs, though some were apparently fertilized, which did not develop beyond early cleavage. Adult worms transferred to axenic cultures soon ceased to copulate, and the females produced only a few eggs.

Following Weinstein and Jones' work with *N. braziliensis,* several workers have attempted the axenic culture of gastrointestinal nematodes from domestic animals (see Table II). The infective larvae of *Haemonchus contortus* and *Trichostrongylus colubriformis* (ovine) can develop and molt to the fourth stage in balanced salt solutions (Stoll, 1940; Silverman *et al.,* 1962; Sommerville, 1966) and nutrients may perhaps generally not be required for this phase of growth (Silverman, 1965a).

Further development requires complex, poorly defined media, including heat-labile tissue extracts. Leland cultured several species from the third larval stage to the adult stage (see Table V) and *Cooperia punctata* (Leland, 1962, 1963) from the egg to the egg-producing adult stage in axenic culture. Leland's medium contained serum, sodium caseinate, cystine, liver extract, vitamins, and antibiotics in a balanced salt solution in which the pH was adjusted to 7.2–7.3 with $NaHCO_3$ or CO_2. Douvres (1960b, 1962a,b) found the addition of extracts from bovine intestinal tissue to an already complex medium (trypticase, vitamins, yeast extract, glucose, serum, rabbit embryo, and pig liver extracts) stimulated the growth of *Oesophagostomum radiatum.* Douvres (1966) also reported that in *O. dentatum* the development of some of the third-stage larvae in culture to fourth or fifth-stage worms inhibited the development of the remaining third-stage larvae. The removal of the developing worms,

TABLE V
AXENIC CULTURES OF GASTROINTESTINAL PARASITES FROM SHEEP, CATTLE, AND PIGS IN COMPLEX MEDIA[a]

Species	Host from which derived	Most advanced stage reached in culture	Authors
Haemonchus contortus	Sheep	4th larval	Glaser and Stoll (1938b)
Haemonchus contortus	Sheep	Adult	Silverman (1959)
Ostertagia species	Sheep	Adult	Silverman (1959)
Hyostrongylus rubidus	Pig	4th larval	Diamond and Douvres (1960, 1962)
Oesophagostomum quadrispinulatum	Pig	4th larval	Diamond and Douvres (1960, 1962)
Oesophagostomum radiatum	Sheep	4th molt	Douvres (1960a,b)
Haemonchus contortus	Sheep	Early 5th	Douvres (1960a,b)
Haemonchus contortus	Cattle	4th larval	Douvres (1960a,b)
Ostertagia ostertagi	Cattle	4th larval	Douvres (1960a,b)
Cooperia species	Cattle	4th larval	Douvres (1960a,b)
Oesophagostomum radiatum	Cattle	Late 4th	Douvres (1960a,b, 1962a,b)
Cooperia punctata	Cattle	Early 5th	Douvres and Alicata (1962)
Cooperia punctata	Cattle	Adult, sperm, infertile eggs	Leland (1961, 1962, 1963)
Cooperia oncophora	Cattle	Adult	Leland (1961, 1962, 1963)
Ostertagia ostertagi	Cattle	Adult, sperm, infertile eggs	Leland (1961, 1962, 1963)
Oesophagostomum radiatum	Cattle	4th larval	Leland (1961, 1962, 1963)
Haemonchus contortus	Cattle	4th larval	Leland (1961, 1962, 1963)
Trichostrongylus axei	Cattle	3rd larval	Leland (1961, 1962, 1963)
Ostertagia circumcincta	Sheep	Adult, infertile eggs	Leland (1961, 1962, 1963)
Ostertagia trifurcata	Sheep	Adult	Leland (1961, 1962, 1963)
Ostertagia species	Sheep	Adult	Leland (1961, 1962, 1963)
Oesophagostomum species	Sheep	4th larval	Leland (1961, 1962, 1963)
Nematodirus species	Sheep	4th larval	Leland (1961, 1962, 1963)
Cooperia pectinata	Cattle (?)	Adult, infertile eggs	Leland (1965a)
Oesophagostomum quadrispinulatum	Pig	Adult	Leland (1965b)
Hyostrongylus rubidus	Pig	Adult, infertile eggs	Leland (1965c)
Oesophagostomum dentatum	Pig	Adult	Douvres (1966)
Haemonchus contortus	Sheep	Early 5th	Hansen *et al.* (1966)

[a] Cultures were initiated with third stage (infective) larvae, axenized with sodium hypochlorite.

or the transfer of the undeveloped worms to fresh media, stimulated the development of the latter. This observation may be significant because other workers have noted that only a small proportion of the third-stage larvae generally show development in culture, and because of the importance of retarded larval development in many natural infections.

Leland's and Weinstein and Jones' successful cultures of *C. punctata* and *N. braziliensis* from egg to adult in unbroken axenic culture show that no stimuli other than those fortuitously present in the cultures were required to initiate the change from free-living to parasitic phases of the life cycle.

In species such as *C. punctata,* which are infective when ingested, exsheathment, marking the transition from the free-living to the parasitic phase, follows specific stimuli received by the worm in the host's gut leading to the secretion of enzymes by the worms (discussed by Rogers and Sommerville, 1963; Rogers, 1966). In *N. braziliensis* the ruptured sheath is lost when the third-stage larva penetrates the host's skin. Neither the lack of the appropriate stimulus for natural exsheathment, nor the artificial removal of the sheath by sodium hypochlorite when larvae from fecal cultures are used, apparently prevents development. Leland and Wallace (1966) showed that the parasitic stages of *C. punctata* complete development normally when transferred surgically from axenic cultures to the rabbit's intestine and produce fertile eggs. Conditions in culture do not, therefore, permanently impair the capacity of the worms to produce fertile eggs, even though these have not yet been produced in culture. Females, separated from males in the fourth stage, produce infertile eggs in culture (Leland, 1962). Unmated *N. braziliensis* and *N. dubius* also produce eggs *in vitro* (Weinstein and Jones, 1959; Sommerville and Weinstein, 1964).

The life cycle of *Stephanurus dentatus,* a parasite of pigs, differs sharply from those discussed above. The infective larvae develop to the fifth stage in the liver and then migrate to the kidney or perirenal fat, from where eggs are passed in the urine. The third-stage larvae have been cultured to the early fourth stage in cell-free culture (Douvres and Tromba, 1962) and to the late fourth stage on pig kidney cell monolayers cultured in a complex medium (Douvres *et al.,* 1966).

The swine lungworms, *Metastrongylus apri* and *M. pudendotectus,* have been cultured from infective larvae (obtained by digestion from earthworms) to the fifth stage (Sen and Kelly, 1960). *Dictyocaulus viviparous* (the cattle lungworm), taken from fecal cultures, has been cultured from the third to fourth stage in salt solutions by Silverman *et al.* (1962).

3. *Ascaris suum*

In *Ascaris suum*, the large roundworm of pigs, the egg, which is passed in the pig's feces, is the infective stage. Embryonated eggs, when ingested by the pig, release second-stage larvae in the small intestine which migrate via the portal circulation, liver, lungs, and trachea before returning to the intestine and completing their development.

Pitts and Ball (1953, 1955) attempted to culture the second-stage larvae from artificially hatched eggs in a variety of axenic culture media. CEE in saline or with serum gave best results, the larvae surviving for many days, showing some growth and signs of an approaching molt. *A. suum* eggs hatch in mice and the larvae complete their migration through the tissues, though they fail to become established in the intestines. Newly hatched larvae from mice did not survive as well in axenic culture as those from eggs hatched artificially, though older larvae from the liver of lungs apparently showed better survival and limited growth (Pitts, 1960).

Cleeland and Laurence (1962) induced sterilized eggs of *A. suum* to hatch by agitation with a magnetic stirrer. The larvae, when cultured in a tissue culture medium (199) supplemented with 20% bovine serum survived for long periods, showed some growth and differentiation, and molted.

In further experiments, the effect of different gas atmospheres was investigated (Cleeland, 1963). Best results were obtained when concentrated suspensions of larvae were cultured under semianaerobic conditions. Only a small proportion of the larvae grew appreciably, but these, when isolated and cultured individually in media gassed with 5% CO_2, 95% N_2, showed further growth, equal to that during the visceral migration in the host, though much less rapid.

Douvres and Tromba, in work unpublished at the time of writing, reported at the 41st meeting of the American Society of Parasitologists (1966) that more rapid growth to the fourth stage was obtained in culture media together with swine kidney cells.

4. *Filarioidea*

Adult Filarioidea live in the tissues of vertebrates and release early larval stages, or eggs, into the host's tissues. The larvae, or microfilariae, develop to the third larval stage in blood-sucking arthropods. The infective larvae escape from the head or mouth parts of the arthropod when it feeds, to infect susceptible vertebrate hosts.

Earl (1959) has reviewed earlier attempts to culture Filarioidea and experimented with *Dirofilaria immitis*, a parasite of dogs transmitted by mosquitoes. Earl found that the microfilariae from dog's blood would survive for several weeks in a tissue culture medium (199) supplemented with dog, horse, or cow serum, but for only a few days in unsupplemented medium. The presence of dog kidney cells in the medium did not enhance survival.

The adults also survived for several weeks in tissue culture media (199 or Eagle's) supplemented with horse serum, but ceased ovulation after a few days. They produced a lot of acid, necessitating repeated changes of the medium. An atmosphere of N_2 was unfavorable.

Taylor (1960) maintained the adults of *Litomosoides carinii* (a parasite of the cotton rat transmitted by mites) alive for 23 days in a tissue culture medium (Parker 199) plus rat serum. Microfilaria were produced for 18 days. She maintained *D. immitis* for 8 days, but microfilaria were produced for only two. Taylor (1960) also maintained the microfilaria of *D. immitis* (dog), *Loa loa* (human) and *Wuchereria bancrofti* (human) alive for 10–14 days and the infective larvae from *D. immitis* alive for 9 days in culture. Sawyer and Weinstein (1963a,b) and Weinstein (1963) studied the microfilaria of *D. immitis, W. bancrofti,* and *L. carinii* in culture. They reported the development of the microfilaria to the "sausage" stage in a tissue culture medium (NCTC) plus serum representing an early stage in the development in the arthropod host. Yoeli *et al.* (1964) reported some metamorphosis, growth, and molting by *D. immitis* infective larvae from mosquitoes in a tissue culture medium (1066) supplemented with dog serum and normal blood. Sawyer (1963) and Sawyer and Weinstein (1965) also observed the fourth molt of *D. immitis* in culture (NCTC plus human serum) starting with infective third stage larvae from the mosquito.

Wood and Suitor (1966) reported the growth and development of the microfilaria of *Macanema formosiana,* a parasite of monkeys, to an advanced stage in an insect cell culture medium. Multiplying insect cells originally derived from the mosquito *Aedes aegypti,* which had been maintained in culture for several years, were an essential component of the system (which is therefore a monoxenic culture). Both filaria and insect cells developed without apparent mutual interference. In the filaria, mouth, esophagus, mid-gut and nerve ring differentiated, and a molt was completed. The worms had therefore completed the second larval stage and possibly reached the third infective stage, though their capacity to infect the vertebrate host has not yet been tested.

5. *Trichinella spiralis*

The adults of *Trichinella spiralis* live in the intestine of many different mammals. The viviparous females release second-stage larvae which migrate to the musculature where they become encapsulated in the host's tissues. When eaten by a susceptible mammal, the encapsulated larvae complete their development in the intestine.

Berntzen (1962a, 1965, 1966) has cultured the infective larvae, obtained from experimentally infected rats by digestion of the muscles, to the adult stage in axenic culture. The adults, after molting once in culture (fourth molt), released second-stage larvae (the first molt, completed in the egg) into the culture. Berntzen's methods were adapted from those previously developed for the axenic culture of cestodes (Berntzen, 1962b).

The composition of the gas phase proved critical, a mixture of 85% N_2, 10% O_2 and 5% CO_2, giving optimal results, and even brief exposures to media equilibrated with air were deleterious. Reducing agents, either cysteine HCl or glutathione at critical concentrations, 0.01 *M*, were necessary. Several different media, all very complex, gave satisfactory results. Chick embryo homogenate was used in earlier studies, but in his latest work, Berntzen (1966) used a very complex chemically defined medium supplemented with human serum fractions. In earlier experiments (Berntzen, 1965) the adult worms often failed to escape from the cuticle of their final molt. This prevented copulation and reproduction, and a special desheathing apparatus was designed to facilitate escape. Worms, at the appropriate stage, were allowed to migrate through a silk pad with no deleterious changes in the gas phase. Later (Berntzen, 1966) it was found that when two proteolytic enzymes (co-carboxypeptidases A and B) were added to the culture medium, the desheathing step could be dispensed with. Best results were obtained with a continuous flow of culture medium through the culture vessel.

The digestion procedure for releasing the nematodes from the host tissue was also critical. A satisfactory technique involved the brief successive treatment of minced infected rat muscle with pepsin in HCl, followed by a solution of trypsin, pancreatin, and sodium taurocholate. Larvae collected from the final digest were washed and axenized by treatment with an antibiotic mixture, washed again and transferred to culture media.

Under suboptimal conditions, the immature adult worms often failed to escape from the cuticle of the fourth molt. The adults then molted repeatedly, becoming progressively smaller. Other workers (Kim, 1961, 1962; Meerovitch, 1962, 1965a,b) have reported multiple sheath formation and failure to escape from the sheath in cultures of *T. spiralis*.

The culture apparatus has undergone progressive development. In its latest form it consists of the following four units:

1. A gas mixing and distributing unit supplying the correct gas mixture at the right pressure, humidified, and sterilized by filtration.

2. A medium supply unit including a reservoir (in which the medium could be equilibrated with the gas mixture) and a microperfusion pump controlled by a clock timer.
3. A culture chamber with media and gas inlets and outlets enclosed in an incubator.
4. A collecting unit for the medium which had been passed through the culture chamber.

6. Tylenchoidea

The growth of plant parasitic nematodes under axenic conditions has yet to be achieved, but useful techniques for the monoxenic culture of many species of Tylenchoidea, including plant parasites, have been developed.

Tyler (1933) cultured the root-knot nematode *Meloidogyne* sp. (= *Heterodera marioni*) on tomato seedlings grown from seeds sterilized with calcium hypochlorite in plant nutrient agar under aseptic conditions. The complete life cycle, starting from egg masses sterilized with H_2O_2, was completed in culture. Mountain (1955) used excised roots of corn, tobacco, or clover to culture *Pratylenchus minys* in plant nutrient agar (White's medium). Nematodes, initially sterilized with streptomycin, reproduced in the culture and were maintained by repeated subculture for several months. Tiner (1966) has adapted Mountain's technique to supply axenic *P. penetrans* for *in vitro* research and the development of nematocides. Axenization of tylenchs is probably simplified because bacteria and yeast are not normally ingested.

A most interesting development has been the monoxenic culture of a number of species on proliferating undifferentiated plant callus tissue on nutrient agar. Krusberg (1961) established prolific cultures of *Ditylenchus dipsaci* and *Aphelenchoides ritzemabosi* on alfalfa callus tissue grown on an agar medium containing minerals, sucrose, glycine, B vitamins, 2,4-dichlorophenoxyacetic acid (2,4-D) and coconut milk. Starting with 50 nematodes sterilized with malachite green and streptomycin, tens of thousands of worms were obtained within 2 months. Several other nematodes were tested less extensively with varying results. Dolliver *et al.* (1962) found that *A. ritzemabosi* would reproduce rapidly on callus tissues from a number of different plants. They introduced some variations in Krusberg's technique, including the use of monoxenic cultures on microscope slides with tobacco cells in a liquid medium for the study of the behavior of this nematode. Krusberg and Blichenstaff (1964) introduced further modifications of the medium for maintaining callus culture. They experimented with *D. dipsaci, P. penetrans,* and *P. zeae* on alfalfa tissue on agar, paying particular attention to the action of plant growth regulating substances. Coconut milk is apparently not essential, but 2,4-D increased the yield of all three nematodes. The best conditions for callus tissue are not necessarily the best for the nematode; kinetin, for example, depressed callus growth but produced higher yields of *D. dipsaci* and lower yields of *Pratylenchus*. Other species of nematodes have been cultured by Krusberg's technique (e.g., Khera and Zuckerman, 1962).

Many species of *Tylenchoidea* can be cultured monoxenically on fungi in an appropriate agar medium (e.g., potato, malt extract agar), including both species which normally feed on fungi and those which are facultative feeders on higher plants. Such cultures, which are used routinely for maintaining various species of *Ditylenchus* and *Aphelenchoides* in the laboratory, show that no sharp line can be drawn between plant parasites and free-living nematodes that feed on plant tissue. Faulkner and Darling (1961) showed that monoxenic cultures of *D. destructor* fed and reproduced on 64 different fungi from the 115 tested. It was also cultured on callus tissue from carrot, clover, alfalfa, radish, sweet potato, parsnip, and iris. A study of the pathology of diseased potatoes following inoculation with one or both nematode and fungus showed that the nematode is a primary parasite attacking the potato tuber in the absence of other organisms. Secondary invaders such as fungi advanced the disease.

Dropkin (1966) has drawn up a long list of soil nematodes which have been cultured monoxenically with fungi or plant callus tissue. Such cultures are not well adapted for the direct determination of nutritional requirements, but they have been used to provide uncontaminated nematodes for biochemical studies and for the identification of enzymes, providing indirect information on their nutrition (Dropkin, 1966).

C. Nutrition

Axenic studies have added little as yet to our knowledge of the nutrition of parasitic nematodes. The media which give good growth invariably contain organic materials undefined chemically, which make the determination of nutritional requirements very difficult.

The chemically defined medium used to culture *N. glaseri* contains over 50 different components (Jackson, 1962a,b, 1966) and medium 126, which comes close to being chemically defined, employed by Berntzen (1966) to culture *T. spiralis* through part of its life-cycle, contains over 70. The possibilities for interaction between so many components and the possible existence of significant endogenous reserves, when only part of a life cycle is completed, render the determination of absolute requirements a daunting task. Nonetheless, some progress has been made.

Jackson (1962a), working with *N. glaseri* in a defined medium, found high levels of arginine, histidine, isoleucine, and lysine were required, possibly reflecting the high levels of amino acids in insect hemolymph. B vitamins were required at a high level, though with the exception of folic acid, the requirement for specific vitamins has not been analyzed. Purines and pyrimidines were beneficial. In further experiments (Jackson and Siddiqui, 1965; Jackson, 1966), folic acid was recognized as a requirement for reproduction but not for maturation in a defined medium. Bacterial assay demonstrated a fall in folic acid and folinic acid from cultures in meat infusion broth. The requirement for folic acid was also shown by the effect of its omission from the defined medium and by the effect of two folic acid antagonists, aminopterin and amethopterin, on *N. glaseri* in meat infusion broth.

Some of the complexities of nutritional analysis in complex media are underlined by the observations that *N. glaseri* took up less folic and folinic acid in media deficient in amino acids. Both *N. dutkii* and *N. carpocapsae* took up more of these vitamins than *N. glaseri,* even though they failed to reproduce in the same medium (Jackson and Siddiqui, 1965; Jackson, 1966). Jackson (1966) also reported changes in the concentrations of protein, soluble nitrogen, and amino acid in cultures of *N. glaseri* and *N. dutkii.*

The addition of vitamins has been reported to improve cultures of several gastrointestinal nematodes (Weinstein and Jones, 1956b, 1957a,b, 1959; Douvres, 1962a; Leland, 1963; Silverman *et al.,* 1965), but specific vitamin requirements have not been analyzed. Strongylid nematodes accumulate high levels of vitamin B_{12} (cyanocobalamin) in their coelomocytes both in axenic cultures and on charcoal/feces cultures (Weinstein, 1961, 1965). In *N. braziliensis,* the larval stages, when raised on pure cultures of B_{12}-synthesizing bacteria, developed a red pigment, probably vitamin B_{12}. When a mutant *Escherichia coli* which produces no detectable B_{12} was substituted, little growth and no pigmentation of the larvae resulted unless the culture was supplemented with the vitamin. Further unpublished observations on the accumulation of vitamin B_{12} by both free-living and parasitic stages of *N. braziliensis* in axenic culture were reported at the 41st meeting of the American Society of Parasitologists (1966).

With the exception of the infective third larval stages of some strongylid nematodes, which can develop to the fourth stage in balanced salt solutions, and *N. glaseri,* which has been cultured through one generation in a defined medium, heat-labile tissue extracts or blood fractions have invariably proved necessary for growth in culture. Of the various different preparations, a crude chick embryo homogenate, usually prepared in a balanced salt solution, has had the widest application. The nature of the requirement is unknown, and many different functions can be imagined. Extracts may, for example, fulfill an unidentified nutritional requirement, offer a feeding stimulus, alter the physicochemical nature of the medium, or be the only method so far found of supplying sufficient nutrient in available form.

D. Gas Phase

The control of the concentrations of dissolved O_2 and CO_2 may be crucial for successful axenic culture. Berntzen (1966) has contrasted the optimal gaseous atmosphere for three species of cestodes and the nematode, *T. spiralis*. Each of these intestinal parasites had different requirements. In *T. spiralis*, even brief exposures to air were harmful, the optimal atmosphere being 5% CO_2, 10% O_2, and 85% N_2.

Other workers have used levels of CO_2 above that of atmospheric air and corresponding more closely with those of vertebrate tissue fluids. Weinstein and Jones (1959) gassed their cultures of *N. braziliensis* with 5% CO_2 in air. Leland (1962) adjusted the pH of media used to culture strongylid worms with CO_2 or $NaHCO_3$. Cleeland (1963) showed that semianaerobic conditions favored the initial growth of *A. suum* larvae in culture, but that they later required 5% CO_2, 95% N_2 for further growth. Meerovitch (1965a) found that CO_2 stimulated molting in *T. spiralis*. Many of the culture media that have been used for nematodes include standard tissue culture media in their composition. These media were usually designed to include a CO_2/HCO_3^- buffer which requires gassing with 5% CO_2 for the control of pH (as described in Umbreit *et al.*, 1964). Published results on nematode culture, particularly the many interesting reports appearing as abstracts of papers read at meetings, are frequently devoid of any detail on the control of the gas phase. It has not been possible to include in this review data on the different kinds of culture vessel, procedures for changing media, the concentrations of the worms, or the depth of media used in different experiments, but all these variations will affect the gaseous equilibrium.

Carbon dioxide is known to affect nematodes in several ways. Between pH 5 and 8, CO_2 comes into equilibrium with H_2CO_3, HCO_3^- and $CO_3^=$; in the presence of added $NaHCO_3$, or $KHCO_3$, CO_2 forms a buffer system. Carbon dioxide forms an essential component of the natural stimulus for the exsheathment of the infective larvae of some strongylid nematodes and the hatching of ascarid eggs (reviewed by Rogers and Sommerville, 1963; Rogers, 1966). Carbon dioxide also influences the development of the third-stage larvae of the strongylids, *H. contortus* and *D. viviparous*, to the fourth stage (Silverman *et al.*, 1966). Sommerville

(1966) found that a high level of CO_2 induced *H. contortus* to develop to the fourth stage and molt in a balanced salt solution. Best results were obtained when the solution was in equilibrium with an atmosphere of 40% CO_2, 10% O_2 and 50% N_2, but after the exsheathed third-stage larva had been exposed to CO_2 for many hours, an atmosphere of air could be used without inhibitory affect. Certain tissue extracts can perhaps be substituted for exposure to high CO_2 (Stoll, 1940; Silverman *et al.*, 1962).

The fixation of CO_2 by pyruvate through "malic" enzyme, with the formation of succinate, is important in the intermediary metabolism of several helminths, including the nematodes, *Heterakis gallinae, Trichuris vulpis*, and *A. suum* (discussed by von Brand, 1966). It may be that the reaction is important in other nematodes and requires relatively high levels of CO_2.

Von Brand (1966) has also discussed at some length the importance of O_2 concentration to parasitic worms. It may be presumed that tissue parasites are exposed to relatively high levels of O_2. The quantities of O_2 available to worms in the intestinal lumen of vertebrates and to the free-living larval stages developing in fecal cultures are less clearly defined, but O_2 may be deficient. Nematodes in which respiration has been studied show a capacity to use oxygen when available and to tolerate periods of anerobiosis. The larger intestinal nematodes may be predominantly anaerobic, though *A. suum* muscle has been shown to possess both aerobic and anaerobic pathways for the generation of ATP. However, most experimenters have contrasted anaerobic respiration with respiration in media equilibrated with air. The effects of intermediate levels of O_2 have been little studied and it seems that more information on levels of O_2 in the natural habitats of parasites and the effects of comparable levels on respiration and on growth in culture should now be sought. CO_2 and O_2 electrodes are now commercially available and should facilitate studies.

E. The Applications of Axenic Cultures in the Study of Nematode Parasites

A number of workers have studied the molting of parasitic worms *in vitro*. Kim (1961, 1962), Meerovitch (1965a,b), and Berntzen (1965) studied the molting of the intestinal phase of *T. spiralis*. By combining axenic studies of the intestinal phase with *in vivo* studies of the migratory and muscle phases, Berntzen (1965) showed that *T. spiralis* has the typical nematode sequence of larval stages and molts. The larvae molt once within the egg, twice in the migratory and muscle phases and for the fourth time in the intestinal phase.

Sommerville (1966) made a detailed study of the fourth molt of *H. contortus* and Taylor (1960), Sawyer and Weinstein (1963a,b, 1965), and Yoeli *et al.* (1964) have made use of axenic cultures to observe the fourth molt of *Dirofilaria immitis,* which normally occurs in the tissue of dogs.

The morphogenesis of the early parasitic stages of several nematodes have been studied with axenic cultures, e.g., of microfilariae (Taylor, 1960; Yoeli *et al.,* 1964), *N. braziliensis* (Weinstein and Jones, 1965a), *O. radiatum* (Douvres, 1962a), and *H. contortus* (Sommerville, 1966). Difficulties in obtaining copulation and fertilization in culture have in turn stimulated more detailed investigations of development and behavior *in vivo* of *N. braziliensis* (Weinstein and Jones, 1959) and *N. dubius* (Sommerville and Weinstein, 1964).

Weinstein (1966) studied the deposition of pigment in cell inclusions, believed to be primary lysosomes, in the intestinal cells of *N. braziliensis.* This pigment should not be confused with the red pigment, probably vitamin B_{12}, found in the coelomocytes. The addition of hemoglobin to axenic cultures led to pigment deposition.

The action of serum and tissue extracts from infected animals has been compared with that of noninfected animals on nematodes in culture. Douvres (1962b) studied the action of serum and intestinal tissue extracts on cultures of *O. radiatum.* Sera and intestinal extracts were taken from, (1) a bovine with patent infections, (2) a bovine rendered resistant by repeated infection with the parasite, (3) several uninfected controls. Precipitates formed on the developing third and fourth stage larvae, especially on the cuticle and in the vicinity of the excretory pore, in the presence of intestinal extracts from the infected and resistant animals, but not from the controls. Sera from infected and resistant animals did not produce precipitates, but retarded growth when compared with those in serum from controls.

Sawyer (1963) observed precipitates on third-stage larvae of *D. immitis* when serum from infected dogs was added to cultures, but not with sera from uninfected dogs. Jackson (1961, 1965) observed precipitates when three species of *Neoaplectana* from axenic culture were incubated in sera from rabbits which had been immunized by subcutaneous injections of live worms. Antisera, when added to RLE, did not prevent development and reproduction of *N. glaseri* in culture. Jackson (1961) also studied the action of some anthelminthics on cultures.

Silverman *et al.* (1962), using the guinea pig as a suitable, though unnatural, laboratory host for *D. viviparous, Trichostrongylus colubriformis,* and *Strongylides papillosus,* explored the possibilities of immunizing animals with preparations from these worms and their secretions. Such a system, when further developed, may provide a means of vaccinating domestic animals against helminths. For this purpose, it is desirable to culture the parasitic stages in media free from macromolecular substances which would contaminate antigens released by the worms in culture. Each of these species were cultured from the third infective to fourth larval stage in a salt solution supplemented with liver extract. Silverman *et al.* (1962) found that a variety of antigen preparations, including some prepared from cultured worms, gave clear evidence of protection against infection, though the characterization of the protective antigens will be a task for the future. A fuller discussion of these possibilities has been given by Silverman (1963, 1965b) and Silverman *et al.* (1966).

III. The Culture and Nutrition of Acanthocephala

The Acanthocephala are all parasites with two or more hosts in the life cycle. Adults which are dioecious may be found in the alimentary canal of all classes of vertebrates. Details of structure and life history are discussed by Pratt (Chapter 2, Section II).

A. Culture

Many Acanthocephala remain viable in simple balanced salt solutions for long periods, showing intermittent writhing movements. Several investigators have studied their survival under such conditions. Gettier (1942) found *Neoechinorhynchus emydis*, from the terrapin, lived for up to 20 days in 0.5–0.7% NaCl. Solutions of KCl, $MgCl_2$ and $CaCl_2$ were less satisfactory, but the addition of traces of $CaCl_2$ to 0.5% NaCl prolonged survival. Various balanced salt solutions, based on Ringer's solution, and isotonic with 0.5% NaCl, gave comparable results. The body, which is normally flattened, tends to become turgid and swollen in these solutions, and Van Cleave and Ross (1944) recommended 0.85% NaCl for the maintenance of *N. emydis* because it retains its normal body shape in this solution. The worms recovered their normal form in 0.85% NaCl after being swollen in tap water for short periods. Von Brand (1940) found that *Macracanthorhynchus hirudinaceus*, a parasite of pigs, depletes its endogenous glycogen *in vitro*. Ward (1952) studied endogenous glycogen consumption by *M. hirudinaceus* under aerobic and anaerobic conditions. Groups of worms were incubated for 43 hours in a balanced salt solution, containing penicillin and streptomycin, some in open flasks which were periodically bubbled with O_2 and others in an atmosphere of N_2. From comparison with unincubated controls, it was concluded that more glycogen was consumed anaerobically than aerobically. Dunagan (1963, 1964) found that added glucose, mannose, and galactose reduced glycogen consumption by *Neoechinrhoynchus* species (from turtles) in balanced salt solutions.

Jensen (1952) cultured the fish parasite, *Pomphorhynchus bulbocolli*, from the cystacanth stage, obtained aseptically from its intermediate host, *Hyalella azteci* (Amphipoda), to the adult form. The two media which gave these results were (1) three parts peptone broth (1% peptone, 1% Lab-Lemco, 0.5% NaCl), one part horse serum, one part 1% glucose and; (2) a similar mixture to which one part yeast extract was added. Microbial contaminants, when these became apparent, were suppressed with penicillin and streptomycin.

The worms showed considerable growth and the males appeared mature, but lacked spermatogenesis. The ovary in the female did not produce ovarian balls, a step marking sexual maturity. Sexually mature males and females taken from the intestines of fish and axenized with antibiotics were active in culture, the males everting the copulatory bursa, but failing to copulate. The cystacanths survived without development for long periods, at least 662 days, in horse serum.

Harms (1965) failed to obtain development of the cystacanths of *Octospinifer macilentis* (from *Ostracoda*), another fish parasite, but found that the immature adult stage increased in length when cultured axenically in a medium consisting of peptone, yeast extract, horse serum, glucose, vitamins, and salts.

Nicholas and Grigg (1965) cultured *Moniliformis dubius* from the rat in various media and with different gas mixtures. Little growth was obtained with immature worms. The sexually mature adults survived and continued to move for at least 8 days in Eagle's vertebrate tissue culture medium or bovine serum. Changes in the gas mixtures made no clear differences, but survival seemed better with N_2 or 95% N_2, 5% CO_2 than 75% N_2, 5% CO_2, 20% O_2. The female worms usually continued to shed acanthors into the medium for the full 8 days, but failure of the uterine bell organ, coupled with inhibition of egg development in culture, led to a rapid fall in the number of fully developed acanthors and increasing numbers of undeveloped ones being released into the medium.

B. Hatching of the Acanthor; Activation of the Cystacanth

The cystacanth is a resting stage invested by the arthropod host in a capsule within its hemocoel, and with the proboscis inverted. When ingested by a suitable host, the proboscis is everted, rupturing the capsule where this has not already been digested, and the parasite becomes attached to the intestinal wall by means of active proboscis movements.

It is not known whether activation of the cystacanth is a necessary prerequisite for its subsequent growth in culture. Pepsin in HCl digests the capsule (Crompton, 1964; King, 1965; Graff and Kitzman, 1965), but does not stimulate proboscis movement. Graff and Kitzman (1965) found that cystacanths of *Moniliformis dubius* (from the cockroach) were activated by bile salt and CO_2 in physiological saline at an alkaline pH; O_2 was inhibitory. They also showed that the surgical transfer of the opening of the bile ducts of rats (the definitive host) to a region of the intestine below that normally occupied by the worm prevented the establishment of infection, when cystacanths were administered orally, unless they had previously been activated *in vitro*.

The acanthors of several species which infect terrestrial arthropods can be hatched *in vitro*. In *M. dubius* the acanthors hatched in solutions of several electrolytes stronger than 0.2 *M* at pH of 7.5 or above (Edmonds, 1966). Sucrose solutions were ineffective, but CO_2 stimulated hatching in electrolyte solutions at pH of 6.0 or greater and may be a component of the natural hatching stimulus in the cockroach gut. In other species with terrestrial intermediate hosts, alternately wetting and drying the acanthors, taken directly from the female worm, caused some to hatch, but did not do so when they were taken from the feces (Manter, 1928; Moore, 1942). This process is presumably not physiological. Acanthors which infect aquatic arthropods have not been hatched artificially, but can be obtained for *in vitro* studies from experimentally infected hosts.

It may be concluded that very little work has so far been done on the culture of Acanthocephala, and this, in turn, has meant that very little information on nutrition has so far come from such studies. Much more information on nutrition has come indirectly from biochemical techniques which lie outside the scope of this chapter. There have, for example, been investigations of intermediary metabolism, amino acid permeation, respiration, and the effects of the host's diet on the parasite. For a recent review of the biochemistry of Acanthocephala, the reader is referred to von Brand's textbook on parasite biochemistry (1966).

Acknowledgment

I am grateful to Dr. P. H. Silverman for the facilities of his department and for his comments on the manuscript.

REFERENCES

Berntzen, A. K. (1962a). *J. Parasitol.* **48**, Suppl., 48.
Berntzen, A. K. (1962b). *J. Parasitol.* **48**, 785.
Berntzen, A. K. (1965). *Exptl. Parasitol.* **16**, 74.
Berntzen, A. K. (1966). *Ann. N.Y. Acad. Sci.* **139**, 176.

Buecher, E. J., Jr., Hansen, E., and Yarwood, E. A. (1966). *Proc. Soc. Exptl. Biol. Med.* **121**, 390.
Chaudhuri, C. (1964). Thesis, University of Illinois, Urbana, Illinois.
Cleeland, R. (1963). *J. Parasitol.* **49**, 64.
Cleeland, R., and Laurence, K. A. (1962). *J. Parasitol.* **48**, 35.
Crompton, D. W. T. (1964). Ph.D. Thesis, University of Cambridge.
Cryan, W. S. (1963). *J. Parasitol.* **49**, 351.
Cryan, W. S., Hansen, E., Martin, M., Sayre, F. W., and Yarwood, E. A. (1963). *Nematologica* **9**, 313.
Diamond, L. S., and Douvres, F. W. (1960). *J. Parasitol.* **46**, Suppl., 25.
Diamond, L. S., and Douvres, F. W. (1962). *J. Parasitol.* **48**, 39.
Dolliver, J. S., Hildebrandt, A. C., and Riker, A. J. (1962). *Nematologica* **7**, 294.
Dougherty, E. C. (1949). *Anat. Record* **105**, 532.
Dougherty, E. C. (1960). *In* "Nematology" (J. N. Sasser and W. B. Jenkins, eds.), pp. 297–318. Univ. of North Carolina Press, Chapel Hill, North Carolina.
Dougherty, E. C., and Calhoun, M. G. (1949). *Proc. Helminthol. Soc. Wash., D.C.* **15**, 55.
Dougherty, E. C., and Solberg, B. (1961). *Nature* **192**, 184.
Dougherty, E. C., Hansen, E. L., Nicholas, W. L., Mollett, J. A., and Yarwood, E. A. (1959). *Ann. N.Y. Acad. Sci.* **77**, 176.
Douvres, F. W. (1960a). *J. Parasitol.* **46**, Suppl., 25.
Douvres, F. W. (1960b). *J. Parasitol.* **46**, Suppl., 25.
Douvres, F. W. (1962a). *J. Parasitol.* **48**, 314.
Douvres, F. W. (1962b). *J. Parasitol.* **48**, 852.
Douvres, F. W. (1966). *J. Parasitol.* **52**, 1033.
Douvres, F. W., and Alicata, J. E. (1962). *J. Parasitol.* **48**, Suppl., 35.
Douvres, F. W., and Tromba, F. G. (1962). *J. Parasitol.* **48**, 269.
Douvres, F. W., Tromba, F. G., and Doran, D. J. (1966). *J. Parasitol.* **52**, 875.
Dropkin, V. H. (1966). *Ann. N.Y. Acad. Sci.* **139**, 39.
Dunagan, T. T. (1963). *J. Parasitol.* **49**, Suppl., 18.
Dunagan, T. T. (1964). *Proc. Helminthol. Soc. Wash., D.C.* **31**, 166.
Earl, P. R. (1959). *Ann. N.Y. Acad. Sci.* **77**, 163.
Edmonds, S. J. (1966). *Exptl. Parasitol.* **19**, 216.
Ells, H. A. (1955). Ph.D. Thesis, University of California, Los Angeles, California.
Faulkner, L. R., and Darling, H. M. (1961). *Phytopathology* **55**, 778.
Gettier, D. A. (1942). *Proc. Helminthol. Soc. Wash., D.C.* **9**, 75.
Glaser, R. W. (1931). *Science* **73**, 614.
Glaser, R. W. (1940a). *J. Exptl. Biol.* **84**, 1.
Glaser, R. W. (1940b). *Proc. Soc. Exptl. Biol. Med.* **43**, 512.
Glaser, R. W., and Stoll, N. R. (1938a). *Parasitology* **30**, 324.
Glaser, R. W., and Stoll, N. R. (1938b). *Science* **87**, 259.
Glaser, R. W., McCoy, E. E., and Girth, H. B. (1942). *J. Parasitol.* **28**, 123.
Gochnauer, M. B., and McCoy, E. (1954). *J. Exptl. Zool.* **125**, 377.
Graff, D. J., and Kitzman, W. B. (1965). *J. Parasitol.* **51**, 424.
Hansen, E. L., and Cryan, W. S. (1966a). *Nematologica* **12**, 138.
Hansen, E. L., and Cryan, W. S. (1966b). *Nematologica* **12**, 355.
Hansen, E. L., Yarwood, W. A., Nicholas, W. L., and Sayre, F. W. (1960). *Nematologica* **5**, 27.
Hansen, E. L., Buecher, E. J., Jr., and Yarwood, E. A. (1964). *Nematologica* **10**, 623.

Hansen, E. L., Silverman, P. H., and Buecher, E. J., Jr. (1966). *J. Parasitol.* **52,** 137.
Harms, C. E. (1965). *Exptl. Parasitol.* **17,** 41.
Hieb, W. F., and Sandman, R. P. (1967). *Federation Proc.* **26,** 797.
Hieb, W. F., and Rothstein, M. (1968). *Science* **160,** 778.
Jackson, G. J. (1961). *Exptl. Parasitol.* **11,** 241.
Jackson, G. J. (1962a). *Exptl. Parasitol.* **12,** 25.
Jackson, G. J. (1962b). *Trans. N.Y. Acad. Sci.* **24,** 954.
Jackson, G. J. (1965). *Parasitology* **55,** 571.
Jackson, G. J. (1966). *Ann. N.Y. Acad. Sci.* **139,** 91.
Jackson, G. J., and Siddiqui, W. A. (1965). *J. Parasitol.* **51,** 727.
Jensen, T. (1952). Ph.D. Thesis, University of Minnesota; *Dissertation Abstr.* **12,** 607 (1952).
Jones, M. F., and Weinstein, P. P. (1957). *J. Parasitol.* **43,** Suppl., 46.
Khera, S., and Zuckerman, B. M. (1962). *Nematologica* **8,** 272.
Kim, C. W. (1961). *Am. J. Trop. Med. Hyg.* **10,** 742.
Kim, C. W. (1962). *Am. J. Trop. Med. Hyg.* **11,** 422.
King, D. (1965). M.S. Thesis, University of New South Wales, Australia.
Krusberg, L. R. (1961). *Nematologica* **6,** 181.
Krusberg, L. R., and Blickenstaff, M. L. (1964). *Nematologica* **10,** 145.
Lapage, G. (1933). *Univ. Cambridge, Inst. Animal Pathol. 3rd Rept. Director* p. 237.
Lawrence, J. J. (1948). *Australian J. Exptl. Biol. Med. Sci.* **26,** 1.
Leland, S. E., Jr. (1961). *J. Parasitol.* **47,** Suppl., 21.
Leland, S. E., Jr. (1962). *J. Parasitol.* **48,** Suppl., 35.
Leland, S. E., Jr. (1963). *J. Parasitol.* **49,** 600.
Leland, S. E., Jr. (1965a). *J. Parasitol.* **51,** Suppl., 46.
Leland, S. E., Jr. (1965b). *J. Parasitol.* **51,** Suppl., 47.
Leland, S. E., Jr. (1965c). *J. Parasitol.* **51,** Suppl., 47.
Leland, S. E., Jr., and Wallace, L. J. (1966). *J. Parasitol.* **52,** 280.
Lower, W. R., Hansen, E., and Yarwood, E. H. (1966). *J. Exptl. Zool.* **161,** 29.
McCoy, O. R. (1929a). *Science* **69,** 74.
McCoy, O. R. (1929b). *Am. J. Hyg.* **10,** 140.
Manter, H. W. (1928). *Trans. Am. Microscop. Soc.* **47,** 342.
Meerovitch, E. (1962). *J. Parasitol.* **48,** Suppl., 34.
Meerovitch, E. (1965a). *Can. J. Zool.* **43,** 69.
Meerovitch, E. (1965b). *Can. J. Zool.* **43,** 81.
Moore, D. V. (1942). *J. Parasitol.* **28,** 495.
Mountain, W. B. (1955). *Proc. Helminthol. Soc. Wash., D.C.* **22,** 49.
Myers, R. F., and Krusberg, L. R. (1965). *Phytopathology* **55,** 429.
Nicholas, W. L. (1956). *Nematologica* **1,** 337.
Nicholas, W. L. (1962). *Nematologica* **8,** 99.
Nicholas, W. L., and Grigg, H. (1965). *Exptl. Parasitol.* **16,** 332.
Nicholas, W. L., and Jantunen, R. (1963). *Nematologica* **9,** 332.
Nicholas, W. L., and McEntegart, M. G. (1957). *J. Helminthol.* **31,** 135.
Nicholas, W. L., Dougherty, E. C., and Hansen, E. L. (1959). *Ann. N.Y. Acad. Sci.* **77,** 218.
Nicholas, W. L., Hansen, E., and Dougherty, E. C. (1962). *Nematologica* **8,** 129.
Pertel, R. (1964). *Nematologica* **10,** 343.
Pitts, T. D. (1960). *J. Parasitol.* **46,** Suppl., 24.
Pitts, T. D., and Ball, G. H. (1953). *J. Parasitol.* **39,** Suppl., 42.
Pitts, T. D., and Ball, G. H. (1955). *J. Parasitol.* **47,** Suppl., 47.

Poinar, G. O., and Thomas, G. M. (1966). *Parasitology* **55,** 385.
Read, C. P. (1966). *In* "The Biology of Parasites" (E. J. L. Soulsby, ed.), pp. 101–126. Academic Press, New York.
Rodriguez, J. G. (1966). *Ann. N.Y. Acad. Sci.* **139,** 53.
Rogers, W. P. (1966). *In* "The Biology of Parasites" (E. J. L. Soulsby, ed.), pp. 33–39. Academic Press, New York.
Rogers, W. P., and Sommerville, R. I. (1963). *Advan. Parasitol.* **1,** 109.
Rothstein, M., and Cook, E. (1966). *Comp. Biochem. Physiol.* **17,** 683.
Rothstein, M., and Götz, P. (1968). *Arch. Biochem. Biophys.* **126,** 131.
Sang, J. H. (1962). *Genet. Res.* **3,** 90.
Sawyer, T. K. (1963). *J. Parasitol.* **49,** Suppl., 59.
Sawyer, T. K., and Weinstein, P. P. (1963a). *J. Parasitol.* **49,** 218.
Sawyer, T. K., and Weinstein, P. P. (1963b). *Am. J. Vet. Res.* **24,** 402.
Sawyer, T. K., and Weinstein, P. P. (1965). *J. Parasitol.* **51,** Suppl., 48.
Sayre, F. W., Hansen, E., Starr, T. J., and Yarwood, E. A. (1961). *Nature* **190,** 1116.
Sayre, F. W., Hansen, E. L., and Yarwood, E. A. (1963). *Exptl. Parasitol.* **13,** 98.
Sayre, F. W., Lee, R. T., Sandman, R. P., and Perez-Mendez, G. (1967). *Arch. Biochem. Biophys.* **118,** 58.
Sen, H. G., and Kelly, G. W. (1960). *J. Parasitol.* **46,** Suppl., 24.
Silverman, P. H. (1959). *Nature* **183,** 197.
Silverman, P. H. (1963). *In* "Techniques in Parasitology" (A. E. R. Taylor, ed.), pp. 45–67. Blackwell, Oxford.
Silverman, P. H. (1965a). *Advan. Parasitol.* **3,** 159.
Silverman, P. H. (1965b). *Am. Zoologist* **5,** 153.
Silverman, P. H., Poynter, D., and Podger, K. R. (1962). *J. Parasitol.* **48,** 562.
Silverman, P. H., Hansen, E. L., and Buecher, E., Jr. (1965). *J. Parasitol.* **51,** Suppl., 46.
Silverman, P. H., Alger, N. E., and Hansen, E. L. (1966). *Ann. N.Y. Acad. Sci.* **139,** 124.
Sivapalan, P., and Jenkins, W. R. (1966). *Nematalogica* **12,** 99.
Sommerville, R. I. (1966). *J. Parasitol.* **52,** 127.
Sommerville, R. I., and Weinstein, P. P. (1964). *J. Parasitol.* **50,** 401.
Stoll, N. R. (1940). *Growth* **4,** 383.
Stoll, N. R. (1953). *J. Parasitol.* **39,** 422.
Stoll, N. R. (1954). *J. Parasitol.* **40,** Suppl., 14.
Stoll, N. R. (1956). Proc. *14th Intern. Congr. Zool., Copenhagen, 1955* p. 382.
Stoll, N. R. (1959). *Ann. N.Y. Acad. Sci.* **77,** 126.
Taylor, A. E. R. (1960). *Exptl. Parasitol.* **9,** 113.
Tiner, J. D. (1966). *Ann. N.Y. Acad. Sci.* **139,** 111.
Tomlinson, G. A., and Rothstein, M. (1962). *Biochim. Biophys. Acta* **63,** 465.
Tyler, J. (1933). *Hilgardia* **7,** 373.
Umbreit, W. W., Burris, R. H., and Stauffer, J. F. (1964). "Manometric Techniques," 4th ed. Burgess, Minneapolis, Minnesota.
Van Cleave, H. J., and Ross, E. (1944). J. *Parasitol.* **30,** 369.
von Brand, T. (1940). *J. Parasitol.* **26,** 301.
von Brand, T. (1966). "Biochemistry of Parasites." Academic Press, New York.
Ward, H. L. (1952). *J. Parasitol.* **38,** 493.
Weinstein, P. P. (1953). *Am. J. Hyg.* **58,** 352.
Weinstein, P. P. (1954). *J. Parasitol.* **40,** Suppl., 14.

Weinstein, P. P. (1958). *Am. J. Trop. Med. Hyg.* 7, 1.
Weinstein, P. P. (1961). *J. Parasitol.* **47,** Suppl., 23.
Weinstein, P. P. (1963). *Trans. Roy. Soc. Trop. Med. Hyg.* **57,** 236.
Weinstein, P. P. (1965). *J. Parasitol.* **51,** Suppl., 55.
Weinstein, P. P. (1966). *In* "The Biology of Parasites" (E. J. L. Soulsby, ed.), pp. 143–154. Academic Press, New York.
Weinstein, P. P., and Jones, M. F. (1956a). *J. Parasitol.* **42,** 215.
Weinstein, P. P., and Jones, M. F. (1956b). *J. Parasitol.* **42,** Suppl., 14.
Weinstein, P. P., and Jones, M. F. (1957a). *Am. J. Trop. Med. Hyg.* **6,** 480.
Weinstein, P. P., and Jones, M. F. (1957b). *J. Parasitol.* **43,** Suppl., 45.
Weinstein, P. P., and Jones, M. F. (1959). *Ann. N.Y. Acad. Sci.* **77,** 137.
Winkler, E. J., and Pramer, D. (1961). *Nature* **192,** 472.
Wood, D. E., and Suitor, E. C. (1966). *Nature* **211,** 868.
Yoeli, M., Upmanis, R. S., and Most, H. (1964). *Exptl. Parasitol.* **15,** 325.
Zimmerman, A. (1921). *Rev. Suisse Zool.* **28,** 357.

CHAPTER 5

Carbohydrate and Energy Metabolism of Nematodes and Acanthocephala[1]

Howard J. Saz

I. Introduction

Increasing realization of the importance of comparative biochemical studies has resulted in a corresponding increase in our knowledge of metazoan metabolism. Unfortunately, most of these studies have dealt with a relatively small group of parasitic metazoans for obvious practical reasons. Therefore, any discussion of nematode metabolism is almost limited to a few of the parasitic species, since relatively little information is available pertaining to the overwhelming number of free-living nematodes. What few data have been obtained, however, indicate that many of the free-living nematodes, like their parasitic counterparts, exhibit basic biochemical differences from the classical metabolic patterns established with mammalian tissues.

The nature of these differences may be morphological, physiological, or biochemical. For example, the structure of a somatic muscle cell of *Ascaris lumbricoides* differs dramatically from that of mammalian

[1] Recent studies from the laboratory of the author which are discussed in this chapter were supported by U.S. Public Health Service grants AI-05290 and 5-T1-AI-149.

muscle in that the former consists of a large nucleus containing cell "belly" which is continuous with two processes; a striated fiber and an elongated arm which in turn subdivides terminally as it approaches the nerve cord (De Bell *et al.*, 1963; Rosenbluth, 1965a,b; Reger, 1965). As will be discussed below, the *Ascaris* muscle bundle comprised of these cells differs markedly from mammalian skeletal, smooth, and cardiac muscles, both physiologically and biochemically. Responses of nematodes to various anthelmintic agents, e.g., piperazine and cyanine dyes, also have served to demonstrate differences at the physiological level (Saz and Bueding, 1966).

Investigations of helminth biochemistry have established at least three areas in which these metazoans may differ from mammalian tissues. First, their overall metabolism frequently diverges from classical patterns, resulting in a variety of end products and hence different or altered pathways. This is particularly true of the parasitic nematodes which have been studied to date. Second, differences may occur at the enzyme level and be expressed kinetically or in terms of varying substrate or cofactor specificities; and finally, differences may be determined at the molecular level of protein structure, particularly as manifested by the immunological specificity of the helminth enzymes when compared to mammalian enzymes which catalyze the same reactions (Bueding, 1962).

Perhaps the most obvious difference between the energy metabolism of the host tissues and the parasitic heminths is the lack of a complete oxidation of substrates by the parasites. Carbohydrate substrates are not "burned" completely to carbon dioxide and water. Instead, more reduced end products are excreted, indicating either a lack or, at best, a limited tricarboxylic acid cycle activity for terminal oxidations. In this respect, the carbohydrate metabolism of many helminths more closely resembles that of anaerobic bacteria than that of mammalian tissues. In accord with this, an anaerobic (or possibly a microaerophilic) existence would be required by the larger helminth parasites which dwell in the lumen of the intestine where oxygen tensions are low (von Brand, 1938, 1960a,b). However, even organisms which are bathed in host blood may have no oxygen requirement for their energy metabolism, as is the case with *Schistosoma mansoni* (Bueding *et al.*, 1953).

The major purpose of this chapter is to discuss in some detail the carbohydrate and energy metabolisms of a selected few nematodes for which sufficient information is available to allow for a reasonably rational biochemical evaluation. Attempts will be made to stress the comparative aspects, that is, the biochemical similarities as well as the differences between metazoan and vertebrate animals; particularly at the pathway level. No attempt will be made to compile a comprehensive review of the literature, which has recently been accomplished in a masterful manner (von Brand, 1966). In addition, a complete and useful bibliography on the metabolism of endoparasites has been published (J. C. Smith, 1965). Books and reviews dealing with various aspects of helminth metabolism also have appeared, only a few of which will be cited here (Fairbairn, 1957, 1960; Rogers, 1961; Lee, 1965; von Brand, 1966, Saz and Bueding, 1966).

The nematodes to be discussed have been divided arbitrarily into two groups. First, the apparent obligate aerobes which presumably require oxygen for the normal functioning of their energy metabolism and for survival. Second, organisms which have at least one stage wherein there appears to be no demonstrable oxygen requirement for the operation of energy-producing pathways. Most members of the second group require carbon dioxide for maintenance or metabolism, and accumulate succinate, or compounds presumably arising from succinate, as end products of their carbohydrate metabolism. It should be stressed that this grouping is for convenience and is not meant to have biological or biochemical significance; particularly since the oxygen requirements for all developmental stages of a given nematode are not known.

II. Nematodes Which Are Obligate Aerobes

To determine whether or not an organism requires oxygen for survival appears to be a simple matter. For many helminths, however, this basic question has remained a very difficult one to answer for several reasons. All of the nematodes, cestodes, and trematodes which have been investigated are capable of taking up at least small amounts of oxygen when incubated *in vitro* in air. On the other hand, the physiological significance of this oxygen uptake has been seriously questioned in those worms which normally dwell in essentially anaerobic environments, such as the adult ascarids and tapeworms surviving in the lumen of the intestines (von Brand, 1938; Bueding, 1949a; Bueding *et al.*, 1953; Scheibel and Saz, 1966). In addition, some of the intestinal helminths survive for extended periods of time or can be cultured anaerobically. The presence of oxygen may be detrimental to survival in these incubation systems (Weinland, 1901; Harpur, 1962; Bueding *et al.*, 1960; Berntzen, 1962; Berntzen and Mueller, 1964; Schiller, 1965). More specific reasons for presuming an anaerobic metabolism for some of the helminths will be discussed below, but in no case has it been completely ruled out that a microaerophilic requirement may be necessary for the normal production of one of the future developmental stages of the life cycle.

Although it is difficult to completely rule out a small requirement for oxygen in apparently anaerobic nematodes, it has been too easy for many authors to conclude the occurrence of a tricarboxylic acid cycle (TCA) cycle as a major energy-yielding pathway in a given helminth. The presence of some or even all of the enzymes of the TCA cycle in an organism does not, by itself, allow for this conclusion. These enzymes may be present even in strict anaerobes where they function in a purely biosynthetic capacity (Krebs and Lowenstein, 1960). As has been demonstrated in bacteria (Saz and Krampitz, 1954; Krampitz, 1961), the reactions must be shown to occur as an orderly sequence of events, which account for the complete oxidation of significant quantities of substrates.

A. *Caenorhabditis briggsae*

C. briggsae is a small free-living nematode living in soil, whose adult form ranges from 1 to 1.2 mm. in length. This hermaphroditic worm has been cultured axenically in a defined medium to which a 10% extract of liver was added as well as in other, more complex media (Dougherty *et al.*, 1959; Tomlinson and Rothstein, 1962; Sayre *et al.*, 1963). It has been assumed that *C. briggsae* is an aerobe, since it consumes oxygen and requires a thin layer of medium with a large air interface for growth. Unfortunately, however, more definitive proof of the aerobic requirement does not seem to be available. To the contrary, Rothstein and Tomlinson (1962) have reported that *C. briggsae* survives when incubated under low oxygen tensions and appears to be unharmed for at least 3 days in the presence of 1×10^{-4} *M* cyanide, although death ensues within 24 hours upon incubation with the higher concentration of 1×10^{-3} *M*. In accord with the relative insensitivity to cyanide, the soil nematode does not possess a typical cytochrome electron transport system, but has been reported to contain a pigment of the cytochrome *b* type (Rothstein, 1963).

Only circumstantial evidence is available which indicates that a tricarboxylic acid may be operative in *C. briggsae*. Rothstein and Tomlinson (1962) demonstrated the biosynthesis of both essential and nonessential amino acids when this nematode was incubated with acetate, glucose, or glycine, respectively, containing ^{14}C. The appearance of isotope in the amino acids generally associated with the tricarboxylic acid cycle, such as glutamic and aspartic acids and alanine, indicated the presence of this terminal oxidative pathway, at least as a route for the biosynthesis of amino acids. Unfortunately, it cannot be concluded from these findings that the cycle functions physiologically as a major energy-yielding pathway for terminal respiration. To the contrary, *C. briggsae* does not oxidize exogenous substrates to completion as would be expected via the

TCA cycle, but rather excretes a complex mixture of products even when incubated in air (Rothstein, 1963). The products have not been characterized, but consist of amino acids and large "acidic" as well as "neutral" fractions. There is, therefore, an incomplete utilization of carbohydrate by the nematode which suggests the absence of the tricarboxylic acid cycle or, at most, the presence of a cycle with limited rate. Further studies pertaining to these points are necessary, however, before any definite conclusion can be reached.

Of particular interest are the demonstrations of the presence of isocitrate lyase (isocitratase) and malate synthetase activities not only in *C. briggsae,* but also in three other free-living nematodes, *Panagrellus redivivus, Turbatrix aceti* (vinegar eel), and *Rhabditis anomala* (Rothstein and Mayoh, 1964a,b, 1965, 1966). Isocitrate lyase was first demonstrated in bacteria (Campbell *et al.,* 1953; Saz, 1954; Saz and Hillary, 1956; R. A. Smith and Gunsalus, 1954) as was malate synthetase (Wong and Ajl, 1955). The former enzyme catalyzes the cleavage of isocitrate to form succinate and glyoxylate, while the latter catalyzes the synthesis of malate from glyoxylate and acetyl-CoA. When both of these enzymes act in conjunction with a number of systems from the TCA cycle, then a cyclic pathway may result which allows for the synthesis of C_4 dicarboxylic acids from acetyl-CoA. This pathway, the glyoxylate cycle, was first proposed by Kornberg and Krebs (1957), and has since been found to be present in a number of bacteria, algae, yeast, protozoa, higher plants and possibly in the pupae of the insect, *Eridinea prodenia* (Kornberg and Elsden, 1961; Carpenter and Jaworski, 1962). Later stages of *Eridinea prodenia* appear to be devoid of glyoxylate cycle activity (Levenbook, 1961).

Neither the quantitative significance nor the physiological role of the glyoxylate cycle in these free-living nematodes can be estimated at this time. Glyoxylate is known to be converted into glycine by animal and plant tissues (Weinhouse and Friedmann, 1951; Nakada and Weinhouse, 1953; Weissbach and Sprinson, 1953; Tolbert and Cohan, 1953; Olson, 1954). It has been postulated, therefore, that in bacteria, the isocitrate lyase system may participate in the biosynthesis of glycine by providing glyoxylate (Saz and Hillary, 1956). Evidence in support of this pathway in *C. briggsae* has been reported by Rothstein and Mayoh (1964b). These authors incubated the nematodes with aspartate-4-^{14}C and found that the worms excreted glycine as a major radioactive product. They postulate that aspartate gave rise to oxalacetate which in turn was converted to citrate and isocitrate respectively. Isocitrate lyase would then be expected to hydrolyze the isocitrate leading to the formation of glyoxylate and succinate. Presumably, glyoxylate would give rise to glycine by transamination. Although this pathway for glycine biosynthesis may be necessary for these worms, it does not require the participation of the other member of the glyoxylate cycle, malate synthetase.

It is of interest that one of the original reasons for seeking a glyoxylate cycle type of pathway was to explain how some bacteria and protozoa could grow on media which contained acetate as the sole source of carbon; since no mechanism was known by which acetate or fatty acids could give rise to a net synthesis of carbohydrate or products derived therefrom (amino acids). Rothstein and Mayoh (1966) point out that the "vinegar eel," *Turbatrix aceti,* normally lives in an environment which contains high concentrations of acetic acid. The ability to employ a glyoxylate cycle for the biosynthesis of carbohydrate and amino acids from acetate would be a definite advantage to this nematode. It has been reported, however, that some members of *Turbatrix aceti* cultures remain alive after 6 months without oxygen (Carvalho and Maniero, 1953). If confirmed, these findings would indicate that this pathway is not required for survival. Unfortunately, the advantage offered by the glyoxylate cycle to the other free-living nematodes is not so obvious.

B. *Litomosoides carinii*

Bueding (1949b) and Berl and Bueding (1951) have investigated the metabolism of the adult filarial worm, *Litomosoides carinii,* whose habitat is the pleural cavity of the cotton rat. This organism resides in an aerobic environment and displays the physiological characteristics of an obligate aerobe. For example, the helminths survive for much longer periods in the presence of air than in an atmosphere of nitrogen. When placed under nitrogen, motility is lost within 60 to 90 minutes, while the reintroduction of air rapidly restores motility. In addition, oxygen is consumed upon endogenous incubations; and the rate of uptake is increased by the addition of exogenous glucose, fructose, or mannose. Aerobically, a net polysaccharide synthesis from exogenous glucose occurs; while anaerobically, the polysaccharide content of the worms decreases even in the presence of glucose. This suggests the requirement of oxidative energy for the synthesis of polysaccharides. Fluoroacetate inhibits both respiration and motility of *L. carinii* and, like anaerobiosis, causes an increased rate of glycolysis. Therefore, a Pasteur effect has been demonstrated in this worm; that is, a more rapid utilization of substrate anaerobically than aerobically. The presence of a Pasteur effect indicates that additional energy is being derived per unit of substrate when it is utilized in the presence of oxygen. The mechanism of the Pasteur effect is thought to involve a competition between aerobic and anaerobic pathways for a cofactor which is common to both and present in a limited concentration. As a consequence, aerobic sequences would decrease the rate of glycolysis by limiting the quantity of cofactor available. It would appear, therefore, that helminths which show this effect are aerobes. In accord with these findings, 2×10^{-4} *M* cyanide completely inhibits both respiration and motility of *L. carinii.* Evidence, therefore, appears to show conclusively that the adult worms are obligate aerobes.

In spite of this definite requirement for oxygen, *L. carinii,* like other aerobic parasites examined, does not oxidize completely all of the carbohydrate which it dissimilates. Aerobically, 30–40% of the glucose carbon utilized was accounted for as lactic acid, while an additional 25–35% of the glucose removed by the parasites was recovered as acetic acid (Bueding, 1949b). In air, appreciable quantities of acetylmethylcarbinol also accumulated (Berl and Bueding, 1951). On the other hand, under anaerobic conditions (nitrogen) lactic acid accounted for over 80% of the carbohydrate dissimilated. The remaining carbon was recovered as acetate. Thus, the ratio of acetate to lactate formed was considerably higher in the presence of oxygen. Pyruvate utilization was faster anaerobically, where a classical dismutation reaction occurred. For each mole of substrate disappearing, 0.5 moles each of lactate, acetate, and CO_2 were formed. Under aerobic conditions, however, acetate accumulation exceeded that of lactate. Therefore, in spite of an incomplete oxidation of substrates by this nematode, aerobiosis has a definite effect on the metabolic pathway, causing a shift toward acetate formation. The possibility exists that ATP generation may be coupled to the oxidative decarboxylation of pyruvate to acetate and CO_2, since oxygen is required for survival.

It is of particular interest that even in air, most of the carbohydrate carbons dissimilated are recovered as the fermentation products acetate, lactate, and acetylmethylcarbinol. It would appear, therefore, that in *L. carinii,* the tricarboxylic acid cycle does not serve as a major terminal respiratory pathway. In accord with this supposition, neither cytochrome *c* nor cytochrome oxidase activities could be detected in these filariae (Bueding, 1949b). The high sensitivity of this worm to cyanide, however, indicates the presence of an electron transport system containing heavy metal. Therefore, the nature and characterization of the aerobic processes occurring in this worm remain to be determined.

A number of the cyanine dyes possess potent anthelmintic activities against *L. carinii* (Brooker, 1945, 1954; Brooker and Sweet, 1947; Peters *et al.,* 1949). Low concentrations of these dyes (6.5×10^{-6} *M*) markedly inhibit oxygen consumption by the adult parasite (Bueding, 1949b). It appears that these inhibitors produce their chemotherapeutic effects by inhibiting an enzyme system concerned with oxidative metabolism. Respiration, cytochrome *c*, and cytochrome oxidase activities of homogenates of mammalian tissues are unaffected by concentrations of cyanines which almost completely inhibit oxygen uptake by *Litomosoides*. Thus, although the respiratory systems of the parasite remain to be elucidated, they must be different from those of the host and presumably play no major role in the respiration of mammalian tissues.

In contrast with *L. carinii*, the adult filarial parasite of man, *Wuchereria bancrofti*, is completely unaffected by the presence of cyanine dyes (Peters *et al.*, 1949; D. Santiago-Stevenson, A. D. Welch, J. Oliver-Gonzalez, and L. Peters, unpublished observations). It would appear, therefore, that either *W. bancrofti* is not dependent upon an oxidative metabolism as is the closely related filarial parasite, *L. carinii*, or the oxidative enzymes of the two worms differ from each other. Cyanine dyes inhibit the respiration of another related helminth, *Dracunculus insignis*, a guinea worm which was formerly classified as a filarial worm. Oxygen uptake by this parasite, however, does not appear to be of physiological significance as far as energy production is concerned, and lactic acid accounts for most of the carbohydrate consumed. *D. insignis* exhibits no Pasteur effect, which lends further support to the assumption that its energy is obtained via a sequence of anaerobic reactions (Bueding and Oliver-Gonzalez, 1950).

Von Brand *et al.* (1963) obtained the adult filarial worm, *Dirofilaria uniformis*, from the subcutaneous tissues of laboratory-infected rabbits. The parasites were maintained in a variety of media in an atmosphere of CO_2 enriched air. Under these aerobic conditions, almost all of the carbohydrate consumed could be accounted for by the lactic acid recovered. This filarial worm, like *D. insignis*, appears to carry out an essentially homolactic acid fermentation even in the presence of air. Unfortunately, survival under an anaerobic atmosphere was not discussed. The possibility remains that *D. uniformis*, unlike *L. carinii*, might have no oxygen requirement for survival or motility.

It appears, therefore, that *L. carinii*, *D. insignis*, and *D. uniformis* are similar to each other and to some mammalian tissues in that all form large quantities of lactate. Equally striking, however, are the marked differences between these morphologically closely related helminths particularly regarding oxygen requirements, oxidative enzymes, and sensitivity to cyanines.

C. *Nippostrongylus brasiliensis* (Syn. *N. muris*)

The adult form of this nematode resides in the jejunal mucosa of the rat where an appreciable oxygen tension appears to be present

(Rogers, 1949a). Like all other nematodes studied, *N. brasiliensis* consumes oxygen at a significant rate (Rogers, 1948). In addition, the adult parasite contains a hemoglobin pigment which has a very high affinity for oxygen, combining at lower oxygen tensions than the corresponding mammalian blood pigment. This high affinity for oxygen, however, makes it an unlikely physiological carrier of this gas, since it binds the oxygen too avidly and would probably not dissociate appreciably under physiological conditions (Rogers, 1949b; Davenport, 1949).

Recently, Roberts and Fairbairn (1965) have studied the adult form of *N. brasiliensis* in further detail. These authors reported that the nematode survives *in vitro* for approximately 8 days in an aerobic atmosphere, although it appears to require only very low concentrations of oxygen, surviving equally well in either 1% oxygen or air. Under anaerobic conditions (N_2 or N_2—CO_2), a loss of motility occurs within 6 hours, which is reversible for up to 24 hours upon the readmission of air. However, if the organisms are incubated anaerobically for more than a 24-hour period, motility does not return upon the readmission of air. Therefore, both motility and survival of *N. brasiliensis* adults are dependent upon the presence of at least low concentrations of oxygen.

The endogenous carbohydrate reserves of this parasite are primarily comprised of an approximately equal mixture of glycogen and trehalose, both of which appear to be utilized nonpreferentially (Roberts and Fairbairn, 1965). In agreement with the finding that *N. barsiliensis* adults are aerobes, the utilization of endogenous carbohydrates was twice as rapid anaerobically as compared to the aerobic rate of dissimilation. Thus, a pronounced Pasteur effect appears to be demonstrable which, in turn, would indicate that the nematode can derive additional energy from an aerobic metabolism. The authors, however, do not discuss this possibility, but suggest instead that the high rate of anaerobic utilization should be attributed to the fact that the worms are dying rapidly (Roberts and Fairbairn, 1965). In support of the apparent existence of a Pasteur effect, a considerable cytochrome oxidase activity has been demonstrated.

It is of particular interest that the rate of oxygen uptake by adult *N. brasiliensis* is considerably greater than can be accounted for on the basis of the complete oxidation of the endogenous carbohydrates dissimilated in an equivalent period of time. In addition, reducing the pO_2 to 8 mm. (1%) has no effect on the rate of disappearance of endogenous stores of carbohydrate. It has been proposed (Roberts and Fairbairn, 1965) that the oxygen uptake in excess of the carbohydrate utilized may signify that oxygen is required for some as yet unknown biosynthetic reaction, and would be of little importance in carbohydrate metabolism. It has even been suggested that carbohydrate dissimilation may be primarily an anaerobic process. The possibility remains, however, that some endogenous store other than carbohydrate may be oxidized by the worm and thereby account for the excess of oxygen uptake.

N. brasiliensis, like essentially all other parasitic helminths studied, seems to possess a limited terminal respiratory pathway; this would be expected to result in the accumulation of fermentation products. Unfortunately, such fermentation products as yet have not been reported, presumably because of the very slow rate at which the parasite utilizes exogenous carbohydrate. Rogers (1948) demonstrated that some CO_2 is released upon the aerobic incubation of the adult parasite. A respiratory quotient (R.Q.) of 0.69 was reported. Values obtained for the respiratory quotients of helminths often have been misleading because these organisms oxidize substrates incompletely, resulting in quotients of less than one. It would appear, therefore, that *N. brasiliensis* is not an exception; and the presence of fermentation products would be predicted. However, oxygen must be of more than minor importance in the energy metabolism of this parasite, since this gas is required for motility as well as survival. In addition, the worm exhibits a Pasteur effect and contains cytochrome oxidase.

D. *Trichinella spiralis* Larvae

These nematode larvae are encysted in muscle and presumably have access to oxygen. Stannard *et al.* (1938) were the first to report a detailed study of the respiratory behavior of this parasite. The organisms consume oxygen at a relatively rapid rate and produce carbon dioxide. Anaerobically, CO_2 was produced at a considerably slower rate. The larvae survive for long periods of time under either aerobic or anaerobic atmospheres; but anaerobiosis results in a reversible loss of motility. Similarly, the larvae are sensitive to KCN, concentrations of 10^{-6} M inhibiting oxygen uptake approximately 75%. Incubation with KCN causes the larvae to coil up and lose motility. Removal of the KCN is followed by a complete reversal and return to normal motility. Thus, the findings of Stannard *et al.* (1938) indicated both an anaerobic metabolism which was sufficient for survival of the larvae, and an aerobic metabolism which was required for the motility of the parasites. Subsequently, cytochrome *c*, cytochrome oxidase, and succinoxidase activities were demonstrated to be present, as well as some of the enzyme systems associated with the tricarboxylic acid cycle (Agosin, 1956; Goldberg, 1957). Anaerobic glycolysis in homogenates of *T. spiralis* larvae occurs in accordance with the reactions of the Embden-Meyerhof scheme (Agosin and Aravena, 1959).

Von Brand *et al.* (1952) accounted for the dual aerobic and anaerobic metabolism of the larvae. Employing bacteria-free preparations of the larval parasite, these investigators found glycogen to be present, which was consumed at essentially the same rate either aerobically or anaerobically. Concomitant with glycogen disappearance, volatile fatty acids accumulated in the medium. The rate of total acid production was the same regardless of the presence or absence of oxygen. From these findings, it appears warranted to conclude that carbohydrates are fermented without the intervention of oxygen, and that the energy derived from this fermentation suffices to maintain the parasites alive but in a nonmotile state. Quantitatively, the most significant volatile acid formed was *n*-valeric acid. In addition, acetic acid and an unidentified C_6 acid were determined. Only traces of lactate were detected aerobically or anaerobically. Von Brand *et al.* (1952) proceeded to demonstrate the presence in the larvae of appreciable quantities of lipids which disappeared at a significant rate throughout the incubation period, independent of glycogen utilization, but completely dependent upon the presence of air. It is probable, therefore, that a major portion of the oxygen consumed by the parasite is utilized for the oxidation of lipids, which, in turn, supplies the energy necessary for motility. *T. spiralis* larvae were the first helminths demonstrated to possess such a dual metabolism. A number of other nematode larvae appear to be similar in this respect.

III. Nematodes Which Lack an Oxygen Requirement for Energy Metabolism in at Least One Stage

It is now apparent that although many nematodes exhibit an oxygen requirement, they do not possess completely aerobic pathways for terminal respiration whereby substrates would be quantitatively oxidized to CO_2 and water. Even in an aerobic environment, a variety of fermentation products accumulate, indicating either the absence of a functional energy-yielding tricarboxylic acid cycle, or the presence of a rate-limited cycle. In either instance, the formation of end products would be more rapid than their removal. If specific enzymic sites exist which limit the rate of the tricarboxylic acid cycle sequence in parasitic forms, inhibitors of these rate-limiting reactions might become available which would possess chemotherapeutic activity. It is also of interest that the aerobic helminths may or may not contain cytochrome oxidase, indicating how little is known of electron transport and oxygen activation systems in these aerobes.

In contrast to helminths which require oxygen for survival, an increasing number of worms have been found to survive and metabolize carbohydrates for extensive periods of time under anaerobic conditions. Many of the intestinal helminths would be placed in this category, but apparent anaerobiosis is not limited to these forms. For example, the blood fluke, *Schistosoma mansoni,* contains cytochrome oxidase and consumes oxygen (Bueding and Charms, 1952). Respiration of this trematode is almost completely inhibited by cyanine dyes both *in vivo* and *in vitro*. However, schistosomes which are inhibited by these dyes continue to survive and produce viable eggs (Bueding, 1950; Bueding *et al.,* 1953). Thus, in spite of the availability of oxygen in its natural habitat, the observed oxygen uptake by *S. mansoni* is not an essential requirement for survival and egg production.

It should be pointed out that complete proof of the anaerobic nature of the following nematodes is still lacking, since none of these worms can be cultured through the complete life cycle. Therefore, it is not possible to determine the effects of oxygen lack upon all of the subsequent developmental stages. In certain instances, however, it appears reasonable to conclude that oxygen is not required for the production of energy by the form of the parasite under investigation.

Many of the anaerobic helminths require CO_2 for optimal survival, whether it be obtained exogenously or from metabolic sources. In view of this, it is of importance to realize that commercial preparations of CO_2 or other gas mixtures containing CO_2 are not completely free of oxygen. On the other hand, many laboratory procedures for removing traces of oxygen from gases may also remove CO_2 (alkaline pyrogallol, hot copper, etc.). These factors should be borne in mind upon evaluation of reports on the anaerobic nature of a given organism.

A. *Heterakis gallinae*

This small intestinal nematode which inhabits the cecum of poultry was studied by Glocklin and Fairbairn (1952) and by Fairbairn (1954). The organisms survive for approximately 3 days in a Krebs-Ringer phosphate solution. Survival and motility were the same in air or an atmosphere of pure nitrogen, indicating the anaerobic nature of the parasite. An additional indication that *H. gallinae* is an anaerobe comes from the observation that endogenous carbohydrate reserves were utilized at essentially equal rates aerobically and anaerobically. Thus, no definite Pasteur effect could be demonstrated in this parasite. Fermentation acids accumulated in the medium both aerobically and anaerobically. Propionic and acetic acids, as well as a component identified by chromatography on paper as succinic acid, were quantitatively the most significant fermentation products. Only very small amounts of lactic and pyruvic acids were detected.

In spite of the fact that the cecum probably contains very little oxygen and the organism appears to be an anaerobe, *H. gallinae,* like all other helminths studied, consumes oxygen at an appreciable rate. Glocklin and Fairbairn (1952) demonstrated that oxygen is utilized very efficiently. At the low oxygen tension of 7.6 mm. (1%), the uptake was 43% of that in air. High oxygen tensions (266–380 mm.) were toxic, the parasites becoming moribund within 2 hours. The possibility exists that O_2 toxicity may be a result of H_2O_2 formation via flavin systems, but this does not appear likely since an active catalase is present in *H. gallinae* homogenates (Glocklin and Fairbairn, 1952).

Carbon dioxide was produced endogenously under both aerobic and anaerobic atmospheres, although anaerobically, CO_2 was produced at a somewhat slower rate. Very interesting was the finding of Glocklin and Fairbairn (1952) that in either environment, the addition of CO_2 to incubation media had a carbohydrate sparing effect. That is, in the presence of carbon dioxide, the carbohydrate reserves of the nematodes disappeared more slowly. Subsequently, Fairbairn (1954) demonstrated that radioactive bicarbonate was fixed by *H. gallinae* primarily into the carboxyl carbon of propionate and to a lesser extent into what was tentatively identified as succinic acid. No fixation occurred into acetate. Results of these isotope studies indicate either a fixation of CO_2 into pyruvate followed by reduction of the product formed to propionate, or a reversible CO_2 fixation into propionate to form succinate in which case both succinate and propionate would become labeled. These pioneering studies with *H. gallinae* set the pattern for further investigations of other intestinal helminths, particularly *Ascaris lumbricoides.*

B. *Ascaris lumbricoides*

1. *Adults*

Of all nematodes, *Ascaris* is certainly the most studied because of its size and ready availability. This intestinal parasite, like all other nematodes examined, is capable of taking up oxygen (Laser, 1944). It survives equally well, however, under aerobic or anaerobic conditions (Toru, 1935; Weinland, 1901), indicating that the adult stage obtains its energy via anaerobic pathways of metabolism. In accord with this, the organism is not sensitive to cyanide (Laser, 1944), and neither cytochrome *c* nor cytochrome oxidase activity was detected in *Ascaris* muscle or female reproductive systems (Bueding and Charms, 1952). In addition, the participation of a flavin system for terminal oxidation in indicated by two other facts: first, in the presence of air, hydrogen peroxide is formed (Laser, 1944) which is toxic to the parasite because of a deficiency of catalase (Laser, 1944; Magath, 1918); and second, the rate of oxygen utilization is dependent upon the oxygen tension, being almost negligible at the O_2 tensions found in the physiological environment of adult ascarids (Bueding and Charms; 1952; Harnisch, 1933; Laser, 1944; Rathbone, 1955). A further indication that the parasite is an anaerobe is that, qualitatively at least, the same end products are formed in atmospheres of either air or N_2—CO_2 (Harpur and Waters, 1960). These observations are consistent with the conclusion that this parasite is predominantly anaerobic and, unlike mammalian tissues, possesses a terminal flavin, rather than a cytochrome oxidase system. The oxygen taken up appears to play no physiological role in energy-generating reactions.

A mixture of volatile fatty acids comprises the major "end products" of *Ascaris* fermentations. Positive identification of these acids have been made by Bueding and Yale (1951) and Bueding (1953) as acetic, propionic, butyric, α-methylbutyric, *n*-valeric and *cis*-α-methylcrotonic (tiglic) acids. Saz and Gerzon (1962) have identified the C_6 acid as α-methylvaleric acid. Ellison *et al.* (1960) confirmed these findings with the exception of the C_6 acid which was subsequently confirmed by Whitlock and Strong (1963). Female worms were employed for most of these studies, but Greichus and Greichus (1966) have reported the excretion of the same volatile acids by males. Succinic acid (Bueding and Farrow, 1956) and acetylmethylcarbinol (Saz *et al.*, 1958) have also been identified as products of the anaerobic *Ascaris* fermentation. Quantitatively, the most significant products are α-methylbutyric, α-methylvaleric, and succinic acids. It is of particular interest that both α-methylbutyrate and α-methylvalerate appear to be excreted as the racemic mixture which would indicate either the presence of racemases in the tissues of the helminths, or of one or more unusual nonstereospecific enzyme systems.

Saz and Weil (1962) have proposed the pathway for the formation of succinate and volatile acids by *Ascaris* muscle strips as illustrated in Fig. 1 (See addendum).

This sequence of reactions takes place in the absence of air. All of the reactions from glucose to pyruvate have been shown to occur in *Ascaris* muscle via the Embden-Meyerhof scheme of phosphorylating glycolysis (Bueding and Yale, 1951; Rathbone and Rees, 1954). However, the activity of the enzyme which reduces pyruvate with NADH to form lactate, lactic dehydrogenase, is relatively low. This is reflected in the fact that intact helminths produce, at the most, only trace amounts of lactate as a fermentation product (Bueding and Yale, 1951). On the basis of lactic dehydrogenase activity, however, more lactate might be expected than is actually found, posing an interesting problem in the mechanism for this control. The possibility that phosphoenol pyruvate (PEP) rather than pyruvate serves as the substrate for CO_2 fixation will be discussed. During glycolysis, the oxidation of 3-phosphoglyeraldehyde to 3-phosphoglyceric acid requires the reduction of the electron acceptor nicotinamide-adenine dinucleotide (NAD). Glycolysis cannot continue unless the NADH formed in this reaction is reoxidized, since this cofactor is present in only catalytic quantities. This function is served by lactic dehydrogenase during the anaerobic phase of glycolyzing mammalian tissues as well as in homolactic acid fermenting organisms. *Ascaris,* on the other hand, must rely upon other reactions for the reoxidation of NADH.

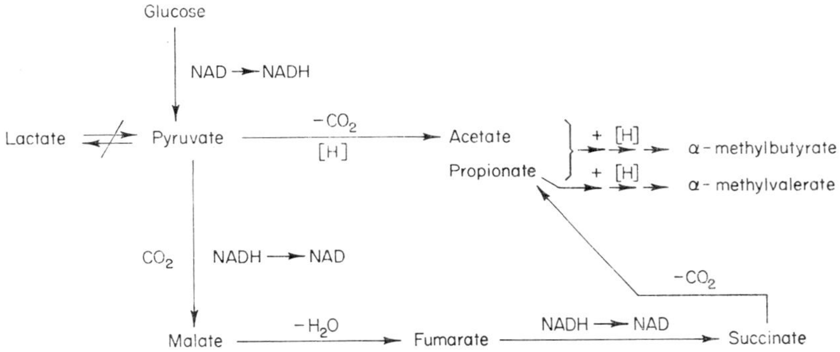

FIG. 1. Proposed pathway for the formation of succinate and volatile acids in *Ascaris* muscle (Saz and Weil, 1962).

Saz and Vidrine (1959) investigated the incorporation of $^{14}CO_2$ and various species of ^{14}C-labeled glucose and lactate into succinate upon incubation of these substrates with *Ascaris* muscle strips in an atmosphere of 95% N_2 and 5% CO_2. The incorporation and distribution of isotope in the recovered succinate was in all cases in accord with the hypothesis that succinate is formed via the fixation of CO_2 into pyruvate, resulting in the formation of a dicarboxylic acid, which is subsequently reduced to succinate. This pathway requires the fixation of one mole of CO_2 for each mole of succinate which accumulates. In these experiments, sufficient amounts of CO_2 were fixed by the *Ascaris* muscle strips to account for the quantities of the dicarboxylic acid which accumulated. As would be predicted by the proposed pathway, malonate inhibited the incorporation of isotope into succinate. The CO_2 requirement has been further substantiated by the finding of Harpur (1962) that the *in vitro* survival time of adult *Ascaris* is markedly increased by the inclusion of $NaHCO_3$ in the medium. Each of the enzyme systems involved in the proposed pathway from pyruvate to succinate has been shown to be present in *Ascaris* muscle.

Whether the initial product of CO_2 fixation into pyruvate is malate or oxalacetate is not clear. A very active malic enzyme linked to NAD has been isolated from *Ascaris* muscle (Saz and Hubbar, 1957); but under the assay conditions employed, the enzyme catalyzed the decarboxylation of malate more rapidly than the reverse CO_2 fixation reaction. Recently, phosphoenolypyruvate carboxykinase activity has been found to be present in homogenates of *Ascaris* muscle (Saz and Lescure, 1968); this enzyme catalyzes the reversible fixation of CO_2 into phosphoenolpyruvate upon the addition of inosine diphosphate (IDP). The kinetics of this enzyme in mammalian tissues, favor the decarboxylation of oxalacetate rather than its formation, and the enzyme is thought to be required for the biosynthesis of phosphoenolpyruvate from pyruvate in glyconeogenesis by coupling with pyruvate carboxylase as follows (Utter and Kurahashi, 1954; Wood and Utter, 1965):

(1) *Pyruvate carboxylase*[2]

$$\text{Pyruvate} + CO_2 + \text{ATP} \xrightleftharpoons{\text{Acetyl-CoA}} \text{OAA} + \text{ADP} + P_i$$

(2) *PEP Carboxykinase*[2]

$$\text{OAA} + \text{ITP} \rightleftharpoons \text{PEP} + CO_2 + \text{IDP}$$

(1) + (2)

$$\text{Pyruvate} + \text{ATP} + \text{ITP} \rightleftharpoons \text{PEP} + \text{ADP} + \text{IDP} + P_i$$

In *Ascaris* metabolism, the pyruvate carboxylase reaction would not be required, since OAA could be formed from the succinate present in the fluids of the helminth. The physiological function of PEP carboxykinase in *Ascaris* muscle may be for glyconeogenesis, since fructose-1, 6-diphosphatase activity is also present (Saz and Lescure, 1968). However, a major role of this enzyme in the fixation of CO_2 by *Ascaris* cannot be ruled out, since a net fixation of CO_2 into PEP is catalyzed by crude *Ascaris* muscle homogenates upon the addition of NADH and IDP (Saz and Lescure, unpublished observations). It appears likely that the unusually high malic dehydrogenase activity of *Ascaris* muscle would serve to pull the PEP carboxykinase reaction in the direction of CO_2 fixation. Such a pathway could account for the absence of lactate production if the PEP formed in glycolysis were converted to OAA, and subsequently succinate, before reaching the level of pyruvate.

Although succinate is one of the major "end products" of fermentation by *Ascaris* muscle strips it is also a precursor of propionate (Saz and Vidrine, 1959), α-methylbutyrate (Saz and Weil, 1960), and α-methylvalerate (Saz and Weil, 1962). Succinate-1-^{14}C is readily decarboxylated

[2] OAA, oxalacetate; P_i orthophosphate, ITP, inosine triphosphate.

by the muscle strips to give rise to $^{14}CO_2$ and propionate-1-^{14}C. Furthermore, by incubation of the nematode muscle preparations with distinctly labeled propionate-^{14}C or acetate-^{14}C, it was determined that the branched-chain carbon skeleton of α-methylbutyrate is formed by a condensation of the carboxyl carbon of an acetate unit with the α-carbon of a propionate unit as follows:

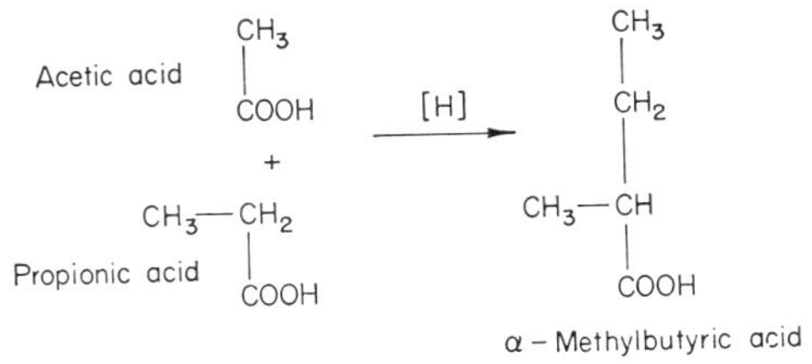

The acetic acid required for the condensation arises from the oxidative decarboxylation of pyruvate (Saz and Weil, 1960).

A similar sequence of reactions is involved in the formation of α-methylvalerate except that the condensation occurs between two propionate units (Saz and Weil, 1962).

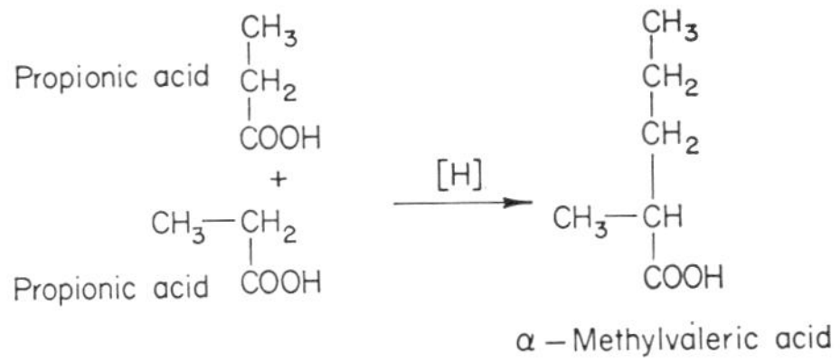

Presumably, the reactions involved in the formation of the branched-chain acids would require the reactants to be in the form of the coenzyme A derivatives.

The proposed sequence of reactions from pyruvate and CO_2 to succinate would provide for the reoxidation of NADH formed during glycolysis in *Ascaris* muscle. It is required by this scheme that the succinic dehydrogenase previously described in *Ascaris* (Bueding and Charms, 1952; Bueding *et al.*, 1955) should function physiologically in the reverse direction, i.e., as a fumaric reductase (Saz and Vidrine, 1959). This was indeed demonstrated to be the case by Kmetec and Bueding (1961). A partially purified particulate fraction obtained from *Ascaris* muscle catalyzed a coupled oxidation-reduction reaction between NADH and fumarate resulting in the formation of succinate. Fumarate reduction by NADH proceeds at a rate which is higher than that of the oxidation of succinate in the presence of either air or methylene blue. This reduction goes to completion because, unlike the corresponding reaction in vertebrate tissues, the reduction of fumarate by the *Ascaris* system is not inhibited by a high molar ratio of succinate to fumarate. The NADH-linked reductions of oxygen or fumarate indicate that in the muscle of *Ascaris*, either component may serve as the terminal electron acceptor. Under physiological conditions, however, fumarate would be expected to be the ultimate electron acceptor. The sequence of electron transport reactions illustrated in Fig. 2 has been proposed to occur in *Ascaris* muscle. Anaerobically, therefore, electrons from NADH are accepted by a flavoprotein and directed to succinic dehydrogenase which, in turn, reduces the terminal electron acceptor, fumarate, to succinate. In the presence of oxygen, succinate or NADH could be oxidized, the

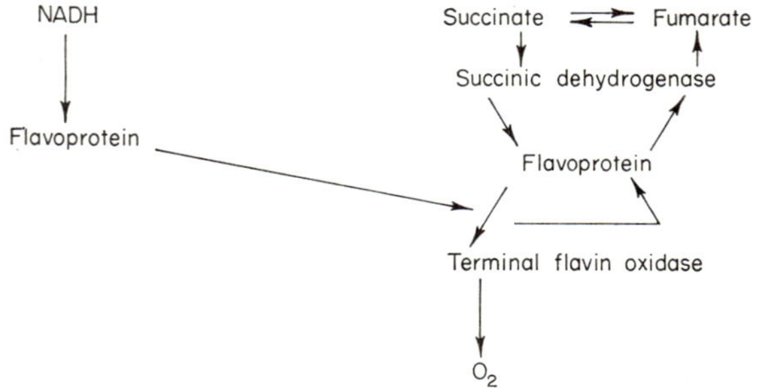

FIG. 2. Proposed pathway for electron transport in *Ascaris* muscle (Kmetec and Bueding, 1961; Bueding, 1962).

electrons passing through the terminal flavin oxidase to oxygen and forming hydrogen peroxide.

The proposed pathway for electron transport not only accounts for the regeneration of NAD from NADH, but also may serve to produce energy-rich phosphate bonds. Chin and Bueding (1954) reported that the dehydrogenation of NADH by *Ascaris* muscle mitochondria was coupled with phosphorylation. On the other hand, no phosphorylation was detected during the oxidation of succinate. It would appear, therefore, that energy-rich phosphate bonds are formed at a site in the proposed electron transport chain above that portion which is used in common by both NADH and succinate, that is, between NADH and flavoprotein, Seidman and Entner (1961) have substantiated the earlier findings and reported phosphorylation coupled to the dehydrogenation of NADH both aerobically and anaerobically in particles prepared from *Ascaris* muscle. Katsume and Obo (1962b,c) and Okamoto *et al.* (1962) have confirmed and extended these findings. Anaerobic succinate formation, as exemplified by the *Ascaris* system, would provide the parasite with a distinct advantage over those tissues which reoxidize NADH via lactic dehydrogenase, since no energy-rich bonds are formed in the latter reaction.

Succinate formation and muscular contraction should be closely associated if the foregoing hypotheses are correct. Piperazine, an effective chemotherapeutic agent in ascariasis, causes a reversible flaccid paralysis of the adult worm. Concurrently, succinate accumulation in both the worm and the incubation media is markedly depressed. The question arises whether the primary effect of piperazine consists of an inhibition of the metabolic formation of succinate, or whether the latter is a consequence of direct action of the drug upon the muscle cell which causes paralysis, resulting in a lower energy requirement of the contractile mechanism and, therefore, in a decreased formation of succinate. In mammalian muscle, for example, exercise initiates a demand for energy which in turn, stimulates glycolysis, resulting in an increased accumulation of lactate. Conversely, little lactate is found in resting muscle. If succinate formation is required by *Ascaris* muscle for the generation of energy in the form of ATP, then increased energy requirements should result in an increased accumulation of succinate, and flaccid paralysis in a decreased accumulation.

To study this problem, Bueding *et al.* (1959) incubated isolated ascarids in a medium with or without piperazine for 24 hours. Muscle strips were obtained from the piperazine-treated and control helminths, and the incorporation of lactate-2-^{14}C into succinate in each was compared. Only muscle strips previously exposed to piperazine were incubated in the presence of the base. Neither the total nor the specific radioactivities of the succinate isolated from control or piperazine-treated muscles differed significantly. Piperazine, therefore, had no direct inhibitory effect upon succinate formation. Presumably, the drug did have a direct paralyzing effect upon the *Ascaris* muscle which resulted in a decreased energy requirement and hence a decreased accumulation of succinate. These findings indicate that succinate formation by *Ascaris* muscle plays an important role in supplying energy for muscular contraction, as would be required by the above mechanisms.

Contrasting with the findings discussed above, Kikuchi *et al.* (1959) and Kikuchi and Ban (1961) presented evidence in support of the existence of cytochrome oxidase, succinic-cytochrome *c* reductase and a cytochrome *b*—like pigment in a particulate fraction from *Ascaris* muscle. Attempts by Kmetec and Bueding (1961) to repeat some of these findings failed. Similarly, attempts of other investigators to demonstrate significant cytochrome oxidase activity in *Ascaris* muscle preparations have had no success (Bueding and Charms, 1952; Rathbone, 1955; Katsume and Obo, 1962a). More recently, Chance and Parsons (1963) examined *Ascaris* mitochondria employing very sensitive techniques and confirmed the earlier findings that no cytochrome oxidase could be detected. Cytochromes of types *b* and *c* were reported by Chance and Parsons (1963) based upon oxidized and reduced spectra of *Ascaris* muscle mitochondria at low temperatures. Unfortunately, the possibility remains that the cytochrome *b* and *c* activities reported by Kikuchi *et al.* (1959) and Chance and Parsons (1963) may be identical with that described by Kmetec and Bueding (1961) which appears to be attributable to *Ascaris* muscle hemoglobins rather than cytochromes and which could not account for the rate at which oxygen is consumed by the particulate preparations.

Oya *et al.* (1962, 1965) have systematically attempted to implicate the tricarboxylic acid cycle as a significant pathway for terminal respiration in *Ascaris.* Ultimately, conclusions were reached which were similar to those of other investigators (Rathbone, 1955; Katsume and Obo 1962a). Oya *et al.* (1965) stated, "The functioning of the tricarboxylic acid cycle in *Ascaris* muscle appeared to be considerably limited on account of the relatively low activity of terminal oxidase system as well as aconitase, condensing enzyme and pyruvate dehydrogenase. Therefore, the physiological role of the cycle might be of a minor importance as compared to the dismutation type reaction." Overall agreement, therefore, appears to have been reached that the TCA cycle, if it exists per se in *Ascaris,* would be of little physiological significance as an energy-producing pathway. As a source of precursors for amino acids and other biosynthetic products, however, the enzyme systems may be of importance to this apparently anaerobic parasite (Krebs and Lowenstein, 1960).

2. *Ascaris Egg and Larval Stages*

In striking contrast to the adult *Ascaris,* the eggs elaborated by these adults have an obligatory requirement for oxygen to initiate and maintain development, although the potential to develop normally remains after extended periods of anaerobiosis (Hallez, 1885; Martin, 1913; Zavadovsky, 1916; Brown, 1928; Kmetec *et al.,* 1963). Unfortunately, most procedures employed to remove oxygen from the incubation environment also removed carbon dioxide; and commercial gas preparations which contain CO_2 are generally contaminated with sufficient quantities of oxygen to allow for some development to occur. In view of the dependence of the adult stage upon CO_2, D. K. Saz and Saz (unpublished observations) incubated *Ascaris* eggs in various stages of development in Brewer anaerobic jars under atmospheres of nitrogen-carbon dioxide mixtures which contained small quantities of hydrogen. Sparking in the presence of catalyst removed the traces of oxygen present without affecting CO_2 concentrations. Under anaerobic atmospheres containing CO_2, no further development of any of the stages occurred, confirming the earlier findings.

Oxygen consumption by *Ascaris* eggs was studied by Fauré-Fremiet (1913), Brown (1928), and Jaskoski (1952); but the most critical evaluation was reported by Passey and Fairbairn (1955) who compared the rates at which oxygen was consumed by the various developmental stages within the egg. Neither cytochrome *c* nor cytochrome oxidase could be detected by the latter investigators. Their findings, however, suggested the presence of these pigments as a terminal electron transport system, since respiration was highly sensitive to carbon monoxide in the dark, to cyanide, and to azide. Cytochrome oxidase and succinoxidase activities of *Ascaris lumbricoides* eggs which had been allowed to develop for 14 days or more, were first reported by Kmetec *et al.* (1963), and the activity of the cytochrome oxidase in the extracts was sufficiently high to account for the rate of oxygen uptake by the intact eggs. Costello *et al.* (1963) confirmed these findings which were subsequently extended (Oya *et al.*, 1963) when the cytochrome oxidase content during development of the eggs was determined. The oxidase was not detectable until after the eight-cell stage whereupon elaboration continued to occur through the 30-day incubation period except for a plateau from approximately 10–20 days. A recent report of Costello *et al.* (1966) suggests the presence of a flavin-mediated terminal oxidase in unembryonated eggs as contrasted with a cytochrome oxidase system in the developing eggs. Therefore, the adult and developing egg forms of this helminth have different terminal oxidases.

Metabolic pathways have been difficult to elucidate in *Ascaris* eggs because of their impermeability to organic compounds. The eggs are assumed to be cleidoic, allowing only gases and water to exchange. The experiments of Hoshino and Suzuki (1956), which indicate the utilization of exogenous carbohydrates, were performed on eggs which had been spawned from nonsterile worms by several overnight incubations at 37°. Increases in the oxygen consumption upon the addition of carbohydrates could be readily explained by the presumed bacterial population present. Passey and Fairbairn (1955) removed the hard parts of the shell either chemically with sodium hypochlorite, or biologically by incubation with chitinolytic bacteria, and demonstrated that such deshelled eggs were still unaffected by solutions of acids, bases, salts, or formaldehyde. Thus, the vitelline membrane is responsible for the permeability characteristics of the eggs. Recently, a technique for inducing an increased permeability by application of hatching fluid has been described (Kaulenas and Fairbairn, 1966).

Lipids appear to comprise over 50% of the protoplasmic solids in *Ascaris* eggs, a significant portion of the triglyceride fraction being comprised of esterified volatile acids (Fairbairn, 1955; Saz and Lescure, 1966). Saponifiable lipids, including the esterified volatile acids, decrease throughout the period of embryonation and postembryonic survival. Passey and Fairbairn (1957) reported that egg carbohydrates, on the other hand, decrease while development occurs to the vermiform stage (10 days), but are completely resynthesized during the next 15 days whereupon the embryo becomes infective. Thorough examination of possible carbon sources available in the eggs which could account for the resynthesis of carbohydrate resulted in the conclusion that lipids were incorporated into glycogen and trehalose (Passey and Fairbairn, 1957). This constituted the first unequivocal demonstration of the relatively rare net synthesis of carbohydrate from lipid in animal tissues. The subsequent postulation of the glyoxylate cycle in other organisms (Kornberg and Krebs, 1957) suggested this pathway as a means for the lipid to carbohydrate interconversion in *Ascaris* eggs. However, enzymes of the glyoxylate cycle have not been demonstrated in any stage of *Ascaris*.

Evidence has been reported suggesting that the volatile fatty acids arising as "end products" of muscle metabolism may be transported to the reproductive tissues, incorporated into the egg lipids, and subsequently serve as precursors for about 20–25% of the resynthesized carbohydrates as illustrated in Fig. 3 (Saz and Lescure, 1966).

Unlike adult *Ascaris* muscle, the reproductive system does not appear capable of the *de novo* synthesis of either α-methylbutyrate or α-methylvalerate in spite of the fact that these branched chain compounds comprise a major portion of the esterified volatile acids in the egg lipids. In addition, low, but significant quantities of α-methylbutyrate-^{14}C and α-methylvalerate-^{14}C are incorporated into the corresponding bound acid fraction upon incubation with intact reproductive tissues. During embryonation, the concentrations of bound α-methylbutyrate and α-methylvalerate decrease to very low levels. These findings indicate that the perienteric fluid serves to supply reproductive tissues with the pentanoic and hexanoic acid isomers, which are subsequently utilized during development.

$^{14}CO_2$ is incorporated rapidly into the small pools of free propionate and succinate within the eggs, indicating a relatively rapid turnover of these acids. However, $^{14}CO_2$ more slowly enters into the bound propionate fraction, which apparently arises from the labeled free propionate pool. Radioactivity from $^{14}CO_2$ is also found in egg glycogen and trehalose demonstrating a capacity for glyconeogenesis. It has been proposed, therefore, that the bound α-methylbutyrate and α-methylvalerate of the neutral lipid fraction are cleaved to form acetate and propionate units in the former case, and two propionate units with the latter compound. The reactions may constitute a reversal of the pathway by which the branched chain acids are synthesized. The formation of acetyl-CoA and propionyl-CoA from α-methylbutyryl-CoA has been demonstrated to take place in pig heart and rat liver (Robinson *et al.*, 1956).

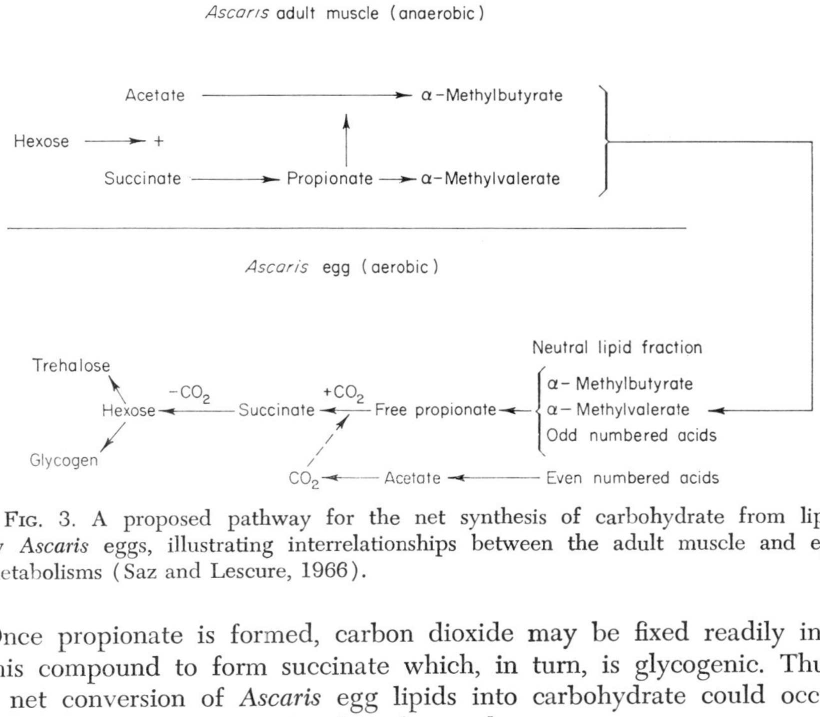

FIG. 3. A proposed pathway for the net synthesis of carbohydrate from lipid by *Ascaris* eggs, illustrating interrelationships between the adult muscle and egg metabolisms (Saz and Lescure, 1966).

Once propionate is formed, carbon dioxide may be fixed readily into this compound to form succinate which, in turn, is glycogenic. Thus, a net conversion of *Ascaris* egg lipids into carbohydrate could occur without the intervention of a glyoxylate pathway.

The hypothesized reactions presume that the larvae are capable of glyconeogenesis from succinate. Recent findings indicate that this is indeed the case (Saz and Lescure, 1968). Hatched larvae have been shown to possess both fructose-1,6-diphosphatase and phosphoenolpyruvate carboxykinase activities. These enzymes are subordinate to the glycolytic scheme and are required for glyconeogenesis. In addition, both hatched larvae and eggs incorporate $^{14}CO_2$ into glycogen. Carbohydrates formed via this pathway would presumably be utilized by the subsequent stages in the life cycle of the parasite. Aerobic respiration in the egg stages would be required according to the postulated pathway both for the dissimilation of lipids, including acetate arising from the breakdown of α-methylbutyrate, and for the presumed cytochrome oxidase-linked oxidation of succinate (Kmetec *et al.*, 1963) to fumarate and thence to oxalacetate. Respiratory carbon dioxide may serve in the formation of succinate from propionate, thereby doing away with the requirement for an external supply of this gas and allowing development of the eggs to occur essentially in its absence; as when the eggs are embryonated in H_2SO_4. Only catalytic amounts of carbon dioxide would be required, since the dicarboxylic acid formed would be decarboxylated subsequently to allow for the formation of triose.

Some of the enzymes of the tricarboxylic acid cycle have been demonstrated in *Ascaris* egg homogenates and mitochondria isolated from eggs which had been incubated for 10 days at 30° (Costello and Brown, 1962; Bloom and Entner, 1965). Neither investigation provided a completely functional tricarboxylic acid cycle preparation, but the latter investigators reported the esterification of inorganic phosphate coupled to the oxidation of malate, α-ketoglutarate, and NADH. Larval mitochondria were different in many respects from the corresponding adult muscle preparations, and more closely resembled the classically described aerobic mammalian mitochondria. For example, mitochondria obtained from eggs catalyzed phosphorylation coupled with NADH oxidation with no added electron acceptor other than O_2; and did not form succinate from malate in the presence of NAD with or without fumarate, indicating that, unlike the adult muscle, a dismutation reaction was not required for the oxidation of malate. These findings would be in accord with the earlier postulations of Kmetec *et al.* (1963) to the effect that a cytochrome-linked terminal respiratory pathway is present in the eggs.

It is also of interest that infective *Ascaris* larvae (second stage) which have been hatched essentially as described by Jaskoski and Colucci (1964), and larvae isolated from infected guinea pig lungs up to 7 days postinfection (third stage) both contain cytochrome oxidase activity which is inhibited by KCN (Saz and Lescure, unpublished observations). Hatched *Ascaris* larvae also appear to require aerobic metabolism for motility, but not for survival as was found first with *Trichinella spiralis* larvae (von Brand *et al.*, 1952). Further indication of the aerobic nature of the third stage lung larvae arises from the finding that endogenous glycogen stores are depleted more rapidly under an anaerobic atmosphere, demonstrating the presence of a Pasteur effect (Saz, Bueding, and Lescure, unpublished observations). Therefore, the change from an aerobic to an anaerobic type of metabolism must occur some time after this third-stage form and before isolation of the adult from the lumen of the intestine.

C. *Trichuris vulpis*

The adult whipworm of the dog, *Trichuris vulpis,* resides in the essentially anaerobic environment of the intestine; but its close proximity to the mucosa, its small size, and the suggestion that the parasite feeds upon tissues and blood cells does not allow the conclusion that this parasite derives its energy solely from a series of anaerobic reactions (von Brand, 1960a,b). Studies by Bueding *et al.* (1960), however, have demonstrated that the adult worms survive for considerably longer periods in an atmosphere of N_2—CO_2 than in either air or O_2—N_2—CO_2 mixtures. Similarly, CO_2 is required for optimal survival, and the rate of glucose uptake was not only highest in the N_2—CO_2 environment, but the higher rate was maintained for more extended periods of time. Elimination of metabolically formed CO_2 from the atmosphere by the use of alkali resulted in a marked reduction in glucose utilization, as well as a reduction in the quantities of the metabolic products formed.

The products accumulating in adult *T. vulpis* fermentations consist of CO_2, a mixture of volatile acids, lactic acid, and succinic acid (Bueding *et al.,* 1960; Kmetec and Bueding, 1965). Formic, acetic, propionic, butyric and *n*-valeric acids in addition to a C_6 acid fraction were identified as the volatile acid components by means of gas chromatography. Lactic acid accounted for no more than 5–10% of the carbohydrate utilized under aerobic or anaerobic conditions, while propionate and succinate each accounted for an additional 10–15%. Only 65% of the total carbohydrate carbon disappearing was recovered, indicating the presence of a considerable quantity of unidentified fermentation products. Similarities become apparent, therefore, between the metabolism of *T. vulpis* and *Ascaris lumbricoides.* Both appear to obtain energy through anaerobic pathways which require CO_2 and form succinate and products derived from succinate.

T. vulpis differs markedly from *Ascaris* in that the former contains succinate-cytochrome *c* reductase, cytochrome oxidase, and catalase activities. In air, the parasite consumes oxygen as do other nematodes; and the activities of both cytochrome *c* and cytochrome oxidase in homogenates are sufficiently high to account for the rate of oxygen uptake by the intact worm (Bueding *et al.,* 1960). Whether or not the cytochrome system plays a physiological role in the essentially anaerobic environment of the parasite cannot be ascertained from these data. The detrimental effects of oxygen on survival of the whipworm should not be the result of H_2O_2 accumulation, since catalase is present and oxygen would presumably be activated by the cytochrome system, which precludes the formation of the peroxide. It would appear, however, that whatever energy the parasite requires for maintenance can be obtained by an anaerobic metabolism.

IV. Acanthocephala

Relatively few biochemical analyses of this group of helminths are available. *Moniliformis dubius* and *Macracanthorhynchus hirudinaceus,* the spiny-headed worms of the rat and pig respectively, have been studied the most. It is somewhat surprising that a number of investigators have concerned themselves primarily with the aerobic oxidative metabolism of these parasites based upon oxygen uptake data. Relatively few experiments have been conducted in an anaerobic environment, despite the fact that information has been accumulating which indicates similarities between the catabolisms of the *Acanthocephala* and *Ascaris.*

The *Acanthocephala* are all intestinal parasites which survive in this relatively anaerobic physiological environment. Like other intestinal helminths, they are capable of consuming oxygen upon *in vitro* incubation (Laurie, 1959). Early reports of von Brand (1940), indicated that the aerobic vs. anaerobic glycogen consumption by *M. hirudinaceus* was almost equal, suggesting the absence of a definite Pasteur effect. Attempts to culture *M. dubius* under a variety of conditions, including atmospheres of N_2 or N_2/CO_2 or $N_2/O_2/CO_2$, have led Nicholas and Grigg (1965) to conclude that, "Eagle's medium and serum, equilibrated with a N_2/CO_2 mixture, gave the best results." Although viable acanthors were produced, they were clearly abnormal in all media employed. Under the N_2/CO_2 atmosphere, however, the worms survived the full 8-day term of the experiment.

Laurie (1957) reported the accumulation of acidic products when *M. dubius* fermented carbohydrates in an anaerobic atmosphere. Unfortunately, the acids were not characterized further. Aerobic fermentations, however, resulted in the accumulation of a mixture of organic acids which had the chromatographic properties of formic, acetic, and lactic acids. The former two constituents were quantitatively the most significant (Laurie, 1959). Subsequent studies indicated the accumulation of small quantities of other components, particularly succinate, within the worm or in body wall preparations (Graff, 1965; Bryant and Nicholas, 1965). *M. dubius* fixed $^{14}CO_2$ primarily into malate, succinate, and aspartate. Isotope was recovered to a lesser extent in fumarate, lactate, and alanine (Graff, 1965). The incorporation of glucose-^{14}C, acetate-^{14}C, and succinate-^{14}C into organic and amino acids has also been investigated (Graff, 1964; Bryant and Nicholas, 1965). Unfortunately, data from these and the previous isotope experiment (Graff, 1965) are expressed in terms of total radioisotope incorporation rather than specific radioactivities. Therefore, relatively little can be concluded regarding pathways since the sizes of the pools containing the isotope are not known. Primarily from the findings that isotope was recovered in aspartate, malate, fumarate, and succinate, but not in glutamate, Bryant and Nicholas (1965) suggested that *M. dubius* forms succinate by a pathway similar to that described previously in *Ascaris*.

Recently, particulate preparations of *M. dubius* were obtained (Bryant and Nicholas, 1966) which possessed succinic dehydrogenase activity when assayed in the presence of methylene blue or phenazine methosulfate; oxidized succinate-^{14}C to fumarate-^{14}C and malate-^{14}C; was only partially inhibited in oxygen uptake by high concentrations of KCN or antimycin A; reduced added cytochrome *c* upon the addition of either NADH or succinate, but could not reoxidize the reduced cytochrome *c* formed; and did not possess demonstrable cytochrome *c* oxidase activity. It appears, therefore, that the electron transport system of this parasite may be similar in several respects to the corresponding system in *Ascaris* which was previously discussed. Further investigations are necessary, however, before any definite conclusions can be reached; particularly in regard to the possible role of fumarate, flavins, and NADH under anaerobic conditions.

The disaccharide, trehalose, was found to be present in a large number of invertebrates, including *M. dubius* (Fairbairn, 1958). Laurie (1959) obtained glycogen synthesis from glucose, fructose, mannose, or maltose in this worm, but failed to demonstrate the synthesis of trehalose. More recently, homogenates of both *Moniliformis* and *Macracanthorhynchus* have been prepared, which synthesized trehalose from glucose-U-^{14}C (Fisher, 1964). UDPG and ATP were required for optimal incorporation of glucose carbons into trehalose, indicating that the mechanism of synthesis was similar to that previously reported in yeast and insects (Cabib and Leloir, 1958; Candy and Kilby, 1961). Acanthocephala, like all other helminths studied, appear to form pyruvate via the Embden-Meyerhof scheme of phosphorylating glycolysis. Some of the enzymes associated with this pathway in addition to isozymes of lactic and malic dehydrogenases have been determined (Dunagan and Scheifinger, 1966a; Dunagan and de Luque, 1966). As might be predicted, it has been concluded that *M. hirudinaceus* does not possess a significant tricarboxylic acid cycle activity (Dunagan and Scheifinger, 1966b). This conclusion was based upon the finding that aconitase and isocitric dehydrogenase activities were insignificant in mitochondrial preparations obtained from the parasite.

V. Summary

Unfortunately, this discussion has been limited to a relatively small number of nematodes. Several generalities, however, appear to be warranted from our present knowledge. All nematodes studied to date are capable of consuming oxygen upon *in vitro* incubation under appropriate conditions. The physiological role of this gas in the metabolism of some helminths is established; in others it is poorly understood; and in still others it is decidedly inhibitory to survival. It can be said that a relatively large group of parasites is similar to *Ascaris* in that there is no oxygen requirement for the normal physiological production of energy in the adult stage. As has been pointed out, this is true not only for a number of nematodes, but also for some cestodes and trematodes in addition to certain of the parasitic protozoa (Scheibel and Saz, 1966; Saz and Bueding, 1966). Regardless of the status of oxygen, however, substrates are not oxidized to completion by any of the helminths studied to date. Products accumulate, indicating either the absence or participation of rate-limited terminal respiratory pathways.

In these same areas where broad similarities exist among the helminths, more detailed analyses show marked differences. Patterns of fermentation products formed are characteristic for each organism in spite of the fact that all appear to dissimilate carbohydrates as far as PEP or pyruvate by the Embden-Meyerhof glycolytic sheme. Differences arise from the pathways for the further utilization of these three carbon acids. Terminal respiratory pathways and particularly mechanisms for electron transport associated with such pathways, are still poorly understood. Considerable variation exists not only from one nematode to another, but also within the developmental stages of a single helminth.

It should be obvious that the metabolic behavior of nematodes cannot be assumed to be the same *in toto* as mammalian tissues. Comparative studies have demonstrated both overall and more subtle differences. Therefore, each organism must be examined as a separate biochemical entity before any reasonable understanding of helminth metabolism can be attained.

REFERENCES

Agosin, M. (1956). *Bol. Chileno Parasitol.* **11**, 46.
Agosin, M., and Aravena, L. C. (1959). *Exptl. Parasitol.* **8**, 10.
Berl, S., and Bueding, E. (1951). *J. Biol. Chem.* **191**, 401.
Berntzen, A. K. (1962)., *J. Parasitol.* **48**, 785.

Berntzen, A. K., and Mueller, J. F. (1964). *J. Parsitol.* **50**, 705.
Bloom, S., and Entner, N. (1965). *Biochim. Biophys. Acta* **99**, 22.
Brooker, L. G. S. (1945). *Frontiers Chem,* **3**, 63–136.
Brooker, L. G. S. (1954). *In* "Theory of the Photographic Process" (C. E. K. Mees, ed.).), pp. 371–429. Macmillan, New York.
Brooker, L. G. S., and Sweet, L. A. (1947). *Science* **105**, 496.
Brown, H. W. (1928). *J. Parasitol.* **14**, 141.
Bryant, C., and Nicholas, W. L. (1965). *Comp. Biochem. Physiol.* **15**, 103.
Bryant, C., and Nicholas, W. L. (1966). *Comp. Biochem. Physiol.* **17**, 825.
Bueding, E. (1949a). *Physiol. Rev.* **29**, 195.
Bueding, E. (1949b). *J. Exptl. Med.* **89**, 107.
Bueding, E. (1950). *J. Gen. Physiol.* **33**, 475.
Bueding, E. (1953). *J. Biol. Chem.* **202**, 505.
Bueding, E. (1962). *Federation Proc.* **21**, 1039.
Bueding, E., and Charms, B. (1952). *J. Biol. Chem.* **196**, 615.
Bueding, E., and Farrow, G. W. (1956). *Exptl. Parasitol.* **5**, 345.
Bueding, E., and Oliver-Gonzalez, J. (1950). *Brit. J. Pharmacol.* **5**, 62.
Bueding, E., and Yale, H. (1951). *J. Biol. Chem.* **193**, 411.
Bueding, E., Peters, L., Koletsky, S., and Moore, D. V. (1953). *Brit. J. Pharmacol.* **8**, 15.
Bueding E., Entner, N., and Farber E. (1955). *Biochim. Biophys. Acta* **18**, 305.
Bueding E., Saz, H. J., and Farrow, G. W. (1959). *Brit. J. Pharmacol.* **14**, 497.
Bueding, E., Kmetec, E., Swartzwelder, C., Abadie, S., and Saz, H. J. (1960). *Biochem. Pharmacol.* **5**, 311.
Cabib, E. and Leloir L. F. (1958). *J. Biol. Chem.* **231**, 259.
Campbell, L. J. R., Smith, R. A., and Eagles, B. A. (1953) *Biochim. Biophys. Acta* **11**, 544.
Candy, D. J., and Kilby, B. A. (1961). *Biochem. J.* **78**, 531.
Carpenter, W. D., and Jaworski, E. G. (1962). *Biochim. Biophys. Acta* **58**, 369.
Carvalho, J. C., and Maniero, J. (1953). *Rev. Inst. Adolfo Lutz* **13**, 83.
Chance, B., and Parsons, D. F. (1963). *Science* **142**, 1176.
Chin, C. H., and Bueding, E. (1954). *Biochim. Biophys. Acta* **13**, 331.
Costello, L. C., and Brown, H. (1962). *Exptl. Parasitol.* **12**, 33.
Costello, L. C., Oya, H., and Smith, W. (1963). *Arch. Biochem. Biophys.* **103**, 345.
Costello, L. C., Smith, W., and Fredericks, W. (1966). *Comp. Biochem. Physiol.* **18**, 217.
Davenport, H. E. (1949). *Proc. Roy. Soc.* **B136**, 271.
De Bell J. T., del Castillo, J., and Sanchez, V. (1963). *J. Cellular Comp. Physiol.* **62**, 159.
Dougherty, E. C., Hansen, E. L., Nicholas, W. L., Mollette, J. A., and Yarwood, E. A. (1959). *Ann. N. Y. Acad. Sci.* **77**, 176.
Dunagan, T. T., and de Luque, O. (1966). *J. Parasitol.* **52**, 727.
Dunagan, T. T., and Scheifinger, C. C. (1966a). *J. Parasitol.* **52**, 730.
Dunagan, T. T., and Scheifinger, C. C. (1966b). *Comp. Biochem. Physiol.* **18**, 663.
Ellison, T., Thomson, W. A. B., and Strong, F. H. (1960). *Arch. Biochem. Biophys.* **91**, 247.
Fairbairn, D. (1954). *Exptl. Parasitol.* **3**, 52.
Fairbairn, D. (1955). *Can. J. Biochem. Physiol.* **33**, 122.

Fairbairn, D. (1957). *Exptl. Parasitol.* **6**, 491.
Fairbairn, D. (1958). *Can. J. Zool.* **36**, 787.
Fairbairn, D. (1960). *In* "Nematology" (J. N. Sasser and W. R. Jenkins, eds.), pp. 267–296. Univ. of North Carolina Press, Chapel Hill, North Carolina.
Fauré-Fremiet, E. (1913). *Arch. Anat. Microscop. Morphol. Exptl.* **15**, 435.
Fisher, F. M., Jr. (1964). *J. Parasitol.* **50**, 803.
Glocklin, V C., and Fairbairn, D. (1952). *J. Cellular Comp. Physiol.* **39**, 341.
Goldberg, E. (1957). *Exptl. Parasitol.* **6**, 367.
Graff, D. J. (1964). *J. Parasitol.* **50**, 230.
Graff, D. J. (1965). *J. Parasitol.* **51**, 72.
Greichus, A., and Greichus, Y. A. (1966). *Exptl. Parasitol.* **19**, 85.
Hallez, P. (1885). *Mem. Soc. Lille* **15**, 1.
Harnisch, O. (1933). *Z. Vergleich. Physiol.* **19**, 310.
Harpur, R. P. (1962). *Can. J. Zool.* **40**, 991.
Harpur, R. P., and Waters, W. R. (1960). *Can. J. Biochem. Physiol.* **38**, 1009.
Hoshino, M., and Suzuki, H. (1956). *Fukushima J. Med. Sci.* **3**, 51.
Jaskoski, B. J. (1952). *Exptl. Parasitol.* **1**, 291.
Jaskoski, B. J., and Colucci, A. V. (1964). *Trans. Am. Microscop. Soc.* **83**, 294.
Katsume, T., and Obo, F. (1962a). *Acta Med. Univ. Kagoshima* **4**, 56.
Katsume, T., and Obo, F. (1962b). *Acta Med. Univ. Kagoshima,* **4**, 66.
Katsume, T., and Obo, F. (1962c). *Acta Med. Univ. Kagoshima.* **4**, 72.
Kaulenas, M., and Fairbairn, D. (1966). *Parasitol.* **52**, 303.
Kikuchi, G., and Ban, S. (1961). *Biochim. Biophys. Acta* **51**, 387.
Kikuchi, G., Ramirez, J., and Barron, E. S. G. (1959). *Biochim. Biophys. Acta* **36**, 335.
Kmetec, E., and Bueding, E. (1961). *J. Biol. Chem.* **236**, 584.
Kmetec, E., and Bueding, E. (1965). *Comp. Biochem. Physiol.* **15**, 271.
Kmetec, E., Beaver, P. C., and Bueding, E. (1963). *Comp. Biochem. Physiol.* **9**, **115**.
Kornberg, H. L., and Elsden, S. R. (1961). *Advan. Enzymol.* **23**, 401.
Kornberg, H. L., and Krebs, H. A. (1957). *Nature* **179**, 988
Krampitz, L. O. (1961). *In* "The Bacteria" (I. C. Gunsalus and R. Y. Stainer, eds.), Vol. 2, pp. 209–256. Academic Press, New York.
Krebs, H. A., and Lowenstein, J. M. (1960). *In* "Metabolic Pathways" (D. M. Greenberg, ed.), 2nd ed., Vol. 1, pp. 129–203. Academic Press, New York.
Laser, H. (1944). *Biochem. J.* **38**, 333.
Laurie, J. S. (1957). *Exptl. Parasitol.* **6**, 245.
Laurie, J. S. (1959). *Exptl. Parasitol.* **8**, 188.
Lee, D. L. (1965). "The Physiology of Nematodes." Freeman, San Francisco, California.
Levenbook, L. (1961). *Arch. Biochem. Biophys.* **92**, 114.
Magath, T. B. (1918). *J. Biol. Chem.* **33**, 395.
Martin, A. (1913). *Ann. Sci. Nat. Zool.* [9] **18**, 1.
Nakada, H. I., and Weinhouse, S. (1953). *J. Biol. Chem.* **204**, 831.
Nicholas, W. L., and Grigg, H. (1965). *Exptl. Parasitol.* **16**, 332.
Okamoto, O., Katsume, T., and Obo, F. (1962). *Acta Med. Univ. Kagoshima.* **4**, 61.
Olson, J. A. (1954). *Nature* **174**, 695.
Oya, H., Kikuchi, G., Hayashi, H., and Bando, T. (1962). *Japan J. Parasitol.* **11**, 291.
Oya, H., Costello, L. C., and Smith, W. N. (1963). *J. Cellular Comp. Physiol.* **62**, 287.

Oya, H., Kikuchi, G., Bando, T., and Hayashi, H. (1965). *Exptl. Parasitol.* **17**, 229.
Passey, R. F., and Fairbairn, D. (1955). *Can. J. Biochem. Physiol.* **33**, 1033.
Passey, R. F., and Fairbairn, D. (1957). *Can. J. Biochem. Physiol.* **35**, 511.
Peters, L., Bueding, E., Valk, A., Higashi, A., and Welch, A. D. (1949). *J. Pharmacol. Exptl. Therap.* **95**, 212.
Rathbone, L. (1955). *Biochem. J.* **61**, 574.
Rathbone, L., and Rees, K. R. (1954). *Biochim. Biophys. Acta* **15**, 126.
Reger, J. F. (1965). *Z. Zellforsch. Mikroskop. Anat.* **67**, 196.
Roberts, L. S., and Fairbairn, D. (1965). *J. Parasitol.* **51**, 129.
Robinson, W. G., Bachhawat, B. K., and Coon, M. J. (1956). *J. Biol. Chem.* **218**, 319.
Rogers, W. P. (1948). *Parasitology* **39**, 105.
Rogers, W. P. (1949a). *Australian J. Sci. Res.* **B2**, 157.
Rogers, W. P. (1949b). *Australian J. Sci. Res.* **B2**, 399.
Rogers, W. P. (1961). "The Nature of Parasitism: The Relationship of Some Metazoan Parasites to Their Hosts." Academic Press, New York.
Rosenbluth, J. (1965a). *J. Cell. Biol.* **25**, 495.
Rosenbluth, J. (1965b). *J. Cell. Biol.* **26**, 579.
Rothstein, M. (1963). *Comp. Biochem. Physiol.* **9**, 51.
Rothstein, M., and Mayoh, H. (1964a). *Arch. Biochem. Biophys.* **108**, 134.
Rothstein, M., and Mayoh, H. (1964b). *Biochem. Biophys. Res. Commun.* **14**, 43.
Rothstein, M., and Mayoh, H. (1965). *Comp. Biochem. Physiol.* **16**, 361.
Rothstein, M., and Mayoh, H. (1966). *Comp. Biochem. Physiol.* **17**, 1181.
Rothstein, M., and Tomlinson, G. (1962). *Biochim. Biophys. Acta* **63**, 471.
Sayre, F. W., Hansen, E. L., and Yarwood, E. A. (1963). *Exptl. Parasitol.* **13**, 98.
Saz, H. J. (1954). *Biochem. J.* **58**, XX.
Saz, H. J., and Bueding, E (1966). *Pharmacol. Rev.* **18**, 871.
Saz, H. J., and Gerzon, K. (1962). *Exptl. Parasitol.* **12**, 204.
Saz, H. J., and Hillary, E. P. (1956). *Biochem. J.* **62**, 563.
Saz, H. J., and Hubbard, J. A. (1957). *J. Biol. Chem.* **225**, 921.
Saz, H. J., and Krampitz, L. O. (1954). *Bacteriol.* **67**, 409.
Saz, H. J., and Lescure, O. L. (1966). *Comp. Biochem. Physiol.* **18**, 845.
Saz, H. J., and Lescure, O. L. (1968). *Comp. Biochem. Physiol.* **22**, 15.
Saz, H. J., and Vidrine, A., Jr. (1959). *J. Biol. Chem.* **234**, 2001.
Saz, H. J., and Weil, A. (1960). *J. Biol. Chem.* **235**, 914.
Saz, H. J., and Weil, A. (1962). *J. Biol. Chem.* **237**, 2053.
Saz, H. J., Vidrine, A., Jr., and Hubbard, J. A. (1958). *Exptl. Parasitol.* **7**, 477.
Scheibel, L. W., and Saz, H. J. (1966). *Comp. Biochem. Physiol.* **18**, 151.
Schiller, E. L. (1965). *J. Parasitol.* **51**, 516.
Seidman, I., and Entner, N. (1961). *J. Biol. Chem.* **236**, 915.
Smith, J. C. (1965). *Exptl. Parasitol.* **16**, 236.
Smith, R. A., and Gunsalus, I. C. (1954). *J. Am. Chem. Soc.* **76**, 5002.
Stannard, J. N., McCoy, O. R., and Latchford, W. B. (1938). *Am. J. Hyg.* **27**, 666.
Tolbert, N. E., and Cohan, M. S. (1953). *J. Biol. Chem.* **204**, 649.
Tomlinson, G., and Rothstein, M. (1962). *Biochim. Biophys. Acta* **63**, 465.
Toru, Y. (1935). *Sci. Rept. Tohoku Univ., Fourth Ser.* **4**, 361.
Utter, M. F., and Kurahashi, K. (1954). *J. Biol. Chem.* **207**, 821.
von Brand, T. (1938). *Biodynamic* **2**, No. 41, 1.
von Brand, T. (1940). *J. Parasitol.* **26**, 301.

von Brand, T. (1960a). *In* "Nematology" (J. N. Sasser and W. R. Jenkins, eds.), pp. 242–248. Univ. of North Carolina Press, Chapel Hill, North Carolina.
von Brand, T. (1960b). *In* "Nematology" (J. N. Sasser and W. R. Jenkins, eds.), pp. 233–241. Univ. of North Carolina Press, Chapel Hill, North Carolina.
von Brand, T. (1966). "Biochemistry of Parasites." Academic Press, New York.
von Brand, T., Weinstein, P. P., Mehlman, B., and Weinbach, E. C. (1952). *Exptl. Parasitol.* **1,** 245.
von Brand, T., Bowman, I. B. R., Weinstein, P. P., and Sawyer, T. K. (1963). *Exptl. Parasitol.* **13,** 128.
Weinhouse, S., and Friedmann, B. (1951). *J. Biol. Chem.* **191,** 707.
Weinland, E. (1901). *Z. Biol.* **42,** 55.
Weissbach, A., and Sprinson, D. B. (1953). *J. Biol. Chem.* **203,** 1023.
Whitlock, B., and Strong, F. M. (1963). *Arch. Biochem. Biophys.* **103,** 459.
Wong, D. T. O., and Ajl, S. J. (1955). *Nature* **176,** 970.
Wood, H. G., and Utter, M. F. (1965). *In* "Essays in Biochemistry" (P. N. Campbell and G. D. Grenville, eds.), Vol. 1, pp. 1–27. Academic Press, New York.
Zavadovsky, M. (1916). *Compt. Rend. Soc. Biol.* **75,** 798.

Addendum

Recently, Bueding and Saz [*Comp. Biochem. Physiol.* (1968) **24,** 511] reported that pyruvate kinase activity was very low in *Ascaris* muscle, indicating that pyruvate could not be formed directly from PEP. Saz and Lescure (43rd Annual Meeting of the American Society of Parasitology, June, 1968) subsequently presented evidence indicating that CO_2 is fixed into PEP to form OAA, which in turn is reduced to malate in the extra-mitochondrial portion of the cell. Malate then may serve as the mitochondrial substrate where a dismutation reaction takes place, one molecule generating mitochondrial NADH by being oxidatively decarboxylated to pyruvate by the NADH-linked malic enzyme; another molecule being reduced to succinate with the generation of ATP.

CHAPTER 6

Lipid Components and Metabolism of Acanthocephala and Nematoda

Donald Fairbairn

I. Introduction

Knowledge of the lipids of the phylum Aschelminthes is limited to the two major classes, the Nematoda and the Acanthocephala. A considerable literature exists in which lipids have been detected or identified histochemically in these helminths (see von Brand, 1952, 1966, for citations). This will be referred to only when it can be connected to chemical observations, or when it appears to be independently valid. While recognizing that such selection is arbitrary to a degree, it is nevertheless true that some tissues of the Aschelminthes are so difficult to fix and stain that results may be difficult to interpret.

Nematodes and acanthocephalans are not closely related, nor have they been equally successful in the evolutionary sense. All adult acanthocephalans live in a vertebrate intestine, with arthropods serving as the secondary host for larval stages. Nematodes, on the other hand, are among the most ubiquitous of animals. Many are parasites of all kinds of plants and animals, and others are free-living and exceedingly diverse in their selection of habitats. Both classes have in common a fairly restricted range of morphological variation, and both are limited to an aquatic or semiaquatic existence, except during periods of lethargus or cryptobiosis.

Until very recently, extensive investigation of the lipids of animals as small and as difficult to obtain as most nematodes and acanthocephalans was extremely difficult owing to limitations in methodology. This situation has now changed greatly, and in addition, methods are available for the *in vitro* cultivation of plant and animal parasites (Krusberg, 1961; Berntzen, 1965; Schiller, 1965) and of free-living nematodes (Rothstein and Cook, 1966). Some benefits have already been derived from such advances, and more can be expected.

In considering the lipids of the Aschelminthes, it has been desirable to compare them with those of the Platyhelminthes whenever possible, for species belonging to these two phyla sometimes live in similar habitats such as the vertebrate intestine, and may have adapted to their environment in similar physiological ways.

II. Nature and Distribution

A. Historical Background

The first reliable estimate of total lipids in a nematode is that of Weinland (1901) who reported that *Ascaris lumbricoides* adults contained between 1.4% and 1.5% of fat. Later, Flury (1912) and Fauré-Fremiet (1913a) published more detailed studies of *Ascaris lumbricoides* and *Parascaris equorum* (= *Ascaris megalocephala,* the horse ascarid), respectively, which were remarkable in their thoroughness and prescience. Both workers independently discovered the unusual glycosides which are quantitatively the most important constituents of the unsaponifiable fraction. Von Kemnitz (1912) was perhaps the first of many investigators to apply histological methods for the demonstration of fats in various tissues. Although scattered isolations and identifications were made during the next 35 years (see von Brand, 1952, for literature citations) the modern era began with the work of Rogers and Lazarus (1949), Fairbairn (1955a, 1956), and Fouquey *et al.* (1957).

B. Total Lipids

Selected values for the total lipids of a number of species are given in Table I. Additional information is to be found in the reviews of Hobson (1948), von Brand (1952), and Fairbairn (1957). The older values in particular may not be entirely reliable. It is clear, however, that most species contain substantial amounts of lipids. The apparent large interspecific variation might be much less if the results were expressed in terms of tissue nitrogen rather than wet or dry weight, for many species such as *Ascaris lumbricoides, Ascaridia galli, Eustrongylides ignotus, Porrocaecum decipiens,* and *Moniliformis dubius* contain glycogen amounting to 20% or even 40% or more of the dry weight (von Brand, 1966). Although the data are too few to permit firm conclusions, it is of interest that the two free-living nematodes which have been examined (*Turbatrix aceti,* the vinegar eelworm, and *Panagrellus redivivus,* a soil organism) and a plant-parasitic genus (*Ditylenchus* spp.) all contain exceptionally large amounts of lipids.

TABLE I
TOTAL LIPIDS IN NEMATODES AND ACANTHOCEPHALANS

Species	Stage	Habitat	% Wet wt.	% Dry wt.	Reference
Nematodes					
Ascaris lumbricoides	Adult ♀	Pig intestine	1.6	8.9	Cavier *et al.* (1958)
	Adult ♂		1.3		
Ascaridia galli	Adult ♀	Fowl intestine	1.34		Monteoliva (1960)
	Adult ♂		1.22		
Dioctophyme renale	Adult	Dog kidney		1	von Brand (1938)
Eustrongylides ignotus	Larva	Fish mesenteries	1.1	4.4	von Brand (1957)
Porrocaecum decipiens	Larva	Fish muscle		3.5	Fairbairn (1958)
Trichinella spiralis	Larva	Vertebrate muscle		5.5	von Brand (1952)
Ditylenchus spp.	Adult	Plant parasites		33	Tracey (1958)
Panagrellus redivivus	Adult	Soil		24	Sivapalan and Jenkins (1966)
Turbatrix aceti	Adult	Vinegar		41	Comenga (1955)
Nippostrongylus brasiliensis	Adult	Rat intestine		11.9	Roberts and Fairbairn (1965)
	Larva	Soil		15.2	Wilson (1965)
Acanthocephalans					
Moniliformis dubius	Adult ♀	Rat intestine	4.2		Beames and Fisher (1964)
	Adult ♂		7.2		
Macracanthorhynchus hirudinaceus	Adult ♀	Pig intestine	0.9		Beames and Fisher (1964)
	Adult ♂		1.7		

C. GROSS COMPOSITION AND DISTRIBUTION IN TISSUES

Most of the available chemical data on gross composition and distribution are summarized in Table II. *Ascaris lumbricoides* is the only species in which details of the distribution in various organs and tissues are known. Hemolymph, cuticle, and muscle contain relatively little lipid, and the reproductive organs (including fertilized eggs) a great deal. In the latter, glycerides are abundant. A notable feature of *Ascaris,*

Ascaridia, and also of the two acanthocephalans is the abundance of the unsaponifiable fraction.

Histochemical evidence indicates that fats are especially abundant in nematodes in the lateral lines, and sometimes in the intestinal cells, particularly of free-living infective larvae (see von Brand, 1966, for

TABLE II

NEUTRAL AND PHOSPHOLIPID COMPOSITION OF ASCHELMINTHES

Species	Total lipids (percent wet weight)	Neutral lipids (percent total lipids): Glycerides	Neutral lipids (percent total lipids): Unsaponifiables	Phospholipids (percent total lipids)
Ascaris lumbricoides[a]				
Adult ♀				
Hemolymph	0.35		5.7	57
Cuticle	0.61	43	16	41
Muscle	0.75	40	11	49
Reproductive organs	6.0	75	22	8.3
Fertilized eggs	17.5	69	22	9.2
Ascaridia galli[b]				
Adult ♀	1.34		22	10
♂	1.22		16	15
Panagrellus redivivus[c]	24			33
Macracanthorhynchus hirudinaceus[d]				
Adult ♀	0.9	32	21	47
♂	1.7	16	38	46
Moniliformis dubius				
Adult ♀	4.2	71	10	19
♂	7.2	83	8	9

[a] All values for *Ascaris* are those of Fairbairn (1955a,b, 1956) except those for hemolymph (Cavier *et al.*, 1958). Cavier *et al.* also give values for male worms. Beames (1965) reported similar values for muscle and cuticle, but found 28.8% of phospholipids, 57% glycerides and 17.9% unsaponifiables in the reproductive organs.

[b] Monteoliva (1960).

[c] Sivapalan and Jenkins (1966).

[d] Beames and Fisher (1964). Values for glycerides are obtained by differences.

details). In acanthocephalans, lipid droplets are commonly observed in the lacunar system, subcuticula, muscles, and reproductive system (e.g., von Brand, 1939; Bullock, 1949). Von Brand (1939) also reported that in *Macracanthorhynchus hirudinaceus* the cuticle and body wall musculature contained 61% of the total lipids and the female reproductive tract 30%, although the concentration in the reproductive organs was higher.

D. Triglycerides and Phospholipids

Flury (1912) clearly established the presence in *Ascaris* lipids of phospholipids, volatile and nonvolatile acids, and glycerol. Rogers and Lazarus (1949) reported that lecithins, cephalins, and sphingomyelins were present in the hemolymph and intestine, and lecithins and cephalins, but no sphingomyelins in the ovaries. Lecithins predominated and the cephalin fraction consisted mainly of phosphatidyl ethanolamine, although phosphatidyl serine was also present. Fairbairn (1955a) and Beames (1964) observed a similar pattern in the entire female reproductive system. Plasmalogens are present in very high concentration in the body wall (30 to 38% of the total phospholipids) and in lower concentrations in the reproductive organs. Although small amounts of phosphatidic acids (or cardiolipins) are also present, no detailed study of these has been made.

Following removal of phospholipids from a total lipid extract by acetone precipitation, the acetone-soluble glyceride contained amounts of volatile acids with 2 to 6 carbon atoms sufficient to account for approximately 50% of the ester linkages (Fairbairn, 1955a). Acetic, pentanoic, and hexanoic acids were most abundant, and were considered to be esterified fermentation acids. They occurred also in the glyceride fraction of muscle, cuticle, and fertilized eggs (Fairbairn, 1955b, 1956), but not in the phospholipids. Beames (1965) established that these volatile acids were identical with the fermentation acids, i.e., acetic, propionic, valeric, α-methylbutyric, and α-methylvaleric acids. Isovaleric acid may also have been present. In the acanthocephalans *Moniliformis dubius* and *Macracanthorhynchus hirudinaceus,* no esterified volatile fatty acids were found (Beames and Fisher, 1964), even though acetic acid is the major fermentation acid produced by *M. dubius* (Laurie, 1958).

A soil nematode, *Panagrellus redivivus,* has been reported to contain an exceptionally large concentration of phospholipids, amounting to 8% of the tissue solids (Sivapalan and Jenkins, 1966), but the method of isolation (precipitation from acetone at $-20°$) suggests that the phospholipids may have been contaminated with triglycerides, which in this species are extremely abundant (Table I). The major phospholipids were phosphatidyl ethanolamine and phosphatidyl choline. Phosphoinositides were also identified, but phosphatidyl serine could not be detected.

Among the acanthocephalans, *Moniliformis dubius* contains about 4 times as much lipid as *Macracanthorhynchus hirudinaceus,* the difference being attributable to triglycerides (Beames and Fisher, 1964). Phospholipids are qualitatively and quantitatively similar to those of *Ascaris,* although sphingomyelins may be absent and phosphoinositides are present. The concentration of plasmalogens is much lower, but analyses were based on the whole worm rather than on the muscle, so that direct comparison is not possible. The total lipid extract contained no bound glucose or galactose, suggesting that cerebrosides were absent. As in higher animals these are normally found in nervous tissue, in which acanthocephalans (and nematodes) are not rich, the observation is not surprising. On the other hand, certain Platyhelminthes such as cestodes, which also have little nervous tissue, contain substantial amounts of cerebrosides (Ginger and Fairbairn, 1966a). On the whole, the qualitative and quantitative features of the phospholipids of Aschelminthes, in so far as they are now known, are similar to those of other invertebrates (Hack *et al.*, 1962), particularly with respect to the preponderance of phosphatidyl choline and the low concentration of phosphatidyl serine. This is true also of the cestode, *Hymenolepis diminuta* (Ginger and Fairbairn, 1966a). Additional study is needed before firm conclusions are possible concerning the occurrence of sphingomyelins, phosphatidic acids, cardiolipids, and lysophospholipids.

E. Fatty Acid Composition of Triglycerides and Phospholipids

Although a few values for the fatty acid composition of triglycerides and phospholipids have appeared (Beames, 1964, 1965; Beames and Fisher, 1964; Sivapalan and Jenkins, 1966), many peaks appearing on gas chromatograms have not been identified, and the nature of the C_{20} and other higher acids in particular remains in doubt. Values for the triglyceride acids (excluding the volatile acids prevalent in *Ascaris* lipids) are given in Table III.

Some 80% of the triglyceride acids are unsaturated, with most of the unsaturation occurring in the C_{18} acids although higher acids also make a significant contribution. The methods used did not permit separation of linoleic ($C_{18:2}$) and linolenic acids ($C_{18:3}$) except in the case of *Panagrellus redivivus*. In that species very little linolenic acid was present, and the amount of linoleic acid greatly exceeded that of oleic acid. Such high concentrations of linoleic acid would be considered to be very unusual in higher animals, but have been observed also in Platyhelminthes such as *Hymenolepis diminuta* (Harrington, 1965; Ginger and Fairbairn, 1966a), and *Spirometra mansonoides* (Meyer *et al.*, 1966). All authors (Table III) reported a number of unknown acids. *Ascaris lumbricoides* and *Panagrellus redivivus* also contain odd-numbered acids and *Ascaris* contains branched chain acids. These may arise from known odd-numbered straight or branched chain fermentation acids such as propionic or α-methylbutyric acids. In *Panagrellus*, which was cultured on an oatmeal-yeast mixture, the authors point out that some of the peculiarities in constituent fatty acids may arise from the unusual nature of the yeast acids. A comparison with *Panagrellus* cultured in an axenic, defined medium (Rothstein and Cook, 1966) is desirable.

TABLE III
TRIGLYCERIDE FATTY ACIDS IN THE ASCHELMINTHES[a]

<table>
<tr><th rowspan="4">No. carbons:
No. double bonds</th><th colspan="4">Percent of total methyl esters recovered</th></tr>
<tr><th colspan="2">Acanthocephalans[b]</th><th colspan="2">Nematodes</th></tr>
<tr><th>M. hirudinaceus adult</th><th>M. dubius adult</th><th>A. lumbricoides[c] ♀ reproductive organs</th><th>P. redivivus[d] mixed culture</th></tr>
<tr><th></th><th></th><th></th><th></th></tr>
<tr><td>12:0</td><td></td><td>0.53</td><td>0.1</td><td>0.1</td></tr>
<tr><td>12:1</td><td></td><td></td><td></td><td>0.4</td></tr>
<tr><td>14:0</td><td>2.75</td><td>5.33</td><td>0.9</td><td>0.5</td></tr>
<tr><td>14:1</td><td></td><td></td><td></td><td>3.1</td></tr>
<tr><td>15:0</td><td></td><td>0.35</td><td>0.3</td><td></td></tr>
<tr><td>16:0</td><td>9.16</td><td>6.86</td><td>13.6</td><td>1.5</td></tr>
<tr><td>16:1</td><td>1.22</td><td>4.02</td><td>0.7</td><td>3.3</td></tr>
<tr><td>17:0</td><td></td><td>0.27</td><td>0.8</td><td></td></tr>
<tr><td>17:1</td><td></td><td></td><td>0.4</td><td></td></tr>
<tr><td>18:0</td><td>7.46</td><td>5.12</td><td>8.2</td><td>0.2</td></tr>
<tr><td>18:1</td><td>42.6</td><td>18.0</td><td>27.1</td><td>15.3</td></tr>
<tr><td>18:2</td><td rowspan="2">28.4</td><td rowspan="2">48.9</td><td rowspan="2">35.5</td><td>32.9</td></tr>
<tr><td>18:3</td><td>0.2</td></tr>
<tr><td>20:0</td><td>1.29</td><td>0.54</td><td>0.6</td><td></td></tr>
<tr><td>20:1</td><td>1.66</td><td>0.59</td><td></td><td></td></tr>
<tr><td>20:2</td><td>1.51</td><td>1.10</td><td></td><td>0.4</td></tr>
<tr><td>20:3</td><td></td><td></td><td></td><td>2.7</td></tr>
<tr><td>20:4</td><td>3.60</td><td>0.38</td><td></td><td>1.9</td></tr>
<tr><td>20:5</td><td></td><td></td><td></td><td>2.3</td></tr>
</table>

[a] Excludes unknowns, trace amounts, and values less than 0.1%.
[b] Beames and Fisher (1964).
[c] Beames (1965).
[d] Sivapalan and Jenkins (1966).

No fatty acid analysis of acanthocephalan phospholipids has appeared. In *Ascaris lumbricoides*, Beames (1964) found that the phospholipid fatty acids of the muscle and reproductive system closely resembled one another. Odd-numbered acids with straight or branched chains were present in significant amounts. Unfortunately, analysis of the C_{20} and higher acids, which are of great importance in phospholipids, was incomplete. Sivapalan and Jenkins (1966) reported that *Panagrellus redivivus* phospholipids contain C_{20} acids with 2 to 5 double bonds, as well as some 30% of unidentified acids, many of which may have originated in the yeast of the culture medium.

As mentioned previously, plasmalogens in *Ascaris lumbricoides* are unusually abundant. It is also of interest that their constituent aldehydes are predominantly unsaturated C_{18} compounds (Beames, 1964) compared with the more usual C_{16} saturated aldehydes of mammals (Gray, 1960).

F. Unsaponifiables

1. *Sterols*

Sterols occur in the Aschelminthes, and it is possible that cholesterol is characteristic. This is certainly the case in *Ascaris lumbricoides,* in which the distribution of cholesterol has been studied (Fairbairn, 1955a,b; 1956) and a definitive identification made (Fairbairn and Jones, 1956). A second sterol which accounted for 25% of the total was thought to be cholestanol, an identification which has since been confirmed using gas chromatographic techniques (Jezyk and Fairbairn, unpublished observation). Earlier workers obtained less definitive evidence for the presence of sterols in a number of helminth parasites (see von Brand, 1966, for review). According to Beames and Fisher (1964), cholesterol (identified by gas chromatography) comprises 50% to 80% of the total unsaponifiables in *Macracanthorhynchus hirudinaceus* and *Moniliformis dubius. M. hirudinaceus* contains 4 distinct unsaponifiable components, and *M. dubius* contains 10. These authors suggested that some of these unknown compounds might be alkali degradation products of cholesterol.

Among the Platyhelminthes, cholesterol has been unequivically identified in the cestodes *Taenia saginata* (Cmelik and Bartl, 1956), *Hymenolepis diminuta* (Fairbairn *et al.,* 1961), *Taenia taeniaeformis,* and *Moniezia expansa* (Thompson *et al.,* 1960). Clearly, present evidence suggests that cholesterol is the commonly occurring sterol in helminth parasites.

2. *Ascarosides*

Flury (1912) and Fauré-Fremiet (1913a) discovered independently that *Ascaris lumbricoides* and *Parascaris equorum* contain large quantities of unsaponifiable matter which is readily isolated by precipitation from nonpolar solvents, melts at 82°, has an empirical formula approximating $C_{32}H_{64}O_4$, and forms a diacetate. It occurs in most tissues of both male and female worms, but is particularly abundant in the ovaries.

In fertilized eggs it accounts for 10% or more of the dry weight (Fairbairn, 1955b) and is the major constituent of the vitelline membrane (see Fairbairn, 1957, for review).

The name ascaryl alcohol (Flury, 1912) was changed to the generic name, ascarosides, by Fouquey *et al.* (1957) when they established that this material as isolated from *Parascaris equorum* unsaponifiables was a mixture of closely related glycosides in which differences were due to variations in the aliphatic aglycone moiety (ascaryl alcohol). Three major fractions, isolated by chromatography on alumina columns, were designated ascarosides A, B, and C respectively. In these the glycone is always 3,6-dideoxy-L-arabinohexose (ascarylose) and is isomeric with a number of important bacterial hexoses (Fouquey *et al.*, 1958, 1959). The aglycones in ascaroside A are C_{23} to C_{27} 2-ols. In ascarosides B and C the aglycone is hentriacontane-2-6-diol, one mole of ascarylose being present in B, and two moles in C. As obtained from the unsaponifiables, the proportions of A, B, and C are 20:67:13 in *P. equorum* and 18:74:8 in *Ascaris lumbricoides* (human strain). In *Parascaris* ovaries the ascarosides occur as the acetate and propionate esters.

Using silica gel columns, Jezyk and Fairbairn (1967a) found that the ascaroside esters of *Ascaris lumbricoides* appear in the neutral lipid fraction and can be separated from other neutral lipids by thin-layer chromatography. Free ascarosides and ascaryl alcohols occur only in traces. The acid moieties of the ascaroside esters are volatile, and are identical with the fermentation acids which also occur in the triglycerides of the neutral lipid fraction. Aglycones were easily obtained from the mixed glycoside esters by transmethylation with methanolic boron trifluoride, following which they were reduced to the corresponding hydrocarbons and identified by means of gas chromatography. Nearly all of these hydrocarbons contain 28 or more carbon atoms, the maximum being 37 and the commonest 29, 31, and 33. Hydrocarbons derived from monols are more abundant than those derived from diols. This is not in agreement with the results of Fouquey *et al.* (1957), but other observations led to the conclusion that much of the so-called ascaroside A (monol) is an artifact derived from ascaroside B (diol) as a result of saponification. The naturally occurring free ascarosides of the ovaries or vitelline membrane are also richest in monols, the proportions of A, B, and C being 67:18:15.

Although ascarosides occur in most *Ascaris* tissues, their only known function is in connection with the vitelline membrane, in which they comprise about 75% of the total solids (Fairbairn and Passey, 1955) and are primarily responsible for the extreme resistance of the eggs to adverse environmental influences (Fairbairn, 1957). Their occurrence in other ascarids, or in less closely related nematodes or acanthocephalans has not been investigated. Recently developed methods for the detection of ascarosides and their esters (Jezyk and Fairbairn, 1967a) should make such a survey possible. Such methods applied to the cestode *Hymenolepis diminuta* (Ginger and Fairbairn, 1966a) suggest that ascarosides do not occur in the Platyhelminthes.

III. Digestion and Absorption. Lipases and Esterases

Nematodes and acanthocephalans differ fundamentally in that the former have a functional digestive tract, whereas the latter rely solely on cuticular absorption. Although cuticular absorption of water, small inorganic molecules, and some organic molecules, such as nematocides, may occur in nematodes (Fairbairn, 1957, 1960) there is no substantial evidence for a general cuticular absorptive function (Roberts and Fairbairn, 1965). It may be significant that all adult acanthocephalans (like all adult cestodes, which also lack a digestive tract) live only in the vertebrate intestine, in which predigested food is available.

Digestion and absorption of lipids in the Aschelminthes has received little attention. As many species of free-living nematodes live on bacteria or other particulate matter, it can perhaps be assumed that they possess a more or less complete array of digestive enzymes, including lipases. Free-living larvae of the hookworm family are also known to feed upon bacteria (McCoy, 1929). Other nematodes are predators, and may puncture and paralyze their prey by means of secretions (Linford, 1937), which may also possess esterase activity (Lee, 1964). Many plant parasitic nematodes possess a stylet which is used to puncture the cell wall and, according to Goffart and Heiling (1962), enzyme-containing secretions are then poured into the cell.

Nematodes parasitic in animals and feeding on intestinal contents, such as the Ascaridae, have access to predigested food, whereas blood and tissue feeders presumably do not. Rogers (1941) found esterases and lipases in the intestine of *Ascaris lumbricoides* and *Strongylus edentatus,* the latter being a tissue feeder in the large intestine of the horse. In a more detailed study (Carpenter, 1953) *Ascaris* lipases and esterases were shown to be most prevalent in the cells of the foregut, although present also in other regions. *Leidynema appendiculata* (Lee, 1958), a nematode living in the cockroach intestine, possesses an intestinal lipase with a pH optimum of 7.0 which hydrolyzes tributyrin but is inactive with respect to olive oil or ethyl butyrate. These observations are of interest, but unfortunately have contributed little to our understanding of digestion. Lee (1962) presented histochemical evidence that the nonspecific esterase of *Ascaris lumbricoides* intestinal cells is present in secretory granules which are discharged into the lumen. The fine structure of these cells also suggests an important secretory function (Sheffield, 1964).

Absorption of lipids by nematodes is not understood, and in acanthocephalans there is indirect evidence for absorption in only one species, *Acanthocephalus renee,* based upon feeding experiments with stained fat (Pflugfelder, 1949). Nevertheless, the high fat content of the Aschelminthes suggests strongly that at least some of the tissue fatty acids are derived from the diet, as they are in other animals. This probability is made more attractive by recent work which indicates clearly that cestodes such as *Hymenolepis diminuta* and *Spirometra mansonoides* not only absorb fatty acids freely (Lumsden and Harrington, 1966; Ginger and Fairbairn, 1966b) but have little or no capacity for their biosynthesis (Meyer *et al.*, 1966; Jacobsen and Fairbairn, 1967).

The presence of a tissue lipase acting on tripalmitin has been reported in infective (free-living) larvae of *Strongyloides ratti* and *Nippostrongylus muris* (= *N. brasiliensis*) (Mandlowitz *et al.*, 1960), and positive histochemical evidence for lipases was obtained in the ovary, testes, and embryonated eggs of the common pinworm, *Enterobius vermicularis* (Marzullo *et al.*, 1957), as well as in the hypodermis and lemnisci of several acanthocephalans (Bullock, 1949). The hemolymph of *Ascaris lumbricoides* apparently contains no lipase (Monteoliva, 1960). Nonspecific esterases, as determined histochemically, are present in cells of the cuticle, excretory canals, hypodermis, muscle, reproductive organs, and other tissues of *Ascaris* (Lee, 1962). Lee observed an enzyme "cap" on fat globules in the hypodermis and muscle. The presence of a specific esterase (cholinesterase) has been reported in nematodes (and in Platyhelminthes) using methods which are now known to be of limited value (see von Brand, 1966, for review). Bueding (1952) developed a specific assay method for the trematode, *Schistosoma mansoni* and applied it to *Ascaris lumbricoides* and the filariid nematode, *Litomosoides carinii,* both of which reacted positively. Cholinesterase has since been identified in *Ascaris* (Lee, 1962) and in other ascarids (Lui *et al.*, 1964) and in plant parasitic nematodes (Rohde, 1960). It is prevalent in the nervous system, but is frequently not confined to it.

IV. Origins and Catabolism

A. Origins

Consideration of the origins of the lipids of the Aschelminthes must take into account not only the fatty acids, sterols or sterol precursors, and glycosides, but also water-soluble components of low molecular weight such as glycerol, inositol, sphingosine, choline, serine, and ethanolamine. Apart from recent work on the glycosides, these origins are almost completely unknown. It can perhaps be assumed, pending actual investigation, that given the proper components, these organisms can synthesize the glycerides and phospholipids whose presence has been established. Such an assumption cannot be applied to the sterols, which in many invertebrates cannot be synthesized *de novo*. The question of origins is of considerable interest from the evolutionary point of view, because the phylum Aschelminthes is extremely diverse ecologically, including a wide variety of free-living and plant and animal parasitic species whose diet may or may not provide most of the lipid constituents, from the simplest to the most complex. Many free-living nematodes, for example, may be expected to possess a biosynthetic capacity for fatty acids and sterols comparable to those of other free-living invertebrates. This may also be true for plant parasitic nematodes, as in these the dietary acids and sterols may differ considerably from animal types. On the other hand, nematodes and acanthocephalans parasitic in animals may rely heavily on their hosts for many lipid components, either through adaptation to such a constant supply, or because their environment may be unsuitable for biosynthesis. As an example of the latter possibility, one might cite the necessity of molecular oxygen for the production of double bonds in higher fatty acids (Erwin and Bloch, 1964) and cholesterol (Bloch, 1965) and the very limited supply of oxygen available in some habitats such as the intestine or bile ducts. For such species, sterols and unsaturated fatty acids may be a dietary requirement.

These speculations are as yet untested in the Aschelminthes, but it may be noted that they are supported by observations on two adult cestodes which live in a similar environment. The abundant fatty acids of the lipids of *Hymenolepis diminuta* and *Spirometra mansonoides*, which inhabit the small intestine of the rat and cat, respectively, are apparently almost exclusively derived from the diet, i.e., from the intestinal contents (Ginger and Fairbairn, 1966b; Meyer *et al.*, 1966; Jacobsen and Fairbairn, 1967). These species cannot synthesize fatty acids *de novo*, although they can lengthen carbon chains beginning with C_{18} acids and can synthesize glycerides and phospho- and glycolipids from absorbed fatty acids. Similar observations were made on the larval stage of *Spirometra mansonoides*, whose environment in the tissues of the mouse and other animals is not so highly deficient in oxygen (Meyer *et al.*, 1966). Thus, although dependence upon the diet for unsaturated acids may be an adaptation to oxygen deficiency, it appears that these cestodes have also lost their capacity for synthesizing saturated fatty acids.

Whatever the origins of the lipids, their turnover in an organism such as *Ascaris lumbricoides* is high. A calculation based on reliable analyses indicates that the daily passage of eggs in this species includes lipids amounting to 1.6% of the total female solids (Fairbairn, 1957).

Jezyk and Fairbairn (1967b) have presented details concerning the biosynthesis of ascarosides. The 3,6-dideoxyglucose (ascarylose) is freely synthesized by *Ascaris lumbricoides* ovaries from either glucose or glucose 1-phosphate by a sequence of reactions which probably involves a nucleotide-diphosphate glucose as an intermediate and NAD and NADPH as coenzymes. Details of ascarylose synthesis by a bacterium, *Pasteurella pseudotuberculosis* are known (Matsuhashi, 1966; Matsuhashi *et al.*, 1966a,b). As free ascarylose does not occur in *Ascaris*, the end product of synthesis is possibly ascarylose-dinucleotidephosphate, which then condenses with the aglycones to form ascarosides as in other glycoside syntheses (Neufeld and Hassid, 1963). The aglycones are synthesized by a Claisen-like reaction between two moles of palmitic or other higher fatty acids, in which the carboxyl carbon of one acid condenses with carbon number 2 of the second acid, with elimination of one mole of carbon dioxide. Thus, two even-numbered fatty acids give rise to odd-numbered carbon chains containing 2_{n-1} carbon atoms. The condensation is analogous to that which occurs in the malonyl CoA pathway for fatty acid biosynthesis. The mechanism by which the condensation product is converted to the characteristic monols and diols is unknown.

B. Catabolism

At present, there is neither positive nor negative evidence concerning β-oxidation of fatty acids in the Aschelminthes. Nor, in plant-parasitic nematodes and acanthocephalans, is there information concerning the "utilization" or disappearance, of tissue lipids.

Free-living infective larvae of the hookworm and related families exhibit a histochemically determined decrease in the tissue lipids during a period of weeks or months when they are not feeding (e.g., Giovannola, 1936). Eventually, lipid losses are associated with loss of infectivity (Engelbrecht and Palm, 1964). Similar observations have been made on infective eggs of *Ascaridia galli* (Elliott, 1954) and the common pinworm *Enterobius vermicularis* (Engelbrecht, 1963). Munnich (1965) reported a gradual loss of fats from infective eggs of *Ascaris lumbricoides* stored in the laboratory for 4 years.

In the only chemical analysis reported thus far, Wilson (1965) found that lipids comprised 15–20% of the tissue solids in infective larvae of *Nippostrongylus brasiliensis*, and disappeared at the rate of 0.9% per day over a period of 8 days at 25°. No glycogen was present, and total carbohydrate did not decrease. Clearly, lipids are utilized extensively during the survival of nonfeeding, infective larvae. This is true also of infective eggs of *Ascaris lumbricoides,* in which the saponifiable lipids decrease to 27% of their initial amount after 53 days at 31° (Fairbairn, 1955b). These eggs contain large amounts of glycogen as well as the disaccharide, trehalose, neither of which is utilized in the postembryonic survival period (Passey and Fairbairn, 1957; Munnich, 1965). Wilson (1965) suggested that decreasing rates of oxygen uptake (Rogers, 1948; Passey and Fairbairn, 1955; Schwabe, 1957; Fernando, 1963) and substrate utilization during survival may be an indication of a physiological adaptation which enables the infective larvae (or eggs) to conserve their stored energy reserves.

Ascaris egg lipids are utilized not only during survival, as noted above, but also during embryonation. Fertilization in the uterus is followed immediately by shell formation, including the very important vitelline membrane (Fairbairn and Passey, 1955; Fairbairn, 1957). In the formation of this membrane about one-half of the total ascaroside esters are de-esterified, the remainder being retained in the egg cytoplasm (Jezyk and Fairbairn, 1967b). The fate of the liberated volatile acids is not known. Following their discharge from the uterus, *Ascaris* eggs will embryonate only if oxygen is present (Fauré-Fremiet, 1913b). During development to the vermiform stage there is a 17% reduction in the total lipids. Only the triglycerides are involved and esterified volatile and nonvolatile acids disappear at nearly equal rates (Fairbairn, 1955b). An accelerated rate of fatty acid decrease accompanies further development to the infective stage (Passey and Fairbairn, 1957) after which lipids continue to be utilized at a very low rate. Utilization of the various nonvolatile acids is nonselective with respect to chain length and degree of unsaturation, but the volatile acids, α-methylbutyric and α-methylvaleric, decrease much more markedly than acetic acid (Saz and Lescure, 1966).

The extensive disappearance of esterified fatty acids during the time when a decreasing rate of oxygen consumption is grossly inadequate to account for their complete oxidation (Fig. 1), and a number of related observations led to the conclusion that much of the fatty acid carbon is converted to glycogen and trehalose (Passey and Fairbairn, 1957). A mechanism for the conversion of fat to carbohydrate in microorganisms and plants by means of the glyoxylate cycle is now well documented (Kornberg and Elsden, 1961) but has not been demonstrated in *Ascaris* eggs or in other animals. Saz and Lescure (1966) presented tentative evidence for the conversion of esterified volatile acids (α-methylbutyric, α-methylvaleric) to glycogen and trehalose during embryonation, by a mechanism involving degradation to propionate followed by carbon dioxide fixation to form succinate, which is glycogenic. Although this mechanism may be valid, the amounts of available esterified volatile acids are sufficient to account for only a small proportion of total carbohydrate synthesis.

Apart from much less detailed studies on a related species, *Ascaris megalocephala* (Fauré-Fremiet, 1913b; Szwejkowska, 1929), the fatty

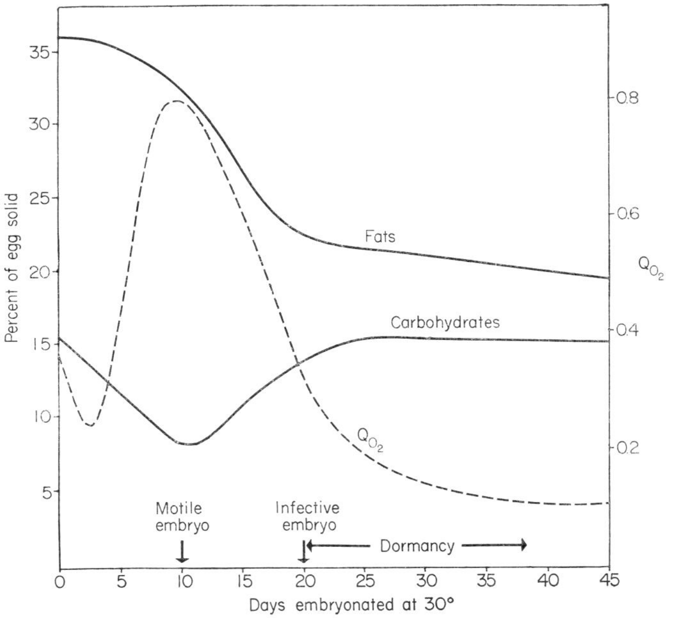

FIG. 1. Oxygen consumption and carbohydrate and lipid utilization during embryonic and postembryonic stages in *Ascaris lumbricoides* eggs. Lipids (triglycerides) are extensively utilized during a period of decreasing respiration and increasing synthesis of carbohydrates (Passey and Fairbairn, 1955, 1957).

acid metabolism of other free-living nematode eggs is unknown. Similarly, no useful general conclusions can be drawn concerning their metabolism in parasitic stages. By use of histochemical methods, Mueller (1929) and Hirsch and Bretschneider (1937) observed a decrease in lipid droplets in surviving cells of *Ascaris* lateral lines and intestine, respectively. On the other hand, von Brand (1934, 1941) found that the total lipids of *Ascaris* remained constant during aerobic *in vitro* starvation, whereas those of *Trichinella spiralis* larvae decreased (von Brand *et al.*, 1952). No decrease occurred when *Trichinella larvae* were incubated anaerobically. The latter observation is not surprising since with the possible exception of rumen ciliates (Gutierrez *et al.*, 1962;

Williams *et al.*, 1963) anaerobic β-oxidation of fatty acids occurs rarely, if at all, in any organism. As the availability of oxygen to parasite varies greatly according to their habitat, there is an essential need for comparative studies on different species.

V. Summary and Conclusions

With the exception of ascarosides, the lipids of Aschelminthes do not differ greatly from those of other animals. They tend to be abundant, particularly in the reproductive organs. Phospholipids may be somewhat unusual in that the concentration of plasmalogens is relatively large and the constituent aldehydes relatively highly unsaturated. As in many other invertebrates, phosphatidyl serine occurs sparingly. Oleic and linoleic acids comprise the majority of the nonvolatile triglyceride acids, but esterified volatile acids are abundant in *Ascaris lumbricoides* and *Parascaris equorum* triglycerides, and in ascaroside esters. The latter are glycoside esters of high molecular weight whose occurrence has been established only in *Ascaris* and *Parascaris*. Cholesterol is the predominant sterol.

It is probable, though unproven, that aerobic free-living and plant-parasitic nematodes obtain their fatty acids from the diet, and also by biosynthesis. In animal-parasitic nematodes and acanthocephalans, the origins of the fatty acids are unknown. Lipases and esterases are present in the nematode intestinal cells, and may be secreted into the lumen. By analogy with certain parasitic platyhelminths, it may be predicted that animal-parasitic species have a very limited capacity for *de novo* synthesis or the formation of unsaturated acids, although chain lengthening may occur. These possibilities have yet to be tested. In contrast, *Ascaris lumbricoides* freely synthesizes the unusual glycone and aglycone moieties of the high molecular weight ascarosides. The glycone (a 3,6-dideoxyhexose) is probably synthesized from glucose by pathways known to occur in bacteria, and the aglycones by a Claisen-type condensation between two moles of higher fatty acids.

As details of lipid chemistry and metabolism become known, they can be expected to contribute importantly to an understanding of the complexities of adaptations to an exceedingly large variety of habitats, including parasitic habitats. The acquisition of such knowledge will require a number of carefully designed comparative investigations.

REFERENCES

Beames, C. G. (1964). *Exptl. Parasitol.* **15**, 387.
Beames, C. G. (1965). *Exptl. Parasitol.* **16**, 291.
Beames, C. G., and Fisher, F. M. (1964). *Comp. Biochem. Physiol.* **13**, 401.

Berntzen, A. K. (1965). *Exptl. Parasitol.* **16**, 74.
Bloch, K. (1965). *Science* **150**, 19.
Bueding, E. (1952). *Brit. J. Pharmacol.* **7**, 563.
Bullock, W. L. (1949). *J. Morphol.* **84**, 201.
Carpenter, M. F. P. (1953). *Dissertation Abstr.* **27**, 11071.
Cavier, R., Savel, J., and Monteoliva, M. (1958). *Bull. Soc. Chim. Biol.* **40**, 177.
Cmelik, S., and Bartl, Z. (1956). *Z. Physiol. Chem.* **305**, 170.
Comenga, M. (1955). *Rev. Espan. Fisiol.* **11**, 181.
Elliott, A. (1954). *Exptl. Parasitol.* **3**, 307.
Engelbrecht, H. (1963). *Z. Parasitenk.* **23**, 384.
Engelbrecht, H., and Palm, V. (1964). *Z. Parasitenk.* **24**, 88.
Erwin, J., and Bloch, K. (1964). *Science* **143**, 1006.
Fairbairn, D. (1955a). *Can. J. Biochem. Physiol.* **33**, 31.
Fairbairn, D. (1955b). *Can. J. Biochem. Physiol.* **33**, 122.
Fairbairn, D. (1956). *Can. J. Biochem. Physiol.* **34**, 39.
Fairbairn, D. (1957). *Exptl. Parasitol.* **6**, 491.
Fairbairn, D. (1958). *Nature* **181**, 1593.
Fairbairn, D. (1960). *In* "Nematology" (J. N. Sasser and W. R. Jenkins, eds.), pp. 267–296. Univ. of North Carolina Press, Chapel Hill, North Carolina.
Fairbairn, D., and Jones, R. N. (1956). *Can. J. Chem.* **34**, 192.
Fairbairn, D., and Passey, B. I. (1955). *Can. J. Biochem. Physiol.* **33**, 130.
Fairbairn, D., Wertheim, G., Harpur, R. P., and Schiller, E. L. (1961). *Exptl. Parasitol.* **11**, 248.
Fauré-Fremiet, E. (1913a). *Arch. Anat. Microscop. Morphol. Exptl.* **15**, 435.
Fauré-Fremiet, E. (1913b). *Compt. Rend. Soc. Biol.* **75**, 90.
Fernando, M. A. (1963). *Exptl. Parasitol.* **13**, 90.
Flury, F. (1912). *Arch. Exptl. Pathol. Pharmakol.* **67**, 275.
Fouquey, C., Polonsky, J., and Lederer, E. (1957). *Bull. Soc. Chim. Biol.* **39**, 101.
Fouquey, C., Polonsky, J., and Lederer, E. (1958). *Bull. Soc. Chim. Biol.* **40**, 315.
Fouquey, C., Polonsky, J., and Lederer, E. (1959.) *Bull. Soc. Chim. France* p. 803.
Ginger, C. D., and Fairbairn, D. (1966a). *J. Parasitol.* **52**, 1086.
Ginger, C. D., and Fairbairn, D. (1966b). *J. Parasitol.* **52**, 1097.
Giovannola, A. (1936). *J. Parasitol.* **22**, 27.
Goffart, H., and Heiling, A. (1962). *Nematologica* **7**, 173.
Gray, G. M. (1960). *Biochem. J.* **77**, 82.
Gutierrez, J., Williams, P. P., Davis, R. E., and Warwick, E. J. (1962). *Appl. Microbiol.* **10**, 548.
Hack, M. H., Gussin, A. E., and Lowe, M. E. (1962). *Comp. Biochem. Physiol.* **5**, 217.
Harrington, G. W. (1965). *Exptl. Parasitol.* **17**, 287.
Hirsch, G. C., and Bretschneider, L. H. (1937). *Cytologia (Tokyo) Fujii Jubilee Vol.* p. 424
Hobson, A. D. (1948). *Parasitology* **38**, 183.
Jacobsen, N. S., and Fairbairn, D. (1967). *J. Parasitol.* **53**, 355.
Jezyk, P. F., and Fairbairn, D. (1967a). *Comp. Biochem. Physiol.* **23**, 691.
Jezyk, P. F., and Fairbairn, D. (1967b). *Comp. Biochem Physiol.* **23**, 707.
Kornberg, H. L., and Elsden, S. R. (1961). *Advan. Enzymol.* **23**, 401.
Krusberg, L. R. (1961). *Nematologica* **6**, 181.
Laurie, J. S. (1958). *Exptl. Parasitol.* **8**, 188.
Lee, D. L. (1958). *Parasitology* **48**, 437.
Lee, D. L. (1962). *Parasitology* **52**, 241.

Lee, D. L. (1964). *Proc. Helminthol. Soc. Wash., D.C.* **31,** 285.
Linford, M. B. (1937). *Proc. Helminthol. Soc. Wash., D.C.* **4,** 41.
Lui, A., Coric, D., and Krvavica, S. (1964). *Vet. Arhiv* **34,** 84.
Lumsden, R. D., and Harrington, G. W. (1966). *J. Parasitol.* **52,** 695.
McCoy, O. R. (1929). *Am. J. Hyg.* **10,** 140.
Mandlowitz, S., Dusanic, D., and Lewert, R. M. (1960). *J. Parasitol.* **46,** 89.
Marzullo, F., Squadrini, F., and Taparelli, F. (1957). *Boll. Soc. Med.-Chir. Modena* **57,** 84.
Matsuhashi, M. (1966). *J. Biol. Chem.* **241,** 4275.
Matsuhashi, S., Matsuhashi, M., Brown, J. G. and Strominger, J. L. (1966a). *J. Biol. Chem.* **241,** 4283.
Matsuhashi, S., Matsuhashi, M., and Strominger, J. L. (1966b). *J. Biol. Chem.* **241,** 4267.
Meyer, F., Kimura, S., and Mueller, J. F. (1966). *J. Biol. Chem.* **241,** 4224.
Monteoliva, M. (1960). *Rev. Iberica Parasitol.* **20,** 573.
Mueller, J. F. (1929). *Z. Zellforsch. Mikroskop. Anat., Abt. Histochem.* **8,** 361.
Munnich, H. (1965). *Z. Parasitenk.* **25,** 231.
Neufeld, E. F., and Hassid, W. Z. (1963). *Advan. Carbohydrate Chem.* **18,** 309.
Passey, R. F., and Fairbairn, D. (1955). *Can. J. Biochem. Physiol.* **33,** 1033.
Passey, R. F., and Fairbairn, D. (1957). *Can. J. Biochem. Physiol.* **35,** 511.
Pflugfelder, O. (1949). *Z. Parasitenk.* **14,** 274.
Roberts, L. S., and Fairbairn, D. (1965). *J. Parasitol.* **51,** 129.
Rogers, W. P. (1941). *J. Helminthol.* **19, 35.**
Rogers, W. P. (1948). *Parasitology* **39,** 105.
Rogers, W. P., and Lazarus, M. (1949). *Parasitology* **39,** 302.
Rohde, R. A. (1960). *Proc. Helminthol. Soc. Wash., D.C.* **27,** 121.
Rothstein, M., and Cook, E. (1966). *Comp. Biochem. Physiol.* **17,** 683.
Saz, H. J., and Lescure, O. L. (1966). *Comp. Biochem. Physiol.* **18,** 845.
Schiller, E. L. (1965). *J. Parasitol.* **51, 516.**
Schwabe, C. W. (1957). *Am. J. Hyg.* **65,** 325.
Sheffield, H. G. (1964). *J. Parasitol.* **50,** 365.
Sivapalan, P., and Jenkins, W. R. (1966). *Proc. Helminthol. Soc. Wash., D.C.* **33,** 149.
Szwejkowska, G. (1929). *Bull. Intern. Acad. Polon. Sci., Classe Sci. Math. Nat.* **BI 4-89.**
Thompson, M. J., Mosettig, E., and von Brand, T. (1960). *Exptl. Parasitol.* **9,** 127.
Tracey, M. V. (1958). *Nematologica* **3,** 179.
von Brand, T. (1934). *Z. Vergleich. Physiol.* **21,** 220.
von Brand, T. (1938). *J. Parasitol.* **24,** 445.
von Brand, T. (1939). *J. Parasitol.* **25,** 329.
von Brand, T. (1941). *Proc. Soc. Exptl. Biol. Med.* **46,** 417.
von Brand, T. (1952). "Chemical Physiology of Endoparasitic Animals." Academic Press, New York.
von Brand, T. (1957). *Z. Tropenmed. Parasitol.* **8,** 21.
von Brand, T. (1966). "Biochemistry of Parasites." Academic Press, New York.
von Brand, T., Weinstein, P. P., Mehlman, B., and Weinbach, E. C. (1952). *Exptl. Parasitol.* **1,** 245.
von Kemnitz, G. (1912). *Arch. Zellforsch.* **7,** 463.
Weinland, E. (1901). *Z. Biol.* **42,** 55.
Williams, P. P., Gutierrez, J., and Davis, R. E. (1963). *Appl. Microbiol.* **11,** 260.
Wilson, P. A. G. (1965). *Exptl. Parasitol.* **16,** 190.

CHAPTER 7

Nitrogenous Components and Their Metabolism: Acanthocephala and Nematoda

W. P. Rogers

I. Introduction

The Acanthocephala form a relatively homogenous group. All species are parasitic, the adults in vertebrates and larval stages in arthropods.

The nitrogen metabolism of this group has not been extensively examined, and even this work has been restricted to adults of a few species.

On the other hand the Nematoda, though many parasitize plants and animals, inhabit a variety of terrestrial and aquatic environments. Research on this group, though extensive, deals largely with parasites of domestic plants and animals, or of species which can be maintained easily in the laboratory. Only one free-living species, *Caenorhabditis briggsae,* has been examined in detail, and this species has been maintained in culture for many generations during which selection of a particular strain may well have occurred.

It should be emphasized, therefore, that the picture of nitrogenous metabolism of both the Acanthocephala and Nematoda presented here is very limited; for the Acanthocephala because so few species have been examined; for the Nematoda because the species which have been examined are probably highly specialized.

An exhaustive review of the literature has not been attempted. In particular, data obtained from analyses, which are not obviously related to our present understanding of nitrogen metabolism, have been summarized briefly.

II. Proteinases in Nematodes

Nematodes have a functional alimentary canal and most interest has been focused on digestive enzymes (Barrington, 1962). Except for proteinases concerned in extracorporeal digestion, however, there is no unequivocal evidence that nematodes produce proteinases for extracellular digestion because extracts of the tissues of the alimentary canal have been used rather than the material actually secreted into the lumen of the gut. For convenience, enzymes in extracts of tissues of the alimentary canal are discussed here under the heading "digestive proteinases." Enzymes in extracts of whole organisms, though they may include digestive proteinases, are discussed as "tissue proteinases." Proteinases which may be concerned in the hatching of eggs and in the molting of juveniles are discussed under the heading "developmental proteinases."

A. Digestive Proteinases

Many nematodes ingest solid material. Indeed some juveniles have been maintained in media in which nutrients must have been obtained from whole bacterial cells (McCoy, 1929a,b; Lapage, 1933). Even parasitic forms which live in the gut ingest solid materials (W. P. Rogers, 1940a). That this material is digested in the intestine is largely inferred, but for some species, histological evidence points strongly in this direction (Hsü, 1938; Hsü and Li, 1940) and the breakdown products of hemoglobin have been found in the intestines of parasites which feed on blood (W. P. Rogers, 1940b; M. H. Smith and Lee, 1963).

Secretory activity in the intestine of *Ascaris lumbricoides* has been described (Carpenter, 1952; Lee, 1962). The results indicate that holocrine secretion, in which parts of cells containing secretory material are passed into the lumen of the intestine where they disintegrate, occurs in the anterior section of the alimentary canal. Merocrine secretion, in which granules coalesce within the cell and pass through the cell boundary into the lumen, has also been described.

Though there is some evidence that cells in the intestine of nematodes may take up solid particles, especially in the posterior part of the intestine (Enigk, 1938), evidence for intracellular digestion is lacking.

Much of the work on digestive enzymes has been carried out using *Ascaris lumbricoides*. The interpretation of the results is difficult because this parasite ingests the intestinal contents of the host and it is possible that the proteinases could have come from the host. However, the pH optima of the digestive enzymes found in the parasite in most cases differed from those of the host. And Carpenter (1952) showed that *Ascaris lumbricoides* could be maintained in saline solution *in vitro* without diminishing its enzymic activity—an unlikely result if the enzymes were of external origin. A study of these proteinases by the method of gel electrophoresis would be profitable.

Extracorporeal digestion is a common feature of nematodes which parasitize plants and it probably occurs also in some predatory nematodes. Secretions which promote the breakdown of connective tissue are also secreted externally by the juvenile stages of parasites which invade the host through the skin. Proteinases which presumably originate in the pharyngeal glands are sometimes secreted by these organisms (Myuge, 1957; Lewert and Lee, 1957). It is possible that the proteinase present in the pharynx of adult *Ancylostoma caninum* may be concerned in extracorporeal digestion (Thorson, 1956).

1. *Endopeptidases*

W. P. Rogers (1941) found that crude extracts of the intestines of *Ascaris lumbricoides* and *Strongylus edentatus* hydrolyzed casein and gelatin at an optimum of pH 6.2. The results obtained with *Ascaris* were confirmed and extended by Savel (1955a) and Carpenter (1952) who also showed that proteolytic activity was not enhanced by cyanide. Studies with frozen sections of the alimentary canal indicated that most of the activity occurred in the anterior part of the intestine. Histological studies showed that this region would have a high secretory activity (Carpenter, 1952).

The proteolytic activity of extracts of the intestine of *Strongylus edentatus* with casein as the substrate, was found to be 12.5 to 40.9 times greater than in extracts of the intestine of *Ascaris lumbricoides* (W. P. Rogers, 1941). This may be related to the feeding habits of these parasites. *Ascaris lumbricoides* lies in the host's intestinal contents and feeds by swallowing fluid and small particles (Fairbairn, 1957) so presumably it usually takes in material which has been partly digested by the host, whereas *Strongylus edentatus* is found firmly attached to the mucosa of the large intestine of the host and ingests blood and solid tissues (W. P. Rogers, 1940a).

TABLE I
ENDO- AND EXOPEPTIDASE ACTIVITY IN INTESTINAL EXTRACTS OF ADULT NEMATODES

Sources of enzyme	Substrates	Optimum pH	Activators	Inhibitors	References
Ascaris lumbricoides	Casein				W. P. Rogers (1941)
	Gelatin				Carpenter (1952)
		6.0–6.2	*not* KCN		
	Albumin				Savel (1955a)
	Hemoglobin				
	Glycylglycine	6.0	Co^{++}	Zn^{++}, Mn^{++}	Carpenter (1952)
	DL-Leucylglycine	6.0	Mn^{++}	cysteine	
	DL-Leucylglycylglycine	6.0	*not* Mn^{++}		
Strongylus edentatus	Casein				
	Gelatin	6.0			W. P. Rogers (1941)
	Albumin				
Leidynema appendiculata	Casein	6.0	*not* KCN		Lee (1958)
	Edestin				
	Glycylglycine				
	DL-Leucylglycine	6.0			
	DL-Leucylglycylglycine				Lee (1958)
	Glycyl-L-proline				

A proteinase similar to that found in *Ascaris* and *Strongylus* occurs in *Leidynema appendiculata*, a parasite of beetles. Lee (1958) found that the proteinase extracted from the intestine of this parasite hydrolyzed casein and edestin at a pH optimum of 6.0. Proteolytic activity was not enhanced by cyanide.

These results indicate that some nematodes which are parasitic in animals produce a digestive proteinase which differs from mammalian trypsin in its pH optimum, and which may differ from tissue cathepsins because it is not activated by cyanide (Table I). In this respect, however, it should be emphasized that these results were obtained with crude extracts which may have contained natural activators.

Proteolytic activity has been detected in media in which plant parasites *Ditylenchus allii*, *D. destructor*, and *Heterodera marioni* had been incubated for 48 hours (Myuge, 1957; Zinovev, 1957), but Krusberg (1960) who used whole homogenates of *D. dipsaci* and *D. triformis* obtained doubtful or negative results with hemoglobin as the substrate. It is possible, then, that the results obtained by Myuge and Zinovev were due to contaminating microorganisms.

A collagenase-like enzyme has been found in secretions of the infective juveniles of *Strongyloides ratti* (Lewert and Lee, 1957) which infects the host via the skin. The enzyme actively hydrolyzed hide powder, azocoll, cartilage, and azocartilage, but its effect on bovine native collagen was slight compared to that of *Clostridium welchii* toxin. The enzyme secreted by *Strongyloides* was similar to collagenase in that (1) it was not inhibited by ovomucoid or soy bean trypsin inhibitor, (2) it was inhibited by cysteine and ethylenediaminetetraacetic acid, Cu^{++}, Hg^{++}, and Co^{++}, and (3) it was activated by Mg^{++} and Ca^{++}.

2. *Exopeptidases*

It is probable that exopeptidases are secreted by pharyngeal glands and intestinal cells of some parasitic nematodes.

The most fruitful work has been done by Carpenter (1952) and Lee (1958). Carpenter used DL-leucylglycine, glycylglycine, DL-leucylglycylglycine and DL-alanylglycine as substrates and concluded that at least four peptidases were present in the alimentary canal of *Ascaris lumbricoides*. The hydrolysis of DL-leucylglycine was activated by Mn^{++} whereas the splitting of DL-leucylglycylglycine was not. Cysteine inhibited the hydrolysis of DL-alanylglycine and the presence of Co^{++} enhanced the hydrolysis of glycylglycine but Zn^{++}, Mn^{++}, and cysteine inhibited. Lee (1958) found that the peptidases which he extracted from the intestines of *Leidynema appendiculata* hydrolyzed a somewhat similar group of substrates (Table I).

It would appear, then, that nematodes have a number of digestive exopeptidases similar to those in mammals.

B. Tissue Proteinases

Proteinases in homogenates of free-living nematodes (*Panagrellus redivivus* and *Turbatrix aceti*), parasites of animals (*Trichuris muris*) and of plants (*Aphelenchoides ritzemabosi, Ditylenchus dipsaci* and *D. triformis*) have been examined (Table II).

With the exception of *D. triformis* and possibly *D. dipsaci* (Krusberg, 1960), these organisms hydrolyzed hemoglobin at low pH. Extracts of *Trichuris muris* attacked only a limited range of substrates. The hydrolysis of casein and hemoglobin by this parasite was enhanced in the presence of Mn^{++}, Co^{++}, and Mg^{++}. Ethylenediaminetetraacetic acid at

TABLE II
ENDO- AND EXOPEPTIDASES IN EXTRACTS OF WHOLE NEMATODES

Sources of enzyme	Substrates	pH	Activity	References
Panagrellus redivivus	Hemoglobin	2.0	+	Miller and Jenkins
(free-living)	Hemoglobin	7.5	+	(1964)
	Glycyl-L-proline	8.0	+	
	Glycylglycylglycine	8.0	+	
Turbatrix aceti	Hemoglobin	2.0	+	Miller and Jenkins
(free-living)	Hemoglobin	7.5	+	(1964)
Aphelencoides ritzemabosi	Hemoglobin	2.0	+	
(parasite of plants)	Hemoglobin	7.5	+	
	Glycyl-L-proline	8.0	+	
	Glycylglycylglycine	8.0	+	
Ditylenchus dipsaci	Hemoglobin	2.0	+	Miller and Jenkins
(parasite of plants)	Hemoglobin	8.0	+	(1964)
	Hemoglobin	7.5	?	Krusberg (1960)
Ditylenchus triformis	Hemoglobin	7.5	−	Krusberg (1960)
(parasite of plants)				
Trichuris muris	Casein	3.7–3.2	+	Nimmo-Smith and
(parasite of animals)	Hemoglobin	3.5	+	Keeling (1960)
	Egg albumin	2.9 and 4.6	−	
	Serum albumin	2.9 and 4.6	−	
	Serum globulin	2.9 and 4.6	−	
	Serum fibrinogen	2.9 and 4.6	−	

concentrations above 4×10^{-4} *M* caused increasing degrees of inhibition. Cysteine also inhibited but cyanide was without effect (Nimmo-Smith and Keeling, 1960).

The enzymes in *Trichuris muris* differ from mammalian cathepsins in several respects but it seems reasonable to suppose that they are concerned in the parasites' own protein metabolism rather than in the invasion of hosts' tissues (Nimmo-Smith and Keeling, 1960).

Activity at pH 6 was found in all the organisms which were examined except *Trichuris muris, Ditylenchus triformis,* and possibly *D. dipsaci*

(Nimmo-Smith and Keeling, 1960; Krusberg, 1960). Exopeptidases were found in all the organisms which were examined for activity of this sort (Miller and Jenkins, 1964; Table II).

C. Developmental Proteinases

In the course of development of nematodes, extracellular structures, e.g., egg shells and the outer cuticle of some juveniles which forms a loose "sheath" or protective covering, are broken down. There is evidence that enzymes are involved in these processes (W. P. Rogers, 1958, 1965) and as proteins are important components of these extracellular structures, proteinases may be concerned in their breakdown.

1. *The Hatching of Eggs*

W. P. Rogers (1958) suggested that a proteinase may be concerned in the hatching of eggs of *Ascaris lumbricoides*. No proteolytic activity could be detected in "hatching fluid" with dialyzed ox serum as the substrate, but amino-nitrogen was released from egg shells which were incubated in the hatching fluid. Most amino nitrogen was released at pH 6 which was the optimum for the chitinase in the hatching fluid, and it is possible that it was the action of this enzyme which caused the release of the amino nitrogen.

2. *Proteinases Concerned in Molting*

Leucine aminopeptidase was found to be responsible for the breakdown of the outer cuticle of "sheaths" of infective juveniles (W. P. Rogers, 1965). The enzymes found in the exsheathing fluids of *Haemonchus contortus* and *Trichostrongylus colubriformis* which had been stimulated to resume development hydrolyzed L-leucinamide and L-leucyl-β-naphthylamide at an optimum of pH 9.5. These enzymes, which required Mn^{++} or Mg^{++} as a cofactor, were similar to mammalian leucine aminopeptidase in their reaction to inhibitors (Table III). However, the leucine aminopeptidases secreted by the two species of juveniles differed from each other and from the leucine aminopeptidase from pigs' kidneys in the range of substrates they attacked (Table IV). Moreover, the ratio of L-leucinamide hydrolyzed/L-leucyl-β-naphthylamide hydrolyzed was much lower (3-17) for the enzymes from the juveniles than for the mammalian enzyme ($>$5000). In other respects the leucine aminopeptidases from the juveniles resembled the mammalian enzyme. Unfortunately their natural substrates, parts of the outer cuticle of the juveniles, which are themselves small in size, are unknown.

TABLE III
THE EFFECT OF INHIBITORS ON LEUCINE AMINOPEPTIDASE OF *Haemonchus contortus*

Inhibitors	Substrate: L-Leucyl β-naphthylamide (%)	L-Leucinamide (%)	Isolated sheaths of *H. contortus* (%)
Ethylenediamine tetracetate (0.01 *M*)	63	100	100
n-Butanol (0.1 *M*)	11	7	0
N-Methylmaleimide	8	8	?[a]
p-Chloromercuribenzoate	0	0	?[a]
Diisopropylfluorophosphate (0.001 *M*)	0	0	0
L-Leucine (0.05 *M*)	28	12	0

[a] The —SH inhibitors caused distortion of isolated sheaths and it was difficult to assess the degree of inhibition.

TABLE IV
SUBSTRATES ATTACKED BY EXSHEATHING FLUIDS AND LEUCINE AMINOPEPTIDASE[a]

Source of enzyme	Substrates: Isolated sheaths of *H. contortus*	Isolated sheaths of *T. colubriformis*	L-Leucinamide	L-Leucyl β-naphthylamide
H. contortus (exsheathing fluid)	+	−	+	+
T. colubriformis (exsheathing fluid)	−	+	+	+
Sus scrofa (purified from kidneys)	−	±	+	+

[a] From Rogers (1965).

III. The Absorption of Nitrogenous Substances in Nematodes

The presence of a functional alimentary canal in nematodes suggests that absorption of nitrogenous substances takes place through the cells of the gut in these animals. Though the cuticle of *Ascaris lumbricoides* is permeable to water, some ions and hydrophobic anthelmintics, it seems unlikely that appreciable amounts of nutrients are absorbed in this way (Fairbairn, 1957, 1960; Cavier and Savel, 1952; W. P. Rogers and Lazarus, 1949). This view is supported by the observations that free-living infective juveniles which did not feed took up dissolved gases, whereas many other small molecules did not penetrate (W. P. Rogers, 1966). Recent work, however, showed that absorption via the gut may be supplemented by uptake via the cuticle in *Ascaridia galli* (Weatherly *et al.*, 1963).

A. The Absorption of Amino Acids

Hankes and Stoner (1956, 1958) and Stoner and Hankes (1955, 1958) found that ^{14}C of glycine-2-^{14}C, DL-tryptophan-2-^{14}C, DL-alanine-2-^{14}C and DL-tyrosine-2-^{14}C appeared in the parasitic infective stage of *Trichinella spiralis* when it was exposed to the amino acids *in vivo* and *in vitro*. There is no doubt that the ^{14}C was absorbed as the amino acids *in vitro* but it is not clear if this was due to absorption via the cuticle or *per os*. Weatherly *et al.* assumed that the amino acids entered via the cuticle.

Decisive results demonstrating the uptake of DL-alanine-^{14}C via the cuticle have been obtained with *Ascaridia galli in vitro* (Weatherly *et al.*, 1963). In these experiments the uptake of the amino acid in normal adult parasites was compared with the uptake in specimens which had been coated with celloidin to prevent entry through the orifices of the alimentary canal. The results showed that absorption through the cuticle accounted for one-third to one-fourth of the alanine which entered the parasites. It should be noted, however, that the total amount of amino acid absorbed by any route in these experiments was probably very small.

A large number of experiments, especially those in which nematodes have been cultivated *in vitro*, show the amino acids are taken up both by free-living and parasitic forms (e.g., see Sayre *et al.*, 1963) but details of the kinetics of the absorption are not yet available.

B. The Absorption of Proteins

The probability that proteinases act on foodstuffs in the alimentary canal of nematodes suggests that proteins are not absorbed as such. The view of Hurlaux (1947) that horse hemoglobin may be absorbed and appear unchanged in the body fluid of *Parascaris equorum* lacks confirmation. Indeed, W. P. Rogers (1940b) and M. H. Smith and Lee (1963) obtained results which suggested that hemoglobin taken into the alimentary canal of nematodes which ingest the hosts' blood is rapidly denatured and the heme group freed.

A factor which promotes the growth of free-living *Caenorhabditis briggsae in vitro* is polypeptide in nature (Sayre *et al.*, 1961, 1963). It is not clear, however, whether this growth factor is actually absorbed by the nematodes. A protein or protein-bound substance enhances the uptake of glucose by *Ancylostoma caninum in vitro* (Fernando and Wong, 1964). Again it is possible that this substance is absorbed as such by the parasites.

C. The Absorption of Other Nitrogenous Substances

Though nematodes require a wide range of nitrogenous nutrients and growth factors, the mechanism by which they are taken up has rarely been examined. As shown by experiments with cyanocobalamin-^{60}Co (Zam *et al.*, 1961), it is probable that most of these substances are absorbed in the gut.

IV. The Absorption of Nitrogenous Substances in Acanthocephala

The lack of an alimentary canal in the Acanthocephala has led to an interest in the mechanisms by which nutrients are absorbed in this group. Studies on the fine structure of the body wall prompted the suggestion that pinocytosis may be involved; other work has shown that amino acids are taken up by active transport.

A. The Structure of the Body Wall

The structure of the body wall of adult *Polymorphus minutus* has been examined by Compton and Lee (1965); Nicholas and Mercer (1965) and Edmonds and Dixon (1966) have studied *Monoliformis dubius*.

Polymorphus is covered by an irregular deposit, the "epicuticle," which stains with uranyl acetate. The cuticle consists of a triple-layered membrane. It is pierced by numerous pores which are the openings of canals which pass, in a regular pattern, through the underlying homogenous layer, giving it a striped appearance. The canals apparently end in thin-walled vesicles underneath the striped layer. Compton and Lee (1965) suggested that the canals have a function in the absorption of nutrients.

Nicholas and Mercer (1965) found that *Moniliformis* contains vacuoles, tentatively identified as pinocytotic vacuoles, in the outer layers. These vacuoles are smaller than the canals and do not penetrate deeply into the tegument (Mercer, personal communication). Edmonds and Dixon (1966) have also described these vacuoles and commented on their possible function.

B. The Uptake of Amino Acids

Rothman and Fisher (1964) showed that adult *Moniliformis dubius* and *Macracanthorhynchus hirudinaceus* concentrated L-methionine against a concentration gradient. Moreover, the uptake of methionine was competitively inhibited by L-leucine, L-isoleucine, L-serine or L-alanine and vice versa. Using results obtained with cestodes, Read *et al.* (1963) demonstrated that the action of a mixture of amino acids on the uptake of a given amino acid would be predicted by the following equation.

$$v = \frac{V}{\frac{(K_t)}{(S)} + 1 + \frac{(K_t)(S_1)}{(K_{t_1})(S)} + \frac{(K_t)(S_2)}{(K_{t_2})(S)} + \cdots + \frac{(K_t)(S_n)}{(K_{t_n})(S)}}$$

where v = the velocity of transport, V = the maximum velocity when the transport loci are saturated, K_t = the transport constant, and S = the substrate concentration and S_1, S_2, . . . S_n = the concentration of the competing amino acids. Rothman and Fisher (1964) used K_i values (the inhibition constants determined experimentally) instead of K_t values in this formula and were able to predict accurately the effect of mixtures of the neutral amino acids on the uptake of methionine in the Acanthocephala. It appears, therefore, that processes governing the uptake of amino acids in the Acanthocephala are similar to those in the Cestoda. These results also show an interesting analogy to those obtained with Ehrlich ascites tumor cells (Christensen, 1960).

Edmonds (1965) has extended these studies to show that the uptake of L-leucine by *Moniliformis dubius* was inhibited in the presence of DL-valine, DL-serine, or DL-methionine.

V. The Excretion of Nitrogenous Substances by Nematodes

Products which from their nature may be regarded as resulting from the catabolism of nitrogenous substances, accumulate in media in which nematodes have been maintained. Though it is doubtful as to which are the most important anatomical routes by which this excretion is achieved, the nature of the major excretory products has been determined for a variety of these organisms in different stages of their development (Weinstein, 1960).

It must be emphasized that many of these experiments have been carried out under abnormal conditions. This applies especially to experiments with adult parasites which have been maintained in simple media in which their survival is limited. Even the physical conditions under which parasites are kept may have profound effects on the excretion of nitrogenous substances (Cavier and Savel, 1954a). It has been possible, however, to use more physiological conditions for studies with juveniles surrounded by normal fluids within egg shells, or with juveniles which do not feed.

Substances found in media in which nematodes have been maintained may include both excretory and secretory substances. For this reason some of the peptides which have been classed as excretory may have been secreted by pharyngeal glands for extracorporeal digestion or to facilitate the invasion of tissue by parasitic forms (see, for example, Lewert, 1958; Mandlowitz *et al.*, 1960; Goffart and Heiling, 1962). Many other secretions, such as those concerned with reproduction and the freeing of eggs, and those produced during molting must also be passed into the medium.

In experiments in which large numbers of small organisms have been used, excretory and secretory products could be contaminated with substances released from dying and disintegrating organisms. This contamination would be small but it might give results which could be confusing.

A. Total Nitrogen Excreted

The total nitrogen excreted by several species of nematodes has been measured, usually under fasting conditions so the animals were not necessarily in a state of nitrogen balance (Fairbairn, 1960). For the nematodes that are parasitic in animals the total nitrogen excreted varied from 0.39 mg./gm. wet weight/24 hours for the large adult parasitic form *Ascaris lumbricoides* (Savel, 1955b), to 2.8 mg./gm. wet weight/24 hours for the small parasitic juvenile of *Trichinella spiralis* (Haskins and Weinstein, 1957). The smaller forms produced relatively large amounts of nitrogenous material (Table V). Thus Fairbairn (1960) calculated that infective juveniles of *Trichinella spiralis* would have to catabolize 12.5% of their total body protein to account for the amount of nitrogen excreted in 24 hours. This represents a very active metabolism even compared to the relatively small parasite like *Nematodirus* (W. P. Rogers, 1952). It is the more remarkable because infective juveniles of *Trichinella spiralis* may be regarded as resting stages. They do not obviously feed on host tissue though they may take in small amounts of nutrients in solution (Hankes and Stoner, 1956, 1958; Stoner and Hankes, 1955, 1958), and they do not grow *in vivo* over long periods of time. Moreover, they differ from the *Nematodirus* adults used in similar experiments because the adults would be expected to have an extensive protein metabolism concerned in the production of eggs and sperm.

The explanation of the unusually high nitrogen catabolism in the juveniles of *Trichinella spiralis* might lie in the method used for preparing the organisms for experiment (von Brand *et al.*, 1952) which involved the digestion of infected tissue in artificial gastric juice at 37°. This process might have acted as a partial stimulus for the resumption of development. If this was so, it would be expected that the initiation of changes in morphology and physiology would require an active metabolism of protein.

This suggestion is supported by results obtained with infective juveniles of *Haemonchus contortus* (W. P. Rogers, 1963). During the first hour after the juveniles had been stimulated to resume development, the total nitrogen passed into the medium was increased by 104% above

TABLE V
TOTAL NITROGEN PRODUCED BY NEMATODES EXPRESSED AS MG./100 GM. WET WEIGHT/24 HOURS

Organism	N_{total}	References
Ditylenchus triformis (adults and juveniles) aerobic	9– 18	Myers and Krusberg (1965)
Ascaris lumbricoides (adults)		Savel (1955a)
aerobic	39	
anaerobic	41	
Ascaridia galli (adults)		
aerobic	29– 41	W. P. Rogers (1952)
anaerobic	32– 42	
Nematodirus spp.[a] (adults)		
aerobic	123–157	W. P. Rogers (1952)
anaerobic	148–193	
Trichinella spiralis (infective juveniles)		
aerobic	280	Haskins and Weinstein (1957)
Haemonchus contortus[b] (infective juveniles)		
aerobic		
stimulated	192	W. P. Rogers (1963)
unstimulated	125	

[a] A mixture of *N. spathiger* and *N. filicollis*.
[b] Sheaths removed artificially: calculated from 1-hour experiments, assuming that wet weight = dry weight × 5.
Experiments with *D. triformis* were conducted at 30°; other experiments at 37°.

that produced by unstimulated organisms. This increase was partly due to breakdown products released by the fracture of the outer cuticle which was the first obvious evidence of the resumption of development. But even when these breakdown products were not present, the increase in the total nitrogen produced due to stimulation was about 50% (Table V).

It is reasonable, therefore, to regard the high total nitrogen excreted by juveniles of *Trichinella spiralis* and *Haemonchus contortus* as being due to special physiological circumstances. But even if these results are ignored, it is clear that nitrogen metabolism of the nematodes shown in Table V, as judged by the total nitrogen excreted, is generally much higher than in most other organisms. For example, the fasting Crustacea studied by Dresel and Moyle (1950) ranged from the low level of the terrestrial *Oniscus asellus,* 0.03 mg./gm. wet weight/24 hours to that of the estuarine *Gammarus zadduchii,* 0.60 mg./gm. wet weight/24 hours.

The reason for the high nitrogen metabolism of nematodes is obscure. But it is notable that the results obtained with *Ditylenchus triformis* indicate a much lower metabolism than in other species of similar size (Table V). This difference may be attributed, at least in part, to the differences in temperature at which the organisms were maintained. Though many nematodes are adapted to live at 37°, it is possible that a highly accelerated nitrogen metabolism is the outcome, and that parasites living in an environment containing plenty of nutrients may be able to maintain this active metabolism successfully.

A comparison of nitrogen excretion of parasites of warm-blooded hosts and free-living forms over a range of temperatures might throw light on this problem.

B. The Nature of Nitrogenous Excretory Products

1. *Ammonia*

Nematodes are predominantly ammonotelic; ammonia may account for 27–71% of the nitrogen excreted by different species (see Tables VI and VII). This might be expected because these organisms live in a watery medium. But the medium in which parasitic forms live may be hypotonic to the parasites' body fluids (Harpur and Popkin, 1965) and the embryos are usually enclosed in an egg shell which is impermeable except to some dissolved gases (Fairbairn and Passey, 1955). It is not surprising, then, that the one species which has been examined in this respect (Cavier and Savel, 1954a) showed a marked capacity for facultative ureotelism.

2. *Urea*

Under normal circumstances *in vitro* adult parasitic nematodes produce appreciable amounts of urea which may contribute 4–15% of the total nitrogen excreted. However, Cavier and Savel (1954a) found that when access to water over the body surface of *Ascaris lumbricoides* was restricted by placing the adult parasite in a U-tube with the free ends in saline so that water could be ingested and ejected through the alimentary canal, the partition of nitrogenous compounds in the excreta was changed. The ammonia nitrogen fell from 70 to 27% and the urea nitrogen rose from 7 to 51%.

TABLE VI
PARTITION OF SOLUBLE NITROGENOUS SUBSTANCES
EXCRETED BY NEMATODES *in Vitro*

Organism	Substances excreted as nitrogen, percent of total nitrogen						References
	Ammonia	Urea	Peptide	Amino acid	Amine	Uric acid	
Ditylenchus triformis (adults and juveniles)							
Aerobic	39	0	+	28	+	0	Myers and Krusberg (1965)
Ascaris lumbricoides (adults)							
Aerobic	69	7	21	+	−	0	Savel (1955a)
Anaerobic	71	6	18	+	−	0	Cavier and Savel (1954a)
In U-tube	27	51	19	−	−	0	
Ascaridia galli (adults)							
Aerobic	56	12	15	+	−	0	W. P. Rogers (1952)
Anaerobic	59	15	15	+	−	0	
Nematodirus spp. (adults)							
Aerobic	42	14	35	+	−	3	
Anaerobic	29	4	35	+	−	4	
Trichinella spiralis (infective juveniles)							
Aerobic	33	0	21	28	7	0	Haskins and Weinstein (1957)

TABLE VII
OCCURRENCE OF SOLUBLE NITROGENOUS SUBSTANCES
IN THE EXCRETA OF NEMATODES

Organism	Ammonia	Urea	Peptide	Amino acid	Amine	Uric acid	References
Ascaris lumbricoides							
Infective, "ovic-juveniles"	−	−	−	−	+	−	Weinstein and Haskins (1955)
Nippostrongylus brasiliensis							
Infective, free-living juveniles	+	0	−	−	+	−	Haskins and Weinstein (1957)
Caenorhabditis briggsae							
Free-living	+	0	−	+	0	0	Rothstein (1963)

Facultative ureotelism of this magnitude is rarely found in any organism. Changes in the nature of the major nitrogenous excretory product are often associated with changes in life cycles which involve, for example, movements from an aquatic to a terrestrial environment, e.g., in *Bufo* (Munro, 1949). Facultative ureotelism as the result of deprivation of water unassociated with specific developmental changes also occurs (H. W. J. Smith, 1930; Delaunay, 1934; Bahl, 1945). The mechanism by which these changes are regulated is unclear, and the marked changes which occur in *Ascaris* indicate that this animal would be useful for further studies.

3. *Uric Acid*

Uric acid is not an important excretory product of nematodes though it seems that *Nematodirus* excretes small amounts (W. P. Rogers, 1952).

4. *Amino Acids and Peptides*

Amino acids and peptides are excreted by free-living and parasitic nematodes, sometimes in such large amounts that they account for 49% of the total nitrogen excreted. This figure was obtained in studies with infective juveniles of *Trichinella spiralis* (Table VI) and, as mentioned earlier, may have been due to special developmental changes occurring in the organisms when they were being examined. Certainly the nitrogenous material passed into the medium by infective juveniles of *Haemonchus contortus* when they were stimulated contained at least one protein, leucine aminopeptidase, which was concerned in developmental changes (W. P. Rogers, 1965). However, the excretion of amino acids in other nematodes indicates that even under normal circumstances these potentially valuable metabolites may be lost in considerable amounts.

The nature of the amino acids excreted by nematodes gives no clear indication of the mechanism involved in their loss. Thus the following amino acids have been found in the excretions of *Ascaris lumbricoides, Ascaridia galli,* and two species of *Nematodirus:* leucine, phenylalanine, alanine, valine, proline, and aspartic and glutamic acids (W. P. Rogers, 1955). The fluids in which *Trichinella spiralis* was incubated lacked aspartic acid, but in addition to those produced by the ascarids, serine, glycine, and methionine were found (Haskins and Weinstein, 1957). A wide range of "essential" and nonessential amino acids was found in the medium in which the free-living nematode *Caenorhabditis briggsae* was incubated (Rothstein, 1963, 1965; Nicholas *et al.*, 1960). In some of these experiments a complete peptone medium was used; moreover, Rothstein carefully examined and rejected the possibility that the amino acids were freed as the result of damage to the worms. It seems that the excretion of amino acids is an authentic physiological feature of nematodes.

The parasites of plants *Ditylenchus triformis*, *D. dipsaci*, *D. myceleophagus*, *Meloidogyne incognita*, and *Pratylenchus penetrans*, maintained in a medium of 1% aqueous glucose, discharged a variety of

TABLE VIII
AMINO ACIDS AND AMIDES FOUND IN THE MEDIUM (1% AQUEOUS GLUCOSE) IN WHICH SURFACE STERILIZED NEMATODES HAD BEEN INCUBATED AT 30° FOR 24 HOURS[a]

Amino acid	*Ditylenchus triformis*	*Ditylenchus dipsaci*	*Ditylenchus myceliophagus*	*Pratylenchus penetrans*	*Meloidogyne incognita*
Alanine	+	+	+	+	+
Arginine	+	+	+	−	+
Asparagine	+	+	+	−	+
Aspartic acid	+	+	+	+	+
Citrulline	±	±	±	−	±
Cysteine	+	+	−	−	−
Cysteic acid	+	+	−	−	−
γ-Aminobutyric acid	+	±	−	−	±
Glutamic acid	+	+	+	+	+
Glutamine	+	+	−	−	+
Glycine	+	+	+	+	+
Histidine	+	−	−	−	+
Hydroxyproline	+	−	−	−	−
Isoleucine or leucine	+	+	+	−	+
Lysine	+	+	+	−	+
Methionine	−	+	+	−	−
Methionine sulfoxide	+	+	+	+	+
Ornithine	+	+	+	+	+
Phenylalanine	+	+	−	−	+
Proline	+	−	−	−	+
Serine	+	+	+	+	+
Threonine	+	+	+	+	+
Tyrosine	+	+	+	−	±
Valine	+	+	+	−	±

[a] From Myers and Krusberg (1965).
[b] Key: + = present; − = not detected; ± = probably present.

amino acids into the medium (Table VIII). Myers and Krusberg (1965), largely on the basis of their work on *D. triformis*, suggested that the synthesis of amino acids and amides might serve for the detoxification of ammonia. In support of this view they have the following evidence:

1. *D. triformis* synthesized amino acids and amides when incubated in media containing acetate-2-^{14}C or glucose-U-^{14}C.

2. *D. triformis* discharged over 20% of its free amino acids and amides in 24 hours while held under essentially fasting conditions, and 21% of the asparagine, 10% of the aspartic acid, 13% of the glutamic acid, 22% of the glutamine, 47% of the glycine and 18% of the ornithine which were labeled with the ^{14}C were excreted.

3. Urea was not found in *D. triformis.*

Results shown in Table VI support this suggestion. Thus juveniles of *Trichinella spiralis* which, like *Ditylenchus triformis,* did not excrete urea, produced relatively large amounts of amino acids whereas the ureotelic forms must have produced relatively small amounts.

A number of nematodes release peptides into the medium. These substances have not been clearly characterized; they may be secretions from the intestine or body surface. W. P. Rogers (1955) found that adult *Ascaris lumbricoides, Ascaridia galli,* and *Nematodirus* spp. each produced three peptides. Judged by R_f values in two different solvents, the peptides produced by these parasites were similar and according to the size of the spots they were produced in the same relative amounts. Moreover, acid hydrolysis of the peptides from each parasite released the same amino acids—leucine, valine, alanine, and glutamic and aspartic acids. The similarity of the peptides produced by each species suggests that they result from the same physiological processes.

5. *Amines*

An interesting feature of nitrogen metabolism in nematodes is the excretion of amines (Haskins and Weinstein, 1957; Weinstein and Haskins, 1955; Myers and Krusberg, 1965). These substances have not been reported as occurring in the excretions of adult parasites except possibly in some parasites of plants, and it is probable that they are not excreted by the free-living form *Caenorhabditis briggsae* (Rothstein, 1963). They are present, however, in appreciable amounts in the excretions of the infective juveniles of *Trichinella spiralis* (7% of the total nitrogen excreted), *Ascaris lumbricoides,* and *Nippostrongylus brasiliensis* (Table VII).

The amines excreted by *Nippostrongylus brasiliensis* were ethylenediamine, cadaverine, ethanolamine, methylamine, propylamine, and butylamine. In addition, juveniles of *Trichinella spiralis* produced ethylamine, amylamine, heptylamine and 1-amino-2-propanol. Allylamine was tentatively identified (Haskins and Weinstein, 1957). The infective juveniles of *Ascaris lumbricoides,* which are enclosed within the egg shell, did not produce amylamine, heptylamine, and ethylenediamine; otherwise, these juveniles produced the same amines as *Trichinella spiralis.* There seems little doubt that amines are produced by these organisms under normal physiological conditions because they were detected in the fluid surrounding the juveniles of *Ascaris lumbricoides* enclosed by the egg shells.

It seems that *Ditylenchus triformis* (a mixture of adults and juveniles) produces a number of secondary aliphatic amines (Myers and Krusberg, 1965). None of these were positively identified. Primary aliphatic amines were not detected.

The origin of amines excreted by nematodes is obscure though it is clear that some of them could be produced by the decarboxylation of the appropriate amino acids.

6. *Other Nitrogenous Substances*

Other possible excretory products such as creatine and allantoin have been sought in media in which nematodes have been maintained. There is no evidence that these substances are produced in appreciable amounts. Myers (1963) and Myers and Krusberg (1965) found amides in the medium in which plant parasitic nematodes had been incubated (Table VIII). Though xanthine has not been identified as an excretory product, it occurs as the "rhabditin" inclusions in the intestinal cells of *Rhabditis strongyloides* (Whittaker, 1963).

C. The Production of Nitrogenous Excretory Products

There is evidence that ammonia and urea are produced by conventional biochemical mechanisms in nematodes but the particular processes which lead to the excretion of amino acids, amines, and peptides have not been examined. Nor has the synthesis of uric acid, which may be excreted by one parasite, been investigated.

1. *The Production of Ammonia*

Ammonia, the immediate end product of the catabolism of nitrogenous substances, is formed by the deamination of a variety of substrates (Cohen and Brown, 1960). A number of nematodes, like some invertebrates and unlike vertebrates (Florkin, 1949), contain urease, so that the ammonia they excrete may be formed directly, or indirectly from urea.

a. Urease in Nematodes. Urease has been found in the parasites of animals—*Ascaridia galli, Nematodirus* spp., *Ascaris lumbricoides,* and, in smaller amounts, in *Haemonchus contortus.* Activity was greatest in homogenates of the intestine of *Ascaridia galli* and *Ascaris lumbricoides* but high activity was also obtained in the perienteric fluid of *Ascaris* (W. P. Rogers, 1952; Savel, 1955b). Optimum activity was obtained at pH 6.0 (Savel, 1955b).

b. Amino Acids and Ammonia Production. The addition of L-amino acids to homogenates of nematodes gave increased production of ammonia, presumably due to the action of L-amino acid oxidases. Thus ammonia in homogenates of *Nematodirus* spp. and *Ascaridia galli* was increased when alanine, aspartic acid, glutamic acid, or arginine was added (W. P. Rogers, 1952). Similar results were obtained with homogenates of the intestine of *Ascaris lumbricoides*. Histidine, leucine, methionine, tyrosine and valine also served as substrates for the production of ammonia (Savel, 1955b). Arginine, aspartic and glutamic acids were

TABLE IX
PRODUCTION OF UREA AND AMMONIA FROM PURINES[a]

Substrate added	Small specimens		Large specimens	
	Ammonia N	Urea N	Ammonia N	Urea N
Homogenates of *Ascaridia galli* (Rogers, 1952)				
None	0.19	0.04	0.16	Trace
Guanine	0.19	0.03	0.10	Trace
Adenine	0.22	0.04	0.22	Trace
Xanthine	1.58	0.05	0.16	Trace
Uric acid	0.37	0.68	0.16	Trace
Allantoin	0.51	0.91	0.16	Trace
Homogenates of the intestine of *Ascaris lumbricoides* (Savel, 1955b)				
None	0.32		0.29	
Guanine	0.34		0.31	
Adenine	0.42		0.37	
Xanthine	0.47		0.32	
Uric acid	0.78		0.36	
Allantoin	0.92		0.34	

[a] Mg./gm. dry weight/hour.

most effectively deaminated by all the species of nematodes which were examined.

c. Purines and Ammonia Production. The addition of some purines and their derivatives to homogenates of *Ascaridia galli* and of the intestine of *Ascaris lumbricoides* gave increased production of ammonia and urea (Table IX). The range of substrates attacked and the amounts of ammonia and urea produced were much greater in immature adults than in fully mature specimens. In immature specimens of *Ascaridia galli* 5′-adenylic acid was most effectively deaminated and urea production was greatest when uric acid or allantoin was added (W. P. Rogers, 1952). *Ascaris lumbricoides* gave similar results (Cavier and Savel,

1954b,c) though presumably the capacity for deaminating adenine and for converting xanthine to uric acid was somewhat greater (Table XI).

No satisfactory explanation of the loss of the capacity to catabolize purines in mature adult nematodes has been advanced.

2. *The Production of Urea*

The increased formation of urea in the presence of added uric acid or allantoin suggests the presence of uricooxidase, allantoinase, and allantoicase in some species of nematodes. There is also evidence that

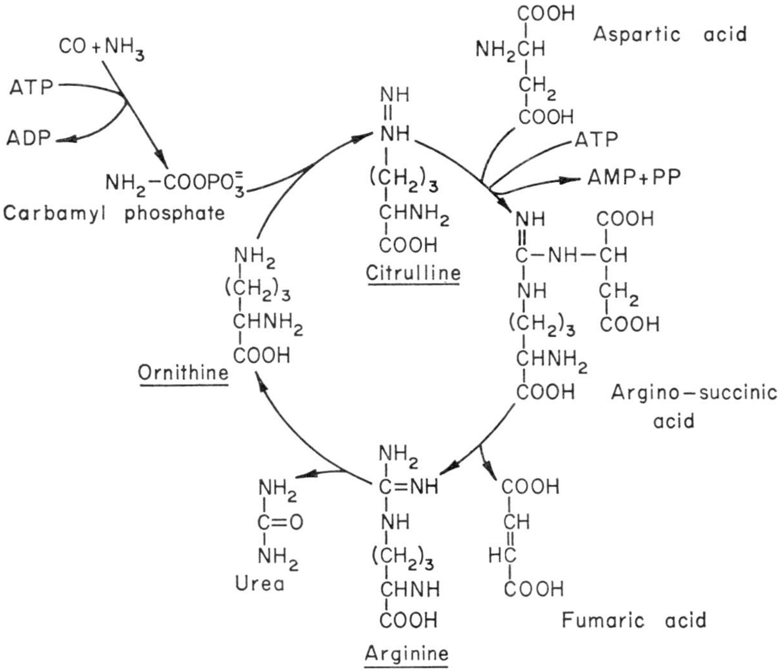

FIG. 1. The ornithine cycle. The compounds which are underlined have been found to increase urea production in some nematodes.

urea is formed by the ornithine cycle of Krebs and Henseleit (W. P. Rogers, 1952; Cavier and Savel, 1954b).

The basic features of this cycle are: the incorporation of ammonia in ornithine to form citrulline; the condensation of citrulline with aspartic acid to form the guanidino group of arginine which is split by arginase to replace the ornithine and to free the urea (Fig. 1). The following observations suggest that a similar biochemical process functions in some nematodes.

(1) Arginase activity has been detected in the tissues of *Ascaridia galli, Nematodirus* spp., and *Ascaris lumbricoides* where activity was greatest in homogenates of the intestine.

(2) As in preparations from vertebrate liver, arginase activity in nematodes was promoted by Co^{++}.

(3) Urea production was increased by the addition of citrulline or ornithine.

Not all species of nematodes produce urea (Tables VI, VII) and it is probable that a functional ornithine cycle is absent from many species. In the others ammonia produced by the deamination of a variety of L-amino acids and amines is probably at least partly metabolized via the ornithine cycle. Transaminations leading to the formation of aspartic acid could also lead to the production of urea.

A relation between the mechanism of energy production and the output of urea in nematodes was suggested by results obtained by W. P. Rogers (1952). Under anaerobic conditions *Nematodirus* spp., which might be expected to be more dependent on aerobic mechanisms, produced less urea. On the other hand, *Ascaridia galli,* which lives in an environment more deficient in oxygen, was not affected by anaerobic conditions.

3. *The Production of Amino Acids*

The relation between excreted amino acids and nutrients in the medium has been examined in free-living and parasitic nematodes (e.g., see Rothstein, 1963, 1965; Myers and Krusberg, 1965).

A remarkable feature of nitrogen metabolism of the nematodes which have been examined in this respect, *Caenorhabditis briggsae* and *Ditylenchus triformis,* is the degree to which externally provided substrates are incorporated in the amino acids which are excreted. Indeed, in some experiments the incorporation of radioactivity from labeled nutrients into the excreted amino acids was greater than in the body components of the organisms. And *Caenorhabditis briggsae,* even when cultured in a complete peptone medium, excreted relatively large amounts of labeled amino acids.

Rothstein and Tomlinson (1962), Rothstein (1963, 1965), and Nicholas *et al.* (1960) have examined the amino acids produced by *Caenorhabditis briggsae* from labeled substrates in the medium. A wide variety of "essential" and nonessential amino acids were secreted (see Section VI).

With *Ditylenchus triformis,* ^{14}C from acetate-2-^{14}C appeared in a number of amino acids and in asparagine and glutamine (see Section I).

^{14}C from labeled glucose was not so freely incorporated into amino acids in the body or into those which were excreted.

Rothstein (1963) has offered an explanation of the low proportion of isotope incorporated into the body compared to that excreted. He suggested that nematodes, lacking discrete metabolic organs such as liver and kidneys, carry out many metabolic processes in the cells of the gut. The constant supply of substrate in the ingesta would be continuously converted into other substances, which, in local excess, would be returned to the lumen of the gut and subsequently excreted. In support of this hypothesis, he pointed out that urease and arginase, and a number of other enzymes, have been found chiefly in the intestine of *Ascaris lumbricoides*. This suggestion implies that there is a low rate of exchange between cells of the intestine, the perienteric fluid, and cells of other tissues.

As pointed out by Myers and Krusberg (1965) both *Caenorhabditis briggsae* and *Ditylenchus triformis* do not excrete urea and so they probably lack the ornithine cycle. The secretion of amino acids in these organisms, the authors suggest, may serve as a means for the detoxification of ammonia. Rothstein (1965) also considered this possibility but made the point that at least some nematodes secrete intermediates as well as "end products" into the medium. As in microorganisms, nematodes may use the intermediates for further processes in metabolism. In this way even the detoxification of ammonia could lead to the production of useful metabolites.

4. *The Production of Amines*

Amonia acid decarboxylases are present in the tissues of nematodes (Cavier and Savel, 1954d). However, amino acids could not serve as substrates for the formation of amines such as ethylenediamine, and amyl-, heptyl- and allylamines which are excreted by some species (Haskins and Weinstein, 1957). Moreover, there is little correlation between amino acid decarboxylases which have been found in nematodes and the nature of amines excreted (see Section VI). It seems reasonable to suppose, then, that mechanisms other than the decarboxylation of amino acids may be responsible for most, if not all, the amines produced by nematodes.

VI. The Metabolism of Amino Acids in Nematodes

The high egg production of nematodes, the rapid growth and anatomical reorganization which takes place at times during their life cycles, imply an extensive metabolism of amino acids and proteins. With the exception of the work by Fairbairn and his co-workers (see the review by Fairbairn, 1960), Cavier and Savel (reviewed by Savel, 1955b), and Rothstein (1963, 1965) little systematic work has been carried out in this area and few generalizations can be made. Results obtained with particular species, however, already indicate unusual features in the metabolism of amino acids in these organisms.

Even the role of amino acids and proteins in the energy metabolism of nematodes is unclear. Though nitrogen catabolism is pronounced under fasting condition, this may not occur under normal circumstances. And, although protein reserves occur in a variety of nematodes, there is no clear evidence that they serve as a source of energy (Fairbairn, 1960) except in infective juveniles of *Nippostrongylus brasiliensis* (Wilson, 1965).

A. Free Amino Acids

The free amino acids in a number of species have been examined, e.g., in juveniles of *Necator americanus* (Desowitz, 1962) and *Nippostrongylus brasiliensis* (Friedman and Kagan, 1958), in various tissues of *Ascaridia galli* (Monteoliva Hernández, 1962), and in the perienteric fluid and ovaries of *Ascaris lumbricoides* (Cavier and Savel, 1954e; Pollak and Fairbairn, 1955a; Salmenkova, 1962; Benediktov, 1962). Some of these results are summarized in Table X.

B. The General Metabolism of Amino Acids

Little information on the regulation of metabolism and the mobilization of amino acids in nematodes is available. Presumably, amino acids are absorbed or synthesized in the intestine and passed into the perienteric fluid. It appears, however, that movement into the perienteric fluid is restricted, at least under some circumstances, in *Caenorhabditis briggsae* (see Section V).

The possibility that the perienteric fluid provides a pool of amino acids on which other tissues may draw needs examination. The perienteric fluid certainly contains a variety of amino acids and presumably tissues bathed by the fluid may take up these amino acids. But there is sound evidence that the organs chiefly concerned with the synthesis of proteins, the ovaries, have a metabolism geared toward the synthesis of amino acids (Pollak and Fairbairn, 1955a,b; Pollak, 1957a,b).

1. Dietary and "Essential" Amino Acids

Many nematodes, especially parasitic and saprophytic species, live in environments which provide an abundant supply of amino acids.

Some species, however, which live in soil or sea- or freshwater may have a more restricted supply, dependent very frequently on the growth of microorganisms in their surroundings.

Evidence for the capacity of nematodes to synthesize amino acids has been obtained from three types of experiment: (1) studies on the

TABLE X
FREE AMINO ACIDS IN NEMATODES

Amino acid	*Ascaridia*[a] females	*Ascaris*[b] ovaries	*Necator*[c] juveniles
Alanine	19.5	17	23.8
β-Alanine	1.4		
Aspartic acid	3.0	2.5	7.6
Cysteine } Cystine }	0.8	<0.5	2.4
β-Aminoisobutyric acid	1.8		
Glutamic acid	6.3	33	11.9
Glycine	6.0	15 (Glycine + Serine)	13.6 (Glycine + Serine)
Serine	8.9		
Tryptophan		trace	
Leucine } Isoleucine }	7.6	1.5	15.0
Histidine	6.8 (Histidine + Lysine + Arginine)	8.2	0.6
Lysine		10	3.9 (Lysine + Arginine)
Arginine		0	
Valine	6.0	1.5	7.2
Proline	9.2	<0.5	+
Taurine	1.3		
Tyrosine	3.0	<0.5	5.7
Phenylalanine	3.0	<0.5	3.7
Threonine	5.3	<0.5	5.1
Methionine	5.7	0	3.8
Glutamine		5.1	
Asparagine		2.5	

[a] Monteoliva Hernández (1962), percent total free amino acid nitrogen.
[b] Pollak and Fairbairn (1955a), percent total free α-amino nitrogen.
[c] Desowitz (1962), percent of free amino acids except proline.

amino acids needed for growth and reproduction in chemically defined media; (2) investigations on the incorporation of ^{14}C from suitable substrates in the media into amino acids in the whole organism; and (3) biochemical studies on the nature of reactions which take place in homogenates. Results obtained from such investigations must, however, be interpreted with care. Thus amino acid requirements can be influenced by the rate of growth, e.g., in a basic medium in which growth is limited by, say, a vitamin requirement, variations in the amounts of amino acids provided may have little effect. Similarly, evidence from tracer studies or from the nature of biochemical processes, though it may indicate a capacity for synthesizing amino acids, does not show whether the amino acids synthesized are available for growth and reproduction, or whether synthesis is rapid enough to meet the needs of the organism.

A further difficulty is that the methods of cultivating nematodes *in vitro* which have been developed usually require complex components, such as liver or chick embryo extract, in the basic medium. Nevertheless, results are already available which give some indication of the probable amino acid requirements.

a. Amino Acids Required in Culture Media. Jackson (1962) found that a chemically defined medium which included seventeen amino acids did not promote development and reproduction of *Neoaplectana glaseri*. One of the requirements for development and reproduction was an increased concentration of the "essential" amino acids—arginine, histidine, isoleucine, leucine, lysine, methionine, phenylalanine, threonine, tryptophan, and valine—in the medium.

In experiments on the axenic cultivation of *Caenorhabditis briggsae* Dougherty *et al.* (1959) found no requirements for the 10 "essential" amino acids when the medium contained 5% "liver protein fraction." But preliminary work indicated that, when the liver protein fraction was reduced to 1%, the omission of any one of the "essential" amino acids reduced development.

b. The Synthesis of Amino Acids in Whole Organisms. The results obtained in nutritional experiments with *Caenorhabditis briggsae* were supported by an examination of the capacity of the nematode to synthesize amino acids from acetate-2-^{14}C provided in the medium (Nicholas *et al.*, 1960). When cultivated in a cell-free medium containing chick embryo extract and autoclaved liver extract the nematode incorporated ^{14}C into the following amino acids: aspartic acid, glutamic acid, alanine, proline, glycine, serine, and cysteine plus cystine (as cysteic acid). The following amino acids were unlabeled: arginine, histidine, lysine, methionine, threonine, tyrosine, valine. A spot on the chromatogram covering phenylalanine, leucine, and isoleucine showed no radioactivity, and as acid hydrolysis was used no tryptophan was detected.

Later work, however, carried out with *Caenorhabditis briggsae* under different conditions suggests that this organism can synthesize "essential" amino acids (Rothstein and Tomlinson, 1962). In these experiments the nematode was maintained in water containing acetate-1-^{14}C, acetate-2-^{14}C, glycine-1-^{14}C, or glycine-2-^{14}C. From the last three substrates the worms not only synthesized a number of "nonessential" amino acids but also threonine, tyrosine, valine, leucine, isoleucine, histidine, and lysine.

The clear-cut nature of these results was somewhat confused by the results obtained when penicillin and streptomycin were added to the medium. In the presence of the antibiotics the organisms failed to synthesize any of the "essential" amino acids from glycine-2-^{14}C, and valine was the only "essential" amino acid synthesized from acetate-2-^{14}C. Extensive tests on media which did not contain antibiotics showed that axenic conditions had been maintained, and Rothstein and Tomlinson concluded that *Caenorhabditis briggsae* can synthesize "essential" amino acids and that the action of the antibiotics led to the suppression of synthesis, except in the case of valine.

Rothstein (1965) has summarized his extensive work on the synthesis of amino acids by *Caenorhabditis briggsae*. Substrates in the medium in which the nematode was cultured produced the amino acid most closely related metabolically. Even when the ^{14}C from the substrate was poorly incorporated into the tissues of the worm, the labeled amino acids appeared in the medium in large amounts. Thus malate-3-^{14}C was readily converted into aspartic acid which appeared in the medium; glycine-2-^{14}C gave serine, pyruvate-1-^{14}C gave alanine and aspartic acid, phenylalanine-3-^{14}C yielded tyrosine, and acetate yielded the amino acids formed from intermediates of the tricarboxylic acid cycle.

The synthesis of a wide variety of free amino acids took place in *Ditylenchus triformis* and *Meloidogyne* spp. when these organisms were incubated in 0.1 *M* phosphate buffer containing acetate-2-^{14}C under conditions in which microbial growth was negligible (Myers and Krusberg, 1965). But tryptophan was the only "essential" amino acid which contained ^{14}C and it was found only in *Meloidogyne* spp. Uniformly labeled glucose was used less efficiently in amino acid synthesis. Thus in *Ditylenchus triformis*, instead of the nine amino acids and glutamine and asparagine labeled in the presence of acetate-2-^{14}C, only alanine, asparagine, aspartic acid, glutamic acid, and serine incorporated ^{14}C from the labeled glucose.

2. *The Metabolism of Amino Acid in Homogenates*

Studies on the metabolism of amino acids in homogenates prepared from the tissues of nematodes showed the presence of many of the classical biochemical processes which lead to the synthesis and breakdown of amino acids. The active carbohydrate metabolism of nematodes and the presence of transaminases indicates a capacity for the synthesis of "glucogenic" amino acids.

Three of the α-keto acids formed in the course of carbohydrate metabolism in mammals, α-ketoglutaric acid, oxaloacetic acid and pyruvic acid, provide the central point for the synthesis of many amino acids (Fig. 2). In the ovaries of *Ascaris lumbricoides* (Pollak and Fairbairn, 1955a,b; Pollak, 1957a,b) the basic processes of amino acid metabolism are somewhat different. It was found that the ovaries had a relatively high content of alanine, and that the reductive amination of pyruvate was an active process whereas glutamic dehydrogenase activity was low. Relatively large amounts of ammonia were present and pyruvate was readily available from an active glycolysis so the production of alanine,

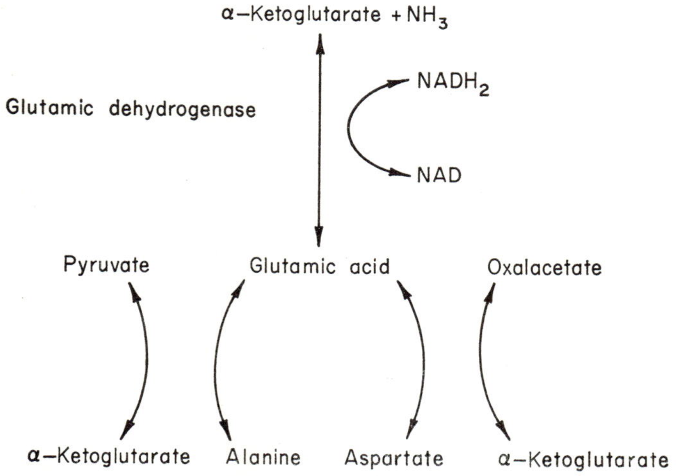

FIG. 2. The metabolic interrelationships of some amino acids in the tissues of mammals.

rather than glutamic acid, was indicated as a central point in amino acid metabolism. Moreover, an unspecific alanine-α-keto acid transaminase was present whereas the corresponding glutamic acid-α-keto acid transaminase was not detected.

These results, and others showing the absence of amino acid oxidases, suggest that anabolic processes predominate in ovaries of *Ascaris lumbricoides,* a conclusion in accord with the view that active protein synthesis would be required for the production of eggs.

a. Transamination. The transformation of amino acids to the analogous α-keto acids, and the reversal of this reaction, plays an important role in the metabolism of mammals. Early work stressed the importance of transaminations involving glutamic acid, aspartic acid, and alanine.

More recent results indicate the ubiquity of transaminations; virtually all of the natural amino acids may take part in enzymic transaminations in mammalian tissues (Meister, 1965).

A number of investigations have disclosed the presence of transaminases in tissues of *Ascaris lumbricoides*. Transaminations involving alanine, valine, and methionine seemed to be most active, on a wet weight basis, in cells of the intestine and least active in the muscles (Savel, 1955b). In ovaries, alanine-glutamic acid, and aspartic acid-glutamic acid transaminases were most active and were freely reversible. Both these systems were found in the muscles, and though, on a wet weight basis, the transaminations were most active in ovaries, the difference was not apparent when results were calculated on the nitrogen contents of the two tissues. Two distinct enzymes were involved in these reactions in ovaries; the optimum for alanine-glutamic acid transaminase occurred at pH 7.8 and that for the aspartic acid-glutamic acid transaminase at pH 7.4 (Pollak and Fairbairn, 1955b).

Of 18 other amino acids tested, which included valine and methionine, only serine and glycine served as amino group donors in the presence of α-ketoglutarate or pyruvate, and even these reactions were weak. A third acid, α-ketovaleric, was transaminated by glutamic acid and valine, but again the reaction was weak (Pollak and Fairbairn, 1955b). A nonspecific alanine-α-keto acid transaminase was also found though the corresponding glutamic acid-α-keto acid transaminase was absent (Pollak, 1957b).

The addition of pyridoxal phosphate to homogenates of ovaries and muscle had no effect on transamination. Pollak and Fairbairn (1955b) surmised, therefore, that the system already contained optimum amounts of the coenzyme. A requirement for pyridoxine has been shown for some nematodes cultured *in vitro* (e.g., see Dougherty *et al.*, 1959).

In homogenates of whole *Ascaris lumbricoides,* aspartic acid gave the greatest yield of glutamic acid. Leucine, valine, arginine, or alanine were somewhat less effective. Cysteine, tryptophan, and lysine were not involved in transaminations leading to the formation of glutamate (Cavier and Savel, 1954d; Savel, 1955b).

In homogenates of whole *Dictyocaulus filaria,* transaminations which produced glutamic acid or alanine with α-ketoglutarate or pyruvate were highly active; transfer to oxaloacetic acid was weak. Thirteen amino acids served as donors of amino nitrogen to α-ketoglutarate; the most active substrates were aspartic acid, alanine, and valine. Glycine, serine, cysteine, and histidine were inactive (Polyakova, 1962). Transaminations involving aspartic or glutamic acids and pyruvate-1-^{14}C with the formation of alanine have been demonstrated in homogenates of the plant-parasitic nematode, *Aphelenchoides ritzemabosi* (Miller and Roberts, 1964).

b. Oxidative Deamination. The addition of L-amino acids to minced tissues of *Ascaridia galli, Nematodirus* spp. (W. P. Rogers, 1952), or *Ascaris lumbricoides* (Cavier and Savel, 1954d) led to increased production of ammonia, presumably due to L-amino acid oxidase. Pollak and Fairbairn (1955b), who used both minces and homogenates of the ovaries of *Ascaris lumbricoides,* found that DL-alanine, DL-aspartic acid, DL-lysine, DL-valine, and DL-phenylalanine had no effect on the oxygen uptake, and considered therefore that neither D- nor L-amino acid oxidases were present. On the other hand, both glutamic acid and glutamine stimulated the consumption of oxygen and it appeared that glutamic dehydrogenase was present.

c. Reductive Amination. The synthesis of alanine plays a key role in amino acid metabolism of the ovaries of *Ascaris lumbricoides* and reductive amination of pyruvate,

$$NH_3 + \text{pyruvate} + NADH_2 \leftrightarrows \text{alanine} + NAD$$

has been studied in some detail (Pollak and Fairbairn, 1955b; Pollak, 1957a,b).

The addition of pyruvate, ammonium chloride, and bicarbonate to homogenates led to a marked synthesis of alanine and aspartic acid which greatly exceeded the decrease in glutamate. In dialyzed preparations under anaerobic conditions alanine formation was dependent upon the presence of $NADH_2$ and ATP; chloramphenicol and ascorbic acid enhanced production. The effect of ascorbic acid could be duplicated by adding large amounts of lactate.

Without chloramphenicol the production of α-amino nitrogen showed two optima—at pH 6.9 and 8.2. When chloramphenicol was present, however, one peak at pH 7.2 was obtained. Pollak (1957b) interpreted these results as indicating that a competitive reaction, which was suppressed by chloramphenicol, caused the depression in the pH curve.

The formation of glutamic acid in the presence of ammonia and α-ketoglutarate in homogenates of *Dictyocaulus filaria* has been demonstrated. When oxaloacetate was added as the substrate alanine was produced rather than aspartic acid (Polyakova, 1962).

d. Decarboxylation. Manometric estimations of the liberation of carbon dioxide from amino acids added to homogenates of *Ascaris lumbricoides* indicated the presence of decarboxylases, which, in descending order of activity, attacked the following substrates: histidine, lysine, ornithine, tryptophan, phenylalanine, tyrosine, and arginine. Alanine, glycine and valine were not attacked (Savel, 1955b). Glutamic acid decarboxylase has been found in *Ascaridia galli* (Monteoliva Hernández *et al.*, 1965).

Though these decarboxylases could provide a number of amines, it is clear that other mechanisms must be involved in producing most the amines excreted by juvenile nematodes (Section V).

C. The Metabolism of Individual Amino Acids

Much of what is known about the metabolism of particular amino acids has been covered in earlier sections. Some amino acids, however, have received special attention.

1. Serine

In *Caenorhabditis briggsae*, ^{14}C from formic acid was readily incorporated into serine. Though glycine-2-^{14}C gave a high yield of serine, serine-1-^{14}C gave little glycine. The mechanism by which this normally reversible one-carbon transfer is controlled is unknown. The results clearly indicate that glycine is formed by a pathway independent of serine (Rothstein and Tomlinson, 1962; Rothstein, 1965).

Rothstein (1965) considered that the formation of pyruvate-3-^{14}C from labeled serine provided the substrate for the production of ^{14}C-labeled glutamic acid, aspartic acid, and alanine.

2. Glycine

Glycine was the most strongly labeled amino acid which was excreted when *Caenorhabditis briggsae* was exposed to aspartate-4-^{14}C (Rothstein, 1963). Some form of tricarboxylic acid cycle functions in *Caenorhabditis briggsae* so Rothstein suggested that the route of synthesis of glycine could be: aspartate-4-^{14}C $\rightarrow$ isocitrate-1-^{14}C (then by the action of isocitric lyase), $\rightarrow$ glyoxylate-1-^{14}C and succinic acid. Transamination of the glyoxalate would then give labeled glycine.

Support for this view was obtained in later experiments (Rothstein and Mayoh, 1964). When *Caenorhabditis briggsae* was incubated with pyruvate-1-^{14}C neither glycine nor serine were significantly labeled. Moreover, the presence of isocitric lyase was demonstrated directly in a number of experiments. This enzyme does not appear to be present in the tissues of other multicellular animals though it has been found in bacteria, in plants, and in a protozoan.

Recent work has shown (Rothstein and Mayoh, 1965) that isocitric acid lyase is present in three other free-living nematodes, *Panagrellus redivivus*, *Turbatrix aceti*, and *Rhabditis anomala*.

Malate synthetase is also present in these three species as well as in *Caenorhabditis briggsae*, hence the complete glyoxalate cycle appears to be present and may be a common feature of free-living nematodes (Rothstein and Mayoh, 1966).

3. *Valine and Lysine*

An interesting feature of the results obtained by Rothstein and Tomlinson (1962) was the high activity in valine and lysine obtained from *Caenorhabditis briggsae* which had been incubated in glucose-1-^{14}C. The authors suggested therefore that a precursor of these amino acids might have been formed before the radiocarbon entered the tricarboxylic acid cycle. Pyruvate, formed by glycolysis, would be a possible precursor.

4. *Glutamic Acid*

γ-Glutamyl hydrazide, an antagonist of glutamine, inhibited the growth and reproduction of *Caenorhabditis briggsae in vitro.* This inhibition could be reversed by adding glutamine in amounts equivalent to that of the antagonist. The inhibition was also overcome by adding nucleotides to the medium, suggesting that glutamine participates in purine synthesis in this nematode (Sayre *et al.*, 1963; see Section XI).

VII. The Metabolism of Amino Acids in Acanthocephala

Information on amino acid metabolism in the Acanthocephala comes largely from the work of Graff (1964) and Bryant and Nicholas (1965). This work was carried out with adults of *Moniliformis dubius* and does not necessarily give a general picture of the metabolism of this group.

The main route of glucose metabolism in *Moniliformis dubius* involves glycolysis (Laurie, 1957, 1959; Read, 1961) and the carboxylation of pyruvate to give malate. From malate, oxaloacetic acid, fumaric, and succinic acids are formed (Bryant and Nicholas, 1965). Citric acid and aconitic acid do not appear to be involved so part of the tricarboxylic acid cycle is relatively inactive (Graff, 1964). Thus, though oxaloacetate and pyruvate are freely available for transamination, α-ketoglutarate may be present in small amounts only.

A. Metabolism of Whole Organisms

Graff exposed whole organisms to glucose-U-^{14}C *in vitro.* Glutamic acid, though present, was not labeled. Alanine and aspartic acid were labeled in approximately equal amounts and presumably transaminations involving oxaloacetate and pyruvate occurred. Serine was slightly labeled and so may have arisen from pyruvate or phosphoenol pyruvate. Glycine was unlabeled.

B. Metabolism in Disrupted Body Wall

In small fragments of body wall, homogenates or sonicates, ^{14}C from glucose-U-^{14}C or succinate-1,4-^{14}C was incorporated into alanine. When acetate-2-^{14}C was the substrate, labeled alanine was obtained from fragments of body wall, but activity was destroyed when the body wall was homogenized or "sonicated." Aspartic acid, and, on one occasion, asparagine, were found to be labeled in preparations which had been exposed to succinate-1,4-^{14}C. Small amounts of labeled glutamic acid were found in some of these experiments; evidently *Moniliformis dubius* has a limited capacity to oxidize succinate (Bryant and Nicholas, 1965).

VIII. Proteins in Nematodes

A number of proteins, especially those of *Ascaris lumbricoides*, have been examined. Interest has centered chiefly around structural proteins of the cuticle and egg shells, but proteins in perienteric fluid and reproductive systems have also received attention. The proteins which have been studied have rarely been isolated, and much of this work has been carried out with histochemical methods (e.g., see Fransden, 1965; Monné, 1955).

Protein synthesis in nematodes has not been examined as such, though the effect of amino acids in the "keeping medium" on the level of proteins in the perienteric fluid has been reported (Cavier and Savel, 1954f).

Much of the work on proteins has been reviewed, e.g., proteins in nematodes generally, by von Brand (1952) and Fairbairn (1960); proteins in *Ascaris* and *Parascaris*, (Savel, 1955a; Fairbairn, 1957); hemoglobins (Lee and Smith, 1965); proteins of egg shells (W. P. Rogers, 1962); proteins of the cuticle (Novelli, 1961); and Chapter 2 in this volume. Except for more recent work, therefore, proteins in nematodes will be discussed only briefly here.

A. Total Protein

The protein content of various stages in the life cycle of *Nippostrongylus brasiliensis* was found to differ to a surprising extent (Moore, 1965). Thus protein in juveniles formed 76% of the dry weight whereas it was 43% in adult worms. The water-soluble total protein of *Ascaris lumbricoides* contains five major fractions and comprises 21% of the dry weight (Kent, 1960).

Guevara Pozo *et al.* (1961) have carried out extensive analyses comparing the proteins of male and female *Ascaridia galli*. The protein content of the female (73% of the total nitrogen) was slightly higher than in males (71%) but the polypeptide content of the male estimated by the method of Cristol (1926) was somewhat higher in males.

The amino acid composition of whole organisms has been reported for several species, e.g., for infective juveniles of *Necator americanus* (Desowitz, 1962) and for adult *Ascaris lumbricoides* (Kanygina, 1952; Kanygina *et al.*, 1962). The total proteins in the various tissues of *Ascaridia galli* have also been analyzed (Monteoliva Hernández *et al.*, 1962).

B. Proteins in Perienteric Fluid

Monteoliva Hernández and Guevara (1960) found that 81.9% of the total nitrogen in perienteric fluid of *Ascaridia galli* could be attributed to protein nitrogen. "Polypeptide" nitrogen comprised a further 13%. Electrophoretic studies revealed four protein fractions. The most mobile fraction, which was analogous to albumin of human serum, comprised 21.9% of the total protein and was the most homogenous fraction. A second fraction, 61.9% of the total protein, was similar to β-globulin. It contained lipoprotein and glucoprotein. The third fraction, 8.4%, had a mobility similar to that of γ-globulin and contained no lipoprotein or glucoprotein. The fraction of lowest mobility contained both of these types of protein. If the fast-moving fraction is regarded as albumin and the remainder as globulin, the albumin/globulin ratio, about 0.25 (Monteoliva Hernández and Guevara, 1960), indicates a marked difference between the perienteric fluid of *Ascaridia galli* and that of *Ascaris lumbricoides* for which the ratio is about 1.4 (Cavier and Savel, 1951; Savel, 1955a).

Proteins form 4.9% of the perienteric fluid of *Ascaris lumbricoides* and 93% of its total nitrogen. The globulins, 2.1% of the protein, have been resolved into α-, β-, and γ-fractions. As in *Ascaridia galli,* polypeptides form a major part (40%) of the nonprotein fraction (Cavier and Savel, 1954e; Savel, 1955a). The presence of low molecular weight proteins in perienteric fluid has also been shown by Giraldo Cardona (1960), who obtained four polypeptide fractions from female *Ascaris,* and three from males.

Amino acid analyses of acid hydrolyzates of *Ascaris* perienteric fluid have been carried out, e.g., by Cavier and Savel (1954e) and by Jaskowski (1963). Glutamic acid, proline, tyrosine, and lysine were most plentiful; arginine, aspartic acid, histidine, tryptophan, phenylalanine, cysteine, methionine, glycine, serine, valine, and leucine were also detected. Jaskowski (1963) also found isoleucine and threonine.

Hemoglobin, which may comprise as much as 2% of the protein in perienteric fluid, has been studied in detail.

C. Hemoglobin

Eighteen species of parasitic nematodes have been found to contain hemoglobins which differ considerably in their physical properties (Lee and Smith, 1965). In *Ascaris lumbricoides* the amounts present vary from individual to individual, with a mean value of about 2% of the total protein in perienteric fluid. It seems probable that the amount of hemoglobin varies in individuals of other species also. Several distinct hemoglobins may be present in the one species. Thus *Tetrameres confusa* contains three hemoglobins (Villela and Ribeiro, 1955a) and *Ascaris lumbricoides* has two or perhaps three of these pigments (Davenport, 1949a; M. H. Smith and Morrison, 1963; Hamada *et al.*, 1963a,b).

1. *Physical Properties*

In the visible region the spectra of nematode hemoglobins conform with the pattern obtained from hemoglobins generally. The pigments in *Ascaris lumbricoides* and *Strongylus* spp., however, differ from most other hemoglobins in that absorption is greater in the β-band than in the α-band. The heme-protein "ratio," estimated on the basis of spectral characteristics, may be appreciably greater than 0.25 (M. H. Smith and Morrison, 1963).

The molecular weights of hemoglobins from perienteric fluid and body wall of *Ascaris lumbricoides* have been examined. M. H. Smith and Morrison (1963) found that the pigment from perienteric fluid had a low molecular weight, 14,000, compared to most other extracellular hemoglobins. These authors also reported a molecular weight of 280,000 for the pigment from body wall. Wittenberg *et al.* (1965) and Okazaki *et al.* (1965) found that the minimum molecular weight per heme unit of this pigment was 40,600, and suggested that there were 8 heme units per molecule.

Isoelectric points of invertebrate hemoglobins generally have relatively low values. The pigments from *Ascaris lumbricoides* with isoelectric points of pH 6.7 (body wall) and 5.0 (perienteric fluid) fit into this pattern; but the isoelectric points for pigments from *Tetrameres confusa* were found to be 8.7 and 6.5 (M. H. Smith and Morrison, 1963; Villela and Ribeiro, 1955b).

The most recent detailed study of nematode hemoglobins has been carried out by M. H. Smith (1963), Gibson and Smith (1965), and Okazaki and Wittenberg (1965) with pigments from *Ascaris lumbricoides*. The rates of reaction of hemoglobin from perienteric fluid with nitrous oxide, oxygen, and carbon monoxide were found to be 4.7×10^6 M^{-1} sec^{-1}, 1.5×10^6 and 1.7×10^5 respectively. The rates for hemoglobin from body wall were about 30% less in each case. Overall equilibrium constants for reactions with oxygen and carbon monoxide were calculated from kinetic data and are shown in Table XI. As shown for hemoglobins of some other nematodes (W. P. Rogers, 1949a), $\log_{10} M$ (where M is the ratio of the equilibrium constants for oxygen and carbon monoxide) had no relation to the "span," i.e., the difference in wavelength between the α-bands of the oxy and carboxy compounds of the hemoglobin. In this respect nematode hemoglobins differ from mammalian hemoglobins (Anson *et al.*, 1924; Theorell, 1934a,b).

All the hemoglobins from nematodes which have been examined in detail have a high affinity for oxygen. Thus the $p_{50}O_2$ for hemoglobin

TABLE XI

EQUILIBRIUM DATA FOR *Ascaris* HEMOGLOBINS AT 37° AND 20°C[a]

	Body wall hemoglobin (mm. Hg)	Perienteric fluid hemoglobin (mm. Hg)
Oxygen		
37°	0.5	0.01
20°	0.11	0.0015
Carbon monoxide		
37°	0.33	
20°	0.13	(27.5°) 0.063
M		
37°	1.5	(27.5°) 0.075
20°	0.8	

[a] From Gibson and Smith (1965). The values were calculated from rate constants, assuming that $n = 1$ in the Hill equation. M is the ratio of the equilibrium constants of oxygen and carbon monoxide.

from perienteric fluid of *Ascaris lumbricoides* at 20° and pH 7.0 was found to be 0.001–0.0038 mm. This result, obtained experimentally by Okazaki and Wittenberg (1965), shows close agreement with the figure 0.0015 mm. calculated from the rates of combination and dissociation of oxygen (Gibson and Smith, 1965). The Hill constant, n, which gives a measure of the heme-heme interaction is probably 1 for pigments from *Camallanus trispinosus, Haemonchus contortus, Nematodirus* spp., and *Nippostrongylus brasiliensis* (Lee and Smith, 1965; Table XII). Hamada (1963), who claimed that there are two hemoglobins in the body wall of *Ascaris lumbricoides* as well as the one in perienteric fluid, gave the values of n as 1.7, 0.7, and 0.7 respectively.

2. *Synthesis*

Nothing is known about the synthesis of the protein moiety of hemoglobins in nematodes. But M. H. Smith and Lee (1963) have shown that *Ascaris lumbricoides* requires an external source of hematin in order to maintain its hemoglobin content (see Section X).

3. *Biological Function of Hemoglobins in Nematodes*

The high affinity of nematode hemoglobins for oxygen suggests that they may function as oxygen carriers (Table XII). The experimental

TABLE XII

THE AFFINITY OF OXYGEN FOR SOME NEMATODE HEMOGLOBINS AND THE OXYGEN TENSION OF THE ENVIRONMENT

Species	$p_{50}O_2$	n	Environment	pO_2
Ascaris lumbricoides				
Body wall	0.49[a]	0.7[b]	Pig, small intestine	8–65[c]
Perienteric fluid	0.001–0.0038[d]	1.7–0.7	(gases)	
Camallanus trispinosus	8[e]	1[f]		
Haemonchus contortus	0.05[g]	1[f]	Sheep, abomasum (near mucosa)	4.3–17.5[g]
Nippostrongylus brasiliensis	0.1[a,g]	1[g]	Rat, small intestine (near mucosa)	4–12.7[g]
Strongylus spp.	<0.01[a]			
Nematodirus spp.	0.05[g]	1[f]	Sheep, small intestine (near mucosa)	8.9–30.2[g]

[a] Davenport (1949a,b).
[b] Hamada (1963).
[c] von Brand (1952).
[d] Okazaki and Wittenberg (1965).
[e] Wharton (1941).
[f] Lee and Smith (1965).
[g] W. P. Rogers (1949a,b).

evidence, however, suggests that this is doubtful under normal conditions for some species (W. P. Rogers, 1962; M. H. Smith and Lee, 1963). It is probable that some pigments may have a function in storing oxygen and releasing it when the pO_2 in the environment is very low. Though the hemoglobin in the body wall of *Ascaris lumbricoides* may function as an oxygen carrier (Gibson and Smith, 1965), the pigment in perienteric fluid does not. The most acceptable hypothesis for the function of this pigment has been put forward by M. H. Smith and Lee (1963). These authors point out that egg production in *Ascaris lumbricoides* must require the provision of large amounts of hematin. Indeed, the eggs produced in one day contain an amount of hematin almost equal to the total amount in the perienteric fluid. They suggest that the hemoglobin in the perienteric fluid serves as a source of hematin compounds which are incorporated into the eggs.

D. Proteins of the Reproductive Systems

The male and female reproductive organs of nematodes contain interesting proteins, some of which appear to have a relationship to the synthesis of new ribonucleic acid (RNA) which appears after fertilization. Though these proteins evoked considerable interest (Fairbairn, 1957, 1960), they have not been studied in recent years. This is unfortunate because the morphology of the ovary, especially in *Ascaris lumbricoides,* seems ideal for studies in changes taking place during the early development of the primary ovocyte and the synthesis of the layers of the egg shell.

1. *Protein in Sperm*

Spermatogonia of *Parascaris equorum* contain granules of protein which fuse to form the acrosome. The protein of the acrosome, which stains gram positive, gradually disappears as the sperm moves toward the female pronucleus in the ovocyte. Small granules of RNA appear around the degenerating acrosome. These granules increase in size and spread further into the cytoplasm together with mitochondria from the sperm. The protein in the acrosome, ascaridine, is thought to have some function in the synthesis of ribonucleoprotein (Filhol, 1937; Panijel, 1947, 1948, 1951; Pasteels, 1948; Panijel and Pasteels, 1951).

Ascaridine has been isolated and examined in detail (Fauré-Fremiet and Filhol, 1937; Panijel, 1950). An electrophoretically homogeneous preparation gave the following amino acids on acid hydrolysis: aspartic and glutamic acids, glycine, alanine, serine, tyrosine, arginine, lysine, phenylalanine, valine, leucine, norleucine and probably hydroxyproline, isoleucine, and threonine. Aspartic acid comprised 30% of the total amino acids, glutamic acid 5%, dibasic acids $<5\%$, aromatic amino acids (after alkaline hydrolysis) about 15%, alanine 5%, glycine 4%, and the leucines, 10%. Purine nitrogen (0.2%), and phosphorus (0.02%) were present. Sulfur was absent.

2. *Proteins in Ovaries*

Proteins comprise 35% of the solid material of the ovaries of *Ascaris lumbricoides* (Pollak and Fairbairn, 1955a). Acid hydrolysis of the protein gave the following amino acids: leucine (12.7% of the total α-amino nitrogen), glutamic acid (10.8%), serine and glycine (10.4%), proline (8.2%), aspartic acid (8.0%), lysine (8.0%), valine (7.8%), alanine (7.1%), histidine (5.8%), phenylalanine (3.4%), threonine (3.1%), and tyrosine (4.1%).

The primary ovocytes of *Parascaris equorum* contain granules which consist largely of three closely related proteins in which proline accounts for more than 20% of the total amino acids. In their solubility characteristics these proteins closely resemble pseudoglobulins or prolamines (Ebel and Colas, 1954). It seems probable that these proteins take part in the formation of one of the layers of the egg shell (Fairbairn, 1957).

3. *Proteins in Egg Shells*

The egg shells of a number of species have been examined, largely by histochemical methods (e.g., see Monné, 1959, 1962, 1963a). At least four layers have been described. The tertiary egg membrane, secreted by the uterus, generally contains protein and polysaccharide. The primary egg membrane contains three layers of which one is predominantly chitin, and the inner one, predominantly lipid. All these may contain protein. A secondary egg envelope, or chorion, is lacking (W. P. Rogers, 1962).

Individual proteins from the various layers have not been isolated, but amino acid analyses of whole eggs and of the various layers have been carried out (e.g., see Krvavica *et al.*, 1964a,b; Kreuzer, 1953; Jaskowski, 1962).

a. The Tertiary Egg Membrane. In most species this is the outermost layer which encloses individual eggs though it may form a jellylike mass in which the eggs are embedded (Chitwood, 1938). The properties of the "jelly-coat" often change after oviposition. Thus the freshly laid eggs of *Ascaris lumbricoides* have a tertiary layer which is colorless and transparent, and is attacked by dilute acids and alkalies and by proteinases. It is believed to contain keratinlike protein, polysaccharide, and bound polyphenol. The eggs collected from feces are brown and the tertiary layer is not readily attacked; the jelly coat is converted into a hard brown "polyphenol-quinone-tanned" cover (Monné and Hönig, 1955a; Monné, 1962).

The outer layer of eggs of *Contracaecum osculatum* consists of acid mucopolysaccharide which is extruded from the eggs themselves. The uterine cells secrete a polyphenol which becomes bound to the protein (Monné, 1963a).

Alkaline and acid hydrolysis of the tertiary layer of *Ascaris* egg shells (Kreuzer, 1953; Jaskowski, 1962) revealed the following amino acids: leucine, valine, alanine, aspartic acid, lysine, proline, histidine, glycine, serine, cystine, glutamic acid, threonine, tyrosine, methionine, tryptophan, phenylalanine, and isoleucine. The presence of arginine, reported by Kreuzer (1953) was not confirmed by Jaskowski (1962).

b. The Primary Egg Envelope. As eggs pass down the oviduct after fertilization, three or possibly four layers are formed by the egg itself. These layers may fuse to some extent, or a layer may be lost entirely so that the egg shells of various species differ considerably (W. P. Rogers, 1962).

i. The outer layer of the primary egg envelope. After the sperm has entered the ovocyte in many species the membrane becomes thickened and lifted slightly from the surface of the egg to form a "fertilization membrane." This layer may become associated in varying degrees with the chitinous layer formed below it. Electron microscopy of egg shells of *Ascaris lumbricoides* (R. A. Rogers, 1956) showed a dense layer about 0.4 μ thick on the outer edge of the primary egg envelope but Rogers did not regard this as being separate from the underlying chitinous material.

There is no doubt that this outer layer contains protein which, like the tertiary layer, changes in its solubility characteristics and becomes more resistant to the action of proteinases as it passes down the oviduct (Monné and Hönig, 1954). However, it remains soluble in sodium hypochlorite and in 10% sodium hydroxide. Hydrolyzates of extracts of eggs which probably contained protein from the chitinous layer as well as protein from the outer layer of the primary envelope contained the following amino acids: leucine, valine, methionine, alanine, lysine, histidine, proline, aspartic acid, arginine, and α-aminobutyric acid (Kreuzer, 1953). On the basis of solubility characteristics and the staining reactions, Monné and Hönig (1954, 1955a) believed the protein of the outer coat to be quinone-tanned, with and without cystine cross links in different species.

ii. The middle, chitinous coat. Chitin forms a major component of the middle coat of egg shells of most nematodes, though in some species it may be confined to small areas or entirely absent (see W. P. Rogers, 1962, for references). Glycogen from the egg is used for the synthesis of glucosamine and chitin (e.g., see Fauré-Fremiet, 1913; Monné, 1963a). Protein from the middle coat of eggs of *Ascaris lumbricoides,* which may be associated with the chitin, contains the following amino acids:

cystine, aspartic acid, glycine, serine, proline, tyrosine, and arginine (Jaskowski, 1962).

A layer of protein may be present on the inner side of the chitinous "hard-shell" of *Ascaris lumbricoides* (Yanagisawa, 1955). It is probable that this layer is formed from protein granules in the primary ovocyte. Certainly these granules move to the surface and become associated with layers of primary egg shell in a number of species (e.g., see Yanagisawa and Ishii, 1954; Monné, 1962, 1963a). In *Contracaeum osculatum* the granules have been described as keratinlike and containing polysaccharide. Monné and Hönig (1955a) suggested that protein in this part of egg shells of a number of species is quinone tanned like that in the outer layer of the primary envelope.

iii. The inner layer of the primary egg envelope. This, in most species of nematodes, is the chief permeability barrier of the nematode egg. Lipid is the predominant component in this layer (Fairbairn, 1960). Protein is also present, at least in the eggs of some species and hydrolyzates of the membrane (*Ascaris lumbricoides*) contain the following amino acids: cysteine, aspartic acid, proline, tyrosine, leucine, isoleucine, tryptophan, phenylalanine, and arginine (Jaskowski, 1962). Bound polyphenol is present in this layer in some species (Monné, 1963a). Jaskowski (1962) regarded the membrane as a lipoprotein complex.

E. Trypsin and Chymotrypsin Inhibitors

What appear to be low molecular weight proteins which inhibit the action of mammalian trypsin and chymotrypsins occur in *Ascaris lumbricoides* (Green, 1955; Peanasky and Laskowski, 1960). There are at

TABLE XIII

Trypsin and Chymotrypsin Inhibitors from *Ascaris lumbricoides*[a]

Source	Molecular weight	Tryptophan	Tyrosine
Trypsin inhibitors			
Body wall	4,650	3.7%	0.3%
Perienteric fluid	7,100	3.5%	0.8%
Chymotrypsin inhibitors			
Body wall	8,760	3.9%	2.3%
Perienteric fluid	12,400	2.1%	0.4%

[a] From Rhodes *et al.* (1963).

least four of these proteins, and the trypsin inhibitors are distinct from the inhibitors of chymotrypsins. The tyrosine and tryptophan content and the molecular weights (calculated on the basis of a 1:1 enzyme-inhibitor complex) for four of the inhibitors are shown in Table XIII (Rhodes *et al.*, 1963). Rola and Puales (1966) who studied the chymotrypsin inhibitors extracted from whole *Ascaris lumbricoides* found that the behavior of a pooled sample on Sephadex columns indicated a molecular weight of about 8,600. One of the chymotrypsin inhibitors from the body wall of *Ascaris* has a single NH_2-terminal amino acid, arginine (Peanasky and Szucs, 1964). All the inhibitors show a remarkable stability to heat and pH changes. They are stable, as inhibitors, in 8 *M* urea.

F. Antigenic Proteins

A number of antigenic proteins and protein-polysaccharide complexes have been found in tissues and secretions of nematodes. Though the serological reactions of these antigens have been extensively examined, the chemical properties have been investigated in detail in a few species only. The methods of density-gradient separation, column chromatography, electrophoresis, and fractional precipitation have been mostly used in purification (e.g., see Kagan *et al.*, 1958; Rose *et al.*, 1964; Sawada *et al.*, 1965; Mills and Kent, 1965; Tanner, 1963).

A number of purified protein-containing antigens from *Ascaris lumbricoides* have been examined (Tormo and Chordi, 1965) and the composition of five water-soluble fractions, separated by electrophoresis, has been determined (Kent, 1960; Table XIV). The antigenic material extracted from *Dirofilaria immitis* contained 2.5% carbohydrate and lysine, arginine, glycine, alanine, glutamic acid, aspartic acid, and valine (Sawada *et al.*, 1965).

Antigenic substances in extracts and secretions of infective juveniles of *Trichinella spiralis* have been examined (e.g., see Kagan, 1960; Sleeman and Muschel, 1961; Tanner and Gregory, 1961; Dusanic and Lewert, 1963; Mills and Kent, 1965). Ten antigenic proteins were found in extracts. One of these was antigenically stable at 100° for 15 minutes but it could be precipitated with trichloracetic acid or ammonium sulfate. Tanner (1963) suggested therefore that this antigen contained polysaccharide as well as protein. An ethanol-soluble antigen from *Trichinella* contained 13–17% carbohydrate which yielded glucose only on hydrolysis (Sleeman, 1961). At least five antigens were found in secretions and excretions. Care was taken to ensure that this material was not contaminated with substances from decaying juveniles. The carbohydrate-protein ratios differed slightly in the different antigens but in general 0.053% (w./v.) was carbohydrate and 0.04% was protein (Mills and Kent, 1965).

TABLE XIV
ANTIGENIC SUBSTANCES IN WATER EXTRACTS OF *Ascaris lumbricoides*[a]

Fraction	Protein content	Amino acids (acid hydrolyzate)	Sugars[b]	Antigenic strength
1	(80%)	Lysine, arginine, glutamic acid, alanine, isoleucine	(20%) Glucose ribose	±
2	(86%)	Lysine, arginine, glutamic acid, valine, tyrosine, alanine, isoleucine	(14%) Glucose ribose	±
3	(99.5%)	Arginine, glutamic acid, alanine, proline, threonine, tyrosine, valine, isoleucine, serine	(None) glucose galactose ribose	2+
4	(23%)	Lysine, histidine, arginine glutamic acid, threonine, valine, leucine	(76.5%) Glucose ribose	4+
5	(37%)	Lysine, arginine, glutamic acid, threonine, valine, leucine	(63%) Glucose ribose	3+

[a] From Kent (1960). The figures within brackets give protein and sugar content determined quantitatively on a dry weight basis. Other results were obtained qualitatively.
[b] Preliminary results.

IX. Proteins of the Acanthocephala

Protein forms 70% of the dry material of *Macracanthorhynchus hirudinaceus* (von Brand, 1939). Apart from histochemical examination, none of this protein has been characterized though the electrophoretic pattern of proteins in perienteric fluid has been examined (Monteoliva Hernández, 1964). Proteins, some of which are like keratins, have been found in the egg shells of *Polymorphus* (Monné and Hönig, 1954, 1955b).

X. Metabolism of Heme Compounds

The presence of hemoglobins and cytochromes in a variety of nematodes indicates that heme compounds are involved in the metabolism of these organisms. A definite requirement for these compounds has not been demonstrated in cultures of nematodes though it is probable that embryo or liver extract, found necessary for most culture media, would contain some of these compounds. It is notable, however, that Jackson (1962) was able to cultivate *Neoaplectana glaseri* in a chemically defined medium which did not contain heme compounds.

The most extensive examination of the metabolism of heme compounds in nematodes has been carried out by M. H. Smith and Lee (1963). These authors showed that the specific hemoglobin in the perienteric fluid of *Ascaris lumbricoides* was increased when horse hemoglobin was included in the medium. Other substrates containing a porphyrin or metalloporphyrin with a vinyl group at position 2 promoted the synthesis of hemoglobin. The addition of ferrous sulfate, ferric sulfate, and albumin to the saline medium used in all their experiments failed to give any detectable synthesis of hemoglobin.

Smith and Lee interpreted their results as showing that hematin cannot be synthesized by *Ascaris lumbricoides,* and that the hemoglobin of the perienteric fluid is synthesized in the cells of the gut. They suggested that the hemoglobin in the perienteric fluid occupies a central position in the metabolism of *Ascaris lumbricoides* and forms the reserve material from which all other hemoproteins are derived.

XI. Nucleic Acids in Nematodes

The distribution of deoxyribonucleic and ribonucleic acid in the tissues of nematodes has been examined by histochemical methods in a variety of nematodes (e.g., see Zarzycki, 1960; Monné, 1963b; Bogoyavlenskii, 1964; Bogoyavlenskii and Drynochkina, 1964; Anya, 1966). RNA has been found in the cortical layer of the cuticle of several species and Anya (1966) suggested that this layer may take part in the synthesis of some of the proteins in this region. The DNA in the intestine of *Meterakis govindi* was found to increase and decrease in phase with reproductive activity (Chen and Chu, 1965).

Biochemical methods have rarely been used. Nigon and Bovet (1955) found that a single spermatid of *Parascaris equorum* contained 2.5×10^{-6} μg. of DNA but immediately after fertilization 290×10^{-6} μg., about 20 times the expected amount, was found. A protein, ascaradine, in the acrosome may be involved in the synthesis of RNA during fertilization (see Section VIII).

Some aspects of nucleic acid metabolism in the reproductive organs of *Ascaris lumbricoides* have been examined by Entner and Gonzalez (1959, 1961). A particulate fraction was found to carry out phosphorylysis of synthetic polynucleotides as well as catalyze exchange reactions between inorganic phosphate and nucleoside diphosphates. Some evidence indicating that this system was capable of synthesizing polynucleotides from nucleoside triphosphate was also obtained. Four specific nucleoside kinases were found in extracts of reproductive organs. Adenosine and guanosine monophosphate kinases were more active than the corresponding kinases for uridine and cytidine monophosphate. The AMP and GMP kinases were partially purified; both required Mg^{++} or Mn^{++} as a cofactor; adenosine triphosphate acted as the relatively specific triphosphate donor. Though the results did not provide evidence of the existence of separate enzymes for the ribo- and deoxyribonucleotides, the kinases for the purine nucleoside monophosphates were significantly more active toward the ribotides than toward the deoxyribotides.

The inhibition of growth and reproduction of *Caenorhabditis briggsae* by γ-glutamyl hydrazide was overcome by adding increased amounts of nucleotides as well as by adding glutamic acid or glutamine (Sayre *et al.*, 1963). These results indicated that glutamine participates in purine ring synthesis in this nematode. Folic acid analogs inhibited reproduction but did not affect growth. Inhibition due to amethopterin was not greatly reduced by adding more thymine to the medium; but nucleotides were completely effective in restoring reproduction, again indicating that the biosynthesis of the purine ring was vulnerable to the action of antagonists. It was not known if the nucleotides were degraded into bases on nucleosides before absorption and incorporation into new nucleotides.

Extracts of *Trichuris muris* contained both ribonuclease and deoxyribonuclease (Nimmo-Smith and Keeling, 1960). The ribonuclease showed a broad peak in activity at pH 5.3–5.8 and an ionic strength (chloride, acetate, or phosphate ions) of between 0.1 and 0.2 was needed for optimal activity. Low concentrations (2.5×10^{-5} M) of Zn^{++}, Cu^{++}, or Mg^{++} caused slight stimulation. The deoxyribonuclease showed a sharp peak in activity at pH 5.2; Mn^{++} did not stimulate and ethylenediaminetetraacetic acid did not inhibit.

ACKNOWLEDGMENTS

I am indebted to Professor M. R. Atkinson who read the manuscript of this paper. Dr. E. Mercer, Dr. M. Rothstein and Dr. P. Weinstein also gave helpful advice for which I am most grateful.

I wish to acknowledge assistance given by grants from the U.S. Public Health Service (A-104093) and from Parke, Davis and Company.

REFERENCES

Anson, M. L., Barcroft, J., Mirsky, A. E., and Oiniuma, S. (1924). *Proc. Roy. Soc.* **B97**, 61.

Anya, A. O. (1966). *Nature* **209**, 827.

Bahl, K. N. (1945). *Biol. Rev.* **22**, 109.

Barrington, E. J. W. (1962). *Advan. Comp. Physiol. Biochem.* **1**, 1065.

Benediktov, I. I. (1962). *Med. Parazitol. i Parazitarn. Bolezni* **31**, 664.

Bogoyavlenskii, Yu. K. (1964). *Tr. Gel'mintol. Lab., Akad. Nauk SSSR* **15**, 38; *Chem. Abstr.* **64**, 14643g (1966).

Bogoyavlenskii, Yu. K., and Drynochkina, Z. V. (1964). *Tr. Gel'mintol. Lab., Akad. Nauk SSSR* **51**, 55; *Chem. Abstr.* **64**, 14643h (1966).

Bryant, C., and Nicholas, W. L. (1965). *Comp. Biochem. Physiol.* **15**, 105.

Carpenter, M. F. P. (1952). *Dissertation Abstr.* **12**, 399.

Cavier, R., and Savel, J. (1951). *Bull. Soc. Chim. Biol.* **33**, 455.

Cavier, R., and Savel, J. (1952). *Compt. Rend.* **234**, 2562.

Cavier, R., and Savel, J. (1954a). *Compt. Rend.* **238**, 2448.

Cavier, R., and Savel, J. (1954b). *Bull. Soc. Chim. Biol.* **36**, 1425.

Cavier, R., and Savel, J. (1954c). *Compt. Rend.* **239**, 205.

Cavier, R., and Savel, J. (1954d). *Bull. Soc. Chim. Biol.* **36**, 1631.

Cavier, R., and Savel, J. (1954e). *Bull. Soc. Chim. Biol.* **36**, 1433.

Cavier, R., and Savel, J. (1954f). *Compt. Rend.* **238**, 2035.

Chen, H.-Y., and Chu, C.-M. (1965). *Acta Biol. Exptl. Sinica* **10**, 213.

Chitwood, B. G. (1938). *Proc. Helminthol. Soc. Wash., D.C.* **5**, 68.

Christensen, H. N. (1960). *Advan. Protein Chem.* **60**, 239.

Cohen, P. P., and Brown, G. W. (1960). *Comp. Biochem.* **2**, 161–244.

Compton, D. W. T., and Lee, D. L. (1965). *Parasitology* **55**, 357.

Cristol, P. (1926). *Bull. Soc. Chim. Biol.* **11**, 93.

Davenport, H. E. (1949a). *Proc. Roy. Soc.* **B136**, 255.

Davenport, H. E. (1949b). *Proc. Roy. Soc.* **B136**, 271.

Delaunay, H. (1934). *Biol. Rev.* **6**, 265.

Desowitz, R. S. (1962). *Trans. Roy. Soc. Trop. Med. Hyg.* **56**, 257.

Dougherty, E. C., Hansen, E. L., Nicholas, W. L., Mollet, J. A., and Yarwood, E. A. (1959). *Ann. N.Y. Acad. Sci.* **77**, Art. 2, 176.

Dresel, E. I. B., and Moyle, V. (1950). *J. Exptl. Biol.* **27**, 210.

Dusanic, D. G., and Lewert, R. M. (1963). *J. Parasitol.* **40**, Suppl., 34.

Ebel, J. P., and Colas, J. (1954). *Compt. Rend. Soc. Biol.* **148**, 1580.

Edmonds, S. J. (1965). *Parasitology* **55**, 337.

Edmonds, S. J., and Dixon, B. R. (1966). *Nature* **209**, 99.

Enigk, K. (1938). *Z. Parasitenk.* **10**, 386.

Entner, N., and Gonzalez, C. (1959). *Biochem. Biophys. Res. Commun.* **1**, 333.

Entner, N., and Gonzalez, C. (1961). *Biochim. Biophys. Acta* **47**, 52.

Fairbairn, D. (1957). *Exptl. Parasitol.* **6**, 491.

Fairbairn, D. (1960). *In* "Nematology" (J. N. Sasser and W. R. Jenkins, eds.), p. 267. Univ. of North Carolina Press, Chapel Hill, North Carolina.
Fairbairn, D., and Passey, B. I. (1955). *Can. J. Biochem. Physiol.* 33, 130.
Fauré-Fremiet, E. (1913). *Arch. Anat. Microscop. Morphol. Exptl.* **15**, 435.
Fauré-Fremiet, E., and Filhol, J. (1937). *J. Chim. Phys.* **34**, 444.
Fernando, M. A., and Wong, H. A. (1964). *Exptl. Parasitol.* **15**, 279.
Filhol, J. (1937). *Arch. Anat. Microscop. Morphol. Exptl.* **39**, 301.
Florkin, M. (1949). "Biochemical Evolution." Academic Press, New York.
Fransden, J. C. (1965). *Trans. Am. Microscop. Soc.* **84**, 154.
Friedman, F., and Kagan, I. G. (1958). *Trans. Am. Microscop. Soc.* **77**, 365.
Gibson, Q. H., and Smith, M. H. (1965). *Proc. Roy. Soc.* **B163**, 206.
Giraldo Cardona, A. J. (1960). *Rev. Iberica Parasitol.* **20**, 425.
Goffart, H., and Heiling, A. (1962). *Nematologica* **7**, 173.
Graff, D. J. (1964). *J. Parasitol.* **50**, 230.
Green, N. M. (1955). *Biochem. J.* **66**, 416.
Guevara Pozo, D., Monteoliva Hernández, M., and Escobar Bueno, C. (1961). *Rev. Iberica Parasitol.* **21**, 3.
Hamada, K. (1963). *Nippon Ika Daigaku Zasshi* **30**, 17.
Hamada, K., Okazaki, T., Shukuya, R., and Kaziro, K. (1963a). *J. Biochem. (Tokyo)* **53**, 471.
Hamada, K., Okazaki, T., Shukuya, R., and Kaziro, K. (1963b). *J. Biochem. (Tokyo)* **53**, 484.
Hankes, L. V., and Stoner, R. D. (1956). *Proc. Soc. Exptl. Biol. Med.* **91**, 443.
Hankes, L. V., and Stoner, R. D. (1958). *Exptl. Parasitol.* **7**, 82.
Harpur, R. P., and Popkin, J. S. (1965). *Can. J. Biochem.* **43**, 1157.
Haskins, W. T., and Weinstein, P. P. (1957). *J. Parasitol.* **43**, 28.
Hsü, H. F. (1938). *Bull. Fan Mem. Inst. Biol., Zool. Ser.* **8**, 121.
Hsü, H. F., and Li, S. Y. (1940). *Chinese Med. J.* **57**, 559.
Hurlaux, R. (1947). *Ann. Sci. Nat., Zool. Biol. Animale* [11] **9**, 155.
Jackson, G. J. (1962). *Exptl. Parasitol.* **12**, 25.
Jaskowski, B. J. (1962). *Exptl. Parasitol.* **12**, 19.
Jaskowski, B. J. (1963). *J. Parasitol.* **49**, Suppl., 50.
Kagan, I. G. (1960). *J. Infect. Diseases* **107**, 65.
Kagan, I. G., Jescka, E. L., and Gentzkow, C. J. (1958). *J. Immunol.* **80**, 400.
Kanygina, K. I. (1952). *Tr. Gel'mintol. Lab., Akad. Nauk SSSR* **6**, 409.
Kanygina, K. I., Smolina, N. I., and Korneichuk, A. A. (1962). *Materialy XI-oi [Odeunadtsatoi] Nauchn. Konf. Prof.-Prepodavat. Sostava. Biol. Fak. Kirg Univ., Funze, Sb.* p. 97.
Kreuzer, L. (1953). *Z. Vergleich. Physiol.* **35**, 13.
Kent, H. N. (1960). *Exptl. Parasitol.* **10**, 313.
Krusberg, L. R. (1960). *Phytopathology* **50**, 9.
Krvavica, S., Maloseja, Z., Wagner, H., and Martincic, T. (1964a). *Vet. Arhiv* **34**, 163.
Krvavica, S., Maloseja, Z., and Lui, A. (1964b). *Vet. Arhiv* **34**, 167.
Lapage, G. (1933). *Nature* **131**, 383.
Laurie, J. S. (1957). *Exptl. Parasitol.* **6**, 245.
Laurie, J. S. (1959). *Exptl. Parasitol.* **8**, 188.
Lee, D. L. (1958). *Parasitology* **48**, 437.
Lee, D. L. (1962). *Parasitology* **52**, 241.
Lee, D. L., and Smith, M. H. (1965). *Exptl. Parasitol.* **16**, 392.

Lewert, R. M. (1958). *Rice Inst. Pam.* **45**, 97.
Lewert, R. M., and Lee, C. L. (1957). *Am. J. Trop. Med. Hyg.* **6**, 473.
McCoy, D. R. (1929a). *Science* **69**, 74.
McCoy, D. R. (1929b). *Proc. Helminthol. Soc. Wash., D.C.* **8**, 11.
Mandlowitz, D., Dusanic, D., and Lewert, R. M. (1960). *J. Parasitol.* **46**, 89.
Meister, A. (1965). "Biochemistry of the Amino Acids," 2nd ed., 2 vols. Academic Press, New York.
Miller, C. W., and Jenkins, W. R. (1964). *Nematologica* **10**, 480.
Miller, C. W., and Roberts, R. N. (1964). *Phytopathology* **54**, 1177.
Mills, C. K., and Kent, H. N. (1965). *Exptl. Parasitol.* **16**, 300.
Monné, L. (1955). *Arkiv Zool.* [2] **7**, 559.
Monné, L. (1959). *Arkiv Zool.* [2] **12**, 99.
Monné, L. (1962). *Arkiv Zool.* [2] **15**, 277.
Monné, L. (1963a). *Z. Parasitenk.* **22**, 475.
Monné, L. (1963b). *Parasitology* **53**, 273.
Monné, L., and Hönig, G. (1954). *Arkiv Zool.* [2] **6**, 559.
Monné, L., and Hönig, G. (1955a). *Arkiv Zool.* [2] **7**, 261.
Monné, L., and Hönig, G. (1955b). *Arkiv Zool.* [2] **7**, 257.
Monteoliva Hernández, M. (1962). *Rev. Iberica Parasitol.* **22**, 255.
Monteoliva Hernández, M. (1964). *Rev. Iberica Parasitol.* **24**, 43.
Monteoliva Hernández, M., and Guevara Pozo, D. (1960). *Libro Homenaje al Dr. Eduardo Caballero y Caballero, Jubileo, 1936–1960* p. 493.
Monteoliva Hernández, M., Escobar Bueno, C., and Guevara Pozo, D. (1962). *Rev. Iberica Parasitol.* **22**, 49.
Monteoliva Hernández, M., Roséro, E. S., Gorge, J. L., and Mayor, F. (1965). *Nature* **205**, 1111.
Moore, D. J. (1965). *Dissertation Abstr.* **26**, 1300.
Munro, A. F. (1949). *Biochem. J.* **54**, 29.
Myers, R. F. (1963). *Phytopathology* **53**, 884.
Myers, R. F., and Krusberg, L. R. (1965). *Phytopathology* **55**, 429.
Myuge, S. G. (1957). *Zool. Zh.* **36**, 620.
Nicholas, W. L., and Mercer, E. H. (1965). *Quart. J. Microscop. Sci.* **106**, 137.
Nicholas, W. L., Dougherty, E. C., Hansen, E. L. Holm-Hansen, O., and Moses, V. (1960). *J. Exptl. Biol.* **37**, 435.
Nigon, V., and Bovet, P. (1955). *Compt. Rend. Soc. Biol.* **149**, 129.
Nimmo-Smith, R. H., and Keeling, J. E. P. (1960). *Exptl. Parasitol.* **10**, 337.
Novelli, A. (1961). *Rivi. Biol. (Perugia)* **54**, 139.
Okazaki, T., and Wittenberg, J. B. (1965). *Biochim. Biophys. Acta* **111**, 503.
Okazaki, T., Briehl, R. W., Wittenberg, J. B., and Wittenberg, B. A. (1965). *Biochim. Biophys. Acta* **111**, 496.
Panijel, J. (1947). *Bull. Soc. Chim. Biol.* **29**, 1098.
Panijel, J. (1948). *Bull. Soc. Chim. Biol.* **30**, 116.
Panijel, J. (1950). *Biochim. Biophys. Acta* **6**, 79.
Panijel, J. (1951). *Experientia* **7**, 100.
Panijel, J., and Pasteels, J. (1951). *Arch. Biol. (Liege)* **62**, 353.
Pasteels, J. (1948). *Arch. Biol. (Liege)* **59**, 405.
Peanasky, R. J., and Laskowski, M. (1960). *Biochim. Biophys Acta* **37**, 167.
Peanasky, R. J., and Szucs, M. M. (1964). *J. Biol. Chem.* **239**, 2525.
Pollak, J. K. (1957a). *Australian J. Sci.* **19**, 208.
Pollak, J. K. (1957b). *Australian J. Biol. Sci.* **4**, 465.

Pollak, J. K., and Fairbairn, D. (1955a). *Can. J. Biochem. Physiol.* **33**, 297.
Pollak, J. K., and Fairbairn, D. (1955b). *Can. J. Biochem. Physiol.* **33**, 307.
Polyakova, D. I. (1962). *Biokhimiya* **27**, 430.
Read, C. P. (1961). *In* "Comparative Physiology of Carbohydrate Metabolism in Heterothermic Animals" (A. W. Martin, ed.), pp. 3–4. Univ. of Washington Press, Seattle, Washington.
Read, C. P., Rothman, A. H., and Simmons, J. E. (1963). *Ann. N.Y. Acad. Sci.* **113**, Art. 1, 154.
Rhodes, M. B., Marsh, C. L. , and Kelly, G. W. (1963). *Exptl. Parasitol.* **13**, 266.
Rola, F. H., and Puales, J. (1966). *Arch. Biochem. Biophys.* **131**, 134.
Rogers, R. A. (1956). *J. Parasitol.* **42**, 97.
Rogers, W. P. (1940a). *J. Helminthol.* **18**, 103.
Rogers, W. P. (1940b). *J. Helminthol.* **18**, 53.
Rogers, W. P. (1941). *J. Helminthol.* **19**, 47.
Rogers, W. P. (1949a). *Australian J. Sci. Res.* **B2**, 287.
Rogers, W. P. (1949b). *Australian J. Sci. Res.* **B2**, 157.
Rogers, W. P. (1952). *Australian J. Sci. Res.* **B5**, 210.
Rogers, W. P. (1955). *Exptl. Parasitol.* **4**, 21.
Rogers, W. P. (1958). *Nature* **181**, 1410.
Rogers, W. P. (1962). "The Nature of Parasitism. The Relationship of Some Metazoan Parasites to Their Hosts." Academic Press, New York.
Rogers, W. P. (1963). *Ann. N.Y. Acad. Sci.* **113**, Art. 1, 208.
Rogers, W. P. (1965). *Comp. Biochem. Physiol.* **14**, 311.
Rogers, W. P. (1966). *Comp. Biochem. Physiol.* **7**, 1103.
Rogers, W. P., and Lazarus, M. (1949). *Parasitology* **39**, 245.
Rose, J. E., Barsden, L. A., and Tromba, F. G. (1964). *J. Parasitol.* **50**, 504.
Rothman, A. M., and Fisher, F. M. (1964). *J. Parasitol.* **50**, 410.
Rothstein, M. (1963). *Comp. Biochem. Physiol.* **9**, 51.
Rothstein, M. (1965). *Comp. Biochem. Physiol.* **14**, 541.
Rothstein, M., and Mayoh, H. (1964). *Arch. Biochem. Biophys.* **108**, 134.
Rothstein, M., and Mayoh, H. (1965). *Comp. Biochem. Physiol.* **16**, 361.
Rothstein, M., and Mayoh, H. (1966). *Comp. Biochem. Physiol.* **17**, 1181.
Rothstein, M., and Tomlinson, G. (1962). *Biochim. Biophys. Acta* **63**, 471.
Salmenkova, E. A. (1962). *Med. Parazitol. i Parazitarn. Bolezni* **31**, 664.
Sayre, F. W., Hansen, E. L., Starr, T. J., and Yarwood, E. A. (1961). *Nature* **190**, 1116.
Sayre, F. W., Hansen, E. L., and Yarwood, E. A. (1963). *Exptl. Parasitol.* **13**, 98.
Savel, J. (1955a). *Rev. Pathol. Gen. Comparee* **55**, 52.
Savel, J. (1955b). *Rev. Pathol. Gen. Comparee* **55**, 213.
Sawada, T., Takei, K., Katamine, D., and Yoskimura, T. (1965). *Japan. J. Exptl. Med.* **35**, 125.
Sleeman, H. K. (1961). *Am. J. Trop. Med. Hyg.* **10**, 834.
Sleeman, H. K., and Muschel, L. H. (1961). *Am. J. Trop. Med. Hyg.* **10**, 821.
Smith, H. W. J. (1930). *J. Biol. Chem.* **88**, 97.
Smith, M. H. (1963). *Biochim. Biophys. Acta* **71**, 370.
Smith, M. H., and Lee, D. L. (1963). *Proc Roy. Soc.* **B157**, 234.
Smith, M. H., and Morrison, M. (1963). *Biochim. Biophys. Acta* **71**, 364.
Stoner, R. D., and Hankes, L. V. (1955). *Exptl. Parasitol.* **4**, 435.
Stoner, R. D., and Hankes, L. V. (1958). *Exptl. Parasitol.* **7**, 145.
Tanner, C. E. (1963). *Exptl. Parasitol.* **14**, 337.

Tanner, C. E., and Gregory, J. (1961). *Can. J. Microbiol.* **7**, 473.
Theorell, H. (1934a). *Biochem. Z.* **268**, 73.
Theorell, H. (1934b). *Biochem. Z.* **268**, 64.
Thorson, R. E. (1956). *J. Parasitol.* **42**, 21.
Tormo, J., and Chordi, A. (1965). *Nature* **205**, 983.
Villela, G. C., and Ribeiro, L. P. (1955a). *Anais Acad. Brasil. Cienc.* **27**, 87.
Villela, G. C., and Ribeiro, L. P. (1955b). *Rev. Brasil. Biol.* **15**, 383.
von Brand, T. (1939). *J. Parasitol.* **25**, 329.
von Brand, T. (1952). "Chemical Physiology of Endoparasitic Animals." Academic Press, New York.
von Brand, T., Weinstein, P. P., Mehlman, B., and Weinbach, E. C. (1952). *Exptl. Parasitol.* **1**, 245.
Weatherly, N. F., Hansen, M. F., and Moses, H. C. (1963). *Exptl. Parasitol.* **14**, 37.
Weinstein, P. P. (1960). *In* "Host Influence on Parasite Physiology" (L. A. Stauber, ed.), pp. 65–92. Rutgers Univ. Press, New Brunswick, New Jersey.
Weinstein, P. P., and Haskins, W. T. (1955). *Exptl. Parasitol.* **4**, 226.
Wharton, G. N. (1941). *J. Parasitol.* **27**, 81.
Whittaker, F. H. (1963). *J. Parasitol.* **49**, Suppl., 40.
Wilson, P. A. G. (1965). *Exptl. Parasitol.* **16**, 190.
Wittenberg, B. A., Okazaki, T., and Wittenberg, J. B. (1965). *Biochim. Biophys. Acta* **111**, 485.
Yanagisawa, T. (1955). *Japan. J. Med. Sci. Biol.* **8**, 379.
Yanagisawa, T., and Ishii, K. (1954). *Japan. J. Med. Sci. Biol.* **7**, 215.
Zam, S. G., Thomas, L. J., and Martin, W. E. (1961). *J. Parasitol.* **47**, Suppl., 15.
Zarzycki, J. (1960). *Weterynaria* **9**, 55.
Zinovev, V. P. (1957). *Zool. Zh.* **36**, 617.

CHAPTER 8

Osmotic and Ionic Regulation in Nematodes

Elizabeth J. Arthur[1] *and Richard C. Sanborn*

I. Introduction

The diversity of habitats and modes of existence exhibited by the nematodes presents a distinct problem in a comparative study of ionic and osmotic regulatory processes within the class. The majority of the nematode species are free-living; these species include marine, fresh-water and soil-dwelling forms. However, experimental investigations for the most part have been limited to a few zooparasitic and phytoparasitic species. Concentration on the parasitic forms primarily is due to health and economic problems resulting from parasitic nematodes. Furthermore, the investigator can obtain large numbers of parasitic nematodes easily; this partially compensates for the difficulty in maintaining larval and adult forms of the parasitic nematodes *in vitro* for prolonged periods of time.

The adoption of a parasitic mode of existence presupposes dependence on the host organism to supply factors essential to the life and well-being of the parasite. This dependence includes adaptation of the parasitic

[1] Supported, in part, By NIH Predoctoral Fellowship No. 5 F1-GM-23, 467.

species to an osmotic and ionic environment regulated to be favorable for the cells and tissues of the host. Moreover, the permeability of the coatings or of the integument enclosing the parasite is frequently modified by the host rather than by the nematode within.

A requirement for a host organism may occur at any or all stages of the nematode life cycle. This life cycle includes four juvenile or

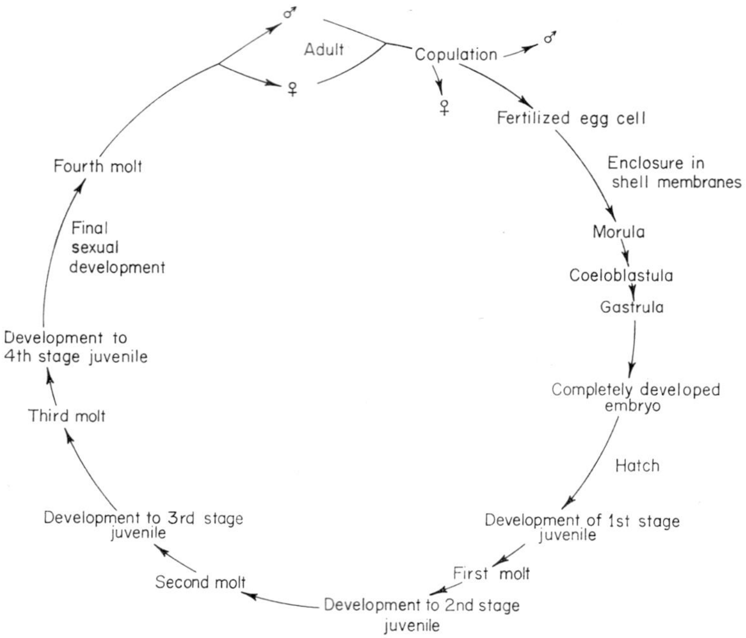

FIG. 1. Nematode life cycle. The stage of development reached prior to release of the egg or embryo by the female is a species variable. For example, *Ascaris lumbricoides* eggs are released from the uterus before development of the fertilized egg occurs, while in *Trichinella spiralis,* the first molt has been completed and the second-stage juvenile is actually hatched *in utero.* (After Hirschmann, 1960; Berntzen, 1965.)

larval stages, isolated from one another by cuticular molts, that intervene between the embryonated egg and the sexually mature adult (see Fig. 1). In the nematode species which have been studied in all stages of the life cycle, embryonated eggs and juveniles often differ from the adult stage and from each other with respect to mechanisms for osmotic and ionic regulation. Because of this, ionic and osmotic regulation in the egg, larval, and adult stages are discussed separately.

II. The Nematode Egg

The membranes surrounding nematode eggs present barriers to the movement of water and most soluble substances, particularly ionizable ones. Thus the eggs, which contain an abundance of available nutrient material, are equipped to survive most adverse environmental conditions encountered prior to the completion of embryonic development.

The fertilized nematode egg is enclosed within three distinct membranes. The outermost layer, containing both protein and carbohydrate components and referred to as the albuminous layer, is a secretion of the parental uterine wall (Chitwood, 1938). Some ovoviviparous species as well as most of the *Rhabditoidea, Tylenchoidea,* and *Strongyloidea* do not form this layer at all (Chitwood and Chitwood, 1950). *Meloidogyne* eggs are deposited embedded in a gelatinous material secreted by the rectal glands, while *Heterodera* eggs are contained within a cyst formed by tanning of the cuticle of the female that produced them (Rogers, 1961; Ellenby, 1946). Both the complex middle coat and the innermost membrane are formed after fertilization by the egg itself (Jacobs and Jones, 1939). The middle coat, or shell proper, has at least two (perhaps three) chemically distinct layers. The outer layer is probably a nonkeratinous scleroprotein (Rogers, 1961), while the inner one contains chitin and protein (Chitwood, 1938; Jacobs and Jones, 1939; Jaskoski, 1962). The innermost, or vitelline membrane, is a lipoprotein. In *Ascaris lumbricoides* eggs, nine amino acids have been identified in the vitelline membrane (Jaskoski, 1962) while the lipid moiety has been shown to consist primarily of ascaroides, the glycosides of ascaryl alcohol (Flury, 1912; Fairbairn, 1955a,b).

The two outer layers of the nematode egg provide structural strength and may be instrumental in preventing desiccation by loss of water. They also prevent the penetration of chemical agents into the perivitelline space (Jaskoski, 1951). For instance, fat solvents cannot penetrate the *Enterobius vermicularis* egg if these coats are intact (Jacobs and Jones, 1939). In *Ascaris lumbricoides* eggs, the scleroprotein layer is dyed by basic vital stains and is permeable to water, but these substances do not penetrate into the perivitelline space once the vitelline membrane has been formed (Wottage, 1937).

The major permeability barrier of the egg shell is the inner or vitelline membrane. Various experimental approaches have been used to demonstrate this fact. Decoating, the removal of the outer layers of the egg shell chemically or by mechanical agitation, can be accomplished while still leaving the vitelline membrane intact. Following 24-hour exposure of decoated *Ascaris lumbricoides* eggs to 2 *N* HCl, 2 *N* HNO_3, 2 *N* NaOH, 0.5 *N* NH_4OH, 3.3 *N* formaldehyde or 4 *N* NaCl, the enclosed embryos remain motile and active. On the other hand, when decoated eggs are hatched by applying pressure lightly on a cover slip, the embryos released are killed almost immediately by exposure to any of these solutions (Fairbairn and Passey, 1955). If the outer layers of the shell are left intact and the eggs heated to about 70°, the vitelline membrane melts. After such treatment, trehalose, a disaccharide found in *Ascaris* perivitelline fluid, diffuses freely through the outer layers and can be detected in the bathing medium (Fairbairn and Passey, 1957). *Heterodera maroni* eggs heated to 70° or washed in absolute alcohol show staining of the enclosed embryo by gentian violet, a basic dye excluded by the normal vitelline membrane (Chitwood, 1938). In partially decoated *Enterobius vemicularis* eggs from which only the outermost albuminous layer has been removed, the embryo stains with neutral red after immersion in fat solvents. Intact eggs are impermeable to this basic vital stain (Jacobs and Jones, 1939). These experiments indicate that the vitelline membrane bars the passage of charged substances into the egg as well as preventing the movement of water-soluble compounds outward.

While the experiments described above imply that the vitelline membrane is impermeable to water as well as to dissolved electrolytes and organic substances, water impermeability is best demonstrated by the behavior of eggs placed in hypotonic or hypertonic solutions. *Meloidogyne arenaria* eggs at the two-celled stage develop to normal, mature larvae in 260 hours at 25° in either distilled water or hypertonic 0.3 *M* NaCl solution. The only difference observed as development proceeds in the two solutions is that the larvae developing in the salt solution fail to move within the egg and do not hatch. Movement of the larvae developing in water begins at 93 hours and continues until they hatch (Dropkin *et al.*, 1958) (see Section III,A). Intact fertilized *Ascaris lumbricoides* or *Trichostrongylus retortaeformis* eggs placed in salt solutions hypotonic to the perivitelline fluid or in distilled water develop to the motile embryo stage in an apparently normal fashion. No swelling of either the eggs or the embryo occurs, indicating that water does not enter the egg under these conditions (Seamster, 1950; Wilson, 1958). Plasmolysis of the embryo does not occur in hypertonic salt solutions until the solutes are present in concentrations osmotically equivalent to 2.6 *M* NaCl, approximately 10 times the concentration of solutes in the *Ascaris* perivitelline fluid (Bataillon, 1901; Zawadowski, 1928; Wilson, 1958). It has been suggested that the plasmolysis observed in extremely hypertonic solutions may be a result of damage to the membrane by the salts, causing it to become permeable (Fauré-Fremiet, 1913). Bataillon's observation that sucrose in any concentration does not cause plasmolysis lends credence to this hypothesis. Since many nematode eggs normally develop in environments where water content is variable, as is the case in agricultural soils or in feces, the barrier to water influx or efflux provided by the vitelline membrane obviously contributes to the survival of the embryo.

The vitelline membrane does allow passage of the respiratory gases, oxygen and carbon dioxide, as well as water vapor and a number of other gases, including hydrogen cyanide, carbon monoxide, and ammonia (Fairbairn, 1957; Seamster, 1950). Permeability to respiratory gases is necessary, since the egg metabolism is aerobic (Costello and Brown, 1962; Costello and Smith, 1964; Bloom and Entner, 1965). Evidence for permeability to oxygen was obtained by Huff (1936) and Jaskoski (1951), both of whom found that normal *Ascaris lumbricoides* eggs consumed a total volume of oxygen of 4 μl per thousand eggs during the 10 days required for development from the single-cell stage to the motile embryo. While Huff reported that removal of the albuminous coat of the eggs resulted in a 5-fold increase in oxygen used, Jaskoski failed to find such an effect. Since a more rapid rate of embryonic development of the decoated eggs which might account for the reported increase in their oxygen consumption was not observed by either author, it seems likely that low-temperature storage of the eggs used by Huff caused the increase which he observed.

Water vapor can enter the embryonating egg through the vitelline membrane and is apparently necessary for normal embryonic development of *Ascaris lumbricoides*. Seamster (1950) reports that in water cultures, fertilized *Ascaris* eggs reach the morula stage in 120 hours, while eggs maintained in air at 31.1° also develop as far as the morula in 120 hours if the relative humidity of the air is 80, 85, or 95%. Development beyond the morula does not occur under these conditions. After 8 days the eggs are desiccated beyond recovery, indicating that loss of water as water vapor can occur. Eggs maintained under conditions of 100% relative humidity at 31.1° develop normally, reaching the motile embryo stage in 200 hours. There is an apparent discrepancy between the permeability of the vitelline membrane to water and to water vapor. The impermeability of the membrane to water may be accounted for by the association of water molecules to form large, loosely bound aggregates or "macromolecules" in the liquid state.

To summarize, during embryonic development the egg shell membranes prevent the movement of charged particles into the perivitelline space. They also prevent passage of water and water-soluble organic compounds either into or out of the egg under the influence of concentration gradients. The exclusion of ionized substances, the prevention of osmotically induced water influx or efflux, and the ease of movement of respiratory gases result in the maintenance of perivitelline conditions that are essentially constant.

III. Larval Stages

A. Hatching and Exsheathment

Hatching requires a change in the permeability of the vitelline membrane to permit osmotic uptake of water as well as movement of inorganic and organic substances into or out of the perivitelline space. This alteration in the osmoregulatory capacity of the membrane may be influenced by the presence of various ions in the external medium. In many nematode species, surface-active agents alter the lipoidal membrane so that it can no longer function as a water-resistant barrier.

The initiation of the hatching process in eggs of nematodes whose first larval stages are free-living seems to be a physiological consequence of the developmental processes of the embyo, since no necessary external stimulus derived from the environment can be detected. On the other hand, eggs of nematodes whose first larval stage is parasitic generally require a stimulus provided by the host organism before hatching can begin. In some zooparasitic species, the infective stage remains ensheathed within the uncast cuticle of the preceding larval stage. This sheath, like the vitelline membrane of the egg, is not permeable to water or solutes. Exsheathment does not occur until an appropriate stimulus provided by the host animal elicits an alteration in the permeability of the sheath so that water and solutes can move through it freely.

Trichostrongylus retortaeformis is a zooparasite whose larvae are free-living in the first stages. The eggs of this nematode do not require any external stimulus in order for hatching to begin. Once the embryo has completed its development, the vitelline membrane becomes permeable to water. Osmotic uptake of water occurs, increasing the hydrostatic pressure within the egg shell, with the result that the enclosed larva loses mobility just before hatching. The actual rupture of the egg shell, releasing the larva, may be completely due to the increased hydrostatic pressure, although the attack on the shell from within by a hatching fluid secreted by the embryo cannot be completely discounted. In study of the hatching process of this nematode *in vitro*, Wilson (1958) found that the normal change in permeability of the vitelline membrane was delayed in the presence of electrolytes. This delay could not be ascribed to the colligative properties of the salts used. Analysis of the depressant effects of 0.05 *N* solutions of CH_3COONa, Na_2SO_4, NaCl, K_2SO_4, KNO_3, KI, Li_2SO_4, $CaCl_2$, and $MgCl_2$ on hatching has led Wilson to claim that the effect of any given electrolyte is controlled by the mobility of its slower ion, regardless of whether this is the anion or the cation. Hence, the effect of the salt on hatching is a consequence of the penetration of the egg membranes by both components of the salt, with specific effects of different ions being relatively unimportant. Apparently the salt stabilizes the vitelline membrane in some manner so that it retains its ability to act as a water-resistant barrier for an extended period of time, delaying the osmotic uptake of water which initiates the hatching process.

The hatching of the eggs of most root-knot nematodes requires the presence of chemical substances or hatching factors which are released by the host plants, although the larvae of some species can emerge in the absence of a hatching factor. *Meloidogyne* spp. show an appreciable hatch in water or dilute salt solutions *in vitro*, and *Heterodera* sp. cysts in soil have been observed to hatch in the absence of host plants. This "spontaneous" hatch of *Heterodera* shows seasonal variations and might be due to the production of hatch-stimulating substances by soil microorganisms (Ellenby, 1963). Furthermore, the influence and specificity of hatching factor on *Heterodera* species is variable. *H. rostochiensis* eggs respond only to diffusates from the root of the normal host plant, the potato. The eggs of *H. schachtii*, the sugar-beet parasite, will hatch in solutions containing root diffusates from a variety of plant species as well as in some inorganic salines (Wallace, 1963). Solutions containing zinc and cadmium ions are more effective than beet-root diffusate solutions in initiating hatching of *H. schachtii* eggs. Solutions containing other cations are less effective and some are highly inhibitory (Wallace, 1956; Clarke and Shepherd, 1965).

The maximum rate of hatching of *H. schachtii* takes place in beet-root diffusates whose concentration is osmotically equivalent to 10^{-2} *M* NaCl. The rate of emergence is reduced in more concentrated solutions of beet-root diffusate. Larval emergence also occurs in 10^{-2} *M* solutions of urea, sucrose, or NaCl, although not to the extent observed in a solution which contains the hatching factor. Increasing the concentration of these solutes above 10^{-2} *M* causes the hatching rate to drop precipitously. Hatching is completely inhibited in 0.65 *M* solutions of urea or sucrose and in 0.42 *M* NaCl. As solute concentration is increased, a marked reduction in the length of the unhatched larvae is observed, indicating osmotic loss of water to the external medium (Wallace, 1956). Mature eggs of *Meloidogyne javanica* and *H. rostochiensis* placed in solutions of NaCl, KCl, $CaCl_2$, or dextrose ranging in concentration from

0.1 to 1.0 *M* (with root diffusate added to the solutions for *H. rostochiensis*) show marked reduction in hatch and an absence of larval movement in all salt solutions whose concentration exceeds 0.2 *M*. Hatching is completely inhibited when salt concentration exceeds 0.3–0.4 *M*. In dextrose solutions, reduction of hatch begins at 0.3 *M*, and inhibition is complete at 1.0 *M*. Transferring *Meloidogyne* eggs to distilled water or *Heterodera* eggs to root diffusate solutions results in the resumption of larval movement and hatching, indicating that the osmotically induced inhibition of hatch is reversible (Dropkin *et al.*, 1958). These results suggest that the osmotic concentration within the shell approximates 0.01 *M* NaCl and is certainly less than 0.2 *M*.

There is a striking parallel between the development of *Meloidogyne* eggs maintained in hypertonic solutions and of *Heterodera* eggs not exposed to root diffusate. Under these conditions, embryonation is either species proceeds to completion, but the larval motion prerequisite to hatching in root-knot nematodes is not observed. *Meloidogyne* larvae begin to move and hatching occurs when the solute concentration is decreased, while the *Heterodera* larvae are activated by the addition of hatching factor to the external medium. In both cases, failure of the egg membranes to admit water, either because of adverse osmotic gradients or due to maintained water impermeability, is involved in hatch inhibition. Solute concentrations in soil water normally are well below the solute concentrations that inhibit hatching of root-knot nematodes *in vitro*. However, drying conditions increase the osmotic pressure of the soil water solution. The permanent wilting point of plants occurs at a soil water free energy of approximately 16 atm., equivalent to the osmotic pressure exerted by 0.7 *M* sucrose or 0.4 *M* NaCl, solutions in which root-knot nematode eggs can develop but will not hatch (Wallace, 1956; Blake, 1961).

The hatching factor for *Heterodera rostochiensis* has been tentatively identified as eclepic acid, a compound containing an unsaturated lactone ring. Eclepic acid is reported to have chemical, physical, and biological properties similar to those of the cardiac glycosides (Ellenby and Gilbert, 1957). *H. rostochiensis* eggs do not hatch *in vitro* in the absence of hatching factor. Tests of the hatching factor in conjunction with the chlorides of Na^{+}, K^{+}, Mg^{++}, and Ca^{++} show that the monovalent ion chlorides reduce hatching while divalent ion chlorides stimulate it. Effects due to osmotic differences between salines were ruled out by adjusting all salines to be osmotically equivalent to 2 m*M* KCl. Further studies of the effects of K^{+} and Ca^{++} ions over a concentration range equivalent to 0.05–100 m*M* KCl indicate that hatching is affected not by the increased cation concentration but by the nature of the cation, with potassium always decreasing hatch and calcium always increasing it by a factor of 2. The depressant effect of potassium can be eliminated by adding 3×10^{-6} *M* quinidine sulfate, a compound known to influence the permeability of cells to potassium (Ellenby and Gilbert, 1958). These observations suggest that the hatching factor promotes ionic changes across the *Heterodera* egg membrane, with a high probability that sodium-potassium transport specifically is affected. The actual mode of action of the hatching factor has not been established. While activity may be associated with the lactone ring, it is also possible that a detergentlike effect may exist. If eclepic acid has the steroidal structure of the cardiac glycosides, the hatching factor may attack the integrity of the lipid membrane of the *Heterodera* egg to alter its osmoregulatory capacity as well as to alter its ability to regulate ionic movement.

In zooparasitic nematodes, the infective stage directly follows a molt in which the membranes of the preceding stage are not cast. Development ceases and is not resumed until the infective larva reaches an appropriate host and is activated (Fairbairn, 1958). Hatching or exsheathment occurs as a result of two separate and distinct events: activation by a stimulus provided by the host animal, and the subsequent release of fluids by the activated nematode larva. This fluid attacks the surrounding membranes to weaken them; the actively moving larva then ruptures the membranes and emerges (Rogers and Sommerville, 1960).

Activation of infective larvae of a wide variety of nematodes is accomplished by a combination of specific conditions encountered in the host animal's alimentary tract. Components of this activating stimulus include a temperature of 37°–38°, and nematode species-specific pH levels and concentrations of CO_2 and undissociated H_2CO_3. Other factors contributing to activation are low oxygen tensions, and the presence of reducing conditions, various host digestive enzymes, and surface-active agents such as bile salts (Rogers, 1958, 1960, 1963).

More detail is known concerning the stimulus conditions necessary to activate infective eggs of *Ascaris lumbricoides* and *Ascaridea galli*. These two species have been studied *in vitro*. To stimulate the eggs of *A. lumbricoides* to hatch in a saline whose temperature is 37°–38° and whose pH is 7.0, brief exposure of the eggs to a pH of 6.0 or lower is required. A carbon dioxide concentration of 5 volume percent and the presence of $0.2–0.3 \times 10^{-3}$ *M* undissociated H_2CO_3 are necessary, as well as nonspecific reducing conditions that may be supplied by cysteine, gluthathione, sodium hydrosulfite, sodium bisulfite, or SO_2. Addition of surface-active agents such as 0.01 *M* sodium taurocholate or 0.001% Tween 80, and NaCl or KCl up to 0.3 *M*, increases the percentage hatching, particularly if stored eggs are used rather than fresh ones (Rogers, 1958). On the other hand, *A. galli* eggs in a physiological saline containing dextrose and $NaHCO_3$, pH 6.8–7.2, show maximum hatching only when trypsin is added to the medium (Hwang, 1960). These observations strongly suggest that the surrounding membrane is permeable to CO_2 or bicarbonate. Bile salts or other detergents would be expected to increase such permeability.

While eggs of either *A. lumbricoides* or *A. galli* hatch 3 hours after the initial exposure to the activating stimulus, the activation of the embryo must occur soon after it encounters the activating stimulus conditions. When *A. lumbricoides* eggs are removed to distilled water after having been exposed to the stimulus for 40 minutes, some of them will hatch. However, the stimulus must act for 90 minutes for optimal results (Fairbairn, 1961). A similar experiment does not appear to have been reported for *A. galli*.

The hatching fluid produced by activated *Ascaris lumbricoides* larvae contains three enzymes, a lipase, a chitinase, and a protease (Rogers, 1958). The lipase acts first, attacking the vitelline membrane to increase its permeability. When *Ascaris* eggs are activated and then refrigerated for 24 hours to prevent completion of the hatching process, analysis of the refrigerated hatching medium indicates the presence of trehalose, a major component of the perivitelline fluid. The detection of trehalose in the medium prior to the rupture of the egg membranes that occurs in hatching proves that the disaccharide has diffused outward through the vitelline membrane because of an increase in its permeability (Fairbairn and Passey, 1957; Fairbairn, 1961). Once the vitelline membrane has become permeable, the chitinase and protease of the hatching fluid diffuse outward to attack the egg shell, hydrolyzing one small area. Larval movement fractures the vitelline membrane and the larva escapes from the shell through this weakened area (Lee, 1965).

Exsheathment of infective larvae follows much the same pattern as hatching of infective eggs. The infective larvae are completely enclosed within one and sometimes two cuticular sheaths. An activating stimulus whose basic components are similar to those required for the activation of infective eggs induces the release of exsheathing fluid by the enclosed larvae. In some species, exposure to digestive enzymes of the host animal is an essential part of the activating stimulus. For example, pepsin is required for *Dictyocaulus viviparus* and *Trichostrongylus columbriformis* larvae to exsheath *in vitro*, while *Haemonchus contortus* larvae will exsheath in a 1% NH_4OH solution saturated with CO_2 (Silverman and Podger, 1964). The digestive enzymes of the host are reported not to digest the sheath. Exsheathment *in vitro* can proceed in distilled water following only 15–30 minutes' exposure of the larvae to the activating stimulus. When infective larvae of *Heamonchus contortus* and *Trichostrongylus axei* enclosed in dialysis sacs are placed in the rumen of sheep, the normal site of ecdysis, larval exsheathment proceeds normally, despite the fact that host enzymes cannot enter the dialysis sacs (Rogers and Sommerville, 1960).

Different nematode species parasitic in the same host animal exsheath in different areas of the host's alimentary tract. Since site specificity is a function of pH and the amounts of dissolved CO_2 and undissociated carbonic acid present, the sheath must be permeable to hydrogen ions and CO_2. *Trichostrongylus axei* requires a pH between 7 and 8, and undissociated carbonic acid in concentrations above 0.5×10^{-3} *M* for maximum exsheathment *in vitro*. *Haemonchus contortus* at the same pH requires 3 times this concentration of H_2CO_3 for exsheathment (Lee, 1965). Sommerville (1957) reports that the ecdysis of *Ostertagia circumcinta*, *H. contortus*, and *T. axei* is initiated in the rumen of the sheep, *T. columbriformis* and *Nematodirus* spp. exsheath in the abomasum, while *Oesophagostomum columbianum* exsheaths in the intestine. While the species normally exsheathing in the more anterior areas of the alimentary tract can also exsheath in more posterior regions, the reverse is not true.

By ligaturing and selective irradiation of activated *Trichostrongylus* larvae, it has been demonstrated that exsheathing fluid is secreted by cells at the base of the esophagus and released through the excretory pore of the larva (Rogers, 1963). This fluid acts on the sheath at an area 19 μ from the anterior end, causing a ring of weakness to appear. The inner cuticle layer is dissolved and the outer layer is split open by the movements of the larva, which then wiggles free (Rogers and Sommerville, 1960).

The activity of the exsheathing fluid produced by stimulated *Haemonchus contortus* larvae is in part due to a leucine aminopeptidase. This enzyme requires Mg^{++} or Mn^{++} as a cofactor, and is unstable in their absence, or below pH 7. The enzyme is species-specific. It has no effect on the sheath of another sheep nematode, *Trichostrongylus columbriformis* (Rogers and Sommerville, 1960; Rogers, 1965).

It has been proposed that the release of hatching and molting fluids is the end result of a neurosecretory process (Sommerville, 1957; Rogers, 1961). With an appropriate stimulus, neurosecretory cells release hormones that activate receptor tissues, stimulating the production or release of hatching or molting fluid. The fact that parasitic nematodes require a stimulus from the host before these processes can begin indicates that in these species, part of the hormonal pathway has been lost. The host stimulus acts indirectly, and the links coupling it with the release of fluids by the nematode larvae are unknown.

Although there is little direct experimental evidence to support these conjectures (however, see Meerovitch, 1965), cells near the nerve ring of *Ascaris lumbricoides* adults show histochemical evidence of neurosecretion. (Gersch and Scheffel, 1958; Davey, 1964). In view of the cell constancy in the nematodes, these cells could have neurosecretory functions in the larval stages as well.

B. Larval Growth and Development *in Vitro*

Short-term studies of regulatory mechanisms in larval nematodes are possible in nonnutrient salines, but carrying either parasitic or free-living forms through all stages of the life cycle *in vitro* requires the use of complex growth media. This suggests that nutritional and physicochemical requirements differ at various stages in the nematode life cycle and offers a fertile field for investigation which is, as yet, relatively untouched (Weinstein and Jones, 1959).

Ascaris lumbricoides larvae are isotonic with 142 m*M* (0.832%) NaCl. They survive for 5½ days in a nonnutrient balanced saline containing 0.8% NaCl, 0.02% KCl, 0.02% $CaCl_2$, and 0.01% $MgCl_2$. Addition of more than 1% glucose to this saline is toxic (Fenwick, 1939). Growth and development of *Ascaris* larvae *in vitro* requires nutrient media and partial to complete anaerobiosis (Cleeland and Laurence, 1962; Cleeland, 1963). Since aerobic conditions are necessary for embryonic development, an abrupt change in metabolic processes occurs at hatching.

Some of the conditions under which three other species of nematodes survive have been defined. *Eustrongylides ignotus* survives longest if the pH of the medium is 7.0, although these larvae can shift the pH of highly acid or alkaline media toward neutrality (von Brand and Simpson, 1942). The free-living soil nematode *Caenorhabditis briggsae* requires various B-vitamins and biotin for growth and development, and aerobic conditions are essential both for egg hatching and normal larval movement (Dougherty *et al.*, 1959; Nicholas *et al.*, 1959, 1962; Nicholas and Jantunen, 1963, 1964). *Trichinella spiralis* larvae require 10% oxygen in order to develop from infective third-stage larvae to sexually competent adults *in vitro*. For normal sheath development in second generation *T. spiralis*, 0.01 *M* cysteine hydrochloride, a reducing agent, must be added to the growth medium (Berntzen, 1965).

C. Osmotic and Ionic Regulation in Nematode Larvae

Most nematodes can tolerate a considerable increase in the osmotic pressure of their body fluid, so hypertonic external conditions present no great regulatory problem. However, hypotonic external conditions often are fatal. Hence, the need for active osmoregulation depends on the environment normally inhabited by the nematode. Those species which are parasites of the animal intestine inhabit a medium in which the ion content fluctuates but normally remains hypertonic to nematode body fluid. In contrast, soil-dwelling nematodes exist in a dilute aqueous solution whose osmotic pressure and ionic composition can vary over a considerable range.

The best-documented case for regulation of water balance in larval nematodes is reported by Weinstein (1952), who worked with third

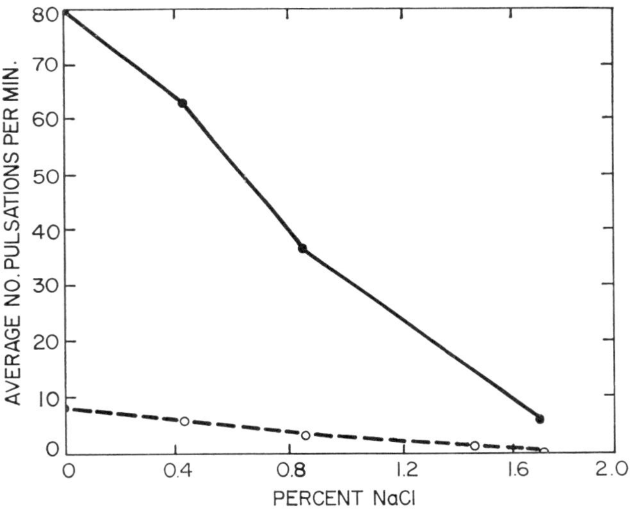

FIG. 2. Regulation of water balance in larval nematodes. ●—● *N. muris* filariform larvae. ○---○ *A. canium* filariform larvae. (After Weinstein, 1952.)

stage larvae of the zooparasites *Nippostrongylus muris* and *Ancylostoma caninum*. In both species, the rate of pulsation of the excretory ampulla, which is the pulsatile portion of the duct connecting the excretory cells with the excretory pore, shows a straight-line inverse relationship with the concentration of NaCl in the external medium over a range of 0–291 m*M* (see Fig. 2). In addition, *N. muris* larvae in sucrose solutions ranging in concentration from 0–438 m*M* exhibit a decrease in pulsation rate with increased solute concentration except for a plateau between 66–248 m*M*. Not only does the rate of pulsation of the ampulla drop to nearly zero in the higher concentrations of either solute, but the percentage of larvae showing pulsations also decreases markedly. Both the pulsation rate and the percentage of larvae showing pulsations rapidly return to the levels observed for distilled water when the larvae are transferred back to that medium.

Regulation of water balance in *N. muris* and *A. caninum* must be fundamentally the same, since the experiments indicate that these larvae adjust to variations in the osmotic pressure of the external medium by a change in the rate of excretion of fluid by the excretory system. The behavior of *N. muris* and *A. caninum* larvae in NaCl solutions differs only quantitatively, with the pulsation rate of the ampulla in *N. muris* being 10 times that of *A. caninum* over the entire range of concentrations tested (Weinstein, 1952). In addition to water and ions, the excretory fluid of *N. muris* larvae displays positive qualitative tests for aliphatic amines, 1,2-dicarboxylic acids, and ammonium salts (Weinstein and Haskins, 1955).

When *Trichinella spiralis* larvae are maintained aerobically for 24 hours in a nonnutritive medium, their nitrogenous excretory products include ammonia, peptides, amines, and free amino acids. Neither urea nor uric acid are excreted under these conditions (Haskins and Weinstein, 1957a,b,c). These observations suggest that *T. spiralis* larvae regulate the composition of the body fluid in a manner similar to that employed by *N. muris* and *A. caninum.*

While *N. muris* and *A. caninum* larvae can regulate water balance actively when they encounter either hypotonic or hypertonic conditions, other larvae are poikilosmotic, absorbing or losing water through the cuticle in response to environmental conditions. *Ascaris* larval life *in vitro* is markedly shortened if the larvae are maintained in hypotonic saline, osmotically equivalent to 120 m*M* NaCl, while in hypertonic salines, osmotically equivalent to 188 m*M* NaCl or more, osmotic effects on life span are less marked (Fenwick, 1939).

Fourth-stage larvae of the soil nematode *Ditylenchus dipsaci* decrease in length in urea solutions ranging in concentration from 0.001 to 1.0 *M*, while death by exosmosis occurs in 1 *M* and 2 *M* solutions. The larvae move less rapidly when solute concentration is greater than 0.1 *M*. Larvae kept in 0.001–0.3 *M* urea solutions for 36 hours regain normal length and activity if they are transferred to distilled water. In sand wetted with either distilled water or urea solutions, larval movement is not affected by osmotic pressure until solute concentration exceeds 0.3 *M*, a concentration considerably above the 0.05 *M* solution concentration normally found in soil water (Blake, 1961). While there are no reported values for isotonicity for soil nematode larvae, maximum emergence of *Heterodera schachtii* larvae occurs in solutions osmotically equivalent to 10 m*M* NaCl (Wallace, 1956, 1963). Thus, it is obvious that soil nematodes can survive in extremely hypertonic conditions, perhaps by osmoregulation.

Very little is known concerning osmotic effects on the activity of nematodes inhabiting environments such as intertidal areas, salt marshes, or estuaries, where the salinity changes frequently. In a study of the marine nematode population of estuarine intertidal mud, Capstick (1959) found that the population density of *Enoplus brevis* decreased with the decrease in water salinity toward the head of the estuary. Thus, it is probable that this nematode cannot osmoregulate to adjust to hypotonic conditions.

Rhabditis dubia, a nematode inhabiting deposits of cow dung, lives in an environment where the osmotic pressure increases as the deposit dries. Several *Rhabditis* generations can develop in one dung deposit. However, as the deposit dries out, a resistant third-stage *dauer* larva is produced. *Dauer* larvae resemble infective larvae of animal parasites in that they are ensheathed and do not feed. The *dauer* larva migrates to the top of the drying dung deposit and is transferred by an insect carrier to freshly deposited dung, where the larva exsheaths and continues development (Lee, 1965).

The abundance or deficiency of certain ions in the environment may affect larval development (Oteifa, 1953; Curtis, 1964; Marks and Sayre, 1964). Both field and laboratory studies have compared the population size of various root-knot nematode species with host-plant levels of potassium, calcium, magnesium, nitrogen, and phosphorus (Dropkin, 1955; Dropkin and King, 1956; Kirkpatrick *et al.,* 1964; Heald and Jenkins, 1963). In addition, attempts have been made to relate the cobalt intake of host animals to the severity of *Haemonchus* infections (Shumard and Herrick, 1954; Downey, 1965). While no clear-cut relationship have been observed, the data indicate that inorganic substances offset parasitic nematode populations in large part through their effect on the metabolism of the host organism. Nevertheless, a direct effect of the ions on larval metabolic processes has not been ruled out.

IV. Adult Nematodes

A. Systems Concerned with Osmotic and Ionic Regulation

The structures most likely to be involved in osmotic and ionic regulation in the larval and adult stages are the body wall, the alimentary canal, and the excretory system. Part of the rationale for believing that all these structures have regulatory functions in adult nematodes is based on morphological and histochemical evidence.

Nematode excretory systems run the structural gamut from a single excretory gland cell with a duct leading to the excretory pore, to a network of lateral tubules and gland cells connected to an excretory sinus which empties through a terminal duct. In some species the system is totally absent (Chitwood and Chitwood, 1950). Only in a few cases has evidence been presented for osmoregulatory activity or even for the excretion of metabolic waste products by this system in adult nematodes. *Rhabditis* spp., *Ditylenchus triformis*, and *Paraphelenchus myceliophthorus* all have been observed to extrude a relatively insoluble material from the excretory pore (Chitwood and Chitwood, 1950; Myers, 1963; Goodey, 1959). Some zooparasitic species excrete administered fluorescent dyes through the excretory system (Beherenz, 1956). Detection of proteolytic enzymes and esterases in the walls of the excretory canals of *Ascaris lumbricoides* indicates that these tissues are metabolically active, but direct evidence for activity in water regulation is lacking (Lee, 1962a,b).

Hypodermal gland cells which open externally through cuticular pores may also be concerned with osmotic or ionic regulation. In species of *Trichuroidea,* which lack an excretory system, these gland cells form the complex structure known as the bacillary band. The apical plasma membrane is elaborately developed to form the lamellar apparatus, similar to membrane structures present in gland cells whose function is ionic and osmotic regulation in other organisms (e.g., the green gland of *Cambarus* and the chloride cells of the seagull). Morphological evidence, therefore, is consistent with a regulatory function for the hypodermal gland cells in the nematodes (Wright, 1963).

The nematode alimentary canal is a straight tube separated into three distinct areas by muscular valves. The epithelium of the intestine proper, the central and longest portion of the alimentary tract, is a single layer of columnar cells. Numerous microvilli are formed on the apical side of each cell by orderly projections of the plasma membrane. Electron micrographs of the apical membrane sometimes show the occurrence of blebbing, suggesting that these cells have secretory as well as absorptive functions. An abundance of mitochondria in the apical region and a well-developed endoplasmic reticulum, together with an abundance of glycogen granules, lipid droplets, and membrane-bound secretory granules, all point toward a high level of metabolic activity in these cells (Kessel *et al.*, 1960; Sheffield, 1964).

The body wall of adult nematodes consists of the cuticle, the hypodermis, and a single layer of obliquely striated, longitudinal muscle cells. The cuticle is an organized secretion of the hypodermis (Bird and Deutsch, 1957; Inglis, 1964). Newly molted 1–2 cm. *Ascaris lumbricoides* adults increase to their final 25–30 cm. length over a period of several months. During this period, the cuticle increases in total thickness in direct proportion to the change in length, with each cuticular component increasing differentially. The hypodermis of the young adult contains abundant endoplasmic reticulum, free ribosomes, many cisternae lying close to the basal cuticular layer, and much RNA. These are all much reduced in the full-sized adult, indicating that active protein synthesis proceeds at a greater rate in young than in older worms. The intimate connection of the hypodermis with all the cuticular layers by a "punctation canal network" permits precursor substances produced in the hypodermis to be transported to the cuticle where final synthesis of cuticular proteins may occur (Inglis, 1964; Watson, 1965a,b).

While the cuticle is a protective covering, it is selectively permeable and metabolically active. A very thin (1000 Å.) osmophilic superficial lipid membrane external to the cuticular cortex has been identified in electron micrographs (Bird and Deutsch, 1957; Watson, 1965a,b). The proteinaceous cortex layer, resistant to collagenase, is not attacked by animal proteolytic enzymes, although plant proteolytic enzymes such as papain digest it (Chitwood, 1938). The middle or matrix layer is hydrolyzed by pepsin or trypsin. The presence of esterases in the matrix layer indicates that it is metabolically active (Lee, 1961). The basal layer of the cuticle is intimately associated with the hypodermis, and is the site of attachment of the muscle cells (Inglis, 1964; Watson, 1965a,b).

In some of the larger nematodes, fibrillar layers are found between the matrix and basal layers. In *Ascaris lumbricoides,* three successive layers of parallel strands of collagenous fibers form a mobile latticework. The three fiber layers are arranged at angles to one another in such a way that a change in tone of the longitudinal muscles to cause a change in body volume does not greatly reduce the internal hydrostatic pressure exerted by the incompressible perivisceral fluid (Harris and Crofton, 1957; Watson, 1965b). In *Ascaris,* the internal fluid pressure is high and varies rhythmically. Harris and Crofton (1957) report the average pressure to be 70 mm. Hg, while Harpur (1964) gives an average value of 120 mm. Hg. The higher figures appears to be more valid. The passage of substances through the body wall is markedly affected by the internal hydrostatic pressure.

B. The Body Fluid: Artificial Salines

The cell-free perivisceral fluid is the extracellular fluid of nematodes. The physical and chemical properties of the perivisceral fluid have been determined only for some of the large zooparasites. Fresh body fluid of *Ascaris lumbricoides* is reddish in color (due to dissolved hemoglobin) and has a distinctive odor. Spectrographic studies of the hemoglobins of *Ascaris lumbricoides* and *Syngamus trachea* prove that the hemoglobin of nematode body fluids is not identical with the hemoglobin of the host (Wittenberg *et al.*, 1965; Kreuger, 1959). The tenacious oxygen binding precludes a respiratory function for the pigment; it is probable that the characteristic hematin compounds of *Ascaris* eggs are derived from the hemoglobin of the perivisceral fluid.

The osmotic pressure of *Ascaris* body fluid is equivalent to 198 m*M* NaCl, its conductivity to 143 m*M* NaCl, and the pH is 6.8. Total solids equal 6.8 gm./ml. of fluid (Hobson *et al.*, 1952a; Rogers, 1945). *Ascaris* body fluid contains a variety of inorganic ions as well as carbohydrates, fats and fatty acids, proteins, polypeptides, free amino acids, enzymes, and metabolic waste products (for a detailed analysis, see Dittmer, 1961). The inorganic anion concentration of the body fluid is less than the cation concentration. The anionic difference is made up by the C_2–C_6 fatty acids, succinic acid, and possibly by 1,2-propanediol-2-phosphate and bicarbonate (Bueding and Farrow, 1956; Fairbairn, 1957; Kubistova and Seth, 1966). Baldwin and Moyle (1947) suggest that the fatty acids may have an osmotic role similar to that of urea in the blood and tissues of elasmobranchs. Rogers' observation (1945) that in *Ascaris* kept 5–6 days in Baldwin's saline, there is a replacement of the fatty acid moiety of the body fluid by chloride speaks to this point.

In the absence of direct analyses, some hints as to the nature of body fluids can be gained from the saline solutions in which physiological activities can be carried out. Table I lists physiological salines which have been used with a variety of nematodes.

Changes in the composition of the bathing solution affect specific metabolic processes and survival of various nematode species *in vitro.* Examples are given below. Potassium stimulates metabolic activity in *Eustrongylus ignotus* and markedly decreases it in *Litomosoides carinii.* In *L. carinii,* metabolic activity also is decreased when the Ca^{++} concentration of the saline is increased (von Brand, 1943; Bueding, 1949). *Ostertagia circumcinta* has a critical requirement for K^+ and Ca^{++}, while Mg^{++} has no effect on its survival (Bueding and Most, 1953). Weinstein (1960) mentions the importance of the Na^+/K^+ ratio and the Mg^{++} dextrose relationship in artificial salines to the survival of nematodes *in vitro.*

Krotov and Kats (1963) have studied the effect of changes in Locke's solution on egg laying by *Ascaris lumbricoides.* They find that the number of eggs laid decreases with time and with lowered temperature. Egg production is increased by the addition of Na^+ and K^+ ions to the medium, as well as by the addition of vegetable fats, butterfat, egg protein, starch, or glucose. Egg laying decreases when Ca^{++} and Mg^{++} are added and when the pH is reduced to 4.0–5.0. Pollack (1957) reports that addition of NH_4Cl to the nonnutrient saline of Baldwin and Moyle prevents the depletion of NH_4^+ ion and nonprotein amino acids, particularly alanine, in the *Ascaris* ovary, indicating that *Ascaris* ovarian tissue is capable of *de novo* synthesis of amino acids if ammonium ion is available from the environment.

Certain ions inhibit or stimulate various enzymically controlled processes. For maximum production of pyruvate in *Ascaris lumbricoides*

TABLE I
COMMONLY USED NEMATODE SALINES

	Constituents (moles/liter)									
Saline	Na	K	Ca	Mg	Cl	SO_4	HCO_3	PO_4	Fatty acid rplt.	Sucrose
Widely used salines										
Kronecker's	172				172					
Locke's	156.4	6	2		164		2.4			11.1[a]
Ascaris salines										
Fenwick	137	3	2	2	148					
Baldwin	136	3	2	0.4	141	0.4				
Baldwin and Moyle	130	24	6	5	170		6			
Harpur and Popkin	168	5			163		10			
Hughes and Arthur	134	16	7	4	53	4	16		95[b]	29
Brading and Caldwell	130	20	6	10	54	35		5	58[c]	82
30% sea water	141	3	3	16	164	8.5				
Ascaris body fluid	129	24.6	5.9	4.9	52.7				122.5	
Pig intestinal fluid	124	27	14	6	61					

[a] Dextrose.
[b] As isethionate and gluconate.
[c] As acetate.

mitochondrial preparations, Mg^{++} is required (Seidman and Entner, 1961). Oxidation of succinate and reduced nicotinamide-adenine dinucleatide (NADH) by *Ascaris* mitochondrial preparations is dependent on the presence of Mn^{++} (Kmetic and Bueding, 1961; Bueding, 1962). Proteolytic activity of *Trichuris muris* with casein or hemoglobin as substrate is stimulated by 0.01 *M* Mg^{++}. *T. muris* ribonuclease is inhibited by Cu^{++} and Zn^{++} at 2.5×10^{-3} *M*. Either ion has a slight stimulatory effect on the ribonuclease at a concentration of 2.5×10^{-5} *M*, while

Mg^{++} at this concentration is 3 times as effective. *T. muris* also contains an acid phosphatase which is stimulated by Mg^{++}, Mn^{++}, and Co^{++}, and inhibited by cyanide, F^-, and MoO_4^{--} (Nimmo-Smith and Keeling, 1960). An inorganic pyrophosphatase of *Ascaridea galli* is inactive in the absence of divalent cations. Mg^{++} is twice as effective as Zn^{++} in activating the enzyme. Other divalent cations are less than one-tenth as effective as Mg^{++} in activating the enzyme (Marsh and Kelley, 1959). Such observations as these suggest that nematode cells or body fluid contain Mn^{++} and cobalt although their presence has not been reported in *Ascaris* body fluid.

C. Cuticular Permeability

The body wall permits free movement of water molecules and is differentially permeable to the other molecules and to ions. The outer cuticular layer with superficial lipid membrane is apparently the major permeability barrier.

When eviscerated cylinders of *Ascaris lumbricoides* body wall are filled with experimental solutions, urea, potassium iodide, and the vital stains methylene blue and neutral red diffuse into the bathing medium readily. Glucose cannot be detected in the external medium unless its concentration in the cylinder is raised to 5% and the incubation time extended to 10 hours (Mueller, 1928). When all muscle and hypodermal tissues are scraped from the cylinders, the cuticle is freely permeable not only to glucose but to a number of other nonelectrolytes (Hobson, 1948). These experiments indicate that the permeability barrier to metabolically required carbohydrates lies in the hypodermis or the muscle layer, not in the cuticle.

Bremner (1961) reports that as copper concentrations of host tissue increase, the copper content of the intestinal parasites *Haemonchus placei* and *Oesophagostomum radiatum* also increases. Both of these nematodes are tissue feeders and are attached to the mucosa of the abomasum. Analysis of the fluid of the abomasum shows that as the copper concentration increases in the fluid, it increases in the nematode tissue. He believes that this indicates that uptake of copper in these two species is through the body wall.

Investigations of the efficacy of antihelminthics provide additional information about cuticular permeability. Phenothiazine enters *Ascaridea galli* through the cuticle. Ligaturing oral and anal openings produces no change in the uptake of phenothiazine, and incubation of drug-treated worms in drug-free saline does not result in a loss of the drug from the tissues (Lazarus and Rogers, 1950). The rate of uptake of 4-hexylresorcinol by *Ascaris lumbricoides* is unaffected by ligaturing. Tissue concentrations of hexylresorcinol increase with time of exposure to drug-containing saline. Transfer of drug-treated worms to drug-free saline does not decrease the tissue concentration of hexylresorcinol (Trim, 1944). Therefore, these compounds diffuse inward through the cuticle, but they can neither move outward by passive diffusion through the cuticle nor be actively expelled.

Uptake of 4-hexylresorcinol is inhibited by the presence of the host's intestinal mucin and bile salts, as well as by certain soaps. However, drug uptake is increased when less than 0.25% sodium oleate is present in the medium (Trim, 1944). Sodium oleate at 0.01–0.2% concentration increases the uptake of hexylresorcinol by *Nippostrongylus muris,* while 0.2–1.0% concentrations are inhibitory (Rogers, 1944). The concentrations of sodium oleate promoting drug uptake are those that would produce minimum tensions at an oil-water interface. Thus, these nonpolar compounds enter the cuticle with some difficulty. Fatty acid soaps are effective in promoting passage through the cuticle in proportion to their effectiveness in solubilizing them in lipids.

Nicotine enters *Ascaris* almost exclusively through the cuticle. The rate of penetration is accelerated by the presence of surface-active agents. Since both cationic and anionic detergents are equally effective in accelerating nicotine uptake, the detergents must act by altering the cuticle itself. Cuticular breakdown by detergent action is further supported by the fact that detergent treatment causes a loss of inorganic phosphate from ligatured *Ascaris* that cannot be prevented by transferring the worms to normal saline (Trim, 1949).

Trim (1949) proposed that these observations could be explained by assuming the presence of an outer, thin, homogeneous lipid layer. Increased permeability caused by the presence of detergent substances would be due to the removal or disorientation of this layer by the lipid-solvent action of the detergent miscelles. The thin external osmophilic layer of the cuticular cortex detected by electron microscopy could be the lipoid layer whose presence Trim postulated (Bird and Deutsch, 1957; Watson, 1965a,b). However, as Rogers (1961) points out, while there is little doubt that lipids are present on the cuticular surface and that they facilitate the entry of fat-soluble substances, this superficial lipid layer is clearly incomplete since it does not present a permeability barrier to the entry of water. There is also the problem of why antihelminthic drugs can enter through the cuticle but do not leave the body when the concentration gradient is reversed. The inhibitory role of host intestinal mucin is not understood. Investigations concerned with the penetration of the cuticle by anionic substances similar to those already carried out with cationic vital dyes and nicotine might provide information which would help to elucidate the question of the differential permeability of the nematode cuticle.

D. Osmoregulation in Adult Nematodes

In addition to the elimination of water from the body across the cuticle, by the alimentary tract and by the excretory system, there are other mechanisms that contribute to osmoregulation. High internal turgor pressure slows the rate of water influx from hypotonic solutions. Internal osmotic pressure can be reduced by excretion of metabolic waste products or of electrolytes from the body fluid, while an adjustment to hypertonic conditions can be made by uptake of ions from the bathing solution.

Rhabditis terrestris, a soil nematode, maintains its internal osmotic pressure higher than that of the environment. The evidence is quite convincing that it does so by active osmoregulation. When *R. terrestris* adults are transferred from isotonic saline, osmotically equivalent to 20 m*M* NaCl, to distilled water, they increase 13% in length within 5 minutes. They are less active during the first 5–10 minutes, and the rhythmic pumping action of the esophageal bulb ceases. Left in distilled water, they return to their original length and activity in 25–30 minutes. In the latter half of the recovery period, considerable amounts of water are expelled through the anus. Worms shrink when placed in hypertonic salines equivalent to 0.34 and 0.51 *M* NaCl. While there is also an initial increase in activity, movement is aberrant and soon ceases completely. Esophageal pumping also ceases and the intestinal length is reduced. Partial recovery of body length and movement occurs after 20–25 hours in 0.34 *M* saline. In 0.51 *M* saline, length recovery is even slower and activity never resumes (Stephenson, 1942).

Osmoregulation in *R. terrestris* is not accomplished by the alimentary canal alone. Water must be actively eliminated through the cuticle also, since recovery in distilled water begins before esophageal pumping is resumed or defecation occurs. Hypertonic external conditions result in an osmotic efflux of water, probably accompanied by the uptake of electrolytes through the cuticle. The hypertrophy of the lateral and postanal organs which sometimes occurs indicates that recovery in hypertonic media involves the slow removal of electrolytes from the body fluid. Since recovery is never complete, these organs apparently are unable to completely restore electrolyte balance as long as external conditions remain extremely hypertonic (Stephenson, 1942).

The body fluids of *Angusticaecum* sp., an intestinal parasite of the tortoise, can remain hypertonic to tapwater for 4 days. On the other hand, when worms are kept in 50% or 100% seawater, the osmotic pressure of the body fluids increases to that of the bathing solution. These observations suggest that *Angusticaecum* lacks a regulatory mechanism adequate to prevent osmotic water loss in hypertonic solutions but that it is capable of regulation in hypotonic media. Regulation is a function of the body wall, not the intestinal epithelium. Ligaturing a worm at both ends and transferring it from tapwater to seawater results in an increase in the osmotic pressure of the body fluids, while worms ligatured in seawater and transferred to tapwater show a decrease in the osmotic pressure of the body fluid. There is no appreciable change in body size in either case. Hence, the change in osmotic pressure of the body fluid is not only due to the passage of water through the cuticle, but also to a slow movement of ions in or out (Panikkar and Sproston, 1941). Since these observations were made using only 14 worms taken from a tortoise that had been dead for 3 days, further studies are desirable.

Hammerschmidtiella diesingi, a parasite of the cockroach hindgut, is subjected to marked changes in osmotic pressure during the cockroach molt. When the thin cuticular lining of the cockroach gut separates from the hypodermis during molting, the nematodes found in it are shrunken and inactive, indicating that the environment is hypertonic. As the molt nears completion, the nematodes become more active and are able to move to their normal position in the gut. These observations suggest that *H. diesingi* possesses active osmoregulatory mechanisms. Experiments *in vitro* bear this out. Both 150 m*M* NaCl or 30% seawater are isotonic for *H. diesingi.* Specimens placed in 200 or 300 m*M* NaCl or 50% seawater shrink. They recover to their original length within 24 hours in these solutions, and they swell when they are transferred back to isotonic saline. Worms taken from molting cockroaches and placed in isotonic saline also swell. However, when the worms are placed in hypertonic sucrose, they decrease in length steadily and no recovery occurs. When specimens of *H. diesingi* are transferred from isotonic salines to distilled water, they swell rapidly and burst. When they are transferred from isotonic saline to hypotonic 50 or 80 m*M* NaCl solutions, they swell. They do not return to original length even after 24 hours in hypotonic salines (Lee, 1960).

Unlike *Augusticaecum, H. diesingi* is capable of osmoregulation in hypertonic salines, while it lacks the ability to adjust to hypotonic conditions. The fact that the nematode returns to its original length in hypertonic saline and fails to do so in hypertonic sucrose indicates that the regulatory mechanism involves uptake of ions from the bathing medium. In the absence of ligaturing studies, the site of the regulatory mechanism in *H. diesingi* cannot be established.

In contrast to the nematodes already discussed, the freshwater nematode *Punctodora ratzeburgensis* returns to its original body length in a matter of seconds when it is transferred to solutions which are either hypotonic or hypertonic to the original medium (Inglis, 1964). This strongly suggests that there is an active osmoregulatory mechanism in this nematode. On the basis of the rapidity of adaptation, it may not be ionic in nature.

Ascaris lumbricoides and *A. megalocephala* are poikilosmotic. Schopfer (1932) found that adult *A. megalocephala* females gained weight in hypotonic salines and lost weight in hypertonic ones. The osmotic pressure of the perivisceral fluid decreased in hypotonic solutions. When diluted horse intestinal fluid was used as the bathing medium, the weight increase of the parasites was approximately proportional to the dilution factor. Hobson *et al.* (1952a) proved that water flux in *Ascaris lumbricoides* occurs across both the cuticle and the alimentary tract. In hypotonic 20% seawater, the weight increase of worms whose oral and anal openings were ligatured was significantly less than that of nonligatured worms. Neither *Ascaris* species is capable of active osmoregulation, since there is no reversal of the osmotically induced swelling or shrinking even if the worms remain in the hypotonic or hypertonic media for days.

E. Ionic Regulation in Adult Nematodes

Selective ion movement has been suggested as an osmoregulatory mechanism by several workers. In his study of the permeability of the *Rhabditis terrestris* cuticle to water, Stephenson (1942) proposed that the adjustment of internal osmotic pressure in hypertonic media required the uptake of osmotically active substances. Subsequent experiments showed that the *R. terrestris* cuticle was not penetrated at the same rate by all cations. When worms were transferred from a balanced isotonic saline, osmotically equivalent to 20 m*M* NaCl, to hypertonic solutions of NaCl, KCl, $CaCl_2$ or $MgCl_2$, osmotically equivalent to 510 m*M* NaCl, shrinking occurred. Reduction in length was unequal in the four salines. Most shrinkage occurred in KCl, followed in order by $MgCl_2$, $CaCl_2$, or balanced saline, while least shrinkage occurred in NaCl (Stephenson, 1944).

In hypertonic saline, either a water efflux, an ion influx, or both could occur. An efflux of water must cause shrinking. This shrinking would be less extensive if ions were able to enter the body to effect more rapid osmotic equilibration. If water efflux is a constant in all equiosmolar solutions, then the greater the reduction in length at equilibrium, the slower the influx of ions from the bathing medium. Stephenson's data show that in *R. terrestris,* the rates of penetration of the cuticle for the four cations tested are $Na^+ > Ca^{++} > Mg^{++} > K^+$.

The small size of *R. terrestris* precludes an analysis of the body fluid to determine quantitative changes in composition resulting from ion movement. *Ascaris lumbricoides,* a larger nematode, is more suitable for studies of the electrolyte content of body fluid under normal and abnormal environmental conditions. The inorganic ion content of the body fluid of *Ascaris* removed from the host intestine immediately after the pig has been slaughtered is given in Table II.

Ascaris body fluid is hypotonic to the intestinal contents of the pig. The osmolality of pig intestinal fluid is 427 mosmoles/liter, and that of fresh *Ascaris* body fluid is 323 mosmoles/liter. When *Ascaris* specimens are placed in saline (330 mosmoles/liter), at 38°, the internal osmotic pressure remains almost unchanged, but the osmotic gradient is reversed within 7 hours, with the body fluid osmolality increasing

TABLE II
INORGANIC CONSTITUENTS OF *Ascaris* BODY FLUID AND HOST INTESTINAL FLUID

Fluid	Constituent (mmoles/liter)					
	Na	K	Ca	Mg	Cl	Total P
Ascaris body fluid	119.2	—	—	—	55.5	12.0
Ascaris body fluid (Hobson *et al.*, 1952)	129	24.6	5.9	4.5	52.7	17.0
Pig intestinal fluid (Hobson *et al.*, 1952)	124	26.7	4.5	5.5	61	24.2

to 353 mosmoles/liter. The reversal is assisted by the ejection of hypotonic feces. Reversal is delayed at 30°, possibly due to physiological closure of the intestine (Hobson *et al.*, 1952a; Harpur and Popkin, 1965).

The increase in body fluid osmolality in saline is accompanied by changes in the ionic composition of the body fluid. Ion movement occurs primarily across the *Ascaris* body wall. The total ionic concentration of the body fluids of ligatured *Ascaris* kept in 30% seawater is equivalent to 142 m*M* NaCl, while that of nonligatured worms is equivalent to 134 m*M* NaCl. These figures indicate that little ion movement occurs across the intestinal epithelium (Hobson *et al.*, 1952a). However, tracer studies using ^{32}P-labeled phosphate prove that phosphate is absorbed almost exclusively by the alimentary tract. While some ^{32}P-labeled phosphate does enter the *Ascaris* cuticle, its passage from the cuticle to other tissues does not occur. Radioactivity is detected in *Nippostrongylus muris,* a tissue feeder, when ^{32}P-labeled phosphate is injected into the bloodstream of the host but not when it is administered to the host *per os*. Therefore phosphate uptake of *N. muris* is also primarily across its intestinal epithelium. Studies of sheep nematodes *Trichostrongylus* sp. and *Haemonchus contortus*, both tissue feeders, indicate that *Trichostrongylus* is capable of phosphate uptake through the cuticle as well as the intestinal epithelium, while *H. contortus* is not. Finally, *Ascaridea galli* shows no evidence of ^{32}P-labeled phosphate uptake by either the cuticle or the intestinal epithelium (Rogers and Lazarus, 1949; Esserman and Sambell, 1951). These observations indicate the problem inherent in generalizing from one nematode species to the entire class.

Determinations of ionic composition of the body fluid of *Ascaris* maintained in various artificial salines indicate that phosphate, normally lower

TABLE III
Ascaris BODY FLUID IONIC COMPOSITION MEASURED AFTER 24–48 HOURS IN SALINE

	Constituent (mmoles/liter)					
Saline used	Na	K	Ca	Mg	Cl	Total P
30% seawater	133	11.5	4.7	4.9	70.5	13.2
Hobson saline A	121	10.4	4.4	4.3	76.0	12.8
Hobson saline B	148.5	15.4	4.6	5.8	78	13.7
Baldwin's	123.4	23.5	—	4.94	59.2	13.2
Saline A	126	20.2	19.7	3.3	162.4	2.4
Saline B	137	38.9	3.2	9.5	200.0	2.4

in perivisceral fluid than in pig intestinal fluid, remains at a fairly constant level even when saline phosphate concentrations are low (see Table III). The Ca^{++} and Mg^{++} concentrations of the body fluid also remain relatively constant regardless of the concentration of these ions in the external medium. While the body fluid Na^{+} and K^{+} concentrations are somewhat variable, the variations do not appear to depend directly on the concentration of the ions in the bathing solution (Rogers, 1945; Hobson *et al.*, 1952b). However, K^{+} concentration of the body fluid is lower in worms which have been kept in saline than it is in fresh body fluid (see Table IV). Recent investigations indicate that the decrease in K^{+} concentration of the perivisceral fluid occurs very soon after transfer to saline (Harpur and Popkin, 1965; Arthur and Sanborn, unpublished). Comparing the K^{+} concentrations measured by Harpur and Popkin and by Arthur and Sanborn with those of Hobson *et al.*

it appears that the initial rapid decrease is followed by a period during which body fluids K^+ remains low but fairly constant. This, in turn, is succeeded by a very gradual rise in the K^+ concentration of the body fluid.

The chloride concentration of *Ascaris* body fluid is slightly less than that of pig intestinal fluid. When worms are kept 5–6 days in Baldwin's saline, there is a replacement of the fatty acid moiety of the body fluid by Cl^-. This increase in the body fluid Cl^- concentration is more pronounced when worms are maintained in phosphate-free media. However, body fluid Cl^- always is much lower than the Cl^- concentration of this

TABLE IV
Ascaris Body Fluid Potassium Concentrations

Investigator and medium used	Maintenance time in medium (hours)	Potassium concentration	
		Medium (mmoles/liter)	Body fluid (mmoles/liter)
Hobson, pig intestinal	—	26.7	24.6
Hobson, 30% seawater	44	3	11.5
	96	3	12.3
Hobson, Saline A	48	20.2	10.4
	96	20.2	12.4
Hobson, Saline B	48	38.9	15.4
	96	38.9	18.7
Rogers, Baldwin's	24	3	23.5
Harpur and Popkin	2–8	5	13.3
Arthur and Sanborn	3–5	24	17.2

saline (Rogers, 1945). Other investigators also report that the Cl^- concentration of the perivisceral fluid is maintained below that of the bathing saline (see Table V).

While the body fluid Cl^- concentration does increase above the normal value when the Cl^- concentration of the bathing saline is high, the change is independent of time. This, together with the fact that the Cl^- concentration of the body fluids is always less than that of the bathing saline, suggests that *Ascaris* regulates Cl^- concentration by active excretion of the ion against a concentration gradient. Experiments in which worms are transferred to a medium isotonic with 30% seawater, but with the Cl^- concentration reduced to 83 m*M* by replacement of NaCl with $NaNO_3$, bear this out. When worms exposed to 30% seawater (166 m*M* Cl^-) to increase the Cl^- concentration of the body fluid to 72 m*M* are transferred to the nitrate-substituted saline, the Cl^- concentration of the body fluid drops from 72 m*M* to 56 m*M* in 48 hours (Hobson *et al.*, 1952a).

The mechanism for regulating Cl^- concentration is located in the body wall, since the body fluid Cl^- concentration of ligatured worms kept in 30% seawater is 68 m*M*, compared with a Cl^- concentration of 78 m*M* in nonligatured worms. When eviscerated cylinders of complete body wall and cuticular cylinders from which all muscle and hypodermal tissue has been removed are filled with perivisceral fluid and kept in 30% seawater for 48 hours, the fluid contained in the body wall cylinders has a total ion concentration equivalent to 128 m*M* NaCl and 64.6 m*M* Cl^-. The fluid contained in the cuticular cylinders has a total ion concentration equivalent to 160 m*M* NaCl and 131 m*M* Cl^- (Hobson *et al.*, 1952a). Thus, the cuticle presents little opposition to the free movement of chloride ions, and the Cl^- regulatory mechanism must lie either in the hypodermis or in the somatic musculature.

TABLE V
Ascaris Body Fluid Chloride Concentrations

	Chloride concentration	
Investigator and medium used	Medium (mmoles/liter)	Body fluid (mmoles/liter)
Rogers, Pig intestinal fluid	61	55.5
Baldwin's saline	143	59.2
Hobson, *et al.* Pig intestinal fluid	61	52.7
20% seawater	111	53
25% seawater	—	65
30% seawater	166	78
35% seawater	—	93
40% seawater	220	108
Saline A	162.4	76.8
Saline B	200.0	80.0
Harpur and Popkin	169	6.4
	69	5.7
Arthur and Sanborn	150	51.4

F. Ionic Conditions and Electrical Activity

Electrical potentials measured across biological membranes are the result of an unequal distribution of ions between intracellular and extracellular fluids. In most muscle and nerve cells studied, the steady-state transmembrane potential of about —90 mV. (inside negative) is due to maintained inequalities in the concentrations of sodium, potassium, and chloride ions. Depolarization of the membrane of excitable cells to cause an action potential or spike is the result of a sudden change in the permeability of the membrane to sodium and potassium. Return to the resting condition requires the expulsion of sodium from the intracellular fluid and the accumulation of potassium in it. The contractile activity of muscle cells is correlated with the changes in electrical potential which result from ion fluxes across the membrane. [Detailed treatment of the subject of biological electrical activity can be found in Hodgkin (1958) and Katz (1966).]

In 30% seawater, *Ascaris* muscle cells have a transmembrane potential of about —30 mV. Spontaneous rhythmic depolarizations occur at frequencies of 1.5 to 7 per second. These depolarizations consist of a slow wave on which a spike potential may be superimposed. The spikes occasionally overshoot zero potential to about +10 mV. (DeBell *et al.*, 1963.) Brading and Caldwell (1964), using a saline more closely resembling the ionic composition of *Ascaris* body fluids than does 30% seawater, report a mean transmembrane potential of —27.5 mV. Using the physiological saline of Hughes and Arthur (Table I), we find a mean transmembrane potential of —28.3 mV. Depolarizing spikes with an overshoot of 20 mV. are common. The spike frequency ranges from 1.5 to 5 per second and the spikes usually occur in a grouped burst rather than singly (Arthur and Sanborn, 1965).

Adding 3.3×10^{-6} *M* acetylcholine to the bathing solution causes depolarization of the muscle cells. Hyperpolarization and loss of spontaneous spike activity is caused by the addition of either γ-aminobutyric acid (GABA) or piperazine (del Castillo *et al.*, 1963a,b; Jarman, 1964). It also has been shown that strips of dorsal or ventral musculature with intact myoneural junctions show maintained contraction when 10^{-6} *M* acetylcholine is present in the bathing solution (Baldwin and Moyle, 1947). Both GABA and piperazine cause the relaxation of such strips (Ash and Tucker, 1966). There is good evidence that acetylcholine and GABA both are neuromuscular transmitter substances in invertebrates while piperazine mimics the activity induced by stimulating *Ascaris* inhibitory nerve fibers (Krotov, 1965; del Castillo *et al.*, 1963b; Goodwin and Vaughan Williams, 1963).

Neuromuscular transmitter substances act to alter the ionic permeability of the postsynaptic membrane, resulting in changes in electrical potential. One can study changes in ionic permeability during activity by observing the effects of changes in external ion concentrations on transmembrane potentials. When the Na^+, K^-, or Cl-content of the bathing saline is changed, the transmembrane potential of the *Ascaris* muscle cell is shifted.

According to del Castillo *et al.* (1964), as the K^+ of an artificial 30% seawater medium is increased from 3 m*M* at the expense of Na^+ with Cl^- kept constant at 175.4 m*M*, there is an initial slight increase in transmembrane potential. When the external K^+ concentration is above 45 m*M*, the membrane becomes increasingly depolarized, with a slope of 23 mV. for a 10-fold change in potassium. On the other hand, when the external Na^+ and Cl^- concentrations are both held low and constant at 0.8 m*M* and 40.4 m*M* respectively, the transmembrane potential decreases almost linearly as K^+ is increased. The slope of the line relating the logarithm of the K^+ concentration with the measured transmembrane potential is approximately 50 mV. When external Cl^- is increased from 40.4 m*M* in a saline lacking Na^+ and containing 45 m*M* K^+, the membrane potential increases with a slope of 59 mV. for a 10-fold change in Cl^-. Finally, a 17 mV. hyperpolarization occurs when Na^+ is removed from the external medium. In this study, the K^+, Na^+ and Cl^- concentrations of the basic media were far from physiological conditions.

Arthur and Sanborn (1965) report that increasing the concentration of either K^+ or Na^+ in Hughes and Arthur saline while holding all other ionic concentrations constant (at the normal level) results in a nonlinear decrease in the transmembrane potential. Increasing the extracellular Cl^- causes a nonlinear increase in the potential. The two curves relating the logarithms of K^+ or Cl^- concentration with measured transmembrane potentials are very nearly inverse mirror images. Changing ion concentrations in the saline causes progressive changes in the spontaneous spiking activity recorded intracellularly as well as changes in the gross contractile activity exhibited by strips of body wall (see Table VI).

Brading and Caldwell (1964) report that a 10-fold increase in the external Cl^- concentration changes the resting potential by —13 mV. Increasing the extracellular K^- concentration 10-fold results in only a 1.5 mV. change in the resting potential, according to their experiments. They suggest that the relative insensitivity of the resting potential of *Ascaris* muscle cells to changes in extracellular Cl^- is due to a stabilizing mechanism operating in the membrane. They state that the lack of sensitivity to changes in concentration of extracellular potassium can be explained by assuming either virtual impermeability of the cell membrane to K^+, which seems unlikely, or extremely effective operation of the stabilizing mechanism. They propose that this mechanism involves either changes in the relative permeability of the membrane to ions or very rapid changes in intracellular Cl^- or K^+.

The latter alternative is not tenable in view of analyses of the intracellular fluid of muscle cells equilibrated in salines of various K^+ and Cl^- concentrations. These analyses show that the intracellular K^+ and Cl^-

concentrations change very little as the K^+ of the saline is progressively increased from 8 to 48 m*M* or as the Cl^- is changed from 26 to 93 m*M* (Arthur and Sanborn, unpublished data).

The *Ascaris* muscle cell membrane apparently has the ability to control intracellular K^+ and Cl^- concentrations within very narrow limits regardless of the fluctuations in ionic content of the extracellular fluid. Active control of the Cl^- content of the body fluid clearly indicates that constancy of body fluid chloride is an important factor for normal function in *Ascaris*. Furthermore, the inverse mirror image character of the curves relating changes in transmembrane potential with increases in extracellular K^+ and Cl^-, together with the other similarities in muscle activity

TABLE VI
RELATIONSHIPS BETWEEN SALINE ION CONCENTRATIONS AND ACTIVITY OF *Ascaris* MUSCLE TISSUE

Ionic concentration		Mean resting potential (mV.)	Spike activity	Gross contractile activity
K (m*M*)	Cl (m*M*)			
2–8	93–148	34–38	Low frequency, single slow waves, or abortive spikes	Maintained relaxation
12–32	47–67	24–33	Normal shape and frequency	Alternate contraction and relaxation
48–69	1.4–26	14–19	Bursts of activity or plateau complexes[a]	Maintained contraction with convulsive writhing
120	—	12	None	—

[a] Plateau complex—prolonged depolarization with superimposed spikes.

(see Table VI), suggest that K^+ and Cl^- contribute equally to the development of the electrical potentials across the *Ascaris* cell membrane. The equilibrium potentials calculated for Cl^- and K^+ are both greater (more negative) than the transmembrane potential, so a Na^+ contribution to the potential cannot be discounted (Arthur and Sanborn, 1965).

On the basis of his observations, del Castillo (del Castillo *et al.*, 1964) suggests that a chloride battery of low internal resistance is largely responsible for the maintenance of the membrane potential. He states that the contribution of potassium is limited by a low membrane conductance to K^+, and that both the chloride and potassium batteries are shunted by a relatively large conductance to sodium ions.

The major point of disagreement in these three studies is with respect to the role of potassium. The conflicting observations make specific statements regarding the relationship between transmembrane ion movement and electrical events essentially meaningless. Much intensive study of the activity of *Ascaris* nerve and muscle cells under carefully controlled conditions is necessary. The use of salines whose ionic composition resembles that of *Ascaris* body fluid is one step in this direction.

V. Summary

Some broad general statements can be made concerning ionic and osmotic regulation in nematodes. Since these generalizations are based on data derived from very few species, most of them parasitic, it cannot be emphasized too strongly that they may not be representative of the entire class.

During the early developmental stages, the embryo is almost completely isolated from its aqueous environment, since the egg membranes permit only gases to enter the perivitelline space. When embryonic development is complete, the embryo causes the surrounding membranes to become permeable to water and water-soluble substances. If the environmental conditions then permit, water enters and increases the pressure exerted by the perivitelline fluid. The egg shell ruptures, and hatching occurs. Under some circumstances salts in the environment may inhibit the hatching process after the membrane has become permeable to them. Thus, once the "biological clock" of the embryo has prepared the embryo for further development by altering the permeability of its sheath, environmental factors determine whether or not that development will ensue.

The situation described above is seen most clearly in free-living species. Some of the parasitic nematodes have lost part or all of the timing mechanism and with it, the self-contained ability to alter the permeability of the membranes surrounding them. In the root-knot nematodes, permeability changes are initiated by hatching factors, chemical compounds produced by the root of host plants. These act directly on the membranes to cause them to become permeable to water. The initial control has, thus, passed from the animal to the environment.

Intestinal parasites seem to have adopted activating stimuli which are so nonspecific that they can only be supposed to act by turning on the "clock." The larva then becomes competent to release the mixture of enzymic fluids which attack the surrounding membranes or integumentary sheaths from within, first to increase their permeability and then to digest them. In essence, then, the initial triggering has passed from the parasite to the host, but the effective change remains the alteration of permeability.

The osmoregulatory ability of larvae or adults is closely related to the usual environment of the species. Nematodes which normally encounter high environmental osmotic pressures commonly are hyperosmotic regulators and hypoosmotic adjusters (Prosser and Brown, 1961). Conversely, nematodes inhabiting environments whose osmotic pressure is likely to be lower than that of the body fluids are hypoosmotic regulators and hyperosmotic adjusters. The inability of intestinal parasite species to osmoregulate suggests that regulatory mechanisms have been lost in adaptation to the parasitic mode of existence, where conditions are essentially constant.

Nematodes which are capable of osmoregulation employ mechanisms which are not unique to the class. These mechanisms include modification of the internal hydrostatic pressure, removal of water and solutes from the body fluid by the excretory system, and movement of water resulting from the transport of electrolytes.

The ability to regulate the concentration of specific electrolytes is manifested at the cellular level by the requirements for specific cations as cofactors. The necessity to maintain specific ion concentrations in perivisceral fluid is most clearly seen in *Ascaris lumbricoides,* which actively regulates the chloride concentration of its body fluid. The correlations between *Ascaris* muscle cell electrical activity, gross muscular movements, and ionic concentrations in the bathing media are evidence for the importance of ionic regulation.

There are differences in the mechanisms employed and in the ability of those nematodes which have been studied to regulate internal osmotic and ionic conditions. Studies of the problem of regulation have not yet been reported for most of the free-living species. Further studies *in vivo* and *in vitro,* with parameters closely controlled, are needed, both to confirm and to expand what is now known. The nematodes are a unique class, an evolutionary "dead end," but since they are able to occupy almost every aqueous ecological niche that exists, knowledge of their osmotic and ionic regulatory ability and the regulatory mechanisms they employ would be a valuable contribution to our understanding of the general problem of homeostasis.

REFERENCES

Arthur, E. J., and Sanborn, R. C. (1965). *Amer. Zoologist* **5**, 207.
Ash, A. S. F., and Tucker, J. F. (1966). *Nature* **209**, 306–307.
Baldwin, E. (1943). *Parasitology* **35**, 89–111.
Baldwin, E., and Moyle, V. (1947). *J. Exptl. Biol.* **23**, 277–291.
Bataillon, E. (1901). *Arch. Entwicklungsmech. Organ.* **11**, 149–185.
Beherenz, K. W. (1956). *Z. Wiss. Zool.* **159**, 129–164.
Berntzen, A. K. (1965). *Exptl. Parasitol.* **16**, 74–106.

Bird, A. F., and Deutsch, K. (1957). *Parasitology* **47**, 319–328.
Blake, C. D. (1961). *Nature* **192**, 144–145.
Bloom, S., and Entner, N. (1965). *Biochim. Biophys. Acta* **99**, 22–31.
Brading, A. F., and Caldwell, P. C. (1964). *J. Physiol.* (*London*) **173**, 36P (abstr.).
Bremner, K. C. (1961). *Australian J. Agr. Res.* **12**, 1188–1199.
Bueding, E. (1949). *J. Exptl. Med.* **89**, 107–130.
Bueding, E. (1962). *Comp. Biochem. Physiol.* **4**, 343–351.
Bueding, E., and Farrow, G. W. (1956). *Exptl. Parasitol.* **5**, 345–349.
Bueding, E., and Most, H. (1953). *Ann. Rev. Microbiol.* **7**, 295–326.
Capstick, C. K. (1959). *J. Animal Ecol.* **28**, 189–210.
Chitwood, B. G. (1938). *Proc. Helminthol. Soc. Wash., D.C.* **5**, 68–75.
Chitwood, B. G., and Chitwood, M. B. (1950). "An Introduction to Nematology." Monumental Printing Co., Baltimore, Maryland.
Clarke, A. J., and Shepherd, A. M. (1965). *Nature* **208**, 502.
Cleeland, R. (1963). *J. Parasitol.* **49**, 64–68.
Cleeland, R., and Laurence, K. A. (1962). *J. Parasitol.* **8**, 35–38.
Costello, L. C., and Brown, H. (1962). *Exptl. Parasitol.* **12**, 33–40.
Costello, L. C., and Smith, W. (1964). *Arch. Biochem. Biophys.* **106**, 23–228.
Curtis, G. J. (1964). *Ann. Appl. Biol.* **54**, 269–280.
Davey, K. G. (1964). *Can. J. Zool.* **42**, 731–734.
DeBell, J. T., del Castillo, J., and Sanchez, V. (1963). *J. Cellular Comp. Physiol.* **62**, 159–178.
del Castillo, J., deMello, W. C., and Morales, T. (1963a). *Arch. Intern. Physiol. Biochim.* **71**, 741–757.
del Castillo, J., Morales, T. A., and Sanchez, V. (1963b). *Nature* **200**, 706–707.
del Castillo, J., deMello, W. C., and Morales, T. (1964). *J. Gen. Physiol.* **48**, 129–139.
Dittmer, D. S., ed. (1961). "Blood and Other Body Fluids." Biol. Handbooks, Washington, D.C.
Dougherty, E. C, Hansen, E. L., Nicholas, W. L., Mollett, J. A., and Yarwood, E. A. (1959). *Ann. N.Y. Acad. Sci.* **77**, 176–217.
Downey, N. E. (1965). *Brit. Vet. J.* **26**, 362–370.
Dropkin, V. H. (1955). *Exptl. Parasitol.* **4**, 282–322.
Dropkin, V. H., and King, R. C. (1956). *Exptl. Parasitol.* **5**, 469–480.
Dropkin, V. H., Martin, G. C., and Johnson, R. W. (1958). *Nematologica* **3**, 115–126.
Ellenby, C. (1946). *Nature* **157**:302–303.
Ellenby, C. (1963). *Nature* **198**, 1110.
Ellenby, C., and Gilbert, A. B. (1957). *Nature* **180**, 1105–1106.
Ellenby, C., and Gilbert, A. B. (1958). *Nature* **182**, 925–926.
Esserman, H. B., and Sambell, P. M. (1951). *Australian J. Sci. Res.* **B4**, 575–580.
Fairbairn, D. (1955a). *Can. J. Biochem. Physiol.* **53**, 122–129.
Fairbairn, D. (1955b). *Can. J. Biochem. Physiol.* **53**, 130–134.
Fairbairn, D. (1957). *Exptl. Parasitol.* **6**, 491–554.
Fairbairn, D. (1958). *Nature* **181**, 1593–1594.
Fairbairn, D. (1960). *In* "Nematology" (J. N. Sasser and W. R. Jenkins, ed.). Univ. of North Carolina Press, Chapel Hill, North Carolina.
Fairbairn, D. (1961). *Can. J. Zool.* **39**, 153–162.
Fairbairn, D., and Passey, B. I. (1955). *Can. J. Biochem. Physiol.* **33**, 130–134.
Fairbairn, D., and Passey, R. F. (1957). *Exptl. Parasitol.* **6**, 566–574.
Fauré-Fremiet, E. (1913). *Arch. Anat. Micr. Morphol. Exptl.* **15**, 435–458.

Fenwick, D. W. (1939). *J. Helminthol.* **17**, 211–228.
Flury, F. (1912). *Arch. Exptl. Pathol. Pharmatol.* **67**, 275–392.
Gersch, M., and Sheffel, H. (1958). *Naturwissenschaften* **45**, 345–346.
Goodey, J. B. (1959). *Nematologica* **4**, 157–159.
Goodwin, L. G., and Vaughan Williams, E. M. (1963). *J. Physiol.* (*London*) **168**, 857–871.
Harpur, R. P. (1964). *Comp. Biochem. Physiol.* **13**, 71–85.
Harpur, R. P., and Popkins, J. S. (1965). *Can. J. Biochem.* **43**, 1157–1169.
Harris, J. E., and Crofton, H. D. (1957). *J. Exptl. Biol.* **34**, 116–130.
Haskins, W. T., and Weinstein, P. P. (1957a). *J. Parasitol.* **43**, 19–24.
Haskins, W. T., and Weinstein, P. P. (1957b). *J. Parasitol.* **43**, 25–27.
Haskins, W. T., and Weinstein, P. P. (1957c). *J. Parasitol.* **43**, 28–32.
Heald, C. M., and Jenkins, W. R. (1963). *Phytopathology* **53**, 877.
Hirschmann, H. (1960). *In* "Nematology" (J. N. Sasser and W. R. Jenkins, eds.). Univ. of North Carolina Press, Chapel Hill, North Carolina.
Hobson, A. D. (1948). *Parasitology* **38**, 183–227.
Hobson, A. D., Stephenson, W., and Beadle, L. C. (1952a). *J. Exptl. Biol.* **29**, 1–21.
Hobson, A. D., Stephenson, W., and Edsen, A. (1952b). *J. Exptl. Biol.* **29**, 22–29.
Hodgkin, A. L. (1958). *Proc. Roy. Soc.* **B148**, 1–37.
Huff, G. C. (1936). *J. Parasitol.* **22**, 455–464.
Hwang, J. C. (1960). *J. Parasitol.* **46**, 27 (abstr.).
Inglis, W. G. (1964). *Proc. Zool. Soc. London* **143**, 465–502.
Jacobs, L., and Jones, M. F. (1939). *Proc. Helminthol. Soc. Wash. D.C.* **6**, 57–60.
Jarman, M. (1964). *J. Physiol.* (*London*) **173**, 35P (abstr.).
Jasjoski, B. J. (1951). *Exptl. Parasitol.* **1**, 291–302.
Jaskoski, B. J. (1962). *Exptl. Parasitol.* **12**, 19–24.
Katz, B. (1966). "Nerve, Muscle and Synapse." McGraw-Hill, New York.
Kessel, R. G., Prestage, J. J., Sekhon, S. S., Smalley, R. L., and Beams, H. W. (1960). *Trans. Am. Microscop. Soc.* **80**, 103–118.
Kirkpatrick, J. D., Mai, W. F., Parker, K. G., and Fisher, E. G. (1964). *Phytopathology* **54**, 706–712.
Kmetic, E., and Bueding, E. (1961). *J. Biol. Chem.* **236**, 584–591.
Krotov, A. I. (1965). *Med. Parazitol. i Parazitarn. Bolezni* **34**, 219–228.
Krotov, A. I., and Kats, K. M. (1963). *Med. Parazitol. i Parazitarn. Bolezni* **32**, 336–338 (abstr.).
Krueger, R. F. (1959). *J. Parasitol.* **45**, 38.
Kubistova, J., and Seth, D. (1966). *Science* **154**, 1461.
Lazarus, M., and Rogers, W. P. (1950). *Nature* **166**, 647–648.
Lee, D. L. (1960). *Parasitology* **50**, 241–246.
Lee, D. L. (1961). *Nature* **192**, 282–283.
Lee, D. L. (1962a). *Parasitology* **52**, 533–536.
Lee, D. L. (1962b). *Parasitology* **52**, 241–260.
Lee, D. L. (1965). "The Physiology of Nematodes." Oliver & Boyd, Edinburgh and London.
Marks, C. F., and Sayer, R. M. (1964). *Nematologica* **10**, 323–327.
Marsh, C. L., and Kelley, G. W. (1959). *Exptl. Parasitol.* **8**, 274–285.
Meerovitch, E. (1965). *Can. J. Zool.* **43**, 81–85.
Mueller, J. F. (1928). *Z. Zellforsch. Mikroskop. Anat.* **8**, 361–403.
Myers, R. F. (1963). *Phytopathology* **53**, 884.

Nicholas, W. L., and Jantunen, R. (1963). *Nematologica* **9,** 332–336.
Nicholas, W. L., and Jantunen, R. (1964). *Nematologica* **10,** 409–418.
Nicholas, W. L., Dougherty, E. H., and Hanson, E. L. (1959). *Ann. N.Y. Acad. Sci.* **77,** 218–236.
Nicholas, W. L., Dougherty, E. C., and Hanson, E. L. (1962). *Nematologica* **8,** 129–135.
Nimmo-Smith, R. H., and Keeling, F. E. D. (1960). *Exptl. Parasitol.* **10,** 337–355.
Oteifa, B. A. (1953). *Phytopathology* **43,** 171–174.
Panikkar, N. F., and Sproston, N. G. (1941). *Parasitology* **33,** 214–223.
Pollack, J. K. (1957). *Australian J. Sci.* **19,** 208–209.
Prosser, C. L., and Brown, F. A., Jr. (1961). "Comparative Animal Physiology," pp. 6–25. Saunders, Philadelphia, Pennsylvania.
Rogers, W. P. (1944). *Parasitology* **36,** 98–108.
Rogers, W. P. (1945). *Parasitology* **36,** 211–218.
Rogers, W. P. (1958). *Nature* **181,** 1410–1411.
Rogers, W. P. (1960). *Proc. Roy. Soc.* **B152,** 367–386.
Rogers, W. P. (1961). "The Nature of Parasitism: The Relationship of Some Metazoan Parasites to their Hosts." Academic Press, New York.
Rogers, W. P. (1963). *Ann. N.Y. Acad. Sci.* **113,** 208–216.
Rogers, W. P. (1965). *Comp. Biochem. Physiol.* **14,** 311–321.
Rogers, W. P., and Lazarus, M. (1949). *Parasitology* **39,** 302–314.
Rogers, W. P., and Sommerville, R. I. (1960). *Parasitology* **50,** 329–348.
Schopfer, W. H. (1932). *Rev. Suisse Zool.* **39,** 59–194.
Seamster, A. P. (1950). *Amer. Midland Naturalist* **43,** 450–470.
Seidman, I., and Entner, N. (1961). *J. Biol. Chem.* **236,** 915–919.
Sheffield, H. G. (1964). *J. Parasitol.* **50,** 365–379.
Shumard, R. F., and Herrick, C. A. (1954). *J. Parasitol.* **40,** 16.
Silverman, P. H., and Podgers, K. R. (1964). *Exptl. Parasitol.* **15,** 314–324.
Smith, M. H., and Lee, D. L. (1963). *Proc. Roy. Soc.* **B157,** 234–257.
Sommerville, R. I. (1957). *Exptl. Parasitol.* **6,** 18–30.
Stephenson, W. (1942). *Parasitology* **34,** 253–265.
Stephenson, W. (1944). *Parasitology* **35,** 167–172.
Trim, A. R. (1944). *Parasitology* **35,** 209–219.
Trim, A. R. (1949). *Parasitology* **39,** 281–290.
von Brand, T. (1943). *Biol. Bull.* **84,** 148–156.
von Brand, T., and Simpson, W. F. (1942). *J. Parasitol.* **30,** 121–129.
Wallace, H. R. (1956). *Ann. Appl. Biol.* **44,** 274–282.
Wallace, H. R. (1963). "The Biology of Plant Parasitic Nematodes." Arnold, London.
Watson, B. D. (1965a). *Quart. J. Microscop. Sci.* **106,** 75–81.
Watson, B. D. (1965b). *Quart. J. Microscop. Sci.* **106,** 83–91.
Weinstein, P. P. (1952). *Exptl. Parasitol.* **1,** 363–376.
Weinstein, P. P. (1960). *Proc. Helminthol. Soc. Wash. D.C.* **27,** 233.
Weinstein, P. P., and Haskins, W. A. (1955). *Exptl. Parasitol.* **4,** 226–243.
Weinstein, P. P., and Jones, M. F. (1959). *Ann. N.Y. Acad. Sci.* **77,** 137–162.
Wilson, P. A. G. (1958). *J. Exptl. Biol.* **35,** 584–601.
Wittenberg, B. A., Okazaki, T., and Wittenberg, J. B. (1965). *Biochim. Biophys. Acta* **111,** 485–511.
Wottge, K. (1937). *Protoplasma* **29,** 31–59.
Wright, K. A. (1963). *J. Morphol.* **112,** 233–245.
Zawadowski, M. M. (1928). *Trans. Lab. Exptl. Biol. Zoo-Park Moscow* **4,** 201–206.

CHAPTER 9

Chemical Aspects of Growth and Development

W. P. Rogers and R. I. Sommerville

I. Introduction

Papers which deal strictly with chemical aspects of growth of nematodes and Acanthocephala are relatively few in number and cover only limited areas of this topic. In writing this review we have interpreted the title somewhat broadly and selected topics where it seems possible to give a more connected narrative, though in doing so we have found it necessary to include papers which deal with biological as well as chemical aspects of growth.

II. Special Features Which Affect Growth

Some features of the organization and life cycles of nematodes and Acanthocephala are unusual, and we shall discuss these in relation to their effect on growth. In addition, many species of nematodes, and all Acanthocephala, are parasitic, and their growth can be modified in a variety of ways by this relationship.

A. Structure and Life Cycles

1. Structure, Form, and Growth

In most nematodes, the juvenile closely resembles the adult, probably because the hydrostatic skeleton and longitudinal muscle system are retained throughout life, and growth must therefore take place within the restrictions laid down by the mechanics of the basic structure. It is perhaps significant that the more obvious departures from the vermiform shape are found in sessile reproductive females like *Tetrameres americana* in birds and *Heterodera rostochiensis* in plants.

The most obvious changes associated with growth are an increase in length and the development of a reproductive system. Some nematodes, like *Heterotylenchus pawlowskyi* in fleas (Kurochkin, 1960), do not increase in length. Others, like *T. americana,* increase in length, but the female also increases in width and becomes globular. Increase in length seems unlikely to take place simply by stretching the cuticle, for it would become weaker, and necessitate a reduction in internal pressure. Most nematodes increase in diameter as they lengthen. This could give rise to an increase in tangential stress so that a thicker cuticle would be required (Crofton, 1966). In *A. lumbricoides* longer worms (which presumably have a greater diameter) have thicker cuticles than shorter ones (Watson, 1965).

2. Size and Habitat

Nematodes are the most numerous of the small metazoa in the sublittoral sediments (Wieser, 1960) and are very common in the soil. Most of these species are 1 to 2 mm. long and relatively few exceed 5 mm. The nematodes which are parasitic in animals are often very large, and it may be significant that in these environments concealment and escape from predators are not necessary.

3. Cell Constancy

Eutely or cell constancy is a characteristic of most organs in nematodes and the Acanthocephala, and the early work on this subject has been reviewed by B. G. Chitwood and Chitwood (1940) and Hyman (1951). Current ideas on cell constancy in nematodes are based largely on studies with *Oxyuris equi, Turbatrix aceti,* and *Rhabditis longicaudata* (cited by Hyman) and *Rhabditis anomala* and *Ascaridia galli* (Wessing, 1953; Frenzen, 1955). Observations show that except for the gonads and some other structures, cell division ceases during the late stages of embryonic development. Subsequent growth is by increase in cell size. Some parasitic nematodes become so large that nuclear division with the formation of syncytial tissues is common. In structures which, like the pharynx, are wholly or partly syncytial (Reger, 1966), the nuclei may be constant, in both number and arrangement, for a given species.

Most of the work on cell constancy was done many years ago, and very little biochemical information on nucleic acid metabolism is available. Intestinal cells of *Caenorhabditis briggsae* have a constant number of giant nuclei which contain both metabolic and genetic DNA, with the former predominating. Although other instances of metabolic DNA are known (Iwamura, 1966), the turnover of DNA in these giant nuclei is exceptionally high (Nonnenmacher-Godet and Dougherty, 1964). The significance of this in terms of the histological peculiarities of *C. briggsae* is not known.

The relationship between cell size and nuclear size in nematodes has not been examined. Some nuclei become bigger as the cell enlarges, but whether this is associated with polyploidy is unknown. In adult ascarids, the "Feulgen positivity" of the nuclei of somatic cells is inversely proportional to nuclear size (Monné, 1963). Why some tissues should increase in size by cell division, and others by cell enlargement is unknown.

4. The Uptake of Nutrients by Acanthocephala

Acanthocephala have neither oral opening nor digestive tract and nutrients enter through the body wall, the structure of which has been described by Lee (1966) and elsewhere in this volume.

The body wall seems to be similar in all species which have so far been examined, and a prominent feature is the presence of many small pores on the cuticle which pass as canals into deeper layers. Particles of carbon and thorium dioxide are taken up by the cuticle of *Moniliformis dubius,* and electron micrographs have revealed these in the canals (Edmonds and Dixon, 1966). These structures are believed to provide a means whereby nutrients are absorbed, but the mechanism is unknown. One suggestion is that a form of pinocytosis could be involved (Nicholas and Mercer, 1965).

5. Plan of the Life Cycle

All nematodes molt, usually four times. The molt may be preceded by a quiescent period or "lethargus," and the interval between successive molts may be a time of extensive and rapid growth. Little is known about the growth processes, or the way in which they are interrupted by molting.

The Acanthocephala do not molt. They are all parasites, and the life cycles always include an intermediate as well as a definitive host. The infective "egg" must be ingested by an arthropod before the larval form can develop. The juvenile is enclosed within membranes, and remains quiescent until contact is made with the definitive host.

6. Hatching, Molting, and Metamorphosis

a. Hatching. The chemistry of hatching is best known for *Ascaris lumbricoides* and will be discussed later in this chapter (Section VI). The hatching process is both chemical and mechanical. Enzymes help in the breakdown of the egg membranes, and the worm becomes very active and pushes against them (Rogers, 1958; Fairbairn, 1960). Accounts of the hatching of other species (D. P. Taylor, 1962; Thomas, 1965; Berntzen, 1965) suggest that both chemical and mechanical factors also operate, although the relative importance of enzymic and muscular activity probably varies from species to species, and the presence of enzymes has been demonstrated unequivocally for only one nematode, *A. lumbricoides,* and one acanthocephalan, *Moniliformis dubius* (Edmonds, 1966).

b. Molting. The onset of molting is commonly associated with a period of inactivity or lethargus, which may last from an hour to several days (Veglia, 1915; Hansen *et al.,* 1964; Hechler and Taylor, 1966). By the time the nematode again becomes active, a new cuticle has been formed and structures associated with the cuticle, such as the lining of the mouth, pharynx, rectum, parts of the excretory duct and tubules associated with the amphids and phasmids, are also replaced. Many species do not feed when molting (Hansen *et al.,* 1964; Bailey, 1967), presumably because the lining of the mouth and esophagus is being renewed.

The significance of the lethargus is not known, but it occurs at a time when metabolic activity must be high, and in *Parascaris equorum, Enoplus communis,* and *Hemicycliophora arenaria* there is an increased oxygen uptake at this time (Hopkins, 1955; Wieser and Kanwisher, 1960; Van Gundy and Stolzy, 1963). The investigations of Harris and Crofton (1957) suggest that a nematode could be immobilized if the pressure in the body cavity exceeded a critical level. This might happen, if, for example, the concentration of osmotically active components in the body cavity increased during molting. It must be emphasized that these ideas are quite speculative: there is almost no information on the physiological changes associated with molting.

Molting of the infective stages of some nematodes is incomplete and the juvenile worm is surrounded by two cuticles. Two cuticles are also found on "*dauer* larvae," a specialized stage for the dispersal of species which live in temporary environments. The term exsheathment is often used to describe the ecdysis of infective stages and *dauer* larvae. Sometimes exsheathment seems to require a special external stimulus (Rogers and Sommerville, 1963), but this has been defined for only a few species and is discussed in Section VII.

c. Metamorphosis. The metamorphic process is primarily one of differential destruction and development of various tissues in a larval organism (Berrill, 1961). A few nematodes undergo a metamorphosis of their mouth parts when they change from a free-living to a parasitic existence (Matsusaki, 1950; Sommerville, 1966), but otherwise metamorphoses are rare in these animals. Although the Acanthocephala do not molt, they undergo a distinct metamorphosis, and the changes are associated with the infection of the intermediate and definitive hosts. A metamorphosis of a parasite which is induced by infection of the host could be an important adaptation which ensures that the infective stage remains in a resting condition until the appropriate host is encountered (Rogers, 1962).

B. The Effect of Parasitism on Growth

Parasitism can be thought of as influencing the growth of nematodes in two ways. First, most parasitic nematodes have a special infective stage, during which developmental processes are temporarily arrested. Second, an established parasite may have its growth pattern modified in a variety of ways by the host.

1. The Infective Stage

The infective stage occupies a special place in the biology of many parasitic nematodes, acting as a "bridge" between two environments. Infective stages may be required to withstand very adverse conditions, in places where the chances of meeting a potential host are high. They are usually provided with energy reserves and are independent of exogenous sources of food, but they do not grow. Developmental processes are suspended until suitable stimuli signal the presence of a potential host. More detailed discussion on the infective stage can be found in Rogers and Sommerville (1963).

2. *Growth After Infection*

a. Enhanced Growth. Many nematodes parasitic in animals are larger than most free-living species, and some, like *Dracunculus medinensis*, may be over 100 cm. long or between 500 and 1000 times the length of many free-living species. We do not know why some parasitic nematodes should be so very large. In the Ascaroidea, increase in body size apparently parallels the course of evolution in this group, but it is quite independent of the size of the host (Osche, 1958).

Sometimes parasitic females become large and round, while the males remain vermiform. Such females are sessile and their shape in this instance may be related to the size of the reproductive organs.

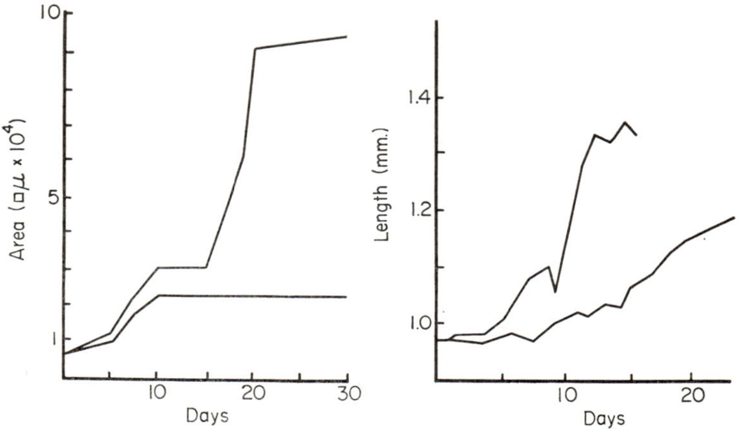

FIG. 1. Growth curves of *Meloidogyne javanica* (left) and *Ditylenchus dipsaci* on plants. In both instances the lower curve represents growth on a resistant plant. Adapted from Burdett *et al.* (1963) and Blake (1962a) respectively.

b. Retarded Growth of Nematodes Parasitizing Plants. All nematodes which parasitize plants seem to have mouth parts which function by piercing and sucking. Plants respond promptly to this invasion in a great variety of ways (Dropkin, 1955; Wallace, 1963; Rohde, 1965) and their response ranges from "immune," through various degrees of resistance to "susceptible" (Blake, 1962a). In susceptible oats, the nematode *Ditylenchus dipsaci* grows readily, cells about it enlarge, and air spaces form between the cells. In a resistant variety the growth of *D. dipsaci* is slower, the final size smaller (Fig. 1), and enlargement and separation of the cells of the plant is almost lacking (Blake, 1962a; Goodey and Hooper, 1962). Similarly, the growth of *Meloidogyne javanica* on resistant varieties of *Prunus persica* is quite different from growth on susceptible species (Fig. 1), sometimes stopping altogether (Burdett *et al.*, 1963).

Little is known about the physiological changes in plant and nematode which lead to retarded development of the parasite. The resistance of potato to two biotypes of *Heterodera rostochiensis* is believed to be dependent upon phenolic glucosides in the root cells, which, if hydrolyzed by β-glucosidase, release polyphenols and so cause cell necrosis. One biotype of *H. rostochiensis* secreted large amounts of β-glucosidase, and failed to infect the potato. But the resistance of the potato was broken by the second biotype which had much lower levels of enzyme activity, so that the glucosides were not hydrolyzed and necroses were not formed. Nematodes of this biotype are therefore able to develop (Wilski and Giebel, 1966). This example shows that if the plant cells in the vicinity of an invading nematode die, or perhaps live but fail to provide the nematode with suitable nutrients, the nematode dies, or at best grows slowly. These reactions are quite different from the way in which animals respond to invasion by nematodes.

c. Retarded Growth of Nematodes Parasitizing Animals

i. Phagocytosis. Nematodes which invade animal tissue can become surrounded by a variety of phagocytic cells, and may become encapsulated. This reaction is not found in plants, but it is particularly important in invertebrates.

In larval mosquitoes, infection by an undescribed species of rhabditid nematode is followed within a few minutes by the deposition of fine particles of melanin over the entire cuticle of the parasite. Blood cells containing particles of melanin accumulate and eventually become deposited on the surface of the nematode (Welch and Bronskill, 1962; Bronskill, 1962), which ceases to grow and ultimately dies. This is only one of a number of examples which could be cited, but the mechanism of these reactions in invertebrates is still largely uninvestigated.

ii. Immunological mechanisms. Possibly all vertebrates except perhaps the most primitive can produce antibody, and if the association of vertebrate host and nematode is close, the parasite can provide an antigenic stimulus. There is an extensive literature on this subject (Urquhart *et al.*, 1963; Kent, 1963; Soulsby, 1963, 1966), but little information is available on the ways in which these immunological mechanisms affect the growth of nematodes.

Frequently the response of the host takes the form of a more or less abrupt elimination of worms, but in other instances growth of nematodes in immune animals is slower than in nonimmune hosts, and sometimes egg production may be impaired (Weinstein, 1955; Soulsby, 1957; Donald *et al.*, 1964; Ogilvie, 1965).

iii. Nonspecific mechanisms. A great variety of nonspecific mechanisms may retard or stop growth of an invading nematode, and although many of these have been examined closely, the way in which they interfere with growth is not understood. Weinstein (1958) has pointed out that retardation of development can occur when *Nippostrongylus brasiliensis* is cultivated in certain media. Addition of growth factors enables development to proceed, but in their absence the worms remain alive and in good condition, although they do not grow (Weinstein and Jones, 1959). Weinstein suggested that some of the factors which influence arrested development may be either a lack of, or imbalance in, certain nutrients.

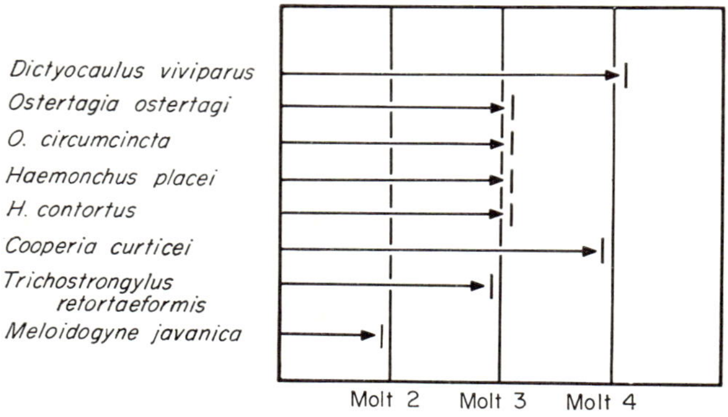

FIG. 2. Inhibition of development of some parasitic nematodes in relation to molting. Sources in order: Michel (1955, 1963), Sommerville (1954), Bremner (1956), Dineen *et al.* (1965), Sommerville (1960), Michel (1952), and Burdett *et al.* (1963).

d. Inhibition of Development in Relation to Molting. Sometimes the development of parasitic nematodes is arrested just before or just after a molt (Fig. 2). Most nematodes stop growing when they molt, and the requirements for growth to restart might be critical, so that the nematode would be more vulnerable to adverse environments when molting than at other times. It might be speculated that in the evolution of the infective stage, this has been turned to the advantage of the parasite, so that growth cannot restart except in a suitable host.

III. Growth Curves and the Measurement of Growth

A. Introduction

Little attention has been given to the quantitative aspects of the growth of nematodes and Acanthocephala. Growth curves of a few species of nematodes have been compiled, and sometimes subjected to mathematical analysis (Scott, 1929a; McCoy, 1930), and allometric growth has been examined, although not extensively (Inglis, 1954; Blake, 1962b; Gysels, 1962). There is a voluminous literature on quantitative aspects of the growth of animals (von Bertalanffy, 1960; Needham, 1964), but investigations on nematodes appear to have contributed little to the general topic.

B. The Measurement of Growth

1. Length as a Measure of Growth

Changes in length are the most common way of measuring growth in nematodes; the method is not without fault.

1. Most nematodes are essentially cylinders, so that a single measurement of length is an obvious way to follow growth. This ignores the problem of shape, although for many species this is of little importance. Different methods (see below) must be used for species which increase significantly in size in other than the longitudinal axis. Some nematodes may not increase in length much during development. Others may cease to increase in length, although internal structures such as the ovaries may continue to do so (Blake, 1962b). In these species, length is not a good measure of the addition of new material to the living system.

2. The length of a worm may depend upon the amount of food in the gut (Townsley *et al.*, 1963) and defecation can be associated with a rapid decrease in volume (Mapes, 1965). The osmotic pressure of the external medium and length of a nematode are related (Lee, 1960; Kampfe, 1962; Anya, 1966a), and at low osmotic pressures nematodes may lengthen by as much as a third.

3. Sometimes the length of nematodes has been measured when the animals have not been fixed (Weinstein, 1955; Haley and Parker, 1961), by straightening them out with needles. More commonly, fixatives are first used, but can give rise to errors from shrinkage which may not be the same for all stages of development (Scott, 1929b; Mackerras and Sandars, 1955).

2. Area and Weight as Measures of Growth

In species like *Tetrameres americana*, females may double their length from time of infection of the host to maturity, but quadruple their diameter (Cram, 1931). A similar pattern of growth is seen in *Meloidogyne javanica*, and growth of this species has been followed by measuring the cross-sectional area (Bird, 1959).

Some nematodes may stop increasing in size, but the growth of internal structures, especially the reproductive tract, may continue. One way in which these changes could be detected is by weighing. Large nematodes have been weighed directly (Mykytowycz and Dudzinski, 1965) or weight can be determined indirectly, using estimates of the specific gravity of the worms (Wieser and Kanwisher, 1960).

C. Growth Curves

In all the growth curves on which the following discussion is based, individual worms, of which there may have been one or many in each

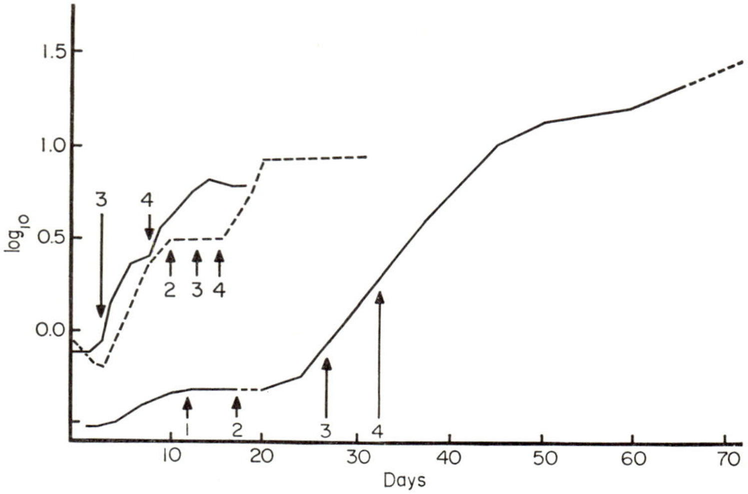

Fig. 3. Growth curves of *Angiostrongylus cantonensis* (lower), *Meloidogyne javanica* (broken line) and *Cooperia curticei* (parasitic stages). Ordinate: *M. javanica*, cross-sectional area; remainder, length. Adapted from Mackerras and Sandars (1955), Bird (1959), and Sommerville (1960) respectively.

age group, were measured once only. The curves are average curves, and will differ from individual curves to a greater or lesser degree, according to the variability of the material. This is unlikely to affect our discussion about the growth curves, because we are concerned with relatively obvious differences, and not with a detailed analysis of growth cycles.

Growth curves of the females of three species are shown in Fig. 3. They are *Angiostrongylus cantonensis*, *Meloidogyne javanica*, and *Cooperia curticei*. The curves are a semilog plot of data adapted from the papers of Mackerras and Sandars (1955), Bird (1959), and Sommerville (1960).

The growth curve of *A. cantonensis* consists of two cycles, and each cycle includes two molts. These growth cycles take place in different places, the first in mollusks, the second in rats. The first cycle of growth is terminated about the time of the first molt and the second molt follows shortly afterward. The third and fourth molts occur early in the second growth cycle, but there is no evidence that growth is stopped at this time. The nematode nearly doubles its length before the first molt, but there is also considerable increase in girth, which of course is not shown in this curve. But more than 90% of the increase in length takes place after the last molt.

These growth cycles and molts are associated with quiescent periods, some of which may be a lethargus, or inactive period associated with molting. But *A. cantonensis* is inactive for all of the first, second, and third stages in the rat, and there is at present no physiological criterion to enable us to distinguish a lethargus from general inactivity.

M. javanica is parasitic in the roots of plants. Shortly after penetration of the root, the female becomes sessile and spherical in shape. The growth curve is based on the cross-sectional area of the various developmental stages, and like that of *A. cantonensis,* it also embodies two growth cycles which are separated by molts. But unlike *A. cantonensis,* the two cycles take place in the same environment. Infection of the plant is accomplished by the second-stage juvenile which hatches from the egg where the first molt took place. After an initial period of 3 days during which it decreases in size, the second-stage juvenile grows rapidly for several days until the onset of the second molt, when growth ceases. The third and fourth molts follow and some 5 days after the second molt, growth is resumed.

The growth curve of *Cooperia curticei,* a parasite of sheep, is based on measurements of length of the parasitic stages only. Although there are two cycles, there are only two molts, the third and fourth. The first two molts take place before this species infects the sheep, the second of these being an example of exsheathment, because it cannot be completed until the infective stage reaches the gut of the host. It is the loss of the cuticle at exsheathment which slightly reduces the length of the nematode immediately after infection; the process is quite different from the "negative" growth of *M. javanica.* There is little increase in length during the ensuing third stage, but both the third and the fourth molts are followed by a rapid increase in length. The growth curve is flattened before the fourth molt, suggesting that growth slowed down, or stopped, but these worms were not alive when measured, and we do not know whether there was also a lethargus at this time.

The growth curves of the third, fourth, and fifth stages of *C. curticei* are very similar to those of the free-living stages of the parasitic nematode *Ancylostoma caninum* (McCoy, 1930), *Aspiculuris tetraptera* in mice (Anya, 1966b), *Trichostrongylus retortaeformis* in rabbits (Bailey 1967), and *Ditylenchus dipsaci* in oat seedlings (Blake, 1962b).

It has been suggested that the idealized form of the growth curve of nematodes might be represented by a series of five stages or cycles of growth, each cycle being terminated by a molt, when growth is temporarily halted (Rogers, 1962; Rogers and Sommerville, 1963). While there can be little doubt that all nematodes molt, the growth curves of some species depart significantly from this model. There are at least two reasons for this. First, nematodes do not always increase in length between molts, a point well illustrated by the growth curve of *M. javanica*. Second, growth is not invariably interrupted by molting. For example, the growth of *A. cantonensis* did not stop at the third and fourth molts, and the free-living nematode *Caenorhabditis briggsae* is similar in this respect (Hansen *et al.*, 1964). But the growth curve of the vinegar eel-worm *Turbatrix aceti* has well-defined breaks which coincide with the molts (Watson, personal communication), and there is therefore no evidence that this phenomenon is limited to parasitic nematodes. Perhaps failures to detect arrested growth at the time of molting depend upon the techniques employed, as Hansen *et al.* (1964) suggest for *C. briggsae*.

The relationship of the lethargus to molting and interruption of growth is also uncertain. *D. dipsaci* stops increasing in length at the fourth molt and enters a lethargus, but the growth of the ovary continues. Although *C. briggsae* has a lethargus when it molts, there is no interruption to growth as a whole. Whatever physiological significance the lethargus may have, these examples show that it is not incompatible with growth of specific organs, or even of the whole organism.

Evidently the growth of nematodes can be modified in many different ways by molting and by the presence of cycles of growth. We do not understand the relation between molting, cessation of growth and the lethargus, and more information is required before the generalization that nematodes "have growth cycles geared to the periodic molting of an exoskeleton" (Needham, 1964, p. 24) can be accepted.

IV. Factors in the Environment Which Affect Growth

A. Introduction

In this section we shall consider some of the factors which enhance the growth of nematodes and Acanthocephala. Many of the environmental components which affect growth are nutritional, and the requirements for various nutrients have been investigated with the aid of culture techniques in which the growth of the organism is followed in different media. As yet no species of Nematoda or Acanthocephala can be grown for an indefinite number of successive generations in a chemically defined medium. Investigations with the nematode *Neoaplectana glaseri* come closest to this ideal, for this species has been grown as far as the F_1 generation, with limited extension to the F_2 generation when fresh medium is added (Jackson, 1962a,b).

It seems almost certain that further alterations of the composition of this medium will give better growth, but a nutritionally well balanced medium may not be *the* crucial condition for optimal axenic culture (Jackson, 1962a). Requirements for other factors may exist in culture systems, and "duplicating these factors in their correct sequence and in time may perhaps be necessary for the successful cultivation of many parasitic nematodes" (Weinstein and Jones, 1956).

If special stimuli are necessary for growth of parasitic nematodes, it is perhaps surprising that it has been possible to induce any growth *in vitro* without some knowledge of them. But it should be remembered that the growth of most species *in vitro* is very slow, that worms are stunted, and that many species fail to reproduce. The media are no doubt nutritionally inadequate, but also special requirements for growth could either be missing or else be supplied at the wrong time.

Some of the processes for which special factors may be necessary are discussed below. We have not attempted to discuss the strictly nutritional requirements for growth, because these have been discussed in Chapter 3.

B. Requirements for Specific Events in the Life Cycle

1. Exsheathment

Exsheathment has been discussed elsewhere in this chapter. When it is necessary to exsheath an infective juvenile before placing it in nutrient media, this is usually done by immersing the nematodes in a solution of sodium hypochlorite, which attacks the sheath chemically (Lapage, 1933). More physiological methods which might be used, depending on the particular species, include carbon dioxide (Rogers, 1960; A. N. Taylor and Whitlock, 1960a), carbon dioxide and digestive enzymes (Silverman and Podger, 1964) or exposure to high hydrogen ion concentrations (Christie and Charleston, 1965).

Exsheathment of an infective juvenile of a species like *Haemonchus contortus* is induced by carbon dioxide. It seems unlikely that stimulation with carbon dioxide is concerned only with exsheathment. It probably involves changes in some system which governs development generally, because treatment which inhibits exsheathment also reduces infectivity, i.e., it interferes with development (Rogers, 1966a). Although *H. contortus* has been grown to the fourth or fifth stage in complex media under air (Silverman, 1959; Leland, 1963), having first been exsheathed in sodium hypochlorite, this does not disprove the hypothesis, because sufficient carbon dioxide may accumulate locally under these conditions to trigger changes in the systems governing development. Or alternatively, substrates usually synthesized in the worms as a result of the stimulus may be replaced by substrates in the medium which are not normally present in the natural environment.

2. *Molting*

Although a specific stimulus is required to induce the infective stages of some parasitic species to complete the second molt, we have no information about special requirements for earlier or later molts of these species. Probably they are controlled by internal mechanisms which regulate development. *Trichinella spiralis* has attracted a lot of attention in this connection because it undergoes a form of "multiple molting," which seems to be associated with adverse conditions (Berntzen, 1965). The fourth molt of this species may be enhanced by high concentrations of carbon dioxide, together with an extract of the wax moth, *Galleria mellonella,* collected at a stage when high amounts of ecdysone (the molting hormone) were likely to be present (Meerovitch, 1965b).

In the continuous flow culture system used by Berntzen (1965), *T. spiralis* differentiated to the adult stage and molted. The molt was not completed, the cuticle being retained like the sheath of an infective juvenile. A special apparatus was devised which the worms penetrated and so eventually lost their sheaths. In later experiments this was replaced by "a mixture of the enzymes cocarboxypeptidase-A and cocarboxypeptidase-B," which resulted in 100% exsheathment. Although these enzymes were selected after experiments with a "variety of normally-occurring enzymes and hormones found in the intestinal tract" (Berntzen, 1966), we do not know whether this is the system which is effective *in vivo.* However, the important point is that by provision of these enzymes at the correct time, development *in vitro* was able to proceed.

3. *Development of the Third Stage*

In vitro, carbon dioxide is a necessary component in a nonnutrient medium for the development of *H. contortus* through the third to the fourth stage (Sommerville, 1964, 1966; Silverman *et al.,* 1966). During this stage the juveniles do not feed and development does not require the intake of conventional nutrients. Carbon dioxide is required for at least half the period of the third stadium and its concentration is critical. Whatever its role as a physiological trigger, it may also have a direct effect upon metabolism, for example, as an important metabolite in the production of energy (Rogers, 1966a).

Third-stage juveniles of some other species, such as *Nippostrongylus brasiliensis, T. retortaeformis,* and *Nematospiroides dubius* are different from *H. contortus,* and seem to be able to ingest food during the latter part of the third stage both *in vitro* and *in vivo.* But the ingestion of food may not alone be sufficient to induce development to the fourth stage. For example, in *Rhodnius prolixus,* a hemimetabolous insect, the epidermal tissues are normally "dormant." Nutrition alone has very little effect on the epidermal cells, but nutrition together with a hormone (molting hormone) restores their capacity for protein synthesis (Wigglesworth, 1963). The tissues of the infective stage could also be "dormant" and some additional factor may be required if food is to be used. It has already been suggested that this factor could be provided by the stimulus which leads, not only to exsheathment of the infective stage, but also to changes in systems which control development, and there is some evidence in favor of this hypothesis, at least for *Nematospiroides dubius.* And one outstanding feature of nematodes in culture is the lack of uniformity in development. For example, numerous juveniles of *Nippostrongylus brasiliensis* may be in excellent condition after many days in culture, but show little or no differentiation, while others in the same culture have matured (Weinstein and Jones, 1959). This could mean that in these cultures, which were initiated with third-stage juveniles, many individuals are not provided with an effective stimulus to restart the mechanisms which control development.

A heat-stable polypeptide material which has a stimulating effect on the development of third-stage juveniles of *Dictyocaulus viviparus* to the fourth stage has been described (Silverman, 1965; Silverman *et al.*, 1966). The way in which this supplement induces its effect seems to be unknown. But when added as a supplement to a balanced salt solution under a gas phase containing carbon dioxide, it enhanced the conversion of third-stage *H. contortus* to the fourth stage. It seems unlikely that it could function as a conventional nutrient, because third-stage juveniles of *H. contortus* are unable to ingest culture media (Sommerville, 1966), and the amounts of the supplement that are required are small (Silverman *et al.*, 1966).

4. *Feeding*

Roberts and Fairbairn (1965) suggested that serum may provide a feeding stimulus for adult *Ancylostoma caninum.* When this species was incubated in a balanced salt solution containing glucose but no serum the glucose was utilized only in very small amounts. When serum was added significant amounts of the glucose were taken up (Warren *et al.*, 1962; Fernando and Wong, 1964; Roberts and Fairbairn, 1965). There is no doubt, however, that some glucose can be used in the absence of serum (Komiya *et al.*, 1956; Fernando and Wong, 1964). Furthermore, Fernando and Wong have shown that some worms can ingest a medium composed of a balanced salt solution containing a trace of India ink, which serves as a marker. They argued that serum plays an important part in the nutrition of *A. caninum,* though its actual role is unknown.

The survival of *Nippostrongylus brasiliensis* in balanced salt solution was enhanced by the addition of glucose. India ink added to this medium was detected in the intestines of some young adult worms (Sommerville and Weinstein, 1967). These results conflict with those obtained in similar media by Roberts and Fairbairn (1965), who not only found no evidence that glucose in the medium was metabolized, but also noted that the pharyngeal muscles were inactive, and concluded that the worms were not ingesting the medium. The reason for the discrepancy in these results may be in the different methods used for handling the worms. It should be noted, however, that, in contrast to *A. caninum, N. brasiliensis* was not stimulated by serum to increase its use of glucose.

The lumen of the intestine of third-stage juveniles of species like *N. brasiliensis* is small, and it is often impossible to be certain about the presence or absence of ingesta in the very early stages of parasitism. Furthermore, any particles which may be present could have been carried over from the preparasitic stage. If these juvenile worms ingest culture media, the presence of the medium in the intestine can be determined by adding a fluorochrome to the medium and examining the worm for fluorescence under appropriate conditions (Sommerville, 1966; Bailey, 1967). Worms may also be collected from the host at various stages of development and put into similar media. When this was done with third-stage juveniles of *Trichostrongylus retortaeformis* from rabbits, worms which had been in the host for less than 14 hours failed to take up the medium *in vitro,* but those collected at later times did so (Bailey, 1967). Similar results were obtained with third-stage *Nematospiroides dubius* from mice (Sommerville, unpublished observations). If we can assume that only those worms which ingested culture media were ingesting food *in vivo* at the time they were removed from the host, the results suggest that the onset of feeding does not immediately follow infection. Perhaps a special stimulus is needed, but it may act indirectly. As with other events associated with infection by *N. dubius,* feeding may only commence when the systems which govern development generally have been restarted.

5. Mating

Nippostrongylus brasiliensis and *Cooperia punctata* can be grown "from egg to egg" in complex media (Weinstein and Jones, 1959; Leland, 1962) but fertile eggs have not yet been produced by worms which have been reared *in vitro*. The complete cycle has yet to be attained *in vitro* and "the primary limitation appears to be a lack of understanding of the factors that control mating" (Weinstein and Jones, 1959). This could prove a serious drawback, for the efficiency of a particular combination of nutrients to promote growth can be evaluated properly only if the medium can support several successive generations.

Dioecious species like *Turbatrix aceti* and *Panagrellus redivivus* mate readily in various types of culture systems (Hansen and Cryan, 1966). At least three species of parasitic nematodes, *Ancylostoma caninum, Neoplectana glaseri,* and *N. dubius* mate readily *in vitro* (Komiya *et al.*, 1956; Jackson, 1962a; Sommerville and Weinstein, 1964). But the conditions under which different species will mate *in vitro* vary greatly. Some will do so in salt solutions, but others fail to mate in media which are adequate for growth. The onset of sexual maturity is presumably associated with physiological changes which may lead to the production of appropriate stimuli, or to the induction of a receptive state, and so in turn to copulation. It is possible that specific components of the natural environment induce or enhance these physiological changes (Sommerville and Weinstein, 1964). Some nematodes are now known to release sex attractants (Greet, 1964; Jones, 1966). The efficiency or even the release of these substances might be inhibited *in vitro*. Alternatively, a system of sex attractants might break down because the worms are different from normal ones in more subtle ways. For example, the chemical composition of *Nippostrongylus brasiliensis* grown *in vitro* is different from worms grown *in vivo* (Moore, 1965) and this may in turn be reflected in their behavior and reactions to the opposite sex.

Berntzen (1965) has recently described a method for the *in vitro* culture of *Trichinella spiralis*, in which the worms copulated and produced normal juveniles. This is the first time this nematode has developed to full sexual maturity *in vitro*, and we do not know whether it is a species which would copulate readily *in vitro*, like *A. caninum*, or whether some special features of the system are necessary for mating to take place.

C. General Requirements

1. The Gas Phase

Most of the early work on the culture of nematodes employed air as the gas phase, but more recently other gas mixtures have been used. Special stages of the life cycle may require different proportions of gases to induce exsheathment, or to induce early development of the parasitic stage.

Cleeland and Laurence (1962) and Cleeland (1963) showed that the best survival of *Ascaris lumbricoides* was obtained under air in deep tubes. Although under these conditions, fewer than 1% of the worms showed signs of growth, the results suggest that *A. lumbricoides* might require relatively less oxygen and more carbon dioxide than is available in air. Harpur (1962) has used the production of carbon dioxide from minced muscle of *A. lumbricoides* as a measure of the health of worms *in vitro.* Although as he pointed out, it is difficult to relate these findings to survival *in vitro,* there was little doubt that carbon dioxide in the gas phase was helpful, especially when associated with low oxygen tensions. Cleeland (1963) used some gas mixtures which were devoid of oxygen or which probably contained only traces of oxygen, and survival of the worms was relatively poor.

Trichinella spiralis has proved a difficult organism to grow *in vitro,* and most reports show that complete development has not been attained (Kim, 1961, 1962; Tarakanov, 1964; Meerovitch, 1965a). But Berntzen (1965) has developed a system, based on a continuous-flow culture, in which this nematode has been grown from excysted larvae to adults which produced progeny *in vitro.* The type and proportions of the gases used had a profound effect on this species. The most promising mixture was found to be one which contained 85% nitrogen, 5% carbon dioxide, and 10% oxygen, and relatively slight changes in these proportions produced abnormal worms. In a gas mixture composed of 95% air and 5% carbon dioxide, there was a multiple molting, the adults were never found free of the sheath, and there was no production of embryos. Worms decreased in size, although sexual differentiation occurred. Exposure to air for as little as 1 hour, followed by a return to the original gas mixture (85% nitrogen, 5% carbon dioxide, and 10% oxygen), was sufficient to induce multiple molting. These abnormalities were typical of those reported by many investigators who have worked with *T. spiralis in vitro,* and it is evident that for this species at least, composition of the gas phase must be adjusted with care, and within relatively narrow limits.

Detailed information on the effects of the gas phase is not available for other species. Weinstein and Jones (1956) used air in their early work on *Nippostrongylus brasiliensis,* but later (1959) used 5% carbon dioxide in air. In the latter experiments, the worms were very similar to those in rats, although smaller. The improved yields compared with those obtained in their earlier work were probably more closely related to the components in the medium than to the gas phase. Nicholas and Jantunen (1964, 1966) have examined the effect of different gas mixtures on growth and reproduction of the free-living nematode, *Caenorhabditis briggsae.* Using mixtures of carbon dioxide, nitrogen, and oxygen, they found that as the level of oxygen fell below partial pressures of about 30 mm. Hg, the rate of growth decreased, and eventually stopped, at oxygen tensions about 1.2 mm. Hg. Carbon dioxide did not seem to have any very marked effect on either growth or reproduction. These results are consistent with the observations that increased populations of *C. briggsae* can be obtained when the worms are grown in thin films instead of in deep tubes (Hansen and Cryan, 1966). This requirement for oxygen is surprisingly high for a species which is probably associated with putrifying organic matter in the soil (Nicholas and Jantunen, 1966). Perhaps *C. briggsae* has become adapted to survival at relatively high oxygen tensions.

Carbon dioxide is fixed by *Moniliformis dubius* (Acanthocephala) under conditions which differed greatly from those which prevail in complex culture media (Laurie, 1959). Most attempts to grow various species of Acanthocephala *in vitro* have employed air (Dunagan, 1962; Harmes, 1965). Nicholas and Grigg (1965) have used with advantage higher concentrations of carbon dioxide and lower concentrations of oxygen than normally occurs in air, in their work with *N. dubius.* Apart from this, nothing seems to be known of the requirements of Acanthocephala for different gas mixtures *in vitro.*

2. *Temperature*

Parasitic nematodes often require a change in temperature as part of a stimulus for further growth. Microfilariae from a warm-blooded animal continue development inside a mosquito. When cultured *in vitro,* it was found that high temperatures inhibited development, which would only proceed at relatively low temperatures, such as might be encountered within a mosquito (Sawyer and Weinstein, 1963; Wood and Suitor, 1966). An infective stage which is free-living, or lives in a poikilothermic animal, seems to require an increase in temperature before growth will commence. The infective stage of *Phocanema decipiens* lives in the muscles of the Atlantic cod, *Gadus morhua,* and the adult stage lives in seals which eat the cod. When the infective juvenile is held in physiological saline for 3–6 days at 35°, it molts to the preadult, or fourth stage. It appears that the stimulus which induces development is temperature (Townsley *et al.*, 1963). This sort of stimulus is satisfactory if an animal lives in a cold marine environment. Infective stages which live in terrestrial environments, like those of *Haemonchus contortus,* might encounter quite high temperatures while still free-living, so that reliance on temperature alone as a stimulus for growth would be hazardous. Infective juveniles of *H. contortus* require an additional stimulus, provided by carbon dioxide (Rogers, 1960), to induce development, although temperature is still important, because the response only reaches a maximum at about 39° or 40°, which is the body temperature of sheep, the definitive host (Rogers and Sommerville, 1960).

In nutrient media, temperature may still be an important part of the stimulus. The free-living stages of *Nippostrongylus brasiliensis* can grow to the filariform stage in a medium composed of chick embryo homogenate, but at temperatures about 20° to 25°, no further development takes place. If the temperature is raised to 37.5°, the filariform juvenile quickly differentiates to a stage corresponding to that of moderately developed juveniles found in the lung of the rat (Weinstein and Jones, 1957). But few complete the third molt, and those which succeed invariably die. Further development required additional sources of nutrients.

3. *Equilibrium Between Medium and Organism*

In the culture of mammalian cells it has been found that there is an equilibrium between the cellular pool and the culture medium for certain compounds. These may be lost to the medium at a rate which exceeds the synthetic capacity of the cell, but if cells are sufficiently abundant, the concentration of the compound in the medium, and hence in the cell, rises to a level which enables the cell to survive. Thus a need might be found to supplement the medium with such compounds at low population levels, but at higher levels the same compounds might not be required (Eagle and Levintow, 1965).

Rothstein (1965) has suggested that the culture medium for *Caenorhabditis briggsae* becomes a "reservoir for metabolic intermediates, rather than for a fixed group of end products." The nutritional requirements of this and perhaps other nematodes in axenic culture might be related to the numbers present if an equilibrium is established between the organisms and the medium for essential metabolites. It might be worthwhile examining the growth of some species under crowded conditions, because some of the special requirements we have discussed may no longer be a problem under these circumstances.

D. Nutritional Requirements of Acanthocephala in the Host

The growth of *Moniliformis dubius* is arrested if the rat host is placed on a diet deficient in carbohydrate, but is resumed if the diet is restored to normal (Read and Rothman, 1958). From this and related experiments it has been concluded that *M. dubius* requires carbohydrate for growth, and that the only adequate source of this is the host's ingesta. There are many accounts of the effect of the diet of the host on parasites, but usually these give no direct information about the requirements of the parasite. Deficient diets must have extensive effects on the host, and some of these in turn might modify the growth of the parasite, either directly or indirectly, by interfering with its ability to take up nutrients. It is difficult to learn much about growth requirements in such complex systems.

V. Intermediary Metabolism in Relation to Growth

Intermediary metabolism provides the energy and many of the small molecules needed for the synthesis of tissue, and so forms an important part of a discussion on growth and development. However, detailed accounts of intermediary metabolism of the Acanthocephala and Nematoda are given in other parts of this book (Chapters 5, 6, and 7), so only a brief summary is given here.

A. The Nature of Internal Reserves

Reserve materials are found in all stages of the life cycles of nematodes and Acanthocephala. They are especially important in eggs and in the reproductive systems of species which produce large amounts of eggs and sperm. Infective stages which do not feed and *dauer* larvae also often contain large amounts of food reserves, but, as these stages do not grow or undergo morphological changes, energy is required for movement and maintenance only. In some species of parasities, e.g., *Haemonchus contortus,* the early parasitic stages do not feed (Section IV) and early developmental changes are dependent upon internal reserves accumulated before the infective stage is reached (Sommerville, 1964, 1966).

Lipid reserves predominate in the early stages of eggs and in the free-living infective stages of some nematodes. Lipids decrease as the infective stages age, whereas glycogen decreases to a lesser degree (e.g., see Rogers, 1940; Elliott, 1954; Wilson, 1965). During embryonation in the eggs of *Ascaris lumbricoides* there is a quantitatively important conversion of triglyceride reserves to carbohydrate (Passey and Fairbairn, 1957). The biological significance of these changes is unknown.

Carbohydrate is the important reserve material in most stages of the life cycles of nematodes and Acanthocephala (e.g., see Giovannola, 1936). Thus glycogen and trehalose may form as much as 14.8% and 1.1%, respectively, of the fresh tissue of adult *Ascaris lumbricoides* (Harpur, 1963). The amount present in individuals of a given species may vary considerably with the condition of the host, though in general it tends to be greater in species which live in environments deficient in oxygen (see von Brand, 1966, for references). Glycogen is important not only as a source of energy; it provides the substrate for the formation of acetylglucosamine used for the synthesis of chitin in egg shells, and so it is found in considerable amounts in ovaries.

Granules of protein have been found in the ovocytes of some ascarids. However, the concentrations of protein and nonprotein nitrogen do not change during embryonation of the cleidoic eggs of *Ascaris lumbricoides* and it seems unlikely that the protein reserves serve as a source of energy (Passey and Fairbairn, 1957). Indeed, it is probable that this protein is used in the formation of membranes of the egg shells (Yanagisawa, 1955). Granules of protein are present in the intestinal cells of some juvenile stages of nematodes (M. B. Chitwood, 1951). It is uncertain whether this protein is catabolized or whether it becomes incorporated in tissue components. In infective juveniles of *Nippostrongylus brasiliensis* protein is an important source of energy (Wilson, 1965).

B. Intermediary Metabolism

Glycolysis is a major metabolic route for providing energy both in the Acanthocephala and Nematoda (Read, 1961; Graff, 1964; Bryant and Nicholas, 1965; Saz and Hubbard, 1957; Mancilla and Agosin, 1960; Krusberg, 1960). This route provides the mechanism whereby certain hexoses in the external environment as well as the internal reserves of glycogen are used. In species which live in environments deficient in oxygen, glycolysis is the major route by which energy is obtained (Bueding and Farber, 1960). However, a wide variety of terminal anaerobic processes have been described and lactic acid is rarely an important end product (von Brand, 1966).

Some form of the pentose phosphate shunt functions in tissues of nematodes (Entner, 1957; Entner and Gonzalez, 1959; Krusberg, 1960) but it seems unlikely that it is quantitatively important though it may, of course, provide basic substrates for complex biosyntheses.

Features of the tricarboxylic acid cycle and the terminal aerobic processes associated with it vary considerably in different species of nematodes and Acanthocephala (e.g., see von Brand, 1966; Bryant and Nicholas, 1966). Carbon dioxide fixation is a common feature in the metabolism of nematodes (von Brand, 1966) and occurs in the Acanthocephala also (Graff, 1965). This process may be of special importance in the development of the early parasitic stages of *Haemonchus contortus* at a time when external substrates are not used by this organism (Rogers, 1966a). Some lipids and amino acids are metabolized via the tricarboxylic acid cycle in many species of nematodes and Acanthocephala and so may provide energy for growth and development. Even in species in which aerobic processes predominate, parts of the tricarboxylic acid cycle are probably important in providing some amino acids needed for protein synthesis (Chapter 7). In general, nematodes and Acanthocephala (Graff, 1965) are probably similar to most other organisms in their capacity to synthesize nonessential amino acids. Some species of nematodes may even have the capacity to synthesize "essential" amino acids (Rothstein and Tomlinson, 1962; Rothstein, 1965; Myers and Krusberg, 1965).

C. Oxidative Metabolism in Relation to Growth

Oxygen is necessary for the growth of many nematodes, though survival, even for prolonged periods, in the absence of oxygen has been reported (e.g., see Rogers and Sommerville, 1963; Section IV). Anaerobiosis may affect growth of nematodes in several ways. Sufficient energy for syntheses may not be available; basic substrates for synthetic processes may not be formed and the accumulation of toxic end products may interfere with metabolism. Whether one or all of these factors operate to prevent growth under anaerobic conditions is unknown.

VI. The Chemistry of the Hatching of Eggs

The early development of most nematodes takes place within a cleidoic egg shell. As a rule it is the first-stage juvenile which emerges through the egg shell but in some species further development takes place before the eggs hatch. In parasitic species in which the egg is infective, the host evidently supplies a stimulus which causes the egg to hatch, e.g., in *Ascaris lumbricoides* (Fairbairn, 1960). This even applies to certain species of plant parasites such as *Heterodera rostochiensis* where the infective stage is enclosed within the egg shell, and the eggs in turn are enclosed in a cyst formed from the cuticular remains of the adult female. In this species the eggs hatch and the infective stage emerges from the cyst in response to a stimulus supplied by the host plant (see Shepherd, 1962, for references).

Noninfective eggs and eggs of free-living species of nematodes hatch when development has reached the appropriate stage, though environmental conditions, which are less specific than those required for infective eggs, influence the hatching (Wilson, 1958; Loewenberg *et al.,* 1960; Christie, 1962).

In the Acanthocephala the egg forms the first infective stage in the life cycle. It hatches when ingested by the proper intermediate host, which is usually a larval insect, amphipod, or isopod. Evidently the conditions in the gut of the intermediate host induce the hatching of the eggs.

A. The Chemistry of Egg Shells

In general the egg shells of nematodes have an internal membrane of lipid and a "hard shell" of chitin. Proteins form one or possibly two separate layers and are probably present in the lipid and chitinous layers also. The layer of lipid may be absent in some species, e.g., in the eggs of *Strongyloides ratti* which are laid parthogenetically (B. G. Chitwood and Graham, 1940), and it varies in thickness in different species. The amounts of chitin and protein in egg shells of different species also vary considerably (see Rogers, 1962, for references).

In view of these differences in the egg shells of nematodes, the mechanisms of hatching in different species may vary in detail though in most species they must involve the piercing of lipoid, chitinous, and proteinacious layers.

The so-called eggs of the Acanthocephala are really acanthors enclosed in an egg shell. The shell contains four layers, one of which contains chitin, though the amount present varies in different species (von Brand, 1940; Monné and Hönig, 1954; West, 1964). There is no evidence that a membrane of lipid is present; the other layers are chiefly composed of protein (e.g., see Monné and Hönig, 1954).

B. Noninfective Eggs of Nematodes

The ovic juveniles of plant parasitic nematodes have a stylet which pierces the egg membranes and helps the release of the juvenile. Thus mechanical processes are more important in the hatching of these eggs. In general the hatching seems to take place in three stages. After the ovic juvenile becomes active and pharyngeal pulsations have occurred, the shell becomes more flexible. In some species an outward flow of material from the pharyngeal glands can be observed and this secretion may be responsible for changes in the egg membranes. During the second stage, the egg shell is punctured by the stylet of the juvenile. The final stage involves the rupture of the membrane by the movements of the whole juvenile (Dropkin *et al.,* 1958; Rhoades and Linford, 1961; D. P. Taylor, 1961).

The "spontaneous" hatching of eggs of *Trichostrongylus retortaeformis* takes place at the appropriate stage of development. Wilson (1958) regarded the first step in hatching as involving the breakdown of the inner lipoid membrane of the egg. This, he thought, was achieved by the movements of the juvenile within the egg shell in the presence of some emulsifying agent in the fluids of the egg. The breakdown of the lipoid layer would cause the uptake of water so that pressure would be exerted on the "hard" layers of the shell. This might be assisted by a chemical weakening of the shell caused by the secretions of the ovic juvenile.

Wilson was able to inhibit hatching by placing the eggs in dilute solutions of electrolytes; this inhibition was reversible. Thus 0.05 *M* sodium chloride slowed down the rate of hatching but the treated eggs hatched normally when placed in distilled water. The hatching of eggs of the plant parasitic nematodes, *Meloidogyne arenaria* and *Heterodera rostochiensis* was also inhibited by 0.2 *M* solutions of salts and 0.3 *M* solutions of dextrose (Dropkin *et al.,* 1958).

C. Infective Eggs

Examination of the hatching of infective eggs of nematodes and Acanthocephala has shown that several steps are involved in the process. A signal from the host acts as a stimulus for the infective ovic juvenile which then secretes the enzymes which attack the egg shell. Though little is known about the intermediate events in the ovic juvenile which link the action of the stimulus with the secretion of the hatching enzymes, it seems probable that in nematodes the process has some similarity to that concerned with molting (Section VII).

1. The Stimulus from the Host

Eggs of ascarids, especially *Ascaris lumbricoides,* have been examined (e.g., see Rogers, 1958, 1960, 1961; Fairbairn, 1961; Jaskoski and Colucci, 1964). It is clear that dissolved gaseous carbon dioxide (and/or undissociated carbonic acid) at 37° is the major factor required in the stimulus from the host. The optimal concentration of carbon dioxide is somewhat different for different species. Under physiological conditions a strongly reducing potential is needed in the stimulus for the eggs of *Ascaris lumbricoides,* though this appears to be less important for the eggs of *Ascaridia galli.* The pH, apart from its effect on the concentration of dissolved carbon dioxide and on the redox potential, also affects the action of the stimulus. Minor factors which influence the efficiency of the stimulus include the type and concentration of electrolytes and the presence of surface-active agents in the medium.

There is little doubt that the stimulus for the hatching of infective nematode eggs acts as a physiological trigger. Nevertheless, if the normal infective eggs of *Ascaris lumbricoides* are used, the stimulus must be applied for 2 or 3 hours, a much longer period than that needed for the stimulation of exsheathment (Section VII).

The eggs of some plant parasitic nematodes contain infective juveniles, and "root diffusates" obtained from host plants cause the eggs to hatch. It is not known how the active material in root diffusate works. It might have a direct effect on the egg shell or it might stimulate the ovic juvenile to secrete substances which attack the egg shell. The results obtained from a study of synthetic hatching agents on *Heterodera schachtii* led Clarke and Shepherd (1964) to suggest that hatching results from a stimulation of the metabolism of the ovic juvenile. Other workers have suggested that natural hatching factors may also work indirectly (see Shepherd, 1962, for references). However, results from a direct experimental attack on this problem are lacking.

The structure of the active substances in root diffusates has not yet been determined but highly purified preparations which cause the hatching of eggs of *Heterodera rostochiensis* have been obtained (Hartwell *et al.*, 1960; Clarke and Widdowson, 1966).

The chemistry of the hatching of eggs of one species of Acanthocephala, *Moniliformis dubius,* has been examined (Edmonds, 1966). Normally these eggs hatch in the midgut of the host, a cockroach. There is no doubt that a stimulus from the host acts as a physiological trigger. Edmonds found that exposure to a suitable stimulus *in vitro* for about 15 minutes was sufficient to set in train the process of hatching, which took at least 2 hours for completion. Solutions of electrolytes such as sodium or potassium chloride above 0.2 *M*, with small amounts of phosphate or bicarbonate buffer at pH's above 7.5, were effective stimulants. The efficiency of stimulation was improved somewhat if carbon dioxide was present. Solutions of sucrose had no effect, nor did the presence of reducing agents. Solutions of electrolytes above 0.5 *M* killed the acanthors (Edmonds, 1966).

2. *The Breakdown of Egg Shells*

The signal from the host evidently causes changes in ovic juveniles which lead to the release of substances which attack the egg shell. Stimulation may also cause changes in the behavior of ovic juveniles so that they disrupt the weakened egg membranes mechanically. Details of the processes directly involved in the breakdown of egg shells are known for two species, *Ascaris lumbricoides* and *Moniliformis dubius.*

As might be expected from the nature of the membranes of the egg shells of nematodes and Acanthocephala, a chitinase, esterase, and possibly a proteinase are secreted by the stimulated juveniles. The chitinase secreted by the ovic juvenile of *Ascaris lumbricoides* was found to be the most active at pH 6. The esterase, with tributyrin or phenylpropionate as substrate, showed optimum activity at pH 7.0–7.5. This enzyme was only slightly inhibited by 10^{-3} *M* diaminoethanetetraacetic acid, diisopropylfluorophosphate, or ethylmaleimide (Rogers, 1958, 1963).

One of the first changes which occurs in the eggs of *Ascaris lumbricoides* which have been stimulated to hatch is an increase in the permeability of the vitelline membrane (Fairbairn, 1960). The significance of this in the process of hatching is unknown but it may be important because similar changes in permeability also occur in the "spontaneous" hatching of noninfective eggs of *Trichostrongylus retortaeformis* (Wilson, 1958).

As a result of the stimulus, the acanthor within the egg shell of *Moniliformis dubius* releases a chitinase which attacks the egg membranes from within. Thereafter the escape of the acanthor seems to be a mechanical process in which the remaining layers of the egg shell are disrupted by the hooks of the acanthor (Edmonds, 1966).

VII. The Chemistry of Molting and Excystment

The chemistry of molting in nematodes has been examined only in infective or parasitic stages. There is good evidence that exsheathment or molting which precede infection take place as the result of a signal from the host. Much of the work in this field has been concerned with the nature of the signal rather than with its effect on the nematode or the actual processes of exsheathment or molting.

Our knowledge of excystment in the Acanthocephala falls in the same category. Excystment takes place during the process of infection of the vertebrate host, and it is the stimulus for excystment rather than the chemistry of excystment itself which has been studied.

A. Composition of the Nematode Cuticle

This has been discussed elsewhere in this volume (Chapter 2). It is sufficient here to point out that in most nematodes molting involves the loosening of the cuticle from the underlying structures followed by the breakdown of a specific area of the cuticle which allows the escape of the next stage of the life cycle. Though much is known about the structure of the cuticle, the actual substrates attacked during the loosening of the cuticle and the specific breakdown are unknown. However, the fact that proteins predominate both in sheaths and normal cuticles suggests that proteinases would be concerned in exsheathment and molting.

B. The Stimulus from the Host

For a number of species of trichostrongyles which infect the host *per os,* the major component in the signal from the host which sets in train the process of exsheathment is dissolved gaseous carbon dioxide and/or undissociated carbonic acid at 37°. The pH, and, in some species the redox potential, affects the efficiency of the stimulus. The effective concentration of carbon dioxide differs for different species (see Rogers, 1966b, for references).

The presence of salts in the stimulating medium also affects the efficiency of the signal causing exsheathment. A. N. Taylor and Whitlock (1960a,b) and Whitlock (1966) suggested that these salts may act by influencing the rate of formation of carbonic acid and that this is of prime importance in stimulating exsheathment.

There is evidence that a stimulus from the host is required to initiate molting in some nematode parasites of plants (Rhoades and Linford, 1959; Fisher, 1966). Thus it was shown that the fourth molt of *Paratylenchus nanus* took place more frequently when the juveniles had been exposed to root secretions obtained from growing apricot seedlings. Dormant seedlings did not produce active root secretion. Stimulation probably acted as a physiological trigger because exposure to root secretion for 24 hours was sufficient to induce molting, which occurred 8 to 13 days later (Fisher, 1966). As in parasitic nematodes of animals (Sommerville, 1957; Rogers and Sommerville, 1960), Fisher found that ligation restricted molting to the anterior ends of the worms.

The activation of cystacanths of *Moniliformis dubius* has been examined by Graff and Kitzman (1965). Bile salts were found to be an essential component of the stimulus but their efficiency in experiments carried out *in vitro* was greatly enhanced in the presence of carbon dioxide. Reducing agents had no effect but oxygen inhibited the activation.

Whatever the precise mechanism by which carbon dioxide acts in promoting exsheathment or excystment, it is clear that this substance, or a related compound, is of major importance in the stimulus for a variety of parasites, not only nematodes and Acanthocephala, but also for some trematodes and protozoa, which infect the host *per os* (Rogers,

1963). It is interesting to note, therefore, that the signal for exsheathment of the infective juvenile of the nematode, *Nematodirus battus*, does not require carbon dioxide. Hydrochloric acid was found to be the chief component in the stimulus for this species (Christie and Charleston, 1965).

C. THE ACTION OF THE STIMULUS ON NEMATODES

It has been postulated (Rogers, 1965) that the stimulus for exsheathment acts on "receptors" in the anterior region of juveniles. The activated

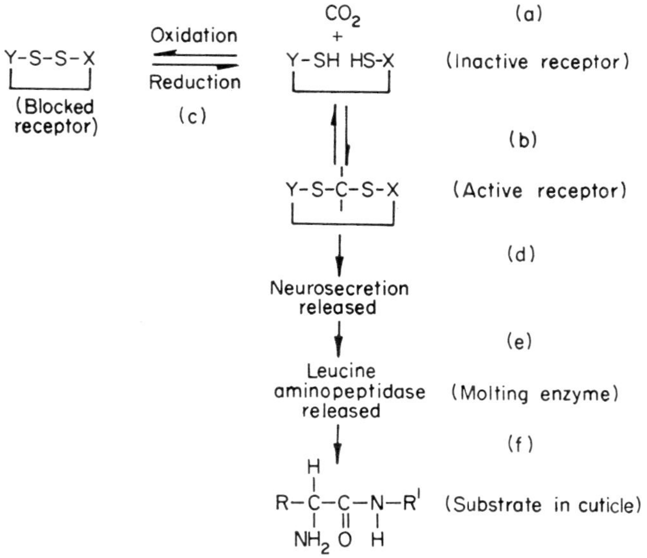

FIG. 4. Hypothetical model for the mechanism of molting in some parasitic nematodes (for details, see text).

"receptor" in turn caused changes in the juvenile which ultimately led to the secretion of leucine aminopeptidase which attacks the sheath (Fig. 4). From the nature of the stimulus and the effect of hydrogen ion concentration and redox potential on its efficiency in a variety of species, it has been suggested (Rogers, 1966c) that the carbon dioxide reacts with adjacent sulfhydryl groups to form an $-S-\overset{|}{\underset{|}{C}}-S-$ link in the activated receptor (Fig. 4b). To a limited extent the hypothesis has been supported by results of experiments with *Haemonchus contortus* and *Trichostrongylus colubriformis* carried out *in vitro*. Thus short exposure to 10^{-3} *N* iodine inhibited the action of carbon dioxide as a stimulus and this inhibition could be reversed by subsequent treatment with hydrogen sulfide (Rogers, 1966c).

It has also been shown that carbon dioxide may act as a physiological trigger for developmental changes as well as for exsheathment (Rogers, 1966a). Thus treatment with iodine prevented infection with infective juveniles of *Nematospiroides dubius* from which the sheaths had been removed by hypochlorite, and infectivity was restored by subsequent exposure to hydrogen sulfide. In these experiments the inhibition of exsheathment would not have prevented infection.

That the neurosecretory system of nematodes is involved in molting has been shown by Davey (1966), who found that the neurosecretory cycle which occurred during the last molt of the nematode parasite of seals, *Phocanema decipiens,* closely paralleled the synthesis and release of leucine aminopeptidase in the excretory gland. Moreover, saline extracts of the anterior ends of worms prepared at the time of maximum neurosecretory activity brought about the synthesis of leucine aminopeptidase in excretory glands *in vitro.*

D. The Breakdown of the Cuticle of Nematodes

The cuticle of nematodes is composed largely of proteins and it seems that a proteinase is the important agent which weakens a specific area of the cuticle and so allows the escape of the next developmental stage (Fig. 4,e,f). The proteinases secreted by infective juveniles of *Haemonchus contortus* and *Trichostrongylus colubriformis* after they have been stimulated have been identified as leucine aminopeptidases with properties similar to those of the mammalian enzyme. However, the leucine aminopeptidase secreted by juveniles of *Haemonchus contortus* did not attack the sheaths of *Trichostrongylus colubriformis* and *vice versa,* and mammalian leucine aminopeptidase had no appreciable effect on the sheaths of either species. All three leucine aminopeptidases, however, did attack L-leucyl-β-naphthylamide and L-leucinamide (Rogers, 1965; this volume, Chapter 7).

It seems probable that leucine aminopeptidase is concerned in molting as well as in exsheathment. Thus Davey and Kan (1967) have obtained histochemical evidence that this enzyme is synthesized and released during the last molt of *Phocanema decipiens.*

ACKNOWLEDGEMENTS

We are grateful to Dr. K. G. Davey and Dr. B. D. Watson for permission to refer to some of their unpublished results.

We wish to acknowledge assistance given by research grants from the U.S. Public

Health Service (Grant No A104093) and Parke, Davis and Co. (W.P.R.); the Australian Meat Research Committee and the Rural Credits Development Fund of the Reserve Bank of Australia (R.I.S.).

REFERENCES

Anya, A. O. (1966a). *Parasitology* **56**, 583.
Anya, A. O. (1966b). *J. Helminthol.* **40**, 261.
Bailey, A. (1967). Ph.D. Thesis, University of Adelaide, South Australia.
Berntzen, A. K. (1965). *Exptl. Parasitol.* **16**, 74.
Berntzen, A. K. (1966). *Ann. N.Y. Acad. Sci.* **139**, 176.
Berrill, N. J. (1961). "Growth Development and Pattern" Freeman, San Francisco, California.
Bird, A. F. (1959). *Nematologica* **4**, 31.
Blake, C. D. (1962a). *Ann. Appl. Biol.* **50**, 703.
Blake, C. D. (1962b). *Ann. Appl. Biol.* **50**, 713.
Bremner, K. C. (1956). *Australian J. Zool.* **4**, 146.
Bronskill, J. F. (1962). *Can. J. Zool.* **40**, 1269.
Bryant, C., and Nicholas, W. L. (1965). *Comp. Biochem. Physiol.* **15**, 103.
Bryant, C., and Nicholas, W. L. (1966). *Comp. Biochem. Physiol.* **17**, 825.
Bueding, E., and Farber, E. (1960). *Comp. Biochem.* **1**, 421–423.
Burdett, J. F., Bird, A. F., and Fisher, J. M. (1963). *Nematologica* **9**, 542.
Chitwood, B. G., and Chitwood, M. B. (1940). "An Introduction to Nematology." M. B. Chitwood, Babylon, New York.
Chitwood, B. G., and Graham, G. L. (1940). *J. Parasitol.* **26**, 183.
Chitwood, M. B. (1951). *J. Parasitol.* **37**, 96.
Christie, M. G. (1962). *Parasitology* **52**, 297.
Christie, M. G., and Charleston, W. A. G. (1965). *Exptl. Parasitol.* **17**, 46.
Clarke, A. J., and Shepherd, A. M. (1964). *Nematologica* **10**, 431.
Clarke, A. J., and Widdowson, E. (1966). *Biochem. J.* **98**, 862.
Cleeland, R. (1963). *J. Parasitol.* **49**, 64.
Cleeland, R., and Laurence, K. A. (1962). *J. Parasitol.* **48**, 35.
Cram, E. B. (1931). *U.S. Dept. Agr., Tech. Bull.* **227**.
Crofton H. D. (1966). "Nematodes." Hutchinson, London.
Davey, K. G. (1966). *Am. Zoologist* **6**, 243.
Davey, K. G., and Kan, S. P. (1967). *Nature* **214**, 737.
Dineen, J. K., Donald, A. D., Wagland, B. W., and Offner, J. (1965). *Parasitology* **55**, 515.
Donald, A. D., Dineen, J. K., Turner, J. H., and Wagland, B. M. (1964). *Parasitology* **54**, 527.
Dropkin, V. H. (1955). *Exptl. Parasitol.* **4**, 282.
Dropkin, V. H., Martin, G. C., and Johnson, R. W. (1958). *Nematologica* **3**, **115**.
Dunagan, T. T. (1962). *Proc. Helminthol. Soc. Wash. D.C.* **29**, 131.
Eagle, H., and Levintow, L. (1965). *In* "Cells and Tissues in Culture" (E. N. Willmer, ed.), Vol. 1, pp. 277–296. Academic Press, New York.
Edmonds, S. J. (1966). *Exptl. Parasitol.* **19**, 216.
Edmonds, S. J., and Dixon, B. R. (1966). *Nature* **209**, 99.
Elliott, A. (1954). *Exptl. Parasitol.* **3**, 307.
Entner, N. (1957). *Arch. Biochem. Biophys.* **71**, 52.
Entner, N., and Gonzalez, C. (1959). *Exptl. Parasitol.* **8**, 471.

Fairbairn, D. (1960). *In* "Host Influence on Parasite Physiology" (L. A. Stauber, ed.), pp. 50–64. Rutgers Univ. Press, New Brunswick.
Fairbairn, D. (1961). *Can. J. Zool.* **39,** 153.
Fernando, M. A., and Wong, H. A. (1964). *Exptl. Parasitol.* **15,** 284.
Fisher, J. M. (1966). *Australian J. Biol. Sci.* **19,** 1073.
Frenzen, K. (1955). *Z. Wiss. Zool.* **158,** 304.
Giovannola, A. (1936). *J. Parasitol.* **3,** 307.
Goodey, J. B., and Hooper, D. J. (1962). *Nematologica* **8,** 33.
Graff, D. J. (1964). *J. Parasitol.* **50,** 230.
Graff, D. J. (1965). *J. Parasitol.* **51,** 72.
Graff, D. J., and Kitzman, W. B. (1965). *J. Parasitol.* **51,** 424.
Greet, D. N. (1964). *Nature* **204,** 96.
Gysels, H. (1962). *Nematologica* **7,** 11.
Haley, A. J., and Parker, J. C. (1961). *J. Parasitol.* **47,** 461.
Hansen, E. L., Buecher, E. J., and Yarwood, E. A. (1964). *Nematologica* **10,** 623.
Hansen, E. L., and Cryan, W. S. (1966). *Nematologica* **12,** 138.
Harmes, C. E. (1965). *Exptl. Parasitol.* **17,** 41.
Harpur, R. P. (1962). *Can. J. Zool.* **40,** 991.
Harpur, R. P. (1963). *Can. J. Biochem. Physiol.* **179,** 435.
Harris, J. E., and Crofton, H. D. (1957). *J. Exptl. Biol.* **34,** 116.
Hartwell, W. V., Dahlstrom, R. V., and Neal, A. L. (1960). *Phytopathology* **50,** 612.
Hechler, H. C., and Taylor, D. P. (1966). *Proc. Helminthol. Soc. Wash., D.C.* **33,** 40.
Hopkins, C. A. (1955). *Trans. Roy. Soc. Trop. Med. Hyg.* **49,** 1.
Hyman, L. H. (1951). "The Invertebrates: Acanthocephala, Aschelminthes and Entoprocta." McGraw-Hill, New York.
Inglis, W. G. (1954). *Nature* **173,** 957.
Iwamura, T. (1966). *Progr. Nucleic Acid Res. Mol. Biol.* **5,** 133.
Jackson, G. J. (1962a). *Exptl. Parasitol.* **12,** 25.
Jackson, G. J. (1962b). *Trans. N.Y. Acad. Sci.* [2] **24,** 954.
Jaskoski, B. J., and Colucci, A. V. (1964). *Trans. Am. Microscop. Soc.* **83,** 294.
Jones, T. P. (1966). *Nematologica* **12,** 518.
Kampfe, L. (1962). *Parasitol. Schriftenreihe, Jena* No. 14, 1–205 (original not seen); abstract in *Helminthol. Abstr.* **31,** 507 (1962).
Kent, N. H. (1963). *Exptl. Parasitol.* **13,** 2.
Kim, C. W. (1961). *Am. J. Trop. Med. Hyg.* **10,** 742.
Kim, C. W. (1962). *Am. J. Trop. Med. Hyg.* **11,** 491.
Komiya, Y., Yasuroaka, K., and Sato, A. (1956). *Japan. J. Med. Sci. Biol.* **9,** 283.
Krusberg, L. R. (1960). *Phytopathology* **50,** 9.
Kurochkin, Y. V. (1960). *Dokl. Akad. Nauk SSSR* **135,** 1281.
Lapage, G. (1933). *Nature* **131,** 583.
Laurie, J. S. (1959). *Exptl. Parasitol.* **8,** 188.
Lee, D. L. (1960). *Parasitology* **50,** 241.
Lee, D. L. (1966). *Advan. Parasitol.* **4,** 187.
Leland, S. E. (1962). *J. Parasitol.* **48,** Suppl., 35.
Leland, S. E. (1963). *J. Parasitol.* **49,** 600.
Loewenberg, J. R., Sullivan, T., and Schuster, M. L. (1960). *Phytopathology* **50,** 215.
McCoy, O. R. (1930). *Am. J. Hyg.* **11,** 413.
Mackerras, M. J., and Sanders, D. F. (1955). *Australian J. Zool.* **3,** 1.

Mancilla, R., and Agosin, M. (1960). *Exptl. Parasitol.* **10**, 43.
Mapes, C. J. (1965). *Parasitology* **55**, 583.
Matsusaki, G. (1950). *Yokohama Med. Bull.* **1**, 111.
Meerovitch, E. (1965a). *Can. J. Zool.* **43**, 69.
Meerovitch, E. (1965b). *Can. J. Zool.* **43**, 81.
Michel, J. F. (1952). *Nature* **169**, 933.
Michel, J. F. (1955). *J. Comp. Pathol.* **65**, 149.
Michel, J. F. (1963). *Parasitology* **53**, 63.
Monné, L. (1963). *Parasitology* **53**, 273.
Monné, L., and Hönig, G. (1954). *Arkiv Zool.* **7**, 261.
Moore, J. D. (1965). *Dissertation Abstr.* **26**, 1300.
Myers, R. F., and Krusberg, L. R. (1965). *Phytopathology* **55**, 429.
Mykytowycz, R., and Dudzinski, M. L. (1965). *Parasitology* **55**, 527.
Needham, A. E. (1964). "The Growth Process in Animals." Pitman, London.
Nicholas, W. L., and Grigg, H. (1965). *Exptl. Parasitol.* **16**, 332.
Nicholas, W. L., and Jantunen, R. (1964). *Nematologica* **10**, 409.
Nicholas, W. L., and Jantunen, R. (1966). *Nematologica* **12**, 328.
Nicholas, W. L., and Mercer, E. H. (1965). *Quart. J. Microscop. Sci.* **106**, 137.
Nonnenmacher-Godet, J., and Dougherty, E. C. (1964). *J. Cell Biol.* **22**, 281.
Ogilvie, B. M. (1965). *Parasitology* **55**, 325.
Osche, G. (1958). *Z. Parasitenk.* **18**, 479.
Passey, R. F., and Fairbairn, D. (1957). *Can. J. Biochem. Physiol.* **35**, 511.
Read, C. P. (1961). *In* "Comparative Physiology of Carbohydrate Metabolism in Heterothermic Animals" (A. W. Martin, ed.), pp. 3–34. Univ. of Washington Press, Seattle, Washington.
Read, C. P., and Rothman, A. H. (1958). *Exptl. Parasitol.* **7**, 191.
Reger, J. F. (1966). *J. Ultrastruct. Res.* **14**, 602.
Rhoades, H L., and Linford, M. B. (1959). *Science* **130**, 146.
Rhoades, H. L., and Linford, M. B. (1961). *Proc. Helminthol. Soc. Wash., D.C.* **28**, 51.
Roberts, L. S., and Fairbairn, D. (1965). *J. Parasitol.* **51**, 129.
Rogers, W. P. (1940). *J. Helminthol.* **18**, 183.
Rogers, W. P. (1958). *Nature* **181**, 1410.
Rogers, W. P. (1960). *Proc. Roy. Soc.* **B152**, 367.
Rogers, W. P. (1961). *J. Helminthol.* **R. T.** Leiper Suppl. p. 151.
Rogers, W. P. (1962). "The Nature of Parasitism: The Relationship of Some Metazoan Parasites to Their Hosts." Academic Press, New York.
Rogers, W. P. (1963). *Ann. N.Y. Acad. Sci.* **113**, 208.
Rogers, W. P. (1965). *Comp. Biochem. Physiol.* **14**, 311.
Rogers, W. P. (1966a). *Exptl. Parasitol.* **19**, 15.
Rogers, W. P. (1966b). *In* "The Biology of Parasites" (E. J. L. Soulsby, ed.), p. 33. Academic Press, New York.
Rogers, W. P. (1966c). *Comp. Biochem. Physiol.* **17**, 1103.
Rogers, W. P., and Sommerville, R. I. (1960). *Parasitology* **50**, 329.
Rogers, W. P., and Sommerville, R. I. (1963). *Advan. Parasitol.* **1**, 109.
Rohde, R. A. (1965). *Phytopathology* **55**, 1159.
Rothstein, M. (1965). *Comp. Biochem. Physiol.* **14**, 541.
Rothstein, M., and Tomlinson, G. (1962). *Biochim. Biophys, Acta* **63**, 471.
Sawyer, T. K., and Weinstein, P. P. (1963). *J. Parasitol.* **49**, 218.
Saz, H. J., and Hubbard, J. A. (1957). *J. Biol. Chem.* **225**, 921.

Scott, J. A. (1929a). *Am. J. Hyg.* **10**, 125.
Scott, J. A. (1929b). *J. Parasitol.* **16**, 54.
Shepherd, A. M. (1962). "The Emergence of Larvae from Cysts in the Genus *Heterodera*," Tech. Commun. No. 32. Commonwealth Bureau of Helminthology, St. Albans, Herts., England.
Silverman, P. H. (1959). *Nature* **183**, 197.
Silverman, P. H. (1965). *Advan. Parasitol.* 3, 159.
Silverman, P. H., and Podger, K. R. (1964). *Exptl. Parasitol.* **15**, 314.
Silverman, P. H., Alger, N. E., and Hansen, E. L. (1966). *Ann. N.Y. Acad. Sci.* **139**, 124.
Sommerville, R. I. (1954). *Australian J. Agr. Res.* **5**, 130.
Sommerville, R. I. (1957). *Exptl. Parasitol.* **6**, 18.
Sommerville, R. I. (1960). *Parasitology* **50**, 261.
Sommerville, R. I. (1964). *Nature* **202**, 316.
Sommerville, R. I. (1966). *J. Parasitol.* **52**, 127.
Sommerville, R. I., and Weinstein, P. P. (1964). *J. Parasitol.* **50**, 401.
Sommerville, R. I., and Weinstein, P. P. (1967). *J. Parasitol.* **53**, 116.
Soulsby, E. J. L. (1957). *Nature* **179**, 783.
Soulsby, E. J. L. (1963). *Ann. N.Y. Acad. Sci.* **113**, 492.
Soulsby, E. J. L., ed. (1966). "The Biology of Parasites," pp. 255–276. Academic Press, New York.
Tarakanov, V. I. (1964). *Veterinariya* **41**, 43 (original not seen); abstract in *Helminthol. Abstr.* **34**, 305 (1965).
Taylor, A. N., and Whitlock, J. H. (1960a). *Cornell Vet.* **50**, 339.
Taylor, A. N., and Whitlock, J. H. (1960b). *Am. J. Vet. Res.* **21**, 318.
Taylor, D. P. (1961). *Proc. Helminthol. Soc. Wash., D.C.* **28**, 60.
Taylor, D. P. (1962). *Proc. Helminthol. Soc. Wash., D.C.* **29**, 52.
Thomas, P. R. (1965). *Nematologica* **11**, 395.
Townsley, P. M., Wight, H. G., Scott, M. A., and Hughes, M. L. (1963). *J. Fisheries Res. Board Can.* **20**, 743.
Urquhart, G. M., Jarrett, W. F. H., and Mulligan, W. (1963). *Advan. Vet. Sci.* **7**, 87.
Van Gundy, S. D., and Stolzy, L. H. (1963). *Nematologica* **9**, 605.
Veglia, F. (1915). Third and Fourth Reports, pp. 347–500. Director of Veterinary Research, Union of South Africa.
von Bertalanffy, L. (1960). *In* "Fundamental Aspects of Normal and Malignant Growth" (W. W. Nowinski, ed.), pp. 137–259. Elsevier, Amsterdam.
von Brand, T. (1940). *J. Parasitol.* **26**, 301.
von Brand, T. (1966). "Biochemistry of Parasites." Academic Press, New York.
Wallace, H. R. (1963). "The Biology of Plant Parasitic Nematodes." Arnold, London.
Warren, L. G., Guevara, A., and Patzrek, D. (1962). *J. Parasitol.* **48**, Suppl., 25.
Watson, B. D. (1965). *Quart. J. Microscop. Sci.* **106**, 83.
Weinstein, P. P. (1955). *Am. J. Trop. Med. Hyg.* **4**, 61.
Weinstein, P. P. (1958). *Am. J. Trop. Med. Hyg.* **7**, 1.
Weinstein, P. P., and Jones, M. F. (1956). *J. Parasitol.* **42**, 215.
Weinstein, P. P., and Jones, M. F. (1957). *Am. J. Trop. Med. Hyg.* **6**, 480.
Weinstein, P. P., and Jones, M. F. (1959). *Ann. N.Y. Acad. Sci.* **77**, 137.
Welch, H. E., and Bronskill, J. F. (1962). *Can. J. Zool.* **40**, 1263.
Wessing, A. (1953). *Zool. Jahrb., Abt. Allgem. Zool. Physiol. Tiere* **73**, 69.
West, A. J. (1954). *J. Parasitol.* **50**, 731.

Whitlock, J. H. (1966). *In* "The Biology of Parasites" (E. J. L. Soulsby, ed.), pp. 185–197. Academic Press, New York.
Wieser, W. (1960). *Limnol. Oceanog.* **5**, 121.
Wieser, W., and Kanwisher, J. (1960). *Z. Vergleich. Physiol.* **43**, 29.
Wigglesworth, V. B. (1963). *J. Exptl. Biol.* **40**, 231.
Wilski, A., and Giebel, J. (1966). *Nematologica* **12**, 219.
Wilson, P. A. G. (1958). *J. Exptl. Biol.* **35**, 584.
Wilson, P. A. G. (1965). *Exptl. Parasitol.* **16**, 190.
Wood, D. E., and Suitor, E. C. (1966). *Nature* **211**, 868.
Yanagisawa, T. (1955). *Japan. J. Med. Sci. Biol.* **7**, 215.

CHAPTER 10

The Pigments of Nematoda and Acanthocephala

Malcolm H. Smith

I. Introduction

The study of pigment materials has been of the greatest importance throughout the animal kingdom, but examples for nematodes and acanthocephalans are generally conspicuous by their absence from standard works (D. L. Fox, 1953; H. M. Fox and Vevers, 1960). Unfortunately this probably reflects lack of adequate study rather than actual absence of many of the pigments which one would expect to occur. It must also be pointed out that the literature on these animals is so scattered that the interested worker is at a considerable disadvantage in trying to ascertain the extent of present knowledge. The following does not claim to be exhaustive, but merely a summary of the better-known facts fitted, as far as is possible, into the wider framework of knowledge of animal pigments.

II. Nematoda

The Nematoda may be divided on a species basis into two roughly equal groups, one parasitic and the other nonparasitic. The former vary in size from microscopic to large, while the latter are all very small, generally microscopic. Such pigments as occur in nematodes are almost all known from parasitic forms, partly because of the ease of handling the larger species and partly because of the obvious color of many of these worms, even those of microscopic dimensions. The most prominent of these pigments is undoubtedly hemoglobin, the Nematoda being one of the invertebrate groups in which it is very common. The main source of information on hemoglobin in nematodes is the recent review by Lee and Smith (1965), from which the present article inevitably borrows heavily. The interested reader is referred there for further discussion of the role of hemoglobin in parasitic animals, as well as for further information, references, etc., to conclusions merely summarized here. (For the reader aware of older controversies, but not the conclusions of later work, it is emphasized that there is no reliable evidence that any parasite absorbs hemoglobin directly from the host without causing it to undergo some further change; whereas there is considerable evidence that any hemoglobin found in the tissues of a parasite is specific to the worm itself.)

At least one other group of heme pigments occurs in nematodes, namely the cytochromes, but convincing evidence for the presence of other hemoproteins, or for hematin itself, appears to be lacking. Reports of other pigments are very few and it appears probable that in general all pigments, with the exception of hemoglobin, occur in quite small amounts. There do not appear to be any pigments which are characteristic of nematodes alone.

A. Distribution of Hemoglobin

Lee and Smith (1965) listed 18 species of nematode in which hemoglobin is definitely known, and a further 13 related species in which its presence was judged possible from published descriptions of the color of each worm. However, only three species have been definitely reported to have no hemoglobin. Ronald (1960) has recently reported the presence of hemoglobin in a further species, *Terranova (Porrocaecum) decipiens.* The available evidence therefore suggests that hemoglobin is common, possibly widespread, among parasitic nematodes. One would welcome further information in order to establish whether or not a strong red color could be due to some other pigment, possibly a carotenoid. There is no evidence in favor of such a possibility at present.

One is also curious about the three species in which hemoglobin is reported to be absent. These worms, *Cruzia testudinis, Falcaustra officine,* and *Oxyuris equi,* are all members of the order Oxyuroidea. Apart from variation by genus, one also has to take into account the possibility of species variation. The reported colors of the parasitic larvae of the insect *Gasterophilus* vary by species from blood red to pale yellow, though only one, described as red, is definitely known to contain hemoglobin. Similarly, only one species of the parasitic nematode *Tetrameres* is known to contain hemoglobin, but three other species are also described as blood red. *Tetrameres* is of further interest in this respect because it is reported (Villela and Ribeiro, 1955a,b) that only the female contains hemoglobin. Cases are also known in other species where the pigment has been reported in either the larvae or the adults, but not both. Finally, it should be noted that there can be considerable variation in color even within a species, and even in the same individual, in certain cases (see Section II,E). There are obviously a number of possibilities to be explored before one can be confident that hemoglobin is completely absent, or present at all times, at least where the Nematoda are concerned.

The species which have been mentioned above are all parasitic, and until quite recently it seemed that hemoglobin was restricted to such species. It is now known that it occurs in the head region of the free-living adult of the insect-parasitizing nematode *Mermis* (Ellenby, 1964). As well as this possibly aberrant case, Ellenby and Smith (1966) have described the occurrence of the pigment in the free-living marine nematode *Enoplus*. They found it "in abundance" in *E. brevis* and "to a lesser extent" in *E. communis*. It occurred in the pharynx of both sexes, in the "copulatory muscles" of the male, and in lesser quantity in the hypodermal chord. The authors point out that it is surprising that hemoglobin has not previously been reported from this widespread genus, which may be due to excessive reliance upon fixed specimens. They think that hemoglobin may eventually prove to be common in free-living nematodes, particularly in the hypodermal chord (hemoglobin occurs in this tissue both in the free-living stage of *Mermis*, and in the parasite *Ascaris*).

It seems likely, therefore, that no distinction need be made between parasitic and nonparasitic forms in this respect. On the other hand, many nematodes, of both types, do not appear to be strongly colored. This need not necessarily rule out the presence of hemoglobin—the protozoan *Paramecium* may also appear almost colorless, even in mass, and yet contain appreciable hemoglobin (Smith *et al.*, 1962). The classification unfortunately gives no help on any aspect of hemoglobin distribution, particularly now that the distinction between parasitic and nonparasitic forms is going, and in any case there is at present no generally accepted scheme, e.g., Rothschild (1961) gives two alternative ways of classifying the Nematoda, one of which tends to separate parasitic and nonparasitic forms, and the other to intermix them.

B. Localization of Hemoglobin

Knowledge of the exact location of pigments is so essential to the proper understanding of their formation and function that lack of it is a great handicap in discussing the nematode hemoglobins. Table I shows the limited extent of our present knowledge. The data in the table should be regarded as indicative only, since in most cases confirmation of the observation is lacking. As far as one can attempt to generalize, it appears that a number of nematodes contain hemoglobin in both the perienteric fluid and the body wall. Two of these nematodes, *Nippostrongylus* and *Ascaris,* have been looked at in more detail. In *Nippostrongylus* the fluid-filled layer of the cuticle gives a staining reaction which is probably due to hemoglobin, but the larvae, which gave a similar reaction, show no hemoglobin spectrum and probably contain cytochromes only (Lee and Smith, 1965). Unfortunately, more detailed

TABLE I
OCCURRENCE OF SOME NEMATODE HEMOGLOBINS

Nematode	Stage of life cycle and sex	Gross location
Ascaris	Male and female adult	Body wall Perienteric fluid
Dioctophyme	Female adult	Cuticle Perienteric fluid
Eustrongyloides	Larvae	Perienteric fluid
Nippostrongylus	Larvae	No hemoglobin?
	Male and female adult	Body wall Cuticle
Spirocerca	Adult	Perienteric fluid
Strongylus	Adult	Perienteric fluid
Terranova	Larvae	
Tetrameres	Adult male	No hemoglobin
	Adult female	Body wall Perienteric fluid
Toxocara	Adult	Body wall Perienteric fluid
Trichinella	Larvae	

investigation of the adult is hampered by its small size and the proximity of the perienteric fluid. In the case of *Ascaris* a detailed investigation (Smith and Lee, 1963) showed that in the body wall the greatest amount of hemoglobin occurred in the hypodermis, especially in the lateral lines and nerve ring. However, it was also present in all other tissues except the cuticle and the wall of the excretory canal. Although it is known that the hemoglobin of the perienteric fluid is different from that in the body wall, it is not known for certain whether the latter is a single pigment only, although that hemoglobin which can be extracted readily from the tissues does appear to be homogeneous.

In most cases there appears to be no obvious distinction between the hemoglobin content of the adult male and adult female, with the exception of *Tetrameres,* mentioned in the previous section. Here it is reported (Villela and Ribeiro, 1955a,b) that only the female contains the pigment. Further information on this or any similar case would be most useful, since it appears to be without parallel in other animal groups.

On the other hand, there are certainly parallels for variations in pigment content at different stages in the life cycle. Three cases have been reported in which nematode larvae contain hemoglobin, but not, apparently, the adult. These are *Eustrongyloides* and *Trichinella,* and *Terranova* (Ronald, 1960). In the light of the general comments made in Section II,A it would appear to be necessary to make quite certain that the pigment is absent from the adults, and not merely present in lower concentration. A detailed investigation of any change is long overdue, since there is apparently a complete lack of information about the metabolism of hemoglobin in gametes or larvae.

C. Properties of the Hemoglobins

The available physicochemical data on nematode hemoglobins were reviewed previously (Lee and Smith, 1965). Apart from that on *Ascaris* hemoglobins, most of the available data are spectroscopic and relate only to the visible region of the spectrum. It was only very recently, therefore, that it was found that the ratio of the maxima at 400 and 280 mμ. was twice as high in one of the *Ascaris* hemoglobins as in mammalian hemoglobins. This had, however, been foreshadowed by the findings of Scheler and Schneiderat (1959) and Antonini *et al.* (1962) on annelid pigments. The two annelid pigments and that of *Ascaris,* the perienteric fluid hemoglobin, are all of high molecular weight and occur free in solution. Data for other invertebrate pigments are not yet available, but the implication appears to be that this is general for those nematode and annelid hemoglobins which are found in solution rather than in tissues. It is a result, not of any unusual spectral property of the hematin group, but of the fact that the equivalent weight of the protein portion is higher than in previously known hemoglobins, resulting in enhanced absorption at 280 mμ. relative to the hematin absorption at 400 mμ. Wittenberg and co-workers (B. A. Wittenberg *et al.,* 1965b; Okazaki *et al.,* 1965) have shown that the *Ascaris* hemoglobin has an equivalent weight of 40,000, rather than the 18,000 or so which is more usual. The molecule as a whole contains eight hematin groups, giving a molecular weight of 320,000. The annelid pigments have a similar equivalent weight but molecular weights of 3×10^6.

The body wall hemoglobin of *Ascaris* has more recently been analyzed in similar detail by Okazaki *et al.* (1967). They find that the spectral properties are similar, suggesting that the generalization made earlier will also apply to tissue pigments, at least among the nematodes. The molecular and equivalent weights are the same at about 36,000. [An earlier report (Smith and Morrison, 1963) that the molecular weight was 14,000 now appears to have been in error because of the presence of glycogen in the preparation. This could alter the viscosity sufficiently to introduce a large error into the calculation of the sedimentation coefficient, upon which the estimate of the molecular weight was based.] Despite this similarity to the subunit of the perienteric fluid hemoglobin, the two are quite different in a number of respects. The authors confirm the report of Hamada *et al.* (1962) that a further hemoglobin component is extracted. This they found to be present in quite small amount, whereas Hamada *et al.* found comparable amounts of their two components. It seems that the possibility of different tissue hemoglobins, e.g., in nerve tissue and musculature, cannot be ruled out for the present, despite the conclusion of B. A. Wittenberg *et al.* (1965c) that there is no evidence for any such separation in certain annelid and molluscan species.

Fortunately there has been very satisfactory agreement about a more fundamental property of the other, perienteric fluid hemoglobin, namely its reactions with gases. The equilibrium constants of the reactions with oxygen and carbon monoxide were measured directly by Okazaki and Wittenberg (1965), and agree well with those calculated from the ratio of the on and off velocity constants measured by Smith and Gibson (1965). The latter workers have given a comprehensive description of the kinetics of the gas reactions of the two *Ascaris* hemoglobins which, with other evidence (see also B. A. Wittenberg *et al.*, 1965a), shows that invertebrate hemoglobins have kinetic constants which are generally similar to those of mammalian hemoglobins, variations of about 10-fold being possible. An exception, however, is the decomposition of oxyhemoglobin, the observed rates varying by up to 100-fold compared with mammalian hemoglobins, and only in the sense of giving a slower reaction. In other words, although there may be significant differences between hemoglobins, resulting from variations in other reactions, it is variation in the rate of breakdown of the oxygenated form which can lead to really dramatic changes. It is this feature of their reactivity which is responsible for the extremely high oxygen affinities of many invertebrate hemoglobins. Both *Ascaris* hemoglobins share this feature, being half saturated with oxygen at approximately 0.1 mm. (body wall) and 0.001 mm. (perienteric fluid) at 20°. The latter figure represents the highest oxygen affinity of any known hemoglobin. Where the data are available, it appears that several nematode hemoglobins have a high oxygen affinity, shown either by the long half-life of the oxy form during deoxygenation, or by the relatively low affinity for carbon monoxide as compared with mammalian hemoglobin. These comprise *Dioctophyme, Haemonchus, Nematodirus, Nippostrongylus, Strongylus,* and *Tetrameres,* in addition to *Ascaris.*

There seems to be little else of significance known about nematode hemoglobins at present, but it should certainly be mentioned that in addition to their other findings, Wittenberg and his collaborators have published amino acid analyses of the two *Ascaris* hemoglobins (B. A. Wittenberg *et al.*, 1965b; Okazaki *et al.*, 1967). The amino acid compositions and sequences of proteins in general are being widely studied in the hope of expanding our knowledge of phylogenetic relationships. A few invertebrate hemoglobins have now been analyzed, though so far only in terms of composition, and the *Ascaris* values are the only ones available for nematodes. Two dangers seem particularly important in attempting to evaluate such results. One is the possibility of revision, which may be quite substantial, to published results; the other is the difficulty of pinpointing significant differences at this rather gross level. A start can be made by comparing the frequency of occurrence of a particular amino acid as a percentage of the total number of residues (Smith, 1966). This rationalizes comparison of proteins of different molecular weight and residue composition, but is of limited value when, as is the case here, there are only small differences in percentage composition and the available data are too few for the use of statistical analysis. With these reservations the analyses now published seem to suggest (1) that *Ascaris* body wall hemoglobin resembles mammalian myoglobin quite closely, and certainly more so than mammalian hemoglobin, (2) that it resembles myoglobin more than it resembles tissue hemoglobins from mollusks and a polychaete annelid, which resemble each other and mammalian hemoglobin reasonably well, (3) that the perienteric fluid hemoglobin does not particularly resemble any of these hemoglobins but is most like the body wall pigment. To the extent that it can be relied upon, this evidence therefore seems to confirm existing assumptions that the nematodes cannot be clearly related to other known groups, and that hemoglobins of the same kind as that in the *Ascaris* perienteric fluid are considerably different from the more familiar "myoglobin" and "hemoglobin" types. It is unlikely that sequence studies will materially affect these conclusions, but more information about hemoglobins from species other than *Ascaris* would be of considerable value.

D. Functions of the Hemoglobins

Although the significance of mammalian myoglobin is readily accepted, there has always been a reluctance to accept that nematode hemoglobins could have a similar function or metabolic importance. This appears to derive from two mistaken impressions, (1) that parasites possessing tissue hemoglobin had obtained it by absorption of host hemoglobin, and that this was merely acting as food material; and (2) that gut-dwelling parasites could not have any use for hemoglobin because the surroundings were completely anaerobic. The former thesis can no longer be supported by any reliable evidence, and clearly has little relevance to the body structure and metabolic activities of most parasitic nematodes. The latter thesis ignored, or failed to confirm with modern methods, available data on the oxygen content of the environment of the worm, and equally failed to take account of the high oxygen affinities of nematode hemoglobins. It now seems reasonably certain that the oxygen content of the small intestine, the habitat of most of the best-studied parasitic nematodes, ranges from that of venous blood at the mucosa to zero at the center (where bacterial action keeps the contents completely anaerobic). The effective oxygen content will therefore vary with the size and normal position of the worm, and with the mixing of intestinal contents produced by movements of the worm or host. Indeed it may well be not the partial pressure of oxygen which counts, this being essentially a static or average concept, but rather the total quantity reaching the worm in a given time. An example may be seen in the case of the larvae of the insect *Gasterophilus* (Tatchell, 1960) which live in the stomach of the horse and maintain an aerobic existence there. They do this by utilizing a modified tracheal system and a hemoglobin with high oxygen affinity to take advantage of chance contacts with air bubbles in the stomach contents. Although the measured partial pressure of oxygen will vary from zero to full saturation, this is averaged out by the parasite to yield a level which is high enough to maintain the necessary metabolic processes.

It seems to be reasonable to suppose that the hemoglobin in the body wall of nematodes functions in the same way as myoglobin in mammalian muscle, i.e., for the transfer of oxygen to the inner tissues, with possibly a secondary function as an oxygen store for short-term deficiency. It has been established by recent work that solutions of hemoglobin can transfer appreciably more oxygen across a thin layer than can be accounted for by gas diffusion alone. Earlier theories had suggested some sort of gradient from oxyhemoglobin externally to deoxyhemoglobin internally, but it now seems (J. B. Wittenberg, 1966;

Wyman, 1966) that the effect is due to random movement of the pigment molecules. Together with the long half-life of the oxy form, this would have the effect of smoothing out both spatial and temporal variations in the oxygen supply to the inner tissues. Known values for nematode hemoglobins indicate that several intracellular pigments could function in this way at partial pressures of oxygen of 2 mm. or less. *Haemonchus* and *Nematodirus* hemoglobins are half saturated at 0.05 mm., *Nippostrongylus* at 0.1 mm. and *Ascaris* at 0.5 mm. (all at 37°). In several other cases it has been shown that contained hemoglobin can be deoxygenated *in situ* when the surroundings are made anaerobic. Not all of the latter are intracellular pigments, however. Unfortunately, further data to confirm the functioning of perienteric fluid hemoglobin as a normal oxygen-combining pigment are not available. The perienteric fluid is not a circulating fluid, being merely contained within the body of the worm as if in a sac, and subject only to random mixing produced by movements of the body wall and of the organs suspended within it. In the only two cases where the reactivity of perienteric fluid hemoglobin has been studied quantitatively it has been found that the rate of dissociation of the oxy form is so slow that a normal function seems most improbable. In both these cases, *Ascaris* and *Strongylus*, it has been reported that the worm is not able to deoxygenate the pigment *in situ*. For the case of *Ascaris*, which they studied in some detail, Smith and Lee (1963) put forward the alternative hypothesis that the perienteric fluid hemoglobin represents a pool of hematin intended for transfer into the gametes of the parasite. This theory, based on observations of the metabolism of the worm, is the only one so far presented which can explain the marked variation in hemoglobin content which can occur between individuals of *Ascaris* and some other nematodes. Direct confirmation is still lacking, however, and it is not clear whether the eggs contain hemoglobin as such, or a stock of hematin in some other form and possibly for some other purpose, e.g., cytochrome formation.

E. Synthesis and Degradation of Hemoglobin

The only reported studies on the synthesis of hemoglobin in nematodes appear to be those of Smith and Lee (1963) on *Ascaris*. They found that the worm showed a marked increase in the hemoglobin content of the perienteric fluid when kept in a medium containing a supply of hematin, preferably a solution of hemoglobin. The hematin could be either as hemoglobin or as hematin, and further experiments showed that the worm could apparently utilize any porphyrin nucleus containing a vinyl group at one of the same positions (side chain 2) as in protohematin, i.e., as in the commonly occurring hematin of hemoglobin, etc. Free porphyrin and chlorophyll (a magnesium hematin) could be utilized, but iron salts were without effect, suggesting that the worm had an essential requirement for the porphyrin nucleus, while retaining the ability to insert iron into the molecule.

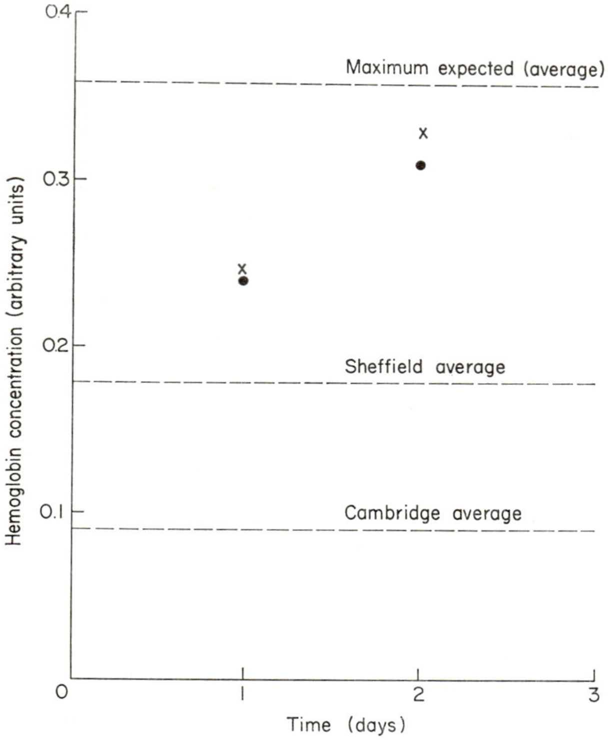

FIG. 1. Change in concentration of perienteric fluid hemoglobin in *Ascaris lumbricoides* during incubation in the presence of horse hemoglobin.

In unpublished studies, I have followed the rate of synthesis of hemoglobin in *Ascaris lumbricoides* with time. The results of two separate experiments, carried out as described in Smith and Lee (1963), are shown in Fig. 1. In these and other experiments where incubation was continued for several days, there always appeared to be a limiting value on the total concentration of hemoglobin which could be reached. The simplest explanation of such a situation would be that a simple equilibrium system exists, so that increase in hemoglobin concentration would eventually inhibit further synthesis. If this were so, then the results would be expected to fall close to a rectangular hyperbola. In fact they appear to be in better agreement with a linear increase in the amount of hemoglobin. This would require a more elaborate hypothesis to explain, and further work will be required to establish whether or not this is actually the case.

As shown on the figure, the worms at Sheffield had on average an appreciably greater concentration of hemoglobin to begin with than was found in earlier experiments at Cambridge. This can be correlated with the fact that pigs in the Sheffield area tend to be fed with a higher proportion of swill and other meat-containing waste than in the Cambridge area, where there is a greater emphasis on grain feeding. I believe that this is further confirmation of the suggestion of Smith and Lee (1963) that the observed color of an individual *Ascaris* depends on the existing balance between the intake of suitable hematin precursors, and its output in the gametes. Pig feed containing meat wastes will obviously be a richer source of these precursors than grain products. I wish to thank Mr. G. D. Rigby, of Sheffield Estate Farm, for providing me with the information on pig feeding practices.

In view of the dramatic results shown in Fig. 1, it was of some interest to determine whether or not the rate of synthesis at the ribosome level was normal. I am grateful to Dr. John Sykes for carrying out the investigation on which the following information is based. If *Ascaris* guts are homogenized (by hand in a glass homogenizer, or by disruption in a Hughes' bacterial press), and the homogenate clarified by brief centrifugation (25,000 g. for 10 minutes), a crude extract of guts may be obtained together with an insoluble stratified debris fraction. In the case of the Hughes' press preparations the extract is green. When the extraction is carried out in either distilled water or 0.01 *M* tris–0.001 *M* magnesium acetate buffer, pH 7.5, the crude extract resolves into three components when centrifuged at 50,740 r.p.m. in the Spinco analytical ultracentrifuge. There are two small fast components (uncorrected S values 17 and 11) and a large slower component (uncorrected S value 5) which accounts for 85% of the total material in the extract.

In chemical analysis the color reactions for protein, DNA, and RNA did not always behave as expected in terms of rate of color development and color finally obtained, and there is consequently some doubt about the absolute values calculated. It is not known whether the unusual results are due to true differences in the *Ascaris* materials, or simply to interference due to side reactions. The latter seems most likely. Unusually large amounts of lipid material were encountered at some stages. It appeared that reasonable values for the total content per gram wet weight of the gut in protein, RNA, and DNA would be 70 mg., 5 mg., and 0.1 mg., respectively. These values may not be accurate for the reasons stated. However, the relative amounts estimated on studying the distribution between the debris fraction and the crude extract are presumed to be fairly reliable. This indicated that 90% of the protein, 40% of the RNA, and all of the DNA remained with the debris fraction. This pattern of distribution would not be expected for a cell system without a defined nuclear body and with ribosomes distributed freely within the cytoplasm, e.g., bacteria. The fact that all of the DNA remained with the debris fraction and that a comparatively poor extraction of RNA was obtained suggest a resemblance to the typical animal cell with defined nucleus and ribosomes attached to an endoplasmic reticulum. This is supported by the stratification observed in the debris fraction, by the absence of conventional ribosome species in solvents favoring their extraction, and by an examination of electron micrographs published by Sheffield (1964). From the latter it appears that the endoplasmic reticulum in *Ascaris* guts is poorly developed and possibly not very extensive. Electron-dense granules, presumably ribosomes, are attached to this structure. These appear to be about 125 Å. in diameter but their numbers are low compared with bacteria and higher animals. It is possible that the 17 S particle seen in the crude extract may be the presumed *Ascaris* ribosome. It also seems fairly clear that in this preliminary work the ribosomes were not brought completely into solution. At the same time, if one assumes that 80% of the total RNA and an equal weight of protein are to be found in these ribosomes (a proportion which is found to be approximately true in other organisms), then roughly 10% of the wet weight of the gut would consist of ribosomes. This is a very low value compared with active tissues such as replicating bacteria, and appears to agree with their apparent sparsity as judged from the electron micrographs.

The value of 10% is also in satisfactory agreement with calculations of the initial rate of synthesis from Fig. 1, since it would correspond to the formation of about 1 mg. of hemoglobin per milligram of ribosomes per hour. Other values are known to range from 0.04 mg. for hemoglobin synthesis in rabbit reticulocytes, through 0.5 mg. protein in isolated pea ribosomes, to 3.5 mg. protein in replicating *Escherichia coli*. In view of the various uncertainties in the numerical values, it is not possible to decide whether the *Ascaris* ribosomes are especially active, but at least their activity is comparable to that of other organisms. The rapid increase in hemoglobin content in the enrichment experiments could therefore be no more than a consequence of the fact that (for reasons not yet clarified) the worm has no way of utilizing the synthesized pigment. The immediate cause of this is probably a cessation of gamete formation following removal of the worm from the pig.

It is not clear to what extent these results reflect the normal course of synthesis by the ribosomes, which must presumably involve other *Ascaris* proteins. Further investigation will require special treatment of the guts to ensure complete breakdown of the endoplasmic reticulum and release of the ribosomes from it, as well as modification to the well-established procedures for quantitating RNA, DNA, and protein.

One approach which would certainly yield further information would be a fractionation of homogenized guts by a technique such as density gradient centrifugation, since in the above work it was found that the insoluble material was stratified into at least three zones. A further observation which has some bearing on cell inclusions in the gut was the finding that guts homogenized in a Hughes' press yielded a green supernatant, whereas in a hand homogenizer one gets an essentially colorless supernatant. The green pigment of *Ascaris* guts, which is referred to in further detail in the following section, can be extracted readily by acid methanol, but not by water or dilute HCl. The Hughes' press is normally expected to break open organelles of about the size of mitochondria or larger, whereas hand homogenization should leave even nuclei undamaged. There is therefore a possibility that the pigment is present in cellular inclusions at least as large as mitochondria and protected by a lipid membrane. There are several unidentified organelles in Sheffield's micrographs which could answer to this description.

The site of synthesis of the *Ascaris* hemoglobins is not known with any certainty, but it seems likely that the perienteric fluid hemoglobin is synthesized in the gut. In histological studies, Smith and Lee (1963) found that there was evidence for the occurrence of hematin or hemoglobin in the gut, mainly at the basal side (adjacent to the perienteric fluid). They therefore suggested that hematin obtained from food material in the gut of the worm is absorbed, converted to a different form, and then reconverted to hematin for combination with globin and discharge into the perienteric fluid. The green pigment mentioned above could be the intermediate form of hematin. There is no evidence that the perienteric fluid contains body wall hemoglobin, or that this undergoes any discernible changes in concentration. It seems more likely that this is formed elsewhere than in the gut, possibly in the actual cells of the body wall, and possibly by the more usual mechanism involving total synthesis of hematin from simple precursors (this appears to be the manner in which myoglobin is synthesized in higher animals).

Evidence for the degradation of either hemoglobin is lacking at present. There is little convincing evidence for specific excretory organs in the worm and it is generally assumed that catabolic products pass out via the gut or cuticle. If it is correct to assume that the perienteric fluid hemoglobin is destined exclusively for the gametes, then one would have only breakdown products of the body wall hemoglobin to consider. These might be quite difficult to detect since, by analogy with higher animals, the pigment probably turns over quite slowly, and also because it represents a relatively small amount of hematin in relation to the total body weight. The normally expected breakdown product of hematin would be some kind of bile pigment. Normal bile pigments appear to be absent, or present only in very low concentration, in the gut of *Ascaris,* but there is a substantial amount of the green pigment, which appears to be related to known bile pigments. As mentioned above, Smith and Lee (1963) suggested that this might be related to the synthesis of the perienteric fluid hemoglobin, rather than to degradative processes. It also seems that the quantity of this material is too great to be accounted for by breakdown of the body wall hemoglobin alone.

F. Bile Pigments

The reported presence of bile pigments in the gut of *Ascaris* (Fauré-Fremiet, 1913) is obviously of considerable interest in the light of the foregoing discussion. The washed guts of living *Ascaris* vary in color from golden-brown to dark green, the color being generally uniform throughout the length of the gut in each individual. The color appears to correlate with the hemoglobin content of the perienteric fluid, the redder worms having greener guts. It also seems that worms which have been damaged or which appear to be less healthy than the rest have the yellowest guts. In the very reddest worms, as in the enrichment experiments, the guts are generally dark brown rather than green. This could be due to contained hematin or hemoglobin but there is no evidence for this at present. In all cases the colors referred to are those of the washed guts, and there is no question of contamination by gut contents.

The nature of the pigment or pigments in the gut is obviously of importance in theories of hemoglobin formation, and the yellow to green variation, coupled with the reported presence of bile pigments, suggests a change between verdin and rubin, as in the change from yellow bilirubin to green biliverdin with mammalian bile pigments. This interpretation has been impossible to sustain, however, and I am doubtful that the washed guts normally contain any appreciable amount of typical bile pigment. The true nature of the pigment remains uncertain and the following observations refer to work in progress at the time of writing.

The pigment cannot be extracted (from guts broken down in a hand homogenizer) into water or 6 *N* HCl, but is readily and completely extracted into 5% (v./v.) concentrated HCl in methanol. The color of the extract varies from golden yellow to grass green, depending on the original color of the guts. Guts which have been kept for about 24 hours become brown even if they were green to begin with, and give a very yellow extract, but on longer keeping there is a reversion toward the green color. In all cases the extract quite quickly becomes green on standing. Neither type of extract shows fluorescence or gives positive Gmelin or Schlesinger reactions (the latter are two standard tests for bile pigments). The green material is not extractable into ether or chloroform, even on neutralizing, nor does it give a positive reaction for iron after decolorizing with hydrogen peroxide. These results would seem to indicate that porphyrins or chlorins are not present (no fluorescence), nor hematin (no iron), nor bile pigments (negative tests). On the other hand, chromatography in a system recommended for vertebrate bile pigments (Garay *et al.*, 1965; Noir *et al.*, 1965) showed that the green material ran with an R_f like biliverdin, while the yellow pigment ran with an R_f similar to that of bilirubin diglucuronide. The R_f of the green pigment in other solvents and on thin layer plates has so far always been similar to that of hematin and porphyrins. The green pigment can be formed from the yellow one by heating or on standing. It can be changed back by the action of borohydride, but not by dithionite (reducing agents). The former reaction was at first thought to be the anticipated reduction of a rubin, giving a verdin. It now appears that this was not so, but was a result of the neutralizing effect of the borate produced in the reaction (thereby explaining why dithionite was without effect). The same change is produced by the addition of sufficient sodium hydroxide to neutralize the HCl present.

The latter observations offer the possibility of a change analogous to that from bilirubin diglucuronide to biliverdin, the presence of HCl tending to hydrolyze some combined group. Even so, the properties of the resulting material are quite different from those of mammalian bile pigments. Apart from the general nonreactivity in standard tests, the absorption spectra are unusual. The yellow form has a smoothly increasing absorption from 500 mμ. to lower wavelengths, with no definite peak apparent down to 350 mμ. The green form has a peak at 394–397 mμ. and a broad band with maximum at approximately 680 mμ. The 394–397 mμ. peak is quite sharp but not very pronounced, there being much greater absorption at lower wavelengths still (peak to trough ratio, 394–397:384 is about 1.2:1). The 680 mμ. peak is much less intense than that at 394–397 mμ., the ratio between them varying between about 4 to 8.5 in traces taken so far. The yellow form is spectrally quite different from bilirubin, which has a single absorption band centered at 454 mμ. The green form is very reminiscent of known biliverdin pigments, but differs in that these generally have peaks at 380 mμ. or less, with the 380:680 mμ. ratio being never more than 4:1. The most interesting of the known pigments in the present context are two isolated by Tixier (1953) from two species of the snaillike mollusk, *Haliotis*. These resemble the *Ascaris* pigment very closely in the oxidized (green) form, but are both reduced by dithionite to a bilirubinlike material. They share with the *Ascaris* pigment the inability to give a fluorescent zinc complex (Schlesinger reaction) but do give a Gmelin-type reaction.

There has also been a recent revival of interest in verdohemochromes, green compounds which have been thought to represent models of an intermediate stage in the breakdown of hematin to biliverdin during hemoglobin catabolism. The structure of these compounds is not yet established, and arguments have been presented in favor of both open and closed ring structures (Levin, 1966). Spectral data indicate some resemblance to the green form of the *Ascaris* pigment. The verdohemochromes contain pyridine, which presumably does not occur *in vivo,* but it is possible that it could be replaced by some other nitrogenous material, possibly an amino acid or peptide.

The interpretation of these data is difficult and it is probable that some of the results with *Ascaris* material are in error due to interference by impurities in the extract (Tixier isolated his pigments chromatographically before carrying out the chemical tests). On balance it seems difficult to avoid the conclusion that some form of tetrapyrrole is present. This is probably a new form of open-chain material, related to both biliverdin and the *Haliotis* pigments, rather than a modified ring structure of the porphyrin type. In the light of the results discussed here it is interesting to observe that Fauré-Fremiet (1913) obtained a positive Gmelin reaction only when the pigment (from *Parascaris*) was extracted into amyl alcohol; whereas Flury (1912), working with both *Parascaris* and *Ascaris* and presumably not employing this variation of the technique, was unable to obtain any positive reaction.

G. Other Pigments

As pointed out in the introduction, little is known of the pigments of nematodes, apart from hemoglobin. The related respiratory pigment chlorocruorin occurs only in a small group of annelids and is not known elsewhere. The other (chemically unrelated) respiratory pigments, hemocyanin and hemerythrin, are not known in nematodes.

1. Cytochromes

Ascaris was one of the animals in which Keilin (1925) reported the presence of cytochrome in his original papers on this group of pigments. He never followed up this observation in detail, and later reports were often contradictory. It appears from the studies of Kikuchi and co-workers (e.g., Kikuchi and Ban, 1961) that the explanation for this is related to the fact that there is little cytochrome *c*, the main component being of the *b* type. Much difficulty also seems to have been caused by the uncritical use of materials, such as mammalian cytochrome oxidase, which need not necessarily be expected to react satisfactorily with *Ascaris* cytochromes. Some spectrophotometric studies on intact mitochondria of *Ascaris* in relation to their cytochrome content have been reported by Chance and Parsons (1963) but a detailed study of the electron transport system in nematodes generally still remains to be carried out. It would be of the greatest value in understanding the metabolism of these animals, not least in relation to their oxygen requirements.

2. Porphyrins

In a recent authoritative review on porphyrins, Rimington and Kennedy (1962) present only one reference to the presence of free porphyrins in nematodes. This is a statement by Lederer (1940) to the effect that the dog parasite *Dioctophyme renale* (= *Eustronglyus gigas*) contains proto- and coproporphyrins. The original observation is reported to have been made by Aducco (1888), author of a comprehensive paper (1889) on the hemoglobin of this parasite (apparently the only study which has ever been presented on the hemoglobin of *Dioctophyme*, one of the largest known parasites—Aducco was able to obtain 25 ml. of perienteric fluid from one individual). I have examined both papers and am unable to trace any mention of porphyrins. The 1888 paper appears to be an almost identical presentation of that of 1889. It therefore appears that there is at present no good evidence for the presence of appreciable amounts of free porphyrins in nematodes.

3. Flavins

Von Brand (1952), reviewing the properties of endoparasites, refers to the findings of Gourevitch (1937). This worker reported a flavin content of about 1 mg. per 100 gm. wet weight in the tissues of *Parascaris*, a value comparable with those of vertebrates.

During water extraction of whole homogenized tissues of *Ascaris* I have found contamination of the hemoglobins by a yellow component which has a bright green fluorescence. This can be readily separated by passage through CM-cellulose, to which the hemoglobins adhere. A similar material can be extracted from the tissues subsequent to water extraction, by using acetone or methanol. No attempt was made to characterize these pigments further, or to remove the gut pigment. It is assumed that the green fluorescence is due mainly to a flavin component, possibly associated with protein *in vivo*.

4. *Melanins*

I have found only one very recent report of melanin in nematodes. Croll (1966) suggests that the reddish-brown eye spots of *Enoplus communis* very probably contain granules of melanin. Similar coloration in other species, particularly in the outer tissues, is very likely to indicate the presence of melanin as well.

5. *Miscellaneous*

There seems to be no report as yet of carotenoids in nematodes, nor of less likely pigments such as ommochromes or pteridines. One might expect that eventually more colored materials must be found, if only because of the presence of enzymes with colored prosthetic groups. It is now known that two enzymes (one lyase and one isomerase) are characterized by possession of a vitamin B_{12} derivative as a prosthetic group. It is therefore of some interest that Weinstein found (1963) that the absorption spectrum of the coelomocytes of *Nippostrongylus* (?) resembled that of the vitamin. The function of these bodies, dark brown in *Ascaris*, has never been satisfactorily explained, although Hurlaux (1947) found that those of *Parascaris* contained a number of oxidative enzymes. It has also been suggested that these bodies may have some kind of catabolic function.

III. Acanthocephala

Knowledge of pigments in Acanthocephala is severely limited, but clearly they occur only in small amounts in general, if they are present at all. Hyman (1951) reports that these worms lack intrinsic coloration but frequently show red, orange, yellow, or brown colors resulting from food material absorbed from the host. While it is probably true that this can occur, it seems likely that it is too ready an explanation of Acanthocephalan coloring to be applicable in all cases. Von Brand (1952) rebuts a similar argument by pointing out that red *Pomphorhyncus* can be found associated with colorless *Acanthocephalus* or *Echinorhyncus*, while the juvenile of *Polymorphus minutus* forms orange-red fats or oils in the colorless body fluid of its host, *Gammarus*. He further lists a total of fourteen species from nine genera in which he believes that the observed color is intrinsic to the worm. The colors range from white and black to red, orange, yellow, or brown. A further species, *Arhythmorhyncus comptus*, probably contains carotene (Van Cleave and Rausch, 1950), as also *Polymorphus minutus* (D. W. T. Crompton, personal communication). These observations suggest that in most cases the coloration may be due to carotenoids. It would be of some interest to have further information on *Bolbosoma* and *Pomphorhyncus* which are reported to be red, in view of the fact that animal carotenoids generally have a yellowish character, and that hemoglobin is at present unknown in this group.

I do not know of any other reports on pigments in Acanthocephala. In a recent examination of *Neoechinorhyncus* spp. from *Mugil cephalus*, the worms appearing white to the naked eye, I found no evidence for the presence of any colored material, not even cytochromes. Irrigation with benzidine and peroxide, a test for hematin or hematin derivatives, did, however, give a strong positive reaction in the lemnisci. The reaction appeared to start at the anterior end and move down the organ, as though the reagents were being ingested. Although the lemnisci have the appearance of digestive sacs and lead from the base of the proboscis, there is in fact no known means of communication with the exterior in Acanthocephala by any observable channel. It is presumed that food materials and catabolic products pass through the body wall as in tapeworms. The fact that a strong reaction was obtained at all is also of interest. The lemnisci appear brownish under the microscope, and the absence of clearly defined absorption bands leaves the presence of hematin itself as the most likely cause. Further comment can only be speculative, but one wonders if there could be any connection with reproductive processes, analogous to the suggested transfer of hematin into the gametes of *Ascaris*.

REFERENCES

Aducco, V. (1888). *Atti Accad. Nazl. Lincei, Rend., Classe Sci. Fis. Mat. Nat.* **4**, Part 2, **187** and 213.

Aducco, V. (1889). *Arch. Ital. Biol.* **11**, 52.

Antonini, E., Rossi-Fanelli, A., and Caputo, A. (1962). *Arch. Biochem. Biophys.* **97**, 343.

Chance, B., and Parsons, D. F. (1963). *Science* **142**, 1176.

Croll, N. A. (1966). *Nature* **211**, 859.

Ellenby, C. (1964). *Nature* **202**, 615.

Ellenby, C., and Smith, L. (1966). *Nature* **210**, 1372.

Fauré-Fremiet, E. (1913). *Arch. Anat. Microscop. Morphol. Exptl.* **15**, 435.

Flury, F. (1912). *Arch. Exptl. Pathol. Pharmakol.* **67**, 275.

Fox, D. L. (1953). "Animal Biochromes and Structural Colours." Cambridge Univ. Press, London and New York.

Fox, H. M., and Vevers, G. (1960). "The Nature of Animal Colours." Sidgwick & Jackson, London.

Garay, E. R., Noir, B., and Royer, M. (1965). *Biochim. Biophys. Acta* **100,** 411.
Gourevitch, M. A. (1937). *Bull. Soc. Chim. Biol.* **19,** 125.
Hamada, K., Okazaki, T., Shukuya, R., and Kaziro, K. (1962). *J. Biochem. (Tokyo)* **52,** 290.
Hurlaux, R. (1947). *Ann. Sci. Nat., Zool. Biol. Animale* [11] **9,** 155.
Hyman, L. H. (1951). "The Invertebrates," Vol. III. McGraw-Hill, New York.
Keilin, D. (1925). *Proc. Roy. Soc.* **B98,** 312.
Kikuchi, G., and Ban, S. (1961). *Biochim. Biophys. Acta* **51,** 387.
Lederer, E. (1940). *Biol. Rev.* **15,** 273.
Lee, D. L., and Smith, M. H. (1965). *Exptl. Parasitol.* **16,** 392.
Levin, E. Y. (1966). *Biochemistry* **5,** 2845.
Noir, B. A., Garay, E. R., and Royer, M. (1965). *Biochim. Biophys. Acta* **100,** 403.
Okazaki, T., and Wittenberg, J. B. (1965). *Biochim. Biophys. Acta* **111,** 503.
Okazaki, T., Briehl, R. W., Wittenberg, J. B., and Wittenberg, B. A. (1965). *Biochim. Biophys. Acta* **111,** 496.
Okazaki, T., Wittenberg, B. A., Briehl, R. W., and Wittenberg J. B. (1967). *Biochim. Biophys. Acta* **140,** 258.
Rimington, C., and Kennedy, G. Y. (1962). In "Comparatie Biochemistry" (M. Florkin and H. S. Mason, eds.), Vol. 4, pp. 577–614. Academic Press, New York.
Ronald, K. (1960). *Can. J. Zool.* **38,** 623.
Rothschild, Lord. (1961). "A Classification of Living Animals." Longmans, Green, New York.
Scheler, W., and Schneiderat, L. (1959). *Acta Biol. Med. Ger.* **2,** 588.
Sheffield, H. G. (1964). *J. Parasitol.* **50,** 365.
Smith, M. H. (1966). *J. Theoret. Biol.* **13,** 261.
Smith, M. H., and Gibson, Q. H. (1965). *Proc. Roy. Soc.* **B163,** 206.
Smith, M. H., and Lee, D. L. (1963). *Proc. Roy. Soc.* **B157,** 234.
Smith, M. H., and Morrison, M. (1963). *Biochim. Biophys. Acta* **71,** 364.
Smith, M. H., George, P., and Preer, J. R., Jr. (1962). *Arch. Biochem. Biophys.* **99,** 313.
Tatchell, R. J. (1960). *Parasitology* **50,** 481.
Tixier, R. (1953). *Mem. Museum Natl. Hist. Nat. (Paris)* **A5,** 41.
Van Cleave, H. J., and Rausch, R. L. (1950). *J. Parasitol.* **36,** 278.
Villela, G. G., and Ribeiro, L. P. (1955a). *Anais Acad. Brasil. Cienc.* **27,** 87.
Villela, G. G., and Ribeiro, L. P. (1955b). *Rev. Brasil. Biol.* **15,** 383.
von Brand, T. (1952). "Chemical Physiology of Endoparasitic Animals." Academic Press, New York.
Weinstein, P. (1963). Referred to in Dr. Weinstein's lecture to the British Society for Parasitology, Aberystwyth meeting; the spectrum is published in an article by B. L. Strehler, "Construction and Application of a Double-beam Recording Microspectrophotometer using the Cary Model 14 Spectrophotometer," printed in the Applied Physics Corporation magazine, *Carygraph* (Vol. 3 or earlier).
Wittenberg, B. A., Brunori, M., Antonini, E., Wittenberg, J. B., and Wyman, J. (1965a). *Arch. Biochem. Biophys.* **111,** 576.
Wittenberg, B. A., Okazaki, T., and Wittenberg, J. B. (1965b). *Biochim. Biophys. Acta* **111,** 485.
Wittenberg, B. A., Briehl, R. W., and Wittenberg, J. B. (1965c). *Biochem. J.* **96,** 363.
Wittenberg, J. B. (1966). *J. Biol. Chem.* **241,** 104.
Wyman, J. (1966). *J. Biol. Chem.* **241,** 115.

CHAPTER 11

Pharmacology of Nematoda

J. del Castillo

I. Introduction

A. Development of Anthelmintic Drugs

The word helminth (from the Greek term for worm) is still employed in medical literature with reference to a number of wormlike organisms parasitic to man and domestic animals. These are classified, somewhat arbitrarily, according to their shape (e.g., roundworms, tapeworms, flukes, hair-snakes, spiny-headed worms, etc.). However, in spite of their common wormy appearance, helminths are a very heterogeneous group of animals representing no less than four zoological phyla (Platyhelminthes, Aschelminthes, Acanthocephala, and Annelida).

Since ancient times, man has correlated the ingestion of certain natural products with the elimination of intestinal parasites. A branch of pharmacology was born in this way. Indeed, the recognition of the anthelmintic properties of some compounds can be said to be the true beginning of chemotherapy.

The great development of the morphological sciences during the nineteenth century led to the accumulation of a vast body of fine anatomical detail on many helminths. Yet, information on their chemistry and physiology was scanty and research on anthelmintic drugs continued to be based on purely empirical grounds. It was only during the second decade of the present century that Trendelenburg (1916) used annelid preparations (earthworm) as models for studying anthelmintic drugs. In the 1920's, Portuguese investigators (Toscano-Rico, Rebello, Gomes da Costa) employed segments of *Ascaris,* suspended in a chamber and attached to a kymograph, to record the effects of anthelmintic drugs on the rhythmic contractions exhibited by the worms (see Toscano-Rico, 1926). Also during the twenties, Hall studied quantitatively for the first time the effects of drugs on parasitized animals, counting both the number of worms expelled as a result of treatment and those which remained in the host (see Hall, 1928).

In the 1940's, Baldwin (1943) and Baldwin and Moyle (1947, 1949) worked with segments of *Ascaris* cut open along the lateral lines, obtaining strips of muscle tissue with an intact nerve cord. In this fashion, the muscle cells were accessible without delay to drugs added to the bath.

During the last two decades research into the biochemistry and physiology of helminths has proceeded at a growing rate, providing for the first time an adequate basis for the understanding of drug actions. However, work of a physiological nature has been concentrated on the larger species, particularly on *Ascaris* which is rapidly becoming a useful laboratory animal for the investigation of some basic problems of biology.

Several reviews on the pharmacology of anthelmintic drugs have appeared within the last few years. Some were written from a strictly medical viewpoint, i.e., regarding the drugs as tools to kill parasitic helminths or, preferably, to dislodge them from their habitat in the body of their host. Those of Mansour (1964) and Saz and Bueding (1966) include excellent summaries of recent biochemical studies showing how the action of some drugs can be traced to their interference with biochemical and physiological processes (see also Bueding and Swartzwelder, 1957; Brown, 1960; Broome, 1962; Bueding, 1962; Hawking, 1963; Standen, 1963).

The present chapter, limited to the pharmacology of Nematoda, attempts to draw a picture, as far as this is possible, of the selective action of drugs on the different tissues and systems of these animals. Drugs are not regarded merely as a part of the medical and veterinarian armamentarium, but as tools for the exploration of interesting types of invertebrate organisms.

In principle, the possibilities of pharmacological interference with the cells, tissues, organs, and functions of nematodes are the same as in any other group of living organisms. Until recently, however, nematode pharmacology has been oriented exclusively toward the development of new and more potent anthelmintics; for these reasons, some of the best-known drugs are those which impair the ability of the worms to maintain their position in the body of the host. Prominent among them are the compounds which interfere with the motor coordination and metabolic processes.

Emphasis, perhaps somewhat undue, has been put on the drugs acting on the neuromuscular system. This can be justified on the grounds that this is the only area of nematode biology with which the author of this chapter has had direct experience. For the same reason an introductory section on the motor system of nematodes has been included.

Some other anthelmintic drugs, for example the phenols, may be described as general (or nonspecific) protoplasm poisons; i.e., drugs which combine with and destroy living matter in general, therefore showing little or no cellular or tissular specificity. Finally, a few drugs are known to affect the gonads and the larvae, or early developmental stages, of the parasitic nematodes.

B. The Motor System of Nematodes

1. *Anatomy*

The diagrams of Fig. 1 illustrate schematically the main anatomical features of a nematode organism as exemplified by *Ascaris*. Diagram A represents a cross section of the worm, which may be regarded as composed of four concentric layers. From the outside they are: cuticle, hypodermis, somatic muscle layer, and the pseudocoelom enclosing the digestive and reproductive systems.

The cuticle, relatively thick, is formed by two sets of inextensible spiral fibrils diagonally crossed. The imbrication of these two sets of fibers forms a basketwork; that is, a system of minute parallelograms which may be distorted by shearing with a consequent change in the length of their diagonals (Harris and Crofton, 1957). This structure seems to be responsible for the fact that the body of nematodes is extensible, being able to change both its length and diameter.

The syncytial layer called hypodermis (see Fig. 1A) protrudes inwardly at four points forming longitudinal ridges: two of them, shaped as troughs or canals, hold the main dorsal and ventral nerve cords; the other two are the lateral lines, believed to be of a secretory nature. Smaller ridges are seen also in association with the subdorsal and subventral nerve cords.

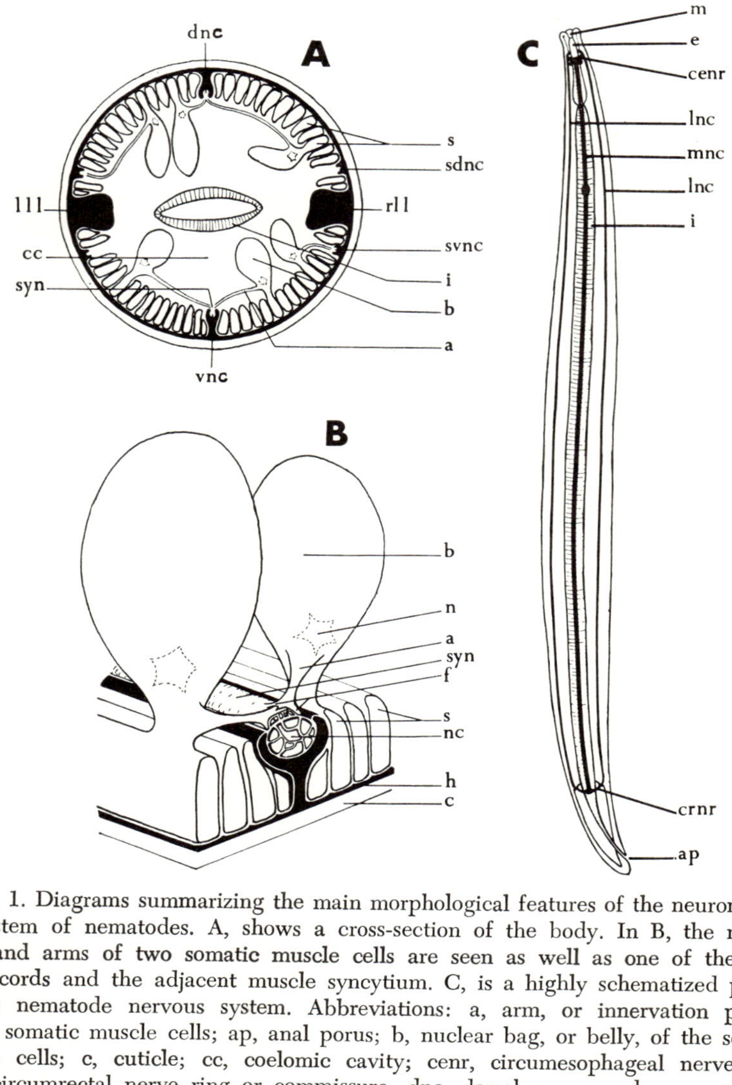

FIG. 1. Diagrams summarizing the main morphological features of the neuromuscular system of nematodes. A, shows a cross-section of the body. In B, the nuclear bags and arms of two somatic muscle cells are seen as well as one of the main nerve cords and the adjacent muscle syncytium. C, is a highly schematized picture of the nematode nervous system. Abbreviations: a, arm, or innervation process of the somatic muscle cells; ap, anal porus; b, nuclear bag, or belly, of the somatic muscle cells; c, cuticle; cc, coelomic cavity; cenr, circumesophageal nerve ring; crnr, circumrectal nerve ring or commissure; dnc, dorsal nerve cord; e, esophagus; f, fingers or terminal branches of the muscle arms; h, hypodermis; i, intestine; lll, left lateral line; lnc, lateral nerve cord; m, mouth; mnc, main nerve cord; nc, nerve cord; rll, right lateral line; s, muscle spindles or contractile part of the muscle cell; sdnc, subdorsal nerve cord; svnc, subventral nerve cord; syn, muscle syncytium.

The third layer is formed by the somatic musculature. Each individual mononucleated muscle cell (see Fig. 1B) has three parts: (1) an elongated spindle, parallel to the longitudinal axis of the worm and only contractile region of the muscle cell; (2) a relatively large bag containing the nucleus, which balloons into the body cavity or pseudocoelom; (3) a prolongation, the so-called arm or innervation process, which joins each muscle cell to either the dorsal or ventral nerve cord. As the diagram shows, the muscle layer is interrupted by the lateral lines which divide it into two (dorsal and ventral) halves or muscle fields.

The innermost cylinder in the body cavity is the digestive tube, which consists of a pharynx or esophagus and the simple intestine. The pharynx, the only motile part of the alimentary canal, is one of the most characteristic features of nematode anatomy. In spite of its length (about 1 cm.) and diameter (up to 1.5 mm.) in *Ascaris,* it may be regarded as a single, syncytial muscle cell (del Castillo and Morales, 1967a,b). It is continued by the intestine, a long, flaccid, and flattened tube formed by a single layer of epithelial cells.

The space between the somatic muscle layer and the intestine, the so-called pseudocoelom, is filled with a liquid—the perienteric or pseudocoelomic fluid—which serves as a carrier of metabolites and oxygen (see its composition in del Castillo and Morales, 1967a).

Drawing C of Fig. 1 shows a diagram of the nematode nervous system which consists of a main ganglion, the periesophageal nerve ring, from which the main dorsal and ventral nerve cords, as well as the accessory subventral and subdorsal ones, depart toward the posterior end of the worm. Thin nerves also are directed toward the anterior end. Some of them form the so-called pharyngeal-sympathetic system which innervates the esophagus.

2. *Neuromuscular Relationships*

Nematodes are unique with respect to the morphological and functional relationships between nerve and muscle. Although in 1866, Schneider recognized that in these animals the nerves do not branch out to the muscles but, rather, it is the muscle which sends branches or fibers to the nervous system, this theory was only proved by electrophysiological and electron microscopical work carried out in the 1960's. DeBell (1965) has summarized the views and ideas held in the intervening 100 years.

Recent observations have established that the arms, or innervation processes, of the muscle cells subdivide upon reaching the vicinity of the nerve cord into a number of terminal arborizations or "fingers" which intertwine with those from adjacent arms, forming a longitudinal band or strip of tissue parallel and in close contact to the nerve cord called the muscle syncytium.

The microanatomy of this band of tissue has been studied with the electron microscope by Hinz (1963) and Auber-Thomay (1964) in the horse ascarid, *Parascaris equorum* (former *Ascaris megalocephala*), and by Rosenbluth (1963, 1965b) in the *Ascaris lumbricoides* of the pig.

Electrophysiological experiments (DeBell *et al.*, 1963) indicate that all the muscle cells are electrically interconnected at this level. Indeed, it can be shown that a fraction of the current injected into the nuclear bag of any muscle cell is first conducted toward the syncytium along its arm and then flows back from the syncytial region along the arms of adjacent muscle cells. The observations of Rosenbluth (1965a) showing the existence of tight junctions between adjacent fingers (i.e., areas where the surface membranes of neighboring cells are fused together with obliteration of the extracellular space between them) provide a morphological basis for such intercellular spread of current. In fact, the electron microscopical demonstration of tight junctions is usually associated with a low electrical resistance between contiguous cytoplasms.

With the electron microscope, neuromuscular synapses are seen to be established between the nerve cord fibers and the muscle syncytium. In the *Ascaris* of the pig the hypodermal tissue, which forms the trough where the nerve cord fibers are embedded, covers at some points the entire nerve cord, but at other sites nerve fibers come into close contact with the syncytium. It is in these areas that neuromuscular junctions are formed. They are characterized by a synaptic gap some 500 Å. wide, and the presence of very large mitochondria and clusters of synaptic vesicles within the presynaptic cytoplasm.

In *Parascaris equorum* the regions of contact between the nerve cord fibers and the fingers of the muscle arms are somewhat different from those in *Ascaris*. The surface of the muscle syncytium facing the nerve cord emits small projections, or prolongations, protruding toward the nerve fibers which have at this point an indented surface. Mitochondria and synaptic vesicles have also been described by Hinz (1963), and Auber-Thomay (1964) has observed tight junctions similar to those described by Rosenbluth.

3. *Body Movement*

An adult *Ascaris* immersed in a large volume of saline solution (30% v./v. seawater which is isotonic with *Ascaris* tissues) at 37° remains almost motionless, usually curled into a wide loop. Only the anterior end performs small lateral, seemingly exploratory, movements. However, in the same worm is enclosed in a glass tube (ca. ½ inch inside diameter) filled with warm saline, its whole body, presumably stimulated by contact with the hard surface, performs vigorous undulatory movements which result in a rather fast progression.

Since the contractile parts of all somatic muscle cells are oriented in the same direction (i.e., parallel to the longitudinal axis of the worm) and there are no circular muscles, the locomotive movements of nematodes take place principally on the dorsoventral plane. Such undulations are due to the progression, along the length of the body, of two parallel and antagonistic waves of contraction and relaxation in the dorsal and ventral fields of the somatic muscle layer.

The aformentioned process of locomotion, which has been called gliding, is aided in some instances by bristles or cuticular excrescences with which the worm obtains purchase on surfaces. Other types of progression (see Stauffer, 1924; Hyman, 1951; Gray and Lissmann, 1964) include swimming, which is essentially another type of undulatory movement; crawling, i.e., alternate lengthening and shortening of the body; and thrashing, in which the ends of the body are brought together and then are straightened suddenly.

Besides the various movements in the dorsoventral plane, nematodes also are able to perform some movements in the horizontal plane. These seem to be produced by the simultaneous contraction of muscle cells close to one of the lateral lines, on both the dorsal and ventral muscle fields, due to the synchronous activity of the subdorsal and subventral nerve cords of the same side of the body (see del Castillo *et al.*, 1967).

Electrophysiological recording with intracellular microelectrodes shows that the contraction of the somatic muscle of *Ascaris* is elicited by rhythmic repetitive action potentials which occur at frequencies of up to 12 per second (del Castillo *et al.*, 1963).

The mechanism of generation of these potentials is different from those which control vertebrate skeletal muscle. In the latter the action potentials are initiated by neural action; i.e., they are produced by the one-to-one transmission of nerve impulses across neuromuscular junctions. In *Ascaris*, however, as in the vertebrate heart, the action potentials are of myogenic origin, arising from the oscillatory activity of the membrane of the syncytial region of the muscle. Depolarization of this region results in the continuous firing of spikes at a rate determined by the prevailing membrane potential (del Castillo *et al.*, 1967).

The role played by the nerve cord fibers appears to be a modulatory one. They control the level of membrane potential and the rate of firing of the spikes. Such regulation is accomplished by two types of synapses: excitatory and inhibitory.

The stimulation of excitatory neuromuscular synapses results in muscle depolarization and an increased frequency of firing. These effects seem to be mediated by acetylcholine or a related choline ester (del Castillo *et al.*, 1963).

The other synapses are inhibitory (Goodwin and Vaughan Williams, 1963). Their activation causes a hyperpolarization of the syncytial region and a decreased frequency or complete suppression of the rhythmic action potentials (del Castillo *et al.*, 1967). The chemical transmitter released at these junctions is likely to be γ-aminobutyric acid or a pharmacologically similar compound. The action potentials disappear altogether when the resting potential increases above 40 mV.

The undulatory movements of adult nematodes can, therefore, be visualized as due to the simultaneous propagation, in the dorsal and ventral muscle fields, of waves of depolarization and hyperpolarization produced by nervous control.

The shortening of the contractile portion of the muscle cells, which are longitudinally arranged, tends to shorten the body and, therefore, to reduce its volume. However, since the liquid constituents of the body are not compressible, the muscle contraction will increase the internal hydrostatic pressure of the worm. The mean value of this pressure in *Ascaris* is, according to Harris and Crofton (1957), as high as 70 mm. Hg, showing rhythmical fluctuations from 16 mm. Hg up to 225 mm. Hg!

4. *The Esophageal Muscle Cell*

As mentioned, the only motile region of the nematode digestive system is the esophagus or pharynx, an organ which may be regarded as a single syncytial muscle cell of tubular shape in which the myofibrils are oriented radially between the external and internal surfaces. Their contraction transforms the lumen, which at rest is an almost virtual triradiate cavity, into a large triangular canal with a consequent development of negative pressure and the suction of liquid and suspended food particles.

Upon relaxation of the myofibrils, the material taken in by the esophagus is pumped into the intestine, a rather formidable task in view of the high hydrostatic pressure existing within the body of the worm (Harris and Crofton, 1957). Since the intestine is formed by a single layer of epithelial cells, it will tend to collapse and can be filled only by a pumping organ capable of generating pressures even higher than those which prevail in the pseudocoelom.

The nematode esophagus is unique as a pump insofar as the ejection of its contents is concerned. The performance of external mechanical work is accomplished not during its contraction, but upon relaxation.

This feature has interesting electrophysiological correlates. Thus, the action potentials recorded from the *Ascaris* esophagus (del Castillo *et al.*, 1964b; del Castillo and Morales, 1967a,b) show two regenerative, all-or-none, phases: a depolarizing one which leads to a positive overshoot of about 15 mV., and a faster repolarizing phase which leads to an "undershoot" or hyperpolarizing spike of up to 100 mV. Such a fast and intense hyperpolarization appears to be necessary for the synchronous and efficient relaxation of the cell.

The triradiate shape of the lumen and the high internal pressures of the nematode body suggest that the entire esophagus may act as a self-sealing valve. The pseudocoelomic pressures which would cause the expulsion of intestinal contents through the esophagus could also act upon the outer surface of the organ, thereby pressing the three sections of the wall together and closing the lumen (Mapes, 1965). Furthermore, the esophageal tissue usually projects into the beginning of the intestine, forming an esophago-intestinal valve.

II. Drugs Acting on the Nervous System

1. *Santonine*

a. Chemistry. Santonine [1] is an essential vegetable oil extracted from the plant *Artemisa cina,* the Levant worm-seed, and related species. Chemically, it is a terpene, an unsaturated hydrocarbon, odorless and colorless, though it becomes yellow upon exposure to light.

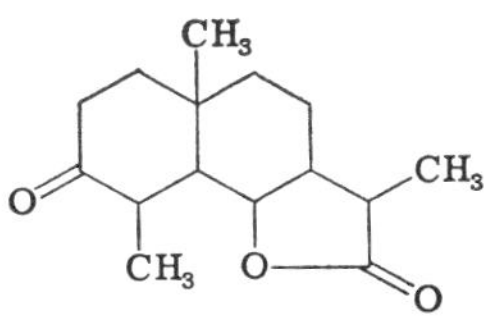

(1)

b. Pharmacology. Santonine, even at concentrations as high as 0.1% does not kill *Ascaris* (Lamson and Brown, 1936), but favors elimination of the worms. Baldwin (1943) demonstrated that santonine in concentrations as low as 1:50,000 paralyzes anterior preparations of *Ascaris* (i.e., in *vitro* preparations which include the anterior end of the worm). In contrast to this inhibitory effect, intermediate preparations of *Ascaris* were stimulated by the drug at concentrations 10 times higher. These results were interpreted as indicating that santonine exerts a paralyzing action on the nerve centers of the anterior part of the worm.

This hypothesis was supported by Goodwin (1958) who suggested that santonine interferes with the coordinating nerve impulses which originate at the main nerve ring. Therefore, the worm can be expelled by giving a purgative, since apparently it is no longer able to maintain its position within the host.

c. Uses. In spite of such observations, santonine was used against *Ascaris* with only doubtful results (dose: from 0.6 to 0.2 gm. combined with calomel—0.2, 0.3 gm.—followed by a saline purgative within 3 hours). According to Hall and Augustine (1929) the *Ascaris* removal rate achieved with santonine was only about 27%.

d. Effects on Host. When santonine is administered, it also affects the mammalian nervous system, causing headache, vertigo, hallucinations, convulsions, etc. (Desoille, 1937).

2. Oil of Chenopodium (Ascaridol)

a. Chemistry. The oil of chenopodium is extracted from the American worm-seed, *Chenopodium ambrosoides.* Its active principle (60–80% of the oil) is ascaridol [2], a liquid organic peroxide, colorless and volatile, with a very pungent odor.

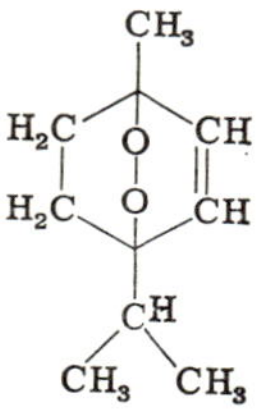

(2)

b. Pharmacology. The actions of ascaridol on nematodes have not been investigated properly. This drug is included somewhat arbitrarily in this section on the basis of botanical and toxicological relationships.

c. Uses. It is a much more efficient ascaricide than santonine (83% reduction rate). However, it is no longer used because of its high toxicity.

d. Effects on Host. It exerts very marked toxic actions on the human central nervous system (ataxia, vertigo, hallucinations, deafness and even blindness). Furthermore, it is extremely irritant to skin and mucous membranes and produces marked gastrointestinal disurbances.

III. Drugs Acting on the Neuromuscular System

A. Compounds Which Activate Depolarizing Receptors

1. Acetylcholine

a. Chemistry. This is the acetic ester of choline [3], a crystalline, highly hygroscopic substance.

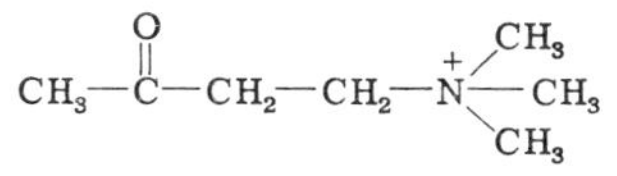

(3)

Widely distributed in nature, acetylcholine plays important physiological roles in many species. It acts as the neuromuscular transmitter in vertebrate skeletal muscle, ganglionic junctions of the autonomic nervous system, as well as in many central nervous synapses and the neuroeffector junctions of the parasympathetic system.

b. Pharmacology. Baldwin and Moyle (1947) found that *Ascaris* muscle is stimulated by acetylcholine at concentrations of 10^{-6} (w./v.) or higher. This was confirmed by Norton and DeBeer (1957) who showed that such an effect is blocked by *d*-tubocurarine.

Mellanby (1955) demonstrated that acetylcholine or a related choline ester is a constituent of *Ascaris* tissues, which also possess a specific cholinesterase (Bueding, 1952; Lee, 1962). Mellanby's observations suggested that acetylcholine is concentrated in the nervous tissue since fragments of *Ascaris* tissue which include the "cephalic" nerve ring contain about 15 times more acetylcholine per unit weight than the rest of the body.

The mechanism of the stimulating action of acetylcholine on *Ascaris* muscle was studied by del Castillo *et al.* (1963), who observed that the addition of this compound to the surrounding saline brings about a marked reduction in the muscle resting potential. The actual depolarization elicited by a given concentration of acetylcholine varied from cell to cell, but it was statistically higher in the nuclear bags over the nerve cords which are joined to the muscle syncytium by short arms. In experiments in which acetylcholine was applied by microelectrophoresis to discrete areas of the cell membrane, no changes in resting potential could be detected even when relatively large amounts were delivered to the nuclear bags and spindles. However, depolarizations were elicited consistently when it was released from micropipet tips placed in the vicinity of the muscle syncytium, indicating that receptors for acetylcholine are restricted to this region of the muscle cells.

The effects of both anticholinesterases and *d*-tubocurarine (see below) indicate that small amounts of acetylcholine are continuously released from the nerve cord. This process maintains the membrane potential of the muscle syncytium at a level in which rhythmic action potentials are generated (del Castillo *et al.*, 1963). Electrical noise, presumably of synaptic origin, occurs in the syncytial region of *Ascaris* muscle (del Castillo *et al.*, 1967). Discrete depolarizing noise components, which resemble the miniature end plate potentials recorded from the neuromuscular junctions of vertebrate muscle, have been recorded with microelectrodes inserted in the syncytium. The possibility exists that they are due to spontaneous acetylcholine liberation from the nerve cord fibers. Yet, further experimentation is needed to establish their actual significance.

c. Uses. None as an anthelmintic.

2. *Bephenium*

a. Chemistry. Bephenium, a quaternary ammonium compound, is the benzyl-dimethyl-2-phenoxyethylammonium [4]; a pale, crystalline substance sparingly soluble in water.

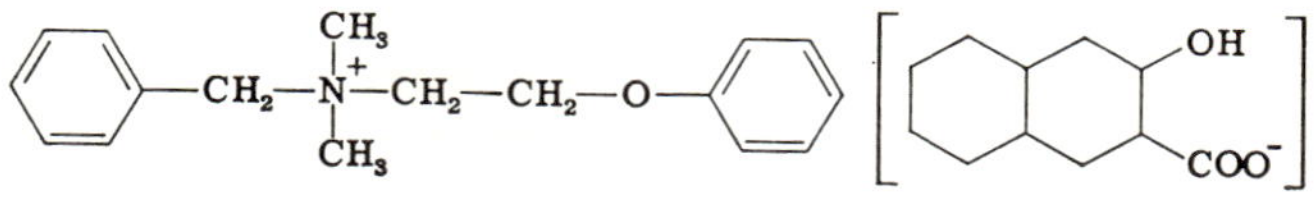

(4)

b. Pharmacology. Bephenium salts are particularly effective against the nematodes which inhabit the small intestine. They cause a contraction of nematode muscle like acetylcholine. A concentration of bephenium of 0.25 μg./ml. has an effect equivalent to a 1 μg./ml. concentration of acetylcholine (Broome, 1962).

c. Uses. This drug is valuable in mixed *Ascaris*-hookworm infections, although it is less effective than piperazine against *Ascaris* (dose: 5 gm. of the salt containing about 2.5 gm. of the base). A single dose is actually sufficient against *Ascaris, Trichostrongylus orientalis,* and both species of human hookworms. It lacks activity, however, against *Strongyloides stercoralis* and has only moderate effectiveness against *Trichuris trichiura.*

d. Effects on Host. It may cause nausea, abdominal pain, vomiting, etc., although it is not very toxic at therapeutic dosages. Such a lack of toxicity is probably due to the fact that bephenium is very poorly absorbed from the intestine.

3. *Thenium*

a. Chemistry. This drug, related to bephenium, is another quarternary nitrogen base [5]. It is crystalline, colorless, slightly soluble in water, and has a bitter taste.

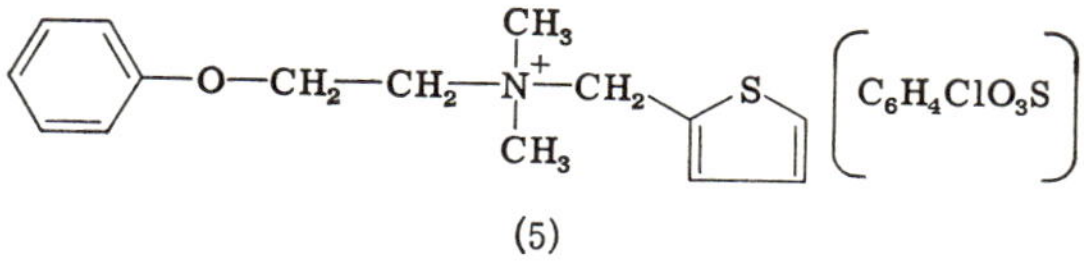

(5)

b. Pharmacology. Not studied in detail but presumably similar to that of bephenium.

c. Uses. In screening tests, this compound had a similar order of activity to that of bephenium against *N. muris* in the rat, but was more active against *A. caninum* in the dog.

4. *Methyridine*

a. Chemistry. Methyridine, 2-(2-methoxyethyl)-pyridine [6], is characterized by colorlessness, a sweet smell, and water solubility.

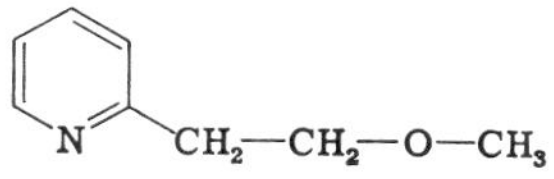

(6)

b. Pharmacology. Methyridine causes a rapid paralysis of rhythmic movements in exposed *Ascaris* preparations, an effect which is reversed by *d*-tubocurarine and piperazine but not by acetylcholine (Broome, 1962). These observations suggested that methyridine has a depolarizing effect on muscle, similar to that exerted by acetylcholine, an action which has been confirmed in this laboratory (unpublished observations). It has been hypothesized, therefore, that the effect of methyridine on nematode muscle is similar to the block caused by decamethonium on vertebrate neuromuscular junctions.

c. Uses. Methyridine was introduced in veterinary medicine for the treatment of gastrointestinal nematodes in sheep and cattle. It is more efficient parenterally than orally, reaching all parts of the gut following its subcutaneous administration (Broome, 1961) (dose: 1 ml. of 90% solution by 10 pounds of body weight, Maximal amount per site of injection: 40 ml.).

d. Effects on Host. It may produce local necrosis, even though it is well tolerated on the whole.

B. Anticholinesterasic Agents

1. *Physostigmine and Neostigmine*

a. Chemistry. Physostigmine [7] is the alkaloid from *Physostigma venenosum* or Calabar bean,

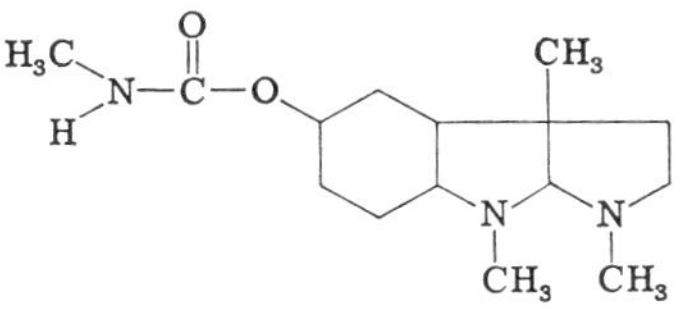

(7)

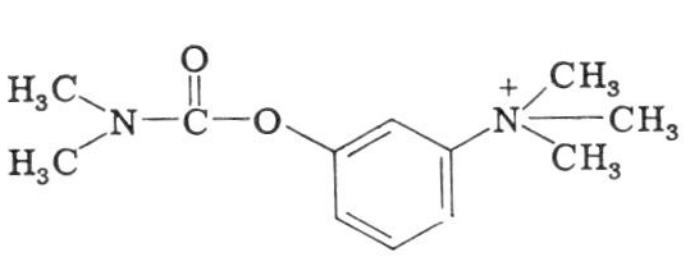

(8)

whereas neostigmine [8], a synthetic compound, differs from physostigmine by having a quaternary nitrogen.

b. Pharmacology. Norton and DeBeer (1957) reported that physostigmine salicylate at concentrations of 10^{-5} (w./v.) markedly enhanced the sensitivity of *Ascaris* muscle to acetylcholine. In the same year, Krotov (1957) showed that the activity of acetylcholine is lost following incubation with *Ascaris* tissues. These results are not surprising since we know today, from Bueding's (1952) and Lee's observations (1962), that the somatic musculature of *Ascaris* contains appreciable amounts of cholinesterase located particularly, according to Lee, in the syncytial region and the terminal part of the arms or innervation processes. These, as mentioned above, are the regions where the acetylcholine receptors seem to be located.

Del Castillo *et al.* (1963), working with neostigmine bromide, confirmed Norton and DeBeer's observations. They showed also that neostigmine causes a depolarization of the syncytial membrane accompanied by an increase in the frequency of the rhythmic action potentials. These effects could be explained by a direct action of the drug; however, the fact that treatment of the same preparations with *d*-tubocurarine increases the resting potential (see below) suggests that the depolarizing action of neostigmine may be due to a potentiating influence upon small amounts of acetylcholine spontaneously released from the nerve cord.

c. Uses. None as anthelmintic.

2. *Organophosphorus Compounds*

Several potent anticholinesterasic compounds, originally developed as insecticides, have proved to be valuable as anthelmintics.

a. Chemistry. These compounds have as a common chemical feature the fact that four of the valences of the phosphorus atom are saturated by double bonds to O or S and linked through an —O— to lower alkyl or directly to amino or alkylamino groups. Three representative compounds and their formulas are:

Asuntol or 3-chloro-4-methyl-umbelliferone-*O*-*O*-diethylthiophosphate [9].

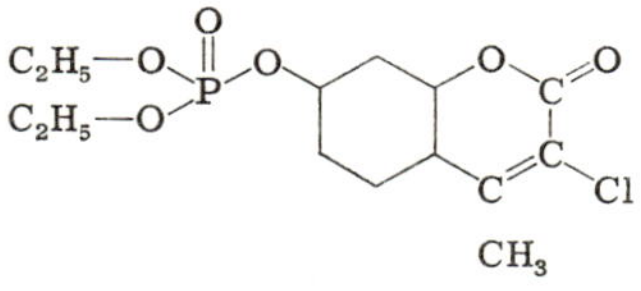

(9)

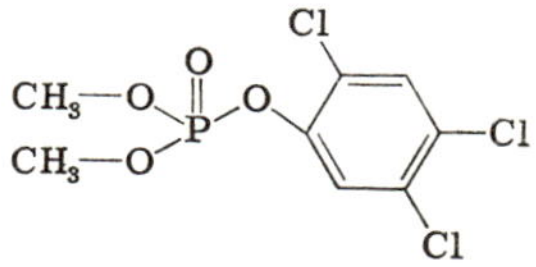

(10)

Dow ET/57 or *O*-*O*-dimethyl-*O*-(2,4,5-trichlorophenyl) phosphorothiate [10].

Neguvon or *O,O*-dimethyl-1-hydroxy-2-trichloromethylphosphonate [11] Bayer L13/59.

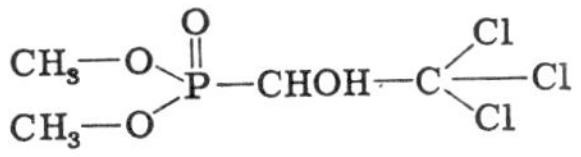

(11)

b. Pharmacology. Although the effectiveness of these compounds as insecticides is due to their anticholinesterasic actions, there is no direct evidence, as yet, that their value against parasitic nematodes is a result of their enzyme-inhibiting properties. Standen (1963) has written a very comprehensive review of the work done on these compounds.

c. Uses. These drugs are used only in veterinary medicine for the treatment of nematode infections of sheep and cattle.

d. Effects on Host. In general, the toxicity of organophosphorus compounds is high and is directly related to their anticholinesterasic properties.

C. Drugs Which Block Depolarizing Receptors

1. *d-Tubocurarine*

a. Chemistry. *d*-Tubocurarine [12] is one of the alkaloids of *Chondrodendron tomentosum,* a plant used by South American Indians for the preparation of the arrow poison called curare.

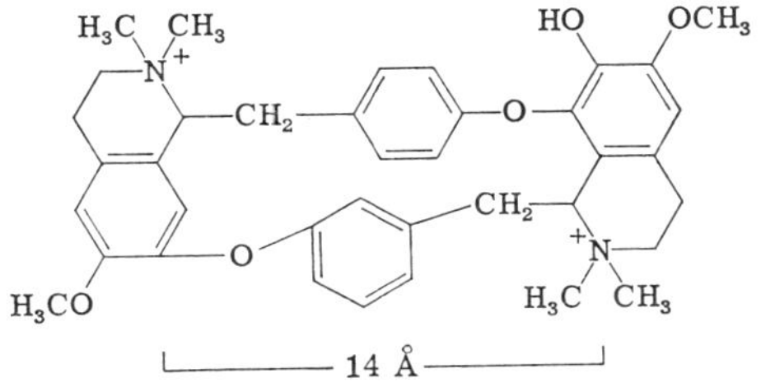

(12)

There are many other compounds, both natural and synthetic, with pharmacological actions similar to that of *d*-tubocurarine. All of them are characterized by having two quaternary nitrogen groups separated by a distance of 14 Å.

b. Pharmacology. *d*-Tubocurarine competes with acetylcholine for the postsynaptic receptors at the neuromuscular junctions of vertebrate skeletal muscle. It combines with such receptors, blocking the actions of both naturally liberated and applied acetylcholine (see del Castillo, 1961).

A similar antagonism between *d*-tubocurarine and acetylcholine on *Ascaris* muscle was reported by Norton and DeBeer (1957) and confirmed by del Castillo *et al.* (1936) with electrophysiological techniques. All the effects of acetylcholine on isolated *Ascaris* preparations were prevented by previous treatment with *d*-tubocurarine. A concentration of this drug of 10^{-5} (w./v.) totally blocked the effects of 10^{-6} (w./v.) acetylcholine.

However, an important difference was observed between the effects of *d*-tubocurarine on vertebrate and nematode muscle. Whereas in the former a block of the neurally initiated muscle impulses is produced, the rhythmic action potentials recorded from *Ascaris* muscle cells are not blocked by *d*-tubocurarine even at concentrations as high as 10^{-3} (w./v.). This lack of action on the generation of muscle action potentials may be regarded as evidence for their myogenic nature.

In most *Ascaris* preparations, *d*-tubocurarine causes an increase in the resting potential of muscle and, therefore, a decrease in the frequency and amplitude of the spontaneous spikes. This can be explained as due to a blocking action on the small concentrations of acetylcholine which are normally released from the nerve cord.

c. Uses. No applications as an anthelmintic.

D. Compounds Which Activate Hyperpolarizing Receptors

1. *Piperazine*

a. Chemistry. Piperazine or diethylendiamine [13] is used as such or in form of its citrate, phosphate, adipate, tartrate, etc., salts.

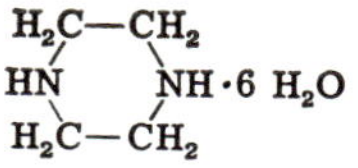

(13)

b. Pharmacology. The efficacy of piperazine for the treatment of *Ascaris* infections was first observed by Boismare, a pharmacist of Rouen, France, who noticed this effect on patients who received piperazine as an uricosuric and antirheumatic agent. This observation was confirmed by Fayard (1949). Soon afterwards, piperazine was firmly established as the drug of choice for the treatment of a number of intestinal nematodes.

Goodwin and Standen (1954) noticed that the worms expelled following treatment with piperazine were narcotized, motionless and relaxed, but still alive. Indeed, tone and motility reappeared upon immersion in Ringer's solution at 37°. These effects were confirmed *in vitro* by

Poynter (1955) using piperazine adipate against *Parascaris equorum* and by Standen (1955) who tested the effect of piperazine salts against the *Ascaris* of the pig. Standen observed that the worms were narcotized if placed in a solution (1:200–1:560) of these salts, but recovered in less than 3 hours in a drug-free medium, even if exposed to piperazine for 48 hours.

Norton and DeBeer (1957) showed that the stimulating action of acetylcholine on *Ascaris* muscle was blocked by piperazine, an observation which suggested that piperazine had a curarelike effect on the neuromuscular junctions. However, subsequent work (DeBell *et al.*, 1963; del Castillo *et al.*, 1963) indicating that the action potentials of *Ascaris* muscle are of myogenic origin led to a reinvestigation of the effects of piperazine with electrical techniques.

The most obvious effect of this drug was to increase the resting potential of the somatic muscle of *Ascaris*. In fact, at concentrations of between 10^{-5} and 10^{-3} (w./v.) the potential across the muscle surface membrane increased with a slope of about 8 mV. per 10-fold increase in concentration. The resting potential was raised from its average value of about 30 mV. to about 45 mV. This was accompanied by a flaccid paralysis of the muscle, as the rhythmic action potentials which elicited its contraction disappeared when the resting potential increased above 40 mV.

c. Site of Action. The hyperpolarization produced by piperazine was greater in the nuclear bags close to the nerve cord (joined to the syncytium by short arms). This suggested that the potential change is maximal at the syncytial region, a possibility which was explored by electrophoretic application of piperazine from micropipets. In these experiments, the distribution of the piperazine receptors was seen to overlap with that of acetylcholine receptors. No changes in the membrane potential could be detected when the drug was applied to the nuclear bags, spindles, and the accessible region of the muscle arms. However, hyperpolarizations were always produced when the drug was delivered in the region of the muscle syncytium.

Concentrations of piperazine of 10^{-2} (w./v.) and higher have a depolarizing effect on *Ascaris* muscle; i.e., decrease the resting potential instead of increasing it. This effect, however, is produced over the entire surface of the cell and is unrelated to the activation of specific synaptic receptors.

Piperazine does not interfere with the function of inhibitory nerve fibers and synapses. On the contrary, the suppression of electrical activity which follows inhibitory stimulation was usually lengthened by piperazine. In fact, the changes induced by this drug on the muscle cells of *Ascaris* are similar to those produced by the stimulation of inhibitory nerve cord fibers and have been interpreted as due to the activation of the hyperpolarizing neuromuscular receptors.

d. The Membrane Change Induced by Piperazine. The resting potential of excitable cells is due basically to the difference in the concentrations of K^+ and Cl^- ions between the cytoplasm and the external solution. If such ionic gradients across the surface membrane are kept constant, the exact value of the resting potential will be proportional to the permeability of the membrane to such ions.

Therefore, in the absence of changes in the ionic composition of the outside solution or the cytoplasm, an increase in the resting potential of an excitable cell can be brought about by an enhancement of the permeability of its membrane to either K^+ or Cl^- ions. The relative importance of these two ions in the piperazine-induced hyperpolarization was explored by changing their concentrations in the surrounding solution. Whereas variations in the external K^+ concentration were ineffective, changes in the concentration of extracellular Cl^- ions markedly influenced the action of piperazine. Indeed, when the extracellular chloride was reduced, the hyperpolarizing effect of piperazine was considerably decreased. The conclusion was drawn that the activation of the inhibitory receptors by the drug increases the permeability of the syncytial membrane to chloride ions (see also del Castillo *et al.* 1964d).

As mentioned above, this change seemed to take place exclusively at the postsynaptic inhibitory receptors. In fact, the activation of these receptors appeared to result in an increased permeability of the membrane to Cl^- ions.

e. Biochemical Effects. Piperazine causes a decrease in the formation of succinic acid, a metabolic product of the worm which originates from the reduction of fumaric acid by reduced nicotinamide adenine dinucleotide; fumaric acid, in turn, rises from malic acid, through a reaction catalyzed by a fumarase (Saz and Hubbard, 1957) whereas malate results from the fixation of CO_2 into pyruvate. The reaction from fumaric to succinic acid which is coupled to phosphorylation supplies energy for the muscle (see Saz and Bueding, 1966, for an up-to-date review on the biochemistry of *Ascaris*).

Ascaris produces large amounts of succinate in both aerobic and anaerobic conditions (Bueding and Farrow, 1956; Bueding *et al.*, 1959); but if *Ascaris* tissues are incubated with paralyzing concentrations of piperazine, succinate production is greatly reduced (Bueding *et al.*, 1959). The lack of succinate production seems to be a result, rather than a cause, of the piperazine-induced paralysis, being due to the lowered energy transformation in the flaccid muscle (Bueding, 1962).

f. Uses. Piperazine is the preferred drug for treating ascariasis. Among its advantages are high potency, simplicity of treatment, almost complete absence of toxicity, low cost, palatability, etc. A single 4-gm. dose is effective in 80% of the cases (Goodwin and Standen, 1954).

Piperazine is also extremely useful for the treatment of *Enterobius vermicularis* (dose: 0.5–2.0 gm. daily for 7 days).

g. Effects on Host. Apart from infrequent cases of urticaria, nausea, and vomiting, some minor neurological symptoms have been reported occasionally.

2. *Hetrazan (diethylcarbamazine)*

a. Chemistry. Hetrazan, the 1-diethyl carbamoyl-4-methyl-piperazine citrate [14], is a compound highly soluble in water in which it yields rather acid solutions (a concentration of 1% has a pH of 4.1).

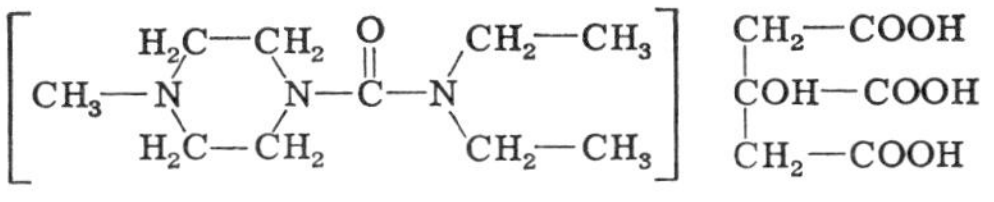

(14)

b. Pharmacology. Hetrazan is extremely effective in the treatment of filarial infections, including Bancroft's filariasis and onchocercosis. Its mechanism of action is unknown. It seems reasonable to assume that a paralysis of the parasites might be involved, as a result of the piperazine ring in its molecule. However, substitutions in this ring are known to reduce the anthelmintic activity of piperazine (Brown *et al.,* 1954). Furthermore, the diethylcarbamoyl group may introduce new, still unknown properties. Thus, Broome (1962) has reported the unexpected observation that diethylcarbamazine increases the muscle tone and spontaneous electrical activity of *Ascaris* preparations. Also, in contrast with piperazine, hetrazan, even at concentrations of 250 μg./ml., does not interfere with the action of acetylcholine on such preparations.

c. Uses. It is the drug of choice for the treatment of filariasis (dose: 2 mg./kg., orally, 3 times daily for 2–3 weeks). It is also useful for the treatment of the lungworm of cattle, *Dictyocaulus viviparus* (dose: 22 ml. of a 10% solution per 100 kg., i.m. repeated three successive days).

d. Effects on Host. The main toxic reactions so far reported consist of fever, headaches, nausea, joint pains and swelling of lymph glands, effects believed to be caused by the proteins released upon the disintegration of the dead microfilaria. Apart from these effects, hetrazan is said to have an effect on the blood pressure similar to that of epinephrine.

3. Other Piperazine Derivatives

A number of piperazine derivatives have been introduced in medicine to take advantage of the outstanding anthelmintic properties of this molecule, but only a few will be mentioned here. For a review on structure-activity relationship of various piperazine derivatives, see Craig and Tate (1961).

a. Chemistry. Tricofenol piperazine or bis-(2,4,5-trichlorophenol)-piperazine [15] is a compound that contains 82% of 2,3,5-trichlorophenol and 18% of piperazine.

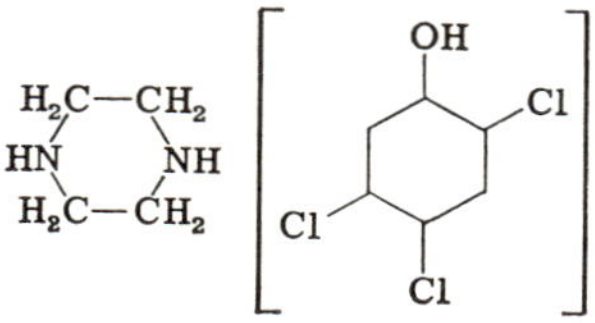

(15)

Piperazine carbon disulfide complex is a polymer composed of piperazine-1-carbodithiotic acid (HN—  —N-CS-SH) and piperazine 1,4-dicarbodithiotic acid (HS-SC-N—  —N-CS-SH).

Polymethylene piperazine [16] and 1,4,bis(D-glucosyl)piperazine [17] are other derivatives.

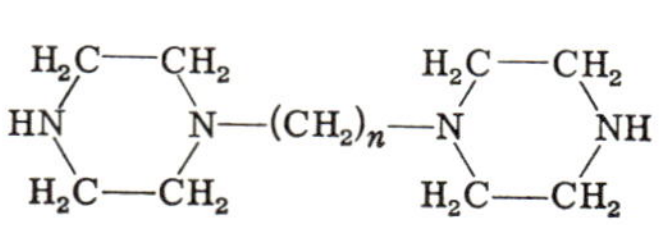

(16)

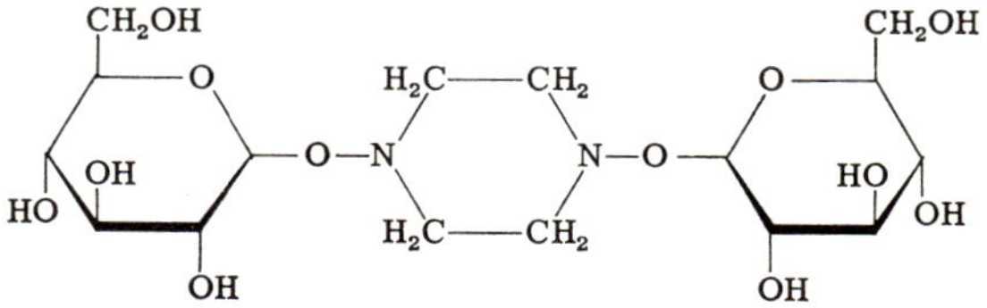

(17)

b. Pharmacology. The activity of all these compounds is largely due to the piperazine moiety of their molecules. Tricofenol piperazine combines this paralyzing action with the necrotizing action of its phenolic moiety. The piperazine carbon disulfide complex decomposes in the stomach, liberating carbon disulfide.

c. Uses. Tricofenol piperazine, acting both on *Ascaris* and hookworm (*Necator americanus*) is particularly useful for the treatment of mixed infections (dose: 50 mg. on an empty stomach). Piperazine carbon disulfide complex has a 100% effectiveness on the *Ascaris* of the pig (dose: 110 mg./kg.). It is also effective on *Oesophagostomum dentatum,* on the nematode infections of cattle (dose: 440 mg./kg.), and on the ascariasis of the horse (dose: 75 mg./kg.).

Polymethylene piperazine is effective against *Parascaris* and the small strongyles of the horse, but is less effective in oxyurasis.

d. Effects on Host. Tricofenol piperazine is very irritating to the oral mucosa.

4. *γ-Aminobutyric Acid (GABA)*

a. Chemistry. GABA ($NH_2CH_2—CH_2—CH_2—COOH$), is a compound, derived from glutamic acid, which occurs naturally in large amounts in vertebrate brain, as well as in the nerve system of some invertebrates.

b. Pharmacology. GABA is believed to function as an inhibitory synaptic transmitter in vertebrate brain. Its action on *Ascaris* somatic muscle was studied by del Castillo *et al.* (1964a). At a concentration of 10^{-7} *M* it was seen to exert a small, but significant ($p < 0.0002$) depolarizing action. However, when its concentration was increased to 10^{-5} *M* the average resting potential increased by about 15 mV. This was accompanied by an inhibition of the spontaneous rhythmic action potentials.

A comparison of the hyperpolarizing actions of GABA and piperazine is shown in Fig. 2. Curve B shows the resting potential as a function of the piperazine concentration (based on the data of del Castillo *et al.,* 1964a), each point being the average of 100 measurements. Curve A shows the effect of GABA (each point being the mean of 120 cells). It is apparent that the maximal levels of membrane potential induced by both compounds are almost equal, about 46 mV. However, the concentration of piperazine necessary to reach such hyperpolarization is about two orders of magnitude higher than that of GABA.

In spite of such marked effects, GABA does not paralyze intact *Ascaris in vitro* (del Castillo and Morales, unpublished). It can be concluded, therefore, that it does not penetrate into the worm or it is metabolized before reaching the muscle cells.

c. Uses. None as an anthelmintic, although it may serve as a starting molecule for the development of new anthelmintic compounds.

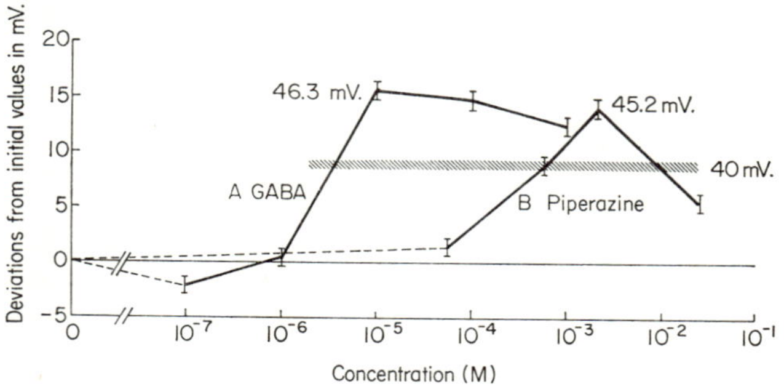

FIG. 2. Graph showing the effect of γ-aminobutyric acid (GABA) and piperazine on the membrane potential of *Ascaris* muscle cells. Ordinates represent average membrane potential in mV., expressed as the difference from the initial values in the absence of drugs. Abscissae: concentration of the drugs in moles. Each point in curve *A* (GABA) is the average membrane potential of 120 cells (six preparations), while the points in curve *B* (piperazine) are the average of 100 measurements (five preparations). The lines drawn through the points are equal to 2 × the S.E. of the mean.

IV. Miscellaneous Drugs Which May Interfere with Excitation Processes

1. *Phenothiazine*

a. Chemistry. Phenothiazine [18] is a light greenish-yellow compound obtained by fusing diphenylamine with sulfur in presence of iodine.

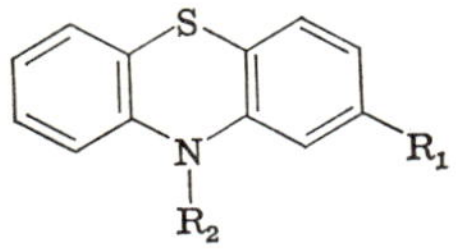

(18)

It is insoluble in water, but extremely soluble in organic solvents and mineral oil. A number of derivatives can be obtained by substitution of the hydrogens at positions R_1 and R_2.

Phenothiazine is one of the most versatile molecules in pharmacology, being the parent compound of at least a dozen compounds widely used in medicine at present (antihistaminics, antispasmodics, hypotensives, tranquilizers, and antiemetics). Two of these derivatives have had some limited applications as anthelmintics; these are:

Promethazine (Phenergan), where $R_1 = H$ and $R_2 =$ 2-dimethyl-aminopropane [19].

Pyrathiazine which is the *N*-(β-pyrrolidyl ethyl)phenothiazine [20] where R_1 = H and R_2 = β-pyrrolidyn-ethyl.

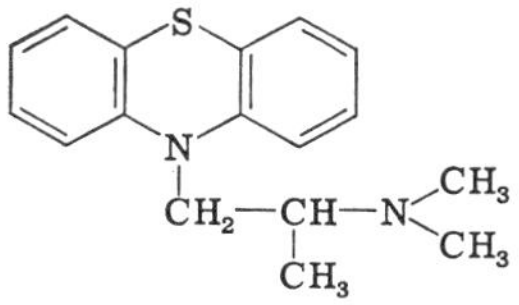

(19)

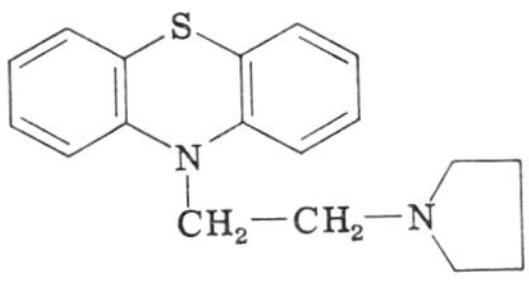

(20)

b. Pharmacology. The anthelmintic properties of phenothiazine were first described by Harwood *et al.* (1938). It appears that this drug is not changed into an active form by metabolic processes, acting as such by a still unknown mechanism. However, Watkins (1958) suggested that it causes a paralysis of the helminth's musculature, as well as the failure of the reproductive system, since phenothiazine inhibits egg-laying both *in vitro* and *in vivo*.

Although the parasites expelled following the administration of phenothiazine continue to live, an observation that supports the idea that they may be paralyzed, Baldwin and Moyle (1949) were unable to demonstrate any effect of this compound on *Ascaris* preparations.

Using phenothiazine labeled with ^{35}S, Lazarus and Rogers (1950) demonstrated that it is absorbed directly through the cuticle, and Knapp *et al.* (1959) related the rate of penetration of this drug to its effectiveness.

Phenothiazine has been shown to inhibit the activity of serum cholinesterase (Collier and Allen, 1942). Although a direct action of this compound on the cholinesterase of *Ascaris* has not been demonstrated, some organic phosphorus compounds which have an anticholinesterasic activity potentiate the anthelmintic action of phenothiazine (Kingsbury, 1961).

c. Uses. Phenothiazine is one of the most widely used anthelmintics against intestinal nematodes in sheep, cattle, and horses. It is also the only anthelmintic drug known to keep on acting outside the host. Indeed, strongyloid worms are not able to grow during their free life in the dung of animals which receive small daily doses of phenothiazine. It has also been used in human medicine against *Enterobius vermicularis* (dose: 7 gm. in 4 days for adults). Promethazine has also a high degree of activity against the parasite but the drug is not recommended because of violent hallucinatory effects observed at the effective dosages (Avery, 1956).

An inverse relationship has been found to exist between the average size of the phenothiazine particles and its effectiveness against some species of nematodes parasitic in sheep. In fact, the efficacy of a given dose would depend upon its total surface area; thus, greatest efficiency was found when the size of the particle was 1–2 μm. No activity was observed if the particles were as large as 140 μm.

d. Effects on Host. In high doses phenothiazine is toxic, producing hepatitis, hemolytic anemia, albuminuria, and hematuria. After treatment with phenothiazine, cattle often develop keratitis due to photosensitization. Promethazine, a very good anthistaminic, is not used as such because of the considerable sedation it produces. Furthermore, it exerts hallucinatory effects at high dosages.

2. *Cyanacetohydrazine and Cyanacetohydrazide*

a. Chemistry. The hydrazine has the formula CN—CH_2—CO—NH—NH_2, and the hydrazide CN—CH_2—CO—NH—NH—CO_2Bu.

b. Pharmacology. Following the injection of cyanacetohydrazine into tracheotomized calves and lambs experimentally infected with lungworms (*Dictyocaulus viviparus* in the former and *D. filaria* and *Metastrongylus apri* in the latter) the worms were expelled alive.

c. Uses. Cyanacetohydrazine is effective against lungworms both by oral and subcutaneous routes of administration. Cyanacetohydrazide, a derivative of the former compound, is effective for both lungworms and tuberculosis.

3. *Tetrachloroethylene*

a. Chemistry. Tetrachloroethylene [21] is a colorless, volatile liquid insoluble in water, but very soluble in fat.

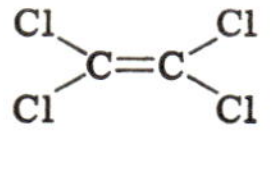

(21)

b. Pharmacology. In the absence of direct experimental evidence it has been suggested that tetrachloroethylene, the same as other halogenated hydrocarbons, acts by becoming dissolved in lipid deposits existing in the muscle cells, particularly in the arms or innervation processes (Mueller, 1929), thus interfering with muscle function.

c. Uses. During the last three decades, tetrachloroethylene has been the drug of choice for the treatment of hookworm, replacing carbon tetrachloride (dose: 5 ml. in a single dose removes 92% of worms; children, 0.05–0.06 ml./pound up to 5 ml. Several treatments are usually needed).

d. Effects on Host. It is less hepatotoxic than carbon tetrachloride.

4. *Hexylresorcinol*

a. Chemistry. This drug is a dihydric phenol [22], only slightly soluble in water (1:2,000), but soluble in alcohol, ether, and chloroform. It has a sharp astringent taste.

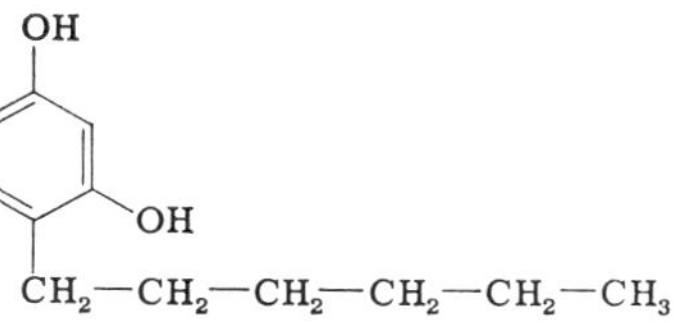

(22)

b. Pharmacology. This drug produces blisters and necrosis in the parasites because of its phenolic nature. It also changes the permeability of the body wall to phosphorus-containing compounds which escape from the worm. Furthermore, it is said to paralyze muscle.

c. Uses. It is of some value for the treatment of combined infections of hookworm and *Ascaris* (dose: 1 gm. for adults and children over 10. Give on empty stomach and follow with cathartic after 2–4 hours).

d. Effects on Host. Local irritation of oral mucosa.

V. Drugs Which Interfere with Nutritional and Metabolic Processes

A. Cyanine Dyes

These compounds are characterized by a resonating amidinium ion system in which a quaternary nitrogen is linked to a tertiary nitrogen by a resonant chain of an uneven number of carbon atoms with alternating single and double bonds (Brooker, 1945, 1954), i.e.:

$$N-(-\overset{|}{C}=\overset{|}{C}-)-C=\overset{+}{N} \leftrightarrow \overset{+}{N}=(\overset{|}{C}-\overset{|}{C}=)=C-N$$

When the nitrogen atoms are part of two heterocyclic rings, the dyes are known as cyanines. If only one nitrogen acts as a member of a ring, the dyes are known as styryls. Brooker and Sweet (1947) found that cyanine dyes have a high activity against *Litomosoides carinii*. This activity is not due to the particular end rings, but to the resonating amidinium ion itself.

1. *Dithiazanine iodide*

a. Chemistry. Dithiazanine is the 3-3′-diethylthiadicarbocyanine iodide, a structurally symmetrical blue-green cyanine dye with the following structural formula [23]:

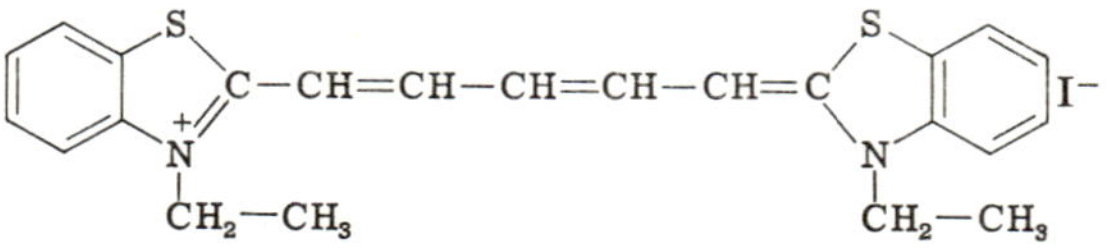

(23)

b. Pharmacology. The biochemical actions of dithiazanine have been thoroughly investigated by Bueding *et al.* (1960; see also Bueding, 1962) on the canine whipworm, *Trichuris vulpis.* This compound was shown to cause a reduction in the motility of the worm preceded by an inhibition of glucose utilization and production of lactic acid as well as an increase in the rate of glyconeogenesis. Concentrations of dithiazanine lower than those which affect motility do inhibit the rate of glucose utilization. This indicates that the biochemical change is not secondary to the decreased motor activity.

However, diathiazanine is believed not to influence the activity of any of the enzymes involved in glycolysis, since it leads to a lowered concentration of glucose within the worms, instead of an increase. This observation suggests that dithiazanine interferes with the availability of sugar from outside sources.

The metabolic effects of dithiazanine appear to be irreversible because if worms are exposed to this drug for 24 hours and then are transferred to a drug-free medium, they continue to show a reduced uptake of glucose.

c. Uses. Dithiazanine was introduced for the treatment of intestinal nematode infections by Swartzwelder *et al.* (1957). It has an outstanding activity against *Trichuris trichiura* and *Strongyloides stercoralis,* parasites not susceptible to other synthetic anthelmintics. It is also active against *Ascaris lumbricoides* and *Enterobius vermicularis,* although not as effectively as piperazine against the former, or pyrvinium against the latter. Variable results have been obtained against *Necator* (adult dose: 200 mg., 3 times daily for 3 days; children's dose: 100 mg., 2 or 3 times daily for 5 days).

d. Effects on Host. Unfortunately, dithiazanine exerts a highly irritant action on the digestive mucosae of the patient. Extreme caution should be exerted if the intestinal mucosa is inflamed. Violent gastritis has often been reported. As a result, production of this drug has been discontinued (see Maldonado, 1965).

2. *Pyrvinium Pamoate*

a. Chemistry. This is the 6-dimethylamino-2-(2,5-dimethyl-1-phenyl-3-pyrryl)-vinyl-1-methyl-quinolinium pamoate [24]; a red, crystalline cyanine dye, slightly soluble in water.

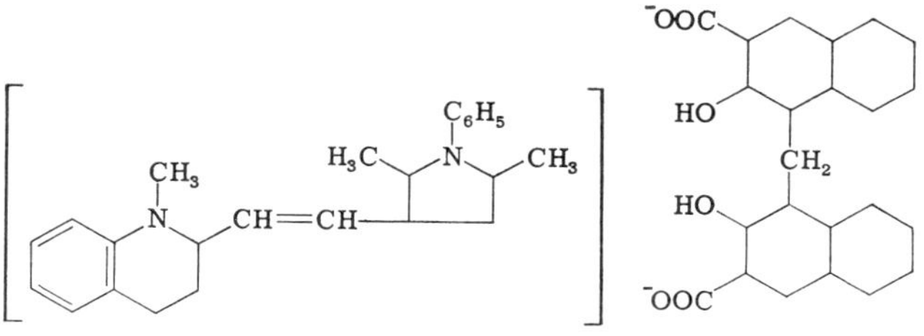

(24)

b. Pharmacology. Its actions are similar to that of the chloride salt. Is very poorly reabsorbed from the intestine.

c. Uses. It is highly active for the treatment of pinworms (enterobiasis) and was developed on the basis of its action against analogous oxyurids in the mouse. It succeeded pyrvinium chloride, having the advantage of not causing nausea; it is also effective against *Ascaris, T. trichura,* and hookworms on dogs (dose: 5 mg./kg., single dose; cure rate, 95%).

d. Effects on Host. The precursor of pyrvinium pamoate, pyrvinium chloride, produces gastrointestinal irritation, accompanied by nausea and vomiting. Some kidney and liver damage was observed in experimental animals. These effects appear to be absent if the drug is given in its pamoate form.

B. Other Metabolic Drugs

1. *Stilbazium Chloride*

a. Chemistry. This is the 2,6-bis(*p*-pyrrolidino-styryl)pyridine ethiodide [25].

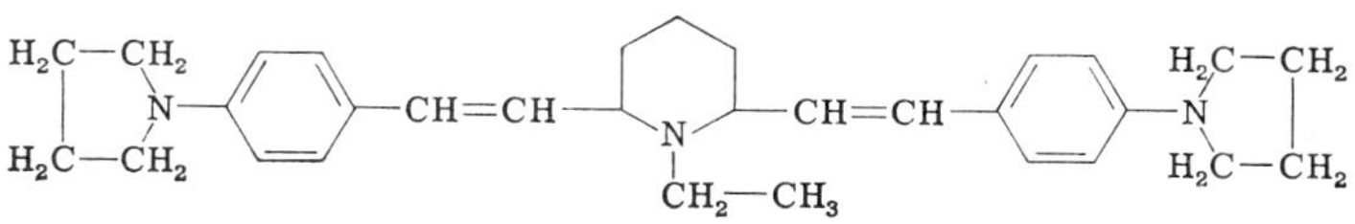

(25)

b. Pharmacology. Its actions may be similar to those of other cyanine dyes which interfere with the transport of glucose into the parasite.

c. Uses. It is highly effective against *Trichuris trichiura.* According to Maldonado (1965) the parasites can be almost totally eradicated by the administration of two 250-mg. doses.

d. Effects on Host. Vomiting and abdominal pain, usually not severe.

VI. Compounds Which Interfere with Reproduction and the Larval Stages

1. *Thiabendazole*

a. Chemistry. This is the 2-(4-thiazolyl) benzimidazole [26].

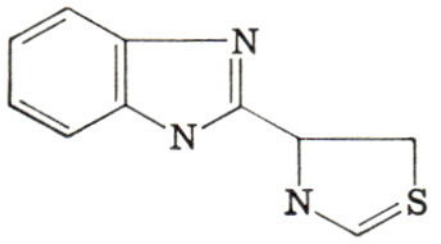

(26)

b. Pharmacology. This compound is a member of a chemical series said to have larvicidal properties *in vitro* at extremely low concentrations (see Standen, 1963). It has been claimed also that this drug inhibits the production of eggs by worms.

c. Uses. In the sheep, a single oral dose of 50 mg./kg. removes more than 95% of *Trichostrongylus, Cooperia,* and other gastrointestinal nematodes. It is active also against hookworm and whipworm, and it is said to affect the tissue migratory stages of *Ascaris* and *Stephanurus* (Standen, 1963). Furthermore, thiabendazole is probably the only chemical agent that may have some effect against *Trichinella spiralis,* since it has produced considerable clinical improvement in experimental trials.

2. *Arsenamide*

a. Chemistry. This is the (*p*-carbamylphenylarsylenedithio)-diacetic acid [27].

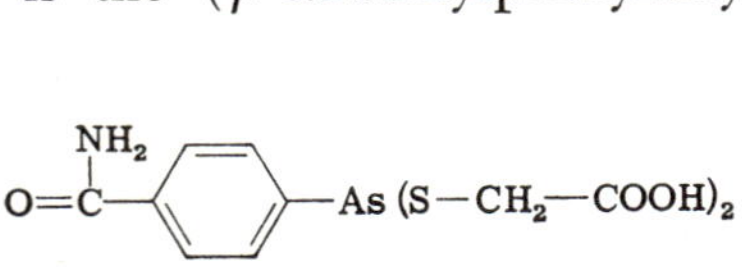

(27)

b. Pharmacology. It has a lethal action on the microfilariae of *Wunchereria bancrofti in vitro,* at concentrations of 0.8 μg./ml. (Hawking, 1950).

c. Uses. It has been used intravenously for the treatment of patients infected with *W. bancrofti* (dose: 1 μg./kg. daily for 15 days).

d. Effect on Host. It may produce very serious toxic effects on the liver; for this reason it has been replaced by hetrazan.

3. *Mel W*

a. Chemistry. This is the pentylthiarsaphenylmelamine [28].

b. Pharmacology. At low doses, it produces an inhibition of cellular

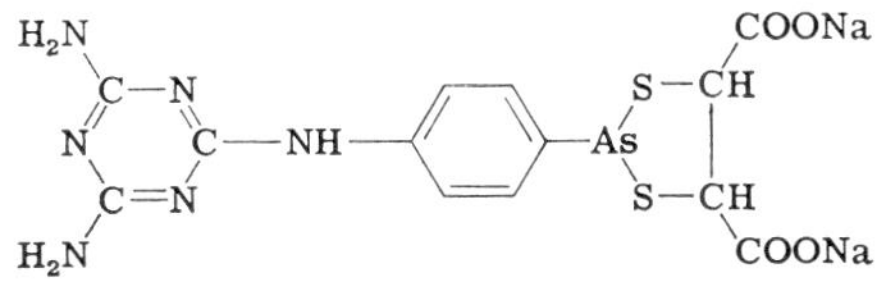

(28)

multiplication in the early development stages of intrauterine filarial embryos at a time when they are also affected by other antimitotic agents.

4. *BIQ 20 and BAC 20*

a. Chemistry. These compounds are members of a series of bis-isoquinolinium compounds synthesized by Austin *et al.* (1959). BIQ 20 is the eicosane 1:20-bis-isoquinolinium iodide [29], and BAC is the eicosane 1:20-bis-4-aminocinnolinium iodide.

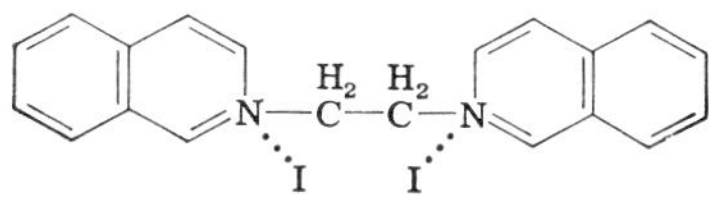

(29)

b. Pharmacology. These compounds were tested upon *Litomosoides* in cotton rats. At concentrations of 0.1 μg./ml. they kill adult worms within 24 hours. Their action seems to be exerted on the gonads and developing embryos (Hawking, 1963).

c. Uses. Still in an experimental stage.

VII. Nonspecific Drugs and Protoplasm Poisons

A. PHENOLS

1. *Thymol*

a. Chemistry. Thymol is the methyl-isopropylphenol ($C_{10}H_{15}O$) [30], a monohydric phenol obtained from several species of plants. It is a colorless, translucent crystalline compound with a pungent odor. It is sparsely soluble in water, but highly soluble in alcohol, chloroform, ether, and oils.

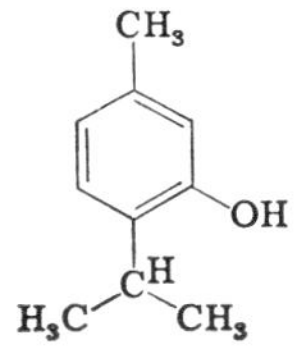

(30)

b. Pharmacology. Phenols exert a coagulant action on proteins and, therefore, protoplasm. Small organisms are particularly susceptible to this action. Thymol is almost totally absorbed from the digestive tract; particularly if fat is present.

c. Uses. It was used for the treatment of hookworm since 1879 (Caius and Mhaskar, 1919) (dose: 2–4 gm., single dose) but it is no longer employed because of its relatively high toxicity.

d. Effects on Host. It has definite toxic properties on humans, irritating mucous membranes and depressing the central nervous system, where it causes hypothermia, vertigo, headache, tinnitis, collapse, etc. It also has toxic effects on the kidney (albuminuria, etc.).

2. *β-Naphtol*

a. Chemistry. This is the β-hydroxynaphthalene [31], a monohydric phenol. Whitish and with a slight phenolic odor, it darkens upon exposure to light. It is slightly soluble in water, but highly soluble in alcohol, ether, chloroform, glycerine and oil.

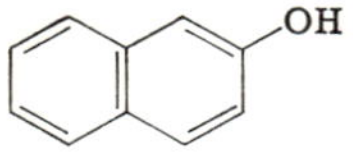

(31)

b. Pharmacology. See thymol, above.

c. Uses. It was employed, since 1904, for the treatment of infection with hookworm. However, it is less efficient than thymol and much less efficient than carbon tetrachloride or tetrachloroethylene. It is no longer used.

d. Effects on Host. It is a rather toxic compound for the mammalian organism, behaving as a strong irritant of the gastrointestinal mucosa. It produces also central nervous symptoms similar to those caused by thymol.

3. *1-Bromo-2-Naphthol*

a. Chemistry. [32]

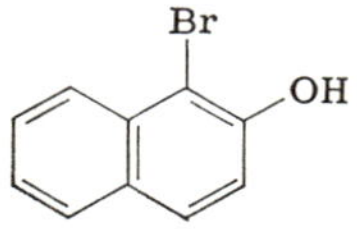

(32)

b. Pharmacology. It is reported to be the more active and less toxic of a series of halogenated naphthols tested against *Rhabdias bufonis* (Miura, 1954).

c. Uses. It has been employed for the treatment of *Ascaris, Trichuris,* and hookworm.

d. Effects on Host. Has a necrotizing effect on the intestinal mucosa.

4. Other Phenols

Other phenols which have some applications as anthelmintics include the 2,6-diiodo-4-nitrophenol and the tricofenol piperazine.

B. GENTIAN VIOLET

a. Chemistry. This dye, also called *Gentian violet medicinal, methyl violet,* or *crystal violet,* is the hexamethylpararosanilin [33], but usually contains some amounts of the tetramethyl and pentamethyl derivatives; for this reason it is also termed methylrosaniline chloride (U.S.P.).

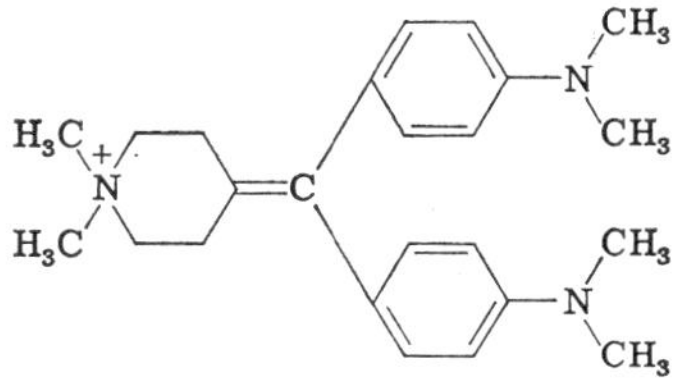

(33)

b. Pharmacology. It is said to combine with the protoplasm of bacteria, fungi, and helminths.

c. Uses. Gentian violet was initially used both as a mild antiseptic and for the treatment of fungal infections of the skin. Later it was introduced as the specific drug for the treatment of the Chinese liver fluke (*Clonorchis sinensis*). It is also very effective for the treatment of strongyloidiasis and oxyurasis (adult dose: 60 mg., 3 times a day before meals for 10 days; children's dose: 10 mg. a day per each year of age).

d. Effects on Host. It may cause nausea, vomiting, diarrhea and abdominal pain; but these symptoms usually subside when dosage is reduced, or the treatment discontinued.

VIII. Proteolytic Enzymes

Although intestinal helminths are highly resistant to digestive enzymes, they are susceptible to the lytic action of several proteolytic enzymes synthesized by tropical plants. The anthelmintic properties of these products were recognized long ago; some of them have been used empirically with relative success for several centuries. The most important are: the pineapple (*Ananassa sativa*) juice, which contains the enzyme bromelin (see Beger and Asenjo, 1939; Asenjo, 1940). The papaya (*Carica papaya*) juice which contains the enzyme papain (see Hassler, 1928; Berger and Asenjo, 1940). The fruit of the *Bromelia pinguina* which contains the enzyme pinguinain (see Asenjo and Fernandez, 1945) and the plants *Ficus glabrata* (syn. F. *laurifolia*) and *Ficus diliaria*, also natives of Central America, which contain an active principle called ficin, also a proteolytic enzyme (see Robbins, 1930).

The latter enzyme has been widely used in some countries in form of the so-called "leche de higuerón" or "higuerolatex"; a white, creamy, viscous liquid which is preserved by adding 1% of sodium benzoate. Ficin is said to be more potent than papain while having no harmful effects on either the gastric or intestinal mucosa. It acts on several species of intestinal helminths, but is particularly effective in the treatment of infection by *Trichocephalus* where it has an 85% effectiveness (see Caldwell and Caldwell, 1929) (dose: 60 ml. on an empty stomach). It is effective also against hookworms and *Ascaris*.

Due to the labile nature of the enzyme, need for refrigeration, etc., the use of higuerolatex is restricted to the areas where the *Ficus* trees grow.

ACKNOWLEDGMENTS

The original work described in this chapter was supported by USPHS Grants Nos. NB-02021 and NB-05235. Mr. T. Morales has kindly provided the drawings of Fig. 1.

REFERENCES

Asenjo, C. F. (1940). *J. Am. Pharm. Assoc.* **29**, 8–10.

Asenjo, C. F., and de Fernandez, M. del C. C. (1945). *J. Agr. Univ. Puerto Rico* **29**, 35–46.

Auber-Thomay, M. (1964). *J. Microscopie* **3**, 105.

Austin, W. C., Lunts, L. H. G., Potter, M. D., and Taylor, E. P. (1959). *J. Pharm. Pharmacol.* **11**, 80–93.

Avery, J. L. (1956). *J. Am. Med. Assoc.* **161**, 681–683.

Baldwin, E. (1943). *Parasitology* **35**, 89–111.

Baldwin, E., and Moyle, V. (1947). *J. Exptl. Biol.* **23**, 277–291.

Baldwin, E., and Moyle, V. (1949). *Brit. J. Pharmacol.* **4**, 145–152.

Berger, J., and Asenjo, C. F. (1939). *Science* **90**, 299–300.

Berger, J., and Asenjo, C. F. (1940). *Science* **91**, 387–388.

Brooker, L. G. S. (1945). *Frontiers Chem.* **3**, 63–136.

Brooker, L. G. S. (1954). *In* "Theory of the Photographic Process" (C. E. K. Mees, ed.), pp. 371–429. Macmillan, New York.

Brooker, L. G. S., and Sweet, L. A. (1947). *Science* **105**, 496.

Broome, A. W. J. (1961). *Brit. J. Pharmacol.* **17**, 327–338.

Broome, A. W. J. (1962). *In* "Drugs, Parasites and Hosts" (L. G. Goodwin and R. H. Nimmo-Smith, eds.), pp. 43–61. Little, Brown, Boston, Massachusetts.

Brown, H. W. (1960). *Clin. Pharmacol. Therap.* **1**, 87–103.

Brown, H. W., Chan, K. F., and Hussey, K. L. (1954). *Am. J. Trop. Med. Hyg.* **3**, 504–510.

Bueding, E. (1952). *Brit. J. Pharmacol.* **7**, 563–566.

Bueding, E. (1962). *In* "Drugs, Parasites and Hosts" (L. G. Goodwin and R. H. Nimmo-Smith, eds.), pp. 15–28. Little, Brown, Boston, Massachusetts.

Bueding, E., and Farrow, G. W. (1956). *Exptl. Parasitol.* **5**, 345–349.

Bueding, E., and Swartzwelder, C. (1957). *Pharmacol. Rev.* **9**, 329–365.

Bueding, E., Saz, H. J., and Farrow, G. W. (1959). *Brit. J. Pharmacol.* **14**, 497–500.

Bueding, E., Kmetec, E., Swartzwelder, C., Abadie, S., and Saz, H. J. (1960). *Biochem. Pharmacol.* **5**, 311–322.

Caius, J. F., and Mhaskar, K. S. (1919). *Indian J. Med. Res.* **7**, 429–463.

Caldwell, F. C., and Caldwell, E. L. (1929). *Am. J. Trop. Med.* **9**, 471–482.

Collier, H. B., and Allen, D. E. (1942). *Can. J. Res.* **D20**, 283–290.

Craig, J. C., and Tate, M. E. (1961). *Fortschr. Arzneimittelforsch.* **3**, 75–150.

DeBell, J. T. (1965). *Quart. Rev. Biol.* **40**, 233–251.

DeBell, J. T., del Castillo, J., and Sanchez, V. (1963). *J. Cellular Comp. Phyisol.* **62**, 159–179.

del Castillo, J. (1961). *Res. Publ., Assoc. Res. Nervous Mental Disease* **38**, 90–143.

del Castillo, J., and Morales, T. (1967a). *J. Gen. Physiol.* **50**, No. 3, 603–629.

del Castillo, J., and Morales, T. (1967b). *J. Gen. Physiol.* **50**, No. 3, 631–645.

del Castillo, J., de Mello, W. C., and Morales, T. (1963). *Arch. Intern. Physiol. Biochim.* **71**, 741–757.

del Castillo, J., de Mello, W. C., and Morales, T. (1964a). *Experientia* **20**, 1–5.

del Castillo, J., de Mello, W. C., and Morales, T. (1964b). *Nature* **203**, 530–531.

del Castillo, J., de Mello, W. C., and Morales, T. (1964c). *Brit. J. Pharmacol.* **22**, 463–477.

del Castillo, J., de Mello, W. C., and Morales, T. (1964d). *J. Gen. Physiol.* **48**, 129–140.

del Castillo, J., de Mello, W. C., and Morales, T. (1967). *J. Exptl. Biol.* **46**, 263–279.

Desoille, H. (1937). *Rev. Oto-Neuro-Oftal., Buenos Aires* **15**, 170–173.

Fayard, C. Thèse pour le Doctorat de Medicine, 66 pp. Thesis, Paris [quoted from *Semaine Hop.* **35**, 1778 (1949)].

Goodwin, L. G. (1958). *Brit. J. Pharmacol.* **13**, 197–201.

Goodwin, L. G., and Standen, O. D. (1954). *Brit. Med. J.* **II**, 1332.

Goodwin, L. G., and Vaughan Williams, E. M. (1963). *J. Physiol.* (*London*) **166**, 44P.

Gray, J., and Lissmann, H. W. (1964). *J. Exptl. Biol.* **41**, 135–154.

Hall, M. C. (1928). *In* "De Lamar Lectures, 1927–28. Johns Hopkins University, School of Hygiene and Public Health," pp. 1–23. Williams & Wilkins, Baltimore, Maryland.

Hall, M. C., and Augustine, D. L. (1929). *Am. J. Hyg.* **9**, 585–628.

Harris, J. E., and Crofton, H. D. (1957). *J. Exptl. Biol.* **34**, 116–130.

Harwood, P. D., Jerstad, A. C., and Swanson, L. E. (1938). *J. Parasitol.* **24**, Sect. 2, 16.

Hassler, E. (1928). *Arch. Schiffs- u. Tropen-Hyg.* **32**, 409–410.

Hawking, F. (1950). *Trans. Roy. Soc. Trop. Med. Hyg.* **44**, 153–186.

Hawking, F. (1963). *Exptl. Chemotherapy* **1**, 893–912.

Hinz, E. (1963). *Protoplasma* **56**, 202.

Hyman, L. H. (1951). "The Invertebrates," Vol. III. McGraw-Hill, New York.

Kingsbury, P. A. (1961). *Res. Vet. Sci.* **2**, 265–271.
Knapp, E. E., Folse, D. S., Moser, H. C., and McFarland, R. H. (1959). *Vet. Med.* **54**, 536–538.
Krotov, A. I. (1957). *Byul. Eksperim. Biol. i Med.* **43**, 95–97.
Lamson, P. D., and Brown, H. W. (1936). *Am. J. Hyg.* **23**, 85–103.
Lazarus, M., and Rogers, W. P. (1950). *Nature* **166**, 647–648.
Lee, D. L. (1962). *Parasitology* **52**, 241–260.
Maldonado, J. F. (1965). "Helmintiasis del hombre en América." Editorial Científico-Médica, Barcelona.
Mansour, T. E. (1964). *Advan. Pharmacol.* **3**, 129–165.
Mapes, C. J. (1965). *Parasitology* **55**, 269–284.
Mellanby, H. (1955). *Parasitology* **45**, 287–294.
Miura, K. (1954). *Japan J. Med. Sci. Biol.* **7**, 265–273.
Mueller, J. F. (1929). *Z. Zellforsch. Milkroskop. Anat.* **8**, 361–403.
Norton, S., and DeBeer, E. J. (1957). *Am. J. Trop. Med. Hyg.* **6**, 898–905.
Poynter, D. (1955). *Vet. Record.* **67**, 159–162.
Robbins, B. H. (1930). *J. Biol. Chem.* **87**, 251–257.
Rosenbluth, J. (1963). *J. Cell Biol.* **19**, 82A.
Rosenbluth, J. (1965a). *J. Cell Biol.* **25**, 494–515.
Rosenbluth, J. (1965b). *J. Cell Biol.* **26**, 579–591.
Saz, H. J., and Bueding, E. (1966). *Pharmacol. Rev.* **18**, 871–894.
Saz, H. J., and Hubbard, J. A. (1957). *J. Biol. Chem.* **225**, 921–933.
Standen, O. D. (1955). *Brit. Med. J.* **II**, 20–22.
Standen, O. D. (1963). *Exptl. Chemotherapy* **1**, 701–892.
Stauffer, H. (1924). *Zool. Jahrb. Abt. Systematik (Oekol), Geograph. Biol.* **49**, 119–130.
Swartzwelder, J. C., Frye, W. W., Muhleisen, P. J., Miller, J. H., Lampert, R., Anthony, S. O., Peña Chavarria, A., Abadie, S. H., and Sappenfield, R. W. (1957). *J. Am. Med. Assoc.* **165**, 2063–2067.
Toscano-Rico, J. (1926). *Compt. Rend. Soc. Biol.* **94**, 918–920.
Trendelenburg, P. (1916). *Arch. Exptl. Pathol. Pharmakol.* **79**, 190–217.
Watkins, T. I. (1958). *J. Pharm. Pharmacol.* **10**, 209–227.

CHAPTER 12

Chemistry of Nematodes in Relation to Serological Diagnosis

José Oliver-González

I. Introduction

Recent trends on investigations dealing with immunological processes in the animal host have strongly indicated that a better knowledge is required of the chemistry of the parasite. This is chiefly in relation to the isolation of chemical constituents which serve as antigens to be used for serological diagnosis and for immunization against infection. There are excellent publications on related topics such as the nature of parasitism (Rogers, 1961), biochemical properties (von Brand, 1966), and biology of parasites (Soulsby, 1966).

The following review intends to inform on the nature of antigens that have been prepared from nematodes, and their value as agents for serological diagnosis. The lack of correlation between chemical approaches for isolation of antigens and the procurement of the specific substance which immunizes the host is also emphasized.

II. Antigen Extractions

Developmental stages of nematodes, eggs, larvae, and adults are known to be very complex, due to the multiplicity of chemical constituents possessed by each stage. Host response to these substances causes additional complications, since upon infection or immunization, antibody formation varies according to the nature of, and concentration of antigen.

It has been a common procedure to make crude extracts of parasitic stages which contain a large number of antigens, each acting independently; and some not related to diagnosis or immunity. Very little progress has been made in the field of isolating specific antigens for both of these purposes, chiefly because such substances require elaborate chemical procedures, and contaminants are usually present.

Among the first attempts to isolate chemical antigenic fractions from nematodes were those by Melcher and Campbell (1942) and Melcher (1943), using *Trichinella spiralis*. These investigators isolated a polysaccharide and an acid-soluble protein which served as antigens in serological and immunological work. The acid-soluble protein has been found to be very sensitive in the bentonite flocculation test and in the precipitin test (Kagan, 1960) for the diagnosis of this infection. The same type of procedure as described by Melcher has been used for preparation of acid-soluble proteins for the diagnosis of other parasitic infections. Other attempts for chemical preparation of specific antigens were made by Sleeman and Musehel (1961). These investigators extracted an ethanol-soluble glycoprotein with 75% protein and 15% carbohydrate. An ethanol-insoluble fraction was also prepared which has a nucleoprotein averaging 60% ribonucleic acid and 14% protein. The ethanol-soluble glycoprotein is very specific for the serological diagnosis of trichinosis while the insoluble nucleoprotein lacked specificity, but was very sensitive.

Trichinella spiralis larvae contain a large number of substances which behave as antigens and react with antisera prepared in rabbits by injecting whole larval material. Tanner *et al.* (1961) observed 11 bands of precipitate using buffered saline larvae extracts against sera from infected rabbits. Of the 11 electrophoretically distinct bands, four moved like human gamma globulins; one each like $beta_1$ and $beta_2$ globulins; two like $alpha_1$ and two like $alpha_2$ globulins. The presence of 11 antigens in saline extracts of trichina larvae illustrates the contrast between chemical complexity on one hand, and the small number of these substances that have been isolated, on the other.

The parasitic nematode most utilized for antigenic studies is the pig roundworm, *Ascaris suum*. Extracts from the whole worms as well as from the isolated tissues have led to preparations which react against sera from infected or artificially immunized animals. Saline extracts made from each of the tissues result in extraction of antigens common to all tissues. The A_2 isoagglutinogen factor which inhibits the α_2 isoagglutinin in human sera of groups O and B is present in all. Oliver-González, (1958) and Kagan *et al.* (1958) using serum-agar double diffusion studies, reported that polysaccharide-protein fractions, prepared with formamide from embryonated and nonembryonated egg, and from cuticle, yielded antigens immunologically less complex than antigens produced by other methods. Using this method of extraction one component was found specific in the cuticle and a second in unembryonated egg.

Soulsby (1956) also studied the reactions of whole *Ascaris* worms against homologous whole ascaris antiserum using double diffusion in agar gels and observed nine bands of precipitate. Extracts of intestine produced fourteen, and cuticle produced nine bands. This illustrates the complexity of such studies in which undescribed antigens are used for immunization, and crude ones are used for testing. Antibodies develop primarily against major antigens; thus no reactions will be observed against minor antigens.

Kent (1960) extracted females of *Ascaris suum* with triple distilled water at 4° after defatting with absolute alcohol and ether at 70°. Six different proteins were separated by column chromatography and agar gel electrophoresis. This represents the first instance in which single purified antigens have been prepared from nematode species. The water-soluble proteins extracted appeared to be glycoproteins with variable amounts of carbohydrates, with the antigenicity related to the glycoprotein. Two of these antigens had diagnostic value in detecting infections as early as 10 days prior to the presence of eggs in stool.

Although progress has been made in the isolation of purified antigens from *Ascaris*, these have not been established as good diagnostic agents. The cross reactivity between antigens from *Ascaris* and other nematodes makes this problem extremely difficult.

Isolation of purified antigens from filarial organisms has been attempted only on few occasions. Serological work has been performed using group, not species-specific, antigens. Crude saline extracts from the dog filarid *Dirofilaria immitis* have been used extensively for the diagnosis of other filarial infections in humans (Kagan, 1963). Acid-soluble fractions of delipidized adult worms have been used for the bentonite and hemagglutination tests to diagnose infections with *Aconthochlionema perstans*. These antigens cross reacted with sera from patients with schistosomiasis, ascariasis, and trichinosis (Kagan *et al.*, 1963).

Sawada and Takei (1965) prepared an antigen from *Dirofilaria immitis* using Sephadex G-100 gel filtration, CM-cellulose chromatography, and DEAE Sephadex-AS column filtration. The finally purified antigen was a protein with 2.5% carbohydrate. This antigen was used for intradermal test on proved cases of filariasis and found to be very reactive. Tada and Kawashima (1964) also prepared a purified antigen from *Dirofilaria immitis* which consisted of polypeptides. This antigen, like Sawada's, was also active and specific used as a skin testing antigen for the diagnosis of infections with *Wuchereria bancrofti*. Although the antigens prepared by Sawada and Tada gave positive reactions in patients with infections with *Wuchereria bancrofti*, further studies are required to establish specificity in view of cross reactions with other nematodes.

The egg and larval forms of most of the parasitic nematodes, except *Trichinella spiralis*, have been used very little for antigen extraction. This is mainly due to the small number of larvae present or difficulties in collection.

Adult and larval stages of a given species have antigenic constituents common to both, although recent investigations have indicated the presence of stage-specific antigens. No comparative studies have been made between the various developmental stages of parasites, but results of some investigations indicate antigenic variation between them. Thus the serological reactions observed in rabbits infected with *Ascaris suum* are distinct when hatching fluid from eggs and antigen prepared from adult worms are used (Soulsby, 1963).

Saline extracts from microfilariae have also been used as antigens. The extracts elicit skin reactions in individuals harboring microfilariae, but these results need further evaluation.

With very few exceptions, no specific antigens have been isolated either from microfilariae of any species or from any larval stage from nematodes. In view of qualitative antibody changes observed during infection, specific antigens should be isolated from those stages which sensitize the host.

III. Antigens from Excretions and Secretions

The excretions and secretions from live nematodes are easily obtained by suspending the worms in adequate fluid media. These may be concentrated by lyophilization or precipitation with mild chemical agents. The importance of these antigens is widely accepted today, both in relation to diagnosis and immunization (Thorson, 1962). A critical analysis of the value of these antigens as compared with antigens from the whole worm is very necessary. Enzymes are also included in the realm of secretions, therefore the number of substances secreted by the parasites become very numerous. Isolation of specific substances from excretions and secretions involves the same difficulties as isolation from the entire worm.

Incubation in immune sera of whole organisms, either eggs or larvae, results in antigen-antibody reactions leading to the formation of precipitate around the egg membrane or around the orifices and inside the intestine of the larvae. This reaction may be also classified as involving secretions and excretions from the parasitic stages combining with antibody in the immune serum. These tests are highly specific and play an important role in the serodiagnosis of infections with nematodes (Anderson, 1963).

IV. Conclusion

Study of the chemical structure of nematodes is very important since during infection of the host some of the constituents serve as antigens which immunize the host. Attempts to isolate these antigens as they occur or as they affect the host have been very few. Antigenic substances of great interest could probably be encountered during the initial period of infection when antibodies have not yet developed.

Chemical extraction of antigens has not followed a regular pattern which should take into consideration the stage or the structure which reacts with the host. That is, if a larval form is found within the host, special attention should be given to the isolation of substances from this particular stage. This will probably be more advantageous in the procurement of diagnostic antigens possessing more sensitivity and specificity.

REFERENCES

Anderson, R. I. (1963). *Am. J. Hyg. Monograph. Ser.* **22**, 97–112.

Kagan, I. G. (1960). *J. Infect. Diseases* **107**, No. 1, 65–93.

Kagan, I. G. (1963). *J. Parasitol.* **49**, No. 5, 773–798.

Kagan, I. G., Jeska, E. J., and Gentzkow, C. J. (1958). *J. Immunol.* **80**, No. 5, 400–406.

Kagan, I. G., Norman, L., and Allair, D. S. (1963). *Am. J. Trop. Med. Hyg.* **12**, No. 4, 548–555.

Kent, N. H. (1960). *Exptl. Parasitol.* **10**, No. 3, 313–323.

Melcher, L. R. (1943). *J. Infect. Diseases* **73**, 31–39.

Melcher, L. R., and Campbell, D. H. (1942). *Science* **96**, 431–432.

Oliver González, J. (1958). *Rice Inst. Pam.* **45**, No. 1, 141–150.

Rogers, W. P. (1961). "The Nature of Parasitism: The Relationship of Some Metazoan Parasites to their Hosts." Academic Press, New York.

Sawada, T., and Takei, K. (1965). *Japan. J. Exptl. Med.* **35**, 125–132.

Sleeman, K. H., and Musehel, L. H. (1961). *Am. J. Trop. Med. Hyg.* **10**, No. 6, 821–833.

Soulsby, E. J. L. (1956). *Trans. Roy. Soc. Trop. Med. Hyg.* **51**, 9–10.

Soulsby, E. J. L. (1963). *Am. J. Hyg. Monograph. Ser.* **22**, 47–59.

Soulsby, E. J. L., ed. (1966). "The Biology of Parasites," pp. 255–273. Academic Press, New York.

Tada, I., and Kawashima, K. (1964). *Japan. J. Parasitol.* **13**, 472–474.

Tanner, C. E., and Gregory, J. (1961). *Can. J. Microbiol.* **7**, 473–481.

Thorson, R. E. (1962). *Am. J. Hyg. Monograph. Ser.* **22**, 63–67.

von Brand, T. (1966). "Biochemistry of Parasites." Academic Press, New York.

CHAPTER 13

Chemical Ecology of Acanthocephala and Nematoda

Alan F. Bird and H. R. Wallace

I. Introduction

Ecology is the study of whole living organisms as they occur in nature, and any attempt to explain how individuals or species or groups of species do certain things involves the study of the relationships between their physiology and behavior, the nature of their environment, and what they actually do in their natural surroundings. The environment of an animal can be divided into various components—weather, food, other animals and organisms causing disease, a place in which to live (Andrewartha and Birch, 1954)—or into biotic and physical factors (Allee *et al.*, 1949). Whichever approach is used, the solution of an ecological problem depends on an explanation of how the environment exerts its influence on the animal. Furthermore, what an animal does in nature is the result of many interacting factors, and experience can only decide which of these are the most important and therefore the ones to study first.

Chemical ecology is one facet of the whole problem; physical ecology is just as important; in fact, the two are inevitably interlocked. We consider, however, that in spite of this limitation, it is worthwhile to consider chemical ecology on its own because there are many situations in which the most useful questions involve a chemical factor.

II. Acanthocephala

A. The Influence of Chemical Factors in the Environment Outside the Host

The Acanthocephala are exclusively endoparasitic, requiring "an arthropod as an intermediate host and a vertebrate as a definitive host" (Harms, 1965). Only the egg is exposed to the external environment. These eggs, which are produced in large numbers (Kates, 1944), only hatch in the gut of the invertebrate intermediate host, and chemical factors in the environment outside the host do not appear to influence hatching.

The eggs of some species, however, are more resistant to desiccation than those of others and these differences may be related to the ecology of the hosts. For instance, the eggs of *Macracanthorhynchus hirudinaceus* and *Moniliformis dubius* (Manter, 1928; Kates, 1942; Moore, 1942; Edmonds, 1966) which hatch in terrestrial insects and whose definitive hosts are terrestrial mammals, are much more resistant to desiccation than those of *Polymorphus minutus* (Hynes and Nicholas, 1957) which hatch in the shrimp *Gammarus* and whose definitive hosts are water birds.

Chemical factors in the environment outside the host may influence the duration of viability of an egg but information on this topic is lacking.

B. The Influence of Chemical Factors in the Environment Inside the Host

The eggs are swallowed by the intermediate host and the embryo or acanthor hatches in its gut. This process begins a few minutes after ingestion in *Polymorphus minutus* (Hynes and Nicholas, 1957) but takes several hours in *Moniliformis dubius* (Edmonds, 1966).

Until recently (Edmonds, 1966) all attempts to induce hatching *in vitro* by chemical means have been unsuccessful. Hynes and Nicholas (1957) attempted to induce hatching of eggs of *Polymorphus minutus* in mashed-up portions of *Gammarus* gut, while Harms (1965) tried to induce hatching of eggs of *Octospinifer macilentis* by treatment with various proteolytic enzymes. Edmonds has succeeded in obtaining hatching of embryonated eggs of *Moniliformis dubius in vitro* at rates which correspond with those found *in vivo*. His experiments show that hatching is "triggered" by solutions of certain electrolytes with a molarity greater than 0.2 and a pH greater than 7.5. This effect is enhanced by CO_2 but not by reducing agents. During the process of hatching, small amounts of a chitinase are released by the embryo. The stimulus appears to be ionic rather than osmotic because 0.1–0.8 *M* sucrose at pH 8.0 does not initiate hatching.

Edmonds attempts to relate his *in vitro* conditions with those existing in the cockroach gut and although he states that "no satisfactory answer to the question can be given at present," he shows that the chemical state of his *in vitro* hatching media is similar in many respects to that which can occur in the host's gut. For instance, although no information is available concerning the concentration of electrolytes in the gut of the cockroach, their concentration in the hemolymph has been determined by a number of workers (Tobias, 1948; Van Asperen and Van Esch, 1956) and the electrolyte concentration can be calculated from "Ringer" solutions used by workers investigating the physiology of the cockroach alimentary tract (Griffiths and Tauber, 1943; Roeder, 1953). In all these experiments the molarity of the electrolytes was between 0.18 and 0.26 *M*. Similarly, the pH levels required for *in vitro* hatching fall into the physiological range (Glaser, 1925; Day and Powning, 1949; Roeder, 1953) found in the insect gut, particularly in the presence of CO_2 when hatching occurs at a pH of 6.0 (Edmonds, 1966). No information is available on the amount of CO_2 in the gut of insects.

It appears that the ingested embryonated egg of *Moniliformis dubius* responds to a number of chemical factors in its new environment and that these are responsible for the "triggering" of the hatching process and may be involved in the movement of the embryo or acanthor through the tissue of the intestinal wall. However, nothing is known about the chemical factors which influence the wandering of the acanthor. In *Moniliformis dubius* this process takes over a week (Moore, 1946) and it is almost 2 weeks from the time of infection before the embryos come to lie in the body cavity and a further 5 weeks before the cystacanths are formed. During this period there is considerable reorganization, the acanthor loses its rostellar hooks and body spines and the characteristic proboscis is formed. This process has been described and illustrated for *Moniliformis dubius* by Moore (1946). Again, nothing is known of the chemical forces which must influence the extensive reorganization and growth which take place and lead to the formation of the cystacanth. At this point further metabolism in response to chemical factors in the arthropod environment appears to cease.

Hynes and Nicholas (1958) have studied the resistance of species of *Gammarus* to infection by *Polymorphus minutus*. They found that it was quite easy to infect shrimps that already contained cystacanths so that there does not seem to be an acquisition of immunity by shrimps. However, failure to develop is common when eggs are fed to a *Gam-*

marus belonging to a species that differs from the one that it originally developed in. The mechanism of this particular type of resistance is unkown, but it is clear that the effects of the invertebrate chemical environment are far from being transient and unspecific. The reaction takes place at an early stage of development within the shrimp. If the parasite survives to develop as far as the acanthella stage it appears to become immune to any further host reaction.

Further development is dependent on the intermediate host being ingested by the definitive host. It is clear from the experiments of Graff and Kitzman (1965) that excretory or secretory products of the definitive host are important in the activation and excystment of the cystacanth. They found that infection with *Moniliformis dubius* did not take place in rats in which the bile ducts had been surgically transplanted to the apex of the cecum and that bile salts were obligatory for the activation of cystacanths *in vitro*. Furthermore, activation was more rapid at high pH (8.5) and was markedly stimulated by treatment with CO_2. Reducing agents, however, did not have any significant effect. Exposure to gastric secretions is not essential for the development of *Moniliformis dubius* as normal development takes place in cystacanths injected directly into the jejunum. The mechanism by which the activation of cystacanths by bile salts is enhanced by CO_2 remains unknown.

The places where acanthocephalans live in the host's intestine are extremely restricted. They are found in the intestine in what Burlingame and Chandler (1941) call the "zone of viability." They are not found in the stomach, presumably because the cystacanth is not activated at the low pH of this environment. Using intact rats Graff and Kitzman (1965) have found that it takes about 1½ hours for larvae of *Moniliformis dubius* to become attached in the proximal 5 inches of the small intestine.

Kates (1944) in his work on *Macracanthorhynchus hirudinaceus* in its definite host, the pig, has studied the localization of these parasites in the gut and has found that practically all specimens occur in the jejunum between 23 and 43 feet from the pylorus. These worms are attached to the intestinal wall by means of their proboscis hooks and nodules are formed in these areas. The nodules appear to be resorbed after the parasite has become detached since they can no longer be detected macroscopically (Kates, 1944). Whether or not these nodules, which appear to be made of necrotic tissue (Weinberg and Romanovitch, 1907), are induced by some exudation from the proboscis or are merely a response to mechanical stimuli is not known. Kates (1944) found that there were more nodules than worms and suggested that these parasites were able to change their position in the small intestine. This new attachment was always further down the intestine in the direction of peristalsis.

Whether or not the Acanthocephala evoke an immune response in the definitive host has not been established in many forms. However, in *Moniliformis dubius* (Burlingame and Chandler, 1940, 1941) it has been shown that generally "the susceptibility of rats to secondary infection was proportional to the susceptibility to primary infections." These workers concluded from their experiments that competition for feeding sites in the intestinal environment led to worms of the superimposed infection "being crowded out of the more favorable regions" and forced to develop in regions which have an adverse effect even in primary infections. This inhibition of secondary infection was not considered to be an immune response on the part of the host. A somewhat similar state of affairs has been shown to occur with *Polymorphus minutus* in the duck (Nicholas and Hynes, 1958) where worms of a superimposed infection will develop normally together with those of a previous infection, although the later infection tends to be inhibited "perhaps because of competition for suitable sites of attachment."

Thus from the evidence that has accumulated from studies of *Moniliformis dubius* and *Polymorphus minutus* it appears that resistance is of an ecological rather than an immunological type.

An example of direct competition between two species with a mutual habitat has been described by Holmes (1961), who determined the effects of crowded single species infections and concurrent infections on the cestode *Hymenolepis diminuta* and the acanthocephalan *Moniliformis dubius* and suggested that this interaction might be analogous to a "single generation" adaptation of the hypothetical model of Andrewartha and Birch (1954, pp. 424–427) illustrating one way in which two nonpredators utilizing the same resources reduce the area of each others' distribution with respect to some other component of environment.

C. Summary

The chemical environment of Acanthocephala may be divided into two parts, the external environment and the host environment.

The only part of the life cycle that is exposed to the first type of environment is the egg. Knowledge of the eggs' response to a natural chemical environment is limited to a few observations on viability which do no more than point to the need for further research on this topic.

The host environment can be divided into that for the intermediate host and that for the definitive host. As has been mentioned, the former is always the gut and body cavity of an arthropod and the latter the gut of a vertebrate. Some information on the chemical nature of both of these environments is available and chemical substances which are known to be part of these environments have been shown to be important both in the hatching of the egg and of the cystacanth in *Moniliformis dubius,* namely, solutions of electrolytes with a molarity greater than 0.2 and a pH greater than 7.5 in the first instance and bile salts with a high pH in the second instance. In both cases hatching is stimulated by CO_2.

There is no evidence in either case of the host developing an immune response to the parasite.

III. Nematoda

A. The Influence of Chemical Factors in the Environment Outside the Host

Nematodes have three basic features that distinguish them from each other: (1) whether they are parasites or not, (2) the kind of hosts they parasitize, (3) the type of habitat they live in. Any differences between species are probably related to these features and in discussing the chemical ecology of nematodes it is our intention to define these differences. Unfortunately there is so little information on the behavior of nematodes in the environment outside the host that a satisfactory assessment is impossible. There is virtually nothing known of the nonparasitic nematodes and information on the ecology of nematodes in environments other than soil is also sparse. However, we can at least define the problem, state the known facts, and speculate on the situation.

1. *Hatching*

The hatching of nematodes' eggs in the soil does not always follow immediately after the embryo has developed into a larva, even when the physical and chemical environment is suitable. In the plant parasites of the genus *Heterodera,* a chemical stimulus in the form of a root secretion is needed to trigger or accelerate the hatching process (Shepherd, 1962a). The eggs are enclosed in a cyst, the body of the female, attached to the root. In fallow soil or under a nonhost crop, the hatch proceeds at a slow rate dependent on physical factors in the soil so that after 1 year the population declines by about 50%. When host plants are grown in the soil, however, the population decreases more rapidly; in *H. rostochiensis* about 20–30% of the eggs hatch in the first 7 weeks after planting potatoes, and up to 70% after 11 weeks (Shepherd, 1962a). The hatching responses of *Heterodera* spp. to different plant species vary widely. *H. rostochiensis,* for example, only responds to a few species in the Solanaceae; *H. schachtii* is stimulated by a wide range of plants, some of which are nonhosts; *H. avenae* has shown no response *in vitro* to diffusates of any host or nonhost; eggs of *H. göttingiana* are not stimulated *in vitro* by host root exudates but in soil there is an increased rate of hatch when pea roots are present (Shepherd, 1963). Although there are obvious differences in the hatching responses of *Heterodera* spp., there may be a basic pattern from which some idea can be obtained of the mechanism of chemical stimulation by root exudates.

It has been suggested (Rogers, 1961) that the free-living phase of parasitic nematodes is a resting stage when the infective larva, although active in the soil, neither feeds nor develops; a stimulus from the host may be necessary to restart the dormant timing mechanism in the parasite. Alternatively, the infective larva may fail to synthesize some essential component of the system and if the host can supply this missing component then development can proceed. Such hypotheses are useful models because they suggest experiments to test them.

The identity of the hatching factor from roots is still unknown in spite of more than 20 years of research by numerous workers. It is probably a monobasic carboxylic acid with an equivalent weight of about 220, a formula approximating $C_{11}H_{16}O_4$ and contains no benzene rings (Shepherd, 1962b). Studies on the hatching of *H. schachtii* in a wide variety of synthetic agents suggest that the hatching response is probably not associated with the availability of a common metabolite or vitamin but with some sort of interaction with the electron transport system, possibly by influencing oxidative phosphorylation or by-passing a rate-limiting step in the electron transport chain (Clarke and Shepherd, 1964). On the other hand, there is evidence to suggest that the factor is consumed during the hatching process as the level of hatch is more closely related to amount of the hatching factor than to concentration (Viglierchio, 1962, 1963).

Whatever the mechanism, it is likely that there is a change in the metabolism of the developing larva. The stimulus may not operate at the time when the larva escapes from the egg shell as the respiration of larvae before and after hatching are similar (Sembdner *et al.*, 1961; Wallace, unpublished). The possibility that the change in metabolism may be associated with the first molt in the egg is worth consideration because the phenomenon would then be similar to the stimulation of molting in *Paratylenchus* spp. by root exudates.

Studies on the hatching of another plant parasitic nematode, *Meloidogyne javanica*, show a similar dependence on the presence of plant roots for increasing the rate of hatch, but the chain of events is quite different (Wallace, 1966a). Eggs of *M. javanica* in water *in vitro* and in sterile soil hatch completely and rapidly. In nonsterile soil the rate of hatch is reduced, however, even when aeration, moisture content, and temperature are optimal. When tomatoes are grown in the soil the rate of hatch again increases. Thus hatching in *M. javanica* is inhibited in nonsterile soil and stimulated in the presence of roots. The eggs of *Meloidogyne* spp. respond to root exudates *in vitro* but the rate of hatch is only a little higher than that in water (Viglierchio and Lownsbery, 1960).

It seems likely that microorganisms are the source of the inhibitor and root exudates the stimulant. It is known that volatile fatty acids, produced directly by bacteria (Johnston, 1957) or during decomposition of plant residues, are more toxic to some plant parasitic nematodes than to nonparasites. Thus decomposing plant residues of rye (*Secale cereale*) and timothy (*Phleum pratense*) produce chemical factors that are toxic to *Meloidogyne incognita* and *Pratylenchus penetrans* but they have no such effect on nonparasitic nematodes. One of the nematocidal components in the residues was identified as butyric acid, which is toxic to plant nematodes (Sayre *et al.*, 1965). Similarly, the metabolic products of the bacterium *Clostridium butyricum* include a mixture of formic, acetic, proprionic, and butyric acids which immobilize the plant nematode *Tylenchorhynchus martini* (Banage and Visser, 1965). It seems probable therefore that nematodes in the soil are constantly exposed to chemicals that are either toxic or inhibitory in some way (Seinhorst, 1950; Luc, 1961). The degree of inhibition might be expected to vary with the density and type of bacteria and plant residues in the soil. The reduced rate of hatch of *H. schachtii* in soils containing high organic content was at first thought to be caused by lack of oxygen (Wallace, 1956a) but it seems more likely that chemical inhibition is the primary cause. What the chemicals inhibit is not known but it is possible that, like hatching factors from roots, it is the metabolism of the nematode that is affected, just as larvae of *M. javanica* and *T. semipenetrans* retain their infectivity longer in soil than in water because activity and consumption of food reserves are inhibited in soil, possibly by some chemical (Van Gundy *et al.*, 1967). Plant roots themselves may produce hatching inhibitors. *Nicotiana* spp. produce a chemical in their root exudates that inhibit hatching without being toxic to hatched larvae (Sembdner and Schreiber, 1960).

In general the hatching process in plant nematodes is probably governed by the combined effect of inhibitors and stimuli. Whether nonparasitic and animal-parasitic nematodes are so affected is not yet known; it is possible that their hatch is controlled by internal hormonal regulation associated with development (Rogers, 1961). Inhibition of hatching of plant nematodes appears to vary in its persistence. In some species like *M. javanica* and, to a lesser extent *H. schachtii,* inhibition soon disappears because the eggs when removed from soil hatch readily in water. In fact, the level of water hatch may indicate the degree of persistence of inhibition. While there is little doubt that the hatching stimulus is in root exudates, it is possible that microorganisms in the rhizosphere stimulate hatching. Thus the metabolites of some fungi in soil appear to have such an influence on the eggs of *H. rostochiensis* (Giebel, 1963) and microorganisms may be responsible for high hatching rate in some soil leachates (Ellenby, 1963).

This interpretation of events may soon be out of date of course as new evidence accumulates, but at least it raises some useful questions: (1) Are nonparasitic and animal-parasitic nematodes influenced by stimuli from root exudates or by microbial inhibitors? (2) What is the source, and identity of the chemicals that inhibit hatch? (3) Do microorganisms stimulate hatch? More information is needed on the hatching responses of nematodes in soil with and without plant roots and observations on species other than *Heterodera* and *Meloidogyne* spp. are lacking. And when the hatching factors have been identified there is great scope for the biochemist to study the nature of stimulation.

As mentioned previously, the hatching of nematode eggs proceeds in soil in the absence of a hatching stimulus from the root although it is slow and protracted. In *H. rostochiensis,* for example, as many as 50% of the eggs may hatch over a growing season in fallow soil. The significance of the root hatching stimulus lies in the production of a high rate of hatch over a period when roots are growing near the cyst (Winslow, 1955; Wallace, 1959). The rate of hatch in the soil, whether influenced by root exudates or not, is influenced by other factors in the soil of which moisture stress (Dropkin *et al.,* 1958; Wallace, 1966b), temperature, and aeration (Shepherd, 1962a) are particularly important. There are two chemical factors which also exert a major influence: toxic principles released from growing roots and the gaseous composition of the soil atmosphere. Some plants such as *Tagetes* produce chemicals (Uhlenbroeck and Bijloo, 1958, 1959, 1960) that reduce numbers in such species as *Pratylenchus* and *Tylenchorhynchus* (Oostenbrink *et al.,* 1957). The toxin does not affect all species of plant nematodes and which particular stage of the nematode is most susceptible is not known. It has only a slight effect on populations of *H. rostochiensis* for example, and there is no toxic effect on the eggs (Omidvar, 1961). However, there is little doubt that such plants have an effective mechanism for resisting nematode attack. Some varieties of asparagus also release a glycoside which is toxic to a variety of plant nematode species (Rohde and Jenkins, 1958) and it is likely that there are many other species of plant with these properties, in particular wild plants growing in areas that have not been disturbed by agriculture.

The gaseous composition of the soil atmosphere has a profound effect on hatch because nematodes are very responsive to lack of oxygen (see reviews by Rogers, 1961; Rogers and Sommerville, 1963; Van Gundy, 1965; Wallace, 1961a, 1963). The number of galls of *M. javanica* on tomato roots is very few when the oxygen falls to 3.5%, probably because of reduced hatching rate and larval infectivity (Van Gundy and Stolzy, 1961). Similarly, the hatching of eggs of *T. semipenetrans* is inhibited when the rate of diffusion of oxygen is reduced to a low level (Van Gundy *et al.*, 1962). Eggs of *Caenorhabditis briggsae* survive several days under anaerobic conditions but only hatch when aerated (Nicholas and Jantunen, 1964). It is sometimes difficult to assess whether inhibition of hatching in badly aerated soils is caused by lack of oxygen or an excess of CO_2 (Gillard *et al.*, 1958); information is needed on the effects of various proportions of these gases because CO_2 may not be just an inhibitor (Onions, 1955; Wallace, 1959), at low concentrations it may be stimulatory. The fact that nematodes are attracted by CO_2 (Klingler, 1959), that CO_2 is the stimulus in exsheathing of some animal parasitic nematodes (Rogers and Sommerville, 1963) and a common fluctuating constituent of the soil and rhizosphere make it a good candidate for further study.

The rate of hatch of nematodes may be influenced by the sort of ions in the soil water. Chlorides of monovalent ions appear to inhibit hatch of *H. rostochiensis* while chlorides of divalent ions are stimulatory (Ellenby and Gilbert, 1958); zinc and cadmium salts also depress hatch (Robinson, 1956). The proportions and concentrations of cations in solution may influence the rate of hatching, possibly by modifying the effect of the hatching factors in root exudates (Robinson and Neal, 1959). Studies on the hatch of nematode eggs at different pH (Ellenby, 1946; Fenwick, 1951; Robinson and Neal, 1956; Loewenberg *et al.*, 1960; Wallace, 1966a) are somewhat contradictory but it can be concluded that while there is an optimum pH at about 6 to 7 in some species there is also a high degree of tolerance over a wide range. This tolerance is also reflected in studies on the effect of pH on survival (Van Gundy and Thomason, 1962; Millan, 1962), infection (Godfrey and Hagan, 1933; Johnson, 1962), and reproduction (Van Gundy and Martin, 1961). So it is likely that pH is of only secondary importance in soil. There is evidence that some species have a maximum rate of hatch at osmotic potentials of about 0.2 atm. (Wallace, 1956b) and that they are strongly inhibited at about 10 atm. In animal feces the concentration of electrolytes may reach a level where hatch is inhibited and so reduce the numbers of larvae entering into a drying environment (Wilson, 1958). In general, nematodes tolerate wide changes in osmotic potential (Stephenson, 1947; Osche, 1952; Kirkpatrick and Van Gundy, 1966).

2. *Molting*

Apart from studies on the plant nematode *Paratylenchus* there is little information on the factors influencing molting in the environment outside the plant or on whether an external stimulus is needed at all.

Paratylenchus spp. are ectoparasites of plants living in the soil for their whole life history. As with other nematodes, there are four molts, four larval and one adult stage. The molt of the fourth preadult stage is stimulated by root secretions so there is an interruption in the life cycle unless a host root is present (Rhoades and Linford, 1959, 1961a; Fisher, 1966). There are similarities here with hatching described in the previous section in that both eggs and fourth-stage larvae are quiescent and require stimulation by root exudates. This response may conserve food reserves until a stimulus is received from the host root that changes the metabolism of the larva in such a way that molting can proceed. Molting in some ectoparasitic plant nematodes may be inhibited by such environmental factors as low oxygen concentration (Van Gundy and Stolzy, 1963) and possibly by chemicals released from microorganisms.

3. *Motility and Orientation*

After hatching, nematodes move about in the soil along pathways that are probably determined by a complex of stimuli. Gradients of moisture and temperature influence movement (Wallace, 1961b) and a variety of chemical gradients arising from numerous sources have an equally important effect (A. F. Bird, 1959; Oteifa and Elgindi, 1961). To analyze the factors that cause a nematode to move in a given direction in soil would be an almost impossible task unless there was detailed information on the stimuli present. However, if the stimuli are controlled under experimental conditions, then it is possible to determine the kinds and relative strengths of stimuli. Larvae of *H. rostochiensis,* for example, move to the wet end of a moisture gradient but this directed movement may be counteracted by plant roots placed at the dry end of the gradient (Wallace, 1960). The point made here is that lack of a directional response may be caused by the absence of some potent stimulus or by the interaction of numerous stimuli acting in different directions. In fact, the expression "random movement" so often used in nematode literature may be merely a description of "events in a system so complex that the number of interrelated variables exceeds the practical possibility of resolution" (Beament, 1960). The evidence that nematodes show directional movement toward roots has been reviewed by Klingler (1965) but there is little information on how they do this and there appears to be little specificity between host and parasite (Shibuya, 1952; Shepherd, 1959). It is sometimes assumed that the paired amphids at the anterior end of nematodes are sense organs and their structure certainly suggests that they have this function. But whether the nematode moves toward a root by detecting differences in concentration on each side of the head, increasing its rate of turning or changing its speed in response to concentration is not known. *Ditylenchus dipsaci* increases speed and rate of turning in the rhizosphere and so tends to aggregate there (Blake, 1962). In an electrical gradient *D. dipsaci* tends to move to the cathode by a combination of increased turning and direct movement (Jones, 1960).

It is possible that true taxes occur, for the bending waves formed during movement may enable the nematode to sample the environment at the extremes of the amplitude of each wave (Croll, 1967). Such behavior would overcome the objection that there is little chance that a nematode could detect differences in intensity of a stimulus over the very small distance between its bilateral sense organs. Croll's experiments with two nonparasitic species, *Panagrellus redivivus* and *Chromadorina viridis,* suggest that orientation is by a tropotaxis in which there is simultaneous comparison of intensities on either side of the nematode rather than by a klinotaxis where each side of the organism is alternately stimulated and comparison of intensities is successive in time.

The fact that nematodes aggregate around a root tip gives no information on orientation; they might have arrived there by chance and then stayed at the root because of an attractant in the rhizosphere or at the root surface or because they were trapped there (Sandstedt and Schuster, 1962). The same nematodes on an agar surface will collect in a drop of water because once in the drop they cannot escape; this is aggregation, not orientation. However, these factors have been taken into account in experiments demonstrating the orientation of *Meloidogyne javanica* to clover roots (A. F. Bird, 1962b).

In an environment like soil, root secretions diffuse out into the soil water, forming concentration gradients whose range will depend on the rate of diffusion of the chemical, its rate of breakdown, and the rate at which it is produced. All three of these factors will vary enormously with changes in soil moisture, temperature, biological activity, growth of the roots, etc. Furthermore, the tortuosity of the channels in soil, drainage of water, and root distribution and density will have a big effect on the chemical gradients. Under such complex conditions it is doubtful whether nematodes could have any efficient means of orientation beyond 4 or 5 mm. from the root and anyway it is doubtful whether effective chemical gradients extend for much more than 1 cm. from the root (Klingler, 1963). In view of these facts, it is likely that the mechanism whereby nematodes reach roots is determined by soil conditions and the presence of a distinct gradient. Under one set of conditions many nematodes may reach roots by chance (Kühn, 1959; Sanstedt *et al.,* 1961) while at the other extreme a high proportion show distinct directional movements (A. F. Bird, 1962b).

The importance of microorganisms in the rhizosphere on the orientation of nematodes cannot be overlooked. While there is good evidence that root exudates do attract nematodes, there is equally good evidence that a variety of other factors is also attractive, including bacteria (Bergman and Van Duuren, 1959a,b; Henderson and Katznelson, 1961).

Chemicals may also repel (Thorne 1951); presumably the mechanisms involved in repulsion are similar to those for attraction except that the nematode moves down the concentration gradient away from the source. Nematodes are repelled by certain bacteria (Bergman and Van Duuren, 1959b) and fungi (Townshend, 1964) as well as roots that have stopped growing (Wieser, 1955, 1956) although this may also be the result of microorganisms that are associated with dying tissues (Wallace, 1966a).

Some chemicals increase the speed of nematodes whereas others slow them down. Larvae of *M. javanica* in the presence of plant roots lose their body contents at a faster rate than larvae in the absence of roots, probably because the nematodes move quicker in the presence of a root stimulus (Van Gundy *et al.*, 1967). The relationship of speed of movement to aggregation at the root is complex. With no orientation increased speed would increase the rate at which nematodes reached the root. On the other hand, a decreased speed near the root surface would tend to increase the rate of aggregation (Rohde, 1960).

4. Physiology

Whereas some chemicals trigger and direct the nematodes' activities there are other chemicals that determine the rate of the various activities. Root exudates, for example, may stimulate an egg to hatch but only if the concentration of oxygen around the egg is sufficiently high.

Aeration is particularly important for nematodes as most species appear to be inactive in its absence and their activity increases as oxygen tension increases (Rogers, 1961; Rogers and Sommerville, 1963; Wallace, 1963; Van Gundy, 1965). There are differences between species, however, in the level of oxygen concentration at which respiration is affected.

It is not possible to relate such differences to habitat or type of life cycle although it might be expected that nematodes living in the muddy bed of a stagnant pool would be able to respire at lower oxygen tensions than nematodes living in running streams. Larvae of *Meloidogyne* sp. cannot survive in soil flooded for longer than 6 months whereas there are several species that can live for as long as 2 years (G. W. Bird and Jenkins, 1965). On the other hand, adaptation to a particular environment may be related to ability to survive low oxygen conditions (Wieser and Kanwisher, 1961). In the soil, oxygen and CO_2 ratios vary widely with depth, season, and soil moisture. After heavy rains some soils may be waterlogged and the available oxygen would be very low. Following drainage and entry of air into the soil pores, however, the nematode would be in an environment of increased oxygen tension. So nematodes in soil probably fluctuate between periods of quiescence and activity determined by fluctuations in soil aeration. Whether a species can persist in such an environment then depends on its ability to survive the anaerobic phases.

Fluctuations in the ratios of oxygen and CO_2 in the environment again raise the question of whether the inactivity of nematodes is depressed by lack of oxygen or excess CO_2. The relative importance of CO_2 is not known although it is an important factor in several nematode activities. It is possible that there is an optimum CO_2 concentration for nematodes and that CO_2 has other influences than just depressing respiration. For example, the infective larvae of some animal parasitic nematodes tend to adopt an erect position on the surface of soil or herbage and wave about in the air. This behavior is apparently elicited by CO_2, which must reach a certain concentration before waving occurs (Sasa *et al.*, 1960).

The relationship between the nutrition of nematodes and chemicals in the environment is difficult to discuss because there is no evidence to suggest that nonparasitic and animal-parasitic nematodes feed on anything but living matter—bacteria, fungi etc.—and the protein of living organisms may even be essential (Hirschmann, 1952). Work on the *in vitro* culture of these nematodes permits some of the essential nutritional factors to be stated (Rothstein and Cook, 1966) but these are not specific enough to contribute much to an ecological discussion. Whether plant parasitic nematodes feed at all outside the host is open to question although larvae of *H. rostochiensis* may absorb amino acids and glucose exuded from the plant roots (Sembdner *et al.*, 1961). Diffusion of ions into the nematode may also influence survival of some other activity such as hatching (Ellenby and Gilbert, 1958), growth, motility, or reproduction (Dolliver *et al.*, 1962).

The variety of names given to the nonparasitic nematodes, free-living, saprozooic, saprophytic, microphagous etc., emphasizes how little is known of their feeding habits or of their chemical requirements. When nematodes can grow and develop in a chemically defined medium *in vitro* it may be possible to relate the distribution and numbers of different species in soil to the food available there and in fact to assess whether food quantitatively or qualitatively is ever in such short supply as to be a limiting factor. For similar reasons the influence of nutrition on the sex ratio of nonparasitic nematodes is little understood although observations on parasitic species suggest that the sex of the nematode may be reversed by lack of food or by some change in quantity or quality of essential factors (Oteifa, 1953; A. F. Bird, 1960; Kämpfe and Kerstan, 1964).

When a soil or similar medium such as feces dries out by evaporation, the concentration of salts increases. The subsequent increase in osmotic pressure tends to remove water from the nematode which may resist by virtue of an impermeable cuticle or by some active means of water retention. Ability to resist desiccation varies between species but is most obvious in those which have prolonged exposure to dry conditions during their life cycle. In an environment such as soil there are two factors that tend to desiccate the nematode, suction and osmotic pressure (Collis-George and Blake, 1959). Of the two, osmotic pressure is the least important for it seldom rises above 10 atm. in soil and nematodes can survive such conditions (Machmer, 1958; Blake, 1961; Wallace, 1966b). In feces (Osche, 1952) and in arid soils on the other hand, the osmotic pressure of the solution may be much higher but even here nematodes can survive the fluctuations.

Salinity becomes an important factor where marine and freshwater habitats meet. Differences in tolerance determine the concentration and distribution of nematodes in river estuaries (Capstick, 1959) and even species in the same genus show marked differences in their preferred salinity. *Panagrolaimus rigidus* occurs commonly in soils and has a much lower tolerance to high salinities than *Panagrolaimus salinus*, an inhabitant of salt marshes (Everard, 1960). Whether the limiting factor in these situations is osmotic pressure or toxicity of particular chemicals in the salt water is not known.

5. *Infection*

For present purposes infection may be defined as the passage of the parasite from the external environment into the host. Hatching, orientation, movement, and penetration can therefore be considered as a sequence of phases in the infection process. Although these aspects have already been discussed, it is useful to consider them together to see if any relationships exist.

The influence of chemicals on plant parasitic nematodes outside the host appears to aid survival by increasing the chance of infection. Chemicals do this in two ways, by inhibiting activity at a time when the environment is unsuitable for infection and by stimulating the dormant nematode when it is suitable. Examples of inhibition by high osmotic pressure and microorganisms and stimulation by root exudates have already been mentioned and this is the most important feature in the chemical ecology of nematodes outside the host. A brief description of the influence of chemicals on infectivity will emphasize this point. As soon as the eggs are laid they are exposed to the environment outside the host. Embryonic development proceeds but chemicals in the soil cause the first interruption in the life cycle by inhibiting hatching. As soon as a plant grows in the vicinity of the egg, chemicals exuded from the roots stimulate the larvae in some way so that hatching occurs providing there are no inadequacies in aeration, osmotic pressure, pH, temperature, etc. Moreover, the general environmental requirements for rapid hatching are similar to those for rapid movement so the larva tends to hatch into an environment conducive for rapid dispersal. The range of activity of root exudates is of the order of a centimeter so stimulation of hatching occurs not just in the presence of roots but when roots are near, i.e., from about 10 to 20 times the nematodes' own length. The chemical gradient from root to egg may now be considered as an extension of the root itself for the larva orientates in the direction of the root and moves along a track which in total length is shorter than that taken by a nematode moving without a directional response. Furthermore, some of the chemical components of the root exudate may cause the nematode to move faster. The general effect of all these events is to reduce the time spent by the larva in moving about in the environment outside the root and so increase the chance of survival and infection because the nematode has a limited supply of energy reserves in the form of fat and protein which it cannot replenish outside the host; if these are used up before the root is reached there is less chance of the nematode penetrating the host root (Wallace, 1966a; Van Gundy *et al.*, 1967).

In ectoparasitic plant nematodes like *Paratylenchus* spp., the interruption of the life cycle at the final molt and its subsequent stimulation by root exudates may also be an adaptation for increased survival but only if the fourth-stage larva has attributes that confer greater advantage than the other stages. It is not clear what these attributes are but it is possibly an increased resistance to desiccation or low temperatures or an ability to conserve its energy reserves during times when the environment does not allow efficient movement.

It is perhaps significant that the production and diffusion of root exudates in soil reach a maximum in conditions that are also conducive for rapid hatch and larval movement (Shepherd and Wallace, 1959).

There are clearly close relationships between plant nematode, host, and environment and the link between any two is often of a chemical nature. Microorganisms may play a more intimate role in the life of nematodes than has hitherto been thought. Chemicals excreted by the nematode could provide a substrate around the body for bacterial multiplication (Rothstein, 1965), in fact there would be a similar ecological situation as in the rhizosphere except that the source of chemical nutrients is a nematode instead of a root. As well as providing a source of food such as essential amino acids, the excreted products of bacteria could influence many aspects of nematode behavior. The possible importance of the intestinal fauna of nematodes, particularly those without stylets, cannot be excluded either, so while it is sometimes proper to study the physiology and behavior of nematodes under sterile conditions, we should not delude ourselves that by removing bacteria we are necessarily measuring the nematodes' natural functions.

Chemical stimuli are probably of great importance in the nonparasitic nematodes. Opposite sexes of *Panagrolaimus rigidus* are attracted to each other by a chemical that can diffuse through a dialysis membrane (Greet, 1964). *Rhabditis oxycerca* aggregates around several species of fungi and actinomycetes (Katznelson and Henderson, 1962). The fungus-feeding nematodes *Aphelenchus avenae* and *Bursaphelenchus fungivorus* are attracted to fungus species (Townshend, 1964). In view of the high concentrations of such soil organisms in the rhizosphere it is probable that nematodes are attracted there not just by root exudates but by the excretions of microorganisms (Katznelson and Henderson, 1962). Ammonia, for example, is produced by ammonifying bacteria in the rhizosphere and strongly attracts *Rhabditis oxycerca* (Katznelson and Henderson, 1963). Although there is little information on the influence of chemicals on the free-living stages of animal parasitic nematodes, it seems likely that they closely resemble the nonparasitic nematodes in their behavior and requirements in an environment like soil or feces. Both groups of nematodes feed on bacteria and might therefore be expected to respond to bacterial excretory substances. Having reached the infective stage, the larvae of animal parasitic nematodes do not feed and the next stage in the life cycle is migration to the host's environment, usually above soil level, and subsequent invasion of the host.

There is little evidence of a chemical relationship between the animal host and the infective stage of its nematode parasites. The reasons for this appear obvious; the animal host does not always share the same environment as its parasite for any length of time and when they do occur together in such an environment as soil the animal host is usually much more mobile than the parasite so that a chemical link between them cannot be established. Whether animal nemtodes show directional movements to a static animal host such as a dormant insect larva or a pupa deserves consideration.

B. The Influence of Chemical Factors in the Environment Inside the Host

The Nematoda, unlike the Acanthocephala, have established themselves in various organs of many different species of animals as well as in the leaves, stems, and roots of plants. It is clear that to attempt to consider the chemical factors in all these structures would be impossible within the confines of this chapter. Instead we will restrict discussion to those few parasites which have been studied in some detail either because of their availability or for their economic importance.

1. The Host Environment

Animals may become infected passively by swallowing eggs or larvae during feeding or, as in the case of all plants, by active penetration into their tissues. In all plant parasites and some animal parasites this is achieved with the aid of a buccal stylet. There are, however, a number of tissue-penetrating nematodes which do not have stylets and rely on mucular activity and the secretion of enzymes for their penetrating ability (Lewert, 1958). These forms, which include *Strongyloides, Ancylostoma,* and *Nippostrongylus,* have been shown to contain enzymes which cause alterations to the glycoprotein connective tissue barriers of the host as they penetrate.

It seems highly probable that the stylet-bearing forms also secrete enzymes as they migrate through tissues. However, as yet, there is little experimental evidence to demonstrate this, beyond reports of observations that have been made on the flow of secretions from the dorsal esophageal gland during stylet activity (Linford, 1937, 1942; Linford and Oliveira, 1937; Rhoades and Linford, 1961b; Anderson, 1964; Poinar and Doncaster, 1965).

Information on the chemical environment of animal hosts has been summarized by Smyth (1962). It is clear from his observations that it is a much more hazardous environment than that found in the tissues of plants so it is not surprising that we do not find host-parasite relationships in animals resembling the complex giant cell or syncytial systems found, for instance, in the sedentary endoparasitic members of the family Heteroderidae.

The vertebrate alimentary canal serves as an environment for many species of parasitic nematodes. It undergoes regular changes related to the feeding habits of the host; its pH ranges from 1.5 to 8.4; it contains a variety of enzymes; it is almost devoid of oxygen and is churned by peristaltic movement. However, it has proved to be extremely difficult to obtain precise physiological measurements of gas tensions and oxidation-reduction potentials at the mucosal surface of the gut where the majority of intestinal nematodes is found. The lumen of the gut is essentially anaerobic but oxygen tensions close to the mucosa are greater (Rogers, 1949) and it is possible that conditions here may resemble those of the intercellular spaces of the host more than the intestinal lumen (Smyth, 1962). Because of this it is referred to as the paramucosal lumen (Read, 1950). CO_2 is found throughout the gut and the contents of the intestine are often saturated with it. Its role as a mechanism for stimulating physiological activity in both eggs and larvae of a number of intestinal nematodes has been well established (Rogers, 1960; Fairbairn, 1961). However, it appears that in some intestinal nematodes it may not have this stimulating activity; for instance, it does not appear to influence the exsheathing of infective larvae of *Nematodirus battus* (Christie and Charleston, 1965) or the hatching of eggs of *Aspiculuris tetraptera* (Anya, 1966).

The variation in hydrogen ion concentration in the alimentary tract of vertebrates, particularly mammals, is well known. The pH of the mouth is close to neutral but becomes strongly acid in the stomach. In this region in sheep the pH is 1.05–3.6 while in cattle it is 2.0–4.1 (Smyth, 1962). In the duodenum the pH is close to neutral and becomes more alkaline in the large intestine. The chemicals which are released into its alimentary tract by the host include hydrochloric acid, which apparently acts as an exsheathing stimulus in *Nematodirus battus* (Christie and Charleston, 1965), numerous enzymes which hydrolyze carbohydrate, fat, and protein, and bile which itself contains a complexity of chemicals. The influence that many of these chemicals have on nematode parasites is not clearly understood because of the technical difficulties associated with obtaining measurements *in vivo* which are truly physiological.

The majority of intestinal parasites are found in the duodenum, ileum, cecum, and large intestine. However, the ecological distribution of individual species and the manner in which this distribution is influenced by other species living in the same region of the intestine has been studied by only a few workers.

The nematodes *Trichostrongylus vitrinus* and *Trichostrongylus colubriformis* are found in their greatest numbers in the first few feet of the small intestine, both occurring in the same part of the gut and neither apparently influencing each other's abundance or distribution. Thus it is possible for closely related nonpredatory species to live crowded together in a restricted space (Andrewartha and Birch, 1954). This does not apply, however, to *Trichostrongylus axei* whose presence will greatly reduce the numbers of either *Haemonchus contortus* or *Ostertagia circumcincta* in the abomasum of sheep (Turner *et al.*, 1962). The linear distribution of nematodes along the small intestine in three different host species has been examined by Sommerville (1963) whose observations indicate that in these hosts the order of succession of species along the intestine appears to be the same and that there is a normal frequency distribution of species even in an unusual host. These findings support those of Tetley (1937) who, in addition, observed that virile adult populations did not migrate away from their area of infection.

An examination of the population ecology of a nematode species has been made by Schad (1963) using the tortoise *Testudo graeca* which contains at least ten species of oxyurids belonging to the genus *Tachygonetria*. Both linear and radial distribution within the colons of six tortoises were studied for eight species of the nematode. Linear overlaps were greatly decreased by differences in radial distribution. However, several broad interspecific distributional overlaps remained which may be accounted for by differences in feeding habit and interspecific seasonal variations in abundance depending on season and age of host. The effect of host response on the distribution of nematodes in the gut will be considered later in this chapter.

Other components of the animal host environment whose chemical composition undergoes fluctuations and to which nematode parasites must adapt themselves for at least part of some life cycles, are blood and muscles. The chemical composition of blood, which constitutes the environment for forms such as microfilarias, undergoes changes in amino acid and glucose composition after the host has fed. Also, oxygen tensions vary considerably in arterial and venous blood. However, both pH and CO_2 content remain stable in blood whereas in muscles, in which forms such as *Trichinella spiralis* encyst, there is considerable fluctuation in pH and CO_2 content due to lactic acid production. This also influences osmotic pressure and oxygen consumption.

Nematodes parasitize many different species of animals. The evolution of their host-parasite relationships in vertebrate animals has been reviewed by Cameron (1964).

The host environment of plants is broadly restricted to two types of tissue—stems and leaves on the one hand and roots on the other. Plant parasitic nematodes may be either ectoparasites which feed on the surface or endoparasites which enter the host tissues. Changes in the chemical environment of the host will naturally have the greatest influence on endoparasitic forms.

There are three principal genera which attack the aerial parts of plants, namely *Aphelenchoides, Ditylenchus,* and *Anguina.* All other major host-parasite relationships that have been described involve interactions between nematodes and plant roots.

It has been observed (Linford, 1941; Powell and Moore, 1961; A. F. Bird, 1962a) that an endoparasitic genus such as *Meloidogyne,* which has a wide host range, is capable of developing just as readily in stems and leaves as it does in roots although it never does this under natural conditions so that the chemical environment of the host in stem and leaves cannot be very different from that in roots.

Work on the physiology of plant roots has been reviewed by Burström (1965). Knowledge of this subject is limited due to the inaccessibility of roots so that experiments are often confined to the laboratory where many of the facets of root ecology cannot be duplicated and so cannot be accepted as valid for what normally takes place under natural conditions.

Roots depend upon leaves for their supply of carbohydrates and hormonal compounds but take up nitrogen and synthesize protein themselves. Although fluctuations in carbohydrate content occur, all the evidence indicates that under normal conditions the amount of carbohydrate in roots is more than adequate for metabolic purposes. Fibrous roots contain more protein nitrogen than leaves and can be regarded as substrates with a higher protein to carbohydrate ratio than other plant parts. Furthermore, roots contain protochlorophyll and if illuminated are capable of developing chloroplasts and photosynthesizing.

The gas phase of root intercellular spaces is at atmospheric pressure containing 78% nitrogen but more CO_2 and less oxygen than air. Fadeel (1963) has measured the amount of CO_2 in fibrous roots and found 8% in the active zone behind the root tip and 5% in the mature part. leaving 13–16% oxygen in the intercellular spaces which connect through the stem and the stomata and hence with the free atmosphere. Thus even in waterlogged or poorly aerated soils plant roots offer endoparasitic nematodes an aerobic environment and what appears to be a more important factor in the chemical environment of nematode parasites, namely a high CO_2 content.

Plant cell sap is buffered to a pH value of 6–6.5 and appears to be more or less independent of external changes in pH (Burström, 1945).

Areas of rapid growth such as are found in meristematic tissues offer a rich chemical environment lacking in mechanical barriers, so that it is not surprising that the endoparasitic nematodes of plants are attracted to these areas and by their influence maintain this type of chemical environment. This results in a host-parasite relationship which is more complex than that found in zooparasitic nematodes.

2. *Host-Parasite Relationships*

When a parasitic nematode enters the tissues of a nonresistant host the host responds by providing the parasite with an environment which permits it to feed, grow, and reproduce. However, in nature, larvae and eggs are often ingested by animal hosts that are resistant and larvae often enter roots of similarly resistant plant hosts and in each case, of course, development of the parasite is retarded or does not take place.

The chemical factors which influence the parasite's environment are perhaps best considered first from the point of view of the host's response to the parasite and second from that of the parasite to the host.

a. The Response of the Host to the Parasite. Both plant and animal hosts respond to parasites by means of defense mechanisms (Horsfall and Dimond, 1959; Smyth, 1962; Sprent, 1963). These may be tissue reactions which are usually localized at the site of parasite invasion or various types of resistance such as natural resistance which in turn can be split into either physiological resistance or behavioral resistance (Read, 1958). Another type of resistance which is restricted to animal hosts is acquired resistance or immunity, which is the result of the host reacting to chemicals, usually proteins with a molecular weight in excess of 10,000 produced by the nematode and known as antigens. Little is known about nematode antigens (Kent, 1963) but it would appear (Silverman, 1965) that they are released during the growth of the parasite as it penetrates, molts, and matures in the host. Their presence at oral and excretory pore orifices has been demonstrated, using fluorescent antibody techniques, in nematode parasites of both animals and plants (Jackson, 1960; Taffs and Voller, 1963; A. F. Bird, 1964). Antigens stimulate antibody formation and this appears to take place in reticuloendothelial cells, particularly in the lymph nodes. The method by which antibodies are formed is controversial and beyond the scope of this chapter. They are usually found in the gamma-globulin fraction of serum and their most characteristic feature is their ability to combine with the homologous antigen and render it less toxic to the host.

The first example of acquired resistance to nematodes by an animal host under natural conditions was that described by Stoll (1928) for

Haemonchus contortus in sheep. Subsequently, it has been suggested by Soulsby *et al.* (1959) that the antigen initiating this self-cure reaction is the exsheathing fluid that is produced at the third ecdysis.

There is now some evidence (Connan, 1966) that developing resistance in the host can influence the distribution of a nematode population along the gut.

An aspect of natural resistance not found in plants is phagocytosis in which foreign bodies are ingested by large cells called macrophages which are found in great numbers in the spleen but which are capable of migrating throughout the body and increasing in numbers and phagocytic activity in response to antibodies. Behavioral resistance is also only found where animal hosts are involved. For instance, a nematode may have a wide host range and be able to parasitize widely divergent species of hosts but some of these hosts may, because of differences in ecology, never come in contact with the parasite and so exhibit behavioral resistance.

Physiological resistance is found in both plants and animals and an interesting aspect of it, namely the effects of various chemicals such as hormones and growth substances on nematode growth and development, will be discussed later in this chapter.

At the cell level the host's chemical defenses respond to invasion by parasitic nematodes in a variety of ways. In animals (Sprent, 1963) an inflammatory reaction takes place which involves a flow of fluid or exudation and the arrival of phagocytes. If it is a first infection the larva will usually leave the inflammatory reaction in its wake and move on through the cells to reach the bloodstream. If, however, the host has acquired resistance from a previous infection, the inflammatory exudate will contain antibodies which will block the nematode's orifices, slow it down and permit the phagocytes to engulf it.

In plants (Mountain and Patrick, 1959; Van Gundy and Kirkpatrick, 1964) the cellular defense mechanisms often consist of a necrotic reaction at the infection point and the formation of wound periderm around this lesion. As a result of these host-parasite relationships chemicals may be liberated which cause further damage to the host or inhibit or kill the parasites in resistant hosts.

Sometimes host tissues both in animals and plants react in an exaggerated way to nematode invasion and this is called hypersensitivity. Although this reaction is derived from animal pathology where its mechanism is akin to that of acquired resistance, this term is nevertheless now used by plant pathologists.

The host's responses to chemical exudations from both animal and plant parasitic nematodes at the tissue and organ level are many and varied and have been reviewed extensively (Seinhorst, 1961; Smyth, 1962; Krusberg, 1963; Rogers and Sommerville, 1963; Mountain, 1965).

To give some idea of the changes which can take place when nonresistant hosts are parasitized, we have selected two nematodes, one a plant parasite and the other an animal parasite whose host response has been studied in some detail. These are *Nippostrongylus brasiliensis* (Symons and Fairbairn, 1963) which parasitizes the jejunum of the rat and *Meloidogyne javanica* (A. F. Bird, 1961) which is found in the roots of various species of plants.

In some respects both hosts appear to respond in a similar way to the chemical exudations from their respective parasites. For instance, there is considerable hyperplasia and hypertrophy which leads to swelling of the intestine and galling of the root. There is also a considerable increase in RNA, which suggests an increased protein synthesis.

However, *Nippostrongylus brasiliensis* is eliminated from the intestine within 21 days whereas *Meloidogyne javanica* remains in the root for at least several months, feeding and laying eggs unharassed by an increasing host resistance, and so it is not surprising that there are also considerable differences in the responses of these two hosts. Thus while infected rat intestines contain half the amount of maltase, leucine aminopeptidase and alkaline phosphatase in their brush borders that are found in uninfected rats, and there are changes in the shape of the villi (Symons and Fairbairn, 1963), these are nevertheless still readily recognized as such; whereas cells from the stele of the tomato root (A. F. Bird, 1961) are transformed into a new kind of tissue so that in this case the host tissue reacts to form a type of neoplasm or tumor which is commonly known as a giant cell or syncytium. It differs from normal tumors (A. F. Bird, 1962a) in that it is not capable of unlimited proliferation and requires a continuous stimulus from the nematode; otherwise it will atrophy.

The structure and biochemistry of this particular host-parasite relationship has been investigated more thoroughly than any other nematode host-parasite relationship (A. F. Bird, 1961, 1962a; Owens and Specht, 1964, 1966; Rubinstein and Owens, 1964; Owens and Bottino, 1966; Owens and Rubinstein, 1966) and in view of the fact that many of the initial cell responses, such as enlargement of nuclei and nucleoli and increase in density and protein content of the cytoplasm are characteristic of cell response to many different species of plant parasitic nematodes, a closer look at this particular host response is warranted.

The larvae of *Meloidogyne javanica* do not appear to exude any chemicals such as enzymes or growth substances before they are stimulated by a root (A. F. Bird, 1966). However, it seems highly probable that they do this on their way into the root and at the feeding site, judging from the flow of secretions from the dorsal esophageal gland as mentioned above. However, there is no information available as yet on the chemical nature of these exudations which in the case of the family Heteroderidae, to which *Meloidogyne javanica* belongs, are responsible for the initiation, development, and maintenance of giant cells. Giant cell formation (Fig. 1), which takes place in the vicinity of the nematode's head, involves cell wall breakdown accompanied by the thickening of a surrounding giant cell wall and an increase in the density and area of cytoplasm, followed by nuclear multiplication by synchronous mitosis within a single giant cell.

Several giant cells are formed (Fig. 1) and they surround the head of the nematode and are perforated by its spear during stimulation and feeding. The giant cell wall is thick, has an irregular surface and similar polysaccharide components to normal cell walls although it has more polyhexuronates and less cellulose. The cytoplasm contains mitochondria, Golgi bodies, proplastids, a dense endoplasmic reticulum, and much RNA. The nuclei and nucleoli are large and it has been demonstrated by microautoradiographic techniques that the developing giant cell is a region of intense nucleic acid biosynthesis and contains 88% more RNA and 71% more DNA than normal root tissue. The chemical changes that *Meloidogyne* induces in the tomato root have been examined by Owens and Specht (1966) who compared the chemical components of normal tomato roots with those of galls and found that carbohydrates and pectins decreased by 36% and cellulose and lignin by 31%. All the other chemical components measured increased, namely, hemicelluloses 36%, organic acids 67%, free amino acids 304%, protein 80%, nucleotides 29%, lipids 154%, and minerals, 4%.

The respiration of galls is similar to that of adjacent tissues or uninfected roots (A. F. Bird and Millerd, 1962; Owens and Rubinstein, 1966). Using tracer techniques Owens and Rubinstein (1966) have shown that intermediary metabolism associated with protein and nucleic acid synthesis is accelerated in galls, so that in this particular instance, exudations from a nematode induce protein and nucleic acid synthesis in plant cells which ceases if the nematode is injured or dies (A. F. Bird, 1962a). The chemical nature of the inducers that the nematode injects into the cell is unknown. Their action on the cell appears to be influenced by antimetabolites, for example 5-bromo-2′-deoxycytidine (A. F. Bird and McGuire, 1966), which presumably influences the thymidine incorporation into DNA described by Rubinstein and Owens (1964) in giant cells and leads to inhibition of nematode development (A. F. Bird and McGuire, 1966).

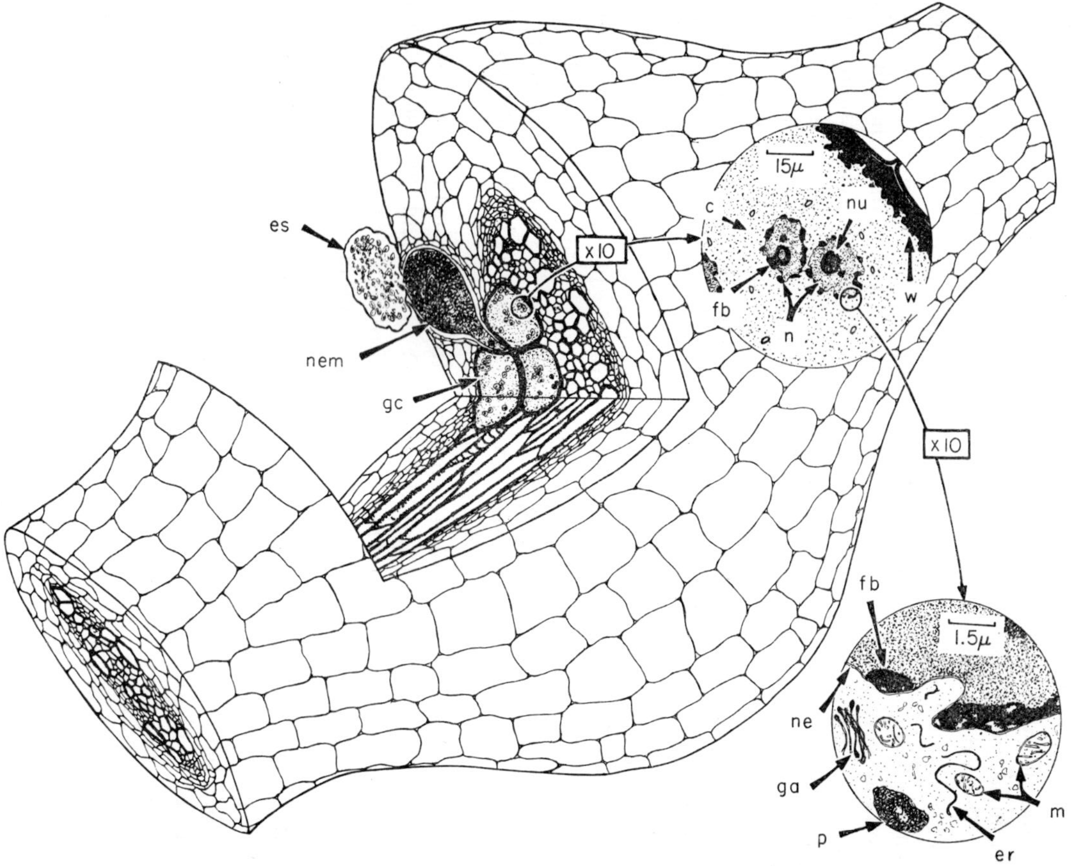

FIG. 1. Semidiagrammatic scale drawing of a single infection in a young tomato root, showing the adult female nematode (nem) with its egg sac (es) protruding out of the gall. Three giant cells (gc) are depicted with thickened cell walls (w) and large irregular nuclei (n) surrounded by nuclear envelopes (ne) and containing Feulgen-positive bodies (fb) and nucleoli (nu). The cytoplasm (c) is dense and contains mitochondria (m), proplastids (p), endoplasmic reticulum (er), and Golgi apparatus (ga). A. F. Bird (1961) reprinted by permission of the Rockefeller University Press from *The Journal of Biophysical and Biochemical Cytology*, Vol. 11, No. 3, pp. 701–715.

Symons and Fairbairn (1963) concluded from their experiments with *Nippostrongylus brasiliensis* in the rat that there was no evidence to support the hypothesis that this nematode secreted antienzymes. The literature on anticoagulants, enzymes, and antienzymes has been reviewed recently (von Brand, 1966). However, information on the host cell response to these various chemicals emitted by nematodes is still sparse.

It is particularly interesting to note here that host production of enzyme inhibitors may not be restricted to animals, as Myuge (1964) has reported that plants parasitized by members of the Heteroderidae produced substances which could inactivate proteases produced by these nematodes. The chemical nature of these particular enzyme inactivators is unknown.

b. The Response of the Parasite to the Host. In many cases, due to the natural resistance of the host, there is no development or growth and the nematode dies, often because of chemicals secreted by the host. In plants death may be due to the release of nematoxic substances from cells surrounding or in the path of the nematode, or in the case of animals it may be caused by the formation of antibodies.

Some forms of natural resistance, particularly in plants, vary in different varieties of the one species. An example of this can be seen in a comparison of the rates of growth of *Meloidogyne javanica* in five different varieties of the peach *Prunus persica* (Burdett *et al.*, 1963). Two of the varieties were nonresistant and the nematodes grew and reproduced normally on them. Two other varieties inhibited the growth of the nematodes and prevented them from reproducing, while the fifth variety was intermediate in its effect. As yet there is very little information available about the biochemistry of this type of host-parasite relationship. This is interesting because the genus *Meloidogyne* has a very wide host range which involves many different families of plants so that these effects must involve metabolic pathways common to many plants. Furthermore, differences in nematode growth rate may be obtained in the same variety of host plants by growing the plant in a medium that is deficient in a particular element (A. F. Bird, 1960) and nutritional stress also has a pronounced effect on the sex ratio (McClure and Viglierchio, 1966).

Some chemicals which are involved in normal physiological processes within the host are capable of influencing the growth of the nematode, for instance progesterone produced by pregnant mice influences the numbers and size of the nematode *Amplicaecum robertsi* (Dobson, 1966) and the synthetic plant growth substance 2,4-dichlorophenoxyacetic acid affects the reproduction of *Aphelenchoides ritzemabosi* (Webster and Lowe, 1966).

The responses of nematode parasites to the chemical composition of the host environment are thus many and varied. The chemical relationships between the parasite and the host are complex and information about them is restricted to work that has been done on a few species only.

C. Summary

As in the Acanthocephala, the Nematoda may exist in either an external environment or a host environment. The difference is, of course, that nematodes often complete their whole life cycle in the external environment. Despite these differences it is clear that there are some chemicals which exert a profound influence on the ecology of nematodes in either of these habitats. Perhaps the most important of these is carbon dioxide which can stimulate or "trigger" hatching and exsheathment in some forms and influence orientation toward the host in others.

The exact role of this chemical on the receptor mechanisms of the nematode is not known as yet and information on the more complex chemicals such as neurosecretory substances, enzymes, and hormones which are thought to be released as a result of chemical stimuli is practically nonexistent. Oxygen is of course important and its effects become pronounced under conditions of oxygen lack.

The influence of chemicals released by microorganisms associated with nematodes will be a fruitful field for research as it may contribute to a better understanding of the behavior, distribution, and physiology of nematodes. We consider, in fact, that microorganisms are an integral part of the nematode in its natural environment and to study the nematode in isolation may not always be a valid approach.

Most chemicals which may influence the ecology of nematodes exert their influences together with many others and it is almost impossible to ascertain the individual physiological role that these substances play in normal environments. Data of any kind on factors in the microenvironment around the parasite are lacking.

The chemical substances that are concerned in nematode host-parasite relationships in both animals and plants may have more affinities with each other than have previously been assumed.

REFERENCES

Allee, W. C., Emerson, A. E., Park, O., Park, T., and Schmidt, K. P. (1949). "Principles of Animal Ecology." Saunders, Philadelphia, Pennsylvania.

Anderson, R. V. (1964). *Phytopathology* **54**, 1121.

Andrewartha, H. G., and Birch, L. C. (1954). "The Distribution and Abundance of Animals." Univ. of Chicago Press, Chicago, Illinois.

Anya, A. O. (1966). *Parasitology* **56**, 733.

Banage, W. B., and Visser, S. A. (1965). *Nematologica* **11**, 255.
Beament, J. W. L. (1960). *Symp. Soc. Exptl. Biol.* **14**, 83–101.
Bergman, B. H. H., and Van Duuren, A. J. (1959a). *Mededel. Inst. Rationele Suikerproduct.* **29**, 1.
Bergman, B. H. H., and Van Duuren, A. J. (1959b). *Mededel. Inst. Rationele Suikerproduct.* **29**, 27.
Bird, A. F. (1959). *Nematologica* **4**, 322.
Bird, A. F. (1960). *Nematologica* **5**, 78.
Bird, A. F. (1961). *J. Cell Biol.* **11**, 701.
Bird, A. F. (1962a). *Nematologica* **8**, 1.
Bird, A. F. (1962b). *Nematologica* **8**, 275.
Bird, A. F. (1964). *Exptl. Parasitol.* **15**, 350.
Bird, A. F. (1966). *Nematologica* **12**, 471.
Bird, A. F., and McGuire, R. J. (1966). *Nematologica* **12**, 637.
Bird, A. F., and Millerd, A. (1962). *Nematologica* **8**, 261.
Bird, G. W., and Jenkins, W. R. (1965). *Plant Disease Reptr.* **49**, 517.
Blake, C. D. (1961). *Nature* **192**, 144.
Blake, C. D. (1962). *Nematologica* **8**, 177.
Burdett, J. F., Bird, A. F., and Fisher, J. M. (1963). *Nematologica* **9**, 542.
Burlingame, P. L., and Chandler, A. C. (1940). *J. Parasitol.* **26**, 18 (Abstr.)
Burlingame, P. L., and Chandler, A. C. (1941). *Am. J. Hyg.* **33**, Sect. D, 1.
Burström, H. (1945). *Arkiv Botan.* **32A**, 1.
Burström, H. G. (1965). *In* "Ecology of Soil-borne Plant Pathogens" (K. F. Baker and W. C. Snyder, eds.), p. 154. Univ. of California Press, Berkeley, California.
Cameron, T. W. M. (1964). *Advan. Parasitol.* **2**, 1.
Capstick, C. K. (1959). *J. Animal Ecol.* **28**, 189.
Christie, M. G., and Charleston, W. A. G. (1965). *Exptl. Parasitol.* **17**, 46.
Clarke, A. J., and Shepherd, A. M. (1964). *Nematologica* **10**, 431.
Collis-George, N., and Blake, C. D. (1959). *Australian J. Biol. Sci.* **12**, 247.
Connan, R. M. (1966). *Parasitology* **56**, 521.
Croll, N. A. (1967). *Nematologica* **13**, 17.
Day, M. F., and Powning, R. F. (1949). *Australian J. Sci. Res.* **B2**, 175.
Dobson, C. (1966). *Parasitology* **56**, 417.
Dolliver, J. S., Hildebrandt, A. C., and Riker, A. J. (1962). *Nematologica* **7**, 294.
Dropkin, V. H., Martin, G. C., and Johnson, R. W. (1958). *Nematologica* **3**, 115.
Edmonds, S. J. (1966). *Exptl. Parasitol.* **19**, 216.
Ellenby, C. (1946). *Nature* **157**, 451.
Ellenby, C. (1963). *Nature* **198**, 1110.
Ellenby, C., and Gilbert, A. B. (1958). *Nature* **182**, 925.
Everard, C. O. R. (1960). *Ann. Mag. Nat. Hist.* **3**, 53.
Fadeel, A. A. (1963). *Physiol. Plantarum* **16**, 870.
Fairbairn, D. (1961). *Can. J. Zool.* **39**, 153.
Fenwick, D. W. (1951). *J. Helminthol.* **25**, 37.
Fisher, J. M. (1966). *Australian J. Biol. Sci.* **19**, 1073.
Giebel, J. (1963). *Biul. Inst. Ochrony Roslin Poznan* **21**, 157.
Gillard, A., D'Herde, J., and Van den Brande, J. (1958). *Mededel. Landbouwhogeschool Opzoekingssta. Staat Gent* **23**, 689.
Glaser, R. W. (1925). *J. Gen. Physiol.* **7**, 599.
Godfrey, G. H., and Hagan, H. R. (1933). *Soil Sci.* **35**, 175.
Graff, D. J., and Kitzman, W. B. (1965). *J. Parasitol.* **51**, 424.
Greet, D. N. (1964). *Nature* **204**, 96.

Griffiths, J. T., and Tauber, O. E. (1943). *J. Gen. Physiol.* **26**, 541.
Harms, C. E. (1965). *J. Parasitol.* **51**, 286.
Henderson, V. E., and Katznelson, H. (1961). *Can. J. Microbiol.* **7**, 163.
Hirschmann, H. (1952). *Zool. Jahrb. Abt. Systematik, Okol. Geograph. Tiere* **81**, 313.
Holmes, J. C. (1961). *J. Parasitol.* **47**, 209
Horsfall, J. G., and Dimond, A. E. (1959). "Plant Pathology," Vol. 1. Academic Press, New York.
Hynes, H. B. N., and Nicholas, W. L. (1957). *Ann. Trop. Med. Parasitol.* **51**, 380.
Hynes, H. B. N., and Nicholas, W. L. (1958). *Ann. Trop. Med. Parasitol.* **52**, 376.
Jackson, G. J. (1960). *J. Infect. Diseases* **106**, 20.
Johnson, L. F. (1962). *Phytopathology* **52**, 410.
Johnston, T. (1957). *Phytopathology* **47**, 525.
Jones, F. G. W. (1960). *Mededel. Landbouwhogeschool Opzoekingssta. Staat Gent* **25**, 1009.
Kämpfe, L., and Kerstan, U. (1964). *Nematologica* **10**, 388.
Kates, K. C. (1942). *J. Agr. Res.* **64**, 93.
Kates, K. C. (1944). *Am. J. Vet. Res* **5**, 166.
Katznelson, H., and Henderson, V. E. (1962). *Can. J. Microbiol.* **8**, 875.
Katznelson, H., and Henderson, V. E. (1963). *Nature* **198**, 907.
Kent, N. H. (1963). *Exptl. Parasitol.* **13**, 45.
Kirkpatrick, J. D., and Van Gundy, S. D. (1966). *Nematologica* **12**, 93.
Klingler, J. (1959). *Mitt. Schweiz. Entomol. Ges.* **32**, 311.
Klingler, J. (1963). *Nematologica* **9**, 185.
Klingler, J. (1965). *Nematologica* **11**, 4.
Krusberg, L. R. (1963). *Ann. Rev. Phytopathol.* **1**, 219.
Kühn, H. (1959). *Nematologica* **4**, 165.
Lewert, R. M. (1958). *Rice Inst. Pam.* **45**, 97.
Linford, M. B. (1937). *Phytopathology* **27**, 824.
Linford, M. B. (1941). *Phytopathology* **31**, 634.
Linford, M. B. (1942). *Phytopathology* **32**, 580.
Linford, M. B., and Oliveira, J. M. (1937). *Science* **85**, 295.
Loewenberg, J. R., Sullivan, T., and Schuster, M. L. (1960). *Phytopathology* **50**, 215.
Luc, M. (1961). *Nematologica* **6**, 95.
Machmer, J. H. (1958). *J. Rio Grande Valley Hort. Soc.* **12**, 57.
McClure, M. A., and Viglierchio, D. R. (1966). *Nematologica* **12**, 248.
Manter, H. W. (1928). *Trans. Am. Microscop. Soc.* **47**, 342.
Millan, F. J. (1962). *Nematologica* **7**, 10.
Moore, D. V. (1942). *J. Parasitol.* **28**, 495.
Moore, D. V. (1946). *J. Parasitol.* **32**, 257.
Mountain, W. B. (1965). *In* "Ecology of Soil-borne Plant Pathogens" (K. F. Baker and W. C. Snyder, eds.), p. 285. Univ. of California Press, Berkeley, California.
Mountain, W. B., and Patrick, Z. A. (1959). *Can. J. Botany* **37**, 459.
Myuge, S. G. (1964). "Parasitic Nematodes of Plants." "Kolos" Publ. House, Moscow.
Nicholas, W. L., and Hynes, H. B. N. (1958). *Ann. Trop. Med. Parasitol.* **52**, 36.
Nicholas, W. L., and Jantunen, R. (1964). *Nematologica* **10**, 409.
Omidvar, A. M. (1961). *Nematologica* **6**, 123.
Onions, T. G. (1955). *Quart. J. Microscop. Sci.* **96**, 495.
Oostenbrink, M., Kuiper, K., and s'Jacob, J. J. (1957). *Nematologica Suppl.* 2, **424**.

Osche, G. (1952). *Z. Morpholo. Oekol. Tiere* **41**, 54.
Oteifa, B. A. (1953). *Phytopathology* **43**, 171.
Oteifa, B. A. and Elgindi, D. M. (1961). *Plant Disease Reptr.* **45**, 928.
Owens, R. G., and Bottino, R. F. (1966). *Contrib. Boyce Thompson Inst.* **23**, 171.
Owens, R. G., and Rubinstein, J. H. (1966). *Contrib. Boyce Thompson Inst.* **23**, 199.
Owens, R. G., and Specht, H. N. (1964). *Contrib. Boyce Thompson Inst.* **22**, 471.
Owens, R. G., and Specht, H. N. (1966). *Contrib. Boyce Thompson Inst.* **23**, 181.
Poinar, G. O., and Doncaster, C. C. (1965). *Nematologica* **11**, 73.
Powell, N. T., and Moore, E. L. (1961). *Phytopathology* **51**, 201.
Read, C. P. (1950). *Rice Inst. Pam.* **37**, 1.
Read, C. P. (1958). *Rice Inst. Pam.* **45**, 36.
Rhoades, H. L., and Linford, M. B. (1959). *Science* **130**, 1476.
Rhoades, H. L., and Linford, M. B. (1961a). *Proc. Helminthol. Soc. Wash., D.C.* **28**, 51.
Rhoades, H. L., and Linford, M. B. (1961b). *Proc. Helminthol. Soc. Wash., D.C.* **28**, 185.
Robinson, T. (1956). *Dissertation Abstr.* **16**, 1335.
Robinson, T., and Neal, A. L. (1956). *Phytopathology* **46**, 665.
Robinson, T., and Neal, A. L. (1959). *Proc. Helminthol. Soc. Wash. D.C.* **26**, 60.
Roeder, K. D. (1953). "Insect Physiology." Wiley, New York.
Rogers, W. P. (1949). *Australian J. Biol. Sci.* **2**, 157.
Rogers, W. P. (1960). *Proc. Roy. Soc.* **B152**, 367.
Rogers, W. P. (1961). "The Nature of Parasitism: The Relationship of Some Metazoan Parasites to their Hosts." Academic Press, New York.
Rogers, W. P., and Sommerville, R. I. (1963). *Advan. Parasitol.* **1**, 109–177.
Rohde, R. A. (1960). *Proc. Helminthol. Soc. Wash., D.C.* **27**, 160.
Rohde, R. A., and Jenkins, W. R. (1958). *Maryland, Univ., Agr. Expt. Sta., Bull.* **A-97.**
Rothstein, M. (1965). *Comp. Biochem. Physiol.* **14**, 541.
Rothstein, M., and Cook, E. (1966). *Comp. Biochem. Physiol.* **17**, 683.
Rubinstein, J. H., and Owens, R. G. (1964). *Contrib. Boyce Thompson Inst.* **22**, 491.
Sandstedt, R., and Schuster, M. L. (1962). *Phytopathology* **52**, 174.
Sandstedt, R., Sullivan, T., and Schuster, M. L. (1961). *Nematologica* **6**, 261.
Sasa, M., Shirasaka, R., Tanaka, H., Miura, A., Yamamoto, H., and Katahira, K. (1960). *Japan. J. Exptl. Med.* **30**, 433.
Sayre, R. M., Patrick, Z. A., and Thorpe, H. J. (1965). *Nematologica* **11**, 263.
Schad, G. A. (1963). *Nature* **198**, 404.
Seinhorst, J. W. (1950). *Tijdschr. Plantenziekten* **56**, 289.
Seinhorst, J. W. (1961). *Ann. Rev. Microbiol.* **15**, 177.
Sembdner, G., and Schreiber, K. (1960). *Nematologica* Suppl. **2**, 127.
Sembdner, G., Osske, G., and Schreiber, K. (1961). *Biol. Zentr.* **80**, 551.
Shepherd, A. M. (1959). *Nematologica* **4**, 253.
Shepherd, A. M. (1962a). "The Emergence of Larvae from Cysts in the Genus *Heterodera*," Tech. Commun. No. 32. Commonwealth Bureau of Helminthology, St. Albans, England.
Shepherd, A. M. (1962b). *Nature* **196**, 391.
Shepherd, A M. (1963). *Nematologica* **9**, 143.
Shepherd, A. M., and Wallace, H. R. (1959). *Nematologica* **4**, 227.
Shibuya, M. (1952). *Mem. Fac. Agr., Kagoshima Univ.* **1**, 1.
Silverman, P. H. (1965). *Am. Zoologist* **5**, 153.

Smyth, J. D. (1962). "Introduction to Animal Parasitology." English Univ. Press, London.
Sommerville, R. I. (1963). *J. Parasitol.* **49**, 593.
Soulsby, E. J. L., Sommerville, R. I., and Stewart, D. F. (1959). *Nature* **183**, 553.
Sprent, J. F. A. (1963). "Parasitism." Univ. of Queensland Press, Brisbane.
Stephenson, W. (1947). *Parasitology* **38**, 116.
Stoll, N. (1928). *J. Parasitol.* **15**, 147.
Symons, L. E. A., and Fairbairn, D. (1963). *Exptl. Parasitol.* **13**, 284.
Taffs, L. F., and Voller, A. (1963). *Trans. Roy. Soc. Trop. Med. Hyg.* **57**, 353.
Tetley, J. H. (1937). *New Zealand J. Sci. Technol.* **18**, 805.
Thorne, G. (1951). *Proc. Helminthol. Soc. Wash. D.C.* **18**, 18.
Tobias, J. M. (1948). *J. Cellular Comp. Physiol.* **31**, 125.
Townshend, J. L. (1964). *Can. J. Microbiol.* **10**, 727.
Turner, J. H., Kates, K. C., and Wilson, G. I. (1962). *Proc. Helminthol. Soc. Wash. D.C.* **29**, 210.
Uhlenbroek, J. H., and Bijloo, J. D. (1958). *Rec. Trav. Chim.* **77**, 1004.
Uhlenbroek, J. H., and Bijloo, J. D. (1959). *Rec. Trav. Chim.* **78**, 382.
Uhlenbroek, J. H., and Bijloo, J. D. (1960). *Rec. Trav. Chim.* **79**, 1181.
Van Asperen, K., and Van Esch, I. (1956). *Arch. Neerl. Zool.* **11**, 342.
Van Gundy, S. D. (1965). *Ann. Rev. Phytopathol.* **3**, 43.
Van Gundy, S. D., and Kirkpatrick, J. D. (1964). *Phytopathology* **54**, 419.
Van Gundy, S. D., and Martin, J. P. (1961). *Phytopathology* **51**, 146.
Van Gundy, S. D., and Stolzy, L. H. (1961). *Science* **134**, 665.
Van Gundy, S. D., and Stolzy, L. H. (1963). *Nematologica* **9**, 605.
Van Gundy, S. D., and Thomason, I. J. (1962). *Phytopathology* **52**, 366.
Van Gundy, S. D., Stolzy, L. H., Szuszkiewicz, T. E., and Rackham, R. L. (1962). *Phytopathology* **52**, 628.
Van Gundy, S. D., Bird, A. F., and Wallace, H. R. (1967). *Phytopathology* **57**, 559.
Vigilierchio, D. R. (1962). *J. Parsitol.* **48**, 53.
Viglierchio, D. R. (1963). *Proc. Helminthol. Soc. Wash., D.C.* **30**, 195.
Viglierchio, D. R., and Lownsbery, B. F. (1960). *Nematologica* **5**, 153.
von Brand, T. (1966). "Biochemistry of Parasites." Academic Press, New York.
Wallace, H. R. (1956a). *Ann. Appl. Biol.* **44**, 57.
Wallace, H. R. (1956b). *Ann. Appl. Biol.* **44**, 274.
Wallace, H. R. (1959). *Nematologica* **4**, 245.
Wallace, H. R. (1960). *Ann. Appl. Biol.* **48**, 107.
Wallace, H. R. (1961a). *Helminthol. Abstr.* **30**, 1.
Wallace, H. R. (1961b). *Nematologica* **6**, 222.
Wallace, H. R. (1963). "The Biology of Plant Parasitic Nematodes." Arnold, London.
Wallace, H. R. (1966a). *Proc. Roy. Soc.* **B164**, 592.
Wallace, H. R. (1966b). *Nematologica* **2**, 57.
Webster, J. M., and Lowe, D. (1966). *Parasitology* **56**, 313.
Weinberg, M., and Romanovitch (1907). *Ann. Inst. Pasteur* **21**, 960.
Wieser, W. (1955). *Proc. Helminthol. Soc. Wash., D.C.* **22**, 106.
Wieser, W. (1956). *Proc. Helminthol. Soc. Wash., D.C.* **23**, 59.
Wieser, W., and Kanwisher, J. (1961). *Limnol. Oceanog.* **6**, 262.
Wilson, P. A. G. (1958). *J. Exptl. Biol.* **35**, 584.
Winslow, R. D. (1955). *Ann. Appl. Biol.* **43**, 19.

CHAPTER 14

Gastrotricha, Kinorhyncha, Rotatoria, Kamptozoa, Nematomorpha, Nemertina, Priapuloidea

Ragnar Fänge

I. Introduction

Hyman (1951) classified Rotifera, Gastrotricha, Kinorhyncha, Priapulida, and Nematomorpha within the phylum Aschelminthes while Kamptozoa (Entoprocta) and Nemertina are usually referred to phyla of their own. The priapulids, nematomorphs, and nemerteans may reach considerable body sizes. On the other hand, the rotifers, gastrotrichs, kinorhynchs, kamptozoans, are microscopic or almost microscopic in size.

The small size causes difficulties for the investigator but also offers some advantages. Culture methods and electron microscopic, microchemical, immunological and electrophysiological techniques have opened new ways to investigate physiology and chemistry of small-sized living objects. So far, however, such techniques have been used only in a few investigations on the animals mentioned above.

II. Gastrotricha and Kinorhyncha (Echinoderida)

These bottom-living forms measure 50–1500 μ in length. (Hyman, 1951). Gastrotrichs are found in salt and freshwater, but kinorhynchs are exclusively marine. Gastrotrichs have external cilia, but kinorhynchs have cilia only in their protonephridia. Freshwater gastrotrichs (Chaetonotoidea) but not marine gastrotrichs have protonephridia. The protonephridia consist of pairwise joined flame-cell tubules (Brandenburg, 1962). They probably eliminate water taken up osmotically.

III. Rotatoria

A. General

Rotifers are microscopic animals, but they may reach sizes which make them visible to the naked eye. A much-investigated species, *Epiphanes* (*Hydatina*) *sena*, measures 100–500 μ (Spemann, 1924) and females of *Asplanchna* may have a body length of 1000 μ (Birky, 1964). Rotifers live in fresh and salt water and are especially abundant in ponds, ditches, and lakes. About 1500 species are known. Some forms tolerate extreme environments and have been found in alkaline lakes (Worley, 1929), at 40–50° in hot springs (Wesenberg-Lund, 1933), or in arctic or antarctic waters. Many forms undoubtedly are ecologically much specialized and difficult to cultivate, but others like the planctonic *Asplanchna* may be kept in culture (Dougherty, 1963; Birky, 1964; Birky and Field, 1966).

The body of rotifers is covered by a cuticle, sometimes superficially resembling that of small arthropods. Chitin, however, probably is absent, because hard parts dissolve in KOH solutions and sodium hypochlorite (Hyman, 1951). The cuticle is not regenerated after its enzymic removal (Koehler, 1965). The poor regenerative capacity of rotifers may be a consequence of the eutely (Nathan and Laderman, 1959).

Typical of rotifers is their "wheel-organ," a ciliated trochal disc in the anterior end. According to the direction of propagation of the metachronal waves of the cilia of the wheel-organ, Knight-Jones (1954) was able to distinguish two different systematic groups of rotifers. However, in Collotheca (Floscularia), reckoned as one of the groups, no ciliary activity occurs. The cilia of rotifers have recently been examined electron microscopically. In analogy with the splitting of phosphate from adenosine triphosphate (ATP) due to enzymic activity of the myosin of striated muscle fibers, the cilia contain components which break down ATP (Lansing and Lamy, 1961). Studies of the ultrastructure of cilia using methods for demonstrations of ATPase activity have shown that cilia operate by a "push-pull" mechanism.

B. Eutely (Cell Constancy)

In adult rotifers each organ contains a constant, species-specific number of nuclei. However, this cell constancy is not absolute. In *Asplanchna* the number of nuclei in the gastric gland and vitellarium is influenced by environmental factors, such as the diet (Birky and Field, 1966). The rigidity of differentiation as manifested by the eutely make rotifers animals of choice for studies of chemical and physical factors with affect growth and differentiation.

C. Reproduction and Duration of Life

Ordinary individuals are amictic females. Their reproductive system consists of: (1) the germinarium—a group of ovocytes, (2) the vitellarium—nurse cells synthesizing yolk and cytoplasm, and (3) the uterus. By production of diploid parthenogenetic eggs the amictic females form clones of new females. Males arise under adverse external conditions by parthenogenetic development of haploid eggs produced by special mictic females. If haploid eggs are fertilized, they transform into large, thick-walled, diploid resting eggs. The latter, after a dormant period, often lasting many months, develop into ordinary amictic females. The main function of sexuality in rotifers seems to be the production of resting eggs (Gilbert, 1963). The ordinary asexual reproduction is rapid. In cultures of *Asplanchna* the number of amictic females doubles within 15 hours (Birky, 1964). The mictic eggs are resistant against external influences, a fact which is utilized in the culture technique. Axenization is obtained by washing the resting eggs in combinations of antibiotics or in a 0.005–0.01% hypochlorite solution.

In culture rotifers live 2–3 weeks only, although in anabiotic condition or as resting eggs life may persist for years. After a period of growth and active life rotifers begin to show signs of age degeneration. Egg-laying stops, the animals get sluggish, the tissues shrink and accumulate pigments. The movements of the pharyngeal cilia are the last signs of life to persist. The short life span and the relative easiness with which rotifers are kept in culture make them suitable objects for gerontological research (Spemann, 1924; Lansing, 1947; Comfort, 1964).

D. Cyclomorphosis

Planctonic rotifers undergo seasonal morphological changes. With successive parthenogenetic generations spines may disappear or change in length. *Asplanchna priodonta* increases its body length from May to July. The armor of *Brachionus* and *Keratella* changes its shape.

Wesenberg-Lund (1900) thought that planctonic animals tend to enlarge their surface as a response to decreasing density of the water, as this is warmed up in the summer. Whitney (1916) reported transformations of *Brachionus pala* into forms resembling another species by addition of sodium silicate. However, in spite of several theories and experimental work, the problem of the cause of cyclomorphosis in rotifers has not been cleared up. Pourriot (1964) brought evidence that the rotifers themselves release substances which induce morphological changes. Such substances may even affect other species. Gilberg (1966) discovered that the predaceous *Asplanchna* liberates a factor which causes another species, *Brachionus*, to develop protective spines. The *Asplanchna* factor probably is a protein, which acts on the development of eggs of *Brachionus* already during oogenesis.

E. Excretion and Osmoregulation

As a result of the large surface volume ratio large amounts of water diffuse into freshwater rotifers and have to be eliminated. It is thus not astonishing that these animals have well-developed protonephridia which empty into pulsatory contractile bladders. The ultrastructure of the protonephridia was studied by Braun *et al.* (1966) in *Asplanchna*, and by Mattern and Daniel (1966) in *Philodina*. In the latter species a 200 Å. thick membrane separates the flame cell lumen from the pseudocoelomic cavity. Observations on *Asplanchna* have shown that the contractile bladder in 15–30 minutes evacuates a volume of fluid equal to the volume of the whole animal (Wesenberg-Lund, 1933). According to Pontin (1966) the normal shape of *Asplanchna* is upheld by osmotic turgor, the body fluid being isoosmotic to a 100–155 mosmolar solution of NaCl. The activity of the vibratile flames and the contractile bladder are directly related to the ionic concentration of the external medium. The rate of beat of the ciliary flames and the output from the bladder decrease in media of high ionic concentrations. The chemical composition of the body fluid is unknown.

F. Resistance Against Desiccation and Extreme Temperatures

The ability of certain microscopic animals to survive drying up was observed already by Leeuwenhoek. The phenomenon was of such general interest that in 1859 the Société de Biologie (Paris) appointed a commission to deal with the matter. As its motto the commission had the Hamlet quotation "to be or not to be, that is the question." It was confirmed that desiccated rotifers, and tardigrades, revive after addition of water (Broca, 1861). Desiccated rotifers for brief periods withstand temperatures above 100°, and their resistance to low temperatures appears unlimited. Rotifers resist dry periods both as resting eggs and by passing into a contracted anabiotic stage of little or no metabolism (see Hyman, 1951, for references). Reversible cytological changes accompanying desiccation in *Philodina* were noticed by Hickernell (1917).

G. PHARMACOLOGY

The small size gives rotifers a relatively enormous body surface through which substances may enter. As a consequence of this they are extremely sensitive to certain toxic agents, and drugs to be tested can be added simply to the surrounding water.

Marriott *et al.* (1948) obtained narcosis with cocaine 1:700 but not with codeine, barbitone, or phenobarbitone. Donner (1962) recommends chloretone (10^{-3} to 10^{-2}), strychnine nitrate (10^{-4} or weaker) or eucaine hydrochloride (2×10^{-3}) for narcosis during microscopic observations. Eucaine hydrochloride may be used in alcoholic solution (Bullough, 1954).

Kreiskott (1958, 1965) used *Epiphanes* as a research object for distinguishing between general toxic effects on cytoplasm and specific toxic effects on mitoses. This was possible because in adult rotifers cell division occurs only in developing eggs. Colchicine had low general toxicity but effectively inhibited mitoses of eggs. Megaphen (chlorpromazine) inhibited the egg deposition but had no effect on mitoses. Pharmacological effects on the neuromuscular system have been studied in connection with egg deposition, because this requires a strong contraction of body wall muscles. Low temperature or Megaphen ($1\text{–}2 \times 10^{-6}$ gm./ml.) blocks the contraction necessary for egg laying in *Epiphanes senta* (Kreiskott, 1958). Lindner and Goldman (1964) investigated the influence of cholinergic and anticholinergic drugs on egg deposition.

IV. Kamptozoa (Entoprocta)

With the exception of the freshwater genus *Urnatella*, the kamptozoans live in marine or brackish water. They are tiny solitary or colonial animals with a bell-shaped body on the top of a contractile stalk. They have a horseshoe-like intestine, a primary coelom and protonephridia. The rim of the bell is surrounded by ciliated tentacles. The cuticle of *Barentia* does not contain chitin (Hyman, 1966). The gut has secretory cells and its content is alkaline (Hyman, 1951).

The rather well-developed muscles, among them retractors of the tentacles, are innervated from a ganglionated nervous system. Kummel (1964) examined the protonephridia of *Urnatella* electron microscopically. In addition to ordinary cilia the flame cells contain microvillae lacking the 9 + 2 structure. Microvillae also occur in flame cells of trematodes and priapulids.

The fertilized eggs after complete cleavage gastrulate and develop into trochophore larvae. Further differentiation is small, and kamptozoans accordingly have been characterized as sexually mature trochophorans. Regenerative capacity is relatively good (Nasonov, 1926; *Arthropodaria*).

V. Nematomorpha (Gordiacea)

Nematomorpha (horse-hair worms) are long, thin wormlike creatures. Most forms (family Gordiidae and Chordodidae) live in freshwater, but an aberrant marine genus (*Nectonema*) exists. The body is surrounded by a thick two-layered cuticle of unknown composition. The body wall has a longitudinal muscle layer. A ventral nerve cord is found in the epidermis. The interior of the body is filled by a mesenchymal tissue. Special respiratory, circulatory, and excretory systems are lacking and the intestine is rudimentary.

The presence of an armored proboscis in the free-living larvae gives some resemblance to kinorhynchs and priapulids. The developing animals are first parasitic but end up as free-living sexually mature individuals. Arthropods such as *Dytiscus marginalis* and myriapods serve as hosts during the parasitic stage. The marine forms parasitize crustaceans.

The nutritional physiology is interesting, as the animals probably absorb nutrition through their body surface, being unable to feed by the mouth (Hyman, 1951).

VI. Nemertina (Ribbonworms)

The soft and nonsegmented body has a ciliated surface. Many forms are brightly red or yellow due to skin pigments. Melanin, ommochrome, and porphyrins were demonstrated in *Lineus ruber* (Vernet, 1966). In several forms the brain ganglia and the longitudinal nerve cords contain red pigment. A muscular proboscis may be withdrawn into a tube-shaped cavity. Most of the about 600 species are marine, but a limited number are terrestrial or live in freshwater. Sexes are separate. Embryonic development is direct or indirect. There are no special respiratory organs. A record of literature on nemertines has been published (Friedrich, 1965).

A. Neuromuscular System

The body muscles are formed by longitudinal and circular nonstriated fibers. As a result of muscular activity most forms may vary enormously in length. The body wall muscles of *Cerebratulus* are contracted by acetylcholine although only in high concentration (5×10^{-5} mg./ml.) after treatment with eserine (Smith *et al.*, 1940). An unusually high activity of cholinesterase was found in *Prostoma rubrum*; 510–899 mg. acetylcholine were hydrolyzed by 100 mg. tissue per hour (Kamemoto, 1957). The central nervous system consists of two pairs of well-developed brain ganglia and two longitudinal nerve cords (Bullock and Horridge, 1965).

B. Circulatory System

Nemertines are among the lowest invertebrates which have a closed vascular system. The blood is circulated by pulsatory vessels. They form three longitudinal trunks anastomosing by numerous smaller vessels and blood lacunes (Martin and Johansen, 1965). The blood usually is colorless, but in marine forms like *Drepanophorus*, *Barlasia*, and *Euborlasia*, hemoglobin-containing erythrocytes may occur in addition to ameboid cells.

Rose-colored blood cells occur in *Tubulanus* and *Amphiporus* (Böhmig, 1933). In *Lineus fuscoviridis* Ohuye (1942) described erythrocytes and various types of leukocytes: eosinophilic, basophilic, spindle-shaped. The erythrocytes give positive nadi- and peroxidase reactions. In *Polia sanguirubra* the plasma contains hemoglobin. *Lineus* has an unusually high content of iron. It is not known whether this is due to hemoglobin (Jones *et al.*, 1935).

C. Excretion and Osmoregulation

Nemertines typically are provided with a pair of branched protonephridia. Terrestrial forms (*Geonemertes*) have thousands of flame cells distributed within the parenchyma of the body. Alkaline phosphatase activity has been demonstrated in the walls of the terminal ducts. Dilution of the body fluids with water stimulates the flame cells and causes the protonephridia to increase their output (Pantin, 1947). In marine species glandlike aggregations of flame cells are situated in the wall of large blood vessels. The arrangement may remind one of the structure of the pronephros of the vertebrate *Myxine* (Böhmig, 1933; Nawitski, 1931).

D. Venom Production

All nemertines are carnivorous. Jennings (1962) thinks that *Lineus* uses the acid digestive fluid to kill its food. Several authors assume that nemertines paralyze their preys by specific venoms. Stylets and rhabdits of the proboscis have been thought to function in association with deposition of venom. Actual observations are few and vague. However, one can distinguish armed nemertines, the proboscis of which has stylets and "venom glands," and unarmed nemertines with naked proboscis. In armed nemertines (*Amphiporus lactifloreus, Drepanophorus crassus*) Bacq (1936, 1937) found a nicotinelike alkaloid, which he called amphiporine. Examination of extracts from 1000 specimens of *Amphiphorus* led to the conclusion that amphiporine is not nicotine but a closely related substance (King, 1939). The venom is not localized in the proboscis nor in the "venom glands," but extracts from all parts of the body are toxic. Amphiporine excites striated muscle of the frog and first excites and later paralyzes the autonomic ganglia of the cat. In *Lineus* Bacq demonstrated another substance, "nemertinine," which has in common with amphiporine the property of exciting crab nerves, but in contrary to amphiporine it does not affect striated muscle of the frog. When a *Lineus lacteus*, weighing 0.15 gm. is extracted with 4 ml. of seawater, the extract diluted 10,000 times still excites the nerve of the isolated leg of the crab *Carcinus maenas* (Bacq, personal communication).

E. Digestive System

Lineus feeds on polychaetes, which may exceed it in size. Digestion is extremely rapid (Gontcharoff, 1961). In *Lineus ruber* the gut contains an endopeptidase (cathepsin C) acting at low pH, and the foregut is provided with acidophile cells probably responsible for production of acid. The cells contain carbonic anhydrase, and production of acid is blocked by the carbonic anhydrase inhibitor Diamox (Jennings, 1962). In *Geonemertes chalicophora* digestion is partly intracellular (von Graff, 1879, cited by Jordan, 1913). Nemertines survive starving for months but decrease in size due to involution of both number and sizes of cells. Effects of starvation have been studied mainly on *Stichostemma* and *Lineus* (Böhmig, 1933).

F. Endocrinology

Scharrer (1941) emphasized the unique structure of the cerebral organs. These contain neurons and gland cells grouped closely together.

Histochemical signs of neurosecretion have been demonstrated (Gabe, 1966). The cerebral organs are thought to function as chemoreceptors (Gontscharoff, 1961), but possibly they serve endocrine functions as well.

G. Regeneration

Nemertines, especially the family Lineidae have a well-developed capacity of regeneration. When irritated, several forms autotomize (Gontscharoff, 1961). Later they probably regenerate missing parts.

VII. Priapuloidea

This group, included in the phylum Aschelminthes by Hyman (1951), could as well be placed in a phylum of its own. The group has half a dozen species widely distributed over the world (Morina and Starobogatov, 1961). Priapulids have a voluminous body cavity, which functions as a hydrostatic skeleton during burrowing (Friedrich and Langeloh, 1936). They are usually considered as pseudocoelomates (Lang, 1953), but Zinkin (1955) and Shapeero (1961) report the existence of an extremely thin and probably discontinuous epithelial peritoneum. This would indicate resemblances with coelomates. According to Zinkin (1955) the body cavity arises as a schizocoel in a mass of mesoderm cells aggregated within the original blastocoel. The embryonic development, however, is imperfectly known. The larval stages remind one both of larval gordiids and of adult gastrotrichs and kinorhynchs. If priapulids are grouped among aschelminths (Hyman, 1951), it ought to be remarked that they are separated from gastrotrichs and kinorhynchs by their large body size. *Priapulus caudatus* may weight 10 gm. or more and *Halicryptus spinulosus* may weight up to 0.5 gm. (own observations).

The priapulids are sexually dioecious, the gonadal products being given off by the urogenital ducts. The ovaries hang freely into the coelomic fluid by which substances are transported to and from the ovocytes. A monograph on ultrastructural aspects of oogenesis in *Priapulus caudatus* has recently been published (Nörrevang, 1965). The body of priapulids is surrounded by a chitinous cuticle which is shed periodically by a molting process (Carlisle, 1959; Shapeero, 1962). It is not known whether the molting is under endocrine control, but this seems plausible.

A. Neuromuscular System

1. *Organization*

Well-developed muscle layers are found in the body wall and the pharynx. Muscle fibers further form short and long proboscis retractor muscles and participate in the walls of the caudal appendage the intestine and the urogenital system. The muscle fibers of the body wall are tube-shaped with a granulated cytoplasm. Body wall muscles and retractor muscles are striated (Lang, 1953; Fänge, unpublished). The nervous system consists of a pharyngeal ring and a ventral longitudinal cord situated within the ectoderm. From the ventral cord transversal branches are given off to the body wall muscles (Bullock and Horridge, 1965).

Electronmicroscopically no cross-striation is noticed in the body wall muscles (Mattisson, 1968, personal communication). Probably a false striation is caused by wrinkling of the surface of the muscles or in some other way.

2. *Pharmacological Effects*

Below are presented results from a few observations on effects of drugs on isolated muscle preparations (Fänge, unpublished). The experiments were made on the longitudinal body wall muscle and the intestine of *Priapulus caudatus.* The preparations were suspended in aerated seawater with a salinity of 32‰. Kymograph recordings were

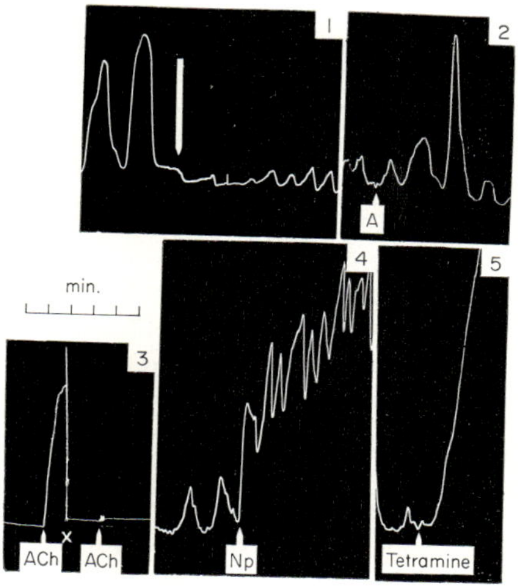

FIG. 1. *Priapulus caudatus.* Kymograph records of isolated muscle preparations. (1) Ventral body wall. At the arrow bubbling of air was started. (2) Same preparation 5 minutes later. At A epinephrine in 5×10^{-5} gm./ml. seawater. (3) Dorsal body wall. Two successive additions of acetylcholine (ACh) to a final concentration of 6×10^{-8} gm./ml. seawater; ×, change of saline. (4) Ventral body wall. 0.2 ml. of a tissue homogenate containing tetramine (salivary gland of *Neptunea antiqua*). (5) Same preparation as (4). Tetramine 10^{-4} gm./ml. seawater.

made at 10°. Preparations of body wall which included the ventral nerve cord showed spontaneous activity. The amplitudes of contractions decreased during aeration (Fig. 1[1]). Acetylcholine in concentrations of 10^{-8} to 10^{-5} gm./ml. seawater caused rise of tonus of the body wall muscle and the intestine. After acetylcholine the preparations were insensitive to new doses of acetylcholine for several minutes (Fig. 1[3]). Responses to acetylcholine were greatly enhanced by pretreatment with eserine at 10^{-5} gm./ml.

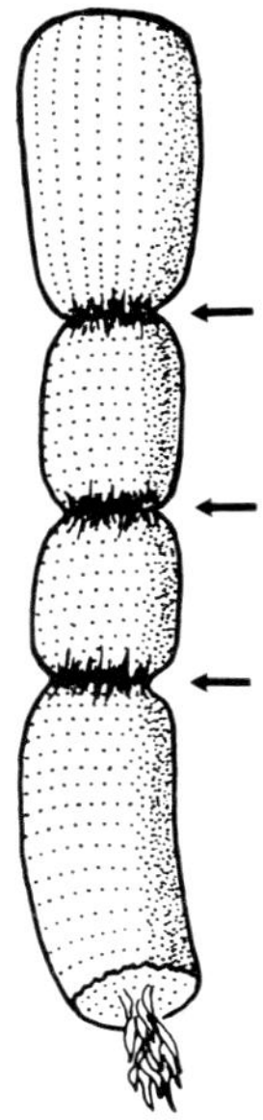

FIG. 2. *Priapulus caudatus*. Response to mechanical stimulation (arrows) in an animal treated by M.S. 222.

Atropine in 10^{-4} gm./ml. seawater weakly depressed the response to acetylcholine. Tetramine (10^{-4} gm./ml.) as well as an organ extract containing this substance caused sustained contraction. The extract also stimulated spontaneous activity of the ventral body wall (Fig. 1[4,5]). Epinephrine and norepinephrine usually had no effects on the body wall muscle, but epinephrine possibly stimulated spontaneous activity (Fig. 1[2]). Epinephrine in 10^{-4} gm./ml. seawater had a weak relaxing effect on the intestine.

5-Hydroxytryptamine caused contraction of the intestine, but did not affect the body wall muscles. γ-Aminobutyric acid (GABA) in 2×10^{-6} gm./ml. seawater contracted the body wall muscle. The effect was additive to that of acetylcholine. The relatively high sensitivity to acetylcholine and eserine of priapulid muscles indicates cholinergic control. The results obtained with epinephrine do not allow any definite conclusions. The vague response observed in Fig. 1[2] could have been due to an unspecific stimulation of sensory nerve elements.

The following observation was made on narcotized specimens of *Priapulus:* M.S. 222 in 10^{-3} gm./ml. seawater within 30 minutes caused immobilization. However, the animals still reacted with local constrictions of the circular body wall muscles, when the skin was pinched (Fig. 2). The constrictions appeared rapidly but lasted several minutes. Longitudinal conduction of impulses in the ventral cord were blocked by the narcosis, as the narcotized animals did not show normal escape movements.

B. Coelomic Fluid

1. *Coelomocytes*

The body cavity takes about 50% of the body volume. It is filled by a pinkish milky fluid containing numerous free cells. The coelomocytes are of two kinds: erythrocytes and amebocytes or leukocytes (Fänge, 1950; Fänge and Åkesson, 1951).

The erythrocytes are biconvex, disc-shaped round cells with a central nucleus and the cytoplasm contains numerous vacuoles and refringent granules (Fig. 3). They have a higher specific gravity than the amebocytes and sediment easily. Details of hematopoiesis are unknown. The erythrocytes carry an iron-containing protein resembling hemerythrin. This is the respiratory pigment found in the coelomocytes of sipunculids and the brachiopod, *Lingula* (Ghiretti, 1962).

The amebocytes are highly movable and phagocytic. They ingest India ink particles and carmine introduced into the body cavity and concentrate neutral red within intracellular vacuoles (Moltschanoff, 1908; Fänge, unpublished observations).

2. *Clot Formation*

In samples of coelomic fluid of *Priapulus,* agglutinating amebocytes rapidly form clotlike masses floating at the surface. The agglutination of amebocytes is accelerated if the samples are shaken or centrifuged for a few minutes. The purely cellular clotting mechanism of *Priapulus* very much resembles processes known to occur in the blood or coelomic fluids of many other invertebrates, crustaceans for example (author's observations).

FIG. 3. *Priapulus caudatus*. Living coelomocytes. The majority of the cells are erythrocytes but four amebocytes are seen. The latter are irregular and smaller than the erythrocytes. Their cytoplasm is granulated. Dark field. Microphotograph by Dr. A. Nörrevang. ×900 Reichert Zetopan.

3. *Preparations of Oxyhemerythrin*

Relative pure samples of erythrocytes are obtained by centrifugations and repeated washings in isoosmotic saline (filtered seawater). Solutions of oxyhemerythrin obtained by hemolysis of erythrocytes with distilled water show a weak absorption band at 495–500 mμ. The hemerythrin sometimes occurs as intracellular crystals. These are birefringent in polarized light and do not dissolve in distilled water. Solutions of oxyhemerythrin are irreversibly deoxygenated by addition of traces of $Na_2S_2O_4$ (Fig. 4A,B). Addition of concentrated sulfuric acid to solutions of oxyhemerythrin produces a red solution with strong green fluorescence (Fänge and Åkesson, 1951). The solution has an absorption maximum at about 520 mμ (Fig. 4C).

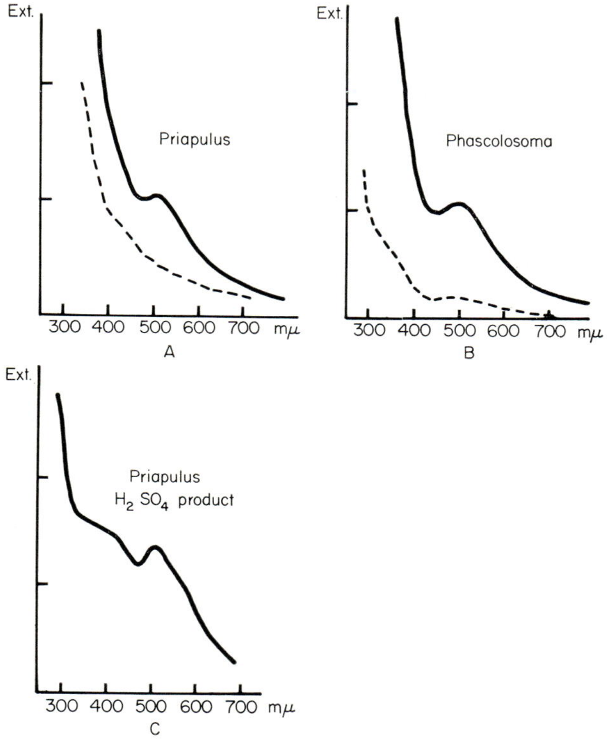

FIG. 4. A. *Priapulus caudatus.* Absorption spectra of oxyhemerythrin (thick line) and deoxygenated hemerythrin (dashed line). B. For comparison: *Phascolosoma vulgare,* similar absorption spectra. C. *Priapulus caudatus.* Absorption spectrum obtained after addition of concentrated sulfuric acid to a solution of oxyhemerythrin. Original curves.

4. Composition of the Coelomic Plasma

The coelomic plasma of *Priapulus* contains 20–40 mg./100 ml. glucose. Examination by starch gel electrophoresis shows two to four protein fractions in the coelomic plasma of *Halicryptus* and *Priapulus.* The plasma proteins from the two species differ from each other and from hemerythrin in rates of migration. Cathode-migrating protein components were found in both species. By immunodiffusion technique further protein components were characterized (B. Holstein, personal communication; author's unpublished observations).

Probably the coelomic fluid of priapulids to a certain extent functionally resembles the blood of more highly differentiated animals. In its composition it might resemble the fluid of the pseudocoel of aschelminth-like nematodes. Undoubtedly the coelomic fluid serves a transport function. The body muscles, the urogenital system, the skin, and the nervous system are supplied by necessary substances via this fluid.

C. Respiration

Main uptake of oxygen occurs through the skin, but the caudal appendage of *Priapulus* may aid in respiratory exchange. The presence of hemerythrin-containing erythrocytes increases the oxygen-binding capacity of the coelomic fluid to about 2 times that of seawater (Fänge and Mattisson, 1961).

D. Osmoregulation and Excretion

Some adaptation to brackish water conditions is obvious in priapulids. In Scandinavia *Priapulus* lives in water with salinities varying approximately between 20 and 33‰ and *Halicryptus spinulosus* is confined to very brackish environments with salinities down to 5–10‰. The very well-developed brightly glittering protonephridia in *Halicryptus* larvae (Lang, 1948b) probably are important water-eliminating structures. Protonephridia also occur in *Priapulus*. The ultrastructure of the protonephridia of the latter form has been examined by Nörrevang (1963b) and Kummel (1964). The wall of the excretory duct shows strong alkaline phosphatase reaction (Nörrevang, 1963a). In contrary to many other excretory organs, the protonephridia of *Priapulus* do not seem to be able to transport certain organic anions. Thus chlorphenol red or fluorescein introduced into the body cavity of *Priapulus* is eliminated by the intestine but not by the protonephridia. *Priapulus* might be an excellent animal for study of protonephridial function, as sampling of urine is possible from the bladderlike enlargements of the urogenital ducts (Fänge, unpublished observations).

E. Digestion

The digestive apparatus consists of the mouth and the extrovertable pharynx with its special muscles, the intestine and the rectum. The intestine is purple-red or red-brown on its surface and contains 0.1–0.5 ml. of a slightly viscous brown fluid; pH of the digestive fluid is about 8.3. Enzymic activity resembling cathepsin, trypsin, and chymotrypsin could be demonstrated (Nilsson and Fänge, 1967). The pigments of the intestinal wall might be a precursor of hemerythrin and the brown pigment of the digestive fluid could be a decomposition product of the same substance. *Priaplus* is carnivorous (Lang, 1948a).

ACKNOWLEDGMENTS

I thank Dr. A. Nörrevang (University of Copenhagen) for the microphotograph (Fig. 3) and Dr. K. Purasjoki (Tvärminne Zool. Station, Finland), Mr. B. Holstein (Göteborg, Sweden), and Dr. B. Swedmark (Zool. Station Kristineberg, Sweden) for material of priapulids. This work was supported in part by the Swedish National Research Board.

REFERENCES

Bacq, Z. M. (1936). *Bull. Acad. Roy. Med. Belg.* [5] **22**, 1072.
Bacq, Z. M. (1937). *Arch. Intern. Physiol.* **44**, 190.
Birky, C. W., Jr. (1964). *J. Exptl. Zool.* **155**, 273.
Birky, C. W., Jr., and Field, B. (1966). *Science* **151**, 585.
Böhmig, L. 1933). *In* "Handbuch der Zoologie" W. G. Kükenthal and T. Krumbach, eds.), Vol. 2, Part 1, No. 13, pp. 1–110. de Gruyter, Berlin.
Brandenburg, J. (1962). *Z. Zellforsch. Mikroskop. Anat.* **57**, 136.
Braun, G., Kümmel, G., and Mangos, J. A. (1966). *Arch. Ges. Physiol.* **289**, 141.
Broca, P. (1861). *Mem. Soc. Biol.* **2**, 1.
Bullock, T. H., and Horridge, G. A. (1965). "Structure and Function in the Nervous System of Invertebrates." Freeman, San Francisco, California.
Bullough, W. S. (1954). "Practical Invertebrate Anatomy," p. 136. Macmillan, New York.
Carlisle, D. (1959). *Arkiv. Zool.* [2] **12**, 79.
Comfort, A. (1964). "Ageing," 2nd ed. Toutledge & Kegan Paul, London.
Donner, J. (1962). "Rädertiere (Rotatorien)." Franckh'sche Verlagshandlung, Stuttgart.
Dougherty, E. C. (1963). *J. Exptl. Zool.* **153**, 183.
Fänge, R. (1950). *Nature* **165**, 613.
Fänge, R., and Åkesson, B. (1951). *Arkiv. Zool.* [2] **3**, 25.
Fänge, R., and Mattisson A. (1961). *Nature* **190**, 1216.
Friedrich, H. (1965). *Veroeffentl. Ubersee-Museum* **3**, 204.
Friedrich, H., and Langelch, H. P. (1936). *Biol. Centr.* **56**, 249.
Gabe, M. (1966). "Neurosecretion" (Transl.). Pergamon Press, Oxford.
Ghiretti, F. (1962). *In* "Oxygenases" (O. Hayaishi, ed.), p. 517. Academic Press, New York.
Gilbert, J. J. (1963). *J. Exptl. Biol.* **40**, 625.
Gilbert, J. J. (1966). *Science* **151**, 1234.
Gontscharoff, M. (1961). *In* "Traité de Zoologie" (P. P. Grassé, ed.), Vol. 4, p. 783. Masson, Paris.
Hickernell, L. M. (1917). *Biol. Bull.* **32**, 344.
Hyman, L. (1951). "The Invertebrates," Vol. III. McGraw-Hill, New York.
Hyman, L. (1966). *Biol. Bull.* **130**, 94.

Jennings, J. B. (1962). *Biol. Bull.* **122**, 63.

Jones, E. J., *et al.* (1935). *J. Exptl. Biol.* **12**, 59.

Jordan, H. (1913). "Vergleichende Physiologie wirbelloser Tiere." Fischer, Jena.

Kamemoto, F. I. (1957). *Science* **125**, 351.

King, H. (1939). *J. Chem. Soc.*, p. 1365.

Knight-Jones, E. W. (1954). *Quart. J. Microscop. Sci.* **95**, 503.

Koehler, J. K. (1965). *Ultrastruct. Res.* **12**, 113.

Kreiskott, H. (1958). *Arch. Hydrobiol.* **54**, 292.

Kreiskott, H. (1965). Ph.D. Thesis, Bonn.

Kummel, G. (1964). *Z. Zellforsch. Mikroskop. Anat.* **62**, 468.

Lang, K. (1948a). *Arkiv. Zool.* **41A**, No. 5, 1.

Lang, K. (1948b). *Arkiv. Zool.* **41A**, No. 9, 1.

Lang, K. (1953). *Arkiv. Zool.* [2] **5**, 321.

Lansing, A. I. (1947). *J. Gerontol.* **2**, 28.

Lansing, A. I., and Lamy, F. (1961). *J. Biophys. Biochem. Cytol.* **9**, 799.

Lindner, L. A., and Goldman, H. (1964). *J. Pharmacol. Exptl. Therap.* **146**, 123.

Marriott, R. H., *et al.* (1948). *Nature* **162**, 189.

Martin, A. W., and Johansen, K. (1965). *In* "Handbook of Physiology" (Am. Physiol. Soc., J. Field, ed.), Sect. 2, Vol. III, Chapter 72, p. 2545.

Mattern, C. F., and Daniel, W. A. (1966). *J. Cell Biol.* **29**, 552.

Moltschanoff, L. A. (1908). *Bull. Acad. Sci. St. Petersbourg* **2**, 957.

Morina, V. V., and Starobogatov, J. I. (1961). *Tr. Inst. Okeanol., Akad. Nauk, SSSR* **46**, 179 (in Russian).

Nasanov, N. (1926). *Trav. Lab. Zool. Sebastopol* [2] No. 5 (cited by Hyman, 1951).

Nathan, H. A., and Laderman, A. D. (1959). *Ann. N.Y. Acad. Sci.* **77**, 96.

Nawitzki, W. (1931). *Zool. Jahrb. Abt. Anat. Ontog. Tiere* **54**, 159.

Nilsson, A., and Fänge, R. (1967). *Comp. Biochem. Physiol.* **22**, 927.

Nörrevang, A. (1963a). *Proc. 16th Intern. Congr. Zool., Washington, D.C.* Vol. 1, p. 33.

Nörrevang, A. (1963b). *Nature* **198**, 700.

Nörrevang, A. (1965). *Vidensk. Medd. Dansk naturhist. Foren. Kiobenhavn.* **128**.

Ohuye, T. (1942). *Sci. Rept. Tohoku Imp. Univ., Fourth Ser.* **17**, 187.

Pantin, C. F. A. (1947). *Quart. J. Microscop. Sci.* **88**, 15.

Pontin, R. M. (1966). *Comp. Biochem. Physiol.* **17**, 1111.

Pourriot, R. (1964). *Bull. Soc. Zool. France* **89**, 555.

Scharrer, B. (1941). *J. Comp. Neurol.* **74**, 109.

Shapeero, W. (1961). *Science* **133**, 879.

Shapeero, W. (1962). *Trans. Am. Microscop. Soc.* **81**, No. 4, 352.

Smith, C. C., Jackson, B., and Prosser, C. L. (1940). *Biol. Bull.* **79**, 377.

Spemann, F. W. (1924). *Biol. Z. Wiss. Zool.* **123**, 1.

Vernet, G. (1966). *Compt. Rend.* **263**, 191.

Wesenberg-Lund, C. (1900). *Biol. Centr.* **20**, 1.

Wesenberg-Lund, C. (1933). *In* "Handbuch der Zoologie" (W. G. Kükenthal and T. Krumbuch, eds.) Vol. 2, Part 1, No. 4, p. 8. de Gruyter, Berlin.

Whitney, D. D. (1916). *Biol. Bull.* **31**, 113.

Worley, L. G. (1929). *Ecology* **10**, 420.

Zinkin, L. N. (1955). *Utschonyie Zapiski (Leningrad)* **110**, 129 (in Russian).

Author Index

Numbers in italics refer to the pages on which the complete references are listed.

D

E

I

J

K

L

M

N

R

S

T

U

V

Subject Index

C

G

H

I

N

O

S

U

Shepherd College Library
3 4550 000010920 0

QP
514
F5.28
v.3

102316

DATE DUE			